Biotechnology: Pharmaceutical Aspects

Biotechnology: Pharmaceutical Aspects

Ronald T. Borchardt and C. Russell Middaugh, Series Editors

Volume I: *Pharmaceutical Profiling in Drug Discovery for Lead Selection*

R.T. Borchardt, E.H. Kerns, C.A. Lipinski, D.R. Thakker, B. Wang

Volume II: *Lyophilization of Biopharmaceuticals*

H.R. Costantino, M.J. Pikal

Volume III: *Methods for Structural Analysis of Protein Pharmaceuticals*

W. Jiskoot, D.J.A. Crommelin

Methods for Structural Analysis of Protein Pharmaceuticals

Wim Jiskoot
Utrecht University, The Netherlands

Daan J.A. Crommelin
Utrecht University, The Netherlands

AAPS PRESS

American Association of Pharmaceutical Scientists
Arlington, VA

Library of Congress Cataloging-in-Publication Data

Jiskoot, Wim.
Methods for structural analysis of protein pharmaceuticals / Wim Jiskoot, Daan J.A. Crommelin.
p. cm. -- (Biotechnology: pharmaceutical aspects ; 3)
Includes index.
ISBN 0-9711767-2-8
1. Protein drugs--Analysis. I. Crommelin, D.J.A. (Daan J. A.) II. Title. III. Biotechnology (Arlington, Va.) ; 3.

RS431.P75J55 2005
615'.1901--dc22

2005041029

Wim Jiskoot
Department of Pharmaceutics
Utrecht Institute for
Pharmaceutical Sciences (UIPS)
Utrecht University
P.O. Box 80082
3508 TB Utrecht
The Netherlands

Daan J.A. Crommelin
Department of Pharmaceutics
Utrecht Institute for
Pharmaceutical Sciences (UIPS)
Utrecht University
P.O. Box 80082
3508 TB Utrecht
The Netherlands

Series Editors:

Ronald T. Borchardt
University of Kansas
Dept. of Pharmaceutical
Chemistry
2095 Constant Avenue
Lawrence, KS 66047

C. Russell Middaugh
University of Kansas
Dept. of Pharmaceutical
Chemistry
2095 Constant Avenue
Lawrence, KS 66047

ISBN: 0-9711767-2-8

Printed in the United States of America.

Executive Director:	John Lisack, Jr.
Project Coordination:	Ericka McIntyre (Custom Editorial Productions Inc.)
Design/Production:	Custom Editorial Productions Inc.
Cover design:	Sitki Kazanci (AAPS)

CONTENTS

PREFACE

Protein pharmaceuticals form a fast-growing category in our arsenal of drugs. New findings in molecular biology and in biotechnology provide more and better opportunities to design new therapeutic agents that are based on endogenous compounds such as cytokines and protein hormones or that are based on the concept of the structure and action of antibodies. At this moment, the FDA has approved one hundred eight biotechnology-based medicines, and the majority of these are pharmaceutical proteins (see http://www.phrma.org/newmedicines/biotech/). Moreover, the pipeline for new drugs shows that the number of pharmaceutical proteins will continue to grow: More than 320 biotechnology-based medicines are currently under development in the United States.

In the coming years, the patents on a number of the first-generation pharmaceutical proteins, such as erythropoietin, will expire. This means that so-called biogenerics may be introduced into the market, just as generic, low-molecular-weight drugs are currently available alongside the products of innovators. However, the discussion on how to deal with biogenerics is still ongoing, as there are a number of distinct differences between pharmaceutical proteins and low-molecular-weight drugs. Low-molecular-weight drug products, such as aspirin, can be characterized in minute detail in terms of chemistry and physics. Biopharmaceutical profiling of such drug products has reached a high level of sophistication that ensures bioequivalence between the innovator's product and the generic product. With pharmaceutical protein products, full characterization of the active compound is much less straightforward, as the high-molecular-weight compounds have, besides their chemical structure (primary structure and post-translational modifications), complex secondary, tertiary, and sometimes even quaternary structures that have to be described and defined in detail and, in the case of biogenerics, have to be mimicked.

The leading issue in this book is, how can protein pharmaceuticals be characterized? What are the techniques currently at our disposal? And, what does each of these techniques from our "toolbox" tell us? This question was brought up in an earlier book edited by us together with Dr. James N. Herron in 1995: *Physical Methods to Characterize Pharmaceutical Proteins* (Plenum, New York, 1995). Since then, ten years have passed and a lot of progress, both in terms of basic insights and in terms of technology, has been made. *Methods for Structural Analysis of Protein Pharmaceuticals* can be considered a follow-up of the 1995 publication. The number of chapters has significantly been expanded and a number of additional techniques have been included.

In this book, the nature of the different analytical techniques and their contribution to define pharmaceutical proteins is described by a group of leading experts. In our 1995 publication, the last paragraph of the preface made the important point that, in general, even in a concerted approach, all the sophisticated techniques described in the different chapters could not fully define the protein structures in their pharmaceutical formulations as can be done with formulations of low-molecular-weight drugs. Ten years later, in spite of all the progress we made, this statement still holds true. The quality of pharmaceutical proteins is based on the combination of a strictly controlled production procedure and specifications based on a set of relevant analytical techniques.

This book deals with analytical techniques that are relevant for characterizing pharmaceutical proteins. Therefore, it should serve the analytical chemist who is responsible for therapeutic protein analysis in the industry. However, the impact of the information goes beyond the analytical chemist. The formulation scientist should also be aware of the principles, the potential, and the limitations of the contents of our analytical toolbox, because the formulation scientist has the responsibility to design a robust formulation of a pharmaceutical protein that can be monitored during production and over (storage) time. Furthermore, our target audience is represented by Ph.D. students, postdoctoral fellows, and research scientists who are working in the academic world and in the pharmaceutical/biotech industry on basic aspects of proteins.

The editors hope that this book will contribute to the rational design of high-quality pharmaceutical protein formulations and to the selection of relevant quality control tests to guarantee product quality when future protein pharmaceuticals are produced and marketed.

Wim Jiskoot and Daan J.A. Crommelin

1

Ultraviolet Absorption Spectroscopy

LISA A. KUELTZO

Department of Pharmaceutical Sciences, University of Colorado Health Sciences Center, Denver, CO

C. RUSSELL MIDDAUGH

Department of Pharmaceutical Chemistry, University of Kansas, Lawrence, KS

History

More than fifty years ago, it was proposed that changes in the ultraviolet absorbance spectrum of a protein could be correlated with its structural perturbations (Beaven and Holiday, 1952). A variety of studies were conducted at that time to explore this potential application, as described in comprehensive reviews by Beaven and Holiday (1952) and Wetlaufer (1962). In the following decades, however, other spectroscopic techniques, including circular dichroism, intrinsic fluorescence, nuclear magnetic resonance (NMR), and Fourier transform infrared (FTIR) spectroscopies, as well as x-ray crystallographic methods, became the methods of choice for characterization of protein structure. Absorbance spectroscopy, although still commonly used in everyday laboratory practices, was primarily relegated to concentration measurements, turbidity assays, and enzymatic activity studies.

At periodic intervals, however, interest in the structural implications of the protein absorption spectrum has been renewed, especially regarding the information contained within the derivative spectrum of the region from 250 to 300 nm—the absorbance region of the three aromatic amino acid residues. It was observed that not only could these signals be employed to quantify the aromatic content of a protein (Balestrieri et al., 1978), but that the difference spectrum (Donovan, 1973), as well as a ratio of the magnitudes of the various derivative peaks

(Ragone et al., 1984), could be used to monitor subtle structural changes, providing information regarding the exposure of the aromatic residues to solvent. Subsequent work has established that more dramatic structural changes (e.g., unfolding) can also easily be detected by these methods. More recently, advances in instrumentation technology, specifically the development of photodiode array detectors, have permitted highly accurate and reproducible measurement of minute shifts in derivative peak positions, which can be directly correlated to small changes in the environment of the aromatic chromophores (Mach et al., 1991b).

Today, interest in ultraviolet absorption spectroscopy as a probe of protein structure is experiencing a minor renaissance. The cost of high-resolution instrumentation has decreased, the availability of photodiode array detectors has expanded the application of absorbance analysis to on-line evaluation of chromatographic separations, and advances in software and algorithm design have increased our ability to obtain the maximum amount of data from spectra. Currently, absorbance spectroscopy offers a fast, nondestructive, high-resolution, and inexpensive alternative to other commonly used spectroscopic techniques, as described in more detail below.

Theoretical Background

Light is traditionally represented as a wave composed of both an electric and a magnetic field component. From a quantum mechanics perspective, light can be described as particles (photons) of energy (E), as defined by Planck's law

$$E = h\nu = hc/\lambda \tag{1}$$

where h is Planck's constant, ν and λ are the frequency and wavelength of the light, respectively, and c is the speed of light. Molecules residing in a ground energy state under normal conditions can be raised to a higher, excited energy state upon interaction with a photon that has an energy equaling the difference between the ground and excited energy states. The molecule can then return to the ground state by emission of the energy by either radiative (as observed in fluorescence and phosphorescence) or nonradiative processes, as seen in electronic absorption. The transition from a ground state to an excited state is customarily defined in terms of the configurations of the molecular orbitals in the ground and excited states; the most commonly observed transitions in protein absorption spectra are the $n \leftrightarrow \pi^*$ and $\pi \leftrightarrow \pi^*$ transitions, where * denotes the excited state.

Each electronic state contains a number of underlying vibrational and rotational energy states. Transition to an excited state can therefore occur at any of a number of potential vibrational energy levels (Figure 1). Collisions between molecules, with solvent molecules under solution conditions, as well as small differences in the absorbance of identical

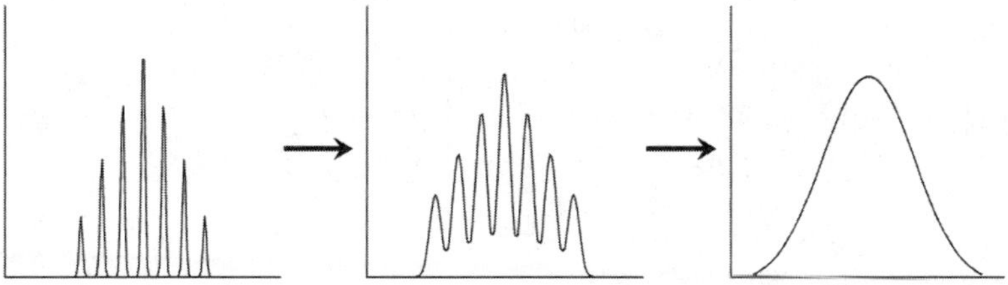

Figure 1 Representation of a vibronic absorption spectrum. The energy of the fine bands is smeared by collisions with other molecules (e.g., solvent molecules), resulting in the traditionally observed broad electronic absorption spectrum.

chromophores under different conditions within the system (e.g., multiple tyrosine residues within a protein), produce a smearing of the energy of the vibronic bands into the broad peak that is normally observed in solution electronic absorption spectra.

Chromophores

The Peptide Group

The most prevalent chromophore in a protein molecule is the amide group of the protein backbone. Its absorbance spectrum contains two main transitions, the first absorbing strongly at 195 nm ($\pi \leftrightarrow \pi^*$) with a second, weaker transition occurring at ~220 nm ($n \leftrightarrow \pi^*$). The peptide group has traditionally been of interest spectroscopically due to the information it can provide regarding the secondary structure content of a protein. Although distinctly different characteristics of the far-UV transition have been observed for model helical, β-sheet, and random coil peptides (Wetlaufer, 1962), interpretation of signals in this region is complicated by the absorbance of common solution components (e.g., inorganic ions and dissolved oxygen) below 200 nm (Wetlaufer, 1962). Additionally, because techniques such as circular dichroism and FTIR spectroscopy possess greater sensitivity to and resolution of secondary structure composition and changes, the application of simple absorption spectroscopic methods to this region has been relatively uninvestigated. Instead, interest has focused on the near-UV region between 250 and 320 nm, which contains the primary absorption from the aromatic amino acid side-chains.

The Aromatic Side-Chains

Tryptophan

The strongest absorbing of the three aromatic amino acids is tryptophan (ε_{280nm} = 5540). As shown in Figure 2, the maximum absorbance of tryptophan (Trp) occurs at ~280 nm, with a less-intense transition observed as a shoulder near 292 nm (both corresponding to $\pi \leftrightarrow \pi^*$

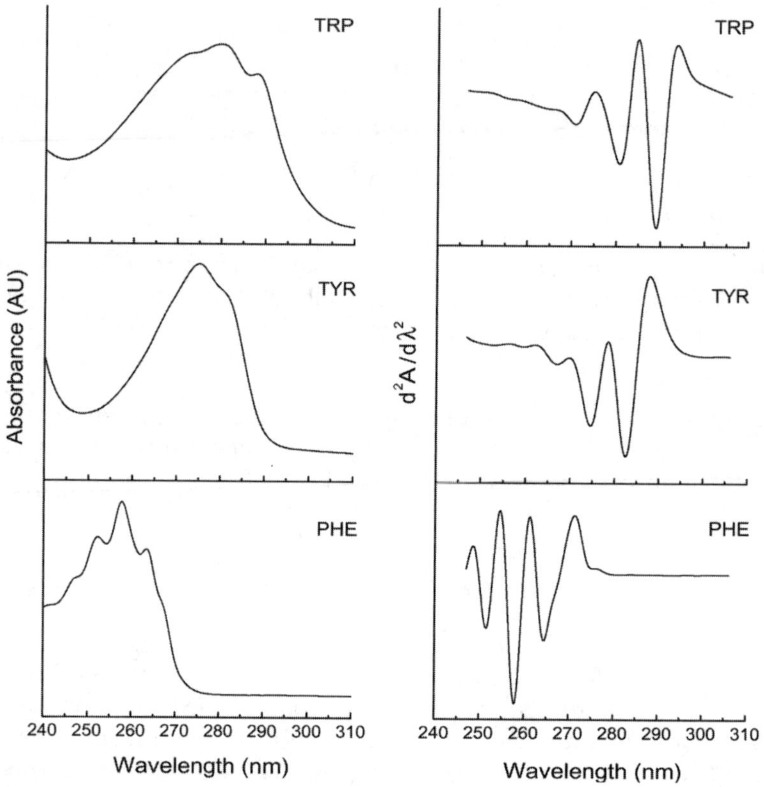

Figure 2 Zero-order and second-derivative absorption spectra of *N*-acetyl-X-ethyl ester derivatives of tryptophan (TRP), tyrosine (TYR), and phenylalanine (PHE). The ordinate axis is presented in arbitrary units.

transitions). The latter is responsible for the shoulder frequently seen in the protein spectrum. Although the fine structure is not well defined in the zero-order spectrum, the derivative shows three distinct minima. Covalent modification of tryptophan residues is occasionally observed, primarily reflecting oxidation of the indole ring (Zhang et al., 2003). Additionally, tryptophan can be modified for investigational purposes (e.g., *N*-bromosuccinimide–induced oxidation) (Wetlaufer, 1962); such modifications change the absorbance properties of the chromophore, and their potential presence always needs to be considered.

Tyrosine

The absorption of tyrosine (Tyr), although less intense (ε_{280} = 1480), overlaps significantly with that of tryptophan, with a maximum absorbance occurring at ~276 nm ($\pi \leftrightarrow \pi^*$). The zero-order spectrum consists of a broad absorbance band with two small shoulders, reflecting the fine structure that is observed once again in the multiple peaks of the derivative spectrum (Figure 2). Of the aromatic amino acids, tyrosine is the most susceptible to structural modifications including nitration, iodination, phosphorylation, and acetylation, as well as ionization of the tyrosine hydroxyl group (see "Effect of pH," below; Creighton, 1993). Such modifications can significantly alter both the intensity and the position of the absorption transitions (Table 1) and complicate spectral interpretation.

Phenylalanine

In contrast to the previous two aromatic side-chains, phenylalanine (Phe) absorbs only weakly (ε_{258nm} = 197) in the 240- to 270-nm region ($\pi \leftrightarrow \pi^*$). Significant fine structure is observed in both the zero order and derivative spectra (Figure 2), with an average of five peaks observable in various environments. In zero-order protein absorption spectra, this fine structure is usually visible only as subtle inflection points (bumps), due to the dominance of the spectrum by the more intense signals of Tyr and Trp.

The Disulfide and Others

Although cysteine does not absorb significantly in the 250- to 300-nm wavelength region, the disulfide-bridged form of the residue, cystine, does show some weak absorbance. The preponderance of disulfide bridges in some proteins can result in a significant contribution to the overall extinction coefficient (see "Protein Concentration Determination and Extinction Coefficients," below). Although methionine and histidine also show some weak absorbance, these residues absorb in the same region as the peptide bond (185 to 220 nm) (Wetlaufer, 1962) and therefore are not usually resolvable in the absorption spectrum of most proteins.

Extrinsic Chromophores

In addition to the intrinsic protein chromophores described above, the absorption spectrum of a protein may include absorbance peaks from a wide variety of other species. Although the absorbance of these extrinsic chromophores is most commonly monitored in the visible wavelength region, in some cases strong and potentially useful signals in the UV or near-UV region are also present. These additional spectral components are sensitive to both the environment as well as the oxidation state of the cofactor and thus can often provide useful information about overall protein conformation, characteristics of the ligand binding site, complex stoichiometry, and reaction kinetics (e.g., cofactor conversion).

The most common extrinsic chromophores are divalent metal cations. Although some cations, such as zinc and magnesium, are optically transparent, others, such as iron, copper, or non-native metals such as cobalt, cadmium, and terbium, possess intense absorbance

Chromophore	λ_{max} 1	λ_{max} 2	λ_{max} 3	λ_{max} 4	ε (λ_{max} 2)
Amide bond	195[a]	220[a]	—	—	—
Tryptophan (in proteins)	—	—	—	—	5540[e]
N-Ac-L-Tryptophan-NH_2	219[b]	280[c]	288[d]	—	5390[f]
Tyrosine (in proteins)	—	—	—	—	1480 (at 280 nm)[e]
N-Ac-L-Tyrosine-NH_2	224[b]	275[d]	[280][f]	—	1390[f]
∈-Iodotyrosine	—	283[c]	—	—	2750[c]
∈-Nitrotyrosine	—	360[c]	—	—	2790[c]
∈-Aminotyrosine	—	275[c]	—	—	1600[c]
O-Acetyltyrosine	—	363[c]	—	—	262[c]
N-Ac-L-Phenylalanine-O-Me	208[b]	252[b]	258[b]	264[b]	197 (at 258 nm)[b]
Cystine	—	[250][b]	[260][b]	[280][b]	134 (at 280 nm)[e]

Brackets denote points of inflection.

[a]van Holde et al. (1998); [b]Fasman (1992); [c]Creighton (1993); [d]Beaven and Holiday (1952); [e]Mach et al. (1992); [f]Mach et al. (1995).

Table 1 Absorption Characteristics of Selected Intrinsic Protein Chromophores

bands (Vergani et al., 2003). These signals can often provide information about the metal binding site, such as binding capacity (Vergani et al., 2003), oxidation state (Favilla et al., 2002), identity of binding site residues, and coordination geometry (Burke et al., 1992). In addition to divalent cations, a variety of enzyme cofactors, such as NADH (Luong and Kirsch, 1997; Piersma et al., 1998) and flavin adenine dinucleotide (FAD; Gadda and Fitzpatrick, 1998), exhibit intense absorbance signals in the near-UV and visible ranges; changes in these signals with oxidation state are often used in the analysis of enzyme kinetics. In a very familiar example, the ferroprotoporphyrin, or heme, group possesses a strong absorbance signal that is frequently monitored in structural studies of heme-containing proteins (e.g., myoglobins, hemoglobins, and various cytochromes), as its absorbance characteristics are often perturbed during functional structural alterations of these proteins.

Instrumentation

A variety of UV-visible spectrophotometers are commercially available, ranging widely in both cost and capability. Regardless of design, however, each spectrophotometer can be broken down into four basic components: the light source, sample compartment, light dispersion device, and the detector. An ideal light source should possess low noise, good intensity over the desired wavelength range, and stability over an extended period of time (Owen, 1996). Most current UV-visible spectrophotometers employ a combination of two sources: a deuterium lamp, possessing good intensity over UV wavelengths, and a tungsten-halogen lamp for the visible region. The instrument then either switches between the lamps during spectral collection, or the two signals are combined into one broad-range light source, depending on the design. Broad-range xenon lamps are also available but are in general a less desirable choice due to their significantly higher noise levels compared to the deuterium-tungsten-halogen combination.

The spectrophotometer sample compartment typically contains one or more aligned cuvette holders. In choosing an instrument, the availability of sample accessories, which expand the experimental capabilities of the instrument, should also be considered. Many systems today offer optional temperature-controlled sample holders (either Peltier-based or water jacketed), multi-cell devices, and stopped-flow and titration accessories. Additionally, some systems are based on a double-beam design, in which the sample and reference scans are collected concurrently from two independent, matched sample cells. Although this design was considered optimal in the past, as it minimizes errors produced by fluctuations in lamp intensity from the time difference between reference and sample spectra collection, the stability of current light sources has made single-beam instruments a viable option.

Scanning Monochromator Based

One of the most significant differences among current spectrophotometer designs is the nature of the light dispersion/detector system. This can be especially crucial in high-resolution derivative absorbance studies (discussed in "High-Resolution Derivative Analysis," below). In a monochromator-based system, the light from the source enters the monochromator through an entrance slit that focuses it on a dispersion device, most commonly a holographic grating, although prisms are still occasionally employed. A narrow band of the dispersed light is then selected by the exit slit of the monochromator and passed through the sample to the detector, most commonly a photomultiplier tube (PMT). In scanning instruments, the grating (or prism) physically moves to produce the desired wavelength range; although these instruments do produce high-quality spectra, the movement of the monochromator can introduce some error into the wavelength reproducibility of repeated measurements.

Diode Array Based

In a diode array instrument, the entire spectrum of the light source is focused on the sample. The light transmitted through the sample passes into the entrance slit of the polychromator and through the dispersion element to the detector. The detector consists of an array of photodiodes set at regular intervals (typically 0.5- to 8-nm apart) that can simultaneously detect the full spectrum of light transmitted (Owen, 1996). Although the resolution of the collected spectrum is limited by the spacing of the photodiodes, simple interpolation algorithms, often included in the spectrophotometer operating software, can mathematically increase the effective resolution of spectra down to 0.01 nm. Offsetting the requirement for interpolation is the very high wavelength reproducibility of the collected data, as the detectors in these systems contain no moving parts.

Data Analysis

Zero-Order Spectra

The raw data collected with a spectrophotometer is referred to as the "zero-order" spectrum. In the case of a protein with intrinsic chromophores, this typically consists of a strong absorbance from the peptide bond at ~200 nm and a weaker, rather broad absorbance between 250 and 300 nm. The weaker absorbance is the additive signal of the overlapping phenylalanine, tyrosine, and tryptophan residues, as described above. The zero-order spectrum has commonly been used for concentration determination, examination of gross spectral characteristics, and detection of major structural changes such as extensive unfolding. Although many of the applications described in this chapter focus on the derivative form of the spectrum, the zero-order spectrum should always be examined before performing other forms of data analysis. This examination provides an estimate of the overall quality of the sample with regard to uniformity compared to previous samples, impurities, as well as the presence of particulates that can result in light scattering, which can distort the pure absorptive spectrum.

Light-Scattering Correction

Interpretation of zero-order absorbance spectra can be significantly complicated by the presence of light scattering. Solute molecules whose size begins to approach 1/20 to 1/50 of the wavelength of the incident light (e.g., protein multimers, aggregates) will elastically scatter the light, preventing it from reaching the detector and artificially inflating absorbance values. The scattering effect can be defined (Timasheff, 1966) as

$$\ln(\text{OD}) = a \ln \lambda + b \qquad (2)$$

When scattering is present, the raw spectrum should be represented as "optical density" (OD) on the ordinate rather than absorbance, as both absorbance and scattering (turbidity) components are included. To obtain an accurate measurement of sample absorbance, the light-scattering contribution, if any, must be eliminated.

The intensity of scattered light is approximately proportional to the inverse of the fourth power of the wavelength of the light in the Rayleigh regime (λ less than 1/20 incident light). Scattering is usually observed in absorption spectra as a negative slope in a region of the spectrum where the sample does not absorb light (greater than 320 nm in most protein samples), as shown in Figure 3. Once detected, scattering can be eliminated by either physical (e.g., removing the larger material by centrifugation or filtration) or mathematical methods. The scattering component within the absorbing region of the spectrum can be extrapolated mathematically through the absorbing region by fitting the nonabsorbing region of the spectrum (e.g., 320 to 500 nm) to Equation 2 and then subtracting the scattering signal to isolate the pure absorbance signal (Figure 3). Most commercial spectrophotometer software packages now include this type of correction in their basic analysis package. Alternatively, the OD contribution at any wavelength can be calculated from the OD values at 320 and 350 nm as follows (Mach and Middaugh, 1993):

$$\text{OD}_{(\lambda)} = 10^{(m+1) * \log(\text{OD } 320) - m \log(\text{OD } 350)} \qquad (3)$$

where $m = 64.32 - 25.67 * \log \lambda$ ($m = 1.5$ at 280 nm). The absorbance component at the desired wavelength can then be calculated. It is important to remember that these corrections take into account only

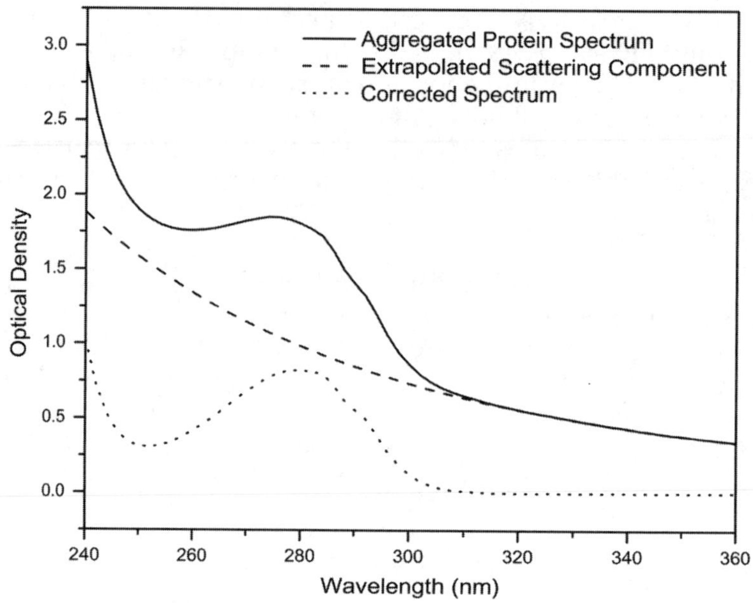

Figure 3 Light-scattering correction of an aggregated protein sample. The light-scattering component is extrapolated through the protein absorbing region (less than 300 nm) and subtracted to obtain the corrected zero-order spectrum.

absorbance and Rayleigh light-scattering components. As particles become larger, absorption flattening, a phenomenon where the chromophores within a particle obscure one another, may also occur (see "Spectral Artifacts and Complications," below) and is not correctable by these methods. Furthermore, correction of scattering by larger particles (e.g., Mie and multiple scattering) requires much more complex methods and is not generally feasible.

Derivative Analysis

Aromatic Amino Acid Quantitation

One of the first applications of derivative absorbance analysis to proteins investigated the quantitation of the aromatic amino acid content, although advances in protein and gene sequencing techniques have made this approach less common in current practices. Calculation of the first through fourth derivatives of a protein's absorption spectrum resolves the overlapping absorbance contributions of Phe, Tyr, and Trp, permitting selective analysis of the individual residues. Balestrieri and colleagues (1978) demonstrated that, once resolved, the intensities of the derivative peaks can be directly correlated to the number of Phe and Trp residues present in a protein. This approach was expanded by Servillo and colleagues (1982), who determined the relationship between the derivative peak intensities and the Tyr/Trp ratio in proteins as follows:

$$r = (Ax + B)/(Cx + 1) \quad (4)$$

where r is the ratio of the peak-to-trough distances between ~283–287 nm and ~290–295 nm, and x is the Tyr/Trp molar ratio. The coefficients A, B, and C are estimated as 0.21, 0.66, and –0.08, respectively, although these values are potentially sensitive to both Tyr/Trp ratio and solvent polarity. When the ratio (x) is calculated during titration of protein samples with model Trp or Tyr containing compounds, the absolute concentrations of these residues can be calculated (Servillo et al., 1982).

High-Resolution Derivative Analysis

It was observed early in the development of protein absorption spectroscopy that the absorption spectrum is sensitive to environmental conditions and as such can be used as a probe of protein structural changes (Beaven and Holiday, 1952). The low resolution of early data only permitted measurement of relatively large shifts (greater than or equal to 1 nm) in the wavelengths of derivative peaks (Ichikawa and Terada, 1979), leading to the more widespread application of difference spectroscopy (discussed below) to study changes in conformation. Advances in instrumentation in the early 1990s (specifically the availability of the diode array detector), however, permitted development of a new approach to determine minute shifts in derivative peak positions (~0.02 nm) with high reproducibility.

In high-resolution derivative analysis, as originally described by Mach and colleagues (1991b), a very high signal-to-noise ratio is required for the absorbance spectrum. The optimal method for collection of high-resolution spectra is to average the signal over an extended length of time (e.g., 30 seconds, compared to a standard collection interval of 0.5 seconds in some instruments), using a diode array–equipped spectrophotometer. This eliminates both transient noise in the spectrum as well as slight but significant differences in absolute peak position that can be introduced with most grating-based detector configurations. Most commonly, the second or fourth derivative of the spectrum is numerically calculated and fitted to a polynomial function. The data in the resulting curve is then interpolated, typically with a final data spacing of 0.01 nm, from which precise peak positions are determined. Although the exact mathematical procedure can differ depending on the statistical analysis software employed, resulting in small differences in calculated spectra, as long as a single program and method is used consistently within a study, most commercially available programs are adequate for this type of analysis. Typical error values for peak measurements using this approach are between 0.01 and 0.03 nm ($n = 3$).

Model studies have shown that shifts in derivative peak positions primarily arise from changes in the polarity of the local environment of the chromophores, with a shift to shorter wavelengths indicating an increase in environment polarity (Mach et al., 1991b; Mach and Middaugh, 1994). Total shifts as small as 0.5 nm and as large as 6.0 nm have been observed for transitions from folded to unfolded protein conformations (Mach et al., 1991a; Lange and Balny, 2002; Kueltzo et al., 2003). Both the intensity as well as the direction of shifts can vary significantly, both between different types of chromophores within a protein as well as among the multiple derivative peaks of an individual chromophore. This illustrates the utility of this particular approach; a typical protein absorption spectrum will provide anywhere between five and eight discrete peaks for detection and interpretation of changes in protein conformation. Because these arise from residues typically dispersed throughout the structure of a protein, they are potentially quite sensitive to a wide variety of protein structural alterations, some quite subtle. Additionally, because calculation of the derivative is not sensitive to broad spectral components, light-scattering effects do not dramatically affect peak position determination (Mach and Middaugh, 1993). Thus, data on aggregated protein systems can also be obtained, at least up to the point of precipitation, although unambiguously interpreting data from such heterogeneous samples can be quite difficult.

The a/b Ratio

The previous discussion concentrates primarily on derivative peak position; peak intensity, however, can also provide useful information about protein conformation. Ragone et al. (1984) proposed a method for determining the exposure of tyrosine residues to bulk solvent based on derivative intensity. Defining the vertical distance between the minimum at ~283 nm and the maximum at ~287 nm as *a* and the vertical distance between the minimum and maximum at ~290 and 295 nm as *b* (Figure 4), the ratio (r) is defined as a/b. (Note this is the same approach employed by this group for the quantitation of Tyr and Trp residues discussed above.) Initial model studies demonstrated that r is relatively insensitive to changes in solvent polarity for Trp models, with an average value of ~0.68. Tyrosine absorption, conversely, appears more sensitive to its environment, producing significant changes in the r value of Tyr/Trp mixtures with changing solvent polarity.

The r values for natively folded (r_n) and unfolded (r_u) proteins have been observed to vary widely from protein to protein (r_n: 0.5–1.5; r_u: 0.7–2.3; Ragone et al., 1984). Changes in r during unfolding transitions can also vary significantly in magnitude (0.03–1.4). These changes depend significantly on both the initial degree of tyrosine exposure, as well as the Tyr/Trp ratio of the protein; no direct correlation can be made between the r_n value of different proteins and the relative

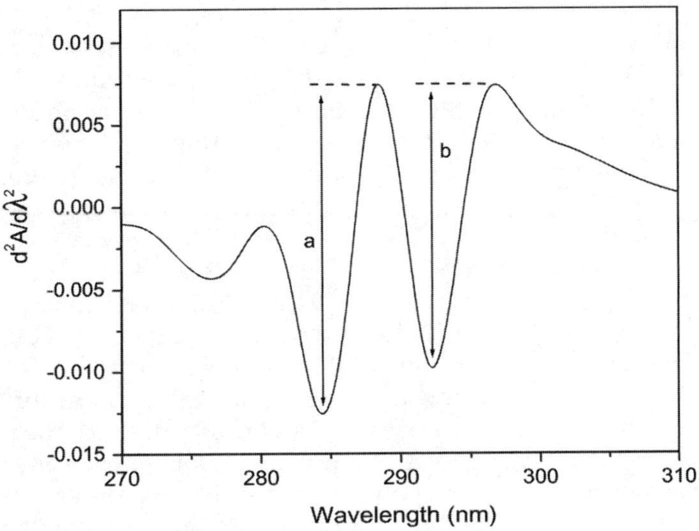

Figure 4 The *a/b* ratio. The ratio (*r*) is calculated from the vertical distances between the primary tyrosine and tryptophan peaks.

exposure of their Tyr residues. The degree of tyrosine exposure (α), however, can be calculated as follows (Ragone et al., 1984):

$$\alpha = (r_n - r_a)/(r_u - r_a) \qquad (5)$$

where r_a is the *a/b* value for a mixture of model amino acids (e.g., N-Ac-Trp-NH_2) in the same ratio as the protein in a solvent of similar polarity to the protein interior (e.g., 20% v/v ethylene glycol). Since its proposal, the *a/b* ratio method has extensively been used to measure tyrosine exposure, providing an indicator of protein conformation change and ligand binding in a variety of proteins including cytochrome c (Fisher and Sligar, 1985), myoglobin (Ragone et al., 1987), and glucose/xylose isomerase (Pawar and Deshpande, 2000).

Difference Spectroscopy

Ultraviolet difference spectroscopy was one of the first absorbance-based techniques developed to detect changes in protein conformation (Wetlaufer, 1962). Difference spectra are generated through the subtraction of a reference spectrum from a sample spectrum, where both the sample and reference solutions contain protein under different conditions (Donovan, 1973). Sample preparation for difference spectroscopy requires that the concentration of the sample and reference solutions are equivalent; if the differences between the spectra are small, higher concentrations may also be required, a potential disadvantage in comparison to other techniques. It is possible to use dual-chambered cells to facilitate solvent perturbation difference spectroscopic studies (see "Effect of Solvents," below). It should be noted that although double-beam instruments were initially preferred for the generation of difference spectra to minimize artifactual differences between spectra resulting from lamp intensity fluctuations and lapses of time between sample collection, current single-beam instruments are both stable and fast enough to eliminate these concerns. Difference spectra in both zero-order (Bailey et al., 1968, Mani et al., 2001) and derivative format ("difference derivative spectra"; Kornblatt et al., 1999) have been calculated.

Although the technique has been used in the investigation of protein conformation (Ragone et al., 1987), ligand binding (Mani et al., 2001), and chromophore microenvironment (Herskovits and Laskowski, 1962), interpretation of difference spectra can be complex. Because difference spectra combine changes in both peak position and intensity, the full implications of multiple changes can be difficult to define. The general rule of thumb, however, is that a negative value in the zero-order difference spectrum indicates a shift of the peak maximum to shorter wavelengths or an increase in the polarity of the chromophore microenvironment, whereas a positive value suggests a shift to longer wavelengths ("red-shift"). Although the spectral region between 250 and 300 nm is most commonly studied with this approach, the tail of the peptide bond absorption peak at 220 nm has also been examined (Bailey et al., 1968). The degree of shift observed in difference spectra has also frequently been correlated to the degree of exposure of the aromatic residues. This application, solvent perturbation spectroscopy, is discussed in detail below (see "Effect of Solvents").

Analysis of Proteins and Protein Structural Changes

Effect of Temperature

A general effect of temperature on absorbance spectra is the sharpening of fine structure bands with decreasing temperature (Wetlaufer, 1962). There have also been observations of the development of new fine structure bands in some chromophore models with changing temperature, for example a new band at 296 nm in protein absorbance spectra at low temperatures (Wetlaufer, 1962). Additionally, the authors have observed new high-energy peaks appearing in the derivative spectra of model amino acids and peptides at higher temperatures (greater than 60°C; Kueltzo et al., 2003). These effects can complicate interpretation of both difference spectra and peak position analyses, as the loss or addition of fine structure components can significantly alter the shape and position of adjacent transitions.

It has also been observed that the position of the absorption bands of the aromatic chromophores changes intrinsically with temperature (Kueltzo et al., 2003). Studies of Phe, Tyr, and Trp derivatives show a curvi-linear increase in peak position with increasing temperatures, with changes as great as 2 nm observed over an 80°C temperature range. Additional studies suggest that this change primarily results from changes in the dielectric constant of the bulk solvent with temperature. As a result, care must be taken in interpreting the conformational significance of continuous (nonabrupt) changes in peak position during temperature unfolding studies.

Effect of pH

The majority of the effects of pH on protein absorption spectra are reflections of pH-induced conformational changes. The exception to this is the ionization of the tyrosine residue, the hydroxyl group of which has a pKa of ~10 (Zubay, 1993). Ionization of this group results in significant changes in the phenolic absorption spectrum, including a redshift of the major absorbance peaks from 222 and 275 nm to 242 and 295 nm, respectively, as well as a significant hyperchromic effect with both peaks (Balestrieri et al., 1980). The fine structure normally observed at longer wavelengths in the derivative spectrum is essentially lost and new fine structure is observed at shorter wavelengths (225 to 260 nm).

Changes in tyrosine absorption spectra can be followed as a function of pH, permitting calculation of the number of titratable tyrosyl groups in the protein (Wetlaufer, 1962). This quantitation may be complicated by changes in the absorbance spectrum in the 295-nm region due to conformational changes at the high pH values required for complete ionization (12.5); if an alternate signal is followed (i.e., the positive peak present in the phenolate absorption spectrum at 260 nm), corrections must be made for contributions from phenylalanine residues as well as ionized thiol groups. In addition to the determination of titratable tyrosine groups, titration experiments have also been used to determine the pKa of individual tyrosine residues in combination with mutation studies (Li et al., 1993), as well as to follow protein unfolding under basic conditions. In the latter case, as the protein unfolds, buried tyrosine residues are exposed to the solvent and ionize, further altering the absorption spectrum (Melo et al., 1997).

Effect of Solvents

It is reasonable to assume that any change in solvent may affect the absorption spectrum of a protein as a direct result of changes in the local environment of the chromophores and/or consequent conformational changes. Early efforts in UV absorptive difference spectroscopy specifically focused on changes in the polarity of the chromophore microenvironment in the absence of conformational change. This approach, initially described by Herskovits and Laskowski (1962) and termed *solvent perturbation spectroscopy*, investigates surface exposed or partially exposed chromophores and their response to the bulk solvent by using mild concentrations of nonaqueous organic solvents or solutes as perturbants.

A typical solvent perturbation experiment involves titration of the protein with up to 20% ethylene glycol or sucrose (Herskovits and Laskowski, 1962; Kosen et al., 1980); both are generally considered (perhaps erroneously) to have little effect on overall protein conformation. The shift in the protein spectrum in the presence of the perturbant is represented as the fraction of the total shift obtained in a completely unfolded protein, in which all of the chromophores are exposed to solvent. The difference spectrum for the unfolded protein is normally obtained through chemical unfolding using agents such as urea or guanidine hydrochloride and disulfide reduction. The choice of perturbing solvent is especially important in this application. Not only must the protein conformation be relatively unaffected by the solvent, but the size of the solvent/solute molecule, and thus its ability to diffuse through the protein matrix, will determine the fraction of the chromophores that are altered (i.e., partially buried as well as fully exposed chromophores).

Cation-π Interactions as a Probe of Protein Dynamics

One powerful use of fluorescence spectroscopy involves the quenching of the fluorescence emission of protein aromatic side-chains (see Chapter 2). In brief, this technique probes protein structural dynamics through the interaction of small neutral and charged compounds (e.g., acrylamide, iodide) with fluorescent side-chains, either through diffusion-limited collision with side-chains or by the formation of nonemittive complexes. Similarly, we have recently found that small cations such as Na^+, Cs^+, and Li^+ can also form complexes with the π-electrons of the three aromatic residues. This causes small shifts in the positions of the absorption peaks of Phe, Tyr, and Trp. Using model compounds, the effect of cation concentration on these shifts has been characterized. When similar experiments are conducted with proteins, it is found that the larger cations (as measured by their dehydrated radii) are less effective in inducing these shifts as side-chains are increasingly buried. Unlike fluorescence quenching, all three aromatic residues can be studied by this method using the high-resolution derivative technique described above. Details of this new technology will be described elsewhere (Ersoy et al., manuscript in preparation).

Empirical Protein Phase Diagrams

The availability of five to seven distinct derivative peaks in the near-UV absorption spectra of proteins has recently provided a powerful but simple new approach to the characterization of their solution behavior. Because the three aromatic residues are dispersed through the three-dimensional structure of most proteins, they potentially provide a series of signals that should be sensitive to a wide variety of both subtle and major structural transitions. In this method, measurements are typically made as a function of solution variables such as temperature, pH, ionic strength, and redox potential. The positions of the five to seven derivative peaks under the different solution conditions are represented as components in a multidimensional vector space. The exact details of this mathematical procedure are provided elsewhere (Kueltzo et al., 2003). In brief, the multidimensional representation can be simplified into a color map, with the "state" or phase of the protein (in this example a dominant structural conformation) represented as a uniform region (Figure 5). This approach provides a rapid method with which one can identify phase boundaries where both physical and/or chemical pathways of degradation are occurring. This information can then be used to develop accelerated stability studies that can, in turn, be used to identify potential stabilizers by high-throughput screening studies. We have found this approach to be both rapid and efficient, and it has successfully been used in collaboration with a number of industrial partners to identify stabilizers of both therapeutic protein and vaccine candidates.

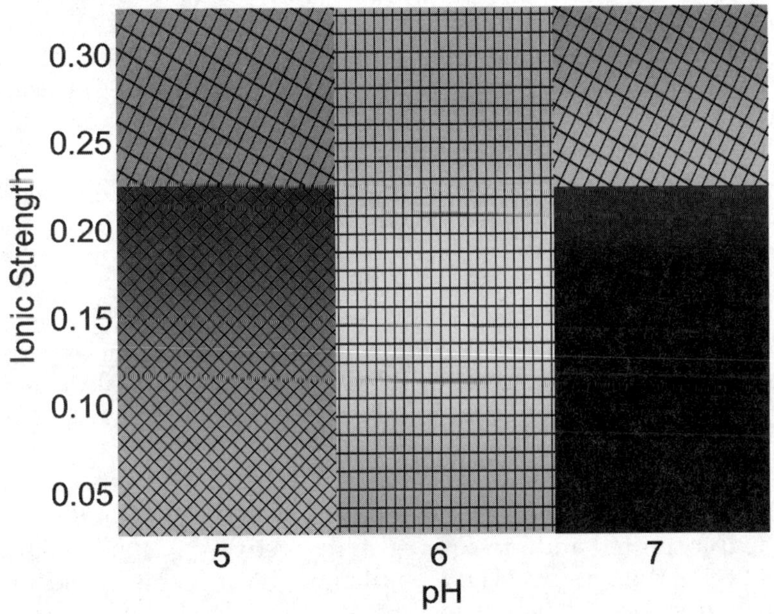

Figure 5 Empirical phase diagram generated from high-resolution derivative absorbance data. Regions of similar pattern denote continuous conformational phases, with shading indicating subtleties within the phase.

Kinetics

All of the approaches described in this chapter are amenable to kinetic use. One of the more common applications of absorbance spectroscopy (actually optical density) to kinetic experiments is the study of protein aggregation. Although, as discussed above, light scattering can be a significant problem in interpreting protein absorbance spectra, it can also be useful as a probe of the protein association state (e.g., oligomerization, aggregation). Many proteins adopt transient aggregation-prone states when undergoing conformational changes, resulting in an increase in overall sample turbidity. The kinetics of such processes can be monitored by following the change in OD in a non-absorbing region of the spectrum. Because Rayleigh scattering is inversely proportional to a high power of the wavelength (i.e., third or fourth power), the strength of the scattered signal greatly increases as wavelength decreases (Figure 3). Thus, measurements are often made in the near-UV (320 to 400 nm) region. The caveat with this method is that little information about the nature of the associated states responsible for the scatter is obtained. This approach has successfully been used in both the detection of protein aggregation events (Kueltzo and Middaugh, 2003) and screening for aggregation inhibiting and/or exacerbating ligands (Volkin et al., 1993; Edwards et al., 2001; Maclean et al., 2002). It is important to note, however, that whether monitoring OD changes or any spectral signal as a function of time, photochemical alteration of the compound can occur. This is especially crucial in longer kinetic studies where the sample is exposed to the light source of the instrument for an extended period of time during spectral collection (Kueltzo et al., 2003). Photochemical modification of the indole group of Trp is especially well-known.

Practical Considerations

Protein Concentration Determination and Extinction Coefficients

The relationship between the intensity of an electronic absorption transition and the concentration of the absorbing sample is simply stated by the well-known Beer-Lambert law

$$A_{(\lambda)} = \log(I_o/I) = \varepsilon(\lambda)cl \tag{6}$$

where A is the absorbance intensity at a specific wavelength, I and I_o are the transmitted and incident light, respectively, c is the molar concentration of the sample, l is the pathlength in centimeters, and ε is the molar extinction coefficient at that wavelength. It should be noted here that extinction coefficients are also commonly reported in the literature as absorbance values for a 1% (w/v) ($A^{1\%}$) or a 1 mg/mL solution ($A^{1mg/mL}$) with a 1-cm pathlength. As a general rule of thumb, the absorbance of a 1 mg/mL protein solution is often near 1 (±0.5). Of course, the actual value varies greatly from protein to protein and is an intrinsic property of each individual protein. To obtain an accurate concentration by absorbance methods, the extinction coefficient must be known. An experimental extinction coefficient can be obtained by extrapolation from alternate concentration determination methods, such as dry weight measurement (Wetlaufer, 1962), amino acid analysis, or colorimetric techniques (Lowry et al., 1951; Bradford, 1976). Alternatively, a surprisingly accurate estimate of the molar extinction coefficient can be calculated from the known protein sequence using the following equation, which is derived from intrinsic extinction coefficients of tryptophan, tyrosine, and cysteine residues obtained from a large experimental dataset of proteins (Mach et al., 1992):

$$\varepsilon_{280} = 5540 n_{\mathrm{Trp}} + 1480 n_{\mathrm{Tyr}} + 134 n_{s\text{-}s} \qquad (7)$$

Here, n is the number of residues or disulfide bonds. This equation applies to proteins in the folded, globular state, but similar analysis has been conducted for unfolded proteins (Gill and von Hippel, 1989), which has the advantage that conformational effects on the extinction coefficients of the individual side-chains are minimized.

Concentration determination using absorbance measurement is the most direct, accurate, and rapid method available at this time, especially compared to colorimetric methods, in which variation in dye binding between the protein sample and the standard can result in a significant deviation from the true concentration (Sapan et al., 1999). To obtain concentrations using absorbance values, however, the sample must be free from both contaminating species as well as light scattering. If a contaminating species absorbs in the same wavelength region as the aromatic amino acids, the two signals must be resolved before calculation. This can be accomplished by using derivative absorbance methods if a derivative peak can be confidently attributed to the protein alone. The concentration can then be extrapolated from a standard curve of derivative intensity at that wavelength *versus* the intensity of the corresponding zero-order spectrum. Alternatively, various other methods of multicomponent analysis are available for this purpose. If particulates are present in the sample, the contribution of light scattering to the signal must be subtracted, preferably by centrifugation or filtration of the sample or, if that is not possible, by the calculation methods outlined above (see "Light-Scattering Correction").

Use in Preformulation/Formulation Studies

The speed and nondestructive nature of absorption spectroscopic techniques coupled with the availability of multiple sample configurations and automated collection software in current instrumentation make absorption spectroscopy a desirable choice for preformulation screening and subsequent formulation optimization. In a typical high-resolution derivative analysis experiment, data on protein tertiary structure (derivative peak positions), tyrosine exposure (the *a/b* ratio), as well as aggregation propensity (OD at 350 nm in the zero-order spectrum) are all collected simultaneously. A series of such experiments as a function of preformulation parameters such as pH, ionic strength, buffer composition, and temperature can be completed in a relatively short time (one to three weeks, depending on the number of parameters examined), permitting selection of a smaller number of promising formulation combinations. As indicated previously, a convenient way to display such data is in the form of an empirical phase diagram (Kueltzo et al., 2003). Absorbance spectroscopy is of further use in formulation and production stages for both stability screening as well as an efficient control for batch-to-batch uniformity using the derivative footprint of the protein.

Use at Very High and Very Low Concentrations

When planning an absorbance spectroscopy experiment, a crucial parameter is protein concentration. A critical factor in determining the concentration of the protein solution is the linearity of the instrument (i.e., the concentration range over which Beer's law is valid). Most current instrumentation has a linear range between 0.05 and 2 absorbance units, although this can vary significantly. Even when Beer's law is obeyed for extreme concentrations, however, there are additional concerns. At very high concentrations, not only is the propensity for aggregation and thus light-scattering interference exacerbated, but absorption flattening may occur (see below), which will also alter the spectral signal. At very low concentrations, the signal may be altered by disproportionate adsorption of the sample to the cuvette. Normally, a small percentage of the sample can be found on the sides of the cuvette; under normal conditions, this does not interfere with the overall sample collection, as the majority of the signal arises from the solution portion of the sample. As concentration decreases, however, the relative percentage of adsorbed protein increases and can dominate the spectrum, resulting in a spectrum that is not representative of the protein in solution. For example, adsorption to the sides of the cuvette may well alter the structure of the protein, producing changes in the extinction coefficient, among other phenomena. Spectra at high concentrations (e.g., greater than 100 mg/mL) can still be obtained by employing sample cells of very short pathlengths. For example, we have been able to obtain excellent spectra of a variety of proteins at concentrations up to 300 mg/mL by using self-constructed cells of 1-μm pathlength. Such cells are easily made using commercially available quartz plates and spacer materials commonly used for infrared spectroscopy.

Spectral Artifacts and Complications

There are a number of spectral artifacts observed in absorbance spectroscopy in addition to those mentioned previously. A particularly important one is the phenomenon of absorption flattening. Absorption flattening can occur in samples at very high concentrations or that are extensively aggregated. In this situation, the chromophores of the protein molecules begin to obscure each other as a result of discontinuous regions of ultrahigh local concentration, preventing an equal amount of light from interacting with each chromophore (Duysens, 1956). This produces both a redshift and a decrease in absorbance intensity in the resulting spectrum, which can be interpreted incorrectly as alterations in the chromophore microenvironment.

Regardless of which of the analytical methods described in this chapter are used, additional solution components that absorb strongly in the 200- to 350-nm region can severely complicate data collection and interpretation. Although calculation of the derivative form of the protein spectrum can eliminate broad absorbance signals and may resolve quite detailed protein signals, high concentrations of additive may still dominate a portion of the absorbance signal, obscuring the protein absorption spectrum. For example, the pharmaceutical preservative benzyl alcohol, at approved bacteriostatic concentrations (0.9% w/v), has been found to obscure the entire protein absorption spectrum, with the exception of the lowest energy Trp derivative minima at 292 nm (Carpenter et al., personal communication). Thus, secondary component absorbance properties and concentrations must carefully be considered in experimental design.

Conclusions

Absorbance spectroscopy is a straightforward, nondestructive, and efficient method for the investigation of protein conformation and stability. Despite being previously eclipsed by other spectroscopic methods, a number of effective applications have been developed in recent decades to investigate both gross protein conformational changes as well as changes in chromophore microenvironments. The inclusion of all three intrinsic chromophores in the data provides a more global picture of protein tertiary structure in comparison to steady-state fluorescence methods, and a wealth of data is obtained from a single spectrum. A number of aspects of the protein absorbance spectrum, however, still remain poorly understood, such as the physical implications of phenylalanine intensity changes and differences in the response of vibrational fine structure within a chromophore to conformational events. Future efforts should continue to expand the utility of this approach to protein conformational studies.

Abbreviations

E	Energy
ε	Molar extinction coefficient
FAD	Flavin adenine dinucleotide
FTIR	Fourier transform infrared
h	Planck's constant
λ	Wavelength
NADH	Nicotinamide adenine dinucleotide
NMR	Nuclear magnetic resonance
OD	Optical density
Phe	Phenylalanine
PMT	Photomultiplier tube
Trp	Tryptophan
Tyr	Tyrosine
UV	Ultraviolet
ν	Frequency

References

1. Bailey JE, Beaven GH, Chignell DA, and Gratzer WB. An Analysis of Perturbations in the Ultraviolet Absorption Spectra of Proteins and Model Compounds. *Eur J Biochem* 1968; 7:8–14
2. Balestrieri C, Colonna G, Giovane A, Irace G, and Servillo L. Second-Derivative Spectroscopy of Proteins. A Method for the Quantitative Determination of Aromatic Amino Acids in Proteins. *Eur J Biochem* 1978; 90:433–440
3. Balestrieri C, Colonna G, Giovane A, Irace G, and Servillo L. Second-Derivative Spectroscopy of Proteins: Studies on Tyrosyl Residues. *Anal Biochem* 1980; 106:49–54
4. Beaven G, and Holiday E. Ultraviolet Absorption Spectra of Proteins and Amino Acids. *Adv Prot Chem* 1952; 7:319–386
5. Bradford MM. A Rapid and Sensitive Method for the Quantitation of Microgram Quantities of Protein Utilizing the Principle of Protein-Dye Binding. *Anal Biochem* 1976; 72:248–254
6. Burke CJ, Sanyal G, Bruner MW, Ryan JA, LaFemina RL, Robbins HL, Zeft AS, Middaugh CR, and Cordingley MG. Structural Implications of Spectroscopic Characterization of a Putative Zinc Finger Peptide from HIV-1 Integrase. *J Biol Chem* 1992; 267:9639–9644
7. Creighton TE. *Proteins: Structures and Molecular Properties*, 2nd ed. New York, NY: W.H. Freeman and Company; 1993. 507p.
8. Donovan JW. Ultraviolet Difference Spectroscopy—New Techniques and Applications. *Methods Enzymol* 1973; 27:497–525

9. Duysens LN. The Flattening of the Absorption Spectrum of Suspensions, as Compared to That of Solutions. *Biochim Biophys Acta* 1956; 19:1–12

10. Edwards K, Kueltzo L, Fisher M, and Middaugh C. Complex Effects of Molecular Chaperones on the Aggregation and Refolding of Fibroblast Growth Factor-1. *Arch Biochem Biophys* 2001; 393:14–21

11. Fasman GD (ed.). *Practical Handbook of Biochemistry and Molecular* Biology, 3rd ed. Cleveland, OH: CRC Press; 1992. 589p.

12. Favilla R, Goldoni M, Del Signore F, Di Muro P, Salvato B, and Beltramini M. Guanidinium Chloride Induced Unfolding of a Hemocyanin Subunit from Carcinus aestuarii. II. Holo Form. *Biochim Biophys Acta* 2002; 1597:51–59

13. Fisher MT, and Sligar SG. Tyrosine Motions in Relation to the Ferric Spin Equilibrium of Cytochrome P-450cam. *Biochemistry* 1985; 24:6696–6701

14. Gadda G, and Fitzpatrick PF. Biochemical and Physical Characterization of the Active FAD-Containing Form of Nitroalkane Oxidase from Fusarium oxysporum. *Biochemistry* 1998; 37:6154–6164

15. Gill SC, and von Hippel PH. Calculation of Protein Extinction Coefficients from Amino Acid Sequence Data. Anal Biochem 1989; 182:319–326.

16. Herskovits TT, and Laskowski M Jr. Location of Chromophoric Residues in Proteins by Solvent Perturbation. I. Tyrosyls in Serum Albumins. *J Biol Chem* 1962; 237:2481–2492

17. Ichikawa T, and Terada H. Estimation of State and Amount of Phenylalanine Residues in Proteins by Second Derivative Spectrophotometry. *Biochim Biophys Acta* 1979; 580:120–128.

18. Kornblatt JA, Kornblatt MJ, Lange R, Mombelli E, and Guillemette JG. The Individual Tyrosines of Proteins: Their Spectra May or May Not Differ from Those in Water or Other Solvents. *Biochim Biophys Acta* 1999; 1431:238–248

19. Kosen PA, Creighton TE, and Blout ER. Ultraviolet Difference Spectroscopy of Intermediates Trapped in Unfolding and Refolding of Bovine Pancreatic Trypsin Inhibitor. *Biochemistry* 1980; 19:4936–4944

20. Kueltzo LA, and Middaugh CR. Structural Characterization of Bovine Granulocyte Colony Stimulating Factor: Effect of Temperature and pH. *J Pharm Sci* 2003; 92:1793–1804

21. Kueltzo LA, Ersoy B, Ralston JP, and Middaugh CR. Derivative Absorbance Spectroscopy and Protein Phase Diagrams as Tools for Comprehensive Protein Characterization: A bGCSF Case Study. *J Pharm Sci* 2003; 92:1805–1820

22. Lange R, and Balny C. UV-Visible Derivative Spectroscopy Under High Pressure. *Biochim Biophys Acta* 2002; 1595:80–93

23. Li YK, Kuliopulos A, Mildvan AS, and Talalay P. Environments and Mechanistic Roles of the Tyrosine Residues of Delta 5-3-ketosteroid Isomerase. *Biochemistry* 1993; 32:1816–1824

24. Lowry OH, Rosenbrough NJ, Farr AL, and Randall RJ. Protein Measurement with the Folin Phenol Agent. *J Biol Chem* 1951; 193:265–275

25. Luong TN, and Kirsch JF. A Continuous Coupled Spectrophotometric Assay for Tyrosine Aminotransferase Activity with Aromatic and Other Nonpolar Amino Acids. *Anal Biochem* 1997; 253:46–49

26. Mach H, and Middaugh CR. Measuring Protein Spectra in the Presence of Light Scattering. *Biotechniques* 1993; 15:240–242

27. Mach H, and Middaugh CR. Simultaneous Monitoring of the Environment of Tryptophan, Tyrosine, and Phenylalanine Residues in Proteins by Near-Ultraviolet Second-Derivative Spectroscopy. *Anal Biochem* 1994; 222:323–331

28. Mach H, Dong Z, Middaugh CR, and Lewis RV. Conformational Stability of Cu,Zn-Superoxide Dismutase, the Apoprotein, and Its Zinc-Substituted Derivatives: Second-Derivative Spectroscopy of Phenylalanine and Tyrosine Residues. *Arch Biochem Biophys* 1991a; 287:41–47

29. Mach H, Thomson JA, Middaugh CR, and Lewis RV. Examination of Phenylalanine Microenvironments in Proteins by Second-Derivative Absorption Spectroscopy. *Arch Biochem Biophys* 1991b; 287:33–40

30. Mach H, Middaugh CR, and Lewis RV. Statistical Determination of the Average Values of the Extinction Coefficients of Tryptophan and Tyrosine in Native Proteins. *Anal Biochem* 1992, 200:74–80

31. Mach H, Volkin DB, Burke CJ, and Middaugh CR. Ultraviolet Absorption Spectroscopy. In: Shirley BA. *Protein Stability and Folding: Theory and Practice*. Vol. 40. Totowa, NJ: Humana Press; 1995:91–114.

32. Maclean DS, Qian Q, and Middaugh CR. Stabilization of Proteins by Low Molecular Weight Multi-ions. *J Pharm Sci* 2002; 91:2220–2229

33. Mani RS, Karimi-Busheri F, Cass CE, and Weinfeld M. Physical Properties of Human Polynucleotide Kinase: Hydrodynamic and Spectroscopic Studies. *Biochemistry* 2001; 40:12967–12973

34. Melo EP, Aires-Barros MR, Costa SM, and Cabral JM. Thermal Unfolding of Proteins at High pH Range Studied by UV Absorbance. *J Biochem Biophys Methods* 1997; 34:45–59

35. Owen T. *Fundamentals of Modern UV-Visible Spectroscopy: A Primer*. Germany: Hewlett Packard Co.; 1996. 142p.

36. Pawar SA, and Deshpande VV. Characterization of Acid-Induced Unfolding Intermediates of Glucose/Xylose Isomerase. *Eur J Biochem* 2000; 267:6331–6338
37. Piersma SR, Visser AJ, de Vries S, and Duine JA. Optical Spectroscopy of Nicotinoprotein Alcohol Dehydrogenase from Amycolatopsis methanolica: A Comparison with Horse Liver Alcohol Dehydrogenase and UDP-Galactose Epimerase. *Biochemistry* 1998; 37:3068–3077
38. Ragone R, Colonna G, Balestrieri C, Servillo L, and Irace G. Determination of Tyrosine Exposure in Proteins by Second-Derivative Spectroscopy. *Biochemistry* 1984; 23:1871–1875
39. Ragone R, Colonna G, Bismuto E, and Irace G. Unfolding Pathway of Myoglobin: Effect of Denaturants on Solvent Accessibility to Tyrosyl Residues Detected by Second-Derivative Spectroscopy. *Biochemistry* 1987; 26:2130–2134
40. Sapan CV, Lundblad RL, and Price NC. Colorimetric Protein Assay Techniques. *Biotechnol Appl Biochem* 1999; 29:99–108
41. Servillo L, Colonna G, Balestrieri C, Ragone R, and Irace G. Simultaneous Determination of Tyrosine and Tryptophan Residues in Proteins by Second-Derivative Spectroscopy. *Anal Biochem* 1982; 126:251–257
42. Timasheff S. Turbidity as a Criterion of Coagulation. *J Colloid Interface Sci* 1966; 21:489–497
43. van Holde KE, Johnson WC, and Ho PS. Absorption Spectroscopy. In: *Principles of Physical Biochemistry*. 1st ed. Upper Saddle River, NJ: Prentice-Hall; 1998:380–417
44. Vergani L, Grattarola M, Dondero F, and Viarengo A. Expression, Purification, and Characterization of Metallothionein-A from Rainbow Trout. *Protein Expr Purif* 2003; 27:338–345
45. Volkin DB, Tsai PK, Dabora JM, Gress JO, Burke CJ, Linhardt RJ, and Middaugh CR. Physical stabilization of acidic fibroblast growth factor by polyanions. *Arch Biochem Biophys* 1993; 300:30–41
46. Wetlaufer D. Ultraviolet Spectra of Proteins and Amino Acids. *Adv Prot Chem* 1962; 17:303–390
47. Zhang Z, Smith DL, and Smith JB. Human Beta-Crystallins Modified by Backbone Cleavage, Deamidation and Oxidation Are Prone to Associate. *Exp Eye Res* 2003; 77:259–272
48. Zubay. *Biochemistry*. 3rd. ed. Dubuque, IA: Wm. C. Brown Publishers; 1993. 1120p.

2

Fluorescence Spectroscopy

WIM JISKOOT

Department of Pharmaceutics, Utrecht Institute for Pharmaceutical Sciences (UIPS), Utrecht University, Utrecht, The Netherlands

ANTONIE J.W.G. VISSER

MicroSpectroscopy Centre, Laboratory of Biochemistry, Wageningen University, Wageningen, and Department of Structural Biology, Vrije Universiteit, Amsterdam, The Netherlands

JAMES N. HERRON

Department of Pharmaceutics and Pharmaceutical Chemistry, University of Utah, Salt Lake City, UT

MARC SUTTER

Department of Pharmaceutics, Utrecht Institute for Pharmaceutical Sciences (UIPS), Utrecht University, Utrecht, The Netherlands

Introduction

The first important steps in the scientific investigation of fluorescence were taken in the nineteenth century. Stokes (1852) observed and published the shift of fluorescence emission to smaller energy relative to the absorption; an effect that is still known as the Stokes shift. Much insight into the physics of fluorescence was gained in the first half of the twentieth century. During this time, important theories for the processes of excitation and emission (Jablonski, 1933), radiationless energy transfer (Förster, 1948), quenching of fluorescence (Stern and Volmer, 1919), and polarization of fluorescence (Perrin, 1926) were developed. As a result of this progress, fluorescence spectroscopy became an established analytical technique. Its applications to biochemical and biophysical systems were pioneered by Weber (1953). Nowadays, fluorescence spectroscopy is an important method for studying proteins and other biological macromolecules. Its high sensitivity, which enables the measurement of very small amounts of (precious) sample material, is a major advantage.

Most proteins contain naturally fluorescent amino acids such as tryptophan or tyrosine. In addition, a large number of fluorescent dyes for probing protein molecules are available. Therefore, fluorescence spectroscopy can readily be applied to proteins. It has great potential to determine structural aspects (e.g., distances, size, shape, state of folding, or ligand binding) and dynamic properties (e.g., local or segmental flexibility) of these molecules. This can be very useful for the characterization of pharmaceutical proteins and for the development of stable formulations, where protein native structure and function are preserved. Fluorescence spectroscopy is indeed increasingly applied for precisely these purposes.

In this chapter, we will introduce the principles of fluorescence and the measurement techniques used for examining proteins in the section "Principles of Fluorescence and Measurement Techniques." With respect to instrumentation, our main focus is on commercially available steady-state and time-resolved spectrofluorometers. For information on other measurement techniques including fluorescence microscopy, confocal techniques, or single molecule fluorescence, the reader is referred to the literature (Lakowicz, 1999; Valeur, 2002). The intrinsic fluorescence of proteins and the use of extrinsic dyes are introduced in the section "Protein Fluorescence." Examples of the application of fluorescence spectroscopy for characterizing pharmaceutical proteins are given in the section "Applications." We have included many example studies that were performed using standard (steady-state) fluorescence instrumentation, which is available in most laboratories. This demonstrates that valuable information on protein structure and function can be obtained from relatively simple experiments. We present a set of practical considerations in the section "Practical Considerations and Experimental Pitfalls" and then end the chapter with the "Conclusions" section. It is our intention to provide the reader with an overview of the possibilities that fluorescence spectroscopy offers for examining pharmaceutical proteins. For more detailed information on either theory or applications, we refer the reader to the numerous literature citations in the text.

Principles of Fluorescence and Measurement Techniques

Fluorescence

The emission of visible or ultraviolet light by typically aromatic molecules in electronically excited states is referred to as luminescence. Depending on the nature of the excited state, luminescence is further categorized into fluorescence and phosphorescence.

The principles of fluorescence and phosphorescence can be illustrated by the Jablonski diagram (Figure 1). In this diagram, four electronic states are depicted for an arbitrary aromatic molecule: the singlet ground state (S_0), two singlet excited states (S_1 and S_2), and the triplet state (T_1). Superimposed on the electronic states are the vibrational levels, which are of much smaller energy. When a photon of sufficient energy, $h\nu$, is absorbed by the molecule, a valence electron is promoted from the ground state to some vibrational level of an excited singlet state. This process departs from the lowest vibrational level of S_0 for almost all molecules that are excited and is extremely rapid, of the order one femtosecond (1 femtosecond = 10^{-15} seconds). The nuclei, having a much greater mass than electrons, are not able to follow quickly any redistribution of charge that may occur within the molecule and are fixed during excitation. This is the Franck-Condon principle. The excited molecules reach the lowest vibrational level of S_1 by vibrational relaxation and internal conversion. The former is the result of redistribution and exchange of vibrational energy, whereas the latter is the transition between two electronically excited states of the same spin multiplicity (e.g., $S_2 \rightarrow S_1$). These radiationless processes take place in the picosecond time range (1 picosecond = 10^{-12} seconds). The lowest vibrational level of S_1 is the starting point for several competing decay processes. Fluorescence emission to the ground state S_0, radiationless decay to S_0 by internal conversion, and transition to the lowest triplet state by intersystem crossing are possible. The duration of fluorescence is usually in the range 1–100 nanoseconds (1 nanosecond = 10^{-9} seconds). It is clear from Figure 1 that, irrespective of the electronic and vibrational energy level reached during excitation, the emitting state is generally the lowest vibrational level of the first excited state, $S_{1,0}$. A few exceptions to this rule exist; for example, azulene and certain metalloporphyrins (Martarano et al., 1976). Figure 1 also indicates the triplet state, which is the starting point for phosphorescence. Once the molecule has reached this state, it will reside there for a relatively long time (between milliseconds and seconds) before it will decay to the ground state. This is due to the spin forbidden excited state singlet-triplet and ground state triplet-singlet transitions involved.

Throughout the rest of this chapter, the focus will be on fluorescence rather than phosphorescence. However, protein phosphorescence can also be observed. Difficulties in the use of phosphorescence for studying protein structure arise from the generally low intensity of phosphorescence in liquid solution, the various possible mechanisms of quenching, notably by oxygen, (see "Quenching of Fluorescence," below) that reduce emission to a greater extent than in the case of fluorescence, and the possible overlap of fluorescence and phosphorescence emission (Papp and Vanderkooi, 1989). According to the Franck-Condon principle, the nuclei are fixed during electronic transitions. This is visualized in Figure 2, which shows Franck-Condon potential curves

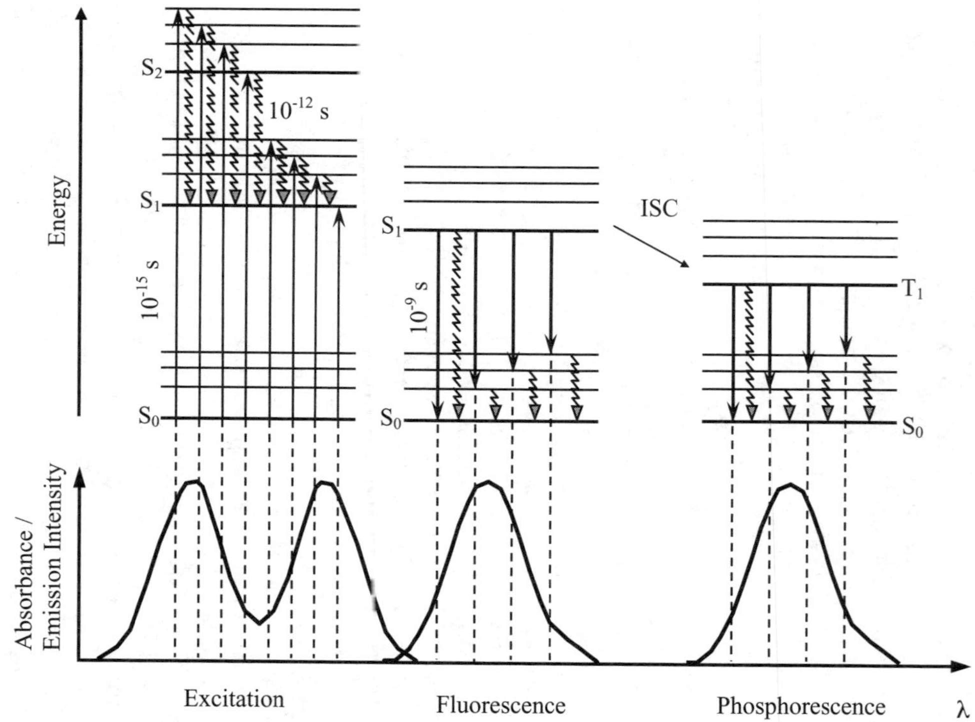

Figure 1 Jablonski diagram of the electronic excitation and relaxation processes within a fluorophore (see section "Fluorescence"). Light absorption and emission are represented by vertical, black arrows. Vibrational relaxation is indicated by the gray, interrupted arrows. The energy difference between S_1 during excitation and emission is caused by solvent relaxation (see section "Emission Spectra"). ISC stands for intersystem crossing to the first excited triplet state.

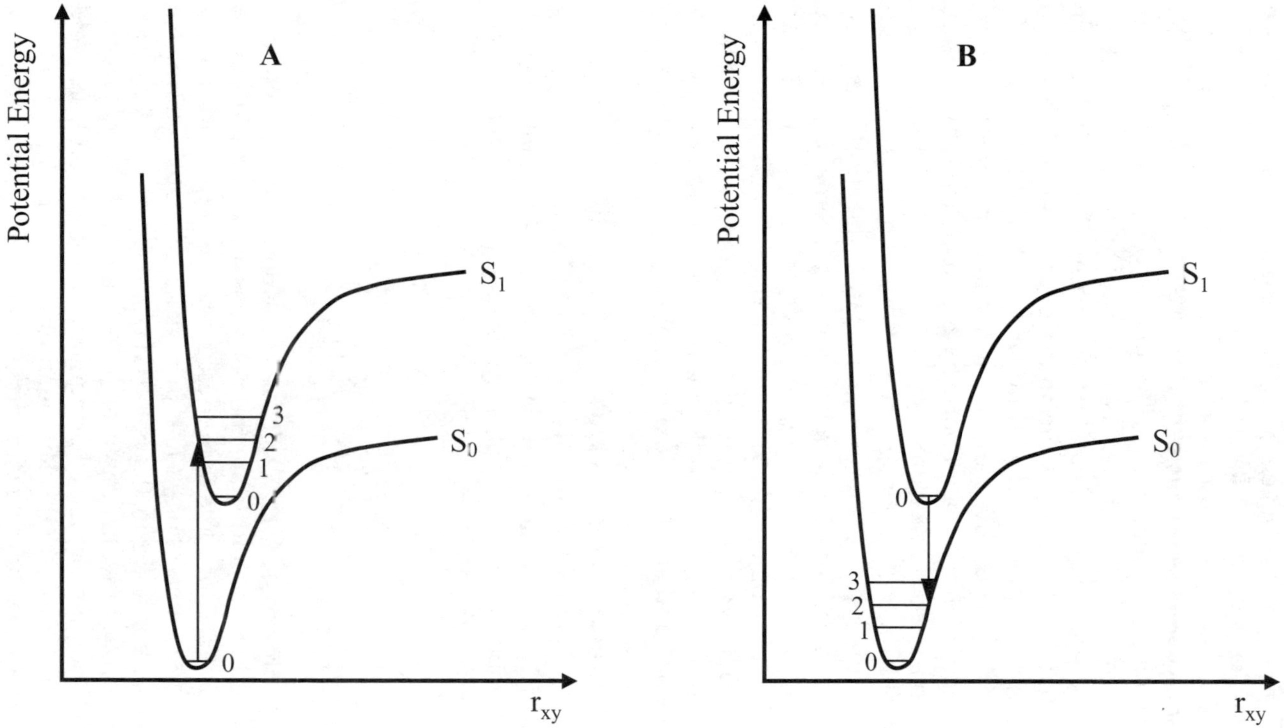

Figure 2 Franck-Condon potential curves for a (hypothetical) two atomic fluorescent molecule, which allows for a two-dimensional diagram that displays the potential energy against the distance between the nuclei (r_{xy}). The S_1 potential curve is slightly shifted to greater bond lengths because electronically excited states normally possess greater antibonding character than electronic ground states. The (A) excitation and (B) emission processes are depicted (see section "Fluorescence").

for the electronic ground state, S_0, and the first excited state, S_1. Before absorption, the molecule is in the lowest vibrational level of S_0, and the nuclei are most probably at their equilibrium separation. Absorption occurs vertically to vibrational levels of S_1 that display the same distance between the nuclei as the ground state. The preferred transition (i.e., the peak of the absorption spectrum) is to the vibrational level that most probably displays the same nuclear distance as the ground state. However, excitation to neighboring vibrational levels in S_1 is also possible. The combination of excitation to higher electronic states and vibrational levels results in an excitation spectrum. The return to the electronic ground state from the lowest vibrational level of S_1 by emission of a photon can be described in a similar way. Also this transition is vertical, and the most probable transition is to a higher vibrational level of S_0. Again, the transition to neighboring vibrational levels may occur, giving rise to an emission spectrum.

It can be seen from Figures 1 and 2 that the amount of energy absorbed may be equal to but is mostly greater than the emitted energy, due to the presence of vibrational relaxation and internal conversion. Therefore, emission occurs at longer, less energetic wavelengths than absorption. This phenomenon was first described by G.G. Stokes in 1852 (Stokes, 1852). Nowadays, the Stokes shift is defined as the difference in wavenumber ($\tilde{\nu}$) between the maximum of the highest wavelength absorption band and the maximum of the emission spectrum.

Quantum Yield and Fluorescence Lifetime

Two important parameters in fluorescence spectroscopy are the quantum yield (Q) and the fluorescence lifetime (τ). The quantum yield, which is the number of emitted photons in relation to the number of absorbed photons, can be expressed as:

$$Q = \frac{\Gamma}{\Gamma + \sum_i k_i} \tag{1}$$

where Γ is the rate constant for radiative relaxation from S_1 to S_0 with emission of fluorescence and k_i are the rate constants for all relaxation processes that do not contribute to fluorescence (e.g., nonradiative relaxation and intersystem crossing). Quantum yields of highly fluorescent molecules (e.g., fluorescein in 0.1 N NaOH or rhodamine) can reach values close to unity. In water, the three fluorescent amino acids tryptophan, tyrosine, and phenylalanine have quantum yields of 0.13, 0.14, and 0.02, respectively (Eftink, 1991). The quantum yield of amino acids in proteins can show considerable variation. Values for tryptophan range from 0.05 to 0.35 in different proteins (Eftink, 1991; Callis and Vivian, 2003). Quantum yields can be estimated by comparison with standards of known quantum yield (Demas and Crosby, 1971; Lakowicz, 1999).

The fluorescence lifetime is the average time that a fluorophore remains in the excited state. It is defined as the reciprocal of the sum of the rate constants for relaxation from S_1 to S_0:

$$\tau = \frac{1}{\Gamma + \sum_i k_i} \tag{2}$$

Fluorescence lifetimes are often in the nanosecond timescale but can range from picoseconds to several hundreds of nanoseconds. A fluorophore in a homogeneous environment is usually expected to have a single fluorescence lifetime. This is not the case for tryptophan in water. The molecule may adopt various conformations, and one of the rotational isomers has a significantly shorter lifetime (0.5 nanosecond) than the others (3.1 nanoseconds) (Szabo and Rayner, 1980). Proteins also often display multiple lifetimes even if they contain a single tryptophan as a fluorophore. Various models have been developed for explaining this heterogeneity (see "Intrinsic Protein Fluorescence," below). In proteins containing multiple fluorophores, the number of fluorescence lifetimes may of course be even larger. The fluorescence lifetime provides a time window for the detection of dynamic processes that occur while the fluorophores are in the excited state. Solvent relaxation (see "Emission Spectra," below), dynamic fluorescence quenching (see "Quenching of Fluorescence," below) and rotational motions of the fluorophores leading to a decrease of fluorescence anisotropy (see "Fluorescence Anisotropy," below) are important examples of such processes.

For fluorophores in the liquid phase, both fluorescence lifetime and quantum yield can be affected by a variety of factors. A simple but important factor is temperature. Generally, the nonradiative relaxation processes are faster at higher temperature, leading to a decrease in quantum yield and fluorescence lifetime. Further factors affecting both parameters include pH, polarity, viscosity, and presence of quenchers.

Spectral Measurements

The energy of the electronic transitions is of the order several electron volts (eV). One eV is equivalent to a wavenumber of about 8000 cm^{-1} (cf., 280 nm $\approx$ 36,000 cm^{-1}), and therefore spectral measurements of fluorescent probe molecules are performed in the ultraviolet and visible region of the electromagnetic spectrum. Three types of spectra—absorption, excitation, and emission—can be recorded. For information on absorption spectra, the reader is referred to Chapter 1 of this book. Excitation and emission spectra will be discussed in the following sections. Spectral measurements are performed with so-called steady-state spectrofluorometers (Lakowicz, 1999; Valeur, 2002). In these experiments, the sample is excited during a time much longer than the fluorescence lifetime, and emission is constantly detected. The term steady-state refers to the equilibrium of fluorophores in the ground state and in the excited state that is reached during the measurement. Fluorescence spectra are either measured as functions of wavelength (λ) in nm or wavenumber ($\tilde{\nu}$) in cm^{-1}. The wavelength scale is more common and will be used throughout this chapter. The wavenumber scale is linear in energy because of the proportionality between energy and wavenumber. Spectral measurements are mostly performed with liquid samples. However, it is also possible to do solid-state fluorescence spectroscopy (Sharma and Kalonia, 2003b).

Excitation Spectra

An excitation spectrum is obtained by measuring the fluorescence emission of a sample at a fixed wavelength, while varying the excitation wavelength. For dilute samples of absorbance less than 0.05, the fluorescence intensity (F) is proportional to the absorption coefficient at each wavelength:

$$F \propto \varepsilon(\lambda) \cdot I_0 \cdot Q \tag{3}$$

Here, $\varepsilon(\lambda)$ is the absorption coefficient at wavelength λ, I_0 the intensity of the incident light, and Q the quantum yield. The quantum yield is usually independent of the excitation wavelength. Consequently, the absorption and excitation spectra of a highly diluted sample should have the same profile but differently scaled y-axes. In reality, the profiles will not exactly match one another because the photon output of the lamp and the efficiency of the excitation monochromator are wavelength dependent. If necessary, excitation spectra can be corrected for nonideality of the spectrofluorometer by the use of a quantum counter, which can be placed behind the excitation monochromator, in front of the reference phototube in some instruments (Lakowicz, 1999).

Emission Spectra

In order to obtain an emission spectrum, the sample is excited with a fixed wavelength, and the fluorescence intensity is recorded as a function of wavelength. Figure 1 shows that the emission spectrum is the mirror image of the longest wavelength band of the absorption spectrum, which can be explained by the similar energy differences between the vibrational levels of the electronic ground state and the first electronic excited state. Its position and profile are independent of the excitation wavelength, as fluorescence emission always starts from the lowest vibrational level of the first singlet excited state (Figure 1). The emission spectrum is shifted to higher wavelengths compared to the absorption spectrum (Stokes shift). However, a close inspection of Figure 1 reveals that there is a small overlap between the absorption and emission spectrum. This exception to the Stokes shift can be explained by the fact that a small number of molecules can reach higher vibrational levels of the singlet ground state at room temperature, if the energy difference between vibrational levels is small. In this case, excitation may start from a higher vibrational level, and the required excitation energy may be lower than the energy released by emission of a photon.

In addition to vibrational relaxation and internal conversion, the magnitude of the Stokes shift can be affected by solvent relaxation. The process of solvent relaxation is depicted in Figure 3. Due to the electronic transitions occurring during absorption, the fluorescent molecule in the excited state may have a different (mostly larger) dipole moment than in the ground state. Sufficiently mobile polar solvent molecules surrounding the fluorophore are able to rearrange to energetically more favorable positions prior to emission, thus lowering the energy of the first electronic excited state. Upon emission, the solvent molecules rapidly rearrange to achieve an energetically favorable interaction with the fluorophore in the singlet ground state. Solvent relaxation explains the increased shift of emission spectra to higher wavelengths (redshift) with increasing solvent polarity.

The fluorescence intensity values of emission spectra are affected by several experimental factors, including the concentration of the fluorophore, the quantum yield, bandpass settings, and instrumental properties, of which the last are usually unknown to the experimenter. For this reason, the intensity values are usually plotted on an arbitrary scale. Also, the emission spectra recorded may be slightly distorted by the wavelength-dependent efficiency of the detection system. In many cases, especially when one simply wants to compare different spectra, these uncorrected or "technical" spectra are sufficient. However, corrected spectra may be necessary in certain cases, for example for the determination of quantum yields. A simple way of obtaining the correction factors is by comparing the measured

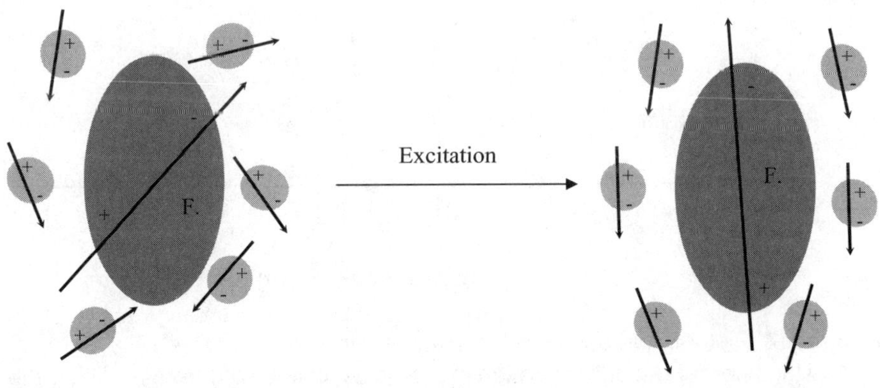

Figure 3 Reorientation of solvent molecules around a fluorophore (F.) in the excited state (see section "Emission Spectra"). Black arrows indicate the dipole axes.

emission spectrum of a reference substance with its known corrected spectrum (Lakowicz, 1999).

Fluorescence Intensity Measurements for Quantitative Purposes
Quantitative fluorescence intensity measurements are usually performed at fixed excitation and emission wavelengths, which can be chosen by the help of excitation and emission spectra. In these types of experiments, it is important to measure highly diluted samples with absorbance below 0.05. In more concentrated samples, inner filter effects lead to absorption of exciting light before it reaches the point of observation (normally the center of the cuvette). Also, absorption of emitted light may occur. Figure 4 shows that deviation from the linear

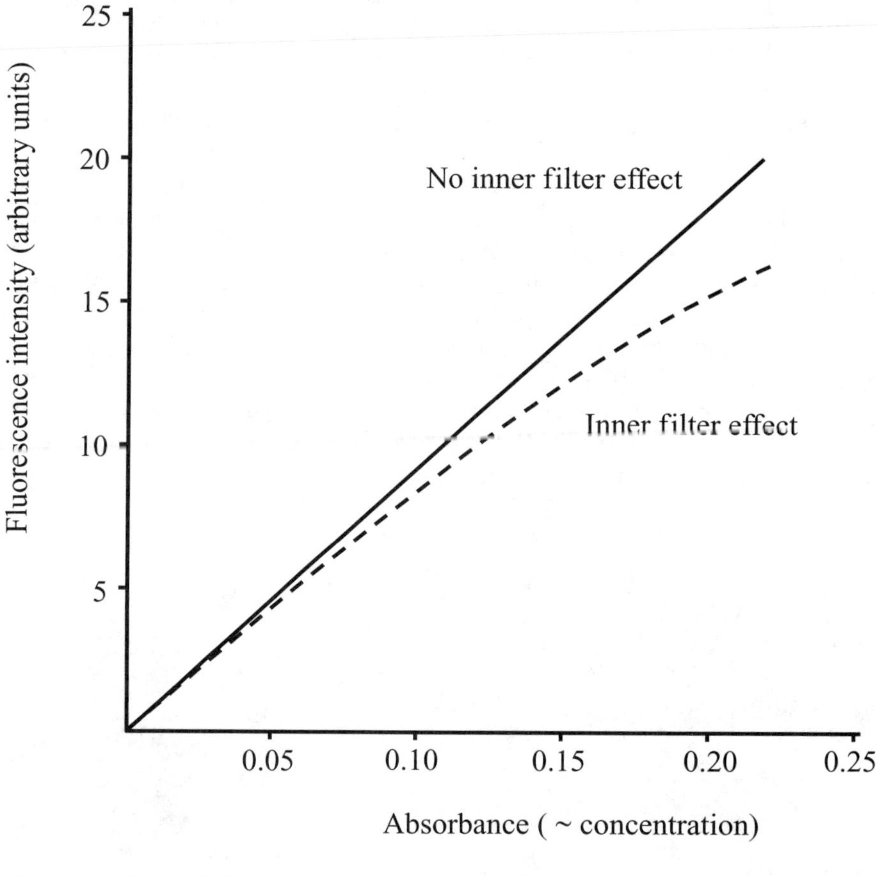

Figure 4 Plot of fluorescence intensity against absorbance of a sample. Increasing fluorophore concentration prevents excitation light from reaching the center of the cuvette, resulting in a decrease of the detected fluorescence intensity (= inner filter effect; see section "Fluorescence Intensity Measurements for Quantitative Purposes").

relationship between fluorescence intensity and absorbance of the sample increases significantly for absorbance values above 0.05.

Fluorescence Lifetime Measurements

If a fluorescent sample is excited with an infinitesimally short pulse of light (δ-pulse), this results in a population, $n(0)$, of molecules in the excited state that decays according to a first-order rate process with a rate constant ($\Gamma + \Sigma_i k_i$) introduced in "Quantum Yield and Fluorescence Lifetimes," above:

$$-\frac{dn(t)}{dt} = \left(\Gamma + \sum_i k_i\right) \cdot n(t) = \frac{1}{\tau} \cdot n(t) \tag{4}$$

Because the number of molecules in the excited state is proportional to the fluorescence intensity (F), Equation 4 can be written as:

$$-\frac{dF(t)}{dt} = \left(\frac{1}{\tau}\right) \cdot F(t) \tag{5}$$

Integration of Equation 5 yields the exponential decay of fluorescence intensity with time, which is depicted in Figure 5:

$$F(t) = F(0) \cdot \exp(-t/\tau) \tag{6}$$

As mentioned in "Quantum Yield and Fluorescence Lifetimes," above, there may be multiple fluorescence lifetimes. In this case, the fluorescence intensity decay with time is commonly expressed as a multi-exponential:

$$F(t) = \sum_i \alpha_i \cdot \exp(-t/\tau_i) \tag{7}$$

In Equation 7, the pre-exponential factors α_i are the amplitudes of the individual fluorescent components, and τ_i are the associated fluorescence lifetimes. Fluorescence intensity decays in proteins can be so complex that they are better described by continuous lifetime distributions (see "Intrinsic Protein Fluorescence," below).

Measurement of the fluorescence lifetime, or generally of the parameters that govern the course of fluorescence intensity with time (see "Fluorescence Anisotropy," below), can be performed with two different techniques that are both widespread: the pulse method and the phase-modulation method.

The pulse method is frequently used in combination with time-correlated single photon counting (TCSPC) (O'Connor and Phillips, 1984; Lakowicz, 1999). The sample is excited with a short pulse of light, and the time between the pulse and detection of the first emitted

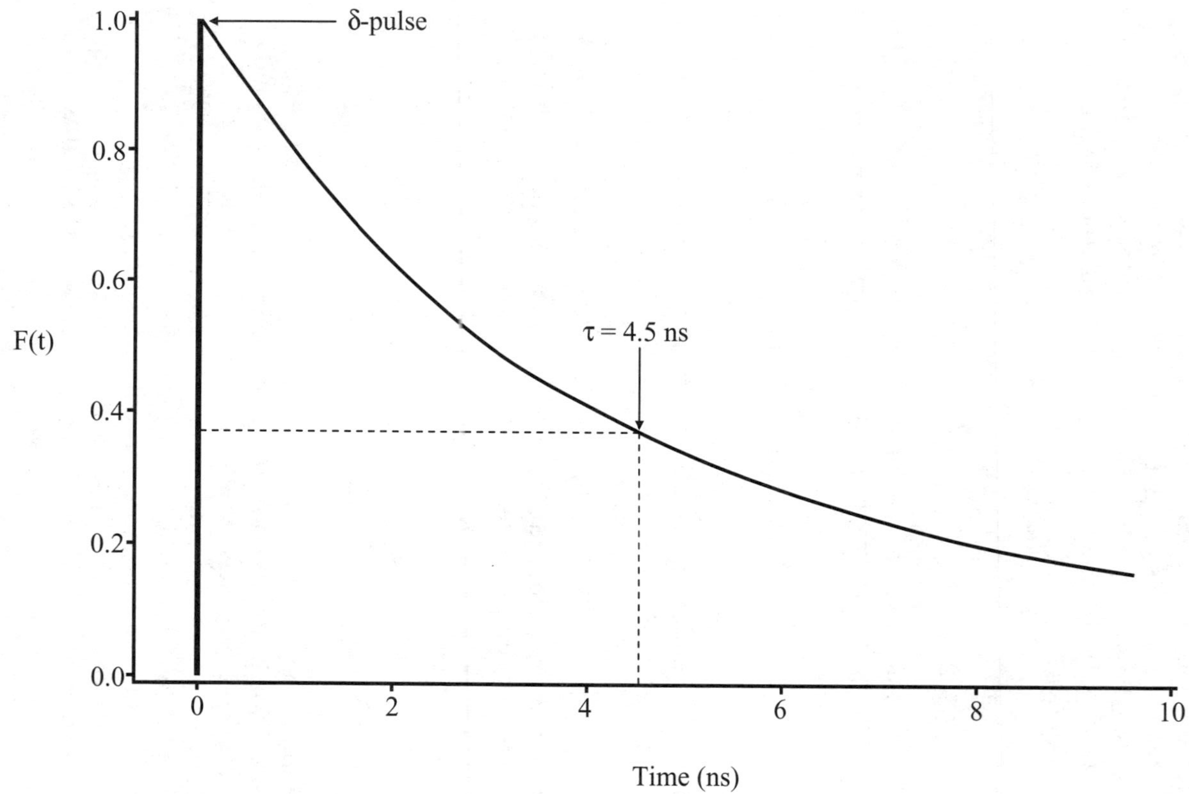

Figure 5 Plot of the decay of fluorescence intensity with time, $F(t)$, after δ-pulse excitation. The lifetime (τ) is indicated as the time, when fluorescence intensity reaches 1/e of its initial value (see Equation 6).

photon is measured. The course of fluorescence intensity with time (Figure 6) is recorded by multiple repetition of this procedure. The resulting signal is a convolution of the decay of fluorescence intensity with time following a δ-pulse with the finite distribution of the excitation pulses of the experiment (i.e., the instrument response function that is also shown in Figure 6). Analysis of the measured fluorescence intensity curves and determination of fluorescence lifetimes are often performed by a nonlinear least-squares method (Grinvald and Steinberg, 1974; Lakowicz, 1999). A model function (e.g., Equation 7) that is assumed to describe the course of fluorescence intensity with time after δ-pulse excitation is convolved with the measured instrument response function. The result of the convolution of Equation 7 with the instrument response function is fitted to the measured fluorescence intensity curves by optimizing the variables α_i and τ_i. The resolution that can be achieved with TCSPC depends much on the light source and detector. Using cavity dumped dye lasers or Ti:sapphire lasers in combination with microchannel plate photomultipliers, instrument response function widths of thirty picoseconds are possible. This allows for the measurement of decay times in the range of ten to twenty picoseconds. Commercially available flashlamps give excitation pulses of about two nanoseconds width. Decay times of several hundred picoseconds can still be resolved when these lamps are used. The major drawback of flashlamps is their low pulse repetition rate (mostly around 50 kHz) compared to lasers (typically 1–4 MHz). This results in longer measurement times of possibly several hours, depending on the sample. During these long measurement times, lamp drifts, and consequently shifts of the instrument response function, are likely to occur.

In the phase-modulation method, the sample is excited by an intensity modulated light source, which can either be a xenon arc lamp or a laser. The excitation is sinusoidally modulated at angular frequency ω (rad/s). The fluorescence is delayed in phase and partially demodulated. The experimental quantities to be determined are the phase difference (ϕ_ω) and the modulation ratio (m_ω) as shown in Figure 7. For a single fluorescence lifetime, phase difference and modulation ratio are related to the lifetime (τ) by:

$$\phi_\omega = \arctan(\omega\tau) \tag{8}$$

and

$$m_\omega = \frac{1}{\sqrt{1 + (\omega\tau)^2}} \tag{9}$$

For the precise determination of fluorescence lifetimes, the phase difference (ϕ_ω) and the modulation ratio (m_ω) are measured as functions of ω. The curves can be analyzed by the method of nonlinear least

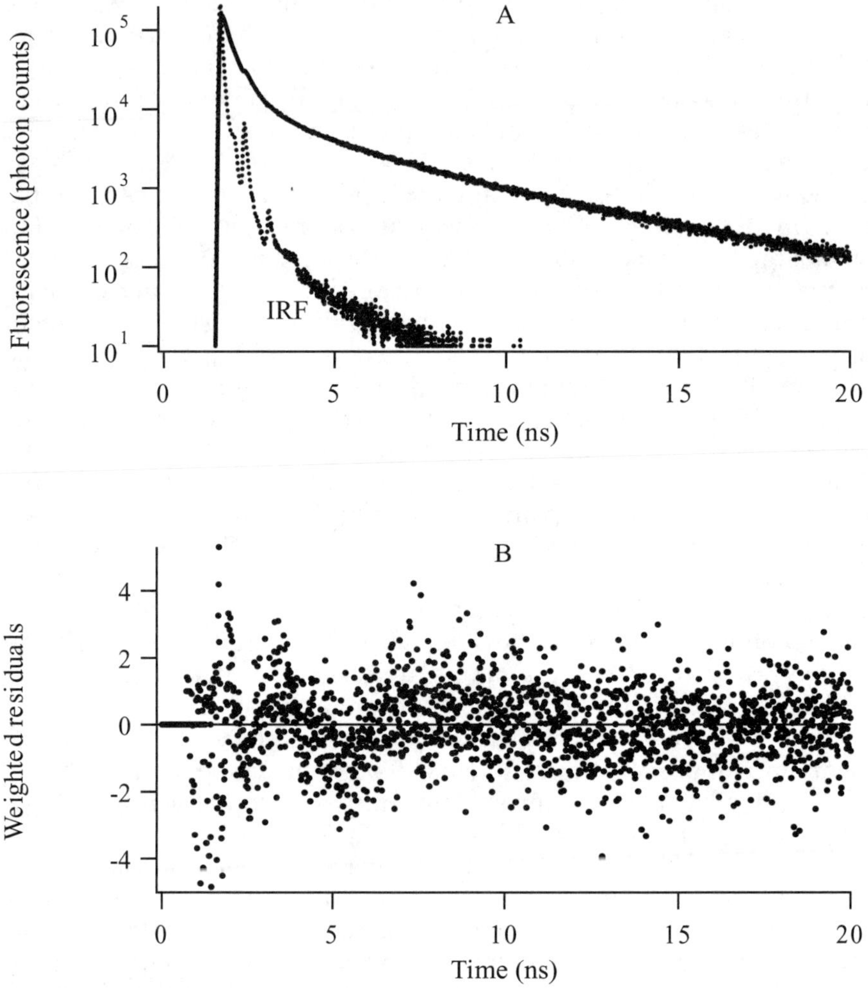

Figure 6 (A) Experimental (dots) and fitted (solid line) fluorescence decay curves of immunity protein Im9 at pH 7.4 and 20°C. The excitation wavelength was 295 nm, and the fluorescence was selected through an interference filter transmitting at 349 nm with half width 5 nm. Also shown is the instrument response function (IRF) obtained with a reference solution of *p*-terphenyl in a mixture of cyclohexane and CCl_4 (3:1 v/v) having a fluorescence lifetime of 15 picoseconds (Visser et al., 1994). The timescale is 11.1 picoseconds per channel, and the analysis was performed over channels 1–2048 of a multichannel analyzer. The fit was according to a four-exponential decay function with the following optimized parameters and standard errors: $\alpha_1 = 0.612 \pm 0.001$; $\tau_1 = 0.115 \pm 0.001$ nanoseconds; $\alpha_2 = 0.346 \pm 0.006$; $\tau_2 = 0.356 \pm 0.003$ nanoseconds; $\alpha_3 = 0.034 \pm 0.001$; $\tau_3 = 1.90 \pm 0.03$ nanoseconds; $\alpha_4 = 0.012 \pm 0.001$; $\tau_4 = 5.46 \pm 0.05$ nanoseconds. The fit criterion $\chi^2 = 1.32$. (B) Weighted residuals between the experimental and fitted data points.

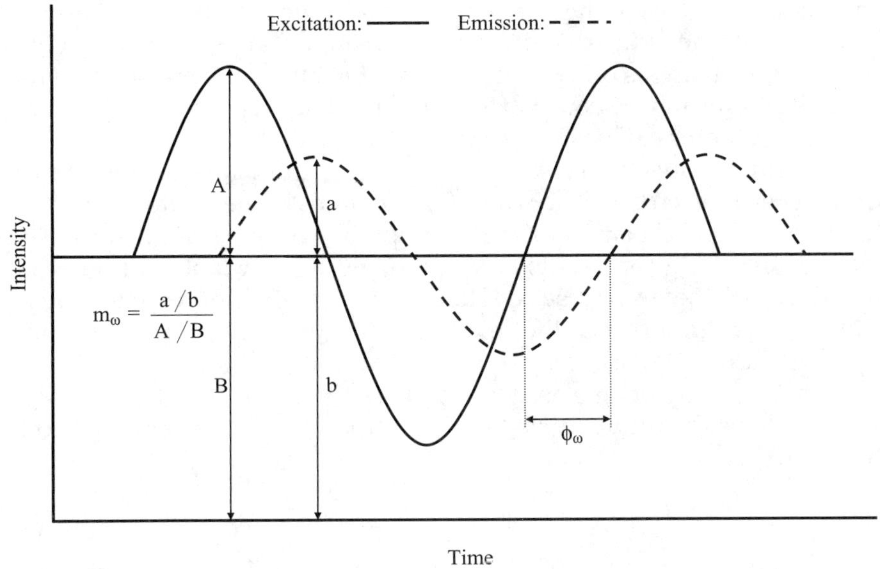

Figure 7 Determination of the phase difference (ϕ_ω) and modulation ratio (m_ω) in the phase-modulation method (see section "Fluorescence Lifetime Measurements").

squares using theoretical expressions of the sine and cosine Fourier transforms of the δ-pulse response (Gratton et al., 1984; Lakowicz et al., 1984; Lakowicz, 1999; Valeur, 2002). The resolution of the phase-modulation method depends on the modulation frequency. Fluorescence lifetimes in the picosecond range can be measured using modulation frequencies between 2 and 10 GHz (Lakowicz, 1993). For lifetimes in the 10-nanosecond range, modulation frequencies between 2 and 200 MHz are suitable.

Quenching of Fluorescence

Generally, quenching of fluorescence refers to all processes that may reduce the emitted fluorescence intensity, including excited state reactions, molecular rearrangements, energy transfer, collisional quenching, and ground state complex formation. We limit the discussion to the last two of these processes, which we refer to as dynamic and static quenching, respectively. For more detailed information on the mechanisms and models of fluorescence quenching, the reader is referred to the literature (Lakowicz, 1999; Valeur, 2002).

Both static and dynamic quenching requires contact between the fluorophore and a quencher molecule. In the case of static quenching, a complex between the fluorophore in the ground state and the quencher molecule is formed, which is usually considered to be nonfluorescent or dark. Consequently, the fluorescence intensity is reduced due to the smaller amount of excitable fluorophores in the sample, but the fluorescence lifetime of the free fluorophores is not changed. This principle of static quenching is questioned by recent results, which show that the quencher-fluorophore complex can still be fluorescent but displays an ultra-short lifetime (Chosrowjan et al., 2003).

Dynamic quenching is based on the diffusion of molecules. Collisions between quenchers and fluorophores in the excited state lead to radiationless relaxation of the fluorophores to the ground state. The rate of quenching represents an additional rate constant for nonradiative decay of the excited state. Therefore, both the fluorescence intensity and the fluorescence lifetime are reduced by dynamic quenching (cf., Equations 1 and 2).

Dynamic quenching is described by the Stern-Volmer equation, which relates the fluorescence intensity (F) to the quencher concentration ([Q]):

$$\frac{F_0}{F} = 1 + k_q\tau_0\,[Q] = 1 + K_D\,[Q] \tag{10}$$

Here, F_0 and τ_0 are the fluorescence intensity and lifetime in the absence of quencher, respectively. The bimolecular quenching constant (k_q) is affected by the frequency of collisions between fluorophores and quenchers and by the efficiency of quenching. It is customary to define the product of k_q and τ_0 as K_D, the Stern-Volmer quenching constant.

For dynamic quenching, the ratio of fluorescence intensities in the absence and presence of quencher equals the ratio of lifetimes in the absence and presence of quencher. Therefore, Equation 10 can also be written as:

$$\frac{\tau_0}{\tau} = 1 + k_q\tau_0\,[Q] = 1 + K_D\,[Q] \tag{11}$$

As for dynamic quenching, there also exists a linear relationship between the ratio F_0/F and the quencher concentration for static quenching:

$$\frac{F_0}{F} = 1 + K_S\,[Q] \tag{12}$$

where K_S is the equilibrium constant (in M^{-1}) for formation of the ground state complex between fluorophore and quencher. Because the ratio τ_0/τ is not changed by static quenching, fluorescence lifetime measurements are a reliable method for distinguishing static and dynamic quenching.

In many fluorescent samples, a combination of dynamic and static quenching may occur. This is accounted for by a modified Stern-Volmer equation:

$$\frac{F_0}{F} = (1 + K_D\,[Q])(1 + K_S\,[Q]) \tag{13}$$

According to the above equation, a plot of F_0/F against [Q] will show an upward curvature. This indication of simultaneous dynamic and static quenching can be confirmed by plotting τ_0/τ against [Q]. Now, in agreement with Equation 11, a linear plot should result.

In some fluorescent samples, it may occur that not all fluorophores are equally accessible to the quencher. In multi-tryptophan proteins, for example, some tryptophan molecules may be buried in the interior of the protein and inaccessible to quenchers. In this case, a plot of F_0/F against [Q] will display a downward curvature. In order to account for the fractional accessibility of fluorophores to collisional quenching, a modification of the Stern-Volmer equation is necessary:

$$\frac{F_0}{\Delta F} = \frac{1}{f_a \cdot K_D \cdot [Q]} + \frac{1}{f_a} \tag{14}$$

Here, $\Delta F = F_0 - F$ and f_a is the fraction of accessible fluorophores. The above equation implies that the Stern-Volmer quenching constant (K_D) is equal for all fluorophores that are accessible to the quencher (Lehrer, 1971). A more complex analysis may be necessary, if there is a distribution of K_D values within the fraction of accessible fluorophores (Lehrer, 1971; Laws and Contino, 1992).

Resonance Energy Transfer

Excited-state energy can be transferred from a donor to an acceptor molecule without the appearance of a photon. This process is referred to as resonance energy transfer (RET), excitation energy transfer (EET), or fluorescence resonance energy transfer (FRET). The last term is misleading because fluorescence is not involved in this process. According to Förster's theory, resonance energy transfer between well-separated, very weakly interacting molecules is enabled by dipole-dipole interactions between the donor and acceptor molecules. Förster derived the following expression for the rate constant of energy transfer (k_T) (Förster, 1948):

$$k_T = \frac{1}{\tau_d^0}\left(\frac{R_0}{R}\right)^6 \tag{15}$$

Here, τ_d^0 is the fluorescence lifetime of the donor in the absence of acceptor, and R is the distance between donor and acceptor. R_0, the so-called Förster distance, corresponds to the distance between donor and acceptor, at which the efficiency of energy transfer is 50%. Typically, R_0 lies between 20 and 60 Å, a range that includes the diameter of most proteins. Therefore, resonance energy transfer can occur within proteins and other biological macromolecules that contain donor-acceptor pairs. R_0 can be expressed as:

$$R_0 = 9.78 \cdot 10^3\,[Q_d^0\, n^{-4}\, J\kappa^2]^{1/6} \tag{16}$$

In the above equation, the Förster distance is given in Å. Q_d^0 is the quantum yield of the donor in the absence of acceptor, and n is the refractive index of the medium. For proteins, n is generally assumed to be 1.4. The overlap integral J (M^{-1} cm^3), which is the degree of overlap between the emission spectrum of the donor and the absorption spectrum of the acceptor, is given by:

$$J = \int_0^\infty F_d(\lambda)\varepsilon_a(\lambda)\lambda^4 d\lambda \tag{17}$$

where $F_d(\lambda)$ and $\varepsilon_a(\lambda)$ are the fluorescence intensity of the donor and the absorption coefficient of the acceptor as functions of wavelength (λ, in cm), respectively. The last coefficient in Equation 16, the so-called orientation factor (κ^2), can range from 0 to 4. Its value depends on the relative position in space of the emission dipole of the donor and the absorption dipole of the acceptor. In a fluorescent sample, a multitude of donor-acceptor pairs are present. If the positions of their transition dipoles randomize prior to energy transfer, κ^2 is 2/3. Because κ^2 cannot be measured, this value is often adopted for donor-acceptor pairs within proteins.

The actual distance between a donor and an acceptor molecule (R) can be determined by measuring the efficiency of energy transfer (E):

$$E = 1 - \frac{\tau_d}{\tau_d^0} = 1 - \frac{Q_d}{Q_d^0} \tag{18}$$

Thus, relative fluorescence lifetime or quantum yield measurements of the donor in the absence and presence of the acceptor will yield the efficiency of energy transfer. This experimentally determined efficiency can be related to the distance R by:

$$E = \frac{k_T}{(\tau_d^0)^{-1} + k_T} = \frac{R_0^6}{R_0^6 + R^6} \tag{19}$$

Figure 8 shows that there is a restricted range of actually measurable distances. Within this range, resonance energy transfer can be used as a spectroscopic ruler to determine the distance between sites in proteins and other macromolecules (Latt et al., 1965; Stryer and Haugland, 1967; Steinberg, 1971; Stryer, 1978; Wu and Brand, 1994; Lakowicz, 1999). In the case of proteins, tyrosine and tryptophan represent an intrinsic donor-acceptor pair. However, according to Equations 16 and 18, one needs to know the donor fluorescence parameters in the absence of acceptor. Tryptophan must therefore be removed from the protein, which may have drastic effects on the structure of the protein. Furthermore, it is not possible to determine absolute distances if resonance energy transfer cannot be related to a

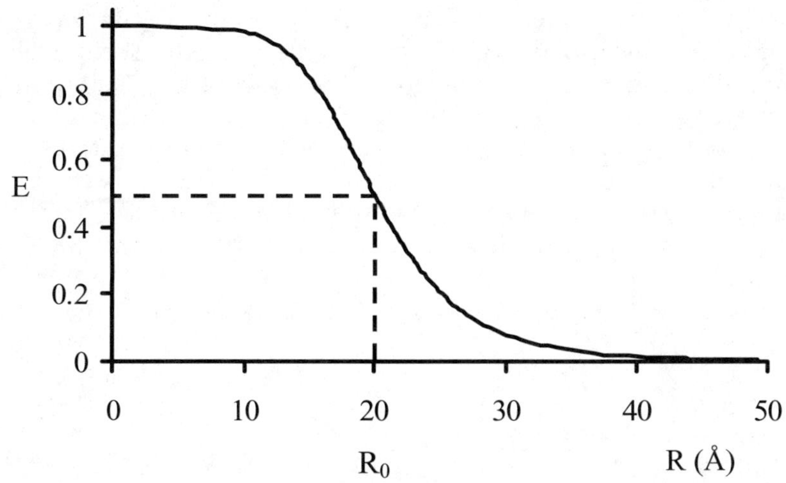

Figure 8 Plot of the quenching efficiency (E) against distance (R). The Förster distance (R_0) is indicated (see section "Resonance Energy Transfer").

specific tyrosine-tryptophan pair. As an alternative, external donors and acceptors may be used that are covalently or noncovalently attached to the protein. Such fluorescent dyes have led to an increased use of resonance energy transfer for obtaining information on protein structure and folding (Wu and Stryer, 1972; Wu et al., 1993; Cai and Schirch, 1996; Taylor et al., 2002). They are also used to study protein function in living cells (Selvin, 2000).

Fluorescence Anisotropy

By fluorescence anisotropy measurements, it is possible to detect rotational motions of fluorophores while they are in the excited state. When the fluorophore is an intrinsic part of the protein or extrinsically attached to it, information about protein hydrodynamic and structural properties can be obtained from such experiments. It is also possible to detect ligand binding by fluorescence anisotropy (see "Fluorescence Lifetime Analysis and Time-Resolved Anisotropy," below). As in ordinary fluorescence, anisotropy measurements can be performed using steady-state or time-resolved instrumentation. Although steady-state measurements are often valuable for detecting changes in protein size or dynamics, time-resolved measurements can give more information, because it is possible to distinguish different rotational modes.

Steady-State Fluorescence Anisotropy

In a fluorescence anisotropy experiment, the sample is excited by linearly polarized light, which results in a preferential excitation of fluorophores with absorption dipoles positioned parallel to the electric field of the incident light (= photoselection). The fluorescence emission (F) is detected at a right angle to the excitation beam through a second polarizer. Measurements with the second polarizer parallel and perpendicular to the position of the excitation polarizer are performed. The measured intensity values are referred to as F_{VV} and F_{VH}, where the subscripts refer to the vertical (V) or horizontal (H) position of the excitation polarizer (left subscript) and the analyzer (right subscript). The intensities F_{VV} and F_{VH} are then used for calculating polarization (p) and anisotropy (r):

$$p = \frac{F_{VV} - G \cdot F_{VH}}{F_{VV} + G \cdot F_{VH}} \tag{20}$$

$$r = \frac{F_{VV} - G \cdot F_{VH}}{F_{VV} + 2G \cdot F_{VH}} \tag{21}$$

Here, the factor G accounts for the different sensitivity of the emission monochromator for vertically and horizontally polarized light and must be determined separately (Lakowicz, 1999). Although polarization and anisotropy are interrelated, the use of anisotropy is generally preferred, because it simplifies many theoretical expressions.

The value of r is affected by several factors (Figure 9). Due to photoselection (Figure 10), the maximum or limiting anisotropy (r_0) in isotropic solutions is 0.4. It is lower if there is a fixed angle (β) between the absorption and emission dipole of the fluorophore:

$$r_0 = 0.4\left(\frac{3\cos^2(\beta) - 1}{2}\right) \tag{22}$$

The angle β depends on the type of fluorophore but may also change with excitation wavelength. For example, a strong variation of r_0 is observed for the amino acid tryptophan at excitation wavelengths between 280 and 300 nm (Valeur and Weber, 1977; Eftink, 1991; Lakowicz, 1999).

A reduction of the steady-state anisotropy below r_0 occurs due to rotational diffusion of the fluorophores during the time of the excited state. For spherical molecules, this is expressed by the Perrin equation:

$$\frac{r_0}{r} = \left(1 + \frac{\tau}{\theta}\right) = 1 + 6D_{rot}\tau \tag{23}$$

Here, θ is the rotational correlation time (in time units), and D_{rot} is the rotational diffusion coefficient (in time^{-1} units). Sometimes, the term rotational relaxation time (ρ) is used, which is related to the rotational correlation time by $\rho = 3\theta$. From Equation 23, it can be seen that anisotropy is a sensitive indicator for rotational motions only when the fluorescence lifetime (τ) and the rotational correlation time

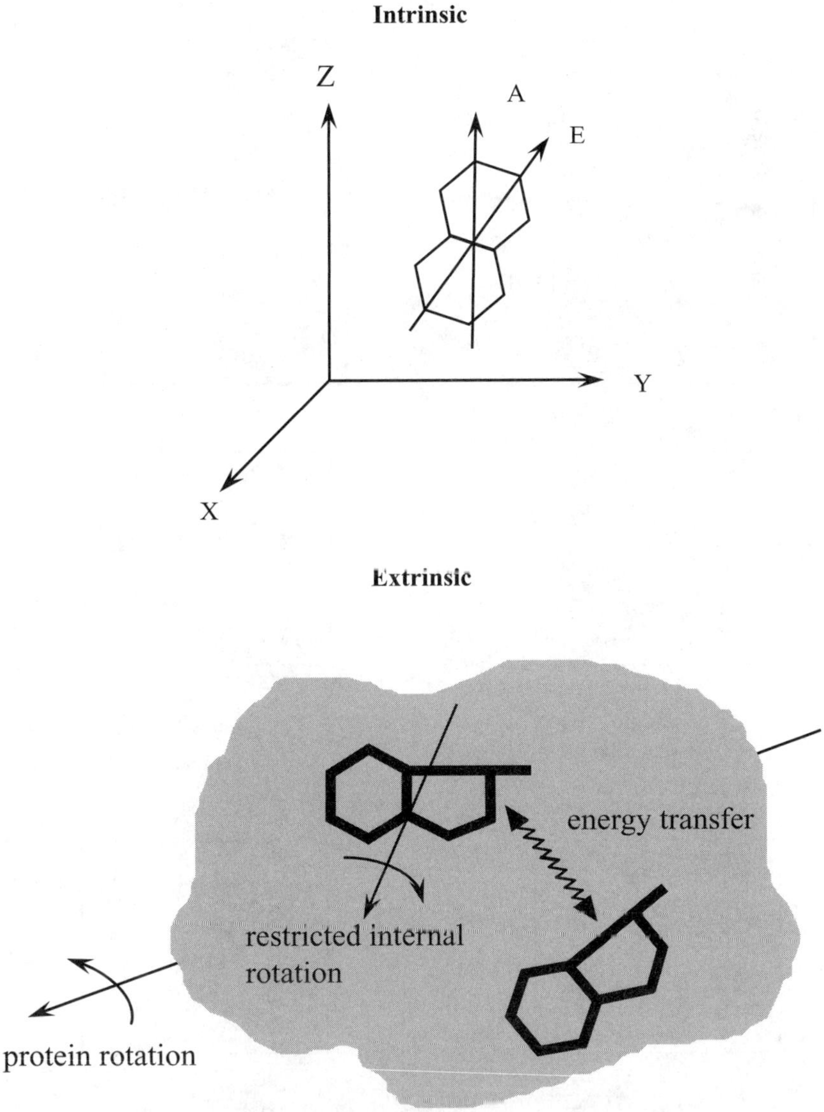

Figure 9 Factors affecting the value of anisotropy (*r*) include the angle between the absorption (*A*) and emission (*E*) dipole, overall protein rotation, restricted internal rotation, and resonance energy transfer.

are in the same order of magnitude. If τ is much greater than θ, the anisotropy is zero, and if τ is much less than θ, the anisotropy equals r_0. Furthermore, it can be seen that for a correct interpretation of anisotropy in terms of rotational motion, r_0 and τ must be known. The determined rotational correlation time can then be used to estimate the hydrodynamic volume of a macromolecule carrying the

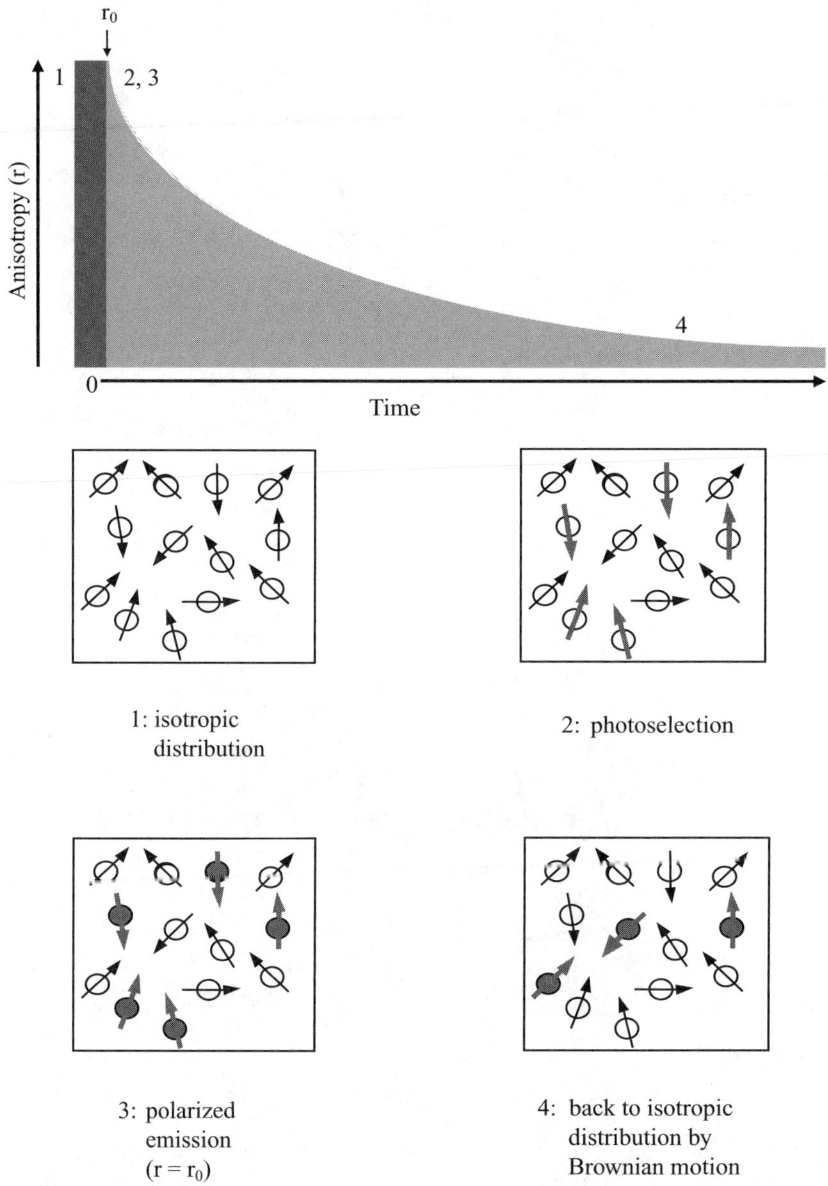

Figure 10 Decay of fluorescence intensity with time after excitation with a light pulse. Only a fraction of the isotropically distributed fluorophores (1) can be excited by linearly polarized light (= photoselection, 2). The fluorophores whose absorption dipoles are parallel to the electric field of the excitation light are most probably excited. Emission is now polarized according to Equation 22; that is, r_0 is reached immediately after pulse excitation (3). Rotational motions of the fluorophores during the time of the excited state lead to a decrease of anisotropy according to Equation 25 (4).

fluorophore, for example, a globular protein, by the Stokes-Einstein equation (Lakowicz, 1999):

$$\theta = \frac{\eta V}{RT} \tag{24}$$

In the above equation, η is the viscosity of the solvent, V is the apparent hydrodynamic volume, R is the gas constant, and T is the absolute temperature. Equations 23 and 24 are valid only for spherical macromolecules, which possess one rotational correlation time. Also, local motions of the fluorophore are not assumed to occur (e.g., the entire motion is due to rotational diffusion of the macromolecule). Many proteins do not fulfill these requirements. Furthermore, in proteins containing several fluorophores, resonance energy transfer (see "Resonance Energy Transfer," above) can cause a reduction of the fluorescence anisotropy. One must also be aware that aggregates can cause an apparent increase of fluorescence anisotropy, because scattered light is vertically polarized (i.e., r = 1). Nevertheless, when all the possible pitfalls are taken into account, steady-state anisotropy measurements can be very useful for detecting changes in protein size or dynamics.

Time-Resolved Fluorescence Anisotropy

In time-resolved fluorescence anisotropy experiments, one measures the angular displacement of fluorophore emission dipoles in real-time. The process of anisotropy decay with time is visualized in Figure 10. Generally, it can be represented by a multiexponential equation:

$$r(t) = r_0 \sum_i \beta_i \exp(-t/\theta_i) \tag{25}$$

where β_i is the fractional amplitude of the associated rotational correlation time θ_i, and $\sum_i \beta_i = 1$. For an irregularly shaped molecule rotating in an isotropic environment, the anisotropy decay is given by a sum of five exponentials. This reduces to three exponentials for molecules of ellipsoidal shape and to one exponential for spheres.

The anisotropy decay of proteins is often affected by rapid reorientation of the fluorophore in addition to the much slower overall rotation of the protein (Figure 10). This is possible for both internal and external fluorophores attached to the protein. There is a simplified equation for the anisotropy decay of a protein exhibiting rapid, internal reorientation and slow overall rotation:

$$r(t) = r_0\{\beta \exp(-t/\theta_{int}) + (1 - \beta)\} \exp(-t/\theta_{prot}) \tag{26}$$

Here, θ_{int} and θ_{prot} are the rotational correlation times for the rapid internal motion and the overall rotation of the protein, respectively. β is the fractional amplitude associated with θ_{int}.

Time domain and frequency domain instruments are both suitable for the measurement of fluorescence anisotropy decays, using similar approaches as for the determination of fluorescence lifetimes (see "Fluorescence Lifetime Measurements," above).

For the analysis of time domain data, two model functions, $F_{VV}^{\text{calc}}(t)$ and $F_{VH}^{\text{calc}}(t)$, are assumed to describe the course of the polarized fluorescence intensities with time upon δ-pulse excitation:

$$F_{VV}^{\text{calc}}(t) = \frac{1}{3}F(t)[1 + 2r(t)] \tag{27}$$

$$F_{VH}^{\text{calc}}(t) = \frac{1}{3}F(t)[1 - r(t)] \tag{28}$$

In the above equation, $F(t)$ represents the decay of the unpolarized fluorescence intensity with time, given by Equation 7. The decay of anisotropy with time, $r(t)$, is given by Equation 25. Simultaneous convolution of Equations 27 and 28 with the corresponding instrument response functions and fitting to the measured polarized intensity curves, $F_{VV}(t)$ and $F_{VH}(t)$, yields the parameters governing fluorescence intensity decay and anisotropy decay.

Depending on the complexity of the model equations (i.e., the number of variables), there can be a substantial correlation between the fitted parameters and a large uncertainty in their values. As for lifetime measurements, this can be reduced by global analysis of several datasets and by fixing certain variables based on either separate measurements or literature values (Beechem et al., 1983; Knutson et al., 1983; Lakowicz, 1999).

Sometimes, the required assumption of a model function may represent a problem of its own. In this case, the maximum entropy method (MEM) can be helpful for determining the distribution of rotational correlation times (and lifetimes) without *a priori* assumptions of the distribution functions (Bastiaens et al., 1992; Brochon, 1994).

In frequency domain experiments, the phase shift at the modulation frequency ω (Δ_ω) and the ratio of the parallel and the perpendicular components of the modulated emission (Λ_ω) are determined:

$$\Delta_\omega = \phi_{VH,\omega} - \phi_{VV,\omega} \tag{29}$$

$$\Lambda_\omega = m_{VV,\omega}/m_{VH,\omega} \tag{30}$$

In the above equations, ϕ and m are the phase difference and modulation ratio, respectively, at given modulation frequency and position of the polarizers. Expected values of Δ_ω ($\Delta_\omega^{\text{calc}}$)and Λ_ω ($\Lambda_\omega^{\text{calc}}$) are calculated using sine and cosine Fourier transforms of the time courses of the individual polarized intensities and are fitted to the measured data by nonlinear regression (Lakowicz, 1999).

Total Internal Reflection Fluorescence

Protein adsorption to surfaces is an important problem in pharmaceutical formulations. Observation of fluorescent molecules at or near interfaces is not possible using conventional fluorometers. Even with front-face illumination, the contribution of the molecules at the interface to the total fluorescence is small compared to the contribution of the molecules in solution. Total internal reflection fluorescence (TIRF), also referred to as evanescent wave spectroscopy (EWS), enables the selective excitation of fluorescent molecules near the interface and observation of their fluorescence. A short introduction to this technique is given here. For more detailed information, the reader is referred to the literature (Hlady et al., 1985, 1999).

The physical basis for TIRF is the total internal reflection of light at a dielectric interface. This is shown in Figure 11 for a solid-liquid

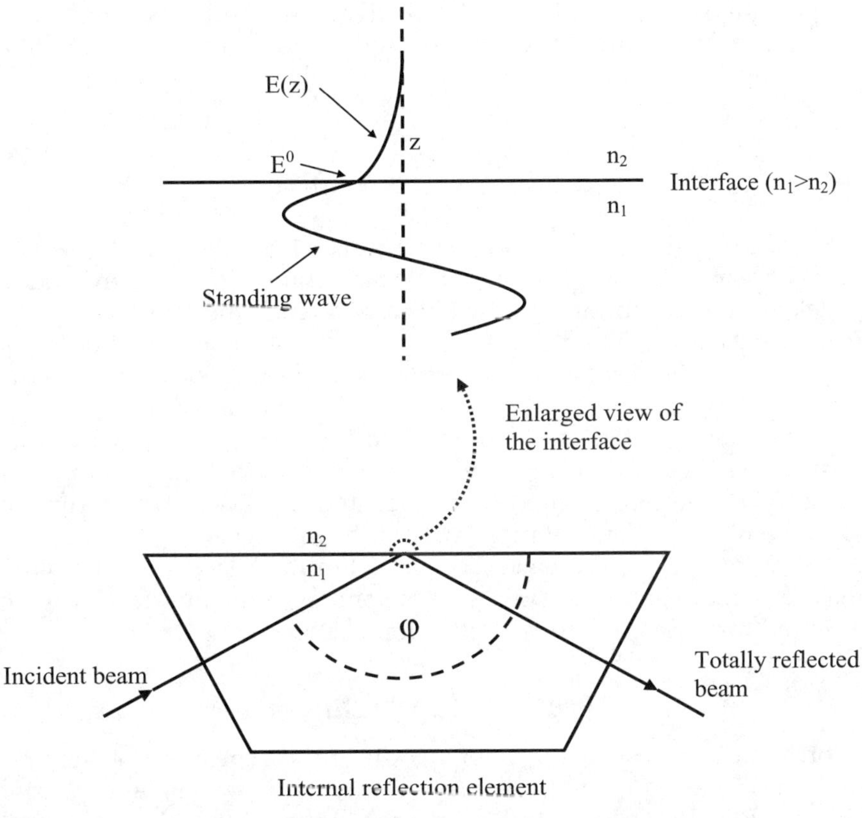

Figure 11 Schematic representation of total internal fluorescence (TIRF) spectroscopy (see section "Total Internal Reflection Fluorescence").

interface. Total reflection of the incident light beam occurs when the incident angle (φ) is larger than the critical angle. Due to constructive interference between the incident and the reflected light, a standing wave is created in the solid. At the interface of two dielectrics, the electric field amplitude of the standing wave has a finite value (E^0). The magnitude of E^0 depends on the incident angle and the refractive indices of the solid (n_1) and the liquid (n_2):

$$E^0 \propto \frac{2 \cos \varphi}{\sqrt{1 - n_2/n_1}} \tag{31}$$

In the optically rarer liquid, an electric field, the evanescent surface wave, $E(z)$, is generated. Its amplitude decreases exponentially with increasing distance from the interface:

$$E(z) = E^0 \exp(-z/d_p) \tag{32}$$

where z is the distance normal to the interface, and d_p is the distance from the interface at which the electric field amplitude decreases to e^{-1} of its interfacial value. The value of d_p depends on the experimental conditions:

$$d_p = \frac{\lambda}{2\pi n_1 \sqrt{\sin^2 \varphi - (n_2/n_1)^2}} \tag{33}$$

In the above equation, λ is the wavelength of the incident light. The depth of penetration is generally of the order one third of the excitation wavelength. Fluorophores in the liquid that reach the interface by diffusion can be excited by the evanescent field. The exponential decay of the evanescent field ensures that excitation of molecules away from the interface is limited.

The internal reflection element used for creating total internal reflection (see Figure 11) is usually a quartz prism. Its liquid facing surface can be chemically modified or coated with a polymer film to affect the physicochemical properties of the interface.

A large part of the fluorescence is emitted back into the solid phase. Therefore, emission of the fluorophores at the interface is again observed through the internal reflection element.

Protein Fluorescence

Fluorescence of protein molecules is due to the presence of the aromatic amino acids tryptophan (Trp, W), tyrosin (Tyr, Y), and phenylalanine (Phe, F). The spectral properties of these intrinsic fluorescent probes as well as their usefulness and limitations for obtaining information on protein structure will be described in this section.

External dyes offer an additional possibility for studying proteins by fluorescence spectroscopy. A great number of different dyes, covering a wide spectral range, are available. Typically, the external probe molecules possess higher absorption coefficients and quantum yields than the aromatic amino acids of the proteins. For many purposes, they must be covalently attached to the proteins. However, information on protein structure can also be obtained from dyes, bound to the proteins by noncovalent interactions.

Intrinsic Protein Fluorescence

Tryptophan, tyrosine, and phenylalanine all absorb light in the ultraviolet region of the spectrum. In water, the absorption maxima are at 280 nm for tryptophan, 275 nm for tyrosine, and 258 nm for phenylalanine. Tryptophan possesses the highest absorption coefficient (5.6×10^3 M^{-1} cm^{-1}), followed by tyrosine (1.4×10^3 M^{-1} cm^{-1}) and phenylalanine (0.2×10^3 M^{-1} cm^{-1}). The emission maxima are around 350 nm for tryptophan, at 304 nm for tyrosine, and at 282 nm for phenylalanine. Quantum yields of tryptophan and tyrosine are 0.13 and 0.14, respectively. Phenylalanine has a very low quantum yield near 0.02 (Demchenko, 1986; Eftink, 1991; Lakowicz, 1999).

Due to its low absorption coefficient and quantum yield, phenylalanine displays a very weak fluorescence. Different functional groups within proteins can further reduce its fluorescence by quenching. Consequently, there are not many studies of phenylalanine fluorescence in proteins (Vanscyoc and Shea, 2001; Vanscyoc et al., 2002).

In most observations of protein fluorescence, excitation wavelengths of 280 nm or higher are chosen. Under these conditions, phenylalanine is no longer excited. If excitation wavelengths near 280 nm are chosen, both tryptophan and tyrosine contribute to fluorescence. One should then be aware of the possibility of resonance energy transfer from tyrosine to tryptophan. This has been used for obtaining information on protein structure (Boteva et al., 1996; Eisenhawer et al., 2001). Furthermore, the hydroxyl group of tyrosine can be deprotonated in an excited state reaction. The resulting tyrosinate has a higher emission maximum and a substantially shorter lifetime than tyrosine (Lakowicz, 1999).

Many proteins contain both tryptophan and tyrosine, and the simultaneous emission of these two fluorophores may also lead to an undesired complication of the fluorescence data. To avoid this, tryptophan can selectively be excited at wavelengths between 295 and 300 nm. Tryptophan shows the strongest fluorescence among the aromatic amino acids. In contrast to tyrosine and phenylalanine, tryptophan emission is highly dependent on the polarity of the environment of the fluorophore. Generally, the emission spectrum is shifted to shorter wavelengths and becomes more structured with decreasing polarity. However, the dynamic ability of polar solvent molecules to undergo dipolar relaxation with the excited state dipole of indole largely affects the extent of the observed Stokes shift (Eftink, 1991; Callis, 1997). In many cases, the contact of water molecules with tryptophan residues of a protein leads to a redshift of tryptophan emission. Furthermore, the wavelength of maximum emission of a tryptophan residue depends on the electric field surrounding it. This field can be affected by the presence of charged amino acid residues of the protein (Vivian and Callis, 2001). All these factors result in a variation of Trp emission spectra for different proteins (Figure 12). With tryptophan fluorescence at femtosecond time resolution, it has been possible to explore the hydration dynamics at the surfaces of two single-tryptophan proteins and relate these observations with a theoretical description of solvation (Pal et al., 2002).

Shifts of the tryptophan emission spectrum are often used to detect structural changes of the protein (Jiskoot et al., 1991a; Tanaka et al., 1997; Metz et al., 2003). However, because the fluorophore is only affected by its immediate environment, it is difficult to estimate the extent of the structural changes. It is also possible to miss structural changes occurring in parts of the protein that contain no tryptophan. In any case, it is advisable to analyze the protein with other analytical techniques and thus get additional structural information.

Besides the dependence of tryptophan emission maxima on the environment, there is also an environmental sensitivity of the fluorescence intensity and quantum yield. In proteins, the quantum yield of tryptophan ranges from almost zero to 0.35. This is due to the varying interaction of tryptophan with quenchers, including water, oxygen, peptide bonds, and a number of amino acid side-chains (MacMahon et al., 1992; Chen et al., 1996, Chen and Barkley, 1998; Pan and Barkley, 2004). Tryptophan fluorescence intensity is sometimes used as an indicator for conformational changes of the protein. Furthermore, there are numerous studies that relate the accessibility of tryptophan to external quenchers (e.g., acrylamide or iodine) with protein conformation (Lakowicz, 1999; Sharma and Kalonia, 2003a).

Because of the various possibilities for quenching of tryptophan fluorescence, there is also a large variation of the fluorescence lifetime. Within proteins, it can range from a few hundred picoseconds to nine nanoseconds. Proteins containing more than one tryptophan usually exhibit multiple lifetimes. A possible approach for determining the origin of the observed lifetimes is the investigation of single tryptophan mutants of the protein (Royer, 1992). However, even single-tryptophan proteins may display heterogeneous fluorescence intensity decays. Various models have been developed to explain this heterogeneity. The rotamer model proposes that three different rotameric states of tryptophan are responsible for multiple lifetimes (Szabo and Rayner, 1980; Adams et al., 2002). Another model assumes a quasi-continuous, dynamical distribution of tryptophan positions, which results in a continuous lifetime

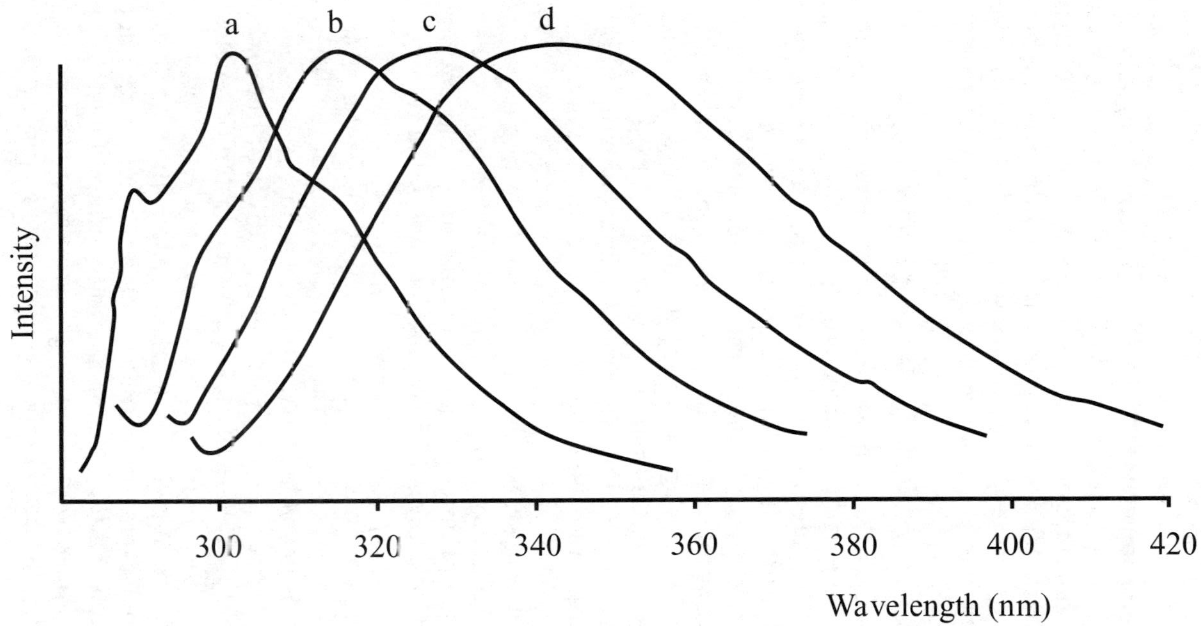

Figure 12 Normalized emission of single tryptophan proteins: a = apoazurin, b = ribonuclease T1, c = nuclease A, d = glucagon (Eftink, 1991).

distribution (Alcala et al., 1987a,b). Further models do not absolutely require the existence of unique conformational states of tryptophan or dynamical motions of the fluorophore. Structural heterogeneities of the proteins are assumed to be responsible for variations in the quenching of tryptophan fluorescence, especially by electron transfer (Petrich et al., 1983; Bajzer and Prendergast, 1993; van Gilst et al., 1994; Callis and Liu, 2004). None of the above models is *a priori* right or wrong. The choice of a model can only be justified by the help of additional experimental techniques or computational data (Moncrieffe et al., 2000). Recently, it has been shown that 5-fluoro-tryptophan built into proteins shows monoexponential fluorescence decay kinetics, which simplifies the interpretation of time-resolved fluorescence data (Broos et al., 2004).

Extrinsic Fluorescent Probes

For most applications, extrinsic fluorescent dyes are covalently attached to the proteins or to molecules that interact with the proteins (e.g., ligands). They are frequently used for resonance energy transfer experiments. Further applications include anisotropy measurements for studying protein-protein or protein-macromolecule interactions, as well as studies of protein structure using environmental sensitive dyes. A large number of dyes are commercially available, and Table 1 lists a selection of them. The choice of a dye depends on the specific application. Appropriate excitation and emission maxima are important for resonance energy transfer experiments. For fluorescence anisotropy measurements, a dye with a suitable lifetime and a high maximum anisotropy (r_0) is preferably chosen. In ligand binding experiments or fluorescence microscopy, a high sensitivity is often required, and dyes with a high absorption coefficient and quantum yield are favorable. Further important aspects of a dye are its photostability, especially in fluorescence microscopy, and its possible pH sensitivity.

Dyes can be conjugated to amino groups within the protein (e.g., the ε-amino group of lysine or the α-amino group of the N-terminus). Alternatively, it is possible to label the thiol group of cystein.

Important types of amino reactive derivatives of fluorophores include succinimidyl esters, sulfonyl halides, and isothiocyanates. These derivatives are linked to the protein by amide, sulfonamide, and thiourea bonds, among which the amide bond is the most stable.

Iodoacetamide and maleimide derivatives of fluorophores are most frequently used for labeling the thiol group of cystein. Because the thiol functional group is not very common in most proteins, this type of labeling often enables the modification of a protein molecule at a specific site.

The conjugation reaction itself varies between the different fluorophore derivatives. Typically, for conjugation of amine groups, it is carried out at a slightly elevated pH between 8 and 9, whereas thiol

Dye	Absorption Maximum (nm)	Absorption Coefficient (M^{-1} cm^{-1})	Emission Maximum (nm)	Remark
2-Dimethylamino-naphtalene -5-sulfonyl chloride (= 2,5-Dansyl-Cl)	275, 363[a]	36,000 (258 nm)[a] 2090 (352 nm)[a]	465[a]	Useful for anisotropy studies because of long lifetime (~30 ns).
1,5-IAEDANS	336[b]	5700[b]	490[b]	Water soluble, photosensitive, long lifetime (greater than 10 ns).
1-Anilino-naphtalene -8-sulfonate (= ANS)	372[c]	7800[c]	480[c]	Noncovalent. Spectral properties depend on solvent polarity. Used for detecting changes in protein conformation.
4,4-Dianilino-1, 1-binaphtyl-5, 5-di-sulfonic acid (= bis-ANS)	395[c]	23,000[c]	500[c]	Noncovalent. Spectral properties depend on solvent polarity. Used for detecting changes in protein conformation.
Cascade blue acetyl azide	409[c]	24,000[c]	558[c]	

Table 1 Extrinsic Fluorescent Dyes

(continued on following page)

Dye	Absorption Maximum (nm)	Absorption Coefficient (M^{-1} cm^{-1})	Emission Maximum (nm)	Remark
Fluorescein	490[d]	93,000[d]	514[d]	Emission intensity is strongly pH dependent.
5,6-Carboxy-fluorescein	492[d]	75,000[d]	517[d]	Emission intensity is strongly pH dependent.
Tetramethyl-rhodamine-5-iso-thiocyanate	543[c]	99,000[c]	571[c]	
Nile red	552[c]	45,000[c]	636[c]	Noncovalent. Spectral properties depend on solvent polarity. Used for detecting changes in protein conformation.
Texas red sulfonyl chloride	587[e]	85,000[e]	602[e]	

[a]Conjugated to IGG.
[b]At pH 8.
[c]In MetOH.
[d]At pH 9.
[e]In chloroform (similar absorption and emission maxima in water).

Table 1 *(continued from previous page)*

conjugation is performed at a pH below 7. In both cases, the protein is initially dissolved in the buffer solution. An excess amount of the reactive fluorophore, dissolved in an organic solvent (e.g., acetonitrile, dimethylformamide, etc.), is added. Conjugation takes place at room temperature. As a final step, it is necessary to remove free dye that has not reacted, either by chromatographic methods or dialysis (Taylor et al., 2002).

In conjugation reactions, it is important to reach an optimal degree of labeling, usually about 1–3 moles of dye per mole of protein. Whereas a too low degree of labeling may simply result in a loss of sensitivity, overlabeling may lead to the quenching of fluorescence and to changes in protein structure and isoelectric point. Furthermore, it is often necessary to know the (preferred) location of labeling. Both degree and location of labeling can be determined by mass spectrometric analysis of the labeled proteins and their digested fractions.

Information on protein structure can also be obtained from extrinsic fluorophores that are not covalently attached to the protein molecule. A prominent example is 8-anilino-1-naphtalene sulfonic acid (ANS). This dye increases its fluorescence intensity substantially with increasing hydrophobicity of its environment. Due to its own hydrophobic nature, ANS can interact with hydrophobic patches on the protein surface. Such patches are more frequently present in partially unfolded than in completely folded proteins. Therefore, ANS is used to detect the intermediate, partially unfolded state of proteins (see "Applications Using Extrinsic Fluorescent Probes," below).

Applications

Applications Using Intrinsic Protein Fluorescence

Spectral Measurements

Changes in the intensity and λ_{max}^{Em} of intrinsic protein fluorescence emission spectra are widely used as indicators for alterations in the conformation of (formulated) proteins. Examples include studies of (1) temperature effects on the structure of bovine granulocyte colony stimulating factor (Kueltzo and Middaugh, 2003), recombinant human interferon-alpha2a (rhIFNα2a) (Sharma and Kalonia, 2003a), adenovirus (Rexroad et al., 2003), meningococcal PorA (Arigita et al., 2004) and PorB (Minetti et al., 1997); (2) pH effects on the conformation of immunoglobulins (Lin et al., 1989; Buchner et al., 1991; Jiskoot et al., 1991a), bovine growth hormone (Holzman et al., 1990), colicin E9 DNase (van den Bremer et al., 2002), rhIFNα2a (Sharma and Kalonia, 2003a), and bacterial toxoids (Paliwal and London, 1996; Xing et al., 1996; Johansen et al., 1998); (3) the influence of organic solvents on the conformation of human growth hormone (Oroszlan et al., 1992; Hovorka et al., 2001); (4) effects of protein-lipid interactions on protein structure (Wang et al., 1997a,b; Koppenhagen et al., 1998; Chen and

Sanyal, 1999; Balasubramanian et al., 2000; van Slooten et al., 2000; Arigita et al., 2003); and (5) the effect of freeze-drying on the structure of a monoclonal antibody (Taschner et al., 2001).

For several immunoglobulins, a dramatic increase in intrinsic Trp fluorescence has been observed at low-pH conditions (Lin et al., 1989; Buchner et al., 1991; Jiskoot et al., 1991a). In these examples, λ_{max}^{Em} was not redshifted, indicating that the acidic environment had not induced complete unfolding. Enhanced Trp fluorescence with no change in λ_{max}^{Em} has also been observed for other proteins under various conditions, for example, for bovine growth hormone at low pH or at intermediate guanidinium hydrochloride (GdnHCl) concentrations (Havel et al., 1988; Holzman et al., 1990) and for interleukin-2 (IL-2) on interaction with liposomal bilayers (Koppenhagen et al., 1998). These observations are usually explained by swelling of the protein molecules, which would result in larger distances between Trps and quenching neighboring groups (Chen and Barkley, 1998) and thus less efficient intramolecular quenching.

Intrinsic Trp fluorescence was used to study the effects of temperature and pH on the structural stability of rhIFNα2a, which contains two Trps (Sharma and Kalonia, 2003a). After a heating/cooling cycle (20–80°C) at pHs between 4.0 and 7.4, the Trp emission showed a drop in intensity and λ_{max}^{Em} had shifted from 336 to 340 nm, indicating that at least one of the two Trps had moved to a relatively polar environment as a result of a conformational change. The non-superimposable heating and cooling curves in emission intensity-temperature plots indicated that this change was irreversible. At lower pH (2.5–3.5) the λ_{max}^{Em} had already shifted to 340 nm at room temperature, and no additional changes were observed after a heating/cooling cycle.

Toxoids are inactivated bacterial toxins used as vaccines. They are usually made by treating a toxin solution with formaldehyde. Based on the shift in λ_{max}^{Em} of the Trp fluorescence of diphtheria toxin and toxoid from 330 nm in the folded state to 350 nm in the unfolded state, it was shown that diphtheria toxoid is more resistant to unfolding by GdnHCl, low pH, and heat than the original toxin (Paliwal and London, 1996). Most likely, formaldehyde-induced cross-links stabilize the folded conformation of the toxoid. Indeed, whereas the midpoint of the denaturation curve (as monitored by the redshift in the fluorescence spectrum) was at about 1.0 M GdnHCl for the toxin, the GdnHCl concentration needed to reach the denaturation midpoint gradually increased from 1.0 to 2.8 M as higher formaldehyde concentrations were used in the inactivation process (Metz et al., 2003).

Xing et al. showed that λ_{max}^{Em} of free tetanus toxoid (excited at 280 nm) shifted from 330.5 nm to 334 nm on incubation for 50 days at 37°C and pH 7.4, indicating partial unfolding (Xing et al., 1996). At pH 2.5 under otherwise the same conditions, shifts in the λ_{max}^{Em} were accompanied by a drastic decrease in intensity and the concomitant appearance of an emission peak around 450 nm, indicative of oxidation of Trp and/or Tyr (Simat and Steinhart, 1998; Arigita et al., 2004). When the release of tetanus toxoid from polylactic-*co*-glycolic acid (PLGA) microspheres was followed (pH 7.4, 37°C), λ_{max}^{Em} had redshifted by 2–3 nm, and a weakly emitting shoulder at ca. 450 nm was seen. These results indicate that acidification of the PLGA matrix during the release stage contributed to the degradation of tetanus toxoid. The appearance of emission peaks at wavelengths above 400 nm indicative of fluorescent amino acid degradation products were also observed in accelerated stability studies of tetanus toxoid (Johansen et al., 1998; Sasiak et al., 2000) and diphtheria toxoid (Johansen et al., 1998) under acidic conditions and in meningococcal outer membrane protein PorA vaccine at pH 7.4 (Arigita et al., 2004).

Spectral information has also been used to probe protein adsorption to solid surfaces by TIRF methodology. Besides quantitation of protein adsorption (Hlady et al., 1986), information on the conformation of proteins in the adsorbed state can be obtained. The spectra of human growth hormone and lysozyme showed broadening of the emission peaks as well as changes in λ_{max}^{Em} on adsorption to a hydrophobic surface, indicating that both proteins had undergone structural changes on adsorption (Buijs and Hlady, 1997).

Solid-state fluorescence emission spectroscopy was used to study the tertiary structure of dried beta-lactoglobulin (βLG) and rhIFNα2a (Sharma and Kalonia, 2003b). The emission spectra of the dried proteins were blueshifted by several nanometers with respect to the solution spectra, most likely as a result of water removal. On the other hand, when βLG was dried in the presence of GdnHCl, redshifts were observed, the extent of which was dependent on the GdnHCl concentration and the residual water content. After precipitation of rhIFNα2a with polyethylene glycol and subsequent vacuum drying, besides the blueshifted λ_{max}^{Em} (from 336 to 332 nm), a shoulder was observed at 352 nm, indicating the presence of unfolded protein species. When the same procedure was followed in the presence of trehalose, but not mannitol, the intensity of the 352-nm shoulder was strongly reduced, indicating that trehalose protected rhIFNα2a against drying-induced unfolding. These results illustrate that, although data interpretation is complicated because changes in both protein structure and polarity of the environment induce spectral shifts, structural information about proteins in the dried state can be obtained via this method.

Most of the above examples deal with multiple-Trp proteins. Spectral information of these proteins therefore reflects the average environment of all Trps. To study the fluorescence behavior of the individual Trps, one can make use of single-Trp mutants (Engelborghs, 2001, 2003). If the protein conformation is not affected by the mutation(s), such mutants can provide useful information about the individual Trps and about local conformational changes near these Trps.

Examples of this approach are studies with single-Trp mutants of diphtheria toxin chains A and B (Wang et al., 1997a), barnase (De Beuckeleer et al., 1999), and E9 colicin DNase (Keeble et al., 2002). Alternatively, selective quenching of accessible Trps by external quenchers can be used to obtain information on the fluorescence properties of buried, nonquenchable Trps (Brewer et al., 1987; Lakowicz, 1999).

Fluorescence Quenching by Collisional Quenchers

In a typical steady-state quenching experiment, the protein is titrated with increasing concentrations of the quencher molecule, and the fluorescence emission is measured at fixed excitation/emission wavelengths. Quenching of Trp fluorescence by oxygen has traditionally been used for probing the conformational dynamics of proteins (Lakowicz and Weber, 1973a,b). Oxygen can deeply penetrate into the hydrophobic interior of proteins and can reach essentially all Trps. In contrast, hydrophilic quenchers such as acrylamide and iodide only reach relatively exposed residues, although acrylamide can diffuse to some extent into the protein interior (Eftink and Ghiron, 1977; Havel et al., 1988).

Fluorescence quenching by acrylamide and iodide is widely employed to probe the localization of Trps in proteins. In one example, the fractional accessibility (f_a in Equation 14) of the Trps of a monoclonal antibody was 100% for acrylamide but less than 75% for iodide (Jiskoot et al., 1991a). The same study revealed that the Stern-Volmer quenching constants (K_D) for both iodide and acrylamide quenching were higher under acidic (pH 3–4) than at alkaline (pH 7.4–10) conditions. A similar increase in the extent of quenching by acrylamide and iodide was observed on acidification of rhINFα2a (Sharma and Kalonia, 2003a).

Koppenhagen et al. reported a dramatic effect of the adsorption of recombinant human interleukin-2 (rhIL2) to liposomes on the quenching by iodide of the single Trp residue in this protein (Koppenhagen et al., 1998). In the absence of liposomes, the fluorescence was only slightly quenched on addition of iodide (see Figure 13). After adsorption of the positively charged protein to negatively charged liposomes, quenching was strongly enhanced. Besides dynamic quenching, a static component was observed, as can be seen from the upward curve in the Stern-Volmer plot (Figure 13).

Enhanced quenching efficiency as seen in the above examples can be explained by an increase in the dynamic flexibility of the protein matrix, which increases the probability that transient channels open and therefore allow a quencher to diffuse more easily into the protein matrix (Eftink and Ghiron, 1977; Eftink and Hagaman, 1986; Somogyi et al., 1986). This would lead to higher values for f_a and/or K_D. It should be noted, however, that K_D is not only dependent on the bicollisional quenching constant (k_q) but also on the fluorescence lifetime (τ_0, see Equation 10). Therefore, a change in K_D (resulting from a structural change in the protein) can result from a change in k_q or τ_0,

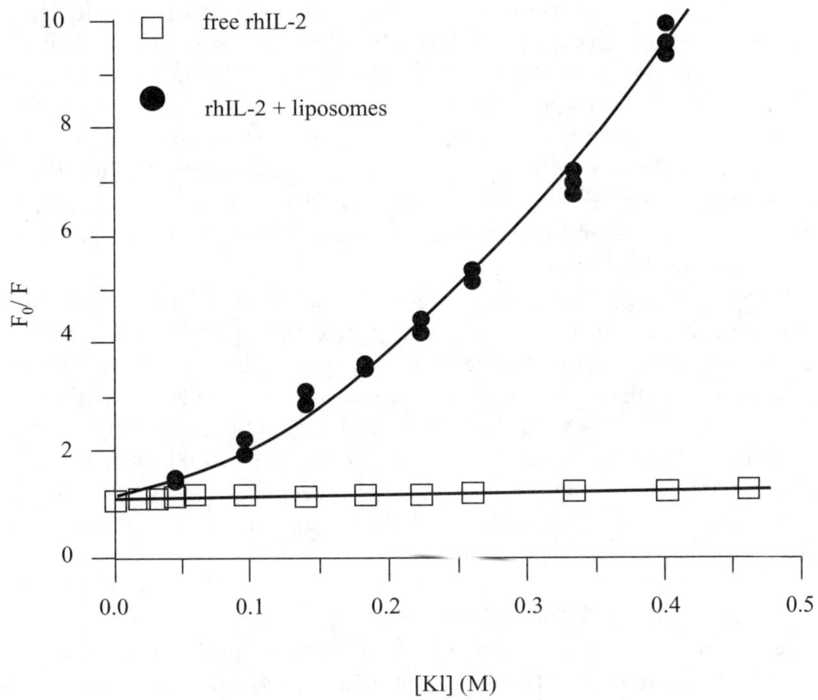

Figure 13 Stern-Volmer plot showing the quenching of rhIL2 Trp fluorescence by iodide. Aliquots of 4 M potassium iodide with 0.2 mM sodium thiosulfate (to prevent oxidation of iodide) were added to 50 µg/mL rhIL2 in the absence or presence of negatively charged liposomes (2 mM phospholipid). The fluorescence intensity was measured using excitation wavelengths of 300 and 335 nm, respectively. F represents the fluorescence intensity corrected for background and dilution. Adapted from Koppenhagen et al. (1998).

or both, which can only be found out by fluorescence lifetime measurements. For instance, in the study by Koppenhagen et al., it was found that the average lifetime of liposome-associated rhIL2 was twice that of free rhIL2 (Koppenhagen et al., 1998), which contributed in part to the observed increase in K_D.

Steady-State Anisotropy

Analogous to quenching, for full interpretation of steady-state anisotropy of protein fluorescence, time-resolved measurements are required (see "Steady-State Fluorescence Anisotropy," above). Still, steady-state anisotropy studies alone can provide useful indications for changes in protein structure (e.g., oligomerization, aggregation, and unfolding) and in some cases can be used to measure ligand binding.

Taschner et al. observed a small but significant decrease in steady-state anisotropy of a monoclonal antibody as a result of freeze-drying, despite a reduction in average lifetime (which by itself would lead to an increase in anisotropy), indicating increased internal flexibility of the Trps (Taschner et al., 2001). A decrease in anisotropy was observed for another monoclonal antibody on acidification from pH 7.4 to 3.0 (Jiskoot et al., 1991a). Because the effect of the pH drop on the fluorescence decay was not studied, one cannot conclude whether the observed change was due to an increase in Trp flexibility or solely to a longer average lifetime (as indicated by the observed increase in fluorescence intensity).

Intrinsic steady-state fluorescence anisotropy can be used to study ligand binding to proteins if the difference between the anisotropy of the free and ligand-bound protein is sufficiently large. Figure 14 shows the increase in anisotropy of lysozyme fluorescence upon titration with heparin (van de Weert, unpublished data). The anisotropy of free lysozyme ($r = 0.098$) increases proportionally with the heparin concentration until a plateau level is reached ($r = 0.119$) above a molar ratio of heparin/lysozyme of 1.0, indicating that lysozyme has one binding site for this ligand.

Fluorescence Lifetime Analysis and Time-Resolved Anisotropy

Cytokines have been formulated in liposomes to improve their efficacy as adjuvant in antitumor therapies. Time-resolved fluorescence has been applied to study the interaction of rhIL2 (Koppenhagen et al., 1998) and rhIFNγ (van Slooten et al., 2000) with liposomal bilayers at low ionic strengths, which are typically used to enhance the encapsulation efficiency via electrostatic interactions between the negatively charged bilayers and the positively charged proteins. Both single-Trp proteins showed multiple exponential fluorescence decay. As compared to free rhIL2, liposome-associated rhIL2 lost one short lifetime component concomitant with an increase of the fractional intensities of the longest lifetime components. This leads to a doubling of the average lifetime (Table 2). This indicates a change in the direct environment of the Trp residue in the adsorbed state. In contrast, the lifetime decay of rhIFNγ remained nearly unchanged after adsorption to preformed liposomes. However, when rhIFNγ was encapsulated in the aqueous compartment of the liposomes, an additional short (0.15 nanosecond) lifetime component was needed to fit the fluorescence decay, and the average lifetime was reduced accordingly (Table 2). Apparently, during the preparation of rhIFNγ-liposomes, stress factors other than bilayer adsorption induced a change in the local Trp environment.

Time-resolved anisotropy measurements of rhIL2 revealed two rotational correlation times (Table 2): a short one indicative of local motion of the Trp residue and a long one ($\theta \sim 79$ nanoseconds). The latter result indicates that the protein formed aggregates under the low ionic strength conditions used, because monomeric rhIL2 has a size of only

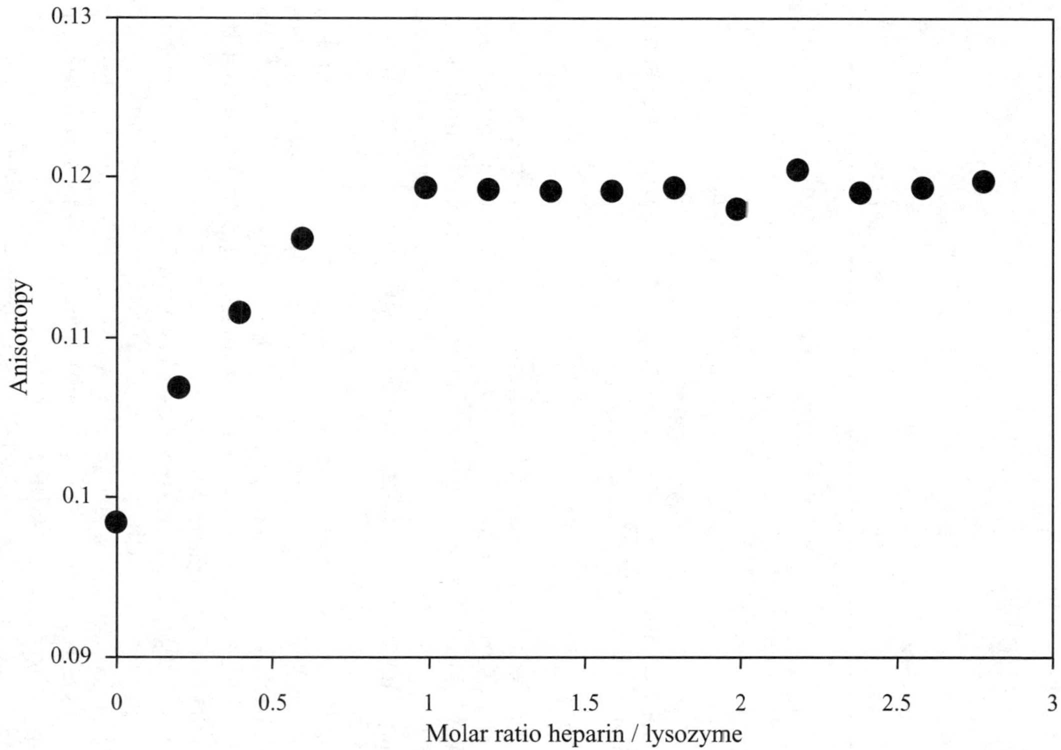

Figure 14 Effect of heparin-lysozyme association on the steady-state anisotropy of lysozyme fluorescence. A solution of lysozyme was titrated with a concentrated solution of the 6-kDa fraction of porcine heparin in 50 mM phosphate buffer, pH 7.4, at 25°C. Excitation at 295 nm, emission at 350 nm. (Unpublished data, kindly provided by Dr. Marco van de Weert.)

Description	$\tau_{average}$ (ns)	θ_1 (ns)	θ_2 (ns)
Free rhIL2	1.0 ± 0.06[b]	79 ± 20[b]	1.2
rhIL2, adsorbed to liposomes	2.0 ± 0.1[b]	132 ± 32[b]	0.9
Free rhIFNγ	3.2	27.6 (24.6–31.0)[c]	—
rhIFNγ, adsorbed to liposomes	3.2	Greater than 100	—
rhIFNγ, desorbed from liposomes	3.3	25.2 (22.9–28.2)[c]	—
rhIFNγ-containing liposomes	2.0	Greater than 100	—

[a]Data of rhIL2 and rhIFNγ were taken from Koppenhagen et al. (1998) and van Slooten et al. (2000), respectively.

[b]Mean ± estimated standard error.

[c]Confidence intervals at the 67% level are shown in parentheses.

Table 2 Time-Resolved Fluorescence Parameters of Free and Liposome-Bound rhIL2 and rhIFNγ[a]

15 kDa with an expected θ of ca. 6 nanoseconds (Koppenhagen et al., 1998). When adsorbed to liposomes, the long rotational correlation time increased, indicating that rhIL2 became even more immobilized. Unlike rhIL2, rhIFNγ exhibits only one rotational correlation time of ca. twenty-five nanoseconds, reflecting the overall motion of the native dimer (van Slooten et al., 2000). This suggests that the single Trp in rhIFNγ is fully immobilized at the nanosecond timescale, although a fast rotation with a minor fractional amplitude (β = 0.006) was observed in an earlier study (Boteva et al., 1996). Analogous to rhIL2, the rhIFNγ molecules were immobilized on adsorption to liposomal bilayers, as reflected by an irresolvably long rotational correlation time (θ much greater than 100 nanoseconds). When rhIFNγ was desorbed from the liposomes by adding sodium chloride, the lifetimes and rotational correlation times were similar to those of untreated rhIFNγ (Table 2), indicating that the conformation was preserved after the adsorption/desorption procedure.

Time-resolved analysis of protein fluorescence is very useful for studying conformational changes induced by ligand binding. Chabbert et al. showed that binding of two peptide ligands to five Trp-containing mutants of calmodulin affected the fluorescence quantum yield and lifetime distribution (Chabbert et al., 1991). The observed effects were attributed to changes in the microenvironment of the Trps as a result of a conformational change in the calmodulin mutants on ligand binding. The five mutants reacted differently, illustrating that conformational changes are not necessarily uniform throughout a protein. Likewise, binding of zinc ion (Zn^{2+}) or immunity protein (Im9) to colicin E9 DNase had distinct effects on its two Trps (Trp22 and Trp58), as shown by studies using single-Trp mutants of this protein (van den Bremer et al., 2004). Whereas Im9 binding resulted in a decrease in the longest lifetime component for Trp22, for Trp58 an opposite effect was observed. Complexation of E9 DNase with Zn^{2+} induced compaction of the E9 DNase structure, as reflected by the decrease in the overall rotational correlation time from 11.5 to 9.5 nanoseconds. Moreover, the local mobility of the Trp residue in either mutant was strongly restricted upon Zn^{2+} or Im9 binding: the fast rotational correlation time of Trp22 completely disappeared, whereas for Trp58 its fractional contribution decreased. Other examples of time-resolved measurements to probe ligand-induced changes are studies of heme-binding to myoglobin (Bismuto et al., 1989; Willis et al., 1990) and calcium binding to parvalbumin (Ferreira, 1989).

Applications Using Extrinsic Fluorescent Probes

Both covalent and noncovalent fluorescent dyes have been used to study diverse aspects of protein structure and function. The surface of partially unfolded proteins generally is more hydrophobic than that of their native counterparts. Consequently, they bind more efficiently to hydrophobic dyes such as ANS, bis-ANS, and nile red, leading to enhanced fluorescence intensity and spectral shifts of the dye. As indicated by increased bis-ANS fluorescence, it was shown that acid-treated or oxidized anti-HSA IgG had undergone partial unfolding (Lin et al., 1991). A similar increase of bis-ANS fluorescence was found after oxidation of HSA (Anraku et al., 2001) and of ANS fluorescence after acid-treatment of mutant interleukin-6 (De Filippis et al., 1996). A recent study has shown that ANS itself may induce structural changes within proteins at low pH (Ali et al., 1999). Such effects have not been observed for another hydrophobicity sensitive dye, nile red, which may therefore be a valuable alternative for ANS (Hendriks et al., 2002).

Particularly for large proteins such as antibodies, the average lifetime of the intrinsic fluorescence is too short to measure accurately the relatively slow rotational motions of these proteins by time-resolved anisotropy. Therefore, fluorescent dyes with longer average lifetimes have been applied for studying the conformational dynamics of large proteins. For instance, dansyl derivatives have been used to study the segmental flexibility of anti-dansyl IgG (Schneider et al., 1988) and IgE (Holowka et al., 1990). Antibody-bound dansyl compounds showed an average lifetime of ca. 27 nanoseconds, long enough to measure both the overall tumbling of the antibodies ($\theta \sim 120$ to 125 nanoseconds) and segmental motions of the antigen-binding fragments (Fab) reflected by a second rotational correlation time ($\theta \sim 15$ to 35 nanoseconds for IgG,

48 nanoseconds for IgE). In another study, the C-terminus of an anti-fluorescein monoclonal Fab was covalently labeled with IAEDANS (Lim et al., 1995). Time-resolved fluorescence studies showed an average lifetime of 15.6 nanoseconds of the label and two rotational correlation times. The longer one (θ_1 ~ 33 nanoseconds) was attributed to global tumbling of the Fab, whereas the shorter one (θ_2 ~ 1.1 nanoseconds) could be a combination of both the local flexibility of the label and segmental flexibility of the constant part of the Fab relative to the antigen-binding domains. The value of the fast rotational component increased significantly (to 1.7 nanoseconds) upon antigen binding, indicating restricted segmental motion of the ligand-bound Fab.

Fluorescent ligands have successfully been applied to study protein-ligand interactions. A well-known example is the anti-fluorescein/fluorescein system (Watt and Voss, 1977). When the hapten (fluorescein) binds to anti-fluorescein antibodies, its fluorescence is strongly quenched through a mechanism most probably involving charge transfer complex formation between the excited fluorescein molecule and two neighboring Trps in the antigen-binding site (Watt and Voss, 1977; Herron et al., 1994), although electrostatic interactions may also play a role (Omelyanenko et al., 1993). Making use of the fluorescence quenching and the increase in fluorescence anisotropy of the bound ligand, this system has been employed to gain insights into, among others, the kinetics (Kranz et al., 1981, 1982) and thermodynamics of ligand-binding (Herron et al., 1986).

For most protein-ligand combinations, the ligand is nonfluorescent. In some cases, however, the ligand can be covalently labeled with a fluorescent dye, and changes in the anisotropy and/or fluorescence intensity of the labeled ligand can then be used to measure binding of the ligand to the protein. Examples are the following antibody-hapten combinations: fluorescein-labeled phenytoin with anti-phenytoin antibody (Bright, 1989), fluorescein-labeled epitope-mimicking peptides with an antibody raised against meningococcal outer membrane protein PorA (Jiskoot et al., 1991b; van den Elsen et al., 1997b), and tetramethylrhodamine-labeled or double-labeled peptides mimicking human chorionic gonadotrophin (hCG) with an anti-hCG antibody (Wei and Herron, 1993; Wei et al., 1994). Such fluorescently labeled peptides can also be used as a tracer to study the interaction of the antibody with the native antigen, as has been described for hCG/anti-hCG monoclonal antibody (Wei and Herron, 1993) and for membrane-bound PorA/anti-PorA monoclonal Fab (van den Elsen et al., 1997a). For more information on the use of fluorescence-based methods to measure equilibrium constants for protein-ligand interactions, the reader is referred to reviews on this topic (Eftink, 1997; Tetin and Hazlett, 2000).

Finally, fluorescent labels have been used for interfacial studies by TIRF, including quantitation (Hlady et al., 1986) and kinetic studies (Hlady and Buijs, 1996; Ho et al., 1996; Buijs and Hlady, 1997) of protein adsorption, as well as conformational analysis of adsorbed proteins (Hlady and Andrade, 1989; Suci and Hlady, 1990). The reader is referred to the literature for more information on this topic (Jiskoot et al., 1995; Hlady et al., 1999).

Practical Considerations and Experimental Pitfalls

Fluorescence measurements require ultrapure materials, including water, buffer components and salts, and cuvettes. It is particularly important to verify the absence of fluorescent impurities. In this respect, it is recommended to take an absorption spectrum (or, if only low protein concentrations are available, a fluorescence excitation spectrum) and a fluorescence emission spectrum of the sample and compare that with a pure reference standard. Note that the protein source itself may contain impurities, such as host-cell derived proteins or degradation products. A purification step may then be needed prior to fluorescence measurements. For verifying the absence of proteinaceous impurities, other techniques such as electrophoresis or chromatography should be used. Several detergents commonly used for the solubilization of membrane proteins are notorious for fluorescent impurities that need to be removed (Dijkstra et al., 1996). In our experience, phospholipids from natural sources may also contain fluorescent impurities.

Measurement of intrinsic protein fluorescence requires quartz cuvettes; for fluorescent dyes that are excited at greater than 320 nm, glass cuvettes suffice. As fluorescence properties are temperature dependent, control of the sample temperature is strongly advised. When time or temperature effects are studied, it is advisable to use cuvettes that can be hermetically closed to avoid evaporation of the solvent. For measurements below room temperature, the sample compartment of the fluorimeter should be flushed with nitrogen to avoid condensation of vapor. Especially when repeated or long-lasting measurements are envisioned, irreversible damage of the fluorophores by the excitation light may occur, leading to a reduction of the fluorescence intensity in time (photobleaching). To minimize photobleaching, exposure times should be kept as short as possible.

Samples should be diluted to optical density values below ca. 0.05 to avoid inner filter effects. If dilution is not an option, one can use cuvettes with shorter pathlengths or use front-face detection geometries (Lakowicz, 1999). The solution in which the protein is dissolved or dispersed should be optically transparent over the entire excitation/emission region. Light scattering (e.g., by protein aggregates) can significantly contribute to the optical density and therefore pose similar problems as high absorbance values. Colloidal protein delivery systems such as liposomes also scatter light. Light scattering by these systems can be limited by using particles with a smaller size or by matching the refractive index of the dispersion medium with that of the particles (Ardhammar et al., 2002).

For quenching studies with iodide or acrylamide, excitation above 290 nm (iodide) or 295 nm (acrylamide) is recommended because these quenchers absorb light at lower wavelengths. If this is not feasible because of too little fluorescence intensity under these conditions, correction for the inner filter effect should be made (Lakowicz, 1999). In addition, for quenching experiments with iodide, one must add an antioxidant (e.g., 0.2 mM sodium thiosulfate) to the stock solution (usually potassium iodide) to prevent oxidation to iodine. In quenching studies with potassium iodide, it is common practice to perform a parallel titration with potassium chloride to exclude ionic-strength effects on the protein. Alternatively, iodide-quenching studies are performed at a fixed high ionic strength by using combinations of potassium iodide and potassium chloride.

One should always subtract the background from the sample signal, even when background fluorescence or light-scattering signals are low. Especially in weakly fluorescent samples, Raman scattering can significantly interfere with the fluorescence signal. The Raman scatter peak in aqueous systems occurs at a wavenumber of ca. 3600 cm^{-1} below the excitation wavenumber. So, when using an excitation wavelength of 280 nm, the Raman peak appears at 311 nm; on excitation at 295 nm, this occurs at 330 nm. Whereas the Raman peak is easily recognized as a symmetrical peak in nonfluorescent background samples, it may appear as a little bump in (uncorrected) protein spectra and as such be misinterpreted. When doing intensity measurements at one single wavelength (e.g., fluorescence quenching, anisotropy, time-resolved studies), one should select an excitation/emission wavelength combination where the fluorescence emission signal does not overlap with the Raman scatter peak. Background subtraction is particularly important in anisotropy studies. As can be seen from Equations 20 and 21, small errors in F_{VV} or F_{VH} cause substantial systematic errors in calculated polarization and anisotropy values. For instance, scattered light is vertically polarized and can substantially contribute to F_{VV} values; uncorrected polarization and anisotropy values will then be an overestimation of the real ones.

Ideally, background measurements are done with exactly the same ingredients as the protein sample, except the protein itself. In some cases, this ideal situation can only be approached. For instance, for liposomal formulations, the presence of protein may affect the size distribution and refractive index of the liposomes and therefore their scattering properties. When dealing with commercial formulations, the exact composition and source of the excipients is often not or only partially known to the experimenter. In such cases, one may consider buffer exchange (e.g., by dialysis).

Importantly, when changing the environmental conditions, such as solution medium or temperature, alterations in fluorescence properties may occur irrespective of conformational changes in the protein.

For instance, when comparing two formulations, the fluorescence intensity may differ just because one formulation contains a salt or buffer species acting as a quencher. Also, addition of miscible organic solvents to aqueous solutions (e.g., ethanol) reduces solvent polarity and thereby induces changes in the emission spectrum. To discriminate direct environmental effects on the fluorescence signal from conformational changes induced by these effects, one should perform control measurements with commercially available Trp and Tyr derivatives, such as *N*-acetyl-L-tryptophanamide and *N*-acetyl-L-tyrosinamide.

Conclusions

Fluorescence spectroscopy can be applied in many different ways for studying proteins. Often, one can make use of the intrinsic fluorophores Tyr and Trp. The environmental sensitivity of Trp emission is useful for detecting protein structural changes and unfolding, and its high r_0 value (greater than 0.2 when excited between 295 and 300 nm) enables fluorescence anisotropy measurements. The tertiary structure of proteins can also be probed by the accessibility of the intrinsic fluorophores to quenchers. Furthermore, the intrinsic fluorophores can be donor-acceptor pairs for resonance energy transfer studies. In addition to the intrinsic fluorophores, external fluorescent dyes can be covalently or noncovalently attached to the proteins. Very often, these dyes are used because of their superior spectral properties (e.g., longer fluorescence lifetimes or higher fluorescence intensities compared to the intrinsic fluorophores).

By fluorescence spectroscopy measurements, it is possible to obtain detailed information on both structural aspects (e.g., distances, size, shape, state of folding, or ligand binding) and dynamic properties (e.g., local or segmental flexibility) of proteins. However, complete interpretation of fluorescence data often requires time-resolved measurements. For example, a change in anisotropy can be caused by a change in fluorescence lifetime or rotational correlation time. In many laboratories, time-resolved instrumentation is not available. This certainly limits the information that can be extracted from fluorescence data. Nevertheless, correctly performed steady-state measurements may be of great value, as demonstrated by the examples in the section "Applications," above. In many cases, it is possible to detect changes in protein structure or dynamics with high sensitivity. This information can be very helpful for finding the optimal formulation of a pharmaceutical protein.

Abbreviations

IAEDANS	5-[[2-[(iodoacetyl)amino]ethyl]amino]naphthalene-1-sulfonic acid
ANS	1-anilinonaphthalene-8-sulfonic acid

bis-ANS	1,1-bis-4-anilinonaphthalene-5,5-sulfonic acid
D_{rot}	Rotational diffusion coefficient
E	Efficiency of energy transfer
F	Fluorescence intensity
Fab	Antigen-binding fragment
GdnHCl	Guanidinium hydrochloride
HSA	Human serum albumin
k_q	Bimolecular quenching constant
K_D	Stern-Volmer quenching constant
MEM	Maximum entropy method
Phe	Phenylalanine
Q	Quantum yield
r	Fluorescence anisotropy
r_0	Maximum (limiting) fluorescence anisotropy
R_0	Förster distance
RET	Resonance energy transfer
rhIFNα2a	Recombinant human interferon-alpha2a
rhIFNγ	Recombinant human interferon-gamma
rhIL2	Recombinant human interleukin-2
TIRF	Total internal reflection fluorescence
Trp	Tryptophan
Trps	Tryptophan residues
TCSPC	Time-correlated single photon counting
Tyr	Tyrosine
βLG	Beta-lactoglobulin
θ	Rotational correlation time
λ	Wavelength
λ_{max}^{Em}	Fluorescence emission maximum
τ	Fluorescence lifetime
$\tilde{\nu}$	Wavenumber

References

1. Adams PD, Chen Y, Ma K, Zagorski MG, Sönnichsen FD, McLaughlin ML, and Barkley MD. Intramolecular Quenching of Tryptophan Fluorescence by the Peptide Bond in Cyclic Hexapeptides. *J Am Chem Soc* 2002; 124:9278–9286
2. Alcala JR, Gratton E, and Prendergast FG. Fluorescence Lifetime Distributions in Proteins. *Biophys J* 1987a; 51:597–604

3. Alcala JR, Gratton E, and Prendergast FG. Interpretation of Fluorescence Decays in Proteins Using Continuous Lifetime Distributions. *Biophys Chem* 1987b; 51:925–936
4. Ali V, Prakash K, Kulkarni S, Ahmad A, Madhusudan KP, and Bhakuni V. 8-Anilino-1-Naphtalene Sulfonic Acid (ANS) Induces Folding of Acid Unfolded Cytochrome C to Molten Globule State as a Result of Electrostatic Interactions. *Biochemistry* 1999; 38:13635–13642
5. Anraku M, Yamasaki K, Maruyama T, Kragh-Hansen U, and Otagiri M. Effect of Oxidative Stress on the Structure and Function of Human Serum Albumin. *Pharm Res* 2001; 18:632–639
6. Ardhammar M, Lincoln P, and Norden B. Invisible Liposomes: Refractive Index Matching with Sucrose Enables Flow Dichroism Assessment of Peptide Orientation in Lipid Vesicle Membrane. *Proc Natl Acad Sci U S A* 2002; 99:15313–15317
7. Arigita C, Kersten GF, Hazendonk T, Hennink WE, Crommelin DJ, and Jiskoot W. Restored Functional Immunogenicity of Purified Meningococcal Pora by Incorporation into Liposomes. *Vaccine* 2003; 21:950–960
8. Arigita C, Jiskoot W, Westdijk J, van Ingen C, Hennink WE, Crommelin DJA, and Kersten GFA. Stability of Mono- and Trivalent Meningococcal Outer Membrane Vesicle Vaccines. *Vaccine* 2004; 22:629–642
9. Bajzer Z, and Prendergast FG. A Model for Multiexponential Tryptophan Fluorescence Intensity Decay in Proteins. *Biophys J* 1993; 65:2313–2323
10. Balasubramanian SV, Bruenn J, and Straubinger RM. Liposomes as Formulation Excipients for Protein Pharmaceuticals: A Model Protein Study. *Pharm Res* 2000; 17:344–350
11. Bastiaens PIH, van Hoek A, Wolkers WF, Brochon JC, and Visser AJWG. Comparison of the Dynamical Structure of Lipoamide Dehydrogenase and Gluthathione Reductase by Time-Resolved Polarized Flavin Fluorescence. *Biochemistry* 1992; 31:7050–7060
12. Beechem JM, Knutson JR, Ross JBA, Turner BW, and Brand L. Global Resolution of Heterogeneous Decay by Phase/Modulation Fluorometry: Mixtures and Proteins. *Biochemistry* 1983; 22:6054–6058
13. Bismuto E, Irace G, and Gratton E. Multiple Conformational States in Myoglobin Revealed by Frequency Domain Fluorometry. *Biochemistry* 1989; 28:1508–1512
14. Boteva R, Zlateva T, Dorovska-Taran V, Visser AJWG, Tsanev R, and Benedetto S. Dissociation Equilibrium of Human Recombinant Interferon Γ. *Biochemistry* 1996; 35:14825–14830
15. Brewer JM, Bastiaens P, and Lee J. Fluorescence Lifetime and Spectral Study of the Acid Expansion of Bovine Serum Albumin. *Biophys Chem* 1987; 28:77–88

16. Bright FV. Multifrequency Phase Fluorescence Study of Hapten-Antibody Complexation. *Anal Chem* 1989; 61:309–313

17. Brochon JC. Maximum Entropy Method of Data Analysis in Time-Resolved Spectroscopy. *Methods Enzymol* 1994; 240:262–311

18. Broos J, Maddalena F, and Hesp BH. In Vivo Synthesized Proteins with Monoexponential Fluorescence Decay Kinetics. *J Am Chem Soc* 2004; 126:22–23

19. Buchner J, Renner M, Lilie H, Hinz HJ, Jaenicke R, Kiefhabel T, and Rudolph R. Alternatively Folded States of an Immunoglobulin. *Biochemistry* 1991; 30:6922–6929

20. Buijs J, and Hlady VV. Adsorption Kinetics, Conformation, and Mobility of the Growth Hormone and Lysozyme on Solid Surfaces, Studied with Tirf. *J Colloid Interface Sci* 1997; 190:171–181

21. Cai K, and Schirch V. Structural Studies on Folding Intermediates of Serine Hydroxymethyltransferase Using Fluorescence Resonance Energy Transfer. *J Biol Chem* 1996; 271:27311–27320

22. Callis PR. ^1l_a and ^1l_b Transitions of Tryptophan: Applications of Theory and Experimental Observations to Fluorescence of Proteins. *Methods Enzymol* 1997; 278:113–150

23. Callis PR, and Vivian JT. Understanding the Variable Fluorescence Quantum Yield of Tryptophan in Proteins Using Qm-Mm Simulations. Quenching by Charge Transfer to the Peptide Backbone. *Chem Phys Lett* 2003; 369:409–414

24. Callis PR, and Liu T. Quantitative Prediction of Fluorescence Quantum Yields for Tryptophan in Proteins. *J Phys Chem B* 2004; 108:4248–4259

25. Chabbert M, Lukas TJ, Watterson DM, Axelsen PH, and Prendergast FG. Fluorescence Analysis of Calmodulin Mutants Containing Tryptophan: Conformational Changes Induced by Calmodulin-Binding Peptides from Myosin Light Chain Kinase and Protein Kinase II. *Biochemistry* 1991; 30:7615–7630

26. Chen Y, and Barkley MD. Toward Understanding Tryptophan Fluorescence in Proteins. *Biochemistry* 1998; 37:9976–9982

27. Chen H, and Sanyal G. Direct Biophysical Characterization of Human Apolipoprotein A-1 in ISCOMs. *J Pharm Sci* 1999; 88:1122–1126

28. Chen Y, Liu B, Yu HT, and Barkley MD. The Peptide Bond Quenches Indole Fluorescence. *J Am Chem Soc* 1996; 118:9271–9278

29. Chosrowjan H, Taniguchi S, Mataga N, Tanaka F, and Visser AJWG. The Stacked Flavine Adenine Dinucleotide Conformation in Water Is Fluorescent on Picosecond Timescale. *Chem Phys Lett* 2003; 378:354–358

30. De Beuckeleer K, Volckaert G, and Engelborghs Y. Time Resolved Fluorescence and Phosphorescence Properties of the Individual Tryptophan Residues of Barnase: Evidence for Protein-Protein Interactions. *Proteins: Structure Function and Genetics* 1999; 36:42–53

31. De Filippis V, de Laureto PP, Toniutti N, and Fontana A. Acid-Induced Molten Globule State of a Fully Active Mutant of Human Interleukin-6. *Biochemistry* 1996; 35:11503–11511
32. Demas JN, and Crosby GA. The Measurement of Photoluminescence Quantum Yields. A Review. *J Phys Chem* 1971; 75:991–1024
33. Demchenko AP. *Ultraviolet Spectroscopy of Proteins.* Berlin: Springer Verlag; 1986. 312p.
34. Dijkstra DS, Broos J, and Robillard GT. Membrane Proteins and Impure Detergents: Procedures to Purify Membrane Proteins to a Degree Suitable for Tryptophan Fluorescence Spectroscopy. *Anal Biochem* 1996; 240:142–147
35. Eftink MR. Fluorescence Methods for Studying Equilibrium Macromolecule-Ligand Interactions. *Methods Enzymol* 1997; 278: 221–257
36. Eftink MR, and Ghiron CA. Exposure of Tryptophanyl Residues and Protein Dynamics. *Biochemistry* 1977; 16:5546–5551
37. Eftink MR, and Hagaman KA. Viscosity Dependence of the Solute Quenching of the Tryptophanyl Fluorescence of Proteins. *Biophys Chem* 1986; 25:277–282
38. Eftink RF. Fluorescence Techniques for Studying Protein Structure. In: Suelter CH. *Protein Structure Determination.* New York: John Wiley & Sons; 1991:127–205
39. Eisenhawer M, Cattarinussi S, Kuhn A, and Vogel H. Fluorescence Resonance Energy Transfer Shows a Close Helix-Helix Distance in the Transmembrane M13 Procoat Protein. *Biochemistry* 2001; 40:12321–12328
40. Engelborghs Y. The Analysis of Time Resolved Protein Fluorescence in Multi-Tryptophan Proteins. *Spectrochimica Acta Part A* 2001; 57:2255–2270
41. Engelborghs Y. Correlating Protein Structure and Protein Fluorescence. *J Fluoresc* 2003; 13:9–16
42. Ferreira ST. Fluorescence Studies of the Conformational Dynamics of Parvalbumin in Solution: Lifetime and Rotational Motions of the Single Tryptophan Residue. *Biochemistry* 1989; 28:10066–10072
43. Förster T. Zwischenmolekulare Energiewanderung Und Fluoreszenz. *Ann Phys - Berlin* 1948; 437:55–75
44. Gratton E, Limkeman M, Lakowicz JR, Maliwal BP, Cherek H, and Lacko G. Resolution of Mixtures of Fluorophores Using Variable-Frequency Phase and Modulation Data. *Biophys J* 1984; 46:479–486
45. Grinvald A, and Steinberg IZ. On the Analysis of Fluorescence Decay Kinetics by the Method of Least Squares. *Anal Biochem* 1974; 59:583–598

46. Havel HA, Kauffman EW, and Elzinga PA. Fluorescence Quenching Studies of Bovine Growth Hormone in Several Conformational States. *Biochim Biophys Acta* 1988; 955:154–163

47. Hendriks J, Gensch T, Hviid L, van der Horst MA, Hellingwerf KJ, and van Thor JJ. Transient Exposure of Hydrophobic Surface in the Photoactive Yellow Protein Monitored with Nile Red. *Biophys J* 2002; 82:1632–1643

48. Herron JN, Kranz DM, Jameson DM, and Voss EW Jr. Thermodynamic Properties of Ligand Binding by Monoclonal Anti-Fluorescyl Antibodies. *Biochemistry* 1986; 25:4602–4609

49. Herron JN, Terry AH, Johnston S, He XM, Guddat LW, Voss EW Jr, and Edmundson AB. High Resolution Structures of the 4-4-20 Fab-Fluorescein Complex in Two Solvent Systems: Effects of Solvent on Structure and Antigen-Binding Affinity. *Biophys J* 1994; 67:2167–2183

50. Hlady V, and Andrade JD. A TIRF Titration Study of 1-Anilinonaphthalene-8-Sulfonate Binding to Silica-Adsorbed Bovine Serum Albumin. *Colloids Surfaces* 1989; 42:85–96

51. Hlady V, Van Wagenen RA, and Andrade JD. Total Internal Reflection Intrinsic Fluorescence (Tirif) Spectroscopy Applied to Protein Adsorption. In: Andrade JD. *Surface and Interfacial Aspects of Biomedical Polymers, Vol. 2, Protein Adsorption*. New York: Plenum Press; 1985, 81–119

52. Hlady V, Reinecke DR, and Andrade JD. Fluorescence of Adsorbed Protein Layers. I. Quantitation of Total Internal Reflection Fluorescence. *J Colloid Interface Sci* 1986; 111:555–569

53. Hlady V, Buijs J, and Jennissen HP. Methods for Studying Protein Adsorption. *Methods Enzymol* 1999; 309:402–429

54. Hlady VV, and Buijs J. Protein Adsorption on Solid Surfaces. *Curr Opin Biotech* 1996; 7:72–77

55. Ho CH, Britt DW, and Hlady V. Human Low Density Lipoprotein and Human Serum Albumin Adsorption onto Model Surfaces Studied by Total Internal Reflection Fluorescence and Scanning Force Microscopy. *J Mol Recognit* 1996; 9:444–455

56. Holowka D, Wensel T, and Baird B. A Nanosecond Fluorescence Depolarization Study on the Segmental Flexibility of Receptor-Bound Immunoglobulin E. *Biochemistry* 1990; 29:4607–4612

57. Holzman TF, Dougherty JJ Jr, Brems DN, and MacKenzie NE. Ph-Induced Conformational States of Bovine Growth Hormone. *Biochemistry* 1990; 29:1255–1261

58. Hovorka SW, Hong J, Cleland JL, and Schoneich C. Metal-Catalyzed Oxidation of Human Growth Hormone: Modulation by Solvent-Induced Changes of Protein Conformation. *J Pharm Sci* 2001; 90:58–69

59. Jablonski A. Efficiency of Anti-Stokes Fluorescence in Dyes. *Nature* 1933; 131:839–840

60. Jiskoot W, Bloemendal M, van Haeringen B, van Grondelle R, Beuvery EC, Herron JN, and Crommelin DJ. Non-Random Conformation of a Mouse Igg2a Monoclonal Antibody at Low Ph. *Eur J Biochem* 1991a; 201:223–232

61. Jiskoot W, Hoogerhout P, Beuvery EC, Herron JN, and Crommelin DJ. Preparation and Application of a Fluorescein-Labeled Peptide for Determining the Affinity Constant of a Monoclonal Antibody-Hapten Complex by Fluorescence Polarization. *Anal Biochem* 1991b; 196:421–426

62. Jiskoot W, Hlady V, Naleway JJ, and Herron JN. Application of Fluorescence Spectroscopy for Determining the Structure and Function of Proteins. *Pharm Biotechnol* 1995; 7:1–63

63. Johansen P, Merkle HP, and Gander B. Physico-Chemical and Antigenic Properties of Tetanus and Diphtheria Toxoids and Steps Towards Improved Stability. *Biochim Biophys Acta* 1998; 1425:425–436

64. Keeble AH, Hemmings AM, James R, Moore GR, and Kleanthous C. Multistep Binding of Transition Metals to the H-N-H Endonuclease Toxin Colicin E9. *Biochemistry* 2002; 41:10234–10244

65. Knutson JR, Beechem JM, and Brand L. Simultaneous Analysis of Multiple Fluorescence Decay Curves: A Global Approach. *Chem Phys Lett* 1983; 102:501–507

66. Koppenhagen FJ, Visser AJ, Herron JN, Storm G, and Crommelin DJ. Interaction of Recombinant Interleukin-2 with Liposomal Bilayers. *J Pharm Sci* 1998; 87:707–714

67. Kranz DM, Herron JN, Giannis DE, and Voss EW Jr. Kinetics and Mechanism of Deuterium Oxide-Induced Fluorescence Enhancement of Fluorescyl Ligand Bound to Specific Heterogeneous and Homogeneous Antibodies. *J Biol Chem* 1981; 256:4433–4438

68. Kranz DM, Herron JN, and Voss EW Jr. Mechanisms of Ligand Binding by Monoclonal Anti-Fluorescyl Antibodies. *J Biol Chem* 1982; 257:6987–6995

69. Kueltzo LA, and Middaugh CR. Structural Characterization of Bovine Granulocyte Colony Stimulating Factor: Effect of Temperature and pH. *J Pharm Sci* 2003; 92:1793–804

70. Lakowicz JR, and Weber G. Quenching of Protein Fluorescence by Oxygen. Detection of Structural Fluctuations in Proteins on the Nanosecond Time Scale. *Biochemistry* 1973a; 12:4171–4179

71. Lakowicz JR, and Weber G. Quenching of Fluorescence by Oxygen. A Probe for Structural Fluctuations in Macromolecules. *Biochemistry* 1973b; 12:4161–4170

72. Lakowicz JR, Lacko G, Cherek H, Gratton E, and Limkeman M. Analysis of Fluorescence Decay Kinetics from Variable-Frequency Phase Shift and Modulation Data. *Biophys J* 1984; 46:463–477

73. Lakowicz JR. Intensity and Anisotropy Decays of [Leu^5] Enkephalin Tyrosyl Fluorescence by 10 GHz Frequency-Domain Fluorometry. *Biophys Chem* 1993; 47:33–40

74. Lakowicz JR. *Principles of Fluorescence Spectroscopy*. New York: Kluwer Academic/Plenum Publishers; 1999. 698p.

75. Latt SA, Cheung HT, and Blout ER. Energy Transfer. A System with Relatively Fixed Donor-Acceptor Separation. *J Am Chem Soc* 1965; 87:995–1003

76. Laws WR, and Contino PB. Fluorescence Quenching Studies: Analysis of Nonlinear Stern-Volmer Data. *Methods Enzymol* 1992; 210:448–463

77. Lehrer SS. Solute Perturbation of Protein Fluorescence. The Quenching of the Tryptophyl Fluorescence of Model Compounds and of Lysozyme by Iodide Ion. *Biochemistry* 1971; 10:3254–3263

78. Lim K, Jameson DM, Gentry CA, and Herron JN. Molecular Dynamics of the Anti-Fluorescein 4-4-20 Antigen-Binding Fragment. 2. Time-Resolved Fluorescence Spectroscopy. *Biochemistry* 1995; 34:6975–6984

79. Lin JN, Andrade JD, and Chang IN. The Influence of Adsorption of Native and Modified Antibodies on Their Activity. *J Immunol Methods* 1989; 125:67–77

80. Lin J-N, Chang IN, Andrade JD, Herron JN, and Christensen DA. Comparison of Site-Specific Coupling Chemistry for Antibody Immobilization on Different Solid Supports. *J Chromatogr* 1991; 542:41–54

81. MacMahon LP, Colucci WJ, McLaughlin ML, and Barkley MD. Deuterium Isotope Effects in Constrained Tryptophan Derivatives: Implications for Tryptophan Photophysics. *J Am Chem Soc* 1992; 114:8442–8448

82. Martarano LA, Wong CP, Horrocks WD, and Ponte Goncalves AM. Luminescence of Yttrium(III), Lutetium(III), and Thorium(IV) Porphyrin Complexes. *J Phys Chem* 1976; 80:2389–2393

83. Metz B, Jiskoot W, Hennink WE, Crommelin DJA, and Kersten GFA. Physicochemical and Immunochemical Techniques Predict the Quality of Diphteria Toxoid Vaccines. *Vaccine* 2003; 22:156–167

84. Minetti CA, Tai JY, Blake MS, Pullen JK, Liang SM, and Remeta DP. Structural and Functional Characterization of a Recombinant Porb Class 2 Protein from Neisseria meningitidis. Conformational Stability and Porin Activity. *J Biol Chem* 1997; 272:10710–10720

85. Moncrieffe MC, Juranic N, Kemple MD, Potter JD, Macura S, and Prendergast FG. Structure-Fluorescence Correlations in a Single Tryptophan Mutant of Carp Parvalbumin: Solution Structure, Backbone and Side-Chain Dynamics. *J Mol Biol* 2000; 297:147–163

86. O'Connor DV, and Phillips D. *Time-Correlated Single Photon Counting*. London: Academic Press; 1984.

87. Omelyanenko VG, Jiskoot W, and Herron JN. Role of Electrostatic Interactions in the Binding of Fluorescein by Anti-Fluorescein Antibody 4-4-20. *Biochemistry* 1993; 32:10423–10429

88. Oroszlan P, Wicar S, Teshima G, Wu SL, Hancock WS, and Karger BL. Conformational Effects in the Reversed-Phase Chromatographic Behavior of Recombinant Human Growth Hormone (Rhgh) and N-Methionyl Recombinant Human Growth Hormone (Met-Hgh). *Anal Chem* 1992; 64:1623–1631

89. Paliwal R, and London E. Comparison of the Conformation, Hydrophobicity, and Model Membrane Interactions of Diphtheria Toxin to Those of Formaldehyde-Treated Toxin (Diphtheria Toxoid): Formaldehyde Stabilization of the Native Conformation Inhibits Changes That Allow Membrane Insertion. *Biochemistry* 1996; 35:2374–2379

90. Pal SK, Peon J, Bagchi B, and Zewail AH. Biological Water: Femtosecond Dynamics of Macromolecular Hydration. *J Phys Chem B* 2002; 106:12376–12395

91. Pan CP, and Barkley MD. Conformational Effects on Tryptophan Fluorescence in Cyclic Hexapeptides. *Biophys J* 2004; 86:3828–3835

92. Papp S, and Vanderkooi JM. Tryptophan Phosphorescence at Room Temperature as a Tool to Study Protein Structure and Dynamics. *Photochem Photobiol* 1989; 49:775–784

93. Perrin F. Polarization of Light of Fluorescence, Average Life of Molecules in the Excited State. *Journal de Physique et le Radium* 1926; 7:390–401

94. Petrich JW, Chang MC, McDonald DB, and Fleming GR. On the Origin of Nonexponential Fluorescence Decay in Tryptophan and Its Derivatives. *J Am Chem Soc* 1983; 105:3824–3832

95. Rexroad J, Wiethoff CM, Green AP, Kierstead TD, Scott MO, and Middaugh CR. Structural Stability of Adenovirus Type 5. *J Pharm Sci* 2003; 92:665–678

96. Royer CA. Investigation of the Structural Determinants of the Intrinsic Fluorescence Emission of the Trp Repressor Using Single Tryptophan Mutants. *Biophys J* 1992; 63:741–750

97. Sasiak AB, Bolgiano B, Crane DT, Hockley DJ, Corbel MJ, and Sesardic D. Comparison of in Vitro and in Vivo Methods to Study Stability of

PLGA Microencapsulated Tetanus Toxoid Vaccines. *Vaccine* 2000; 19:694–705

98. Schneider WP, Wensel TG, Stryer L, and Oi VT. Genetically Engineered Immunoglobulins Reveal Structural Features Controlling Segmental Flexibility. *Proc Natl Acad Sci U S A* 1988; 85:2509–2513
99. Selvin PR. The Renaissance of Fluorescence Resonance Energy Transfer. *Nat Struct Biol* 2000; 7:730–734
100. Sharma VK,and Kalonia DS. Temperature- and pH-Induced Multiple Partially Unfolded States of Recombinant Human Interferon-A2a: Possible Implications in Protein Stability. *Pharm Res* 2003a; 20:1721–1729
101. Sharma VK, and Kalonia DS. Steady-State Tryptophan Fluorescence Spectroscopy Study to Probe Tertiary Structure of Proteins in Solid Powders. *J Pharm Sci* 2003b; 92:890–899
102. Simat TJ, and Steinhart H. Oxidation of Free Tryptophan and Tryptophan Residues in Peptides and Proteins. *J Agr Food Chem* 1998; 46:490–498
103. Somogyi B, Norman JA, and Rosenberg A. Gated Quenching of Intrinsic Fluorescence and Phosphorescence of Globular Proteins. An Extended Model. *Biophys J* 1986; 50:55–61
104. Steinberg IZ. Long-Range Nonradiative Transfer of Electronic Excitation Energy in Proteins and Polypeptides. *Annual Review of Biochemistry* 1971; 40:83–114
105. Stern O, and Volmer M. The Extinction Period of Fluorescence. *Physikalische Zeitschrift* 1919; 20:183–188
106. Stokes GG. On the Change of Refrangibility of Light. *Philos Trans R Soc London* 1852; 142:463–562
107. Stryer L, and Haugland RP. Energy Transfer: A Spectroscopic Ruler. *Proc Natl Acad Sci U S A* 1967; 58:719–726
108. Stryer L. Fluorescence Energy Transfer as a Spectroscopic Ruler. *Annual Review of Biochemistry* 1978; 47:819–846
109. Suci P, and Hlady V. Fluorescence Lifetime Components of Texas Red-Labelled Bovine Serum Albumin: Comparison of Bulk and Adsorbed States. *Colloids Interfaces* 1990; 51:89–104
110. Szabo AG, and Rayner DM. Fluorescence Decay of Tryptophan Conformers in Aqueous Solution. *J Am Chem Soc* 1980; 102:554–562
111. Tanaka N, Nishizawa H, and Kunugi S. Structure of Pressure-Induced Denaturated State of Human Serum Albumin: A Comparison with the Intermediate in Urea-Induced Denaturation. *Biochim Biophys Acta* 1997; 1338:13–20
112. Taschner N, Muller SA, Alumella VR, Goldie KN, Drake AF, Aebi U, and Arvinte T. Modulation of Antigenicity Related to Changes in Antibody Flexibility Upon Lyophilization. *J Mol Biol* 2001; 310:169–179

113. Taylor RM, Lin B, Foubert TR, Burritt JB, Sunner J, and Jesaitis AJ. Cascade Blue as a Donor for Resonance Energy Transfer Studies of Heme-Containing Proteins. *Anal Biochem* 2002; 302:19–27

114. Tetin SY, and Hazlett TL. Optical Spectroscopy in Studies of Antibody-Hapten Interactions. *Methods* 2000; 20:341–361

115. Valeur B, and Weber G. Resolution of the Fluorescence Excitation Spectrum of Indole into the 1la and 1lb Excitation Bands. *Photochem Photobiol* 1977; 25:441–444

116. Valeur B. *Molecular Fluorescence. Principles and Applications*. Weinheim: Wiley-VCH Verlag GmbH; 2002. 387p.

117. van den Bremer ET, Jiskoot W, James R, Moore GR, Kleanthous C, Heck AJ, and Maier CS. Probing Metal Ion Binding and Conformational Properties of the Colicin E9 Endonuclease by Electrospray Ionization Time-of-Flight Mass Spectrometry. *Protein Sci* 2002; 11:1738–1752

118. van den Bremer ETJ, Keeble AH, Visser AJWG, van Hoek A, Kleanthous C, Heck AJR, and Jiskoot W. Ligand Induced Changes in the Conformational Dynamics of a Bacterial Cytotoxic Endonuclease. *Biochemistry* 2004; 43:4347–4355

119. van den Elsen JMH, van Pomeren E, Poolman JT, Wilting J, Herron JN, and Crommelin DJ. On the Interaction between a Bactericidal Antibody and a Pora Epitope of Neisseria Meningitidis in Outer Membrane Vesicles: A Competitive Fluorescence Polarization Immunoassay. *Anal Biochem* 1997a; 247:382–388

120. van den Elsen JMH, van Unen LMA, van Bloois L, Busquets MA, Jiskoot W, Hoogerhout P, Wilting J, Herron JN, and Crommelin DJA. Thermodynamic Analysis of the Interaction between a Bactericidal Antibody and a Pora Epitope of Neisseria meningitidis. *Biochemistry* 1997b; 36:12583–12591

121. van Gilst M, Tang C, Roth A, and Hudson B. Quenchin Interactions and Nonexponential Decay: Tryptophan 138 of Bacteriophage {T}4 Lysozyme. *J Fluoresc* 1994; 4:203–207

122. van Slooten ML, Visser AJ, van Hoek A, Storm G, Crommelin DJ, and Jiskoot W. Conformational Stability of Human Interferon-Gamma on Association with and Dissociation from Liposomes. *J Pharm Sci* 2000; 89:1605–1619

123. Vanscyoc WS, and Shea MA. Phenylalanine Fluorescence Studies of Calcium Binding to N-Domain Fragments of Paramecium Calmodulin Mutants Show Increased Calcium Affinity Correlates with Increased Disorder. *Protein Sci* 2001; 10:1758–1768

124. Vanscyoc WS, Sorensen BR, Rusinova E, Laws WR, Ross JBA, and Shea MA. Calcium Binding to Calmodulin Mutants Monitored by

Domain-Specific Intrinsic Phenylalanine and Tyrosine Fluorescence. *Biophys J* 2002; 83:2767–2780

125. Visser NV, Visser AJWG, Konc T, Kroh P, and van Hoek A. New Reference Compound with Single, Ultrashort Lifetime for Time-Resolved Tryptophan Experiments. *Proc SPIE* 1994; 2137:618–626

126. Vivian JT, and Callis PR. Mechanisms of Tryptophan Fluorescence Shifts in Proteins. *Biophys J* 2001; 80:2093–2109

127. Wang Y, Kachel K, Pablo L, and London E. Use of Trp Mutations to Evaluate the Conformational Behavior and Membrane Insertion of A and B Chains in Whole Diphtheria Toxin. *Biochemistry* 1997a; 36:16300–16308

128. Wang Y, Malenbaum SE, Kachel K, Zhan H, Collier RJ, and London E. Identification of Shallow and Deep Membrane-Penetrating Forms of Diphtheria Toxin T Domain That Are Regulated by Protein Concentration and Bilayer Width. *J Biol Chem* 1997b; 272:25091–25098

129. Watt RM, and Voss EW Jr. Mechanism of Quenching of Fluorescein by Anti-Fluorescein IgG Antibodies. *Immunochemistry* 1977; 14:533–551

130. Weber G. Rotational Brownian Motion and Polarization of the Fluorescence of Solutions. *Adv Protein Chem* 1953; 8:415–459

131. Wei AP, and Herron JN. Use of Synthetic Peptides as Tracer Antigens in Fluorescence Polarization Immunoassays of High Molecular Weight Analytes. *Anal Chem* 1993; 65:3372–3377

132. Wei AP, Blumenthal DK, and Herron JN. Antibody-Mediated Fluorescence Enhancement Based on Shifting the Intramolecular Dimer <-> Monomer Equilibrium of Fluorescent Dyes. *Anal Chem* 1994; 66:1500–1506

133. Willis KJ, Szabo AG, Zuker M, Ridgeway JM, and Alpert B. Fluorescence Decay Kinetics of the Tryptophyl Residues of Myoglobin: Effect of Heme Ligation and Evidence for Discrete Lifetime Components. *Biochemistry* 1990; 29:5270–5275

134. Wu CW, and Stryer L. Proximity Relationships in Rhodopsin. *Proc Natl Acad Sci U S A* 1972; 69:1104–1108

135. Wu PG, and Brand L. Resonance Energy Transfer: Methods and Applications. *Anal Biochem* 1994; 218:1–13

136. Wu PG, James E, and Brand L. Compact Thermally-Denaturated State of a Staphylococcal Nuclease Mutant from Resonance Energy Transfer Measurements. *Biophys Chem* 1993; 48:123–133

137. Xing DK, Crane DT, Bolgiano B, Corbel MJ, Jones C, and Sesardic D. Physicochemical and Immunological Studies on the Stability of Free and Microsphere-Encapsulated Tetanus Toxoid in Vitro. *Vaccine* 1996; 14:1205–1213

3

Circular Dichroism Spectroscopy

MICHAEL BLOEMENDAL

Beth Shalom, Amsterdam, The Netherlands

WIM JISKOOT

Department of Pharmaceutics, Utrecht Institute for Pharmaceutical Sciences (UIPS), Utrecht University, Utrecht, The Netherlands

Introduction

In this chapter, we give a general survey of the application of circular dichroism (CD) spectroscopy for structural studies of proteins. We do not pretend to give here a comprehensive review of the recent literature on CD, and neither will we indulge in extensive theoretical considerations. Rather, we aim to illustrate the possibilities, limitations, potential information to be obtained, and experimental details.

In the section "What Is Circular Dichroism?" of this chapter, we give a global description of the theory and practice of CD spectroscopy. The sections "Far-UV Circular Dichroism" and "Near-UV and Visible Circular Dichroism" are devoted to far-UV and near-UV–visible (UV-Vis) CD, respectively. Far-UV CD is mostly used to estimate the secondary-structure content and distribution. By near-UV–Vis CD, the tertiary structure of proteins and ligand binding to proteins can be studied. Kelly and Price (1997), among many others, reviewed the combined use of far-UV and near-UV–Vis CD to study protein folding. Vibrational CD (VCD), that is, CD with infrared (IR) light, can also be used for monitoring secondary structure and ligand binding. This is discussed in the section "Vibrational Circular Dichroism." Magnetic CD (MCD) has a fundamentally different origin and is beyond the scope of this chapter, but some theoretical and experimental considerations will be given. For a detailed treatise, we refer the reader to Sutherland and Holmqvist (1980), Sharonov (1991), and Cheesman et al. (1991). Finally, in the sections "Circular Dichroism on Oriented Samples" and "Time-Resolved Circular Dichroism," CD on oriented samples and time-resolved CD (TRCD) are discussed, respectively. Kuwajima (1996) has given an extensive review of the latter.

In comparison with the chapter on CD spectroscopy for protein studies that was published in 1995 (Bloemendal and Johnson, 1995), this chapter gives an update of the mentioned techniques and applications. New aspects that are discussed in this chapter are (1) far-UV CD with synchrotron radiation (SRCD), which has higher sensitivity and allows measurements at both higher and lower concentrations of protein than conventional far-UV CD; (2) the analysis of CD spectra of membrane proteins and denatured proteins, the conclusion of which may equally well apply to any measurement in a nonaqueous environment; (3) CD of (pseudo) solid-state samples like films, gels, and liquid crystals; (4) measurement of CD on oriented proteins, which yields information on the orientation of secondary-structure elements with respect to molecular axes; (5) the prediction of secondary-structure segments from far-UV CD spectra; (6) the existence of an interactive Web site for analysis of protein secondary structure from CD spectra; (7) the use of far-UV CD for obtaining information on the distribution of the secondary-structure segments; and (8) the calculation of binding constants and thermodynamic parameters from CD data.

What Is Circular Dichroism?

General Considerations and Brief History

In 1811, Agaro discovered that plane-polarized light is rotated by quartz. Eight years later, Biot showed that optical rotation can also be observed in solution and that its magnitude depends on concentration, pathlength, and wavelength (see Pasteur, 1848; Cotton, 1896; Jirgensons, 1973; Charney, 1979). Pasteur (1848) suggested optical activity to be due to the asymmetry of the molecule; that is, the impossibility to superimpose the mirror image of (part of) a molecule on its original. The asymmetry may be caused by intrinsically asymmetric covalently bound groups or, in case of a macromolecule, by the (asymmetric) conformation. In 1846, Faraday detected that optical activity can also be induced by applying a magnetic field (see Charney, 1979). Magnetically induced optical activity depends on ligand geometry around the chromophore rather than on molecular conformation. Cotton (1896) discovered that optically active materials absorb right- and left-handed circularly polarized light to a different extent.

CD (the difference in absorption between left- and right-handed circularly polarized light) and the related phenomenon optical rotatory dispersion (ORD, the wavelength dependence of the optical rotation) are two manifestations of the optical activity and hence provide information on asymmetries in the molecules under study. Because CD is an absorption difference, it is observed only in an absorption band. ORD, on the other hand, is a refractive index phenomenon, and thus it is measurable at wavelengths remote from the absorption band (Charney, 1979; Bayley, 1980; Cantor and Schimmel, 1980). Because CD detection is more sensitive than ORD, and ORD interpretation involves the weighted sum of the effect of all CD bands, ORD is now rarely used for the study of protein structure. However, when CD cannot be applied (e.g., for secondary-structure determinations at high NaCl concentrations), ORD may be an alternative (Hvidt and Lehrer, 1992).

CD Chromophores and Their Information

CD may arise from inherently asymmetric groups, such as the chiral α-carbon of the peptide bond. CD of such chromophores occurs irrespective of their chemical environment, although the latter may modify the optical properties. In many biological molecules, the asymmetry is not intrinsic but arises from the environment. In these cases, CD provides information on this asymmetric environment.

The prevailing electronic absorptions in proteins are listed in Table 1. In principle, they all yield CD. However, due to the predominance of peptide bonds relative to other chromophores in proteins, the contribution of peptide bonds dominates the far-UV CD spectrum. As the CD band position and intensity of peptide bonds depend on their conformation, the far-UV CD spectrum can be used to determine the secondary-structure content of proteins.

In the near-UV CD region (above 250 nm), only the aromatic residues and cystine absorb light. This part of the spectrum yields information on the tertiary structure. Occasionally, the visible (400–800 nm) part of the CD spectrum can also be used to probe the tertiary structure of proteins (i.e., when colored ligands are present).

Group	Location	$\log \varepsilon_{max}$	Group	Location	$\log \varepsilon_{max}$
Peptide[b]	~210	~2.0	Tyr	274	3.1
	~190	~3.8		222	3.9
His	211	3.8		193	4.7
Cys	250	2.5	Phe	257	2.3
Trp	280	3.7		206	4.0
	219	4.7		188	4.8

[a]From Campbell and Dwek (1984); location in nm.

[b]Additionally, there are absorption bands around 175 and 165 nm, which can only be measured by SRCD (Wallace, 2000a).

Table 1 Some Relevant Electronic Absorption Bands in Proteins[a]

In principle, VCD in the infrared part of the spectrum can also provide information on protein structure and ligand binding. Some important vibrational transitions are listed in Table 2.

Tryptophan is the only amino acid that has a strong MCD signal. Moreover, metal-protein complexes show MCD, which makes this technique very useful for studying structure-function relationships of

Group	Location (in cm^{-1})	Remarks
NH (stretch), α-helix[a]	3300–3290	—
NH (stretch), β-sheet[a]	3300–3280	—
Amide I, α-helix[a]	1660–1650 (1695, 1659,1641)	1695/1641 peaks weak
Amide I, β-sheet[a]	1630 (1682,1634)	1682 peak weak
Amide I, random coil[a]	(1656)	—
Amide I, turn[a]	(1684, 1654, 1633)	1654/1633 peaks weak
Amide II, α-helix[a]	1550–1540 (1546,1516)	1516 peak weak
Amide II, β-sheet[a]	1525–1520 (1510,1530,1550)	1550/1510 peaks weak[c]
Amide II, random coil[a]	1520	—
Saturated aliphatic acids[b]	1725–1700	—
Saturated esters[b]	1750–1720	—
H_2O (stretch)[a]	(3900–3200)	Broad intense band
H_2O (bend)[a]	(1645)	Broad band

[a]From Campbell and Dwek (1984); in parentheses peak positions according to Lee (1985) for amide II and water and to Lee et al. (1990) for amide I. In D_2O, the peak positions are shifted.

[b]From Gans (1980).

[c]1510 band in antiparallel, 1550 in parallel β-sheet.

Table 2 Some Relevant Vibrational Absorption Bands in Proteins

such proteins. The sensitivity and high spectral resolution of MCD (as compared to UV absorption spectroscopy) enables the detection of weak transitions unresolved in the absorption spectrum.

Noteworthy, the position of the CD bands of a chromophore may be and often is shifted with respect to its absorption maximum. This is due to exciton coupling, which in addition to vibrational coupling may also yield multiple CD peaks from a single absorption band (Bailey et al., 1982; Visser et al., 2000).

Parameters and Units

Because CD is an absorption difference, the most straightforward parameter to use in CD is molar absorption difference ($\Delta\varepsilon$). From the Lambert-Beer law, it follows that:

$$\Delta\varepsilon = \Delta A / cL = (A_L - A_R) / cL \tag{1}$$

where A_L and A_R are the absorption of left- and right-handed polarized light, respectively, c is the concentration in mol/liter, L is the pathlength in cm, and $\Delta\varepsilon$ is the molar absorption difference in liter mol^{-1} cm^{-1}. Note that c is the concentration of the chromophores. For secondary-structure determination, this is the concentration of peptide bonds or, by approximation, the concentration of amino acid residues.

An alternative parameter is the molar ellipticity, $[\theta]_\lambda$, which is historically used and originates from the fact that CD causes a linearly polarized light beam to become elliptically polarized after passing through an optically active sample. Molar ellipticity is easily converted to molar absorption difference by (Campbell and Dwek, 1984):

$$[\theta]_\lambda = 3300\ \Delta\varepsilon \tag{2}$$

with $[\theta]_\lambda$ in degrees dL mol^{-1} dm^{-1}.

In MCD, a strong magnetic field in the propagation direction of the light, B_0, is applied to the sample. In first-order approximation, the MCD is proportional to the magnetic field:

$$\Delta\varepsilon_B = \Delta A_B / cLB_0 \tag{3}$$

with the subscript B denoting CD solely caused by the magnetic field, and $\Delta\varepsilon_B$ in L mol^{-1} cm^{-1} T^{-1}.

The intensity of the CD signal is sometimes expressed as the anisotropy ratio or dissymmetry factor, g, which is defined as

$$g = (A_L - A_R) / [0.5(A_L + A_R)] \tag{4}$$

The anisotropy ratio g theoretically depends on the electric and magnetic transition dipole moments and on the rotational and dipole strength (Bormett et al., 1992).

Instrumentation

A conventional CD spectrophotometer essentially comprises a light source (usually a xenon, hydrogen, or deuterium lamp), a monochromator to select light of the desired wavelength, a polarizer to produce plane-polarized light, a photoelastic modulator to convert the plane-polarized light into alternating left- and right-handed circularly polarized light, and a lock-in amplifier to separate the small modulated signal detected by the photomultiplier from the noise. It is possible to measure all wavelengths simultaneously by means of a multichannel detector (Lewis et al., 1992; Nölting et al., 1992).

CD signals are typically small (~10^{-4} of regular absorption). Therefore, to obtain sufficiently high signal-to-noise ratios, very sensitive detectors and bright light sources are needed. Because photoelastic modulator frequencies are in the high micro- or low millisecond range, completely different devices have to be used for ultrafast (pico- to microseconds) TRCD (Lewis et al., 1992).

Commercially available CD spectrophotometers will perform well if they are carefully maintained. For specific purposes (e.g., very far-UV, magnetic, vibrational, ultrafast time-resolved, and solid-state CD), home-designed instruments have been described (Johnson, 1971; Osborne et al., 1973; Brahms and Brahms, 1980; Englinton et al., 1980; Keiderling, 1981; Bokma et al., 1987; Sharonov, 1991; Lewis et al., 1992; Kuroda et al., 2001). SRCD can be measured with dedicated SRCD equipment at synchrotron radiation facilities such as Daresbury (Warrington, UK) and Brookhaven (Upton, NY, USA) (Wallace, 2000a). MCD measurements require a magnet to be incorporated in the spectrometer. This can be a permanent or an electro or superconducting magnet, depending on the application (Sutherland and Holmqvist, 1980; Sharonov, 1991; Lewis et al., 1992).

Possible instrumental artifacts in CD spectrometry, in particular the possible occurrence of circular birefringence, linear birefringence, and linear dichroism, are discussed by Shindo and Nakagawa (1985), Wu et al. (1990), and Lewis et al. (1992). Fortunately, these effects can be neglected for most purposes if the equipment is properly maintained. Regular calibration of a CD spectrophotometer is crucial, because deterioration of the optics causes the relative contribution of the scattered light to increase, resulting in errors in intensity and peak positions. A commonly used and easily available calibration standard is (+)-10-camphorsulfonic acid (CSA). It provides a two-point calibration with $\Delta\varepsilon = 2.36$ and -4.9 at 290.5 nm and 192.5 nm, respectively (Chen and Yang, 1977). An 0.5-mm pathlength cuvette with 1 mg/mL CSA will suffice. As CSA is hygroscopic, one must determine its concentration spectroscopically, using an extinction coefficient of the anhydrous compound of 34.5 at 285 nm (Lowry and French, 1932).

Far-UV Circular Dichroism

Introduction

Extensive reviews on far-UV CD have been written among others by Johnson (1988, 1990), Manning (1989), Woody (1995), and Greenfield (1996). The most widely used application of far-UV CD is estimation of secondary-structure content of proteins. This treatise of far-UV CD will essentially be devoted to understanding of the pitfalls of CD spectrophotometers and the methods to extract secondary-structure information from the measurements, as results are easily misinterpreted.

A major recent development in far-UV CD spectroscopy on proteins is the use of SRCD (Wallace, 2000a,b; Wallace et al., 2004). Therefore, we will devote a separate section to this technique (see "Synchrotron Radiation Circular Dichroism Spectroscopy," below).

By measuring far-UV CD at various conditions, thermodynamic parameters of protein stability can be estimated. However, as this is more often done from near-UV CD measurements, this will be discussed in "Near-UV and Visible Circular Dichroism," below.

What Does It Measure?

Far-UV CD spectra are dominated by the contributions of the peptide bond conformations, although some side chains may also contribute (Table 1). In Table 3, the dihedral angles (see Figure 1) of the most common secondary structures are summarized. It is often overlooked that individual bonds with dihedral angles close to those occurring in a certain type of secondary structure will show spectra resembling that of the particular structure, irrespective of whether an identifiable length of this structure is present. Thus a single residue with (φ,ψ) angles of (–57, –47) will show an α-helical CD spectrum, whereas at least four or five residues are required to recognize an α-helical loop. Oppositely, distorted structures will often be underestimated (Hadden et al., 1994). However, in many cases, the length of a structure as well as its adjacent structures and surroundings do affect the spectra (Goodman et al., 1969; Chen et al., 1972; Chang et al., 1978; Yang et al., 1986;

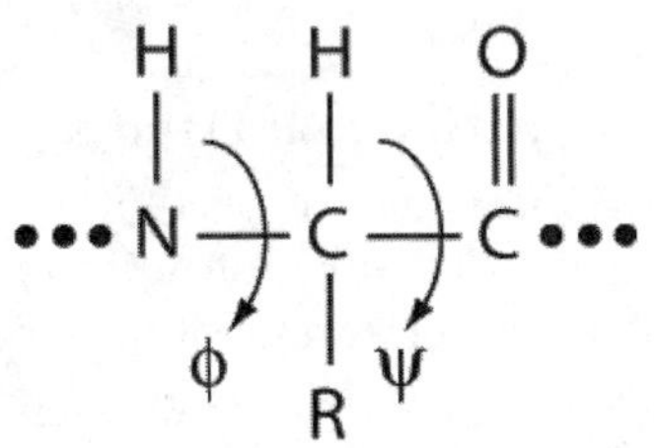

Figure 1 Dihedral angles, according to IUPAC recommendations (*Biochemistry* 1970; 9:3471–3478), shown for φ = 0 degrees and ψ = 0 degrees.

Structure	φ_i	ψ_i	φ_{i+1}	ψ_{i+1}
α-helix	–57	–47		
Parallel β-sheet	–119	+113		
Antiparallel β-sheet	–139	+135		
Collagen	–76	+160		
	–68	+156		
	–67	+168		
β-turn I	–60	–30	–90	0
β-turn I′	+60	+30	+90	0
β-turn II	–60	+120	+80	0
β-turn II′	+60	–120	–80	0
β-turn III	–60	–30	–60	–30
β-turn III′	+60	+30	+60	+30
β-turn V*	–80	+80	–80	+80

[a]All angles transformed to IUPAC-IUB recommendations (*Biochemistry* 1970; 9:3471–3478); *i* denotes the residue, *i*+1 the residue next to *i* (turns are defined by a sequence of residues); all turns according to Lewis et al. (1973), other structures according to Madison and Schellman (1972). For antiparallel supercoiled, antiparallel twisted, and parallel twisted β-sheet, deviating dihedral angles are found (Kubelka and Keiderling, 2001).

Table 3 Dihedral Angles for Peptide Bonds in Different Secondary Structures[a]

Manning 1989; Dousseau and Pezolet, 1990; Kuwajima et al., 1991; Chen and Wallace, 1997a,b; Wallace et al., 2003). Therefore, it has been suggested that far-UV CD can predict not only the overall secondary-structure content but also some higher order structural aspects, like the number of elements of a certain type of secondary structure (Pancoska et al., 1999) and their distribution within the protein (Manavalan and Johnson, 1983; Venyaminov and Vassilenko, 1994; Greenfield, 1996).

Experimental Details

Maintenance of the Optics

An intense light source is needed for low-noise far-UV CD experiments. Unless prevented by a vacuum chamber or very good nitrogen flushing, such sources easily convert oxygen to ozone, which rapidly damages the optics. Ideally, one should monitor possible deterioration of the optics at least once a week by the two-point CSA calibration, for which the ratio of $-\Delta\varepsilon(192.5)/\Delta\varepsilon(290.5)$ should be greater than or equal to 2.0. Replacing the first mirror after the source and cleaning the front surface of the prisms can often renew the optics. CD spectra can be measured down to 170 nm on the newest commercial CD instruments when the equipment is well maintained. Homemade setups have been described by Johnson (1971), Gross and Schnepp (1977), and Toumadje et al. (1992). CD can be measured at even lower wavelengths (in some cases down to 154 nm) by means of synchrotron radiation (Wallace et al., 2003, 2004).

Sample Transparency

The limiting factor in far-UV CD spectroscopy on proteins is usually the transparency of the buffer, as almost any compound absorbs light in this spectral region. Water absorbs essentially all light below 176 nm in a 50-μm pathlength; a 1-mm cell with water has an absorption of 1 at 186 nm. The presence of chloride anion, which is often used in biological buffers, prevents measurements below 205 nm. Fluoride or sulfate salts are more UV-transparant alternatives. Guanidinium hydrochloride and urea absorb strongly below 220 nm. Because some detergents such as Triton tend to oxidize rapidly and form UV-absorbing materials, chaps and octylglucoside at low concentrations are to be preferred as detergents in far-UV CD experiments (Pelton and McLean, 2000). Tables of solvents with their absorption as a function of pathlength and the transparency of buffer additives are given by Johnson (1986, 1990).

In order to preclude artifacts due to light absorption by the buffer, the total absorption of the sample should always be checked before the measurement and be kept below 1 for the entire spectral range studied. In practice, this means that for measurements down to 184 nm, a 0.2-mm pathlength is the upper limit, although better results will be obtained with even shorter pathlengths (e.g., 0.05 or 0.1 mm). Typical protein concentrations for such pathlengths are 0.2 to 1 mg/mL. The use of synchrotron radiation allows for measurements up to 10 mg/mL in 0.01-mm cells and permits less stringent buffer conditions (Wallace et al., 2003).

CD on Limited Amounts of Protein

When only limited amounts of protein are available, CD measurements can be done by adsorbing the protein to quartz or Teflon surfaces. A device to this end and the application for IgG at different conditions is described by Vermeer and Norde (2000). The protein can also be deposited as a film (McPhie, 2001). Alternatively, synchrotron radiation can be used in conventional pathlength cells to measure at lower protein concentrations (Wallace, 2000a). In this way, protein amounts in the range 1–10 μg will suffice to estimate the secondary structure.

Pathlength

When using very-short-pathlength cuvettes, the pathlength has to be checked; it may deviate considerably from the data given by the manufacturer. Errors in pathlength change the apparent $\Delta\varepsilon$, which in turn influences the secondary-structure estimation. From the infrared interference fringes of an empty cell, the pathlength in centimeters (L) may be calculated as follows:

$$L = \Delta n \lambda_1 \lambda_2 / 2(\lambda_1 - \lambda_2) \tag{5}$$

where λ_1 and λ_2 are long and short wavelength interference minima, respectively, and Δn is the number of maxima between them (Bree and Lyons, 1956; Johnson, 1988).

Protein Concentration

Another point of concern is the determination of the protein concentration. As will be discussed in "Estimation of the Secondary-Structure Content," below, reliable estimations of secondary-structure content can be made only when the protein concentration of the sample (and hence $\Delta\varepsilon$) is accurately known. Traditional protein assays (e.g., Lowry et al., 1951; Bradford, 1976; Smith et al., 1985) are only reliable when the specific response of the studied protein is known, as this response varies considerably for different proteins (Lowry et al., 1951; Smith et al., 1985). It is advisable to use specific extinction coefficients at a certain wavelength, provided that these are accurately known (see also Chapter 1). Other reliable methods are quantitative amino-acid analysis combined with peptide-backbone concentration determination by Biuret assay (Goa, 1953) or total nitrogen determination (Lang, 1958).

Solid-State Samples

Special care has to be taken when interpreting CD spectra of solid-state samples in dried thin films. Harada and Kuroda (2002) showed that previous conclusions on conformational changes of the proteins in films were erroneous. The observed changes in the far-UV CD spectra were caused by artifacts arising from macroscopic anisotropy of the samples. The same may occur in gels, micelles, and liquid crystals.

Estimation of Secondary-Structure Content

Introductory Comments

Several methods have been developed to quantify the secondary structure of proteins from their far-UV CD spectra. Many of these have been reviewed by Greenfield (1996). The basic assumption in these methods is that the peptide-backbone spectrum is a linear combination of the spectral contributions of the various types of secondary structure. However, care has to be taken in this simple linear approach. CD spectra of the peptide bond may be affected by contributions of aromatic side chains, disulfide bridges, and other conformations. Aromatic residues, even if they do not have significant far-UV CD themselves, may produce shifts and form changes in the peptide band due to exciton coupling (Kuwajima et al., 1991). Moreover, a certain structure does not necessarily yield a unique CD spectrum, because the

CD spectrum of a chromophore is sensitive to its environment. Hence, the secondary-structure CD spectrum is influenced by the tertiary structure of a protein. Additionally, the α-helix spectrum depends on chainlength (Goodman et al., 1969, Chen et al., 1972) and the vicinity of other helices (Manning, 1989), whereas β-sheets and turns show a range of differently shaped spectra (Chang et al., 1978). In particular, the peak positions for β-sheets show much variety, presumably due to the various types of β-structure like barrels, sheets, propeller, β-helices, and so forth (Wallace et al., 2003), several of them having deviating dihedral angles (Kubelka and Keiderling, 2001). Finally, the polarity of the environment may affect the position and intensity of CD peaks (Chen and Wallace, 1997a,b; Wallace et al., 2003).

Helical Content from a Single Wavelength

The simplest method estimates secondary-structure content from the CD signal at a single wavelength. Although this estimation is very crude, it is still being used. For instance, Dumas et al. (2000), Banerjee and Basu (2002), and Naseem et al. (2003) deduced the fraction of α-helix from the CD signal at 222 nm according to Yang et al. (1986) by:

$$f_H = [\theta]_{222}/(-40{,}000 + 100{,}000/n) \quad (6)$$

where n is the number of residues per protein molecule. Especially when the buffer composition excludes measurement down to low wavelengths, this allows for a crude estimation of the helical content.

Polypeptides with Pure Secondary Structure as Reference

Better results can be obtained when far-UV CD spectra are used. One of the oldest of these approaches uses model polypeptides with assumed pure secondary structure. The spectrum of the unknown structure is then linearly fitted in terms of these model compounds (Greenfield and Fasman, 1969; Brahms and Brahms, 1980).

Comparison with Spectra of Proteins with Known X-ray Structure

More advanced methods use reference proteins with known crystal structure from x-ray diffraction data. If the reference set is large enough, all structures and all specific effects will be present in the set. There will be differently sized α-helices, different types of sheets and turns, and so on. However, when using a large reference set, the number of spectra will exceed the information content of the data. Because the reference CD spectra will be linearly dependent and contain experimental error, the problem is overdetermined, and a large number of solutions will fit the data within that error. Johnson and co workers (Hennessey and Johnson, 1981; Manavalan and Johnson, 1985; Toumadje et al., 1992) have shown that measurements truncated at 200 nm, 190 nm, 178 nm, and 168 nm contain 2, 4, 5, and 6 independent pieces of information, respectively, within the noise level of the experiments. When the wavelength region is extended to 160 nm or below by the use of synchrotron radiation, the information content can be seven (Wallace, 2000a). Effectively, this means that measurements from 200 to 250 nm provide only a reliable percentage of α-helix, which dominates the CD, whereas a description in terms of α-helix, parallel and antiparallel β-sheet, turn, and "other" requires measurements down to at least 184 nm.

The overdetermination problem implies that it is essential to know the exact protein concentration and pathlength of the cuvette, as not only the shape of the spectra (peak positions) but also their absolute magnitude influences the results. In some cases, like in solid samples, measurement of concentration is impracticable. McPhie (2001) solved this problem by so-called g-factor analysis. As Equation 4 shows, g is concentration independent. Hence, when the same cell and sample are used for dichroism and absorption measurement, concentration and pathlength determination is not necessary. Secondary structure can then be deduced from the g-factor spectra by comparison with g-spectra of proteins with known structure. When McPhie applied his method to a test set of 19 test spectra in solution, he obtained a standard deviation of about 0.12 for α-helix, about 0.12 for β-sheet, about 0.10 for turn, 0.15 for remainder, and 0.6 for poly-L-proline II-structure. Although these precisions are clearly less than when the concentrations are known (see below), the g-value provides a useful solution when concentration determination is cumbersome.

Least-Squares Fit with Regularizer (CONTIN)

Several approaches have been used to solve the overdetermination problem. Provencher and Glöckner (1981) used a least-squares fit that includes a regularizer term to stabilize the solution (CONTIN). The effect of this regulation is that solutions emphasizing a particular reference protein are penalized unless they are really mandatory for a good fit. They use two constraints: all secondary-structure fractions found should be positive, and the sum of fractions should be one (Provencher, 1982a,b). A disadvantage of this approach is that although any result looks reasonable, it can lead to unreliable results and even disturb estimation of the α-helix content (Johnson, 1986, 1990; Greenfield, 1996). As shown by Johnson (1986, 1988, 1990), it is better to perform the fits without constraints and use these as a quality check. A solution consistent with the information content of the data rather than the "best solution" given by the program then yields the most reliable result (Provencher, 1982a; Manavalan and Johnson, 1987).

Singular Value Decomposition Combined with Variable Selection

Johnson and co-workers used an alternative approach, namely singular value decomposition (SVD) combined with variable selection (SVD+VS) (Hennessey and Johnson, 1981; Compton and Johnson,

1986; Manavalan and Johnson, 1987). Recently, this approach was further improved. The new algorithm is called CDsstr (Johnson, 1999). In variable selection, sets of CD spectra of the reference proteins are removed from the reference set until certain criteria are satisfied. In this way, conflicting information is eliminated. Criteria for the quality of a solution are that (a) the sum of secondary structures is between 0.9 and 1.1 (preferably 0.95 and 1.05); (b) all fractions are larger than –0.03 (ideally positive), (c) the fit of the reconstituted CD is within the noise level of the experimental data (less than 0.25 $\Delta\varepsilon$ units), (d) the fraction of α-helix must be similar to that obtained using all proteins in the reference set, and (e) a minimum number of reference proteins is used for the fit. Many subsets of the reference proteins will give similar results. Their solutions are averaged.

The Locally Linearized Model

The Gauss-Markoff model developed by Van Stokkum et al. (1990) is a multivariate linear model with noise that provides both the secondary-structure fractions and the estimated real errors in the fractions for a protein of unknown structure. Because there are a lot of nonlinear effects in the relation between secondary structure and CD spectra (e.g., chainlength and environmental dependence of helices and sheets), a locally linearized model using a set of reference proteins is introduced. Whereas variable selection removes conflicting information from the reference set, local linearization synthesizes a minimum reference subset necessary for a proper result from the total reference set. Again, a number of solutions is found. As selection criteria, fractions between –0.05 and 1.05 combined with a sum of fractions as close as possible to unity are chosen. Alternatively, a minimum in the sum of errors of the fractions can be used as the selection standard.

Self-consistent Procedure for Singular Value Decomposition with Variable Selection (SELCON)

Sreerama et al. (1999) have developed a self-consistent procedure for SVD+VS. In this method, the spectrum of the protein being analyzed is included in the reference set. Initially, a guess is made for the structure of that protein, and subsequently the predicted structure is used in the computation until self-consistency is reached. Many of the features of previous approaches have been incorporated in this method, which allows for discrimination between regular and distorted α-helix and between regular and distorted β-sheet. In a later version, Sreerama and Woody (2000) extended the reference set to 48 proteins.

Other Methods

Pancoska and Keiderling (1991) have used the principal component method of factor analysis combined with cluster analysis. Although this method is comparable to SVD, nonlinear effects are not taken into account. This method has the advantage that it does not inherently assume the complete far-UV CD band to be due to secondary-structure contributions.

Perczel et al. (1991, 1992) suggested convex constraint analysis as a superior method for secondary-structure calculation, as it avoids the use of x-ray structures as references. However, for determining the number of basis spectra and for testing the reliability of the method, they used spectra of poly-L-lysine in different conformations. As pointed out above, these spectra may not be representative for real proteins. Moreover, the method is rather sensitive to the set of proteins used in the analysis. Comparison with x-ray structures gave correlation coefficients that are significantly lower than for the other methods discussed here.

Sreerama and Woody (1994) and Merolo et al. (1994) have used neural-network methods to estimate the secondary structure from CD spectra. As noted by Greenfield (1996), some of these methods are very time consuming, whereas the reliability of the secondary-structure predictions is not better in comparison with other methods.

Comparison of the Various Methods

Van Stokkum et al. (1990), Greenfield (1996), Johnson (1999), and Sreerama et al. (2000) have compared the various methods. From the reference set, each protein was successively removed and fitted in terms of the remaining proteins. The results for the secondary structure are compared with that deduced from the x-ray structures. Table 4 summarizes the results. The comparison has to be taken with some caution, as the sets of reference proteins and the wavelength ranges differ, although Sreerama and Woody (2000) showed that the problem of a restricted wavelength range can be at least partly overcome by extending the number of proteins in the reference set. Johnson (1999), Wallace (2000a), and Sreerama and Woody (2000) showed that the quality of the prediction depends strongly on the reference set used. Johnson (1999) improved the reliability of the predictions by limiting the reference set to proteins with an x-ray structure with a resolution better than 2 Å. Wallace (2000a) showed that the inclusion of denatured proteins in the reference set significantly improves the prediction of secondary structure of proteins with a high content of unordered structure. Although an unordered structure in a native protein means that the comparatively stable dihedral angles of the peptide backbone cannot be classified as one of the known structures, whereas a denatured protein is a highly dynamic structure undergoing large-scale fluctuations in conformation, the resulting CD of the two seems to show enough similarity to warrant their treatment as a single category in methods that use a flexible basis set (Wallace, 2000a).

Another complication is that the generally used sets of reference proteins do not include membrane proteins. As shown by Wallace et al. (2003), secondary-structure calculations produce considerably poorer

	Structural Classes[a]							
Method	**H**	**3_{10}**	**A**	**P**	**β-total**	**T**	**O**	**Poly-L**
Linear model[b,c,d]	9	–	16	7	–	11	17	–
LLM-I[b,c,d,e]	9	–	12	8	–	7	9	–
LLM-II[b,c,d,e,f]	7	–	12	7	–	7	8	–
SVD+VS[b,f,g]	6	–	10	8	–	6	9	–
PROV-I[b,c,h]	12	–	21	6	–	11	8	–
PROV-II[b,c,h]	9	–	13	6	–	7	8	–
MLR[i,j]	14	–	–	–	21	16	–	–
LINCOMB[i,k]	8	–	–	–	16	17	–	–
CCA[i,l]	10	–	–	–	18	18	–	–
Neural net[i,m]	10	–	–	–	11	5	–	–
SELCON[i,n]	10	–	–	–	9	9	–	–
SELCON-III[o]	5/5[p]	–	–		8/4[p]	6	12	–
SELCON-III (J)[q]	6	3	–	–	4	4	5	3
SVD-CDsstr[r]	3	3	–	–	4	4	5	3

[a]H = helix, 3_{10} = 3_{10}-helix, A = antiparallel β-sheet, P = parallel β-sheet, β-total = total β-sheet ,T = turn, O = other structure, Poly-L = poly-L-proline II-structure.

[b]Calculated by van Stokkum et al. (1990).

[c] Data are based on 22 protein spectra from 178 to 240 nm (Van Stokkum et al., 1990).

[d]The linear model and LLM use the Gauss-Markoff method (Van Stokkum et al., 1990).

[e]LLM-I and LLM-II are the locally linearized model using the sum of fractions closest to 1 and the same combined with smallest standard deviations in the fractions, respectively, as selection criteria.

[f] Data based on 16 spectra from 168 to 240 nm (Toumadje et al., 1992).

[g]SVD+VS is singular value decomposition with variable selection (Manavalan and Johnson, 1987).

[h]PROV-I and PROV-II are the Provencher method (Provencher and Glöckner, 1981) with the solution suggested by the program and by Manavalan and Johnson (1987), respectively.

[i]From Greenfield (1996).

[j]MLR is multiple linear regression with four model peptides according to Brahms and Brahm (1980).

[k]LINCOMB is a multiple linear regression based on a linear combination of thirty-three proteins with the constraint that the sum of fractions has to be 1 (Yang et al., 1986).

[l]CCA is convex constraint analysis with 17 proteins according to Sreerama and Woody (1994).

[m]Neural net with seventeen proteins according to Sreerama and Woody (1994).

(continued on following page)

(continued from previous page)

[n]SELCON is self-consistent model on thirty-three proteins according to Sreerama and Woody (1994).

[o]SELCON-III is a modified SELCON program. Thirty-seven reference proteins are used (Sreerama et al., 2000).

[p]Regular/distorted structure.

[q]SELCON-III used by Johnson (1999) with his reference set of twenty-two high-resolution proteins.

[r]Modified SVD with reference set of twenty-two high-resolution proteins (Johnson, 1999).

Table 4 Comparison of Root Mean Square Residuals × 100% of the Secondary-Structure Fractions Obtained from Far-UV CD Spectra with Different Methods

results for membrane proteins than for water-soluble proteins, which seems to be caused by altered spectral peak positions due to a different dielectric constant of the membrane environment. Consequently, also for protein studies in nonaqueous solvent systems, secondary-structure estimations by traditional analysis methods have to be taken with caution. For membrane proteins, in contrast to water-soluble proteins, Wallace et al. (2003) also found that there is no correlation between the goodness of the fit and the correctness of the secondary-structure prediction.

Although many of the models give a comparable average quality of the prediction, it is clear from Table 4 that methods with a root mean square deviation larger than 10% for any of the structures have to be rejected. Sreerama and Woody (2000) showed that a slightly modified CONTIN version (CONTIN-LL), SELCON-III, and CDsstr give on the average comparable qualities of the predictions when the same reference sets and same criteria for secondary-structure extraction from the x-ray structures are used. Which method yields the best result depends on the protein and the secondary-structure class. Therefore, they advocate using all three methods in the analysis of the CD spectra. Sreerama et al. (2001) succeeded to slightly improve the quality of the secondary-structure prediction by limiting the reference set to the proteins that belong to the same structural class (see section "Estimation of Structural Class," below).

It should be stressed that the errors in Table 4 are root mean square residuals (i.e., the average of all errors found). This implies that some predictions will be considerably better, but some worse. The Gauss-Markoff model used by Van Stokkum et al. (1990) provides a measure for the uncertainty of a single determination.

In interpreting Table 4, it should also be considered that whereas the reference set contains proteins with an α-helical fraction ranging from 0 to 1, the maximum fraction of antiparallel β-sheet, turn, and other structure is 0.5, and that of parallel sheet only 0.3.

This is not improved in recent reference sets (Sreerama et al., 2000). Thus, Pancoska et al. (1995) noted that relative errors or correlation coefficients have to be used for proper evaluation of the quality of predictions.

A structure present in some proteins that is not included in any of the programs is the collagen structure. This shows a minimum in $[\theta]_\lambda$ of –50,000 at 198 nm and a maximum of +6000 at 220 nm (Campbell and Dwek, 1984).

Availability of Computer Programs and Reference Spectra

Most of the methods for secondary-structure estimation discussed in this section can be obtained from the authors or from Johnson (1999), Greenfield (1996), and/or Sreerama (2000). The computer program SELCON3 can be downloaded from http://lamar.colostate.edu/~sreeram/SELCON3. Different reference sets and computer programs for secondary-structure estimation are available at http://lamar.colostate.edu/~sreeram/CDPro. Lobley and co-workers (2002) have created a user-friendly Web site (DICHROWEB) for the analysis of protein secondary structure from CD (http://www.cryst.bbk.ac.uk/cdweb). The site accepts a variety of input and output formats and units, allows all the methods to use a range of common databases, and permits facile comparisons between the results from different methods using graphics outputs. The site is available via a password-accessible server, which eliminates the problems of installation and software and hardware compatibility for casual users. Data analysis can be done using most of the secondary-structure determination algorithms mentioned above. An on-line user's guide is provided, and links to other CD sites are given.

Prediction of Secondary-Structure Segments

An interesting extension of the information that can be obtained from far-UV (and also vibrational) CD spectra has been given by Pancoska et al. (1996, 1999). They noticed that proteins with more or less identical secondary-structure content can have quite different CD spectra. Although to some extent this can be due to tertiary-structure differences, they attributed it largely to a different distribution of secondary-structure elements. This is illustrated in Figure 2, where two proteins of equal length and secondary-structure content are depicted. However, in protein A there are nine comparatively short helices and three sheet regions, whereas in protein B there is only one sheet region and six larger helices. Moreover, the distribution of the structural segments (and hence the length of the coil sections) is different in the two proteins. Two reasons can be imagined why the CD spectra of the two proteins should be different. First, as discussed above, dipolar and rotational strengths of polypeptide segments depend on their length (Yang et al., 1986; Dousseau and Pezolet, 1990). Second, the residues at the ends of a structural segment will have deviating dihedral angles, the value of which

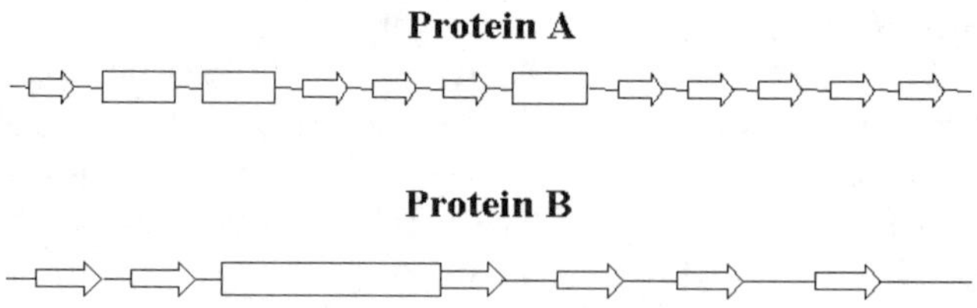

Matrix description	H	E	C	Protein A	H	E	C	Protein B	H	E	C
H	n_H	n_{HE}	n_{HC}	H	3	0	3	H	1	1	0
E	n_{EH}	n_E	n_{EC}	E	0	9	9	E	0	6	6
C	n_{CH}	n_{CE}	n_C	C	3	9	13	C	1	5	7

Figure 2 Schematic derivation of 3 × 3 matrix descriptor to deduce secondary-structure segments from CD spectra for two proteins containing helix (rectangle, H), sheet (arrow, E), and other structure (line, C). The secondary-structure content is equal but the distribution is different. n_{ij} is the number of segments of a certain type of secondary structure for diagonal elements of the matrix and the number of (directional) contacts between structure *i* and *j* for the off-diagonal elements. Based on Pancoska et al. (1996, 1999).

will depend on the adjacent structural segment. The shorter the elements, the larger the contribution of these deviating dihedral angles on the spectra will be. Although the influence of a single distortion on the spectrum may be limited, the cumulative effect of many of them may cause large differences between spectra. In order to take the above-described effects into account, Pancoska et al. (1999) extended the fit procedure of a protein of unknown structure in terms of reference proteins. Whereas the conventional methods in fact have a vector to describe the secondary structure (the elements are the fractions of secondary structure), Pancoska et al. (1996, 1999) used a matrix with integer values on the diagonal describing the number of fragments of a specific structure and the off-diagonal elements of the matrix representing the number of their respective (oriented) interconnections. The corresponding matrices are also given in Figure 2. For a protein with unknown structure, the matrix is solved by a neural-network approach. When Pancoska et al. (1999) tested their method on the far-UV CD and VCD spectra of twenty-three test proteins, they found that the results for far-UV CD were significantly better than those for VCD. In the first case, only 2% of the matrix elements were incorrectly predicted *versus* 6.5% for VCD. The better prediction of far-UV CD is consistent with the observation that VCD is less sensitive to the length of structural elements. The method seems to work better for single domain proteins.

Estimation of Structural Class

It has been suggested that from the far-UV CD spectra also some information on the structural class can be obtained (Manavalan and Johnson, 1983; Venyaminov and Vassilenko, 1994). Proteins are divided in five structural classes, viz. dominantly α-helix (α), separate α-helix and β-sheet domains ($\alpha+\beta$), intermingled α-helix and β-sheet domains ($\alpha\beta$), predominantly β-sheet (β), and mainly unordered (u). The α-helix–containing proteins all have profound bands around 208 and 222 nm. In the α-group, the change of sign of the CD spectrum in the low region (less than 180 nm) takes place below 172 nm, for the $\alpha+\beta$ proteins the 208-nm band is larger than the 222-nm band, whereas in $\alpha\beta$-proteins the opposite is true. The typical α-bands are lacking in β-proteins. Venyaminov and Vassilenko (1994) developed a computer program to determine the protein's structural class from the CD spectrum by cluster analysis. Out of 53 test proteins, 100% of the α- and u-class proteins, 85% of the $\alpha+\beta$-class proteins, and 75% of the $\alpha\beta$-class proteins were correctly identified. However, purely α-helical or β-sheet polypeptides were both identified as $\alpha\beta$ (Greenfield, 1996).

Synchrotron Radiation Circular Dichroism Spectroscopy

In recent years, SRCD has been used in order to improve secondary-structure predictions (Wallace, 2000a,b; Wallace et al., 2003, 2004). Advantages of the use of synchrotron radiation are obvious. The intensity of the radiation in the vacuum ultraviolet (VUV) region is 10^3 as large as that of conventional CD equipment. As a result, data can be collected to much lower wavelengths (in some cases down to 154 nm), which implies that more electronic transitions of the polypeptide backbone group are involved in the spectra and that the information content is higher (Wallace, 2000a). Moreover, the "pure structure" spectra differ more in the region below 190 nm than at higher wavelengths (Wallace, 2000a). In particular, the secondary structure of proteins with a high content of β-sheet, poly-L-proline-II, and/or aperiodic structure is better predicted. Also, the higher intensity means better signal-to-noise levels, which allows measurements at shorter pathlength or with much lower concentration of protein at regular pathlengths, making it a new and valuable tool for structural studies of proteins that are available in limited amounts.

It has been a concern that the intense radiation might affect the chemical integrity and/or secondary structure. However, Orry et al. (2001) showed that the enhanced intensities generally do not alter the spectra and hence that the secondary structure seems to be unaffected. When they used horse myoglobin as a test protein, they found that at least for this protein also, the chemical structure was unchanged during SRCD measurements.

As the light absorption by water is the main restricting factor for far-UV CD measurements, Wallace et al. (2004) have used D_2O as solvent instead of H_2O. When this was combined with the use of CaF_2 instead of quartz cells, data collection was possible to below 154 nm.

SRCD has been used for analysis of secondary structure, for stopped-flow kinetic studies of protein folding (Clarke et al., 1999), and for binding studies (Chang et al., 1999). It may be anticipated that future use will comprise identification of supersecondary features like fold motifs and different types of β-structure barrels, sandwiches, and trefoils (Wallace et al., 2004).

Advantages, Limitations, and Conclusions

Far-UV CD provides a fast method for monitoring the secondary-structure content of proteins. Although it is less accurate than x-ray diffraction or NMR spectroscopy, its relative ease and speed are major advantages. Moreover, it enables the measurement of proteins in solution and allows a fast scanning of the effect of environmental conditions on protein structures, although the dielectric properties of the solvent may influence the spectra significantly (Wallace et al., 2003). Table 4 shows that standard deviations per class (averaged over all reference proteins) become 5% or better with the most recent methods and reference sets (Johnson, 1999; Wallace, 2000a). However, uncertainties in estimations of the secondary structure of individual proteins may be much larger. The Gauss-Markoff method provides a toll to estimate the uncertainties (Van Stokkum et al., 1990). Results may be misleading when the spectra are not measured down to 184 nm at least, although the α-helical content is reasonably reliable when deduced from spectra truncated at 200 nm (Manavalan and Johnson, 1985; Johnson, 1986).

Spectra of "pure" secondary structures deduced by SVD are shown in Figure 3. Greenfield (1996) presented the CD spectrum of the poly-L-proline-II structure. It appears that the spectra for α-helix and antiparallel β-sheet resemble each other to some extent, except for the intensity. Consequently, small antiparallel β-sheet fractions are easily overestimated when much α-helix and little β-sheet is present; conversely, when little α-helix and much β-sheet is present, the latter may be underestimated (Van Stokkum et al., 1990). The poly-L-proline II spectrum and β-turn are similar but opposite in sign (Greenfield, 1996), which may also cause errors in the secondary-structure estimation.

An inherent problem of most methods when deducing secondary structure from CD spectra is that the data are compared with x-ray structures. First, the crystal structure may deviate from the solvent structure (Manning, 1989). Moreover, classification of structures reveal discrepancies of 10% and larger between several investigators (Yang et al, 1986; Wallace et al., 2003). Finally, it may be questioned whether the facility of

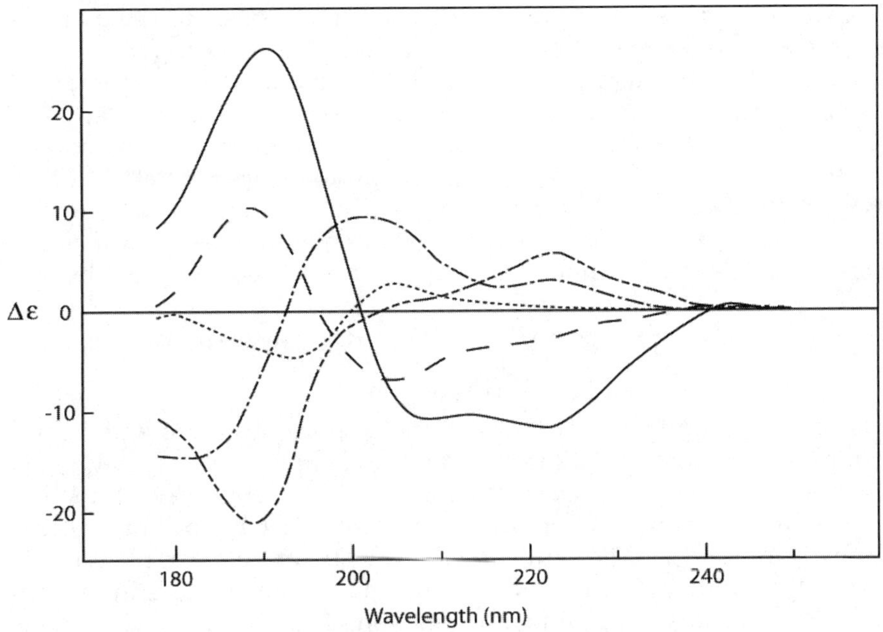

Figure 3 Characteristic circular dichroism spectra for α-helix (solid line), antiparallel β-sheet (dashed line), parallel β-sheet (dash-dotted line), β-turn (short/long-dashed line), and remainder (dotted line). From Compton and Johnson (1986) with permission of the copyright holder (Academic Press).

some proteins to crystallize denotes a structural similarity, thus biasing the reference set of proteins (Pancoska et al., 1991).

Identical secondary-structure content of course does not prove an identical secondary structure. It only shows that there are equivalent amounts of the different types of structure or structural elements at best. A minor problem is the fact that the spectral contributions of mirror-imaged structures (e.g., left- and right-handed turns) may cancel out each other (Chang et al., 1978; Hennessey and Johnson, 1981). A serious point to be considered before selecting far-UV CD for secondary-structure determination is that the choice of the buffers is restricted by their absorption in the far-UV, although on this point synchrotron radiation has significantly enlarged the possibilities of CD spectroscopy.

From far-UV CD spectra, β-sheet structures can be predicted less well than α-helix structure (Pribic et al., 1993; Wallace et al., 2003). Therefore, far-UV CD has been combined with Fourier transfom infrared (FTIR) spectroscopy (Sarver and Kruger, 1991; Pribic et al., 1993; Baumruk et al., 1996) or VCD (Pancoska et al., 1995; Baumruk et al., 1996), as the latter is more reliable for β-sheet determination. Other techniques that allow for relatively quick determination of protein secondary structure are Raman spectroscopy (Chapter 5) and

NMR spectroscopy (Chapter 6). The four mentioned techniques are described and compared in detail by Pelton and McLean (2000).

Far-UV CD spectroscopy has been applied much less on membrane proteins and solid samples than on water-soluble proteins. It appears that the reliability of secondary-structure predictions for membrane proteins is lower than for water-soluble proteins (Wallace et al., 2003). Far-UV spectroscopy on solid samples easily leads to erroneous conclusions (Harada and Kuroda, 2002). CD on oriented samples is further discussed in the section "Circular Dichroism on Oriented Samples," below.

Near-UV and Visible Circular Dichroism

Introduction

Near-UV CD and Vis CD monitor the tertiary structure and ligand binding. Although quantitative structural analysis from near-UV CD is sometimes possible (Scharnagl and Schneider, 1991), the extraction of quantitative information from near-UV–Vis CD spectra is often cumbersome, as will be explained below. Therefore, near-UV–Vis CD is mostly applied to obtain global information on changes in tertiary structure and ligand binding as a function of experimental conditions (e.g., pH, temperature, presence of pharmaceutical excipients). Thus, it has been used for studies on protein-ligand interaction, conformational transitions, protein aggregation, and the kinetics and thermodynamics of these phenomena. Examples are given by Greenfield (1999). Near-UV CD is also a valuable tool for a quick analysis of differences in conformation between related proteins or mutants (Maune et al., 1992). Reviews on near-UV–Vis CD of proteins are given by Strickland (1974), Kahn (1979), Kelly and Price (1997), and Greenfield (1999).

What Does It Measure?

The Coulomb interaction between the asymmetrically positioned transition dipoles of independent chromophores within a molecule forms the quantum mechanical origin of CD. This interaction depends on the orientation of the dipole moments and hence on the relative orientation of the chromophores. Only interactions between nearby chromophores will yield CD, because Coulomb interactions decrease rapidly with distance. Properties of groups showing near-UV CD are given by Strickland (1974). Some information on the effect of binding of metal ions to proteins on the Vis CD spectrum can be found in Sundberg and Martin (1974), Fawcett et al. (1980), and Sigel and Martin (1982).

For phenylalanine and tyrosine only a single band, the so-called ^1L_b, shows intense CD. The CD signal of tryptophan is more complicated, because two of its dipole moments, ^1L_a and ^1L_b, show significant near-UV CD. The shape and intensity of the near-UV CD band of the cystinyl group (S-S) is strongly dependent on the protein conformation (Strickland, 1974).

As a result of charge transfer with aromatic residues, some cations, like Co^{++}, Cd^{++}, Zn^{++}, Hg^{++}, and Cu^{++}, show CD in the near-UV (above 300 nm) and/or the Vis region, when bound to proteins (Beltramini, 1992). This has been used to study conformational changes in the active site of hemocyanins (copper-containing proteins) upon ligand binding (Beltramini et al., 1992), to examine iron binding in iron-sulfur clusters that occur in oxidation/reduction cycles (Vanoni et al., 1992), and to study copper binding to prion protein's octarepeat model peptides (Pappalardo et al., 2004). Other cations like Na^+, K^+, and Ca^{++} have no intrinsic CD, but they may affect far-UV CD and Vis spectra by their influence on the conformation of the proteins. Vis CD can provide structural information on heme-binding proteins (Singh and Wilson, 1990; Eposti et al., 1991).

In order to analyze near-UV–Vis CD in detail, precise peak positions and intensities have to be identified. This can be done by deconvolution of the spectra in Gaussian bands. Commercial software for deconvolution is available from several brands. Visser et al. (2002) have used IGOR-Pro (WaveMetrics) software to deconvolute CD spectra of different fluorescent proteins into two to four absorption bands. In this way, they were able to locate different electronic band positions and to correlate these with electronic transitions and couplings as well as vibrational progression of electronic band positions.

The environment influences the position and sharpness of CD bands. Hydrogen bonding usually results in a redshift, which for tryptophan can be up to 7 nm (Strickland, 1974). Changes in polarity of the environment can lead to shifts in either direction. Hydrophobic environments tend to sharpen peaks. These phenomena are particularly useful to monitor changes in the equilibrium conformation of near-UV–Vis absorbing chromophores and their surroundings. For highly mobile absorbing residues, CD signals from the different orientations will cancel out each other (averaging of asymmetric environments). Thus, Bailey et al. (1982) interpreted the extraordinary intense near-UV CD of epidermolytic toxins as a highly rigid conformation related to the extreme stability of these proteins. Metz et al. (2003) observed an enhancement of the near-UV CD signal of diphtheria toxin after it was inactivated with formaldehyde. This was interpreted as a loss of flexibility of the aromatic residues as a result of the formation of formaldehyde-mediated intramolecular cross-links.

Because of vibrational modes, a single electronic transition may yield multiple CD peaks, which complicates the interpretation of near-UV–Vis CD spectra. Moreover, most proteins contain multiple tyrosines and tryptophans, each of which may have a different environment giving rise to different minima and maxima. As opposite signs of the CD signals decrease the net signal, it has been suggested that the average CD per aromatic residue tends to decline as the number of the residues increases. It is often difficult to separate the effect of mobility and multiple occurrences (Shire et al., 1991). In order to decrease the ambiguity in the interpretation, the use of mutants has been proposed (Craig et al., 1989).

Experimental Details

CD signals in the near-UV region are weak relative to the normal absorption (i.e., $\Delta\varepsilon/\varepsilon$ is small). Therefore, signal-to-noise levels may be low and changes sometimes difficult to detect, although with modern technology, sufficient data collection and proper controls, this no longer provides a major obstacle. In near-UV–Vis CD, unlike in far-UV CD, the problem of buffer interference is virtually absent. Pathlengths are typically between 0.1 and 1.0 cm, depending on the concentration and number of chromophores present; for very low concentrations, even 10-cm cells have been used (Shire et al., 1991). Vis CD bands usually give stronger signals, and measurement of Vis CD is more or less straightforward. The resolution of the CD spectra can sometimes be enhanced by measurements in water/glycerol mixtures at low temperature (Kelly and Price, 1997).

Protein-Ligand Binding

The dissymmetric binding of chromatic ligands to a protein leads to large CD bands. This can be used for ligand-binding studies both in a qualitative and quantitative way. Pappalardo et al. (2004) used CD spectra between 300 and 800 nm to show that copper binds to prion protein's octarepeat model peptides, specifically to the imidazole nitrogen of the histidine residue and the two peptide nitrogens from the subsequent C-terminal glycines.

Near-UV–Vis CD has also been used to calculate thermodynamic parameters for ligand-protein binding. The change in ellipticity as a function of substrate concentration can be used to estimate the affinity constant K. Binding of a ligand at total concentration [L] to n binding sites of a protein with total concentration [P] yields a ligand-protein complex at concentration $[PL_n]$:

$$K = [PL_n]/\{([L] - n[PL_n])([P] - [PL_n])\} \quad (7)$$

The change in ellipticity at a relevant wavelength, θ_i, is directly proportional to $[PL_n]$. When all the protein-binding sites contain ligand, $\theta_i = \theta_{max}$. Hence, $[PL_n] = [P] * (\theta_i/\theta_{max})$ and

$$\theta_i = \theta_{max}\{(1 + K[P]/n + K[P])/(2K[P]) - (\{(1 + K[L]/n + K[P])/(2K[P])\}^2 - [L]/(n[P]))^{1/2}\} \quad (8)$$

By fitting Equation 8 to the raw data, K and n can be estimated.

In some cases, ligand binding leads to a conformational change in the protein. In that case, binding constants can also be obtained for nonchromatic ligands from the CD in the far-UV region or from the near-UV CD of the aromatic side chains.

Thermodynamic Information from CD

Thermodynamic information can be obtained by following changes in the optical activity of a protein as a function of temperature or denaturant. The approach described in this subsection is valid not only for near-UV–Vis CD but also for other types of CD. The appropriate equations can be found in Freire (1995) and Kelly and Price (1997).

The basic idea in deriving thermodynamic parameters from CD is that the change in ellipticity is directly proportional to the change in conformation. If the fraction of folded protein at a certain temperature is α, the equilibrium constant can be written as $K = \alpha/(1-\alpha)$. The free energy of folding ΔG can be written in terms of the enthalpy and entropy of folding (ΔH and ΔS, respectively), the temperature T, and the gas constant R by $\Delta G = RT\ln K = \Delta H - T\Delta S$. Because K at the midpoint of the thermal transition (T_{M}) is one,

$$K = \exp\{(\Delta H/RT)((T/T_M) - 1)\} \qquad (9)$$

$$\alpha = K/(1 + K) \qquad (10)$$

$$\theta_i = (\theta_f - \theta_u)\alpha + \theta_u \qquad (11)$$

where θ_i, θ_f, and θ_u are the observed ellipticity and the ellipticities at fully folded and fully unfolded state, respectively. Measuring θ_i as function of temperature and curve fitting allows calculation of ΔH and ΔS. This approach neglects changes in the heat capacity upon unfolding of the protein. Breslauer (1995), Efting (1995), and Roberts et al. (2003) have discussed more complex mechanisms. Roberts et al. (2003) also deduced kinetic parameters from the CD measurements. Comparable equations for denaturant-induced unfolding can be found in Greenfield (1999).

Advantages, Limitations, and Conclusions

Near-UV–Vis CD is particularly useful for a fast qualitative comparison of the tertiary structure of homologous proteins and mutants, for the assessment of the conformational or aggregational state of a single protein under different conditions, for monitoring protein flexibility or rigidity, and to study ligand and metal-ion binding. As interference by components of the solution medium in this part of the spectrum is usually negligible, near-UV–Vis CD can be very useful as a screening method in formulation design. Moreover, it can be used over a large range of protein concentrations by the use of cells of different pathlengths (0.1–10 cm). Although theoretically a quantitative description of the near-UV CD spectrum of a protein may be obtained, this is rarely done, because the required knowledge about the orientation of the transition dipole moments and their dependence on the direct environment is often not available. When a near-UV–Vis CD spectrum is sensitive to ligand binding, equilibrium constants and thermodynamic quantities can be estimated from changes in the CD signal at a single wavelength.

A drawback of near-UV CD is its intrinsically weak signal intensity. In particular, when the protein contains many chromophores, often the net CD decreases. To overcome the weak intensity of near-UV CD signals, proper controls and abundant signal averaging are important. Furthermore, site-directed mutagenenesis and deconvolution methods can be applied to improve both the sensitivity and the spectral resolution. Vis CD spectra normally show larger signals and are therefore easier to measure.

Vibrational Circular Dichroism

Introduction

Electronic transitions, as measured by UV CD, are restricted in number, often broad and unstructured, overlapping, and mutually interfering. Vibrational transitions absorbing in the infrared, on the other hand, are mostly better resolved and more straightforward to assign. Therefore, IR CD (often called VCD) might be a useful alternative for studying protein structure (Keiderling, 1986; Pancoska et al., 1989; Pancoska and Keiderling, 1991; Gupta and Keiderling, 1992; Baumruk and Keiderling, 1993).

What Does It Measure?

Although any IR absorbing chromophore in an asymmetric environment can in principle be studied with VCD, IR absorption by water molecules and to a lesser extent by CO_2 has limited application. The amide I band (1600–1700 cm^{-1}), which reflects secondary structure, and some ligand bands have been used for most work. The amide II CD band also contains information on secondary structure, but its rather weak and unstructured character (Pancoska et al., 1989) makes it less suitable for quantitative purposes (Gupta and Keiderling, 1992; Baumruk and Keiderling, 1993). Keiderling et al. (1986) and Paterlini et al. (1986) have compared the amide I′ (amide I in D_2O) bands for poly-L-lysine in the α-helix, β-sheet, random coil, and completely denatured form. In the latter case, no VCD signal was observed (Paterlini et al., 1986). For a theoretical description of qualitative relationships between secondary structure and VCD for the amide I and amide II bands, the reader is referred to Snir et al. (1975), Paterlini et al. (1986), and Kubelka and Keiderling (2001). VCD can also be used to study ligand binding (Bormett et al., 1992; Teraoka et al., 1992), as we will discuss in the section "Ligand Binding," below.

Experimental Details

For VCD measurements, one needs an infrared spectrometer that has been modified to produce modulated circularly polarized light and has a fast, highly sensitive IR detector to measure the modulated intensities (Diem et al., 1978; Keiderling, 1981, 1986; Lipp and Nafie, 1984; Pancoska et al., 1989; Bormett et al., 1992; Baumruk and Keiderling, 1993; Pancoska et al., 1995). A sensitivity of $\Delta A/A$ of 10^{-5} has been reported (Keiderling, 1986), which is comparable with that of UV-CD spectrophotometers. However, because IR-CD signals are smaller, the signal-to-noise ratio is worse. The spectral resolution is 10–15 cm^{-1} (Pancoska et al., 1989).

Obviously, problems of ordinary IR spectroscopy on biomolecules (see Chapter 4) are also encountered in VCD. Especially the strong IR absorption of water, the natural solvent for most biomolecules, poses difficulties. In particular, the HOH bend absorption strongly overlaps with the amide I band. Moreover, one should avoid the use of excipients with carbonyl groups, because their signal may interfere with that of the peptide carbonyl groups. Interference by CO_2 can be minimized by thoroughly flushing the spectrometer with dry nitrogen. The spectral interference of amino acid side-chains is more a problem in the amide II than in the amide I band, but the disturbing effect is small, as the transitions are sharp and at well defined frequencies (Pancoska et al., 1995).

To overcome the problem of water absorption, many quantitative measurements on protein secondary structure in solution have been performed in D_2O. However, in heavy water, deuterium exchanges with the protein hydrogens. The rate and extent of this exchange depends on the protein and the environmental conditions (Leichtling and Klotz, 1966; Molday et al., 1972; Delepierre et al., 1987). Therefore, great care must be taken when comparing different systems (Pancoska et al., 1989; Baumruk and Keiderling, 1993), although Baumruk et al. (1996) claimed that the effect of the (partial) deuteration on secondary structure predictions can be neglected in most cases. Deuterium exchanged proteins have been obtained by twice or three times dissolving and lyophilizing the sample in D_2O before final sample preparation. This procedure may be sufficient to counteract differences in rate and extent of exchange but bears the risk of unwanted perturbation of the protein structure by the lyophilization cycles. In 1993, Baumruk and Keiderling were the first to report amide I CD spectra in H_2O, owing to improved instrumentation. Although the signal-to-noise ratio of the measurements is worse than in D_2O, measurements in water avoid interpretative ambiguities due to deuteration.

Importantly, a proper subtraction of the (solvent) water contribution to the spectra requires high concentrations of protein (typically 20–50 mg/mL) and very short pathlengths (6 μm for H_2O, 20–50 μm for D_2O). Usually, CaF_2 windows separated by a Teflon, tin, or Mylar spacer are used as the cell. The use of a flow-in cell, which allows recording baseline and sample spectra without cell demounting, ensures consistency in the optical properties of the cell, like pathlength, stress-induced birefringence, and window orientation (Pancoska et al., 1994).

In order to compensate for uncertainties in pathlength or concentration, the VCD spectrum is normalized to the IR absorbance of the sample by assuming equal absorption at 1650 cm^{-1} for different structures (Pancoska and Keiderling, 1991; Pancoska et al., 1994). However, this procedure was questioned by Lee et al. (1990). Molar absorption coefficients may differ for different structures, and absorption due to other chromophores may also contribute to the absorption at 1650 cm^{-1}. The latter can be solved by peak deconvolution, but it is obvious that the assumptions of the deconvolution may affect the result (Pancoska et al., 1994). However, the effect of pathlength and concentration on secondary-structure prediction seems to be less for VCD than for far-UV CD (Pancoska et al., 1994, 1995).

Artifact signals correlated to the magnitude of the absorption pose an additional problem in VCD (Malon and Keiderling, 1988). Apparently, by subtracting the VCD spectrum of poly-(DL)-lysine measured under identical condition as the sample spectrum, one can solve this problem, as essentially flat base lines are obtained by this procedure (Pancoska et al., 1989). Alternatively, one can reduce the artifact problem to a large extent by adaptation of the experimental setup (Lipp and Nafie, 1984; Malon and Keiderling, 1988).

Estimation of Secondary Structure from Vibrational CD

In order to calculate secondary-structure content from experimentally determined VCD spectra of reference proteins, Pancoska and coworkers have used the principal-component method of factor analysis combined with cluster analysis (Pancosca et al., 1991, 1994, 1995; Pancoska and Keiderling, 1991; Baumruk et al, 1996). The nonlinear effects discussed in the previous section have not been considered explicitly for VCD on proteins. Pancoska et al. (1989, 1991) found that VCD reflects shorter range interactions than far-UV CD, so nonlinear effects may be less important. Indeed, model polypeptides show relatively small chain-length dependence for VCD spectra of α-helices (Yasui et al., 1987). Concentration dependence of the results seems to be less with the factor analysis than with other methods (Pancoska et al., 1995).

Although both far-UV CD and VCD of the amide I band reflect the secondary structure, far-UV CD monitors only dihedral angles, whereas VCD bands also have a through-bond contribution. Therefore, secondary structures obtained with the two techniques may differ.

Secondary structure in H_2O may be deduced from the amide I or amide II band and in D_2O from the amide I′ or amide II′ band. The amide I band is most sensitive to secondary structure, but has a strong overlap with the water absorption band. The amide II band does not have this disadvantage, but shows less variation with secondary structure. In D_2O, the water absorption hardly plays a role, but as described in the previous section, the use of D_2O introduces the ambiguity of completeness of H-D exchange. Pancoska et al. (1995) and Baumruk et al. (1996) showed that the best prediction is obtained by combining the CD of the amide I′ and amide II band.

The quality of the prediction of secondary structure by VCD, by far-UV CD, and by a combination of the two has been compared by Pancoska et al. (1995). They used the principal-component method of factor analysis for predicting secondary-structure content from both VCD and far-UV CD spectra. α-Helix and 3_{10}-helix were combined in a single structural class, as was done for all β-sheet structures. Other categories used were turn, bend (no hydrogen bond, Cα–Cα dihedral angle less than 74 degrees), and other. The results of the comparison are given in Table 5. Comparison of this table with Table 4 is not straightforward, as different test sets were used. From Table 5 it is clear that the prediction by VCD for α-helix is clearly less reliable than by far-UV CD, whereas for β-sheet the prediction is better. This is not surprising, as it has been shown (Pribic et al., 1993) that also FTIR-absorption spectra are more sensitive to β-sheet structure than far-UV CD spectra. Also for turn and other, VCD gives slightly better results, whereas the bend prediction is similar. However, the best results are obtained when the two techniques are combined. Table 5 also shows that whereas the absolute errors for β-sheet, turn, bend, and other structures are comparatively small, the relative errors are considerable.

The real advantage of combining VCD with far-UV CD is found when the prediction of the secondary structure of individual proteins is considered. As Pancoska et al. (1995) showed, only the combination of VCD with far-UV CD consistently yields predictions within 10% for α-helix and β-sheet and within 5.5% on average.

	Structural Classes									
Method	**Helix**		**Sheet**		**Turn**		**Bend**		**Other**	
	σ	σ_{rel}	σ	σ_{rel}	σ	σ_{rel}	σ	σ_{rel}	σ	σ_{rel}
VCD[b]	10	13	7	16	4	27	3	18	3	14
Far-UV	7	9	10	10	4	27	4	21	4	17
VCD[b] + far-UV CD	6	7	7	14	3	19	4	20	3	13

[a]From Pancoska et al. (1995).

[b]VCD of combined amide I′ and amide II band.

Table 5 Comparison of Absolute (σ) and Relative (σ_{rel}) Root Mean Square Residuals × 100 of the Secondary-Structure Fractions Obtained From Infrared and Far-UV CD Spectroscopy[a]

Ligand Binding

An interesting application of VCD has been the study of ligand binding to heme proteins. For this purpose, the ligand-iron stretch vibration between 2100 and 1950 cm^{-1} is measured (Bormett et al., 1992; Teraoka et al., 1992). The theoretical relationship between this signal and the electronic state of the ligand-heme-iron-protein complex is not exactly known (Bormett et al., 1992), but from empirical studies it appears that the VCD signal is affected by the symmetry and constraint in the porphyrin ring, by the electronic state and orientation of the electromagnetic dipole moments of the ligand-iron complex, and by the interaction between the ligand and the heme pocket (Bormett et al. 1992; Teraoka et al., 1992). For instance, azide as the ligand results in strong negative bands, whereas cyanide yields a positive band, and SCN^-, OCN^-, and CO give no significant VCD (Teraoka et al. 1992). Bormett et al. (1992) measured VCD of a series of azide myo- and hemoglobins. The IR absorption spectra in this region show a double transition, which results from the high- and low-spin state iron complexes. Only the low-spin band is reflected in the VCD spectra. The anisotropy ratio, g, (Equation 4) for horse and whale myoglobin is 9.5×10^{-4}. These proteins have identical heme pockets. When the His E7 of the pocket was mutated to glycine or the Val E11 to asparagine, g was almost zero. Replacement of His E7 by glutamine reduced g by 35%. This shows that VCD is highly sensitive to small changes in the weak interactions between the ligand and the heme-pocket residues. All native hemoglobins studied had a g of 8×10^{-4}. The authors suggested that the decrease (compared to the myoglobin value) is caused by structural differences in the proximal side of the heme or by constraint of the porphyrin ring. Also, a mutant monomeric hemoglobin (CTT) had a g of 8×10^{-4}, although the band position was shifted by 20 cm^{-1}. Apparently, only the protein-heme interactions (band shift), but not the ligand-protein interactions (constant g), are affected by the quaternary structure.

Advantages, Limitations, and Conclusions

VCD spectra in general show more differentiation than UV-CD spectra. This is illustrated in Figure 4, where the near-UV CD and VCD spectra of four proteins are compared. Whereas the four proteins have a similar shape of the far-UV CD spectrum, showing a negative band between 200 and 240 nm and a positive one between 180 and 200 nm, they exhibit dissimilar VCD spectra with different signs at comparable wavenumbers. Hence, not only intensity differences, but also sign discrimination can be used for the analysis of VCD.

Advantages of VCD over far-UV CD are the absence of interference by absorbing aromatic residues (Yasui and Keiderling, 1986) and the absence of CD for totally unordered structure (Paterlini et al., 1986). On average, the reliability of secondary-structure prediction from VCD and specifically the reliability of α-helix prediction is less than that

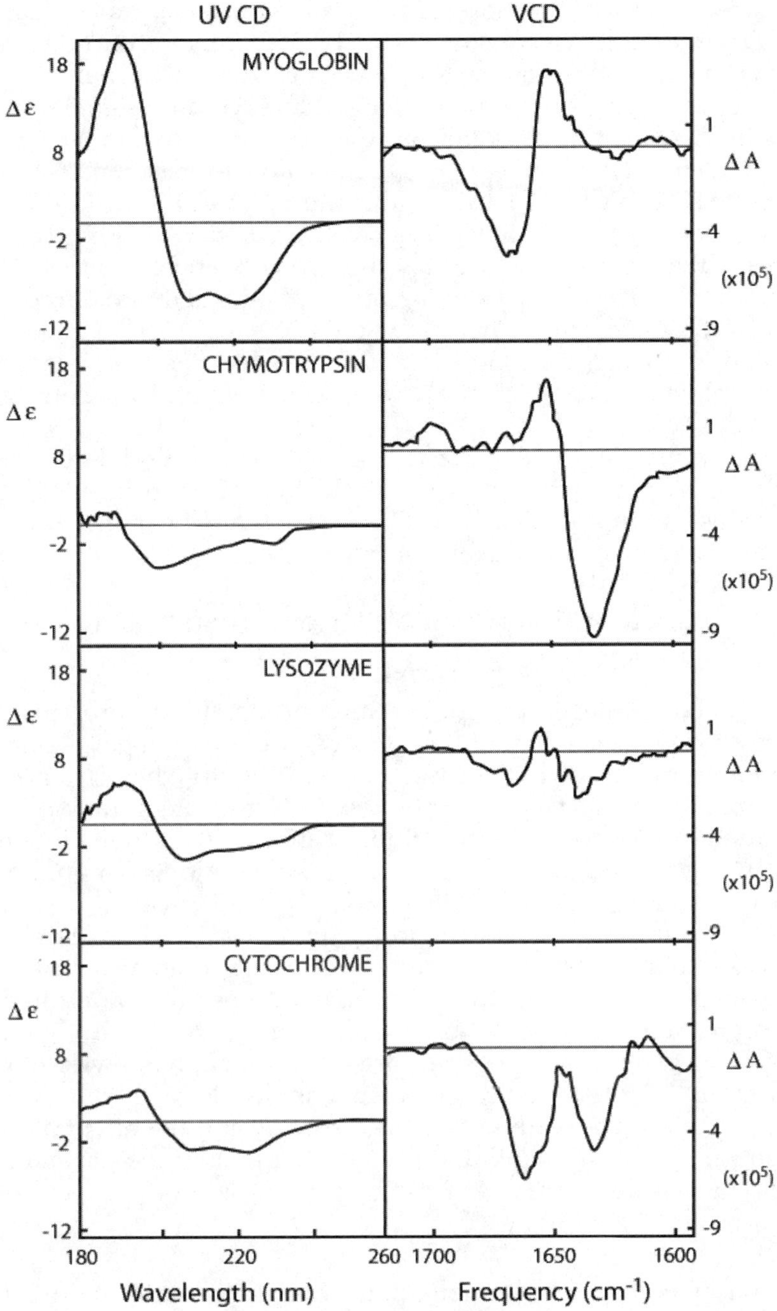

Figure 4 A comparison between UV-CD and VCD spectra of four proteins. From Pancoska et al. (1989), reproduced with permission of the author and the copyright holder (American Chemical Society).

from far-UV CD, and the measurements are more complicated than far-UV CD measurements (Pancoska et al., 1995). It remains uncertain as to what extent spectral differences in VCD are caused by different levels of deuterium exchange in the proteins. Moreover, the signal-to-noise levels in VCD experiments are considerably lower than in far-UV CD measurements. Nevertheless, VCD is a valuable addition to far-UV CD, because in particular the combination of VCD with far-UV CD measurements (see Table 5) yields a better prediction for some types of structure than one of the techniques alone (Pancoska et al. 1995; Baumruk et al., 1996). VCD measurements require high protein concentrations (20–50 mg/mL), whereas far-UV CD is restricted to lower concentrations (~1 mg/mL). So, also in this respect the two techniques are complementary, which may be especially useful for concentration-dependent studies.

The application of VCD for the study of ligand binding is insufficiently developed to give a proper evaluation of the possibilities and limitations, but in specific cases VCD has shown to be highly sensitive to small changes in ligand-receptor interactions.

Circular Dichroism on Oriented Samples

α-Helices

A fundamental question in the research of membrane proteins, like receptor proteins, is whether the α-helices are oriented parallel to the membrane surface, as in the case of an amphipathic helix, or normal to the membrane surface, as in the case of a transmembrane α-helix. This often has been studied by "fluorescence footprinting," in which a tryptophan residue is moved progressively along a peptide sequence (see, e.g., Clayton and Sawyer, 1999). However, a simpler method is applying CD spectroscopy on helical peptides bound to oriented planar phospholipid multilayers. The oriental distribution of the helices with respect to the incident light beam is monitored. In this way, changes on orientation upon changes in the environment or the binding of agents can be measured (Clayton and Sawyer, 2000).

For an oriented axially symmetric molecule with the average angle γ between the main axis of the molecule and the orientation direction (in case of a membrane protein, the membrane normal), the orientational order parameter S can be defined as:

$$S = 1.5\langle\cos^2\gamma - 1\rangle \tag{12}$$

For an α-helix, the helical orientational order parameter, S_h, with respect to the optical propagation axis can be measured from the ratio of the rotational strength of the resolved 207-nm transition in oriented molecules (R_o) to that of the unaligned sample (R_d) by (Bazzi and Woody, 1985; Olah and Huang, 1988):

$$S_h = [1 - (R_o/R_d)] \tag{13}$$

and hence

$$<\sin^2\gamma> = 2(R_o/R_d) \tag{14}$$

The rotational strength is obtained from the area under the CD band at 207 nm. The area can be determined from a Gaussian deconvolution of the far-UV CD spectrum (Bazzi and Woody, 1985; Clayton and Sawyer, 2000) or from the difference CD spectrum between the oriented and the nonoriented sample (Clayton and Sawyer, 2000).

An example is given in Figure 5 and Table 6 for the polypeptide DWLKAFYDKVAEKLKEAF, which is a model for the amphipathic helical regions of apolipoprotein A-I. From the results, it appears that S_h is negative both above and below the transition temperature of the lipid, indicating that the helix lies predominantly parallel to the lipid surface rather than in a transmembrane direction. At higher temperature, the average angle between the membrane normal and the polypeptide is

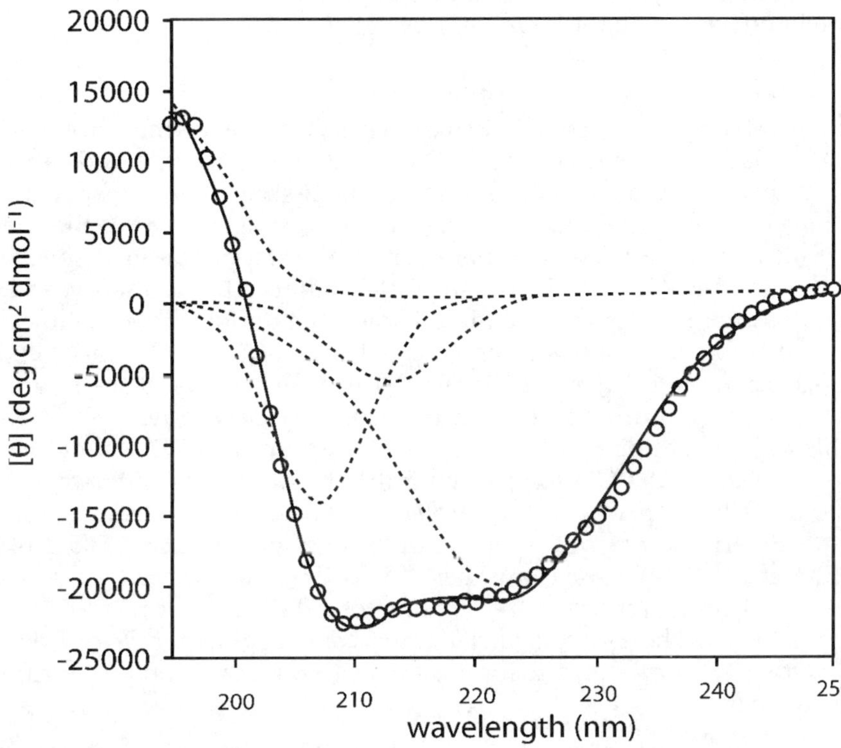

Figure 5 CD spectrum of a model polypetide (A18) in oriented multilayers of dimyristoyl phosphatidylcholine at a molar lipid/peptide ratio of 20:1 at 35°C (circles). The resultant fit to the sum of four Gaussian components is indicated by the solid line and the Gaussian components by the dotted lines. From Clayton and Sawyer (2000), reproduced with permission of the author and the copyright holder (Elsevier Science BV).

Method[a]	15°C (T less than T^m)	35°C (T greater than T^m)
Rotational strength ratio	–0.18	–0.32
Difference ellipticity	–0.20	–0.28

[a]Values determined using the ethanol-treated multilayer spectrum at 35°C as isotropically averaged helix reference spectrum. From Clayton and Sawyer (2000).

Table 6 Helix Order Parameters (S_h) of the Model Peptide 18A in Lipid Multilayers of Dimyristoyl Phosphatidylcholine as Determined by the Rotational Strength Ratio or Difference Ellipticity (See Text)

increased. The results are in accordance with fluorescence and x-ray measurements (Clayton and Sawyer, 2000). Other examples of the determination of the orientation of the amphipathic α-helix at a membrane interface can be found in Hristova et al. (1999).

β-Barrels

Regular β-barrels are also axially symmetric, and hence their orientation can also be monitored from CD measurements on oriented samples. In this case, the mean orientation of the β-strands with respect to the barrel axis is measured. For a barrel of n strands, this gives the so-called shear number, S, which characterizes the stagger of the β-sheets (Tamm and Tatulian, 1997; Marsh, 2000). Although the theory and equations are more complicated in this case, the underlying principles are comparable to those for oriented α-helices.

As explained in "Far-UV Circular Dichroism," above, β-sheets are more reliably quantified by IR than by far-UV spectroscopy. Therefore, in this case VCD appears to be the most appropriate CD technique. Marsh (2000) used VCD for studying the orientation of outer membrane protein OmpA from *Escherichia coli*. From the spectra, he was able to determine by means of the optimized geometrical model of Murzin et al. (1994) a shear number of 9 to 10 (8 strands), a barrel radius of 0.81 nm, an internal free volume of 0.031 nm^3 per residue, an average twist of the β-sheets of 28 degrees, and a coiling angle of 1 degree. Other examples and theoretical considerations can be found in the same paper.

Time-Resolved Circular Dichroism

TRCD allows structural studies of transient intermediates. It has been used to study isomerization (Hasumi, 1980), photoprocesses (Björling et al., 1992; Lewis et al., 1992), and folding intermediates (Labhardt, 1986; Kuwajima et al., 1987, 1991; Elöve et al., 1992; Qin et al., 2001;

Clarke et al., 1999). Clarke et al. (1999) used SRCD with a stopped-flow chamber. An extensive review on TRCD has been written by Kuwajima (1996).

Instrumentation for TRCD has been reviewed by Lewis et al. (1992). When the required time resolution is milliseconds or longer, a conventional CD spectrophotometer supplied with a mixing chamber can be used. Mixing chambers are commercially available or can be homemade (Kuwajima et al., 1987; Qin et al., 2001). Both stopped-flow (Kuwajima et al., 1987, 1991) and continuous-flow (Elöve et al., 1992) experiments have been described. In some cases, measurements are done in ethylene glycol/water mixtures at low temperatures (−77°C) in order to slow down processes. To this end, a special stopped-flow device for high viscosity and low temperature was developed (Qin et al., 2001). When ethylene glycol is added to the sample, it should be realized that in some cases this enhances the α-helical structure in the protein (Hamada et al., 1995, Kuroda et al., 1996, Qin et al., 2001).

The optical pathlength in stopped flow experiments is typically between 1.0 and 10 mm and the dead time 2 milliseconds or longer. In many cases, CD signals have been monitored at single wavelengths, which is time-consuming when a complete spectrum is scanned point by point. Alternatively, a multi-channel detector or a rapid-scanning monochromator can be used. The CD results are subjected to traditional kinetic analyses. Both far- and near-UV–Vis TRCD measurements can be made on these instruments, allowing detailed kinetic studies of secondary and tertiary structural changes.

Stopped-flow SRCD allows measurements in the sub-millisecond range (Wallace, 2000a). Light-induced processes in the nano- and picosecond timescale can be followed by near-UV CD with highly sophisticated TRCD instrumentation (Björling et al., 1992; Lewis et al., 1992). However, these measurements are very sensitive to optical artifacts (Björling et al., 1992; Lewis et al., 1992).

TRCD was used by Hasumi (1980) to elucidate a four-step mechanism for the pH-induced isomerization of horse heart ferricytochrome c. Elöve et al. (1992) defined four distinct stages in the refolding of cytochrome c: less than 4 milliseconds, 4 to 100 milliseconds, 0.1 to 1 second, and greater than 1 second. The first two phases were shown to represent the formation of the secondary structure, whereas in the latter two the tertiary structure is formed.

Kuwajima et al. (1991) used TRCD to demonstrate the existence of various folding intermediates during the transition from the unfolded to the native state of dihydrofolate reductase. They calculated the secondary-structure contents and observed that initially a considerable amount of secondary structure is formed during folding, which rearranges in later stages. Moreover, their studies suggested that unfolding and refolding follow different pathways.

Conclusions

This chapter shows that CD spectroscopy can be used to probe very different levels of protein structure. Far-UV CD and VCD can be applied to obtain information about the secondary structure. Near-UV CD is particularly suitable to monitor the tertiary structure and the conformational flexibility. For measuring protein-ligand interactions, MCD, near-UV–Vis CD, and VCD are available. Although the structural information content of CD is significantly lower than that of NMR spectroscopy and crystallographic x-ray diffraction, it offers a number of advantages, including (1) the speed of the measurements, (2) the relatively low amount of protein needed, (3) the large range of concentrations that can be used, (4) its nondestructivity, (5) the simplicity of the spectra obtained, and (6) the relative ease of their interpretation. Therefore, CD spectroscopy has become a popular tool for the comparison of homologous proteins, the study of a single protein at various conditions, the checking of the integrity of recombinant proteins, the screening of mutants, the estimation of thermodynamic data on protein folding and ligand binding, and the monitoring of structural changes during production, formulation, and storage of proteins. In particular, near-UV CD (and to a lesser extent far-UV CD) spectroscopy, as it is sensitive to relatively small conformational changes and compatible with a large range of commonly used formulation excipients, is particularly useful for the screening of pharmaceutical protein formulations.

Abbreviations

CD	Circular dichroism
CSA	(+)-10-Camphorsulfonic acid
g	Anisotropy ratio
IR	Infrared
MCD	Magnetic circular dichroism
ORD	Optical rotatory dispersion
SRCD	Synchrotron radiation circular dichroism
SVD	Singular value decomposition
SVD+VS	Singular value decomposition with variable selection
TRCD	Time-resolved circular dichroism
UV	Ultraviolet
VCD	Vibrational circular dichroism
Vis	Visible
VUV	Vacuum ultraviolet

$\Delta\varepsilon$	Molar absorption difference
$[\theta]_\lambda$	Molar ellipticity

Acknowledgments

This work was made by updating a review written by the first author and W.C. Johnson, Jr. (Bloemendal and Johnson, 1995), and was financially supported by the ALCON award of H. Bloemendal (1996).

References

1. Bailey JC, Martin SR, and Bayley PM. A Circular Dichroism Study of Epidermolytic Toxins A and B from Staphylococcus aureus. *Biochem J* 1982; 203:775–778
2. Banerjee R, and Basu G. Direct Evidence for Alteration of Unfolding Profile of a Helical Peptide by Far-Ultraviolet Circular Dichroism Aromatic Side-Chain Contribution. *FEBS Lett* 2002; 523:152–156
3. Baumruk V, and Keiderling TA. Vibrational Circular Dichroism of Proteins in H_2O Solution. *J Am Chem Soc* 1993; 115:6939–6942
4. Baumruk V, Pancoska P, and Keiderling TA. Prediction of Secondary Structure Using Statistical Analyses of Electronic and Vibrational Circular Dichroism and Fourier Transform Infrared Spectra of Proteins in H_2O. *J Mol Biol* 1996; 259:774–791
5. Bayley P. Circular Dichroism and Optical Rotation. In: Brown SB. *An Introduction to Spectroscopy for Biochemists*. London: Academic Press; 1980:148–234
6. Bazzi MD, and Woody RW. Oriented Secondary Structure in Integral Membrane Proteins. I. Circular Dichroism and Infrared Spectroscopy of Cytochrome Oxidase in Multilamellar Films. *Biophys J* 1985; 48:957–966
7. Beltramini M, Bubacco L, Salvato B, Casella L, Gulotti M, and Garofani S. The Aromatic Circular Dichroism Spectrum as a Probe for Conformational Changes in the Active Site Environment of Hemocyanins. *Biochim Biophys Acta* 1992; 1120:24–32
8. Björling SC, Zhang C-F, Farrens DL, Song P-S, and Kliger DS. Time-Resolved Circular Dichroism of Native Oat Phytochrome Photointermediates. *J Am Chem Soc* 1992; 114:4581–4588
9. Bloemendal M, and Johnson WC, Jr. Structural Information on Proteins from Circular Dichroism Spectroscopy. Possibilities and Limitations. *Pharm Biotechnol* 1995; 7:65–100
10. Bokma JT, Johnson WC Jr, and Blok J. CD of the Li Salt of DNA in Ethanol/Water Mixtures: Evidence for the B- to C-form Transition in Solution. *Biopolymers* 1987; 26:893–909

11. Bormett RW, Asher SH, Larkin PJ, Gustafson WG, Ragunathan N, Freedman, TB, Nafie LA, Balasubramanian S, Boxer SG, Yu N-T, Gersonde K, Noble RW, Springer BA, and Sligar SG. Selective Examination of Haem Protein Azide Ligand-Distal Globin Interactions by Vibrational Circular Dichroism. *J Am Chem Soc* 1992; 114:6864–6867
12. Bradford MM. A Rapid and Sensitive Method for Quantification of Microgram Quantities of Protein Utilizing the Principle of Protein-Dye Binding. *Anal Biochem* 1976; 72:240–254
13. Brahms S, and Brahms J. Determination of Protein Secondary Structure in Solution by Vacuum Ultraviolet Circular Dichroism. *J Mol Biol* 1980; 138:149–178
14. Bree A, and Lyons LE. The Intensity of Ultraviolet Light Absorption by Monocrystals. Part I. Measurement of Thickness of Thin Crystals by Interferometry. *J Chem Soc (London)* 1956:2658–2662
15. Breslauer KJ. Extracting Thermodynamic Data from Equilibrium Melting Curves for Oligonucleotide Order-Disorder Transitions. *Methods Enzymol* 1995; 259:221–242
16. Campbell ID, and Dwek RA. *Biological Spectroscopy*. Menlo Park, CA: Benjamin/Cummings; 1984:255–277
17. Cantor CR, and Schimmel PR. *Biophysical Chemistry, Volume 2*. New York: Freeman; 1980:409–418
18. Chang CT, Wu C-S, and Yang JT. Circular Dichroism Analysis Of Protein Conformation: Inclusion of the β-turns. *Anal Biochem* 1978; 91:13–31
19. Chang JF, Phillips K, Lundback T, Gstaiger M, Ladbury JE, and Luisi B. Oct-1 POU and Octamer DNA Co-operate to Recognise the Bob-1 Transcription Co-activator via Induced Folding. *J Mol Biol* 1999; 288:941–952
20. Charney E. The Molecular Basis of Optical Activity, Optical Rotatory Dispersion and Circular Dichroism. New York: John Wiley; 1979:1–17 and 323–327
21. Cheesman MR, Greenwood C, and Thomson AJ. Magnetic Circular Dichroism of Hemoproteins. *Adv Inorg Chem* 1991; 36:201–255
22. Chen Y, and Wallace BA. Secondary Solvent Effects on the Circular Dichroism Spectra of Polypetides: Influence of Polarisation Effects on the Far Ultraviolet Spectra of Alamethicin. *Biophys Chem* 1997a; 65:65–74
23. Chen Y, and Wallace BA. Solvent Effects on the Conformation and Far Ultraviolet Circular Dichroism Spectra of Gramicidin A. *Biopolymers* 1997b; 42:771–781
24. Chen Y-H, and Yang JT. Two Point Calibration of Circular Dichrometers with d-10-Camphorsulphonic Acid. *Anal Lett* 1977; 10:1195–1207

25. Chen Y-H, Yang JT, and Martinez H. Determination of the Secondary Structures of Proteins by Circular Dichroism and Optical Rotatory Dispersion. *Biochemistry* 1972; 11:4120–4132

26. Clarke DT, Doig AJ, Stapeley BJ, and Jones GR. *Proc Natl Acad Sci U S A* 1999; 96:7232–7237

27. Clayton AHA, and Sawyer WH. Oriented Circular Dichroism of a Class A Amphipathic Helix in Aligned Phospholipid Multilayers. *Biochim Biophys Acta* 2000; 28:1331–141

28. Compton LA, and Johnson WC Jr. Analysis of Protein Circular Dichroism Spectra for Secondary Structure Using a Simple Matrix Multiplication. *Anal Chem* 1986; 155:155–167

29. Cotton MA. Recherches sur l'absorption et la dispersion de la lumière par les milieux doués du pouvoir rotatoire. *Ann Chim Phys VII* 1896; 8:347–437

30. Craig S, Pain RH, Schmeissner U, Virden R, and Wingfield TP. Determination of the Contributions of Individual Aromatic Residues to the CD Spectrum of IL-1 Using Site Directed Mutagenesis. *Int J Pept Protein Res* 1989; 33:256–262

31. Delepierre M, Dobson CM, Karplus M, Poulsen FM, States DJ, and Wedin RE. Electrostatic Effects and Hydrogen Exchange Behaviour in Proteins: The pH Dependence of Exchange Rates in Lysozyme. *J Mol Biol* 1987; 197:111–130

32. Diem M, Gotkin PJ, Kupfer JM, and Nafie LA. Vibrational circular Dichroism in Amino Acids and Peptides. 2. Simple Alanyl Peptides. *J Am Chem Soc* 1978; 100:5644–5650

33. Dousseau F, and Pezolet M. Determination of the Secondary Structure Content of Proteins in Aqueous Solutions from Their Amide I and Amide II Infrared Bands. Comparison Between Classical and Partial Least-Squares Methods. *Biochemistry* 1990; 29:8771–8779

34. Dumas F, Frank S, Koebnik R, Maillet E, Lustig A, and van Gelder P. Extended Sugar Slide Function for the Periplasmic Coiled Coil Domain of Scr. *J Mol Biol* 2000; 300:687–695

35. Efting MR. Use of Two Multiple Spectroscopic Methods to Monitor Equilibrium Unfolding of Proteins. *Methods Enzymol* 1995; 259:487–512

36. Elöve GA, Chaffotte AF, Roder H, and Goldberg ME. Early Steps in Cytochrome c Folding Probed by Time-Resolved Circular Dichroism and Fluorescence Spectroscopy. *Biochemistry* 1992; 31:6876–6883

37. Englinton DG, Johnson MK, Thomson AJ, Gooding PE, and Greenwood C. Near Infra-Red Magnetic and Natural Circular Dichroism of Cytochrome c Oxidase. *Biochem J* 1980; 191:319–331

38. Eposti MD, Crimi M, Körtner C, Kröger A, and Link T. The Structure of Dihaem Cytochrome b of Fumarate Reductase in Wolinella

Succinogenes: Circular Dichroism and Sequence Analysis Studies. *Biochim Biophys Acta* 1991; 1056:243–249

39. Fawcett TG, Bernarducci EE, Krough-Jespersen K, and Schugar HJ. Charge-Transfer Absorptions of Copper(II)-Imidazole and Copper(II)-Imidazolate Chromophores. *J Am Chem Soc* 1980; 102:2598–2604
40. Freire E. Thermal Denaturation Methods in Study of Protein Folding. *Methods Enzymol* 1995; 259:144–168
41. Gans P. Vibrational Spectroscopy. In: Brown SB. *An Introduction to Spectroscopy for Biochemists*. London: Academic Press; 1980:148–234
42. Goa J. A Micro Biuret Method for Protein Determination. Determination of Total Protein in Cerebrospinal Fluid. *Scand J Clin Lab Invest* 1953; 5:L218–222
43. Goodman M, Verdini AS, Toniolo C, Phillips WD, and Bovey FA. Selective Criteria for the Critical Size for Helix Formation in Oligopeptides. *Proc Natl Acad Sci U S A* 1969; 64:444–450
44. Greenfield NJ. Methods to Estimate the Conformation of Proteins and Polypeptides from Circular Dichroism Data. *Anal Biochem* 1996; 235:1–10
45. Greenfield NJ. Applications of Circular Dichroism in Protein and Peptide Analysis. *Trends Anal Chem* 1999; 18:236–244
46. Greenfield NJ, and Fasman GD. Computed circular dichroism spectra for the evaluation of protein conformation. *Biochemistry* 1969; 8:4108–4116
47. Gross KP, and Schnepp O. Improved Circular Dichroism Instrument in the Vacuum Ultraviolet. *Rev Sci Instrum* 1977; 48:362–363
48. Gupta VP, and Keiderling TA. Vibrational CD of the Amide II Band in Some Model Polypeptides and Proteins. *Biopolymers* 1992; 32:239–248
49. Hadden JM, Bloemendal M, Haris PI, Srai SKS, and Chapman D. Fourier Transform Infrared Spectroscopy and Differential Scanning Calorimetry of Transferrins. *Biochim Biophys Acta* 1994; 1205: 59–67
50. Hamada D, Kuroda Y, Tanaka T, and Goto Y. High Helical Propensity of the Peptide-Fragments Derived from Beta-Lactoglobulin, a Predominantly Beta-Sheet Protein. *J Mol Biol* 1995; 254:737–746
51. Harada T, and Kuroda R. Circular Dichroism Measurements of a Protein in Dried Thin Films. *Chem Lett* 2002:326–327
52. Hasumi H. Kinetic Studies on Isomerization of Ferricytochrome c in Alkaline and Acid pH Ranges by the Circular Dichroism Stopped-Flow Method. *Biochim Biophys Acta* 1980; 626:265–276
53. Hennessey JP Jr, and Johnson WC Jr. Information Content in the Circular Dichroism of Proteins. *Biochemistry* 1981; 20:1085–1094

54. Hristova K, Wimley WC, Mishra VK, Anantharamiah GM, Segrest JP, and White SH. An Amphiphatic α-Helix at a Membrane Interface: A Structural Study Using a Novel X-ray Diffraction Method. *J Mol Biol* 1999; 290:99–117

55. Hvidt S, and Lehrer SS. Thermally Induced Chain Exchange of Frog $\alpha\beta$-Tropomyosins. *Biophys Chem* 1992; 45:51–59

56. Jirgensons B. Optical Activity of Proteins and Other Macromolecules. *Mol Biol Biochem* 1973; 5:1–191

57. Johnson WC Jr. A Circular Dichroism Spectrometer for the Vacuum Ultraviolet. *Rev Sci Instrum* 1971; 42:1283–1286

58. Johnson WC Jr. Extending Circular Dichroism Spectra into the Vacuum UV and Its Application to Proteins. *Photochem Photobiol* 1986; 44:307–313

59. Johnson WC Jr. Secondary Structure of Proteins Through Circular Dichroism Spectroscopy. *Ann Rev Biophys Biophys Chem* 1988; 17:145–166

60. Johnson WC Jr. Protein Secondary Structure and Circular Dichroism: A Practical Guide. *Proteins: Struct Funct Gen* 1990; 7:205–214

61. Johnson WC Jr. Analyzing Protein Circular Dichroism Spectra for Accurate Secondary Structures. *Proteins* 1999; 35:307–312

62. Kahn PC. The Interpretation of Near-Ultraviolet Circular Dichroism. *Methods Enzymol* 1979; 61:339–377

63. Keiderling TA. Vibrational Circular Dichroism. *Appl Spectrosc Rev* 1981; 17:189–226

64. Keiderling TA. Vibrational CD of Biopolymers. *Nature* 1986; 322:851–852

65. Kelly SN, and Price NC. Application of Circular Dichroism to Studies of Protein Folding and Unfolding. *Biochim Biophys Acta* 1997; 1338:161–185

66. Kubelka J, and Keiderling TA. Differentiation of β-sheet-forming Atructures: ab initio-based Simulations of IR Absorption and Vibrational CD for Model Peptide and Protein β-sheets. *J Am Chem Soc* 2001; 123:12048–12058

67. Kuroda Y, Hamada D, Tanaka T, and Goto Y. High Helicity of Peptide Fragments Corresponding to β-strand Regions of β-lactoglobulin Observed by 2D Spectroscopy. *Fold Des* 1996; 1:243–251

68. Kuroda R, Harada T, and Shindo Y. A Solid-State Dedicated Circular Dichroism Spectrophotometer: Development and Application. *Rev Sci Inst* 2001; 72:3802–3810

69. Kuwajima K. The Molten Globule State as a Clue for Understanding the Folding and Cooperativity of Globular Protein Structure. *Proteins: Struct Funct Gen* 1989; 6:87–103

70. Kuwajima K. In: Fasman GD. *Circular Dichroism and the Conformational Analysis of Biomolecules*. New York: Plenum Press; 1996:159–182.

71. Kuwajima K, Yamaya H, Miwa S, Sugai S, and Nagamura T. Rapid Formation of Secondary Structure Framework in Protein Folding Studied by Stopped-Flow Circular Dichroism. *FEBS Lett* 1987; 221:115–118

72. Kuwajima K, Garvey EP, Finn BE, Matthews CR, and Sugai S. Transient Intermediates in the Folding of Dihydrofolate Reductase as Detected by Far-Ultraviolet Circular Dichroism Spectroscopy. *Biochemistry* 1991; 30:7693–7703

73. Labhardt AM. Folding Intermediates Studied by Circular Dichroism. *Methods Enzymol* 1986; 131:126–135

74. Lang CA. The Microdetermination of Kjeldahl Nitrogen in Biological Materials. *Anal Chem* 1958; 30:1692–1694

75. Lee DC. *Infrared Spectroscopic Studies of Biological and Model Membranes*. Ph.D. Thesis. London: Royal Free Hospital School of Medicine; 1985:62, 71

76. Lee DC, Haris PI, Chapman D, and Mitchell RC. Determination of Protein Secondary Structure Using Factor Analysis of Infrared Spectra. *Biochemistry* 1990; 29:9185–9193

77. Leichting BH, and Klotz IM. Catalysis of Hydrogen-Deuterium Exchange in Polypeptides. *Biochemistry* 1966; 5:4026–4037

78. Lewis PN, Momany FA, and Scheraga HA. Chain Reversals in Proteins. *Biochim Biophys Acta* 1973; 303:211–229

79. Lewis JW, Goldbeck RA, Kliger DS, Xie X, Dunn RC, and Simon JD. Time-Resolved Circular Dichroism Spectroscopy: Experiment, Theory, and Applications to Biological Systems. *J Phys Chem* 1992; 96:5243–5254

80. Lipp ED, and Nafie LA. Fourier Transform Infrared Vibrational Circular Dichroism: Improvement in Methodology and Mid-Infrared Spectral Results. *Appl Spectr* 1984; 38:20–26

81. Lobley A, Whitmore L, and Wallace BA. DICHROWEB: An Interactive Website for the Analysis of Protein Secondary Structure from Circular Dichroism Spectra. *Bioinformatics Appl Note* 2002; 18:211–212

82. Lowry TM, and French HS. The Rotatory Dispersive Power of Organic Compounds. Part XX. Rotatory Dispersion and Circular Dichroism of Camphor-B-Sulphonic Acid in the Region of Absorption. *J Chem Soc* 1932:2655–2658

83. Lowry OH, Rosebrough NJ, Farr AL, and Randall RJ. Protein Measurement with the Folin Phenol Reagent. *J Biol Chem* 1951; 193:265–275

84. Madison V, and Schellman J. Optical Activity of Polypeptides and Proteins. *Biopolymers* 1972; 11:1041–1076

85. Malon P, and Keiderling TA. A Solution to the Artifact Problem in Fourier Transform Vibrational Circular Dichroism. *Appl Spectrosc* 1988; 42:32–38

86. Manavalan P, and Johnson WC Jr. Sensitivity of Circular Dichroism to Protein Tertiary Structure Class. *Nature* 1983; 305:831–832

87. Manavalan P, and Johnson WC Jr. Protein Secondary Structure from Circular Dichroism Spectra. *J Biosci* 1985; 8 (Suppl):141–149

88. Manavalan P, and Johnson WC Jr. Variable Selection Method Improves the Prediction of Protein Secondary Structure from Circular Dichroism Spectra. *Anal Biochem* 1987; 67:76–85

89. Manning MC. Underlying Assumptions in the Estimation of Secondary Structure Content in Proteins by Circular Dichroism Spectroscopy - A Critical Review. *J Pharmac Biomed Analysis* 1989; 7:1103–1119

90. Marsh D. Infrared Dichroism of Twisted β-Sheet Barrels. The Structure of E. coli Outer Membrane Proteins. *J Mol Biol* 2000; 297:803–808

91. Maune JF, Beckingham K, Martin SR, and Bayley PM. Circular Dichroism Studies on Calcium Binding in Two Series of Ca^{2+} Binding Site Mutants of Drosophila melanogaster Calmodulin. *Biochemistry* 1992; 31:7779–7786

92. McPhie P. Circular Dichroism Studies on Proteins in Films and in Solution: Estimation of Secondary Structure by *g*-factor Analysis. *Anal Biochem* 2001; 293:109–119

93. Merolo JJ, Andrade MA, Prieto A, and Morán F. Proteinotopic Feature Maps. *Neurocomputing* 1994; 6:443–454

94. Metz B, Jiskoot W, Hennink WE, Crommelin DJA, and Kersten GFA. Physicochemical and Immunochemical Techniques Predict the Quality of Diphtheria Toxoid Vaccines. *Vaccine* 2003; 22:156–167

95. Molday RS, Englander SW, and Kallen RG. Primary Structure Effects on Peptide Group Hydrogen Exchange. *Biochemistry* 1972; 11:150–158

96. Murzin AG, Lesk AM, and Chothia C. Principles Determining the Structure of β-Sheet Barrels in Proteins, I. A Theoretical Analysis. *J Mol Biol* 1994; 236:1369–1381

97. Naseem F, Khan RH, Haq SK, and Naeem A. Characterization of Molten Globule State of Fetuin at Low pH. *Biochim Biophys Acta* 2003; 1649:164–170

98. Nölting B, Jung C, and Snatzke G. Multichannel Circular Dichroism Investigations of the Structural Stability of Bacterial Cytochrome p-450. *Biochim Biophys Acta* 1992; 1100:171–176

99. Olah GA, and Huang H.W. Circular Dichroism of Oriented α-Helices. I. Proof of the Exciton Theory. *J Chem Phys* 1988; 89:2531–2538

100. Orry AJW, Janes RW, Sarra R, Hanlon MR, and Wallace BA. Synchrotron Radiation Circular Dichroism Spectroscopy: Vacuum Ultraviolet Irradiation Does Not Damage Protein Integrity. *J Synchrotron Rad* 2001; 8:1027–1029

101. Osborne GA, Cheng JC, and Stephens PJ. A Near-Infrared Circular Dichroism and Magnetic Circular Dichroism Instrument. *Rev Sci Instrum* 1973; 44:10–15

102. Pancoska P, and Keiderling TA. Systematic Comparison of Statistical Analyses of Electronic and Vibrational Circular Dichroism for Secondary Structure Prediction of Selected Proteins. *Biochemistry* 1991; 30:6885–6895

103. Pancoska P, Yasui SC, and Keiderling TA. Enhanced Sensitivity to Conformation in Various Proteins. Vibrational Circular Dichroism Results. *Biochemistry* 1989; 28:5917–5923

104. Pancoska P, Yasui SC, and Keiderling TA. Structural Analyses of the Vibrational Circular Dichroism of Selected Proteins and Relationship to Secondary Structure. *Biochemistry* 1991; 30:5089–5103

105. Pancoska P, Bitto E, Janota V, and Keiderling TA. Quantitative Analysis of Vibrational Circular Dichroism Spectra of Proteins. *Faraday Disc* 1994; 99:287–310

106. Pancoska P, Bitto E, Janota V, Urbanova M, Gupta VP, and Keiderling TA. Comparison of and Limits of Accuracy for Statistical Analyses of Vibrational and Electronic Circular Dichroism Spectra in Terms of Correlations to and Predictions of Protein Structure. *Protein Sci* 1995; 4:1384–1401

107. Pancoska P, Fabian H, Yoder G, Baumruk V, and Keiderling TA. Protein Structural Segments and Their Interconnections Derived from Optical Spectra. Thermal Unfolding of Ribonuclease T_1 as an Example. *Biochemistry* 1996; 35:13094–13106

108. Pancoska P, Janota V, and Keiderling TA. Novel Matrix Descriptor for Secondary Structure Segments in Proteins: Demonstration of Predictability from Circular Dichroism Spectra. *Anal Biochem* 1999; 267:72–83

109. Pappalardo G, Impellizzeri G, and Campagna T. Copper(II) Binding of Prion Protein's Octarepeat Model Peptides. *Inorg Chim Acta* 2004; 357:185–194

110. Pasteur ML. Sur les relations qui peuvent exister entre la forme cristalline, la composition chimique et le sens de la polarisation rotatoire. *Ann Chim Phys III* 1848; 24:442–459

111. Paterlini MG, Freedman TB, and Nafie LA. Vibrational Circular Dichroism Spectra of Three Conformationally Distinct States and an Unordered State of Poly(-L-Lysine) in Deuterated Aqueous Solution. *Biopolymers* 1986; 25:1751–1765

112. Pelton JT, and McLean LR. Spectroscopic Methods for Analysis of Protein Secondary Structure. *Anal Biochem* 2000; 277:176–176

113. Perczel A, Hollósi M, Tusnády G, and Fasman GD. Convex Constraint Analysis: A Natural Deconvolution of Circular Dichroism Curves of Proteins. *Prot Eng* 1991; 4:669–679

114. Perczel A, Park K, and Fasman GD. Analysis of the Circular Dichroism Spectrum of Proteins Using the Convex Constraint Algorithm: A Practical Guide. *Anal Biochem* 1992; 203:83–93

115. Pribic R, Van Stokkum IHM, Chapman D, Haris PI, and Bloemendal M. Protein Secondary Structure from Fourier Transform Infrared and/or Circular Dichroism Spectra. *Anal Biochem* 1993; 214:336–378

116. Provencher SW. A Constrained Regularization Method for Inverting Data Represented by Linear Algebraic or Integral Equations. *Comput Phys Commun* 1982a, 27:213–227

117. Provencher SW. CONTIN, A General Purpose Constrained Regularization Program for Inverting Noisy Linear Algebraic and Integral Equation. *Comput Phys Commun* 1982b; 27:229–242

118. Provencher SW, and Glöckner J. Estimation of Globular Protein Secondary Structure from Circular Dichroism. *Biochemistry* 1981; 20:33–37

119. Qin Z, Hu D, Shimada L, Nakagawa T, Arai M, Zhou J-M, and Kihara H. Refolding of β-lactoglobulin Studied by Stopped-Flow Circular Dichroism at Subzero Temperatures. *FEBS Lett* 2001; 507:299–302

120. Roberts CJ, Darrington RT, and Whitley MB. Irreversible Aggregation of Recombinant Bovine Granulocyte-Colony Stimulating Factor (bG-CSF) and Implications for Predicting Protein Shelf Life. *J Pharm Sci* 2003; 92:1095–1111

121. Sarver RW, and Kruger WC. An Infrared and Circular Dichroism Combined Approach to the Analysis of Protein Secondary Structure. *Anal Biochem* 1991; 199:61–67

122. Scharnagl C, and Schneider S. UV-Visible Absorption and Circular Dichroism Spectra of the Subunits of C-Phycocyanin. II. A Quantitative Discussion of the Chromophore-Protein and Chromophore-Chromophore Interactions in the β-subunit. *J Photochem Photobiol B Biol* 1991; 8:129–158

123. Sharonov YA. Room Temperature and Low-Temperature Magnetic Circular Dichroism of Hemoproteins and Related Compounds. *Sov Sci Rev D Physicochem Biol* 1991; 10:1–118

124. Shindo Y, and Nakagawa M. Circular Dichroism Measurements. I. Calibration of a Circular Dichroism Spectrometer. *Rev Sci Instrum* 1985; 56:32–39

125. Shire SJ, Holladay LA, and Rinderknecht E. Self-association of Human and Porcine Relaxin as Assessed by Analytical Ultra-Centrifugation and Circular Dichroism. *Biochemistry* 1991; 30:7703–7711

126. Sigel H, and Martin RB. Coordinating Properties of the Amide Bond. Stability and Structure of Metal Ion Complexes of Peptides and Related Ligands. *Chem Rev* 1982; 82:385–426

127. Singh HK, and Wilson MT. Characterization of Haem Disorder in Cytochrome B_5 by Circular Dichroism. *Biochem Soc Trans* 1990; 18:1272–1273

128. Smith PK, Krohn RI, Hermanson GT, Mallia AK, Gartner FH, Provenzano MD, Fujimoto EK, Goeke NM, Olson BJ, and Klenk DC. Measurements of Protein Using Bicinchonic Acid. *Anal Biochem* 1985; 150:76–85

129. Snir J, Frankel RA, and Schellman JA. Optical Activity of Polypeptides in the Infrared. Predicted CD of the Amide I and Amide II Bands. *Biopolymers* 1975; 14:173–196

130. Sreerama N, and Woody RW. Poly(Pro)II Helices in Globular Proteins: Identification and Circular Dichroic Analysis. *Biochemistry* 1994; 33:10022–10025

131. Sreerama N, and Woody RW. Estimation of Protein Secondary Structure from Circular Dichroism Spectra: Comparison of CONTIN, SELCON, and CDSSTR Methods with an Expanded Reference Set. *Anal Biochem* 2000; 287:252–260

132. Sreerama N, Venyaminov SY, and Woody RW. Estimation of the Number of α-helical and β-strand Segments in Protein Using Circular Dichroism Spectra. *Protein Sci* 1999; 8:370–380

133. Sreerama N, Venyaminov SY, and Woody RW. Estimation of Protein Secondary Structure from Circular Dichroism Spectra: Inclusion of Denatured Proteins with Native Proteins in the Analysis. *Anal Biochem* 2000; 287:243–251

134. Sreerama N, Yu S, Venyaminov SY, and Woody RW. Analysis of Protein Circular Dichroism Spectra Based on the Tertiary Structure Classification. *Anal Biochem* 2001; 299:271–274

135. Strickland EH. Aromatic Contribution to Circular Dichroism Spectra of Proteins. *CRC Crit Rev Biochem* 1974; 2:113–175

136. Sundberg RJ, and Martin RB. Interactions of Histidine and Other Imidazole Derivatives with Transition-Metal Ions in Chemical and Biological Systems. *Chem Rev* 1974; 74:471–517

137. Sutherland JC, and Holmqvist B. Magnetic Circular Dichroism of Biological Molecules. *Ann Rev Biophys Bioeng* 1980; 9:293–326

138. Tamm LK, and Tatulian SA. Infrared Spectroscopy of Proteins and Peptides in Lipid Bilayers. *Q Rev Biophys* 1997; 30:365–429

139. Teraoka J, Nakamura K, Nakahara Y, Kyogoku Y, and Sugeta Y. Extraordinarily Intense Vibrational Circular Dichroism of a Metmyoglobin Cyanide Complex. *J Am Chem Soc* 1992; 114:9211–9213

140. Toumadje A, Alcorn SW, and Johnson WC Jr. Extending CD Spectra of Proteins to 168 nm Improves the Analysis for Secondary Structure. *Anal Biochem* 1992; 200:321–331

141. Vanoni MA, Edmondson DE, Zanetti G, and Curti B. Characterization of the Flavins and the Iron-Sulphur Centres of Glutamate Synthase from Azospirillum Brasilense by Absorption, Circular Dichroism, and Electron Paramagnetic Resonance Spectroscopy. *Biochemistry* 1992; 314:4613–4623

142. Van Stokkum IHM, Spoelder HJW, Bloemendal M, Van Grondelle R, and Groen FCA. Estimation of Protein Secondary Structure and Error Analysis from Circular Dichroism Spectra. *Anal Biochem* 1990; 191:110–118

143. Venyaminov SY, and Vassilenko KS. Determination of Protein Tertiary Structure Class from Circular Dichroism Spectra. *Anal Biochem* 1994; 222:176–184

144. Vermeer AWP, and Norde W. CD Spectroscopy of Proteins Adsorbed at Flat Hydrophilic Quartz and Hydrophobic Teflon Surfaces. *J Colloid Interface Sci* 2000; 225:394–397

145. Visser NV, Hink MA, Borst JW, van der Krogt GNM, and Visser AJWG. Circular Dichroism Spectroscopy of Fluorescent Proteins. *FEBS Lett* 2002; 521:31–35

146. Wallace BA. Synchrotron Radiation Circular-Dichroism Spectroscopy as a Tool for Investigating Protein Structures. *J Synchrotron Rad* 2000a; 7:289–295

147. Wallace BA. Conformational Changes by Synchrotron Radiation Circular Dichroism Spectroscopy. *Nature Struct Biol* 2000b; 7:708–709

148. Wallace BA, Janes RW, and Orry A. Secondary Structure content in the VUV Region of SRCD Spectra. *Biophys J* 1999; 80:28a

149. Wallace BA, Lees JR, Orry AJW, Lobley A, and Janes RW. Analyses of Circular Dichroism Spectra of Membrane Proteins. *Protein Sci* 2003; 12:875–884

150. Wallace BA, Wien F, Miles AJ, Lees JG, Vronning SH, Evans P, Wistow GJ, and Slingsby C. Biomedical Applications of Synchrotron Radiation Circular Dichroism Spectroscopy: Identification of Mutant Proteins Associated with Disease and Development of a Reference Database for Fold Motifs. *Faraday Discuss* 2004; 126:237–243

151. Woody RW. Circular Dichroism. *Methods Enzymol* 1995; 246:34–71

152. Wu Y, Huang HW, and Olah GA. Method of Oriented Circular Dichroism. *Biophys J* 1990; 57:796–806

153. Yang JT, Wu C-SC, and Martinez HB. Calculation of Protein Conformation from Circular Dichroism. *Methods Enzymol* 1986; 130:208–269

154. Yasui SC, and Keiderling TA. Vibrational Circular Dichroism of Polypeptides. VI. Polytyrosine α-helical and Random Coil Results. *Biopolymers* 1986; 25:5–15

155. Yasui SC, Keiderling TA, and Katakai R. Vibrational CD on Polypeptides. X. A Study of α-Helical Oligopeptides in Solution. *Biopolymers* 1987; 26:1407–1420

4

Fourier Transform Infrared Spectroscopy

MARCO VAN DE WEERT

Department of Pharmaceutics, The Danish University of Pharmaceutical Sciences, Copenhagen, Denmark

JOACHIM A. HERING AND PARVEZ I. HARIS

Faculty of Health & Life Sciences, De Montfort University, The Gateway, Leicester, United Kingdom

Introduction

In 1950, Elliot and Ambrose (1950) realized that infrared (IR) spectroscopy provides information on the secondary structure of proteins. However, the use of IR in protein structural analysis did not become widespread until the advent of Fourier transform infrared (FTIR) spectroscopy in the early 1980s. The transition from IR to FTIR resulted in markedly improved spectral quality and introduced various mathematical possibilities. The latter significantly increased the information content of the IR spectrum with respect to protein conformation.

It took another fifteen years before FTIR became an accepted technique in the protein formulation process. Further instrumental improvements allowed one to study proteins in their natural environment, H_2O-containing solutions, at acceptably low concentrations. In addition, it was realized that FTIR can be used to analyze protein conformation in *any* physical state and in highly complex matrices. This has made FTIR an important technique in the analysis of lyophilized proteins and proteins entrapped in particulate drug delivery systems or other highly dispersive systems.

Introduction to Infrared Spectroscopy

Infrared spectroscopy is a relatively old technique with an extremely broad application area. The *Handbook of Vibrational Spectroscopy* (Chalmers and Griffiths, 2001), a 5-volume, 4000-page series, deals with most aspects of vibrational spectroscopy, including IR and FTIR. The interested reader is referred to this handbook or other general books on the subject for more detailed background information.

Principle of Infrared Spectroscopy

The atoms within a molecule always oscillate around an equilibrium position, even at zero Kelvin. This results in changes in bond lengths and bond angles. The frequency of these motions is within the infrared region (1–100 μm), and thus, radiation in this energy region can excite vibrational motions. For practical and historical reasons, infrared spectra are often plotted using a wavenumber scale. The wavenumber, $\bar{\nu}$, is the reciprocal of the wavelength in centimeters. Thus, the infrared region ranges from 10 to about 10,000 cm^{-1}.

The vibrational motion of a molecule, even with just three atoms, appears complex on first sight. The motion can be subdivided into 3N–6 fundamental vibrations (3N–5 for linear molecules), where N is the number of atoms. For proteins, this theoretically results in several thousands, if not more, vibrational motions. In addition, these motions may not only be excited to the first energy level but also to higher energy levels. However, for a number of reasons, these vibrational excitations do not all show up individually in a spectrum. First, only vibrations that result in a change in dipole moment can be excited by infrared radiation. For example, during the vibrational motion of N_2, no change in dipole moment occurs. Thus, under normal conditions, no FTIR spectrum of N_2 can be obtained (this is in contrast to Raman spectroscopy; see Chapter 5). Second, some transitions have a very low transition probability compared to others and may thus be too weak to be observed in the spectrum. The transition probability of a vibrational excitation is proportional to the change in dipole moment during the vibration. Third, vibrations may have similar excitation frequencies that cannot be resolved.

A major cause of the low resolving power of IR spectroscopy is the simultaneous excitation of rotational and vibrational motions of the molecule. The rotational excitation is superimposed on the individual vibrational motions and results in a significant band broadening. Only in the gas phase can one sometimes see the fine structure of this rotational excitation spectrum, as discussed further below for water vapor. Intra- and intermolecular hydrogen bonding can also result in band broadening, especially if this hydrogen bonding exists in various different configurations.

In most infrared spectroscopic analyses, one uses the mid-infrared region (ca. 4000–400 cm^{-1}). Within this region, one finds first-order excitations of the vibrations of small parts of the molecule, the so-called functional group vibrations. These vibrations include such vibrations as the stretching of the C–H bonds in a methyl or methylene group, and the stretching of the C=O bonds in, for example, an amide bond. These functional group vibrations are often observed in characteristic regions of the spectrum, allowing an investigator to determine the presence or absence of certain functional groups within the molecule under investigation. In protein analysis, one often confines oneself to the region between 1700 and 1600 cm^{-1}, in which the characteristic absorption band of the C=O stretching of the amide bond can be observed (see the section "Protein Absorption Bands and Protein Structural Analysis," below).

In the region between 1400 and 400 cm^{-1}, the vibrational motions of large sections and of the whole molecule are observed. These vibrations are highly specific for a compound, including its configuration and its physical state. For this reason, it is also known as the *fingerprint* region. For protein analysis by FTIR, it is usually of lesser importance.

Instrumentation: The Interferometer

In contrast to most other absorption-based analytical techniques, FTIR instrumentation does not employ an optical wavelength separation method. Rather, all of the radiation of the source is sent through a so-called interferometer, then impinges on the sample, and subsequently falls onto a detector. The wavelengths are then extracted mathematically.

The function of the interferometer (Figure 1) can be explained most easily by starting out with monochromatic radiation entering the

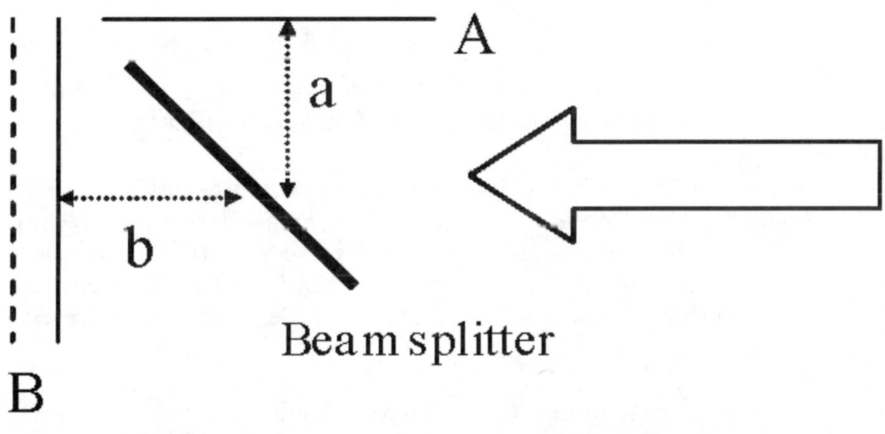

Figure 1 Schematic representation of the Michaelson interferometer used in FTIR spectrometers. Shown are the stationary mirror A, the movable mirror B, and the beam splitter. The arrow shows the radiation beam coming from the course, while a and b indicate the distance from beam splitter to mirror A and B, respectively.

interferometer. With an ideal beam splitter, 50% of the radiation is reflected toward the stationary mirror A, and 50% is allowed to transfer onto the movable mirror B. Subsequent reflection from the mirrors back to the beam splitter results in interference between the two reflected beams at the beam splitter. This interference is constructive or destructive, depending on the pathlength difference (x) from mirror A and B to the beam splitter. If x (= $2a - 2b$) is an integer number (n) of the wavelength of the impinging monochromatic radiation, the interference is completely constructive. If $x = (n + {}^1/_2) \times \lambda$, the interference is completely destructive. Thus, if mirror B is allowed to move, one gets a sinusoidal pattern of the measured intensity as a function of x. This is known as an interferogram.

Obviously, the radiation from an IR source is not monochromatic. The interferogram as a result of all wavelengths combined then becomes a summation over all interferograms of each separate wavelength. As the emitted wavelength range from the source is continuous, this is better described by the integral shown in Equation 1.

$$I(x) = \int_0^\infty I(x, \bar{\nu})d\bar{\nu} = \frac{1}{2}\int_0^\infty I(\bar{\nu})[1 + \cos(2\pi\bar{\nu}x)]d\bar{\nu} \quad (1)$$

in which x is the pathlength difference and $\bar{\nu}$ is the wavenumber (= $1/\lambda$). $I(x)$ is one half of a Fourier pair, the other being $I(\bar{\nu})$.

$$I(\bar{\nu}) = \int_0^\infty I(x, \bar{\nu})dx \quad (2)$$

The two functions $I(x)$ and $I(\bar{\nu})$ can be interconverted using a Fourier transform, hence explaining the name Fourier transform infrared spectroscopy.

The interferometer-based measurement procedure has a number of advantages, most notably a markedly increased sensitivity and wavelength stability compared to the dispersive-type IR instruments. Also, one can acquire a single spectrum within a (few) second(s). However, to reduce the noise level, one often needs to average consecutive scans.

Instrumentation: The Mathematics from Interferogram to Spectrum

The calculation of the Fourier transform of the interferogram is not simple, even with a computer. In addition, there are some complicating factors: Both the number of data points as well as the maximum pathlength distance, X, of the two mirrors in the interferometer are finite.

This transforms the infinite integral into a finite summation. Fortunately, such calculations can be made by approximations known as fast Fourier transforms (FFT). An important feature of the FFT is that the maximal resolution of a Fourier transform of a finite summation is directly proportional to the maximum value of X, as shown in Equation 3:

$$R = \frac{1}{X} \tag{3}$$

In general, protein spectra are acquired at a resolution of 4 cm^{-1}, which is sufficient for most practical purposes.

A typical feature of the interferogram is that at $x \approx X$, the noise in the interferogram is much larger than at $x \approx 0$, simply because the intensity is much smaller. To reduce the influence of this noise, one often employs a *weighting function*, which overemphasizes the intensity measurements at $x \approx 0$. These weighting functions are known as *apodization functions*. These apodization functions do introduce some spectral distortions, most notably, minor side-lobes next to a peak. There has been no formal investigation showing the effect of different apodization functions on protein structural analysis. Van de Weert (unpublished work) found that switching between the so-called triangular and medium Norton-Beer apodization functions does not have a significant effect on protein spectra. These conservative apodization functions are often factory defaults, along with the cosine apodization function.

Mostly disregarded is the influence of *zero-filling* on the spectral characteristics; the parameter setting of this zero-filling is often placed within the advanced parameter setting in the software. In zero-filling, one adds additional zeros in the interferogram to some value beyond the X-value. The apodization function excludes these additional values from being used in the intensity calculations, but these points *are* taken into account for the *resolution* calculation, effectively resulting in *smoothing* of the spectrum. Zero-filling thus results in a mathematical increase in resolution and an increase in the point density of the spectrum. The latter can affect some of the mathematical procedures employed in protein structural data extraction, in which one applies a function over a certain spectral window. Often, one defines in the mathematical function the number of data points rather than the actual spectral window. In general, zero-filling is not required in protein analysis and should be avoided.

Instrumentation: Source, Sample Chamber, and Detector

Besides the interferometer, an FTIR instrument consists of a source, a sample chamber, and a detector. The source usually is SiC-based (known as the globar) and is sometimes water-cooled. These sources provide a high-energy infrared beam, with its maximum intensity around 1500–1800 cm^{-1}. All FTIRs also contain a laser, which is used to determine the exact position of the movable mirror. Furthermore, the laser provides a trigger for data acquisition. For example, at each zero crossing of the laser interferogram, a data point is collected.

The sample chamber is located after the interferometer. The sample chamber should prevent stray light falling onto the detector. Moreover, it should provide a closed chamber that can rapidly be purged of atmospherical interferences. Both water vapor and carbon dioxide have strong absorption bands, and nitrogen gas or dried air is used to reduce this atmospherical interference. In general, the sample chamber can only hold one sample. Therefore, the background and actual samples have to be measured sequentially rather than in parallel.

Two different types of detectors are commonly used within the mid-IR region. An important issue for the detector is its response time, as this determines how fast the movable mirror can move. The most common detector is the deuterated triglycine sulfate (dTGS) detector. It is a very stable detector, but it has a relatively slow response time. A less common detector is the mercury-cadmium-telluride (MCT) detector. It is much more sensitive than the dTGS detector and has a response time that is about ten times faster compared to the dTGS detector. This means that spectra can be acquired much faster. However, the MCT detector has to be cooled with liquid nitrogen, and its linear range is smaller than that of the dTGS. Thus, the dTGS is the preferred detector, unless short analysis times are a crucial issue.

Protein Absorption Bands and Protein Structural Analysis

Protein Absorption Bands and Their Relation to Protein Structure

Amide vibrations are infrared active vibrations based on the amide bond present in proteins. They form the basis for studies related to protein FTIR spectroscopy for the analysis of protein secondary structure. Altogether, up to nine characteristic bands named A, B, I, II, . . . , VII have been identified in the mid-IR region (Bandekar, 1992), and several of these are indicated in Figure 2. A description of the amide modes has been given by various authors, reviewed among others by Krimm and Bandekar (1986). However, due to technical and theoretical limitations, only the amide I, II, and III bands are used to investigate the secondary structure of proteins.

It is assumed that the exact frequencies of the amide I and II absorption are influenced by the strength of any hydrogen bonds involving amide C=O and N–H groups. Each individual secondary structural conformation is associated with a characteristic hydrogen bonding pattern between these groups. Thus, each type of secondary structure gives rise to different frequencies at which the amide bonds

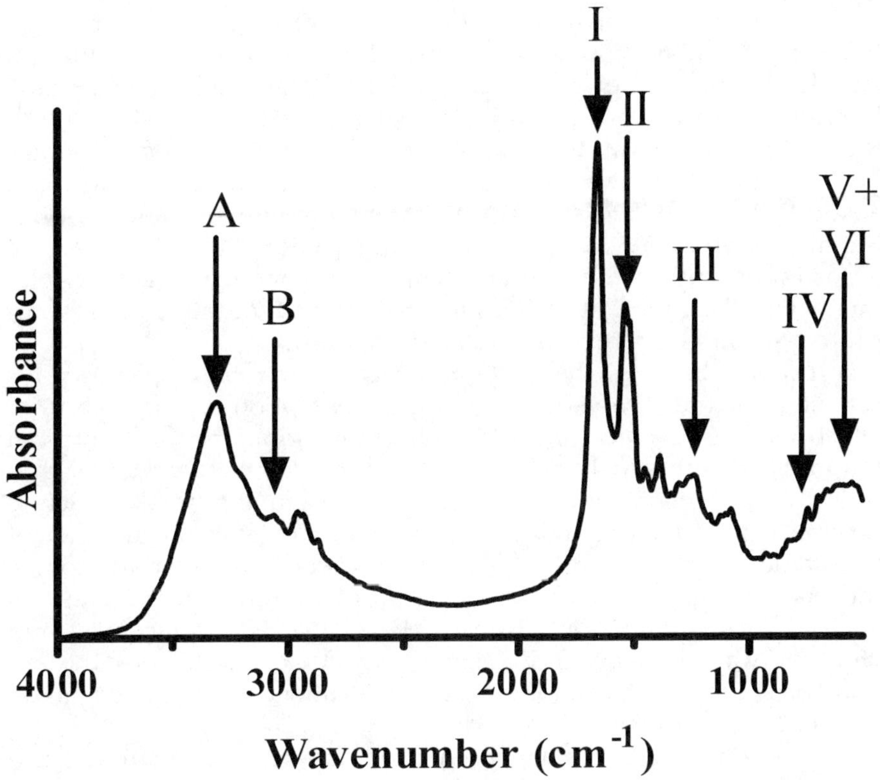

Figure 2 A typical protein IR spectrum, showing the various regions in which vibrational motions of the amide backbone occur, indicated by A, B, I, II, III, IV, V, and VI.

vibrations occur, resulting in characteristic amide I and II vibrations. It is this separation of the amide absorption that forms the basis of protein structure quantification from FTIR spectra of proteins. Correlation between amide frequency and protein secondary structure has been demonstrated using both normal mode calculations (cf. Krimm and Bandekar, 1986) and experimental studies on peptides and proteins (cf. Haris and Chapman, 1992). However, because of the complexity of naturally occurring proteins, most data have been obtained from studies on model compounds with only a single secondary structure. These model compounds ranged from simple amino acid derivatives to large synthetic polypeptides.

Protein structural analysis by FTIR spectroscopy is mainly performed using the amide I band. Amide I absorption is directly related to the backbone conformation with a major contribution from C=O stretching vibration and minor contribution from the C–N stretching vibration. Absorption for this band occurs in the region

1600–1700 cm^{-1}. Other than amides, there are very few other vibrational modes that show an absorption band in this region, thus resulting in few interfering absorptions. The most problematic interference is an absorption band of water within this region (see the section "The Amide I Band and Interference of Water/Water Vapor," below), but also the contribution of amino acid side-chain absorption may be of importance (cf. Venyaminov and Kalnin, 1990; Rahmelow et al., 1998). In the following sections, the characteristic absorptions within the amide I band of secondary structural elements and the different techniques used for quantitative analysis of protein secondary structure will be discussed.

It should be noted that other amide bands have also been used for protein structural analysis. The amide II absorption results both from N–H bending vibration and C–N stretching vibration. The absorption for this band occurs in the region 1500–1600 cm^{-1}. It has been found, empirically, that the amide I absorption is more useful for protein structure determination (cf. Haris and Chapman, 1992). However, the inclusion of the amide II band with the amide I band has been reported to provide improved prediction accuracy by some workers using multivariate data analysis techniques (Dousseau and Pezolet, 1990; Kumosinksi and Unruh, 1994; Rahmelow and Huebner, 1996; Wi et al., 1997; see the section "Structure Identification and Quantification: Pattern Recognition Techniques," below).

The amide II band has been used for monitoring hydrogen-deuterium exchange in proteins. Analysis of the changes in the intensity of the amide II band in ^2H_2O provides a means for determining the replacement kinetics of amide hydrogens by deuterium. Investigation of hydrogen-deuterium exchange using FTIR spectroscopy has been very useful for detecting subtle changes in protein secondary structure. For example, subtle conformational differences between ribonucleases *A* and *S* and between the native and reactive center cleaved forms of α_1-antitrypsin (Haris et al., 1986, 1990) have been shown.

Determination of protein secondary structure has also been demonstrated based on the amide III region (cf. Fu et al., 1994; Schweitzer-Stenner et al., 2002). Absorption for this band occurs in the region 1220–1330 cm^{-1}. Amide III absorption mainly arises from C–N stretching vibrations as well as N–H in-plane bending vibrations, with weak contributions from C–C stretching and C=O in-plane bending vibrations (Bandekar, 1992). The amide III region has been characterized to be a less well defined vibrational mode, with contributions from different vibrations varying between proteins (Krimm and Bandekar, 1986). Additionally, signal contribution arising from amide vibration within the amide III region has been found to be very weak and extensively mixed with CH vibration of the amino acid side chains (Böhm et al., 1992). The advantage of the amide III region is that it is not hampered by OH vibrations of water. However, many other vibrational modes also show an absorption band within the amide III region, thus resulting in many potentially interfering substances. This significantly reduces the applicability of the amide III region for protein structural analysis in protein formulations.

Characteristic Absorption Bands Within the Amide I Region

Many empirical studies have identified regions within the amide I band, that are sensitive to particular secondary structural conformation. This information is used for the *curve fitting* method (described in the section "Structure Identification and Quantification: Curve Fitting," below) to assign resulting bands to protein secondary structure. A collection of such regions reported in the literature is given in Table 1.

Note that there is a difference in the band positions when comparing the characteristic band positions in water and ^2H_2O. This is due to hydrogen-deuterium exchange of the amide hydrogens, resulting in slight changes in vibrational characteristics and subsequent band shifts. For proteins analyzed in ^2H_2O, one generally denotes the amide I region as the amide I′ (prime) region to more clearly distinguish it from proteins analyzed in normal water.

In most cases, good relationships exists between protein secondary structure and respective amide I frequencies. However, this

Structure	Region(s)	Solvent[a] H_2O	Solvent[a] ^2H_2O	Reference
α-helix	1648–1657	x		(Barth and Zscherp, 2002)
	1648–1660	x		(Jackson and Mantsch, 1995)
	1650–1658	x		(Haris, 1999)
	1642–1660		x	(Barth and Zscherp, 2002)
	1650–1658	x	x	(Haris, 2000)
	1648–1657	x	x	(Haris and Chapman, 1994)
3_{10} helix	1662–1666	x		(Arrondo and Goni, 1999; Dong et al., 2000)

[a] Where no solvent has been given, solution in H_2O has been assumed.

Table 1 Empirically Determined, Structure-Sensitive Regions Within the Amide I Band

(continued on following page)

Structure	Region(s)	Solvent[a]		Reference
		H_2O	^2H_2O	
	Near 1645	x		(Arrondo and Goni, 1999)
	1660–1670	x		(Jackson and Mantsch, 1995)
β-sheet	1623–1641	x		(Barth and Zscherp, 2002)
	1674–1695			
	1620–1640	x		(Haris, 2000)
	1670–1695			
	1625–1640	x		(Jackson and Mantsch, 1995)
	1620–1640	x		(Haris and Chapman, 1994)
	1670–1695			
	1620–1640	x		(Haris, 1999)
	1615–1638		x	(Barth and Zscherp, 2002)
	1672–1694			
Turn	1662–1686	x		(Barth and Zscherp, 2002)
	1620–1640	x		(Haris, 2000)
	1650–1695			
	1653–1691		x	(Barth and Zscherp, 2002)
Other (unordered)	1642–1657	x		(Barth and Zscherp, 2002)
	1650–1655	x		(Haris, 2000)
	1660–1670			
	1640–1648	x		(Jackson and Mantsch, 1995)
	1639–1654		x	(Barth and Zscherp, 2002)
	Near 1644		x	(Haris, 2000)

Table 1 *(continued from previous page)*

correlation does not apply to all proteins and peptides. For example, deviations may occur with proteins containing unusual or less common structures. Thus, the α-helix structure of poly-L-lysine in water has been observed at 1638 cm^{-1} and that of bacteriorhodopsin at 1662 cm^{-1} (Haris, 2000), well beyond the region normally assigned to α-helix structure. Moreover, two parallel α-helices wrapped around each other in a left-handed supercoil (coiled-coil structure) give rise to an absorption at 1630 cm^{-1} (Rahmelow et al., 1998), which is the region normally assigned to β-sheet structure. Several conflicting assignments within the amide I region have recently been reviewed (Haris, 2000).

Structure Identification and Quantification: Curve Fitting

Curve fitting, the most widely used method for protein secondary structure quantification, mainly involves curve fitting of the amide I band (cf. Ruegg et al., 1975; Byler and Susi, 1986; Surewicz and Mantsch, 1988; Carrasquillo et al., 2000), and occasionally the amide III band (cf. Singh et al., 1991; Carrasquillo et al., 2000).

The basic principle of the curve fitting procedure is to resolve the original protein spectrum into individual bands that fit the spectrum. However, before the curve fitting procedure can be performed, *band narrowing* techniques need to be applied to obtain estimates of the number and positions of discrete absorption bands that make up the complex amide I band profile. The outcome of the curve fitting procedure is highly dependent on the outcome of these estimates. Most widely used band narrowing techniques are Fourier self-deconvolution (FSD) and derivation. Because amide band assignments are dependent on the outcome of those band narrowing techniques, the latter techniques are briefly discussed in the following.

Band Narrowing Techniques

Most proteins usually contain more than one secondary structure; thus, they give rise to several amide absorption bands. A considerable amount of overlap in terms of width and separation of these absorption bands often results in featureless absorption profiles (Figure 3). This makes it difficult to analyze the frequency of the composite amide maximum and any visible shoulders, which may lead to misinterpretation of spectral shifts (Hübner et al., 1990). Jackson and Mantsch (1995) have therefore argued that deduction of structural parameters from the relatively featureless amide I band alone is of limited use. In an attempt to tackle this problem, a number of mathematical data processing techniques have been developed to visualize those overlapping bands, thus allowing extraction of detailed information from infrared spectra of proteins (Kauppinen et al., 1981a,b; Cameron and Moffat, 1984; Griffiths and Pariente, 1986). These techniques are often referred to as *resolution enhancement* techniques. However, because resolution is an instrumental parameter that cannot be increased after a

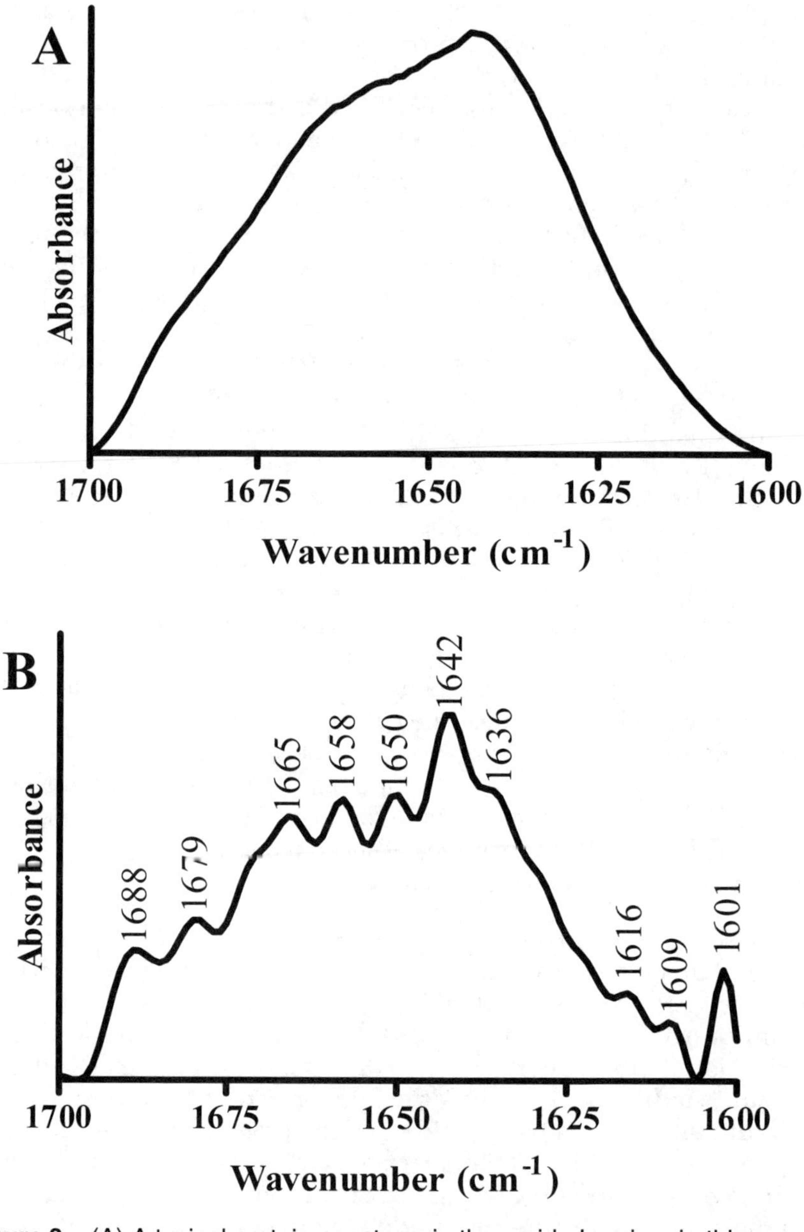

Figure 3 (A) A typical protein spectrum in the amide I region, in this case ribonuclease A. (B) The deconvoluted spectrum (C) The second derivative spectrum of (A), obtained using a second-order Savitzky-Golay derivative. (D) Result of the curve-fitting of (A) using Gaussian peaks based on identification of bands using a combination of FSD and second-derivative analysis. The composite bands are indicated as dotted lines, whereas the original spectrum is shown as a solid line.

(continued on following page)

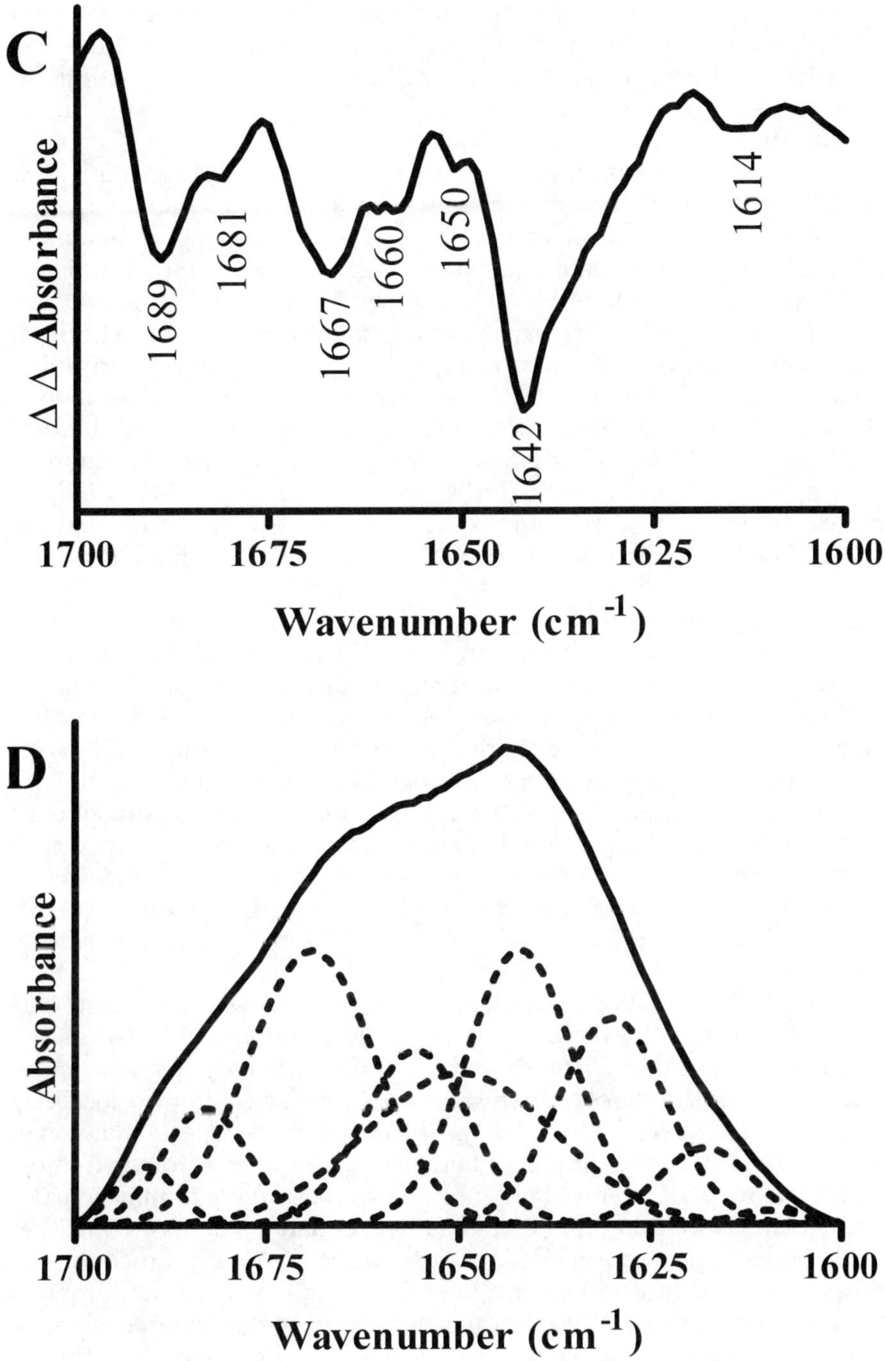

Figure 3 *(continued from previous page)*

spectrum is recorded, these techniques are more correctly referred to as band narrowing techniques. Basically, these techniques are mathematical procedures that are applied to achieve narrower component bands. The two most popular of these techniques are Fourier self-deconvolution and derivation.

Fourier self-deconvolution (Kauppinen et al., 1981a,b) is based on the assumption that in the liquid or solid state, the absorption bands are broadened, or *convoluted*, by a certain function, causing band overlap. As a result, the individual bands cannot be distinguished in the amide envelope. The self-deconvolution procedure uses this function to narrow, or *deconvolute*, the spectrum. Although the exact shape of the convolution function is still not determined, most authors have assumed Lorentzian or Gaussian functions (Byler and Susi, 1986; Dong et al., 1990; Kumosinski and Unruh, 1994). The basic underlying principle of FSD is described elsewhere (Kauppinen et al, 1981a,b; Surewicz and Mantsch, 1988). FSD is mainly controlled by two parameters: The full width at half-height (FWHH) and the *resolution enhancement factor*, *K*. Both have to be determined manually, mainly by trial and error. The number and peak maximum of the resulting component bands is highly dependent on the choice of deconvolution parameters. Different combinations of values for the FWHH and for *K* yield different shapes of the resulting deconvoluted spectrum. For example, if the FWHH is too small, little narrowing is obtained. If, on the other hand, the FWHH is too large, negative side-lobes appear. Because the choice of the deconvolution parameters is subjective, varying results of subsequent curve fitting can be expected for the same protein FTIR spectrum under investigation. Moreover, deconvolution may amplify noise and as a result generate incorrect data if one performs the curve fitting on FSD spectra. An example of FSD of a protein spectrum is shown in Figure 3.

Spectral derivation, an operation similar to deconvolution, was originally realized by the method of Savitsky and Golay (1964). In the area of spectral analysis, the second-order derivative is most widely applied. The Savitsky-Golay method uses information from a localized segment of the spectrum to calculate the derivative at a particular wavelength rather than the difference between adjacent data points. In most cases, this reduces the problem of noise enhancement and may actually apply some smoothing to the data. However, large side-lobes on either side of intense absorption bands are produced, making it difficult to determine the true limits of these absorption bands. A significant problem of spectral derivation is that it does not preserve relative intensities of absorption bands. Relative absorption bands in derivative spectra depend on the width of the absorption in the original spectrum, where narrow absorption bands will be enhanced at the expense of broader bands. Due to these problems, derivative spectra are often merely used to confirm the initial identification of the band positions by deconvolution (Singh, 2000). However, successful application of curve fitting directly based on derivative spectra has been demonstrated (cf. Dong et al., 1990, 1994, 2000; Caughey et al., 1991). Moreover, spectral comparisons have been performed on derivative spectra (see the section "Correlation Coefficient and Area Overlap," below). An example of derivation of a protein spectrum is shown in Figure 3.

Curve Fitting Procedure

A complete description of the curve fitting procedure can be found elsewhere (cf. Singh, 2000). Basically, it involves an iterative process, where a set of Gaussian, Lorentzian, or a mixture of Gaussian/Lorentzian-shaped curves is determined that best fits the original protein spectrum. Often, the band narrowing techniques described above are used to get an initial estimate of the curve fitting parameters, like band height, band width, band position, and the baseline. The best fit is obtained by a root mean square analysis determining the optimal set of curve fitting parameters, after which individual bands are then assigned to protein structure (Table 1). The band area or intensity of each individual band is used to calculate its relative contribution to a particular protein secondary structure in relation to the overall area of the original spectrum. An example of the result of such a curve fitting session is shown in Figure 3.

At the end of the curve fitting method, the area of each fitted band is expressed as a percentage of the total area of the amide envelope under investigation. However, this procedure assumes that molar absorptivities of C=O groups responsible for each of the fitted bands are equal; commonly, no weighting function is applied to each band. However, studies on poly-L-lysine have shown that the molar absorptivities of different secondary structures can in fact vary slightly (Jackson et al., 1989). De Jongh et al. (1996) suggest that sheet structure displays the greatest molar absorptivity. In contrast, others suggest that molar absorptivities are independent of secondary structure (Chirgadze et al., 1973; Chirgadze and Brazhnikov, 1974). Finally, in the curve fitting procedure, one often neglects the contribution of side-chain absorption. It is possible to subtract the contribution of the amino acid side chains from the peptide spectrum, based on the known absorption spectra of the individual amino acids (Venyaminov and Kalnin, 1990), but it is not exactly known if the absorption spectra are the same for the amino acid and the amino acid residue.

Structure Identification and Quantification: Pattern Recognition Techniques

Alternative methods for quantitative analysis of protein secondary structure have been introduced that require fewer assumptions and remove much of the subjectivity inherent in the curve fitting method described above. The basic underlying principle is to extract common features, or *patterns*, associated with fractions of secondary structures from a reference set of protein spectra with well-known and well-defined secondary structure. Hence, these techniques are often referred to as *pattern recognition* techniques. These encompass both multivariate data analysis and neural network methods. Here, the spectral features of an unknown sample are compared to the features extracted from a database of reference spectra, yielding the overall secondary structure of the protein sample. Thus, no band assignments are necessary. Moreover, the contribution of amino acid side-chain absorption and potentially differing molar absorption coefficients are also covered in the comparison. However, problems may arise when the sample protein contains an unusual structure or sequence that is not covered in the database of known proteins or when the protein spectrum is acquired under different experimental settings. Although multivariate data analysis approaches and neural network approaches are different in the actual implementation, their strengths and weaknesses are similar.

Multivariate Data Analysis Techniques

Most pattern recognition–based approaches are multivariate data analysis methods. Here, two sets of data, commonly expressed in two matrices X and Y, are related by regression. Typically, one matrix consists of target secondary structure fractions for proteins in the reference set (Y), and the other matrix consists of the absorption values of these reference proteins for a range of wavenumbers (X). The combined matrices X and Y are known as the calibration set or training set. Based on this calibration set, a multivariate regression model is derived, which can then be used to predict the secondary structure fractions of a new protein based on its infrared spectral data. These methods have been pioneered by applications based on spectral data from other techniques like CD (cf. Chen and Yang, 1971; Bolotina et al., 1980; Pancoska et al., 1991; Bohm et al., 1992; Greenfield and Fasman, 1996) and Raman spectroscopy (cf. Bussian and Sander, 1989; Williams and Dunker, 1981; Williams, 1983; Sane et al., 1999). The most widely used regression methods include principal component analysis (PCA), factor analysis (FA), and singular value decomposition (SVD) (cf. Dousseau and Pezolet, 1990; Lee et al., 1990; Sarver and Krueger, 1991; Pribic et al., 1993; Pancoska, et al., 1995; Rahmelow and Huebner, 1996; Wi et al., 1997; Forato et al., 1998; Baello et al., 2000; Wu et al., 2001).

Neural Network Analysis

Alternative pattern recognition approaches have been applied, which are based on the principle of artificial neural networks. Most neural networks used are so-called feed-forward multi-layer neural networks.

Here, the input values (FTIR spectral data) are "fed" into the input layer nodes. After several calculations in further layers of the neural network, the fractions of protein secondary structure content are the output. In the past, neural networks have been used for secondary structure prediction from amino acid sequences (cf. Moult, 1996; Rost and O'Donoghue, 1997; Baldi et al., 1999, 2000) and CD spectral data (cf. Andrade et al., 1993; Dalmas et al., 1994; Pancoska et al., 1999).

Few neural network approaches have been applied to FTIR spectra of proteins (Pancoska et al., 1996, 1999; Cabiaux et al., 1997; Severcan et al., 2001; Hering et al., 2001, 2002a, 2002b, 2003). Results reported suggest that neural networks are a valid alternative to the multivariate data analysis in making good predictions about the secondary structure of proteins from their FTIR spectra. However, as with all other methods discussed here, the success of neural networks depends very much on the right configuration. It has been shown that the choice of training scheme, type of neural network, and the choice of data pre-processing methods are important design choices in achieving good prediction accuracy (Pancoska et al., 1996, 1999; Cabiaux et al., 1997; Severcan et al., 2001; Hering et al., 2001, 2002a).

Recently, neural network approaches were successfully applied to predict secondary structural features beyond purely average structural content. That is, the number of helix/sheet segments and average helix/sheet length have been determined (Pancoska et al., 1996, 1999, 2001; Hering et al., 2004).

Correlation Coefficient and Area Overlap

In the analysis of protein formulations, the pharmaceutical scientist is sometimes interested only in recording differences. That is, one compares the native structure of the protein to the structure in a selected formulation. This does not require any tedious curve fitting procedures, circumvents any problems with ambiguous band assignments, and prevents any questions whether pattern recognition techniques are applicable. In principle, the methods described below can also be used in various other analytical techniques.

In FTIR analysis, visual comparison of the spectra in the amide I, II, and/or III regions would suffice to detect deviations, if not for the subjectivity of such comparisons: What is a large and/or small difference? Prestrelski et al. (1993) used the correlation coefficient as an objective method to compare spectra of proteins in solution with those of lyophilized proteins. The correlation coefficient, r, is calculated using Equation 4:

$$r = \frac{\sum y_{1i}\, y_{2i}}{\sqrt{\sum y_{1i}^{2} \sum y_{2i}^{2}}} \tag{4}$$

where y_{1i} and y_{2i} are the absorbance values at wavenumber *i* of the two spectra. For identical spectra, this will return the value 1, whereas a value of 0 indicates that the two spectra have nothing in common.

Occasionally, the correlation coefficient calculation results in a low correlation for visually identical spectra. This may be due to, for example, an offset in the baseline. To circumvent this problem, the group of Carpenter and co-workers (Prestrelski et al., 1993) later adopted the area overlap method developed by Heimburg and Marsh (1993). Here, the integrated difference between two area-normalized spectra is computed. Identical spectra yield an area overlap that is identical to the normalized area of the spectra (e.g., 100), whereas completely different spectra will yield an area overlap of 0. An example is shown in Figure 4.

The correlation coefficient and the area overlap can be calculated using raw spectra and resolution-enhanced spectra. The group of Carpenter exclusively uses second-derivative spectra in their calculations. Griebenow et al. (1999) have advised against this and prefer to use Fourier self-deconvolved spectra. Van de Weert et al. (2001) have

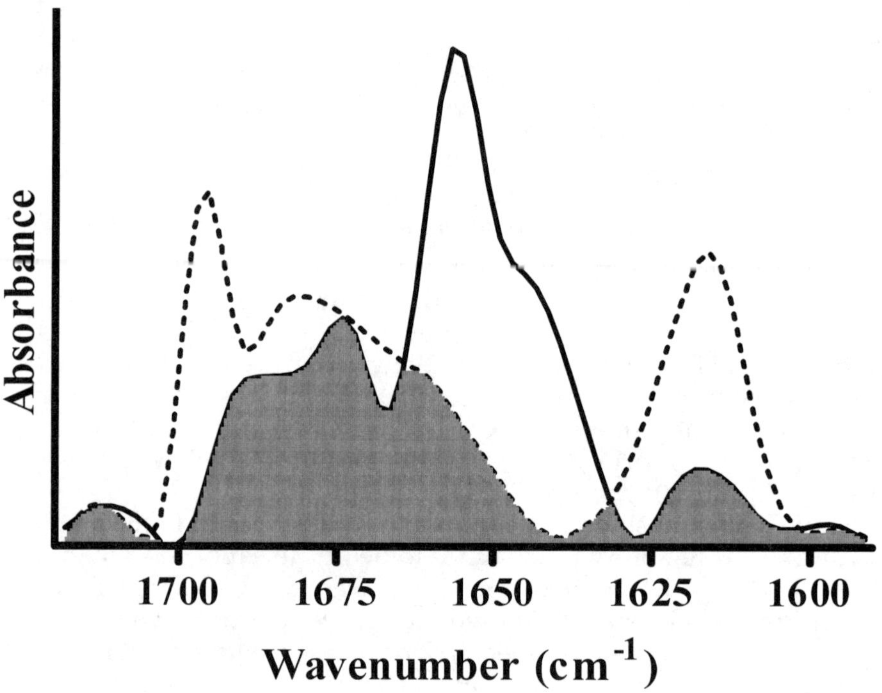

Figure 4 Example of an area overlap. Two inverted, normalized second derivative spectra of lysozyme, in solution (solid line) and dispersed heat aggregates (dotted line), respectively, are shown. The area overlap of the two spectra (55%) is indicated in gray.

used raw spectra in their spectral comparisons using the area overlap method. The choice of method and type of spectra being used typically depends on the personal preferences of the investigator.

It should be stressed that the area overlap and correlation coefficient only indicate whether deviations from the control spectrum to which the sample spectrum is compared have occurred. Any mechanistic insight into any calculated deviations still requires analysis using pattern recognition or curve fitting or by using the two-dimensional correlation analysis described below.

Two-Dimensional Correlation

A relatively new procedure in protein analysis is two-dimensional infrared correlation spectroscopy, reviewed recently by Ozaki et al. (2003). This method can be applied to sets of spectra that are acquired as a function of some type of time-dependent variation (e.g., thermal, mechanical, or chemical perturbations). The data are then subjected to a correlation analysis that indicates whether absorption bands change simultaneously (in phase) or sequentially (out of phase) during application of the external perturbation.

The correlation analysis is usually plotted in two contour maps, in which the wavenumbers are plotted along the x and y axes, while the z axis (the contours) indicates the strength of the correlation. The *synchronous* contour map shows positive and/or negative cross-peaks for structural elements that change simultaneously. For example, in the temperature-induced aggregation of cytochrome *c*, Paquet et al. (2001) observed that a band at 1650 cm^{-1} was negatively correlated to a band at 1616 cm^{-1} in the amide I′ region. This suggested that α-helical structure was transformed into intermolecular β-sheets. They also observed a positive correlation of the absorption band at 1616 cm^{-1} to the band at 1685 cm^{-1}. The latter is the high-energy band of the intermolecular β-sheet vibration, with its main absorption at 1616 cm^{-1}.

The *asynchronous* contour map shows the correlation of bands that change out of phase, thus, sequentially. Paquet et al. (2001) observed a strong negative correlation in the asynchronous contour map between the 1650 and 1616 cm^{-1} absorption bands. This suggests that the aggregation process, as indicated by the 1616 cm^{-1} absorption band, actually precedes the unfolding of the α-helix components. Apparently, the intermolecular β-sheets are initially formed by structural changes of the unordered/undefined structural elements in the protein, which is then followed by unfolding of the α-helical structures.

To our knowledge, two-dimensional correlation analysis has as yet not been applied to any pharmaceutical protein formulations. Yet, time-based perturbations, such as heat-induced aggregation, are routinely applied in infrared analysis of protein formulations. Time-based perturbations are also applied in fluorescence, CD, and ultraviolet-visible

(UV-Vis) spectroscopy, thus two-dimensional correlation may also be used in these analytical techniques. Little quantitative information can be derived using two-dimensional correlation, but it can be used to obtain insights into the structural changes that occur when a stress factor is applied. Computer programs to perform the data analysis are available, such as 2D-Pocha (which can be downloaded from http://science.kwansei.ac.jp/~ozaki/2D-Pocha.htm), Mathcad (e.g., version 7, Mathsoft, Cambridge, MA), or the Win-IR Pro software provided with high-end Digilab FTIR spectrometers (Randolph, MA, formerly known as Bio-Rad).

The Amide I Band and Interference of Water/Water Vapor

For a considerable amount of time, the absorption of water and water vapor mainly within the amide I region (Figure 5) posed a significant problem in protein structural analysis by using FTIR spectroscopy. As a result, many studies were performed in ^2H_2O. However, it is now possible to subtract digitally the overlapping H_2O absorption from the spectrum of the protein solution, provided the protein concentration is high enough (greater than 5–10 mg/mL) and a short pathlength is used (6 μm). An example of subtracting the water signal is shown in Figure 6, where the subtraction is based on obtaining a flat baseline

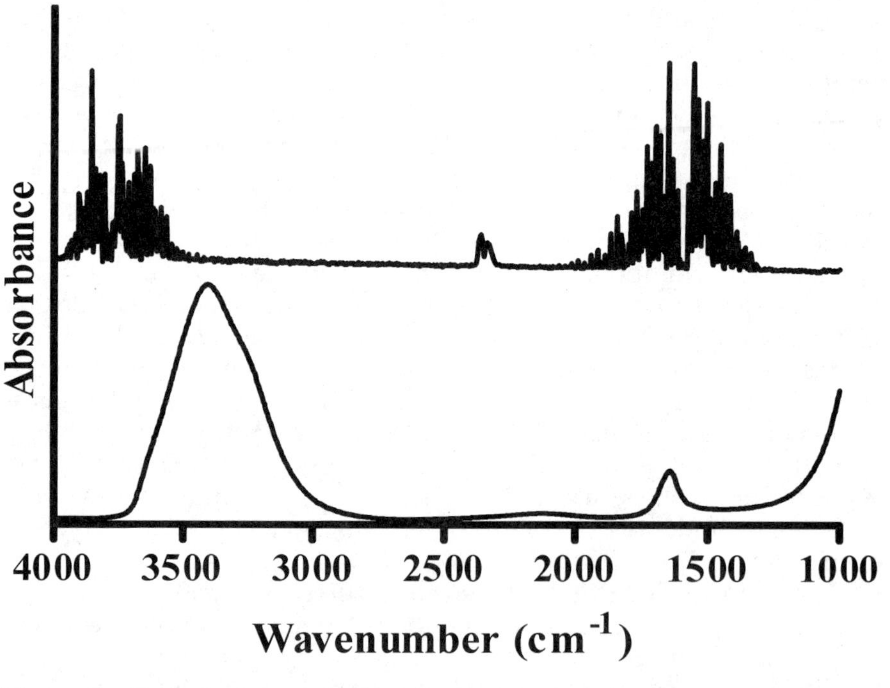

Figure 5 Typical spectra of water (bottom) and water vapor (top), showing their strong absorption within the amide I region.

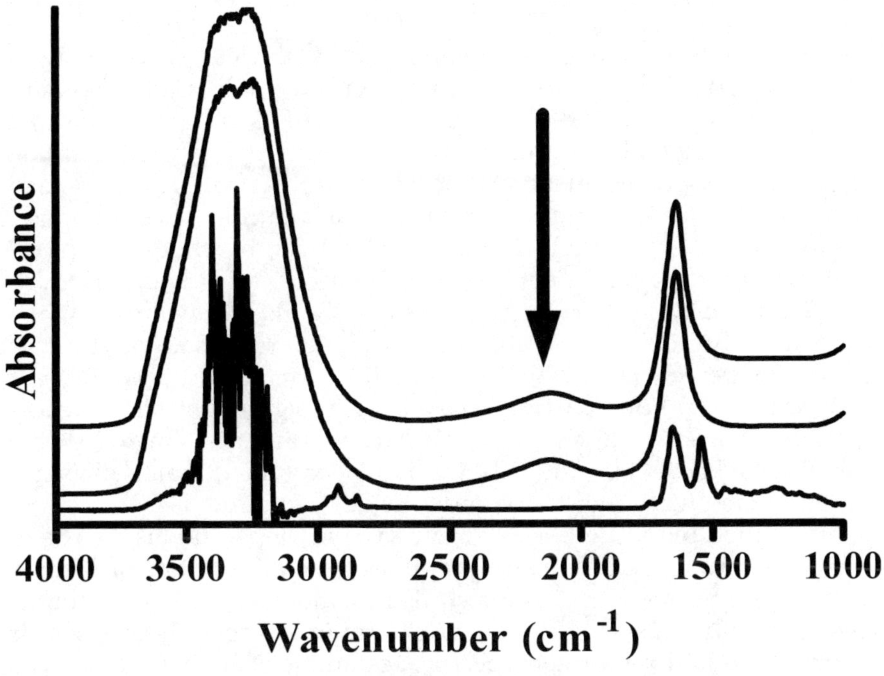

Figure 6 FTIR spectra of, from top to bottom, a protein-containing buffer (lysozyme, 30 mg/mL), the plain buffer solution, and the protein-containing buffer from which the buffer contribution has been subtracted. The latter spectrum has been multiplied by a factor of 20. The arrow points at the region of 1750–2300 cm^{-1}, which has been used to subtract the buffer contribution.

within the region 2300–1750 cm^{-1}. An algorithm for such subtractions is present in the software of most FTIR spectrometers.

Water vapor also absorbs within the amide I region, giving rise to a series of very sharp peaks, arising from excitation of the various rotational motions. In general, purging the sample compartment with dry air or nitrogen will eliminate the water vapor contribution, also improving the signal-to-noise ratio. However, care should be taken that the water vapor contribution is indeed negligible. Band narrowing techniques have a strong bias toward sharp peaks, potentially leading to erroneous band assignments due to residual water vapor contribution.

Sampling Methods

As stated in the "Introduction," FTIR can be used to analyze samples in any physical state. This is partly related to the relatively low scattering power of IR radiation, while sampling methods have been developed that either depend on scattering or do not cause any scattering. Several different sampling methods are being used for protein structural analysis, and below a brief summary is given.

Solid Samples

The most common method to analyze solid samples is co-mixing a small amount of the sample (1–2 mg) with an infrared-transparent salt such as KBr (ca. 200–300 mg). The mixture is compressed into a transparent pellet a few millimeters thick, which is analyzed in transmission mode. Chan et al. (1996) showed that this procedure may lead to minor structural changes of rhDNase, depending on the applied pressure and the sample-to-KBr ratio. However, by choosing appropriate mixing ratios and pressure, these changes can be minimized.

There are two important issues one should be aware of when preparing KBr pellets. First, the moisture level of the sample must be low (a few percent) to obtain good-quality KBr pellets. Second, very large and amorphous particles may lead to significant band distortions, and hard crystalline materials can be difficult to mix with the KBr. The physical stress of grinding hard crystalline materials is not recommended for labile compounds such as proteins.

A method requiring less extensive sample preparation is the use of IR microscopy. Here, one simply applies a small amount of sample under a microscope, through which IR radiation is sent to the sample. This method has been used to show that, in most cases, the use of KBr pellets does not distort protein spectra (Costantino et al., 1997). However, IR microscopy suffers from a poor signal-to-noise ratio and therefore is not the first method to consider during routine analysis.

Souillac et al. (2002) showed that diffuse reflectance (DRIFT) can also be used in the analysis of dry proteins. In DRIFT analysis, the sample is mixed with a crystalline IR-transparent salt such as KBr. The IR radiation is allowed to scatter from this mixture, during which absorption by the sample may occur, and the diffuse-reflected (scattered) radiation is then collimated onto the detector. The authors made no comparison with other solid-state techniques, thus it is unknown whether DRIFT spectra are directly comparable to, for example, KBr spectra. However, the observed trends in protein structural changes upon lyophilization were similar to those observed by others who used KBr pellets in their analysis.

Hydrated Samples, Dispersions, and Solutions

Hydrated and aqueous protein samples cannot be analyzed by any of the methods mentioned above other than IR microscopy. KBr will partially dissolve due to the presence of the relatively large amounts of water. This results in very poor quality KBr pellets, or absence of diffuse reflectance in DRIFT.

Hydrated samples are usually analyzed by using attenuated total reflectance (ATR). In ATR, the IR beam is allowed to reflect within an IR transparent crystal; for example, germanium (Ge) or zinc selenide (ZnSe). At the reflection points, an evanescent wave protrudes from the surface of the crystal. A sample in close contact with the crystal can interact with this evanescent wave, resulting in attenuation of the reflected beam. The effective *pathlength* depends on the refractive indices of the sample and the crystal, the wavenumber, and the reflection angle. Thus, the actual thickness of the sample usually is of no importance, as long as there is close contact to the crystal. However, Jackson and Mantsch (1992) have shown that thick protein samples may show some optical aberrations in ATR analysis, resulting in potentially erroneous data interpretation.

Dispersions and solutions are generally analyzed in transmission mode in a cell. There are two demands on the transmission cell. First, the cell windows should be IR transparent and insoluble in water. The most common window material is calcium fluoride (CaF_2), which has a low wavelength cutoff at ca. 900 cm^{-1}. Second, a pathlength of no more than 8–10 μm is required to allow subtraction of the water signal. Very thin spacers, made of Mylar or tin, are used to achieve these short pathlengths. Recently, special cell windows with a 6- to 7-μm depression have become available (e.g., Confocheck by Bruker Optics (Billerika, MA) and BioCell by BioTools (Wauconda, IL)), which simplifies the loading procedure. This is especially of importance when analyzing dispersions or very viscous samples.

In contrast to H_2O, ^2H_2O does not have a strong absorption in the amide I region. Under these conditions, the pathlength of the cell can be much larger (up to 100 μm). This, in turn, means that much lower protein concentrations can be used. However, the use of ^2H_2O may lead to practical problems, as one has to start with a lyophilized material or has to perform a *buffer-exchange* procedure.

The ATR technique can also be used for dispersions and solutions. Often, low concentrations (1 mg/mL) can be used with the ATR technique, but a larger solvent volume is required (up to 0.5 mL). Moreover, Oberg and Fink (1998) have shown that some protein structural changes occur upon protein adsorption to ATR crystals. These changes appear to be very small and insignificant above concentrations of a few milligrams per milliliter. Van de Weert et al. (2001) showed that at high concentrations, any differences between transmission and ATR protein spectra are mainly related to physical differences between the two acquisition techniques.

Practical Examples

Analysis of Liquid Protein Formulations

FTIR has extensively been used as a rapid method to obtain structural information of proteins in solution. The interested reader is referred to several reviews on this topic (cf. Haris and Chapman, 1992, 1995; Haris, 2000). In the protein formulation process, FTIR spectroscopy on protein solutions is mainly used to study whether the formulation process results in a change in protein structure. Very few studies have been performed on liquid protein formulations as such. In general, the solution spectrum is used to study structural changes during emulsification, lyophilization, and entrapment in particulate delivery systems. FTIR spectroscopy has also been used in the analysis of protein aggregation and protein fibrillation in liquid protein formulations. For example, Nielsen et al. (2001) and Bauer et al. (1994) have shown that insulin and calcitonin, respectively, aggregate through the formation of intermolecular β-sheets, giving rise to two distinct bands at ca. 1625 and 1695 cm^{-1} (ca. 1615–1620 and 1685 cm^{-1} in ^2H_2O).

FTIR spectroscopy has also extensively been used to examine the interaction of proteins with model membranes (cf. Haris and Chapman, 1992; Goormaghtigh et al., 1994). Liposomes are used as a model membrane in these studies. Because liposomes are also used as a drug delivery system, this means that protein structure in such a system may also be analyzed using FTIR spectroscopy, provided the protein concentration is high enough.

FTIR can also be used to investigate complex protein-protein interactions. Haris et al. (1992) developed an FTIR-based approach for investigating protein-protein interaction by uniformly ^{13}C isotope labeling one of the proteins involved in the interaction. The amide I band of the labeled protein shifts toward lower frequency by ca. 45 cm^{-1}. Hence, the amide I bands of the two proteins are sufficiently separated for evaluating structural changes that may accompany the interaction of one protein with the other. This approach has opened up the possibility of investigating, for example, receptor-ligand interaction (cf. Li et al., 1997). Also, conformational differences between the native and reactive-center cleaved forms of α_1-antitrypsin (Haris et al., 1990) have been shown. The spectra indicated a stronger hydrogen bonding within the reactive-center cleaved form of the protein. In agreement, the hydrogen-deuterium exchange within the polypeptide backbone was found to be less for the cleaved form, which was further confirmed by NMR spectroscopy.

Emulsions are a common intermediate in the preparation of polymeric microparticles and are under investigation as potential drug delivery systems as such. For most analytical techniques, the emulsion must first be broken before the protein structure can be analyzed. Unfortunately, these emulsions are often designed to be very stable. For FTIR analysis, the emulsion need not be broken. Jørgensen et al. (2003) have used FTIR to show that some structural changes occur for insulin and growth hormone in a water-in-oil emulsion. The area overlap of second-derivative spectra of insulin was reduced to ca. 93% and that of growth hormone was reduced to ca. 80% when comparing the protein in solution to that entrapped in the emulsion.

Lefèvre and Subirade (2003) have used FTIR to determine the structural changes of β-lactoglobulin as a result of emulsification. By using ^2H_2O as the aqueous phase, it was possible to determine the protein structure at the water-oil interface. Similar to studies in which emulsions were first broken (cf. Van de Weert et al., 2000a), it was shown that the protein at the interface is aggregated through intermolecular β-sheets.

Lyophilized Proteins

Lyophilization of protein formulations is often essential to obtain sufficient stability during shipping and (long-term) storage. However, the lyophilization process itself introduces stress factors that may denature the protein. Until recently, little was known regarding the effect of these factors on protein stability. With FTIR, it is now possible to analyze the protein throughout the lyophilization process: The protein structure can be investigated in solution, the frozen solution, and the solid (lyophilizate). This was most aptly shown by Remmele et al. (1997), who used a homemade ATR cell to investigate the stresses on lysozyme throughout the whole lyophilization process. The described method is of significant benefit in development of the lyophilization process, as detrimental steps can be identified during, rather than after, the lyophilization cycle.

Using FTIR, Carpenter and co-workers have shown that spectral changes occur for many proteins upon lyophilization (reviewed in Carpenter et al., 1998). In addition, these spectral changes correlate to reduced storage stability. For example, Carpenter and co-workers have shown that the storage stability of lyophilized interleukin-2 (Prestrelski et al., 1995) and interleukin-1 receptor antagonist (Chang et al., 1996) is directly related to both the matrix T_g value and the observed spectral changes in the protein spectra upon lyophilization. Costantino et al. (1998) showed an improved stability of human growth hormone when the spectral changes upon lyophilization were minimized. Many other studies on proprietary industrial formulations have been performed but remain as yet unpublished (Carpenter et al., 1998).

The observed spectral changes upon lyophilization are often described as indicative of structural changes of the protein (cf. Prestrelski et al., 1993). This also fits with the contention that successful lyoprotectants replace the water hydrogen bonds to the protein molecule, thus preventing structural changes. However, conflicting reports appear in the literature. Griebenow and co-workers (Griebenow et al., 1999; Al-Azzam et al., 2002) describe the observed spectral changes as indicative of structural changes, with some minor contribution of water removal. In contrast, Forato et al. (1998) claim that the observed spectral changes are due to the removal of water without causing any structural changes. Van de Weert et al. (2001) have noted that both structural changes and removal of water may cause spectral changes, without specifically noting which of the two processes is more important. Finally,

Belton and Gil (1994) proposed a water-sharing mechanism, rather than water-replacement, for successful lyoprotectants. Despite this controversy, the potential of FTIR spectroscopy to predict storage stability of lyophilized proteins remains undisputed.

Proteins in Polymeric Matrices

Polymeric matrices can be used for sustained/controlled release purposes. Because proteins are mainly delivered parenterally and often have a short half-life *in vivo*, such sustained/controlled release systems may significantly improve the therapeutic benefit of these proteins. Also, they could improve patient compliance by reducing the number of injections.

As the use of FTIR in analyzing lyophilized proteins developed, it was realized that FTIR can also be used in analyzing the proteins entrapped in the polymeric matrix. Fu et al. (1999) showed that lysozyme and bovine serum albumin have an altered FTIR spectrum when entrapped in poly(lactic-*co*-glycolic acid) (PLGA) microparticles using an emulsion-based encapsulation method. The spectral changes could be reduced by adding well-known lyoprotectants, resulting in fewer protein aggregates in the released protein. However, the observed spectral changes were bigger than those observed upon mere lyophilization of the protein, suggesting that the encapsulation process caused some additional structural changes of the protein. Van de Weert et al. (2000b) showed that some of the observed spectral changes upon encapsulation of lysozyme may be explained by protein aggregation during the preparation process.

Yang et al. (1999) used FTIR in the analysis of human growth hormone in PLGA microspheres. Here also, well-known lyoprotectants were able to reduce the spectral changes upon encapsulation. In contrast, zinc-complexation of growth hormone was shown not to protect the protein against lyophilization-induced changes but prevented any further changes when the complex was entrapped in the microspheres. The microspheres were also analyzed in the rehydrated state. The data suggested that the lyoprotectants provided moderate protection against protein aggregation upon rehydration, whereas the zinc-complexed growth hormone refolded into its native state. These results have played an important role in development of Nutropin Depot, a sustained release system containing a zinc-growth hormone complex.

In a series of papers, Griebenow and co-workers have further developed the use of FTIR in the formulation design of polymeric sustained/controlled release systems (cf. Pérez et al., 2002 and references therein). They have coined the term *structure-guided protein encapsulation* for this process.

Some investigators have also analyzed protein conformation upon entrapment in polymeric films. Carrasquillo et al. (1999) analyzed the structural changes of human growth hormone upon entrapment in a PLGA film. Very minor spectral changes were observed. Similarly, van de Weert et al. (2002) observed very minor spectral changes of lysozyme in a block co-polymer of poly(ethylene glycol) and poly(butylene terephthalate). Van Dijkhuizen-Radersma et al. (2002) have shown that the nonrelease of calcitonin from a polymer film was due to aggregation of calcitonin during the release process rather than during the preparation of the film.

Conclusions

In the past decade, FTIR has become a standard technique in many protein formulation laboratories. Its biggest asset is the ability to measure a sample in virtually any physical state, thus allowing structural analysis of proteins in solutions, dispersions, and the solid state. Its use in lyophilization development has proved of great importance, but also the development of protein-containing sustained/controlled release systems has benefited from FTIR analysis.

The biggest drawback of FTIR is its limited sensitivity, but this can often be overcome by choosing another sampling method or by altering some of the sample properties. More importantly, some pharmaceutical proteins are formulated at relatively high concentrations (greater than 5 mg/mL). Many analytical techniques are not able to deal with very high protein concentrations, whereas in FTIR the higher concentrations are actually beneficial to the analysis.

Abbreviations

ATR	Attenuated total reflectance
CD	Circular dichroism
DRIFT	Diffuse reflectance
dTGS	Deuterated triglycine sulfate
FA	Factor analysis
FFT	Fast Fourier transform
FSD	Fourier self-deconvolution
FTIR	Fourier transform infrared spectroscopy
IR	Infrared
MCT	Mercury cadmium telluride
PCA	Principal component analysis
PLGA	Poly(lactic-*co*-glycolic acid)
SCOP	Structural classification of proteins
SVD	Singular value decomposition

Acknowledgment

M.v.d.W. acknowledges an Investigational Fellowship of the Alfred Benzon Foundation.

References

1. Al-Azzam W, Pastrana EA, Ferrer Y, Huang Q, Schweitzer-Stenner R, and Griebenow K. Structure of PEG-Modified HRP in Organic Solvents. IR Amide I Spectral Changes upon Protein Hydration Are Largely Caused by Protein Structural Changes and Not by Water Removal per se. *Biophys J* 2002; 83:3637–3651
2. Andrade MA, Chacon P, Merelo JJ, and Moran F. Evaluation of Secondary Structure of Proteins from UV Circular-Dichroism Spectra Using an Unsupervised Learning Neural-Network. *Protein Eng* 1993; 6:383–390
3. Arrondo JLR, and Goni FM. Structure and Dynamics of Membrane Proteins as Studied by Infrared Spectroscopy. *Prog Biophys Mol Biol* 1999; 72:367–405
4. Baello BI, Pancoska P, and Keiderling TA. Enhanced Prediction Accuracy of Protein Secondary Structure Using Hydrogen Exchange Fourier Transform Infrared Spectroscopy. *Anal Biochem* 2000; 280:46–57
5. Baldi P, Brunak S, Frasconi P, Soda G, and Pollastri G. Exploiting the Past and the Future in Protein Secondary Structure Prediction. *Bioinformatics* 1999; 15:937–946
6. Baldi P, Brunak S, Chauvin Y, Andersen CAF, and Nielsen H. Assessing the Accuracy of Prediction Algorithms for Classification: An Overview. *Bioinformatics* 2000; 16:412–424
7. Bandekar J. Amide Modes and Protein Conformation. *Biochim Biophys Acta* 1992; 1120:123–143
8. Barth A, and Zscherp C. What Vibrations Tell Us About Proteins. *Q Rev Biophys* 2002; 35:369–430
9. Bauer HH, Muller M, Goette J, Merkle HP, and Fringeli UP. Interfacial Adsorption and Aggregation Associated Changes in Secondary Structure of Human Calcitonin Monitored by ATR-Spectroscopy. *Biochemistry* 1994; 33:12276–12282
10. Belton PS, and Gil AM. IR and Raman Spectroscopic Studies of the Interaction of Trehalose with Hen Egg White Lysozyme. *Anal Biochem* 1994; 34:957–961
11. Böhm G, Muhr R, and Jaenicke R. Quantitative Analysis of Protein Far UV Circular Dichroism Spectra by Neural Networks. *Protein Eng* 1992; 5:191–195
12. Bolotina IA, Chekhov VO, Lugauskas VY, Finkelshtein AV, and Ptitsyn OB. Determination of the Secondary Structure of Proteins

from the Circular-Dichroism Spectra. 1. Protein Reference Spectra for Alpha Structure, Beta Structure, and Irregular Structure. *Mol Biol* 1980; 14:701–709

13. Bussian BM, and Sander C. How to Determine Protein Secondary Structure in Solution by Raman-Spectroscopy—Practical Guide and Test Case DNase-I. *Biochemistry* 1989; 28:4271–4277
14. Byler DM, and Susi H. Examination of the Secondary Structure of Proteins by Deconvolved FTIR Spectra. *Biopolymers* 1986; 25:469–487
15. Cabiaux V, Oberg KA, Pancoska P, Walz T, Agre P, and Engel A. Secondary Structures Comparison of Aquaporin-1 and Bacteriorhodopsin: A Fourier Transform Infrared Spectroscopy Study of Two-Dimensional Membrane Crystals. *Biophys J* 1997; 73:406–417
16. Cameron DG, and Moffatt DJ. Deconvolution, Derivation and Smoothing of Spectra Using Fourier Transforms. *J Testing Eval* 1984; 12:78–85
17. Carpenter JF, Prestrelski SJ, and Dong A. Application of Infrared Spectroscopy to Development of Stable Lyophilized Protein Formulations. *Eur J Pharm Biopharm* 1998; 45:231–238
18. Carrasquillo KG, Costantino HR, Cordero RA, Hsu CC, and Griebenow K. On the Structural Preservation of Recombinant Human Growth Hormone in a Dried Film of a Synthetic Biodegradable Polymer. *J Pharm Sci* 1999; 88:166–173
19. Carrasquillo KG, Sanchez C, and Griebenow K. Relationship Between Conformational Stability and Lyophilization-Induced Structural Changes in Chymotrypsin. *Biotechnol Appl Biochem* 2000; 31:41–53
20. Caughey BW, Dong A, Bhat KS, Ernst D, Hayes SF, and Caughey WS. Secondary Structure Analysis of the Scrapie-Associated Protein Prp 27-30 in Water by Infrared Spectroscopy. *Biochemistry* 1991; 30:7672–7678
21. Chalmers JM, and Griffiths PR. *Handbook of Vibrational Spectroscopy.* Chichester: John Wiley & Sons; 2001. 4000p.
22. Chan HK, Ongpipattanakul B, and Au-Jeung J. Aggregation of rhDNase Occurred During the Compression of KBr Pellets Used for FTIR Spectroscopy. *Pharm Res* 1996; 13:238–242
23. Chang BS, Beauvais RM, Dong A, and Carpenter JF. Physical Factors Affecting the Storage Stability of Freeze-Dried Interleukin-1 Receptor Antagonist: Glass Transition and Protein Conformation. *Arch Biochem Biophys* 1996; 331:249–258
24. Chen YH, and Yang JT. A New Approach to the Calculation of Secondary Structures of Globular Proteins by Optical Rotary Dispersion and Circular Dichroism. *Biochem Biophys Res Commun* 1971; 44:1285–1291

25. Chirgadze YN, and Brazhnikov EV. Intensities and Other Spectral Parameters of Infrared Amide Bands of Polypeptides in the Alpha-Helical Form. *Biopolymers* 1974; 13:1701–1712

26. Chirgadze YN, Shestopalov BV, and Venyaminov SY. Intensities and Other Spectral Parameters of Infrared Amide Bands of Polypeptides in the Beta- and Random Forms. *Biopolymers* 1973; 12:1337–1351

27. Costantino HR, Shieh L, Klibanov AM, and Langer R. Heterogeneity of Serum Albumin Samples with Respect to Solid-State Aggregation via Thiol-Disulfide Interchange—Implications for Sustained Release from Polymers. *J Controlled Release* 1997; 44:255–261

28. Costantino HR, Carrasquillo KG, Cordero RA, Mumenthaler M, Hsu CC, and Griebenow K. Effects of Excipients on the Stability and Structure of Lyophilized Recombinant Human Growth Hormone. *J Pharm Sci* 1998; 87:1412–1420

29. Dalmas B, Hunter GJ, and Bannister WH. Prediction of Protein Secondary Structure from Circular Dichroism Spectra Using Artificial Neural Network Techniques. *Biochem Mol Biol Int* 1994; 34:17–26

30. De Jongh HHJ, Goormaghtigh E, and Ruysschaert J-M. The Different Molar Absorptivities of the Secondary Structure Types in the Amide I Region: An Attenuated Total Reflection Infrared Study on Globular Proteins. *Anal Biochem* 1996; 242:95–103

31. Dong A, Huang P, and Caughey WS. Protein Secondary Structure from Second Derivative Amide I Infrared Spectra. *Biochemistry* 1990; 29:3303–3308

32. Dong A, Caughey WS, and Du Closs TW. Effects of Calcium, Magnesium and Phosphorylcholine on Secondary Structures of Human C-Reactive Protein and Serum Amyloid P Component Observed by Infrared Spectroscopy. *J Biol Chem* 1994; 269:6424–6430

33. Dong A, Meyer JD, Brown JL, Manning MC, and Carpenter JF. Comparative Fourier Transform Infrared and Circular Dichroism Spectroscopic Analysis of α1-Proteinase Inhibitor and Ovalbumin in Aqueous Solution. *Arch Biochem Biophys* 2000; 383:148–155

34. Dousseau F, and Pezolet M. Determination of the Secondary Structure Content of Proteins in Aqueous Solutions from Their Amide I and Amide II Infrared Bands. Comparison Between Classical and Partial Least-Squares Methods. *Biochemistry* 1990; 29:8771–8779

35. Elliot A, and Ambrose EJ. Structure of Synthetic Polypeptides. *Nature* 1950; 165:921–922

36. Forato LA, Bernades-Filho R, and Colnago LA. Protein Structure in KBr Pellets by Infrared Spectroscopy. *Anal Biochem* 1998; 259:136–141

37. Fu F-N, DeOliveira DB, Trumble WR, and Sarkar HK. Secondary Structure Estimation of Proteins Using the Amide III Region of

Fourier Transform Infrared Spectroscopy: Application to Analyze Calcium-Binding-Induced Structural Changes in Calsequestrin. *Appl Spectrosc* 1994; 48:1432–1441

38. Fu K, Griebenow K, Hsieh L, Klibanov AM, and Langer R. FTIR Characterization of the Secondary Structure of Proteins Encapsulated within PLGA Microspheres. *J Controlled Release* 1999; 58:357–366

39. Goormagtigh E, Cabiaux V, and Ruysschaert J-M. Determination of Soluble and Membrane Protein Structure by Fourier Transform Infrared Spectroscopy. *Subcell Biochem* 1994; 23:329–450

40. Greenfield NJ, and Fasman GD. Computed Circular Dichroism Spectra for the Evaluation of Protein Conformation. *Biochemistry* 1996; 8:4108–4116

41. Griebenow K, Santos AM, and Carrasquillo KG. Secondary Structure of Proteins in the Amorphous Dehydrated State Probed by FTIR Spectroscopy. *Int J Vib Spec* 1999; 3 (1) article 8

42. Griffiths PR, and Pariente GL. Introduction to Spectral Deconvolution. *Trends Anal Chem* 1986; 5:209–215

43. Haris PI. Characterization of Protein Structure and Stability Using Fourier Transform Infrared Spectroscopy. *Pharm Pharmacol Commun* 1999; 4:15–25

44. Haris PI. Fourier Transform Infrared Spectroscopic Studies of Peptides: Potentials and Pitfalls. *Proc ACS Symp Ser* 2000; 750:54–95

45. Haris PI, and Chapman D. Does Fourier-Transform Infrared-Spectroscopy Provide Useful Information on Protein Structures. *Trends Biochem Sci* 1992; 17:328–333

46. Haris PI, and Chapman D. Analysis of Polypeptide and Protein Structures Using Fourier Transform Infrared Spectroscopy. *Proc Microsc Optical Spectrosc Macrosc Techn* 1994; 22:183–202

47. Haris PI, and Chapman D. The Conformational Analysis of Peptides Using Fourier Transform IR Spectroscopy. *Biopolymers* 1995; 37:251–263

48. Haris PI, Lee DC, and Chapman D. A Fourier Transform Infrared Investigation of the Structural Differences Between Ribonuclease A and Ribonuclease S. *Biochim Biophys Acta* 1986; 874:255–265

49. Haris PI, Chapman D, Harrison RA, Smith KF, and Perkins SJ. Conformational Transition Between Native and Reactive Center Cleaved Forms of α_1-Antitrypsin by Fourier Transform Infrared Spectroscopy and Small-Angle Neutron Scattering. *Biochemistry* 1990; 29:1377–1380

50. Haris PI, Robillard GT, Van Dijk AA, and Chapman D. The Potential of ^{13}C and ^{15}N Labelling for Studying Protein-Protein Interaction Using Fourier Transform Infrared Spectroscopy. *Biochemistry* 1992; 31:6279–6284

51. Heimburg T, and Marsh D. Investigation of Secondary and Tertiary Structural Changes of Cytochrome c in Complexes with Anionic Lipids Using Amide Hydrogen Exchange Measurements: An FTIR Study. *Biophys J* 1993; 65:2408–2417

52. Hering JA, Innocent PR, and Haris PI. An Improved Method for Rapid Quantification of Protein Secondary Structure from FTIR Spectra of Proteins. In: Palagi PM, Sanchez J-C, and Stöcklin R. *Proceedings of the Swiss Proteomics Society, 2001 Congress, Functional Proteomics*. Lausanne: FontisMedia Publishers; 2001:128–132

53. Hering JA, Innocent PR, and Haris PI. An Alternative Method for Rapid Quantification of Protein Secondary Structure from FTIR Spectra Using Neural Networks. *Spectroscopy – An Int. J.* 2002a; 16:53–69

54. Hering JA, Innocent PR, and Haris PI. New Approaches for Quantification of Protein Secondary Structure from FTIR Spectra of Proteins. In: Palagi PM, Quadroni M, Rossier J, Sanchez J-C, and Stöcklin R. *Proceedings of the Swiss Proteomics Society, 2002 Congress, Applied Proteomics*. Lausanne: FontisMedia Publishers; 2002b:163–165

55. Hering JA, Innocent PR, and Haris PI. Neuro-fuzzy SCOP Classification for Improved Protein Secondary Structure Prediction. *Proteomics* 2003; 3:1464–1475

56. Hering JA, Innocent PR, and Haris PI. Beyond Average Protein Secondary Structure Content Prediction Using FTIR Spectroscopy. *Appl Bioinf* 2004; 3:9–20

57. Hübner W, Mantsch HH, and Casal HL. Beware of Frequency Shifts. *Appl Spectrosc* 1990; 44:732–734

58. Jackson M, and Mantsch HH. Artifacts Associated with the Determination of Protein Secondary Structure by ATR-IR Spectroscopy. *Appl Spectrosc* 1992; 46:699–701

59. Jackson M, and Mantsch HH. The Use and Misuse of FTIR Spectroscopy in the Determination of Protein-Structure. *Crit Rev Biochem Mol Biol* 1995; 30:95–120

60. Jackson M, Haris PI, and Chapman D. Conformational Transitions in Poly(L-Lysine)—Studies Using Fourier-Transform Infrared-Spectroscopy. *Biochim Biophys Acta* 1989; 998:75–79

61. Jørgensen L, Vermehren C, Bjerregaard S, and Froekjaer S. Secondary Structure Alterations in Insulin and Growth Hormone Water-in-Oil Emulsions. *Int J Pharm* 2003; 254:7–10

62. Kauppinen JK, Moffatt DJ, Mantsch HH, and Cameron DG. Fourier Self-Deconvolution: A Method for Resolving Intrinsically Overlapped Bands. *Appl Spectrosc* 1981a; 35:271–276

63. Kauppinen JK, Moffatt DJ, Mantsch HH, and Cameron DG. Fourier Transforms in the Computation of Self-Deconvoluted and First-Order

Derivative Spectra of Overlapped Band Contours. *Anal Chem* 1981b; 53:1454–1457

64. Krimm S, and Bandekar J. Vibrational Spectroscopy and Conformation of Peptides, Polypeptides and Proteins. *Adv Protein Chem* 1986; 38:181–364

65. Kumosinski TF, and Unruh JJ. Global-Secondary-Structure Analysis of Proteins in Solution—Resolution-Enhanced Deconvolution Fourier-Transform Infrared-Spectroscopy in Water. *Proc Mol Modeling* 1994; 576:71–98

66. Lee DC, Haris PI, Chapman D, and Mitchell RC. Determination of Protein Secondary Structure Using Factor-Analysis of Infrared-Spectra. *Biochemistry* 1990; 29:9185–9193

67. Lefèvre T, and Subirade M. Formation of Intermolecular β-Sheet Structures: A Phenomenon Relevant to Protein Film Structure at Oil-Water Interfaces of Emulsions. *J Colloid Interface Sci* 2003; 263:59–67

68. Li T, Horan T, Osslund T, Stearns G, and Arakawa T. Conformational Changes in G-CSF/Receptor Complex as Investigated by Isotope-Edited FTIR Spectroscopy. *Biochemistry* 1997; 36, 8849–8857

69. Moult J. The Current State of the Art in Protein Structure Prediction. *Curr Opin Biotechnol* 1996; 7:422–427

70. Nielsen L, Frokjaer S, Carpenter JF, and Brange J. Studies of the Structure of Insulin Fibrils by Fourier Transform Infrared (FTIR) Spectroscopy and Electron Microscopy. *J Pharm Sci* 2001; 90:29–37

71. Oberg KA, and Fink AL. A New Attenuated Total Reflectance Fourier Transform Infrared Spectroscopy Method for the Study of Proteins in Solution. *Anal Biochem* 1998; 256:92–106

72. Ozaki Y, Murayama K, Wu Y, and Czarnik-Matusewicz B. Two-Dimensional Infrared Correlation Spectroscopy Studies on Secondary Structures and Hydrogen Bondings of Side Chains of Proteins. *Spectroscopy – An Int J* 2003; 17:79–100

73. Pancoska P, Yasui SC, and Keiderling TA. Statistical-Analyses of the Vibrational Circular-Dichroism of Selected Proteins and Relationship to Secondary Structures. *Biochemistry* 1991; 30:5089–5103

74. Pancoska P, Bitto E, Janota V, Urbanova M, Gupta VP, and Keiderling TA. Comparison of and Limits of Accuracy for Statistical Analyses of Vibrational and Electronic Circular Dichroism Spectra in Terms of Correlations to and Predictions of Protein Secondary Structure. *Protein Sci* 1995; 4:1384–1401

75. Pancoska P, Fabian H, Yoder C, Baumruk V, and Keiderling TA. Protein Structural Segments and Their Interconnections Derived from Optical Spectra. Thermal Unfolding of Ribonuclease T-1 as an Example. *Biochemistry* 1996; 35:13094–13106

76. Pancoska P, Janota V, and Keiderling TA. Novel Matrix Descriptor for Secondary Structure Segments in Proteins: Demonstration of Predictability from Circular Dichroism Spectra. *Anal Biochem* 1999; 267:72–83

77. Pancoska P, Janota V, and Nesetril J. Novel Matrix Descriptor for Determination of the Connectivity of Secondary Structure Segments in Proteins. Analysis of General Properties Using Graph Theory. *Discrete Math* 2001; 235:399–423

78. Paquet MJ, Laviolette M, Pezolet M, and Auger M. Two-Dimensional Infrared Correlation Spectroscopy Study of the Aggregation of Cytochrome c In the Presence of Dimyristoylphosphatidylglycerol. *Biophys J* 2001; 81:305–312

79. Pérez C, Castellanos IJ, Costantino HR, Al-Azzam W, and Griebenow K. Recent Trends in Stabilizing Protein Structure Upon Encapsulation and Release from Bioerodible Polymers. *J Pharm Pharmacol* 2002; 54:301–313

80. Prestrelski SJ, Tedeschi N, Arakawa T, and Carpenter JF. Dehydration-Induced Conformational Transitions in Proteins and Their Inhibition by Stabilizers. *Biophys J* 1993; 65:661–671

81. Prestrelski SJ, Pikal KA, and Arakawa T. Optimization of Lyophilization Conditions for Recombinant Human Interleukin-2 by Dried-State Conformational Analysis Using Fourier Transform Infrared Spectroscopy. *Pharm Res* 1995; 12:1250–1259

82. Pribic R, Van Stokkum IHM, Chapman D, Haris PI, and Bloemendal M. Protein Secondary Structure from Fourier Transform Infrared and/or Circular Dichroism Spectra. *Anal Biochem* 1993; 214:366–378

83. Rahmelow K, and Huebner W. Secondary Structure Determination of Proteins in Aqueous Solution by Infrared Spectroscopy: A Comparison of Multivariate Data Analysis Methods. *Anal Biochem* 1996; 241:5–13

84. Rahmelow K, Huebner W, and Ackermann T. Infrared Absorbances of Protein Side Chains. *Anal Biochem* 1998; 257:1–11

85. Remmele RL, Stushnoff C, and Carpenter JF. Real-time in situ Monitoring of Lysozyme During Lyophilization Using Infrared Spectroscopy: Dehydration Stress in the Presence of Sucrose. *Pharm Res* 1997; 14:1548–1555

86. Rost B, and O'Donoghue S. Sisyphus and Prediction of Protein Structure. *CABIOS* 1997; 13:345–356

87. Ruegg M, Metzger V, and Susi H. Computer Analysis of Characteristic Infrared Bands of Globular Proteins. *Biopolymers* 1975; 14:1465–1471

88. Sane SU, Cramer SM, and Przybycien TM. A Holistic Approach to Protein Secondary Structure Characterization Using Amide I Band Raman Spectroscopy. *Anal Biochem* 1999; 269:255–272

89. Sarver RW, and Krueger WC. Protein Secondary Structure from Fourier Transform Infrared Spectroscopy: A Database Analysis. *Anal Biochem* 1991; 194:89–100

90. Savitsky A, and Golay MJE. Smoothing and Differentiation of Data by Simplified Least Squares Procedures. *Anal Chem* 1964; 36:1627–1639

91. Schweitzer-Stenner R, Eker F, Huang Q, Griebenow K, Mosz P, and Kozlowski P. Structure Analysis of Dipeptides in Water by Exploring and Utilizing the Structural Sensitivity of Amide III by Polarized Visible Raman, FTIR-Spectroscopy and DFT Based Normal Coordinate Analysis. *J Phys Chem B* 2002; 106:4294–4304

92. Severcan M, Severcan F, and Haris PI. Estimation of Protein Secondary Structure from FTIR Spectra Using Neural Networks. *J Mol Struct* 2001; 565:383–387

93. Singh BR. Basic Aspects of the Technique and Applications of Infrared Spectroscopy of Peptides and Proteins. *Proc ACS Symp Ser* 2000; 750:2–37

94. Singh BR, Fuller MP, and Da Gupta BR. Botulinum Neurotoxin Type A: Structure And Interaction with the Micellar Concentration of SDS Determined by FT-IR Spectroscopy. *J Protein Chem* 1991; 10:637–649

95. Souillac PO, Middaugh CR, and Rytting JH. Investigation of Protein/Carbohydrate Interactions in the Dried State. 2. Diffuse Reflectance Studies. *Int J Pharm* 2002; 235:207–218

96. Surewicz WK, and Mantsch HH. New Insight into Protein Secondary Structure from Resolution-Enhanced Infrared Spectra. *Biochim Biophys Acta* 1988; 952:115–130

97. Van de Weert M, Hoechstetter J, Hennink WE, and Crommelin DJA. The Effect of a Water/Organic Solvent Interface on the Structural Stability of Lysozyme. *J Control Release* 2000a; 68:351–359

98. Van de Weert M, van 't Hof R, van der Weerd J, Heeren RMA, Posthuma G, Hennink WE, and Crommelin DJA. Lysozyme Distribution and Conformation in a Biodegradable Polymer Matrix as Determined by FTIR Techniques. *J Controlled Release* 2000b; 68:31–40

99. Van de Weert M, Haris PI, Hennink WE, and Crommelin DJA. Fourier Transform Infrared Spectrometric Analysis of Protein Conformation: Effect of Sampling Method and Stress Factors. *Anal Biochem* 2001; 297:160–169

100. Van de Weert M, van Dijkhuizen-Radersma R, Bezemer JM, Hennink WE, and Crommelin DJA. Reversible Aggregation of Lysozyme in a Biodegradable Amphiphilic Multiblock Copolymer. *Eur J Pharm Biopharm* 2002; 54:89–93

101. Van Dijkhuizen-Radersma R, Nicolas HM, van de Weert M, Blom M, de Groot K, Bezemer JM. Stability Aspects of Salmon Calcitonin

Entrapped in Poly(Ether-Ester) Sustained Release Systems. *Int J Pharm* 2002; 248:229–237

102. Venyaminov SY, and Kalnin NN. Quantitative IR Spectrophotometry of Peptide Compounds in Water (H_2O) Solutions. I. Spectral Parameters of Amino Acid Residue Absorption Bands. *Biopolymers* 1990; 30:1243–1257
103. Wi S, Pancoska P, and Keiderling TA. Predictions of Protein Secondary Structures Using Factor Analysis on Fourier Transform Infrared Spectra: Effect of Fourier Self-Deconvolution of the Amide I and Amide II Bands. *Biospectroscopy* 1997; 4:93–106
104. Williams RW. Estimation of Protein Secondary Structure from the Laser Raman Amide-I Spectrum. *J Mol Biol* 1983; 166:581–603
105. Williams RW, and Dunker AK. Determination of the Secondary Structure of Proteins from the Amide-I Band of the Laser Raman-Spectrum. *J Mol Biol* 1981; 152:783–813
106. Wu YQ, Murayama K, and Ozaki Y. Two-Dimensional Infrared Spectroscopy and Principle Component Analysis Studies of the Secondary Structure and Kinetics of Hydrogen-Deuterium Exchange of Human Serum Albumin. *J Phys Chem B* 2001; 105:6251–6259
107. Yang TH, Dong A, Meyer J, Johnson OL, Cleland JL, Carpenter JF. Use of Infrared Spectroscopy to Assess Secondary Structure of Human Growth Hormone Within Biodegradable Microspheres. *J Pharm Sci* 1999; 88:161–165

5

Raman Spectroscopy

OLE FAURSKOV NIELSEN

Department of Chemistry, Physical Chemistry, University of Copenhagen, Copenhagen, Denmark

Introduction

Raman and Krishnan (1928) discovered that scattering of monochromatic electromagnetic radiation by molecular liquids can take place with a change of frequency. Since then, this effect has been called Raman spectroscopy. Bands characteristic of the molecules studied are shifted in frequency relative to the monochromatic, exciting frequency. In the early days, an atomic emission line was used as a light source. However, it was very difficult to focus the light on a small sample. The appearance of the laser in the mid 1960s introduced the second era of Raman spectroscopy and made Raman spectroscopy into a very efficient tool in studies of molecular structures and dynamics. However, in studies of biologically interesting molecules, fluorescence from inevitable impurities was an ever-competing process to a Raman spectrum. In the mid 1980s, Raman spectroscopy in the near-infrared (NIR) region became possible (Chase, 1986; Hirschfeld and Chase, 1986). By this technique, fluorescence was avoided, and Raman spectroscopy became an efficient tool in studies of biomolecules. This third era of Raman spectroscopy will be described with a particular emphasis on proteins. Basic Raman theory, instrumentation, and sample handling are included in this chapter.

Theoretical Background

The Electromagnetic Spectrum and Raman Spectroscopy

Electromagnetic radiation is used in various studies of biomolecular structures and dynamics. Figure 1 shows a survey of the electromagnetic spectrum. The wavelengths (λ) are given in meters (1 m = 10^9 nm = 10^{10} Å)

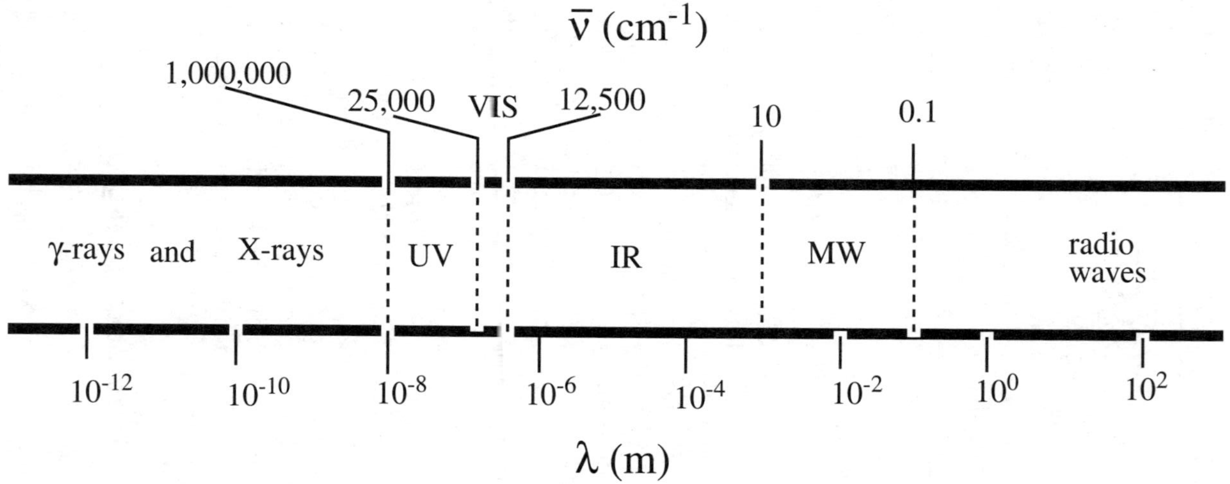

Figure 1 Electromagnetic spectrum. MW, microwave; IR, infrared; Vis, visible; UV, ultraviolet.

together with the corresponding wavenumbers $\bar{\nu}$ (cm^{-1}) (see Chapter 4 on infrared spectroscopy). Radio waves with long wavelengths do not interact with molecules. Molecular dipoles absorb microwaves, and microwave ovens use this effect. Changes in molecular dipoles due to vibrational motions of the nuclei cause absorption of infrared (IR) radiation. Electronic transitions give rise to absorptions in the visible (Vis) and ultraviolet (UV) regions. X-ray wavelengths correspond to distances between the nuclei in molecules (about 1 to 2 Å), and light diffraction in this part of the spectrum yields information about molecular structures. The energy of electromagnetic radiation increases with decreasing wavelengths. In the γ- and x-ray regions, the radiation is so powerful that covalent bonds may be destroyed. Raman spectroscopy is not related to a particular region of the electromagnetic spectrum. In principle, all wavelengths in Figure 1 can be used to excite a Raman spectrum. However, there are limitations and/or advantages by use of a specific wavelength in the electromagnetic spectrum. In Raman studies of biological molecules, the correct wavelength choice is extremely important in order to reach the best result. Thus, throughout this chapter, please keep Figure 1 in mind.

Raman spectroscopists use frequency (or wavenumber) instead of wavelength, because the molecular vibrational frequencies in a Raman spectrum appear in the scattered light as bands shifted with the vibrational frequencies relative to the exciting monochromatic laser frequency. Figure 2 illustrates the Raman effect with a Vis-laser (20,000 cm^{-1} ~ 500 nm). Raman shifts occur to both the high- and low-wavenumber side of the Rayleigh line (same wavenumber as the laser). Thus, the bands in a Raman spectrum shift symmetrically in wavenumber relative to the central Rayleigh line, and the Stokes and anti-Stokes sides appear. The Raman spectrum in Figure 2 shows decreasing Raman shifts to the right in accordance with the IUPAC recommendation (Becker et al., 1981). However Raman spectra in contradiction to this rule are sometimes plotted with increasing Raman shift wavenumbers to the right.

Band intensities for the corresponding Stokes and anti-Stokes bands are larger on the Stokes side of the spectrum (Figure 2), and for nearly all purposes only the Stokes side is shown. All lines in the Stokes side of Figure 2 appear with the same intensity. This is seldom the case in a real Raman spectrum. As shown in the section "Classical Treatment of Raman Spectroscopy" below, the band intensity depends on the change of the laser-induced molecular dipole moment (squared) during the vibration, and each vibrational mode will give different Raman intensities. An IR spectrum gives information about the same molecular vibrations. However, the band intensity in an IR spectrum depends on the change in molecular dipole moment (squared) during

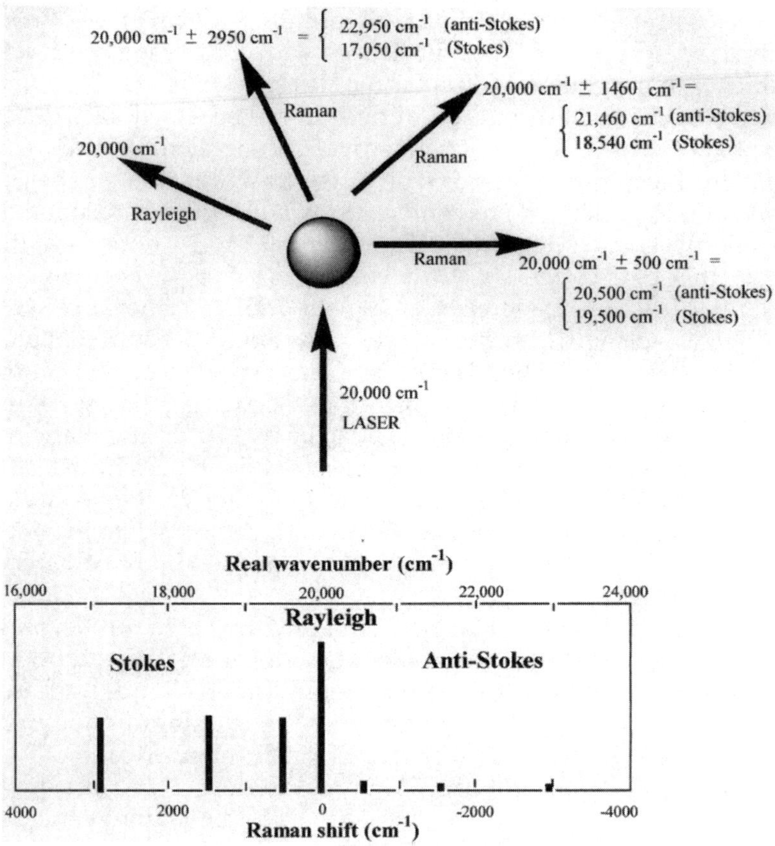

Figure 2 Raman scattering by molecular vibrations with wavenumbers of 2950, 1460, and 500 cm^{-1}. The laser wavelength is 500 nm (20,000 cm^{-1}). The upper part shows the scattering process and the lower part the Raman spectrum. In the Raman spectrum, lines are shown schematically, instead of bands.

the vibration. It often happens that weak bands in Raman are strong in IR. Thus, IR and Raman spectroscopies are complementary techniques in the study of molecular vibrations.

Classical Treatment of Raman Spectroscopy

A classical description of Raman spectroscopy can be made using the diatomic molecule N_2 as an example. The two nitrogen atoms complete harmonic oscillations around the equilibrium distance R_0. The displacement from the equilibrium distance is designated by $\Delta R = R - R_0$. For a harmonic motion, $\Delta R = A\cos(2\pi\nu_q t)$, where ν_q is the harmonic oscillator frequency, t is the time, and A the amplitude of the vibration.

The dipole moment induced (μ_{ind}) in a molecule by an electric field (E) is proportional to the electric field, where the polarizability (α) is the proportionality factor (Equation 1).

$$\mu_{ind} = \alpha E \tag{1}$$

For a harmonically vibrating molecule and a laser with a harmonic oscillating electrical field at frequency ν_L and amplitude E_o, the induced dipole moment is given as:

$$\mu_{ind} = \left[\alpha_0 + \left(\frac{d\alpha}{dR}\right)_{R=R_0} \Delta R\right] E_0 \cos(2\pi\nu_L t)$$

$$= \left[\alpha_0 + \left(\frac{d\alpha}{dR}\right)_{R=R_0} A\cos(2\pi\nu_q t)\right] E_0 \cos(2\pi\nu_L t)$$

$$= \alpha_0 E_0 \cos(2\pi\nu_L t) + \left(\frac{d\alpha}{dR}\right)_{R=R_0}$$

$$E_0 A \frac{1}{2}\{\cos[2\pi(\nu_L + \nu_q)t] + \cos[2\pi(\nu_L - \nu_q)\mathrm{t}]\} \tag{2}$$

Both α_0 and $(d\alpha/dR)_{R=R_0}$ are constants, and the expression for α in Equation 2 is a Taylor expansion in the coordinate ΔR. This expression is a good approximation close to R_0 (Figure 3). The induced dipole, μ_{ind}, acts as an emitter of electromagnetic radiation. The term $\alpha_0 E_0 \cos(2\pi\nu_L t)$ in Equation 2 describes emission of radiation with the laser frequency ν_L; this term gives rise to Rayleigh scattering. The second term describes emission of radiation with the frequencies $\nu_L + \nu_q$ and $\nu_L - \nu_q$. These frequencies correspond to anti-Stokes and Stokes scattering, respectively. In order to observe these two frequencies, the change in polarizability during the vibration $(d\alpha/dR)_{R=R_0}$ must be *non*-zero. In the above treatment, the vector properties of the dipole moment and the electric field have been neglected. In real systems, three components appear for the vectors, and the polarizability is a 3×3 matrix (a tensor with rank 2).

Figure 3 illustrates the change in polarizability during the vibration of N_2. The polarizability α is plotted as a function of the internuclear distance between the two N-atoms. No distance between the two nuclei corresponds to a Si atom, and very large R-values correspond to two *non*-interacting N-atoms. R_0 designates the equilibrium distance between

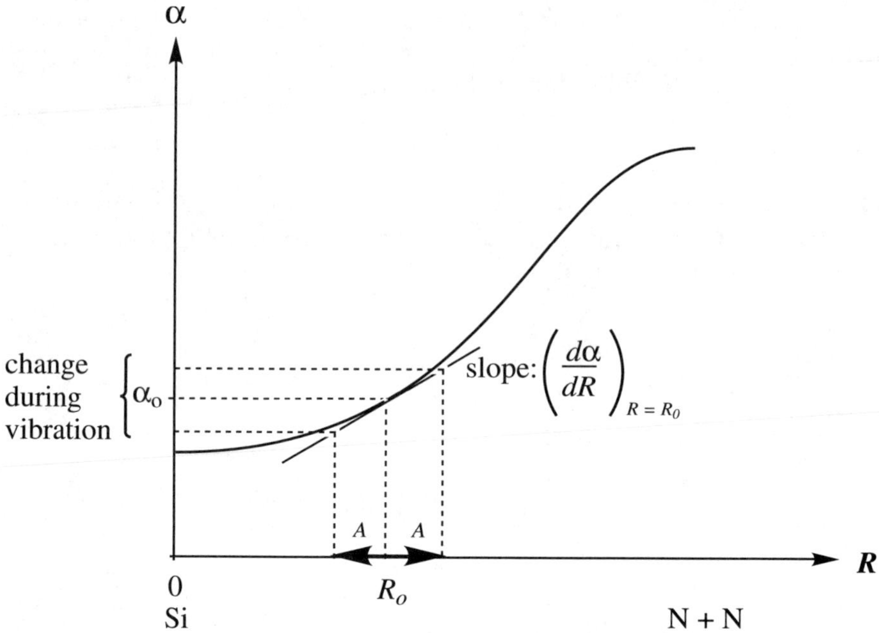

Figure 3 Polarizability of N_2 as function of the internuclear distance, *R*. R_0 is the equilibrium distance; *A* is the amplitude of vibration. The change in polarizability during the vibration is shown. The slope is a good approximation of the polarizability at not-too-large amplitudes. In a real vibration, the amplitude is much smaller relative to the change in polarizability, and it is usually a good approximation to use the slope for the change in polarizability, as is done in Equation 2 in the text.

the two nuclei in the N_2-molecule. The exact polarizability function is unknown, but still Figure 3 illustrates the principle behind the change in polarizability during the molecular vibration. The slope of the tangental straight line at R_0 is $(d\alpha/dR)_{R=R_0}$. At not-too-large vibrational amplitudes, the tangental line is a good approximation to the polarizability curve. This approximation, equivalent to a Taylor expansion, is used for the polarizability in Equation 2. As seen from Figure 3, the N-N stretching vibration in N_2 with frequency ν_q gives a change in the polarizability during the vibration, and frequencies corresponding to $(\nu_L + \nu_q)$ and $(\nu_L - \nu_q)$ are emitted from the molecule according to the second term of Equation 2. The Raman shifts correspond to $\pm\bar{\nu}_q$, where the bar denotes wavenumber units (cm^{-1}). Thus, one band is expected in the Stokes side and one in the anti-Stokes side. For N_2, the wavenumber for the stretching vibration is 2332 cm^{-1}, and bands shifted by this number to either side of the laser wavenumber are expected in the Raman spectrum.

Why begin with a simple molecule like N_2? The reason is that each vibration in a more complex molecule can be treated like the vibration for N_2. A normal coordinate like ΔR for N_2 describes each of the vibrations in the molecule. The number of vibrations for a molecule with N atoms is $3N$-6, thus $3N$-6 bands are expected in the Stokes side. For linear molecules, only $3N$-5 bands are expected. The intensity for each band is given by the change in polarizability (squared) for this vibration. In order for a band to appear in a Raman spectrum, the change in the polarizability must be *non*-zero. This means that the slope of the tangent in Figure 3 must be different from zero. Vibrations with larger slopes give rise to more-intense bands in the Raman spectrum.

IR intensities depend on the change in molecular dipole moments during the vibrations. For N_2 and other homonuclear diatomic molecules, the dipole moment as well as the change in dipole moment are zero during the vibration. Accordingly, N_2 and other homonuclear diatomic molecules do not absorb IR radiation.

Quantum Mechanical Description of the Raman Spectrum

Figure 4 shows potential energy functions for a diatomic molecule in the electronic ground and excited states as a function of internuclear distance. The ground electronic state shows two vibrational energy levels. Vibrational energy levels are also present in the excited electronic state, but for simplicity these are omitted in Figure 4. The diagram shows arrows for electronic transitions causing UV-Vis absorption and fluorescence. An arrow depicts the IR transition corresponding to absorption in the IR spectral region.

Figure 4 also illustrates the Raman effect. The laser arrows indicate the photon energy of the exciting laser. One starts with the molecule in the ground vibrational state and lifts the molecule to the lower of the two virtual states, and the other starts in the vibrationally excited state and lifts the molecule to the upper of the two virtual states. The lifetimes of the virtual states are extremely short, and through emission of radiation, the molecule relaxes to the lower vibrational state, as indicated by the two downward arrows in Figure 4. The Stokes arrow is shorter than the laser arrow, and the anti-Stokes arrow is longer. The arrow lengths correspond to the wavenumber of the respective transitions. An emission also occurs from the virtual states and back to the same vibrational level where the molecule started. This emission, corresponding to the Rayleigh line, is not shown. Figure 4 shows that an anti-Stokes transition starts in the vibrationally excited state and ends up in the ground vibrational state. For the Stokes side, the molecule starts in the ground vibrational state and ends up in the vibrationally excited state. The relative intensities of anti-Stokes and Stokes lines depend on the population in the starting

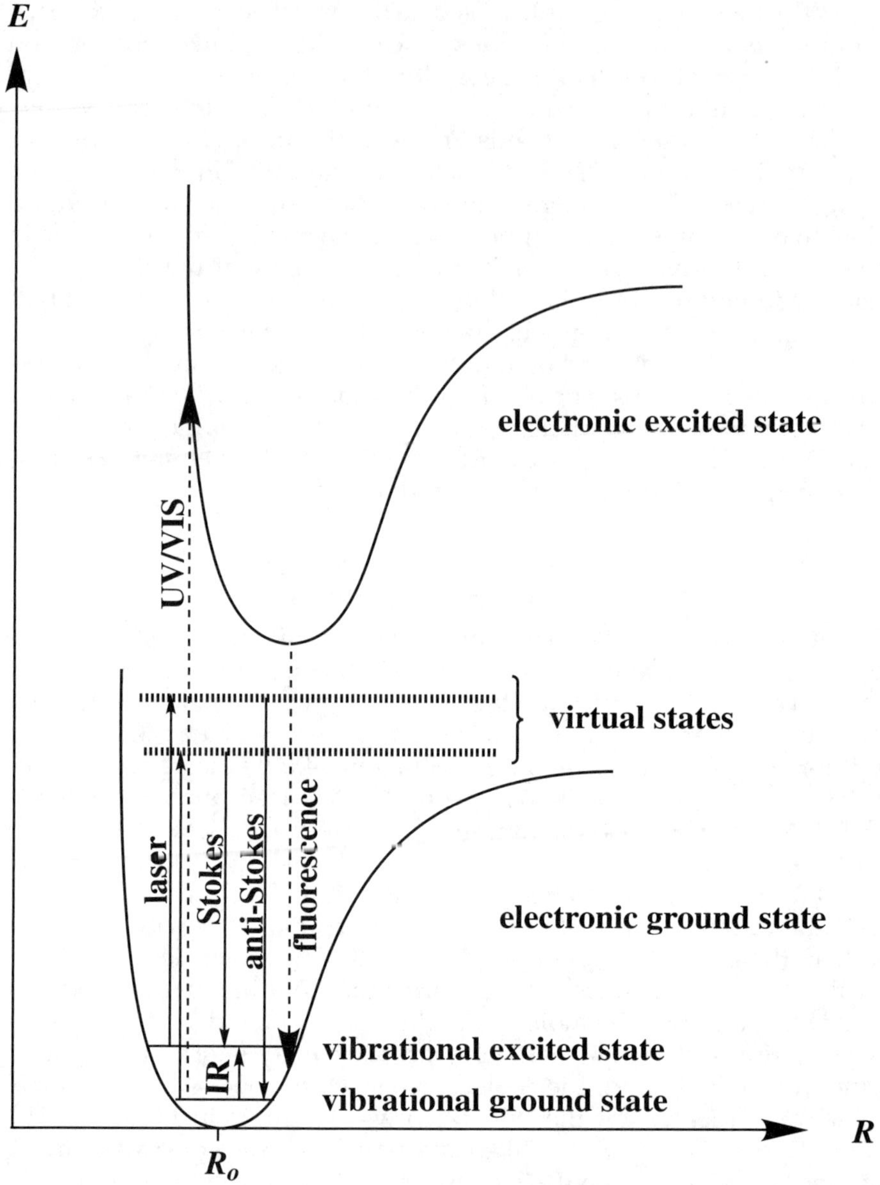

Figure 4 Potential energy function for the ground and electronic excited states of a diatomic molecule. Various spectral transitions giving rise to either the absorption or emission of electromagnetic radiation are shown.

level. Boltzmann's distribution law states that there are more molecules in the lower state. Thus, the anti-Stokes line will be weaker than the Stokes line, and this is the reason for only showing the Stokes side of a

Raman spectrum. However, it should be emphasized that the Boltzmann distribution is temperature dependent, and the relative intensities of anti-Stokes/Stokes pairs depend on the temperature. Thus, a careful determination of the relative intensities in the Stokes and anti-Stokes side of the spectrum can give an estimate of the temperature at exactly the spot of the sample illuminated by the laser. Although it is not so easy to use in practice, this is a tool that can be used to determine whether local heating of the sample in the exciting laser beam takes place.

Raman Spectroscopy *versus* Fluorescence

The virtual levels in Figure 4 are found at energies below the potential energy curve for the electronically excited state, and the electronic state is not excited. The energies of the virtual states increase with increasing laser photon energies (frequencies, wavenumbers). When the laser photon energy becomes high enough to excite the molecule electronically, two possibilities exist: The molecule can return to the ground state by emission of light as either fluorescence or Raman/Rayleigh scattering. However, fluorescence is more intense than Raman scattering. For peptides and proteins, even very small amounts of fluorescent impurities, even of the order 1‰, can hide the Raman spectrum. Figure 5 shows Raman spectra of the same sample of polyglycine (*M* ~15,000, from Sigma Chemical Co., St Louis, MO) with excitation at two different laser wavelengths, a Vis laser at 514.5 nm and a NIR laser at 1064 nm. None of the two lasers can electronically excite polyglycine, but a small amount of an impurity in the sample gives rise to fluorescence when using the laser at 514.5 nm (19,436 cm^{-1}). Excitation at 1064 nm (9398 cm^{-1}) is energetically too low to excite the electronic state of the impurity. Thus, a fluorescence spectrum dominates in Figure 5A, whereas the true Raman spectrum of polyglycine appears in Figure 5B. A careful comparison of Figures 5A and 5B shows that some Raman bands of polyglycine are present in Figure 5A, but they are very weak compared to the strong broad fluorescence band. Fluorescence was for many years the real problem in Raman studies of biochemical and pharmaceutically interesting molecules. For that reason, it was a great advance to be able to record Raman spectra with long-wavelength lasers in the NIR region (Chase, 1986; Hirschfeld and Chase, 1986).

Instrumental

Comprehensive books on Raman spectroscopy describe many instrumental details (Hendra et al., 1991; Schrader, 1995; Lewis and Edwards, 2001). The most important instrumental facts concerning

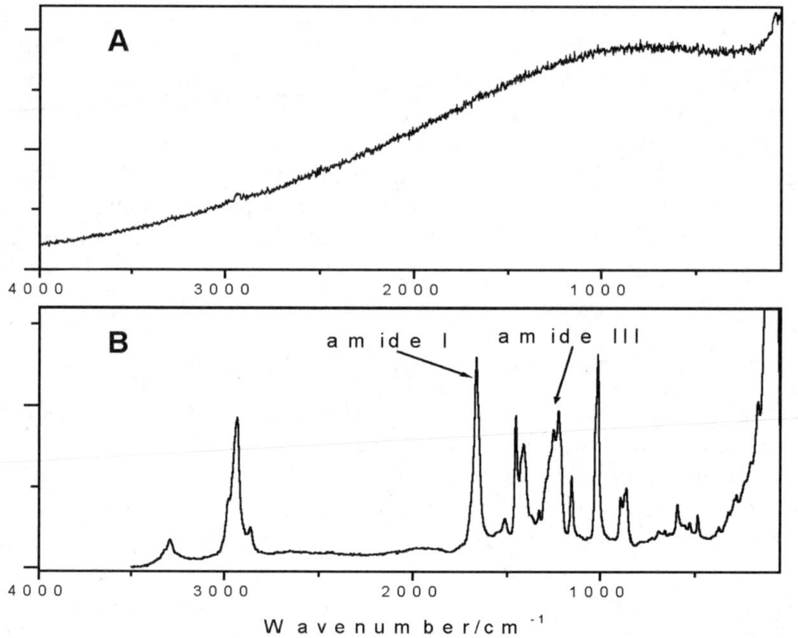

Figure 5 Raman spectra of the same sample of polyglycine obtained with two different laser wavelengths. The spectrum in (A) is obtained with a laser wavelength of 514.5 nm and the spectrum in (B) with a laser wavelength of 1064 nm. A strong broad fluorescence band dominates the spectrum in (A). The Raman spectrum without fluorescence is seen in (B).

protein studies are given. Two different types of instruments are used for Raman spectroscopy: Diffractive instruments and interferometers.

Diffractive Instruments

A monochromator separates different wavelengths of light in diffractive instruments (Figure 6). A lens focuses emitted Raman light on the entrance slit of the monochromator. A grating disperses the light into its spectral elements, and a detector collects the light. A lot of the monochromatic laser light is simply reflected into the instrument, in particular for powdered samples. Thus, an enormous amount of light enters the monochromator with the unshifted laser wavenumber. This tends to give an overall very strong monochromator illumination, which may even be more intense than the Raman scattered light; this scattered light may prevent observation of the Raman spectrum. In order to avoid stray light, a filter cutting off the laser wavelength is placed in the scattered beam before the entrance slit of the instrument (Figure 6). This filter is most often of the notch type (Hirschfeld and

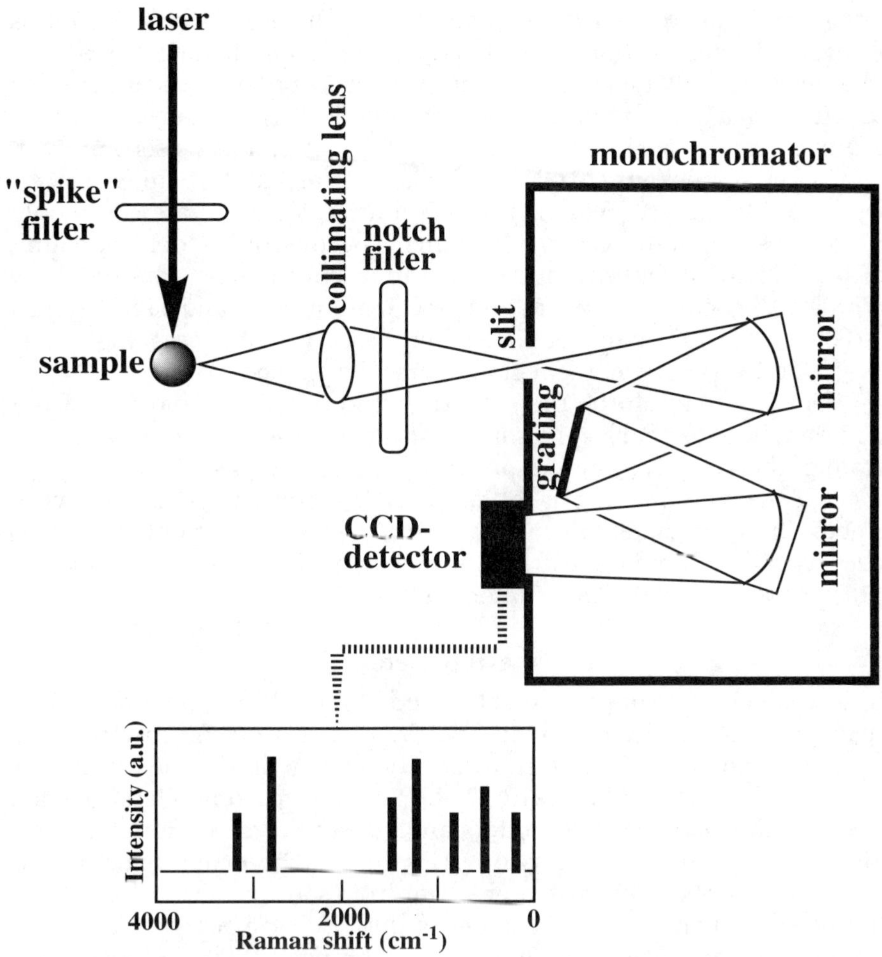

Figure 6 The basic principles of a diffracting laser Raman spectrometer with CCD detection.

Chase, 1986). It is impossible to manufacture a filter that specifically blocks out only the laser wavenumber. Very often, notch filters cut away the spectrum from the Rayleigh line to a Raman shift of 200–300 cm^{-1}. For many purposes, this is not a problem. However, for low-wavenumber studies, it is very important to assure that the notch filter allows low-wavenumber Raman shifts to pass (Colaianni and Nielsen, 1995). Notch filters allowing Raman studies down to ~30 cm^{-1} are now available. Double and triple monochromators can also be used to reduce the stray light (Schrader, 1995; Slater et al., 2001).

Multichannel detectors are today the most commonly used detectors and have replaced photomultiplier tubes (PMT). Charge-coupled device (CCD) detectors have a large number of pixels available. As in Fourier transform infrared (FTIR) spectroscopy (see Chapter 4), the spectrum is recorded during the whole time that the laser is illuminating the sample. This *multiplex* advantage is very important in order to record Raman spectra with a good signal-to-noise ratio. Typically, a Raman spectrum of a protein can be obtained in minutes. CCD detectors lose sensitivity at wavenumbers below ~10,000 cm^{-1} (Slater et al., 2001). Semiconductor lasers at 785 nm (12,739 cm^{-1}) or 835 nm (11,976 cm^{-1}) are commonly used as excitation sources in NIR-Raman spectroscopy. A detection limit of the CCD of 10,000 cm^{-1} indicates a high limit Raman shift of 12,739 minus 10,000 cm^{-1} = 2739 cm^{-1} for the 785-nm laser, and 11,976 minus 10,000 cm^{-1} = 1976 cm^{-1} for the 835-nm laser. In both cases, the O–H, N–H, and C–H stretching regions cannot be studied.

Finally, it should be mentioned that in order to avoid *non*-lasing emission lines as sharp spikes in the spectrum, a *spike* filter is placed in front of the sample, allowing only the laser line to pass.

The wavenumber axis of modern instruments is usually very stable. But of course, wavenumber drift of the instrument should be checked. The Raman spectrum of cyclohexane can be used for this purpose (Mann et al., 2001; Slater et al., 2001).

Interferometers

Interferometer techniques must be used in the NIR region where CCD detectors lack sufficient sensitivity. Chapter 4 on FTIR spectroscopy describes how a Michelson interferometer works. The multiplex advantage is also valid for NIR-FT-Raman spectroscopy (Hendra et al., 1991). Both spike and notch filters are necessary when interferometers are used for Raman spectroscopy (Figure 7). The usual detectors for NIR-FT-absorption spectroscopy are far from sensitive enough to detect the weak Raman signals. An InGaAs detector or a high-performance liquid nitrogen cooled Ge detector can be used (Slater et al., 2001). With an excitation laser at 1064 nm (9398 cm^{-1}), the sensitivity of the Ge detector will only allow detection to a Raman shift of 3500 cm^{-1}, corresponding to the real wavenumber 9398 minus 3500 cm^{-1} = 5898 cm^{-1}. For this reason, the NIR-FT-Raman spectrum in Figure 5B was only recorded to a 3500 cm^{-1} Raman shift.

Scattering Configurations

In a 90-degree scattering configuration, the Raman emitted light is collected perpendicular to the incoming laser beam (Figure 6). In Raman interferometers, a 180-degree scattering configuration is often used, as illustrated in Figure 7.

Intensity Corrections

Different spectrometers have different frequency-dependent intensity transfer functions as a result of the various components in the

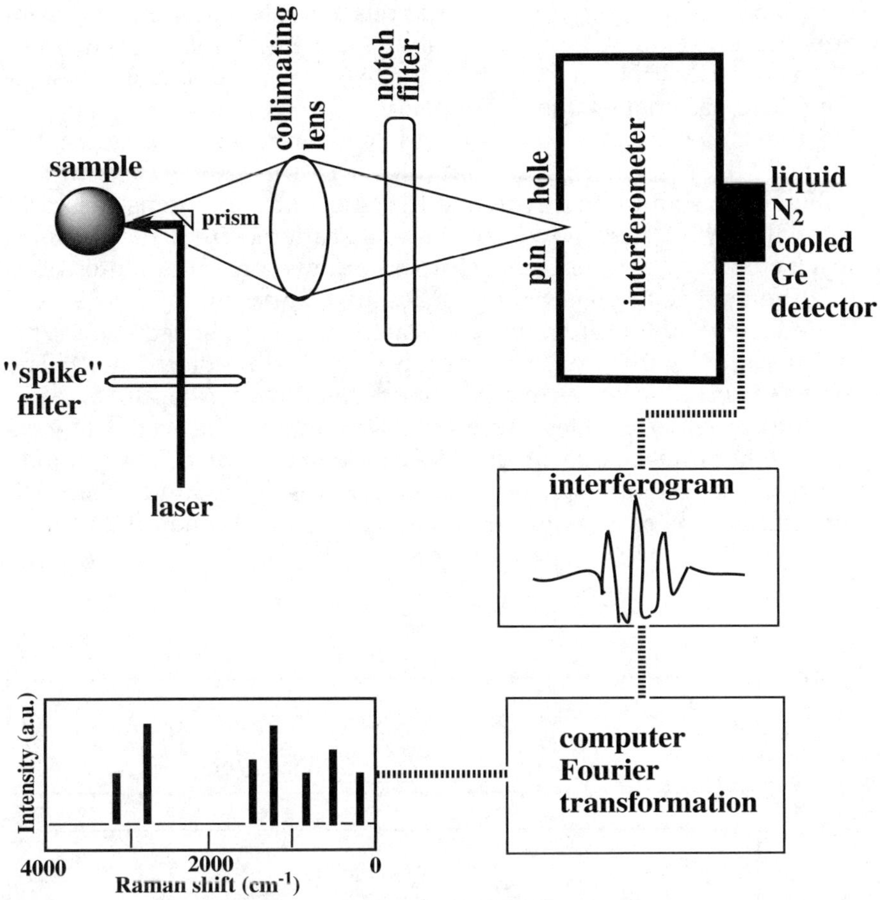

Figure 7 The basic principles of a NIR-FT-Raman Raman instrument (180-degree scattering).

light path. It is not easy to obtain the real intensities (absolute cross sections) from a Raman spectrum (Myers, 1995). Meticulous calibrations can be performed using standard lamps or compounds with known band intensities (Slater et al., 2001). The Raman spectrum of cyclohexane can be used for this purpose (Slater et al., 2001).

For NIR-FT-Raman instruments, a white-light correction can be performed, as shown in Figure 8 (Hendra et al., 1991; Colaianni and Nielsen, 1995; OPUS Spectroscopic Software, 1995). Figure 8A shows the NIR-FT-Raman spectrum of liquid *N*-methylformamide (NMF) and Figure 8B the spectrum from an incandescent lamp. This lamp is a black-body emitter at 3200 K. The white-light spectrum in Figure 8B was divided by a black-body radiation spectrum calculated at 3200 K, and the spectrum in Figure 8A was divided by the resultant spectrum giving the spectrum in Figure 8C. The relative intensities change from Figure 8A to Figure 8C. Figure 8D shows the Vis-Raman spectrum of NMF recorded with laser excitation at 514.5 nm on a dispersive instrument with a very flat frequency-dependent intensity response. The relative intensities in this spectrum are very similar to those in the corrected NIR-FT-Raman spectrum (Figure 8C).

Finally, it should be mentioned that the scattered intensity is proportional to the fourth power of the real scattered frequency. This is also a factor that should be taken into account when Raman intensities with different exciting laser frequencies are compared.

Because of the aforementioned factors, Raman spectra are typically presented with arbitrary intensity units (abbreviated a.u.). The strongest band may be normalized to one, as shown in Figure 8.

Unfortunately, articles presenting Raman spectra do not always state whether intensity corrections have been performed. Thus, it may be difficult to compare spectra from different sources, and it is much more difficult to create reference data archives for Raman than for IR spectra.

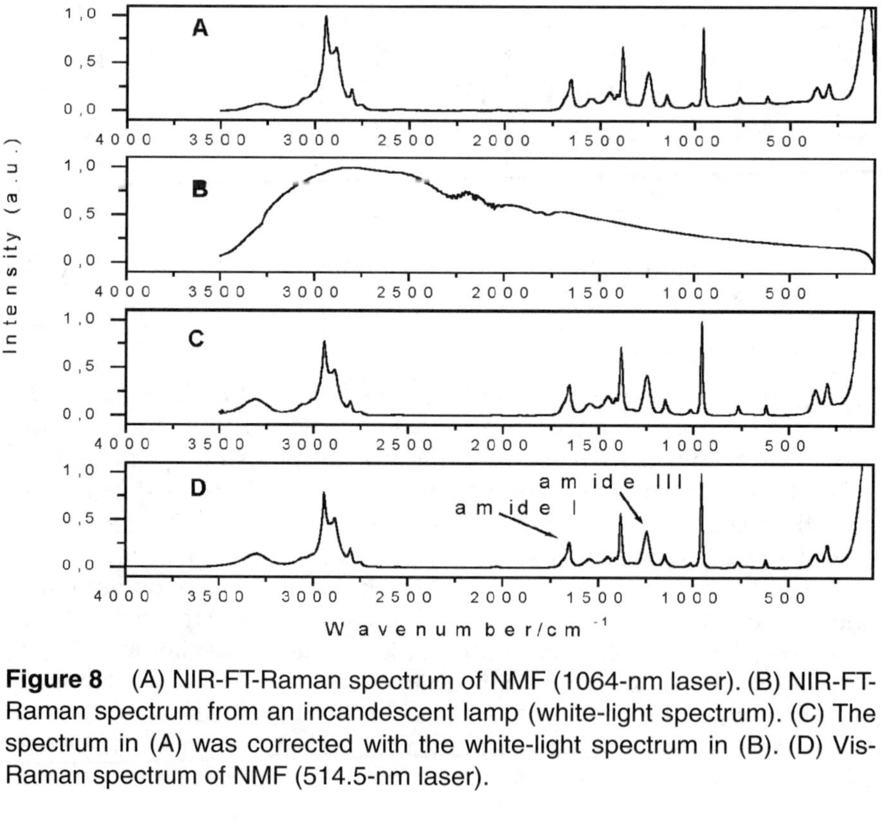

Figure 8 (A) NIR-FT-Raman spectrum of NMF (1064-nm laser). (B) NIR-FT-Raman spectrum from an incandescent lamp (white-light spectrum). (C) The spectrum in (A) was corrected with the white-light spectrum in (B). (D) Vis-Raman spectrum of NMF (514.5-nm laser).

Raman Microscopy

Raman spectra can be obtained through a microscope (Delhay and Dhamelincourt, 1975; Baldwin et al., 2001). The microscope is attached to the Raman instrument by a bundle of optical fibers. In the middle of the bunch, an optical fiber leads the laser through the microscope to the sample, and the optical fibers collect the 180-degree Raman scattered light. The scattered light passes through the optical fibers to the entrance slit of a dispersive instrument (Figure 6) or to the pinhole of an interferometer (Figure 7).

Spectral Manipulations

Background correction, curve fitting, deconvolution, use of second derivatives, principal component analysis (PCA), and the use of neural networks are all methods employed in a general treatment of spectral data (Smith, 1996). These methods are discussed in Chapter 4 on FTIR spectroscopy. However, it should be recalled that Raman intensities are more arbitrary than IR absorbances, and in deriving quantitative results, correction of the Raman intensities should be performed *prior* to any spectral manipulations.

Sample Handling

A real advantage of Raman spectroscopy is that no sample preparation is necessary. For example, it is not necessary to crystallize the samples. Spectra of amorphous freeze-dried samples can be obtained. Spectra of solutions can be recorded, and in the context of biochemicals, it is very important that glass or quartz cells can be used for aqueous solutions. A few milligrams of a solid sample or a few microliters of a liquid sample are sufficient, and when a Raman microscope is used, even smaller amounts can be studied. Figure 9 shows some sample holders. The windowless cell (Brooker et al., 1995) and the solid-state sample holder are for 180-degree scattering. NMR- or ESR-tubes can be used for both 90-degree and 180-degree scattering configurations. However, the spectrum of glass or quartz may appear in the 180-degree scattering configuration. This can be avoided by use of the windowless cell (Brooker et al., 1995).

Characteristic Bands in Raman Spectra of Proteins

Proteins contain a large number (N) of atoms, and the number of fundamental vibrations ($3N$-6) is very large. However, some bands are very intense and dominate the Raman spectrum. In a characterization of large molecules like proteins, only the strong bands are of interest. Vibrations of chemical bonds with many electrons yield a large change in polarizability, $(d\alpha/dR)_{R=R_0}$, and strong Raman bands appear in the

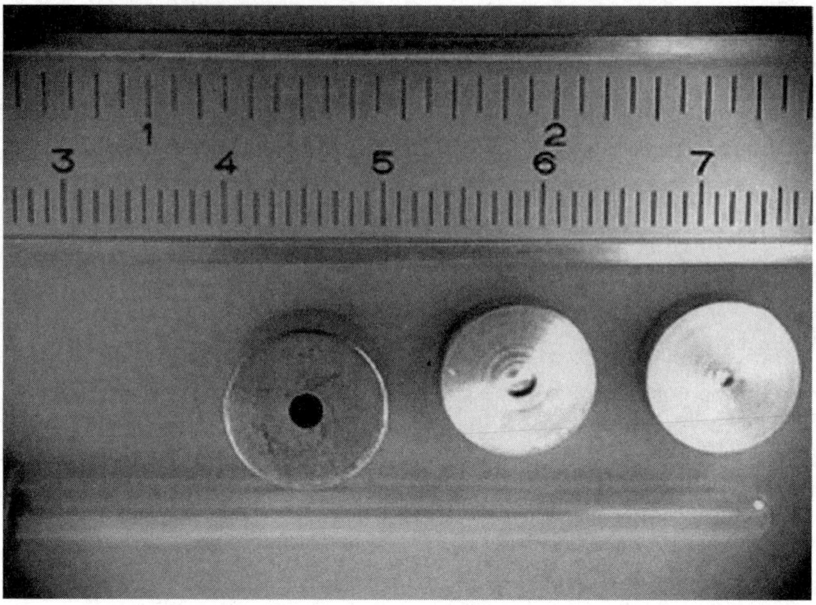

Figure 9 Sample holders for Raman spectroscopy. The scale is in inches (upper scale) and centimeters (lower scale). The three metal holders are made of stainless steel. The holder to the left is a windowless cell for liquids. The liquid is filled in the hole (2-mm diameter). The hole runs through the holder. The two other metal holders are for solid samples. The diameter of the holes are 2 and 1 mm, respectively. The depth of the two holes is approximately 1 mm. The liquid cell in the bottom of the figure is an ESR tube with an inner diameter of 2 mm.

spectrum. Thus the peptide bond, aromatic side chains, and sulfur bridges are expected to show strong Raman intensities. Not quite consistent with this is the fact that C–H stretching and CH_2 bending give rise to relatively strong Raman bands. The NIR-FT-Raman spectrum of solid oxytocin acetate in Figure 10 shows these characteristic features. The bands in the region from 1800 to 2400 cm^{-1} are absorptions from water overtones in the atmosphere. These bands are very often seen in Raman spectra with excitation at 1064 nm.

Amide Bands

Several amide bands are characteristic of peptide group vibrations (see Chapter 4; Tu, 1986; Carter and Edwards, 2001). In Raman, the amide I and III bands are strong. These bands were already seen as characteristic bands in the spectrum of polyglycine (Figure 5), NMF (Figure 8), and oxytocin (Figure 10). The specific positions of the amide I and III bands depend on the secondary structure of the protein. IR and Raman spectra show slightly different spectral regions for the secondary structures.

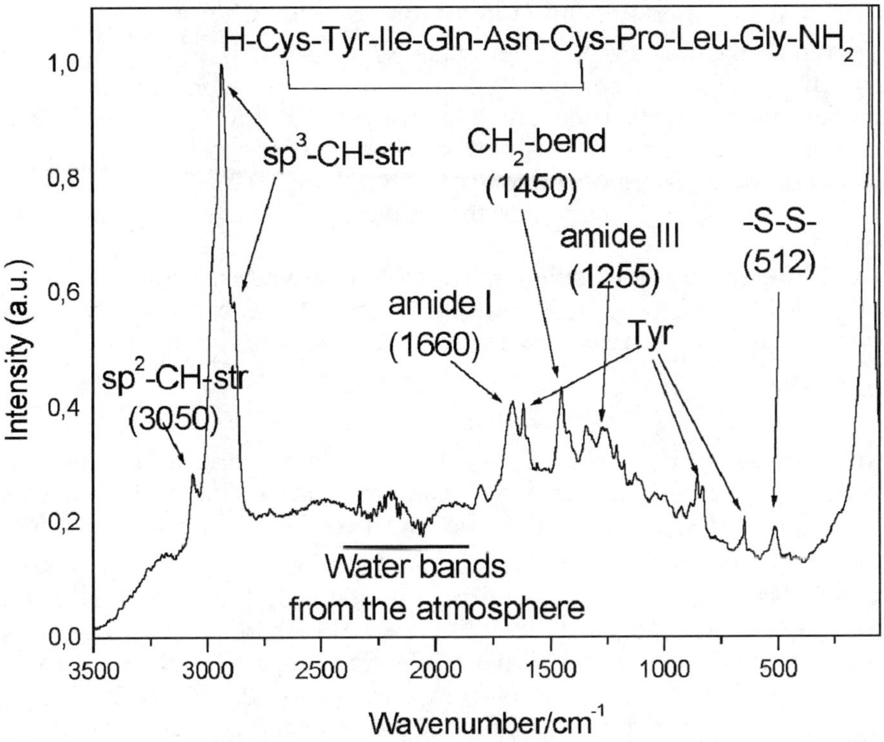

Figure 10 NIR-FT-Raman spectrum of solid oxytocin acetate with excitation at 1064 nm (laser power 280 mW, spectral resolution 6 cm^{-1} after apodization). Five hundred spectra were accumulated, corresponding to a recording time of 15 minutes. Wavenumbers in parentheses are in cm^{-1}. The primary structure is shown at the top. The atmospheric water overtone bands show rotational fine structure and cover the region from 2400 to 1800 cm^{-1}.

Table 1 gives characteristic Raman spectral regions for α-helix, β-sheet, β-turn, and random coil secondary structures (Tu, 1986). According to Table 1, the Raman spectrum of oxytocin (Figure 10) indicates, not surprisingly, a random coil secondary structure

Secondary Structure	Amide I (cm^{-1})	Amide III (cm^{-1})
α-helix	1645–1658	1260–1305
β-sheet	1665–1680	1230–1245
β-turn	1663–1678	1258–1300
Random coil	1660–1665	1242–1255

Table 1 Characteristic Amide Bands in Raman Spectra of Proteins (Tu, 1986).

Aromatic and Heteroaromatic Side Chains

Tyrosine is the only aromatic side chain in oxytocin. Apart from the C–H stretching vibrations around 3050 cm^{-1}, some ring vibrations are marked in Figure 10. The relative intensity of the doublet 850/830 cm^{-1} (Figure 11) shows whether the hydrogen bonding of tyrosine is *buried* or *exposed* (Tu, 1986). The spectrum of oxytocin (Figure 10) reveals that because the band at 850 cm^{-1} is stronger than the one at 830 cm^{-1}, tyrosine is exposed.

Different aromatic side chains show characteristic Raman bands (Tu, 1986). A very significant strong band at 1000 cm^{-1} and a weaker band at 1025 cm^{-1} characterize phenylalanine.

Sulfur Bridges

The S–S stretching vibration is strong in Raman spectra, because the many easily polarized electrons result in a large change in polarizability, $(d\alpha/dR_{S-S})_{R=R_0}$, during the vibration. Due to the large masses involved, the S–S stretching vibration is observed in the region 500–550 cm^{-1}. The exact wavenumber depends on the conformation of the C–S–S–C bridge (Tu, 1986). Newman projections for bonds numbered 1, 2, and 3 in Figure 12 can be either trans or gauche. The conformation around the S–S bond (no. 2) is always gauche. The combinations ggg, tgg, and tgt give rise to different S–S stretching wavenumbers, as shown in Figure 12. Small deviations from the numbers given in Figure 12 occur, but still it is possible to distinguish between the different conformers. Oxytocin in Figure 10 has a S–S stretching vibration at 512 cm^{-1}, and the conformation is ggg for the sulfur bridge.

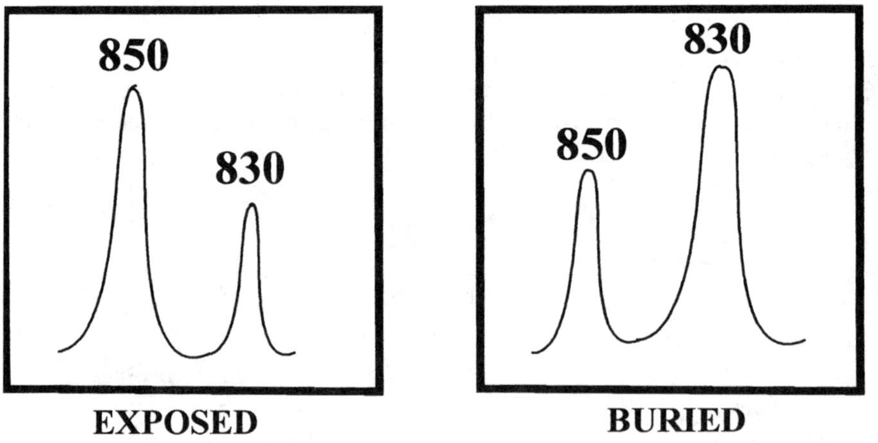

Figure 11 Relative intensities of the doublet for exposed and buried tyrosine.

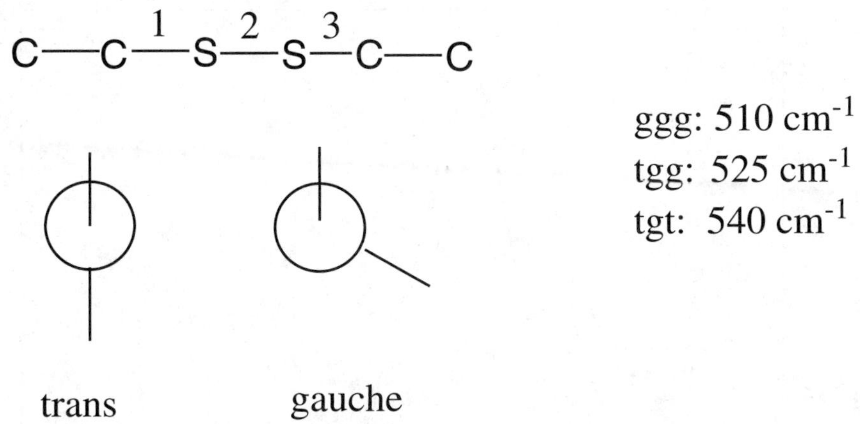

Figure 12 The relation between sulfur bridge conformation (Newman projection) and S–S stretching wavenumbers. For bond nos. 1 and 3, the C–S bonds are either trans (t) or gauche (g). For bond no. 2, the S–S bond is always gauche. A trans conformation corresponds to an angle of 180 degrees between the two bonds in the Newman projection and gauche to all other angles. The symbols ggg, tgg, and tgt correspond to the Newman projection for the three bands numbered consecutively.

Characterization of Proteins by Raman Spectroscopy

The following case studies will show how Raman spectroscopy is used to characterize proteins. The examples will also illustrate some of the problems and limitations in Raman studies of proteins.

Polyglycine

Polyglycine exists in a β-sheet–like structure, I (Lotz, 1974; Moore and Krimm, 1976), and a 3_1-helix structure, II (Crick and Rich, 1955; Dwivedi and Krimm, 1982). It was not stated on the sample from Sigma used in Figure 5 whether it was one of the two structures or a mixture. Figure 13 shows Raman spectra in the amide I region of the two different polyglycine secondary structures I and II together with the sample from Sigma (Figure 5). A comparison of the spectra in Figure 13 reveals that the polyglycine sample from Sigma has the helical structure, II.

Human Growth Hormone

Figure 14 shows the NIR-FT-Raman spectrum of freeze-dried human growth hormone. This protein contains 191 amino acid residues and shows a predominantly α-helical secondary structure (Vos et al., 1992). No limitation regarding the number of amino acid residues exists in

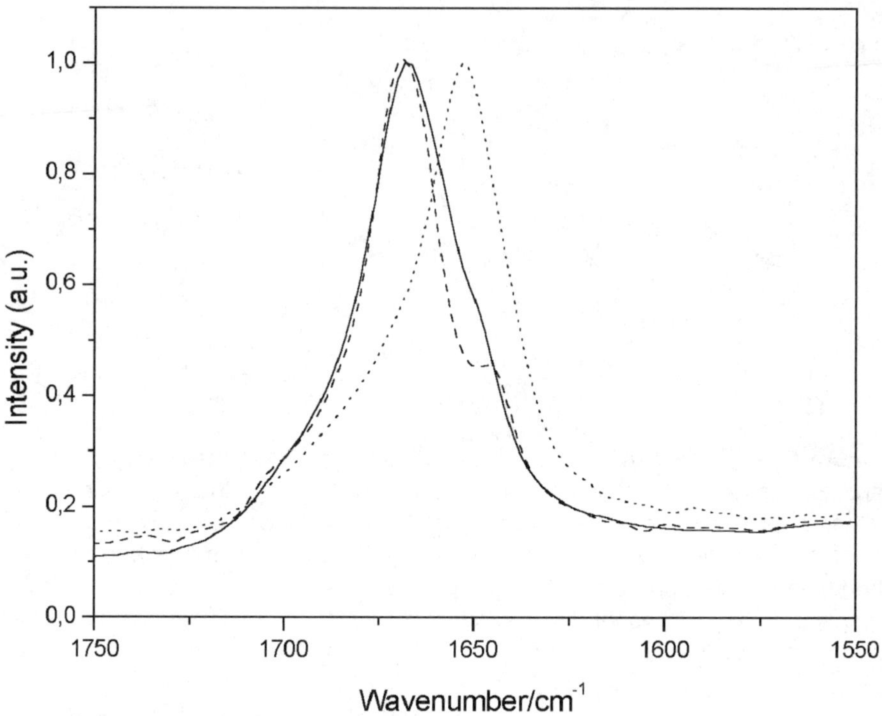

Figure 13 NIR-FT-Raman spectra (1064 nm excitation, ~300 mW, 500 scans) of polyglycine from Sigma (full curve), polyglycine I (broken curve), and polyglycine II (dotted curve).

Raman spectroscopy, and the nice spectrum in Figure 14 also illustrates that spectra of freeze-dried samples are easily obtained. Aromatic side chains show strong Raman bands (Tu, 1986). This is in accordance with the expected large change in polarizability. The two bands at 1000 cm^{-1} (strong) and 1025 cm^{-1} (medium) are characteristic for phenylalanine. The doublet at 850/830 cm^{-1} is from tyrosine. With more tyrosines present, it is still possible to calculate the ratio between buried and exposed from the relative intensities of the doublet (Tu, 1986).

The inset in Figure 14 shows the region 1750–1550 cm^{-1}. Curve fitting was performed (ORIGIN, 1999) using Lorentzian line shapes. The three bands in the region from 1620 to 1580 cm^{-1} are assigned to (hetero)aromatic ring-stretching vibrations. In the amide I region, the very intense band with a maximum at 1654 cm^{-1} shows the dominance of the α-helix secondary structure (Table 1). The band around 1680 cm^{-1} was also seen in the Raman spectrum of frog tropomyosin with a predominanting α-helix content (Nielsen et al., 2000), so it is more uncertain whether this band is assigned to another secondary structure. The amide III region is more complex, and no curve fitting is presented.

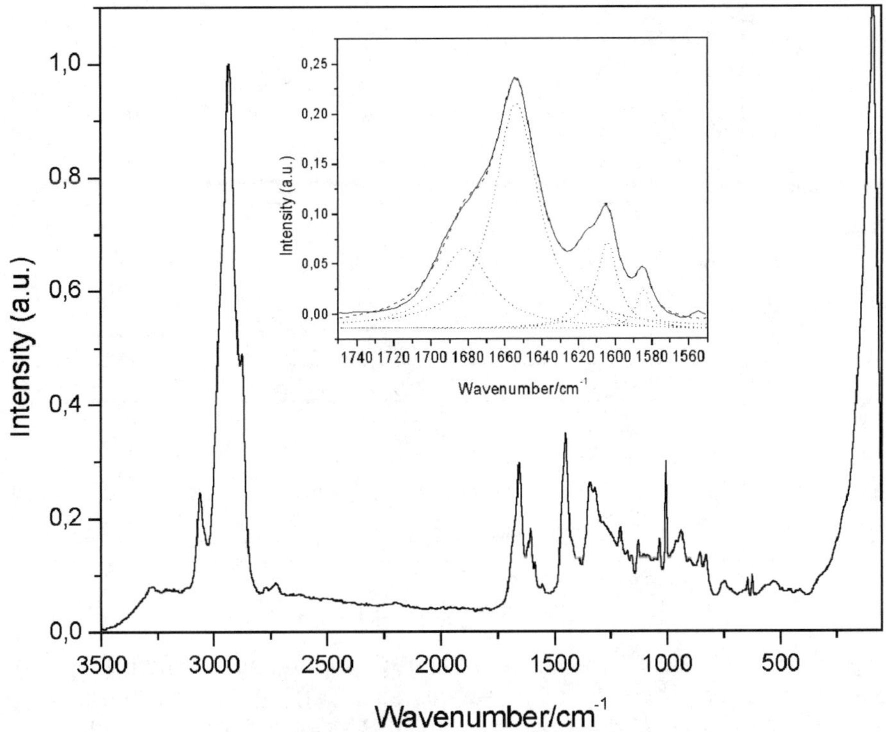

Figure 14 NIR-FT-Raman spectrum of solid human growth hormone (1064-nm excitation, ~125 mW, 500 scans). The inset shows the amide I region: The full curve is the experimental spectrum, the broken curve is the fitted spectrum, and the dotted curves are the individual bands (Lorentzians) found by the curve fitting. Microcal ORIGIN was used for the curve fitting procedure.

At 1064 nm, a laser power of 125 mW was used to obtain the spectrum shown in Figure 14. At higher laser powers, the sample decomposed. The sample was actually burned by the laser. This is a general problem in measuring the Raman spectra of proteins. No general rules can be applied, but it is important to start with very low laser power and record several spectra while increasing the laser power slightly. Although the S/N ratio is poor at low laser powers, it can be checked by comparing spectra recorded with different laser powers, that the laser causes no damage to the sample. Our experience is that the destruction of the sample happens very rapidly when a certain threshold power is reached. Up to this value, no heating and no destruction occur. In order to increase the threshold laser power, it can sometimes help to press the sample as tightly as possible into the sample holder.

Human Hemoglobin

The NIR-FT-Raman spectrum of human hemoglobin is shown in Figure 15A. This spectrum is poor because it was only possible to use a

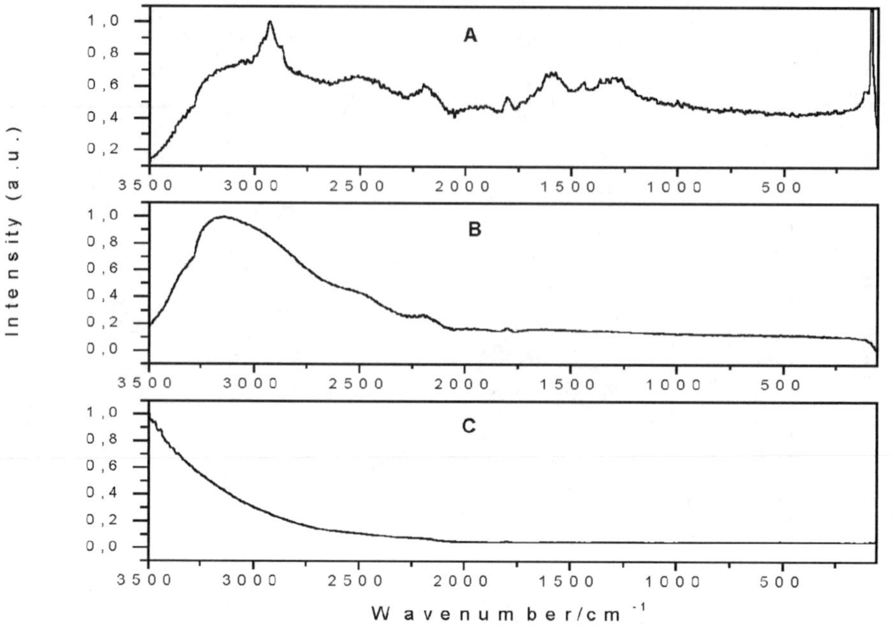

Figure 15 (A) NIR-FT-Raman spectrum of solid human hemoglobin, 500 scans). (B) Spectrum of the same sample as in (A), but the laser power was increased to ~125 mW. (C) The spectrum in (B) is white-light corrected.

laser power of less than 5 mW before the sample was destroyed. Figure 15B shows a spectrum of the decomposed sample recorded with a laser power of 125 mW. A broad band appears with a maximum around 3000 cm^{-1}. This band is due to heating of the sample. A white-light correction for instrumental response was performed in Figure 15C. The curve with a continually decreasing intensity from higher to lower Raman shifts is due to emission from the hot, "burned-off" sample. In general, it is difficult to obtain Raman spectra of strongly colored proteins. Even with excitation in the NIR region at 1064 nm, the molecules absorb the laser light due to electronic transitions (Figure 4), and in many cases the protein will be locally heated and destroyed.

DesB30 Human Insulin

Insulin is a protein containing an A- and a B-chain with 21 and 30 amino acid residues, respectively. Three sulfur bridges are present. DesB30 is an insulin derivative with residue no. 30 (threonine) missing in the B-chain. The NIR-FT-Raman spectrum of DesB30 human insulin is shown in Figure 16. In acidic solutions or in aqueous solutions containing ethanol, DesB30 insulin precipitates in a fibril-like form (Refstrup et al., 2002). The inset of 1750 to 1550 cm^{-1} in Figure 16 shows

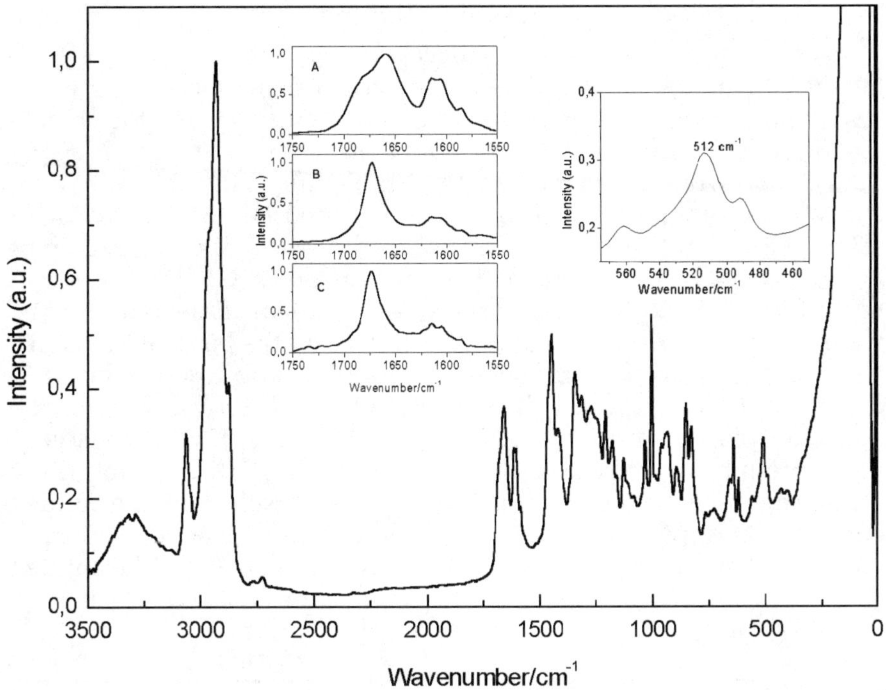

Figure 16 NIR-FT-Raman spectrum (1064 nm, 300 mW, 500 scans) of solid DesB30 human insulin. The inset of 1750 to 1550 cm^{-1} shows in (A) the spectrum of DesB30 insulin, and in (B) and (C) precipitates formed in ethanolic and acidic aqueous solution, respectively. The inset of 575 to 450 cm^{-1} shows the S–S stretching band at 512 cm^{-1}. The figure is reproduced from *Asian Journal of Physics* 2002; 11: 1–8 with permission from ANITA Publications, Ghaziabad, India.

the amide I region for solid DesB30 insulin (Figure 16A) and precipitates formed in (i) an aqueous solution at pH value 1.75 (Figure 16C) and (ii) a 1:1 (w/w) solution of water and 96% ethanol (Figure 16B). The concentration of DesB30 in the solutions prior to precipitation was 2.5% (w/w). The band found at 1655 cm^{-1} for DesB30 insulin in Figure 16A indicates the presence of an α-helix secondary structure (Table 1). In Raman spectra of the precipitates (Figures 16B and 16C), a band emerges at 1673 cm^{-1}, showing the presence of a β-sheet structure (Table 1). The conformation of the sulfur bridges did not seem to change, because the pattern of the S–S stretching vibration, also shown as an inset in Figure 16, was the same for all three samples. The relative intensities of the tyrosine doublet changed slightly, indicating a change from exposed to buried for one of the four tyrosine residues (Refstrup et al., 2002).

Aqueous Solutions

Lysozyme

Figure 17A shows the NIR-FT-Raman spectrum of chicken egg white lysozyme in the solid state. Even in this survey spectrum, the characteristic protein bands mentioned above are easily seen. The NIR-FT-Raman spectrum of water is given in Figure 17B. No white-light corrections were performed. Usually, two intense bands with Raman shifts at 3410 and 3225 cm^{-1} are observed in the stretching region of liquid water (Walrafen, 1964). The rapid drop in Ge-detector sensitivity from 3250 to 3500 cm^{-1} causes the peculiar appearance of the water-stretching vibrations in Figure 17B. Only the band at 3225 cm^{-1} is seen. The weaker water-bending vibration appears around 1650 cm^{-1}. The Raman spectra of 20% (w/w) and 5% (w/w) lysozyme aqueous solutions are shown in Figures 17C and 17D, respectively. Bands from lysozyme are easily recognized in the 20% solution. Only the strongest lysozyme bands are seen in the 5% solution. The amide I band is overlapping with the water-bending band around 1650 cm^{-1}, and even this rather weak band is of the same intensity as the amide I

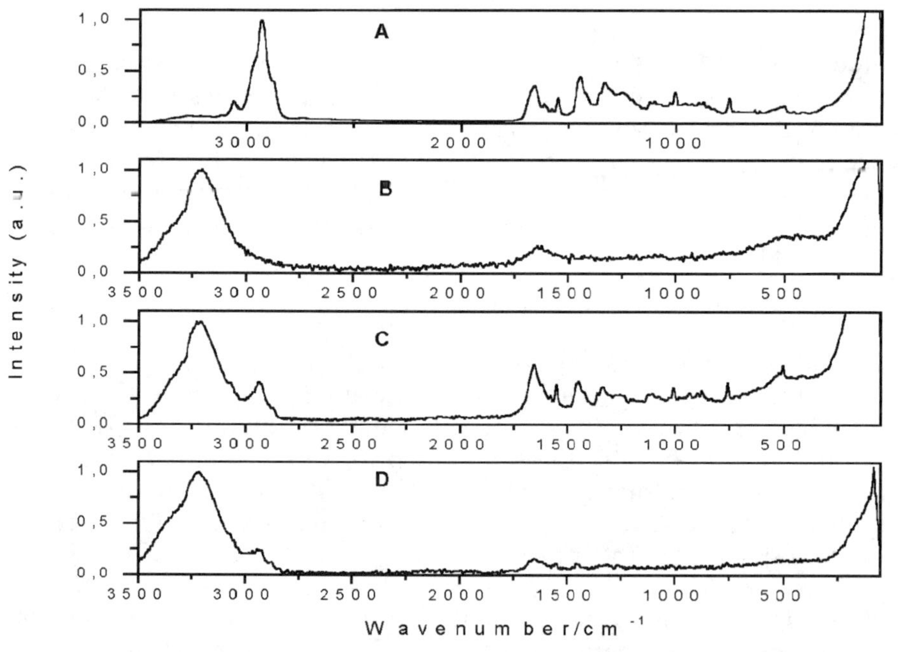

Figure 17 NIR-FT-Raman spectra (1064 nm, 300 mW, 500 scans) of (A) solid lysozyme, (B) liquid water, (C) 20% (w/w) aqueous lysozyme solution, (D) 5% (w/w) aqueous lysozyme solution. The liquid spectra were recorded by use of a windowless cell.

band. Thus, the limit for detection of a protein in aqueous solution is around 1–5% (w/w). Spectra of the solutions were all obtained by use of the windowless cell (Figure 9) to avoid a spectrum of glass in the 180-degree scattering configuration.

γ-Globulin

Figure 18A shows the NIR-FT-Raman spectrum of solid γ-globulin (lyophilized, *M* ~150,000) and a 20% (w/w) aqueous solution (B). The solution spectrum was recorded in the 180-degree scattering configuration using an ESR tube as a cell (Figure 9). A spectrum of the empty cell is included in Figure 18B. Below ~1000 cm^{-1}, the spectrum of glass from the cell is intense, and this part of the spectrum is not useful. This shows the advantage of the windowless cell as described above for lysozyme solutions. However, evaporation of the solvent is a problem for the windowless cell, and this may change the concentration during the course of recording a spectrum. A comparison from 1750 to 1500 cm^{-1} of the solid-state and solution spectrum is performed in the inset

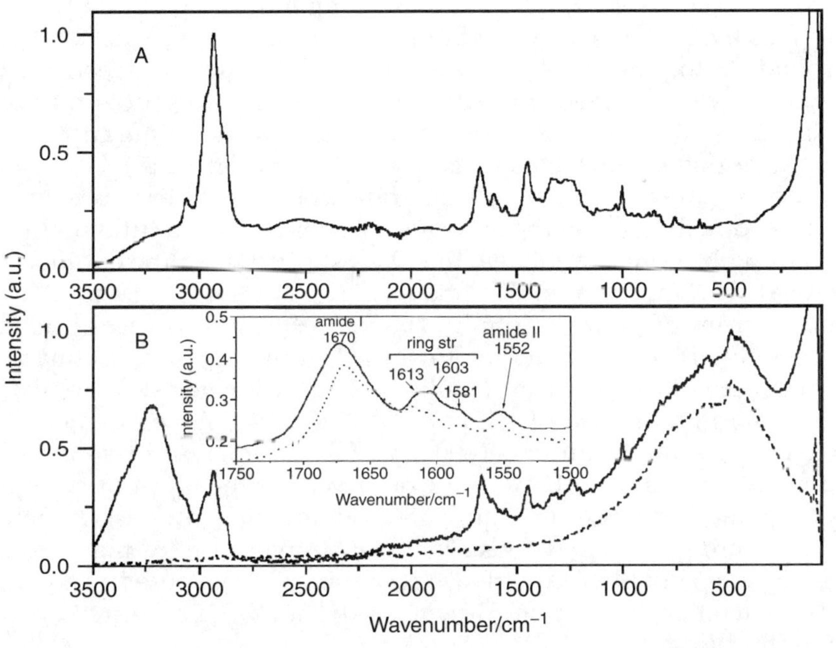

Figure 18 NIR-FT-Raman spectra (1064 nm, 280 mW, 500 scans) of γ-globulin. (A) Solid state, (B) 20% (w/w) aqueous solution (full curve). The dashed curve is recorded from the empty cell (ESR tube). The inset shows a comparison in the region from 1750 cm^{-1} to 1500 cm^{-1} for the solid state (full curve) and the aqueous solution (dotted curve). Wavenumbers for amide I, II, and ring stretchings are shown.

of Figure 18B. The water-bending vibration around 1650 cm^{-1} is too weak in intensity to seriously interfere with the Raman spectrum of γ-globulin. The amide I band in both spectra is observed around 1670 cm^{-1}, indicating a β-sheet secondary structure. The similarity of the amide bands in the two cases confirms that no change in secondary structure occurs in going from the solid state to solution. The band at 1552 cm^{-1} may be assigned to the amide II band. However, this is usually weak in Raman spectra, and the 1552 cm^{-1} band, together with the band at 1581 cm^{-1}, might be assigned to the indole ring in tryptophan (Lord and Yu, 1970). The band at 1613 cm^{-1} may be assigned to the phenyl ring of tyrosine and the band at 1603 cm^{-1} to phenylalanine (Tu, 1986). The ring-stretching vibration at 1603 cm^{-1} shows a different relative intensity in the two spectra, indicating somewhat different surroundings for the phenylalanine ring in the solid state and in solution.

Formulated Proteins

Human Normal Immunoglobulin (Gammanorm)

Gammanorm was obtained in a sealed 10-mL glass ampoule. One milliliter aqueous solution contains 165 mg immunoglobulin human normal, 20 mg glycine, and sodium chloride and sodium acetate corresponding to 2.5 mg of sodium. The NIR-FT-Raman spectrum in Figure 19 was obtained directly from the formulated protein in the sealed glass ampoule. A 180-degree scattering configuration was used as applied in the spectrum of γ-globulin (Figure 18). However, in order to minimize the spectrum of the glass, the laser was focused as far as possible inside the aqueous solution. This is a little tricky to do, but with some experience, the glass spectrum can be avoided in this way. The spectrum of Gammanorm (Figure 19) is very similar to the spectrum of gammaglobulin (Figure 18). In both cases, the intense amide I band is found at 1670 cm^{-1}, showing a predominantly β-sheet secondary structure (Table 1). This is further supported for Gammanorm by the amide III band at 1243 cm^{-1} (Table 1). The ring-stretching vibrations around 1600 cm^{-1} for Gammanorm (see inset in Figure 19) are very similar to those observed for the aqueous solution of γ-globulin (see inset in Figure 18B). Figure 19 shows assignments for the more prominent bands. The relative intensities of the tyrosine 830/850 cm^{-1} doublet show that tyrosine is in the exposed form. The S–S stretching vibration at 510 cm^{-1} indicates ggg-conformations of the sulfur bridges.

Structural changes or degradation of the protein would certainly change the spectrum. Thus, Raman spectroscopy can be used to monitor the stability of formulated proteins even in sealed glass ampoules.

Desmopressin (Minirin)

Desmopressin is a peptide with nine amino acid residues. Minirin, a 200-mg tablet, contains 0.2 mg desmopressin acetate. This corresponds to

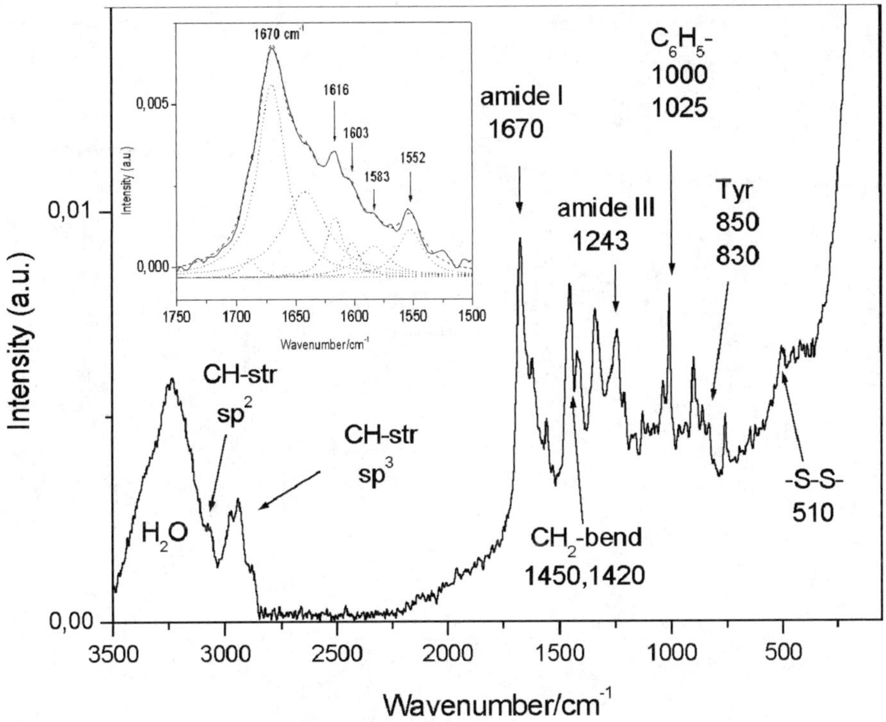

Figure 19 NIR-FT-Raman spectrum (1064 nm, 200 mW, 500 scans) of Gammanorm. The spectrum was obtained directly from the sealed ampoule. The inset shows a curve fit in the amide I, II, and ring-stretching region.

only 0.1% (w/w). The NIR-FT-Raman spectrum of Minirin is shown in Figure 20. Most of the bands in Figure 20 are from compounds used in the formulation. However, even at this very low concentration, it is possible to observe the amide I band from the peptide at 1667 cm^{-1} as seen in the inset of Figure 20. Thus, even in this solid state, it is possible to follow structural changes or degradations of the peptide in formulated form.

Conclusions

Raman spectroscopy is an efficient way to follow secondary structural changes of proteins. In addition, changes in the conformation of sulfur bridges and in the hydrogen bonding of tyrosine are easily monitored. No sample preparation is necessary, and spectra can be obtained directly from crystals or amorphous powders (e.g., freeze-dried samples) and solutions. The water spectrum is rather weak, and aqueous solutions in glass cells and even sealed ampoules can be studied. Raman spectroscopy should be a valuable tool in future studies of protein formulation. Raman spectroscopy was recently used in studies of protein formulation in glycerol (Caliskan et al., 2003) and propellants (Quinn et al., 1999).

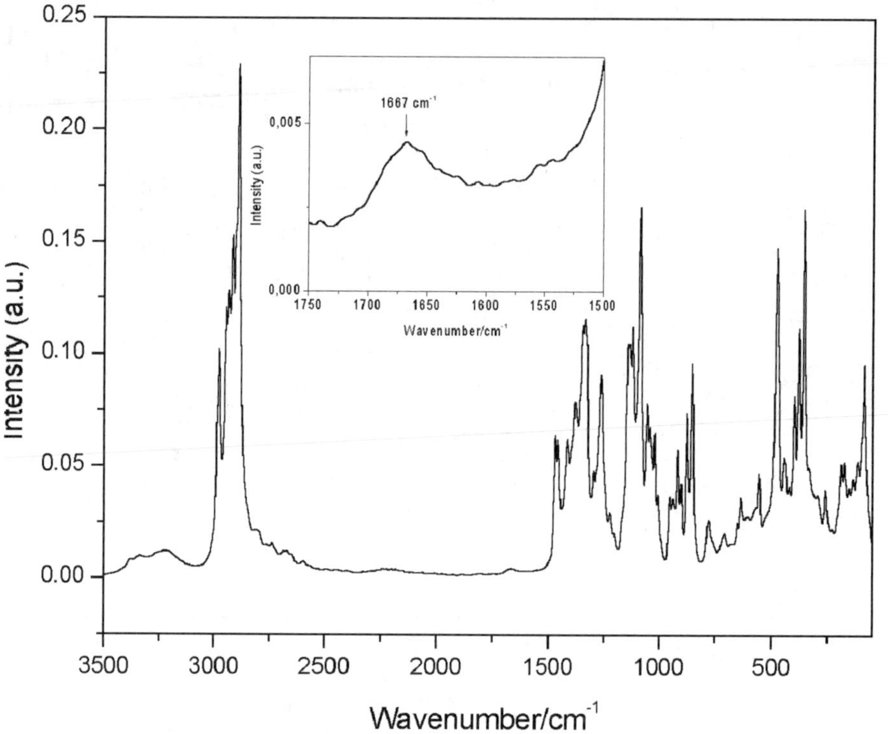

Figure 20 NIR-FT-Raman spectrum (1064 nm, 280 mW, 500 scans) of Minirin, 0.2 mg. The inset shows the amide I region at a higher gain.

Raman optical activity (ROA) measures the difference between spectra obtained for right and left circularly polarized laser light, and the ROA spectra are sensitive to protein secondary structure (Barron et al., 2002). In the future, ROA might also be used for studies of formulated proteins.

The intensity of Raman scattering is rather low, and concentrations in the percent (w/w) region are necessary for solutions. Use of an excitation wavelength in the UV region (Figure 1) can increase the intensity enormously, and protein structures in very dilute solutions can be studied (Asher et al., 2001). Laser wavelengths corresponding to the absorption of electronic transitions in the peptide group are used. According to Figure 4, this should give a fluorescence spectrum similar to that seen in Figure 5 with Vis excitation at 514.5 nm. However, with sufficiently high laser photon energy, the excitation occurs at an even higher energy than shown for fluorescence in Figure 4. In this case, a so-called resonance Raman spectrum appears, and a gain in intensity of a factor of 10^3 can occur for the amide I and III vibrations. Thus, UV-resonance Raman spectroscopy will be a very interesting technique in future studies of protein secondary structure in dilute solutions (Asher et al., 2001). In addition, this might be a profitable method for studying formulated proteins at very low concentrations.

Another very sensitive technique is surface-enhanced Raman spectroscopy (SERS). In biomedical spectroscopy, this is a promising technique that allows detection at the single-molecule level (Kneipp et al., 1999). Thomas (1999) presents novel Raman techniques for the study of protein nucleic acid interactions. However, at present, these techniques are of limited value for protein formulation studies.

Abbreviations

Å	Angstrom
a.u.	Arbitrary units
CD	Circular dichroism
CCD	Charge-coupled device
ESR	Electron spin resonance
FT	Fourier transform
IR	Infrared
IUPAC	International Union of Pure and Applied Chemistry
K	Absolute temperature (Kelvin)
Laser	Light amplification by stimulated emission of radiation
m	Meter
M	Molecular mass
MW	Microwave
NIR	Near infrared
nm	Nanometer
NMF	*N*-Methylformamide
NMR	Nuclear magnetic resonance
PCA	Principal component analysis
PMT	Photomultiplier tube
ROA	Raman optical activity
SERS	Surface-enhanced Raman spectroscopy
UV	Ultraviolet
Vis	Visible

Acknowledgments

Dr. M. Baron, LADIR, Thiais, France, is thanked for giving me a sample of polyglycine I and II. I thank Professor J. Flink and Dr. Susanne Bang, NOVO-NORDISK, Copenhagen, Denmark, for a fruitful collaboration with my previous graduate students C. Johansson and P. Refstrup regarding studies of human growth hormone and DesB30 insulin. I am grateful to Dr. C. Pahl, Octapharma, Holte, Denmark, for allowing me to show the spectrum of Gammanorm and to Ferring AB, Malmø, Sweden, for supplying the Minirin tablets. My technicians I. Blangsted and L. Ryelund are thanked for recording the spectra and for their help with many practical things during the preparation of this chapter. I am grateful to my colleague Associate Professor M.S. Johnson for correcting the English language. The Danish Natural Research Council is thanked for general funding during the period the Raman spectra were recorded (grant no. 51-00-0312).

References

1. Asher SA, Ianoul A, Mix G, Boyden MN, Karnoup A, Diem M, and Schweitzer-Stenner R. Dihedral ψ Angle Dependence of the Amide III Vibration: A Uniquely Sensitive UV Resonance Raman Secondary Stuctural Probe. *J Am Chem Soc* 2001; 123:11775–11781
2. Baldwin KJ, Batchfelder DN, and Webster S. Raman Microscopy: Confocal and Scanning Near-Field. In: Lewis IR. and Edwards HGM. *Handbook of Raman Spectroscopy. From the Research Laboratory to the Process Line.* New York: Basel: Marcel Dekker; 2001:145–190
3. Barron LD, Blanch EW, and Hecht L. Unfolded Proteins Studied by Raman Optical Activity. *Adv Protein Chem* 2002; 62:51–90
4. Becker ED, Durig JR, Harris WC, and Rosasco GJ. Presentation of Raman Spectra in Data Collections. *Pure Appl Chem* 1981; 53:1879–1885
5. Brooker MH, Nielsen OF, and Christensen D. Windowless Cell for Fourier Transform Raman Spectra of Liquids and Aqueous Solutions. *J Raman Spectrosc* 1995; 26:331–334
6. Caliskan G, Kisliuk A, Tsai AM, Soles CL, and Sokolov AP. Protein Dynamics in Viscous Solvents. *J Chem Phys* 2003; 118:4230–4236
7. Carter EA, and Edwards HGM. Biological Applications of Raman Spectroscopy. In: Gremlich H-U, and Yan B. *Infrared and Raman Spectroscopy of Biological Materials.* New York, Basel: Marcel Dekker; 2001:421–475
8. Chase DB. Fourier Transform Raman Spectroscopy. *J Am Chem Soc* 1986; 108:7485–7488
9. Colaianni SEM, and Nielsen OF. Low-Frequency Raman Spectroscopy. *J Mol Struct* 1995; 347:267–284
10. Crick FHC, and Rich A. Structure of Polyglycine II. *Nature* 1955; 176:780–781
11. Delhaye M, and Dhamelincourt P. Raman Microprobe and Microscope with Laser Excitation, *J Raman Spectrosc* 1975; 3:33–43

12. Dwivedi AM, and Krimm S. Vibrational Analysis of Peptides, Polypeptides, and Proteins. XV. Crystalline Polyglycine II. *Biopolymers* 1982; 21:2377–2397

13. Hendra P, Jones C, and Warnes G. *Fourier Transform Raman Spectroscopy. Instrumentation and Chemical Applications*. Chichester, England: Ellis Horwood Limited; 1991. 311 p.

14. Hirschfeld T, and Chase B. FT-Raman Spectroscopy: Development and Justification. *Appl Spectrosc* 1986; 40:133–137

15. Kneipp K, Kneipp H, Itzkan I, Dasari RR, and Feld MS. Surface-enhanced Raman Scattering: A New Tool for Biomedical Spectroscopy. *Curr Sci* 1999; 77:915–924

16. Lewis IR, and Edwards HGM, eds. *Handbook of Raman Spectroscopy. From the Research Laboratory to the Process Line.* New York, Basel: Marcel Dekker; 2001. 1054p.

17. Lord RC, and Yu N-T. Laser-Excited Raman Spectroscopy of Biomolecules. *J Mol Biol* 1970; 50:509–524

18. Lotz B. Crystal Structure of Polyglycine I. *J Mol Biol* 1974; 87:169–180

19. Mann CK, and Vickers TJ. The Quest for Accuracy in Raman Spectra. In: Lewis IR, and Edwards HGM. *Handbook of Raman Spectroscopy. From the Research Laboratory to the Process Line.* New York, Basel: Marcel Dekker; 2001:251–274

20. Moore WH, and Krimm S. Vibrational Analysis of Peptides, Polypeptides, and Proteins. Polyglycine I. *Biopolymers* 1976; 15:2439–2464

21. Myers AB. Excited Electronic State Properties from Ground-State Resonance Raman Intensities. In: Myers AB, and Rizzo TR. *Techniques of Chemistry, Laser Techniques in Chemistry,* Vol. 23. New York: John Wiley & Sons; 1995:325–384

22. Nielsen OF, Johansson C, Christensen DH, Hvidt S, Flink J, Hansen SH, and Poulsen F. Collective Vibrational Effects in Hydrogen Bonded Liquid Amides and Proteins Studied by Isotopic Substitution. *J Mol Struct* 2000; 552:71–80

23. *ORIGIN, version 6*. Northampton, MA: Microcal Software; 1999.

24. *OPUS Spectroscopic Software, version 2.0*. Karlsruhe, Germany: BRUKER Analytische Messtechnique GMBH; 1995:178–182

25. Quinn EA, Forbes RT, Williams AC, Oliver MJ, Mckenzie L, and Purewal TS. Protein Conformational Stability in the Hydrofluoroalkane Propellants Tetrafluoroethane and Heptafluoropropane Analyzed by Fourier Transform Raman Spectroscopy. *Int J Pharm* 1999; 186:31–41

26. Raman CV, and Krishnan KS. A New Type of Secondary Radiation. *Nature* 1928; 121:501

27. Refstrup P, Nielsen OF, and Bang S. Biological Raman Spectroscopy. *Asian J Phys* 2002; 11:1–8

28. Schrader B, ed. *Infrared and Raman Spectroscopy, Methods and Applications.* Weinheim, Germany: VCH Verlagsgesellschaft; New York: VCH Publishers; 1995. 787p.

29. Slater JB, Tedesco JM, Fairchild RC, and Lewis IR. Raman Spectrometry and Its Adaption to the Industrial Environment. In: Lewis IR, and Edwards HGM. *Handbook of Raman Spectroscopy. From the Research Laboratory to the Process Line.* New York, Basel: Marcel Dekker; 2001: 41–144

30. Smith BC. *Fundamentals of Fourier Transform Infrared Spectroscopy*. Boca Raton, FL: CRC Press; 1996. 202p.

31. Thomas GJ Jr. Raman Spectroscopy of Protein and Nucleic Acid Assemblies. *Annu Rev Biophys Biomol Struct* 1999; 28:1–27

32. Tu AT. Peptide Backbone Conformation and Microenvironment of Protein Side Chains. In: Clark RJH, and Hester RE. *Spectroscopy of Biological Systems.* Chichester, England: J. Wiley & Sons; 1986; 13:47–112

33. Vos de AM, Ultsch M, and Kossiakoff AA. Human Growth Hormone and Extracellular Domain of Its Receptor: Crystal Structure of the Complex. *Science* 1992; 255:306–312

34. Walrafen GE. Raman Spectral Studies of Water Structure. *J Chem Phys* 1964; 40:3249–3256

6

Nuclear Magnetic Resonance Spectroscopy

DAVID WISHART

Department of Biological Sciences, Computing Science and Faculty of Pharmaceutical Sciences, University of Alberta, Edmonton, AB, Canada

Introduction

During the past twenty years, nuclear magnetic resonance (NMR) spectroscopy has emerged as one of the most powerful physical methods to study proteins. It combines the strengths of such methods as fluorescence, circular dichroism, and calorimetry for characterizing proteins in solution with the power of x-ray crystallography to perform these characterizations atom by atom. Besides x-ray crystallography, NMR is the only other method available that allows the three-dimensional structure of proteins to be determined to atomic resolution (Wuthrich, 1995). Furthermore, NMR is the only known method that allows protein structures to be determined in solution (i.e., near physiological conditions). However, NMR spectroscopy is not "just another" structure determination technique. Like fluorescence, NMR can provide exquisitely detailed information about protein dynamics spanning an astonishing temporal range from picoseconds to hours (Ishima and Torchia, 2000; Palmer, 2001). NMR can also be used to quantify protein secondary structure and secondary structure changes just like circular dichroism (Wishart and Sykes, 1994a). It can also provide detailed information on a range of protein kinetic parameters including diffusion constants, binding constants, hydrogen exchange rates, and kinetic exchange rates, as well as ligand on-rates and off-rates (Diercks et al., 2001). Furthermore, NMR can be used to monitor protein-ligand interactions and to map these interactions atom by atom (Shuker et al., 1996; Zuiderweg, 2002).

NMR spectroscopy offers protein chemists and physical chemists the possibility to see and quantify an astonishing range of structurally and kinetically useful measurements. What's more, these measurements can be made either globally or locally (residue-by-residue) with remarkable precision. It is because of this enormous range of measurement possibilities that NMR has become increasingly popular in the pharmaceutical and biopharmaceutical industries. For instance, NMR is now routinely used to characterize new protein drug targets, to screen drug libraries, and to assist in rational drug design strategies of new drug leads (Chen and Shapiro, 1999; Diercks et al., 2001). NMR may also be used to monitor protein integrity and to characterize pharmaceutical protein formulations (Jacoby et al., 1996). Given the growing number of applications being discovered for NMR, it is likely that NMR will find its way into many other aspects of drug discovery and development.

In this chapter, we will explore how NMR spectroscopy (particularly solution-state NMR) can be used in the structural analysis of proteins and how these approaches can be applied to study pharmaceutically interesting proteins or protein systems. To facilitate this discussion, we will begin with a review of the general principles of NMR along with a presentation on the key spectral parameters in NMR in terms of what they mean for protein structure. This will be followed by an overview of existing and emerging solution-state NMR techniques for characterizing proteins. Finally, a discussion of the applications of solution-state NMR in pharmaceutical science along with several examples or case studies will be given. Clearly, it is impossible to give a complete summary covering all aspects of protein NMR in a single book chapter. For more detailed or specialized discussions on protein NMR, the reader is invited to check out the following references (Wuthrich, 1986; Cavanagh et al., 1996; Guntert, 1998).

NMR Basics

NMR spectroscopy is a spectroscopic technique that measures the absorbance of radio frequency (RF) radiation that occurs when certain kinds of atomic nuclei (^{1}H, ^{13}C, ^{15}N) are placed under strong magnetic fields. NMR can be performed either on bulk solids (solid-state NMR) or liquid samples (solution-state NMR). To perform a solution-state NMR experiment, a liquid sample (~0.5 mL) containing a pure compound of interest must be placed inside a glass tube that is then placed inside a superconducting magnet. Inside the magnet there is a small saddle-shaped metal coil that surrounds the sample tube. The coil (an RF transceiver) is responsible for sending a microsecond burst of RF radiation to the sample and then collecting the weak, oscillating RF signal that emanates from the sample for several seconds after this initial burst. The resulting RF signal (called a free induction decay, or FID) is amplified, digitally sampled, and then Fourier transformed (FT) to convert the time-dependent oscillations to the frequency domain (Ernst, 1992). The result is an NMR spectrum (Figure 1A), which

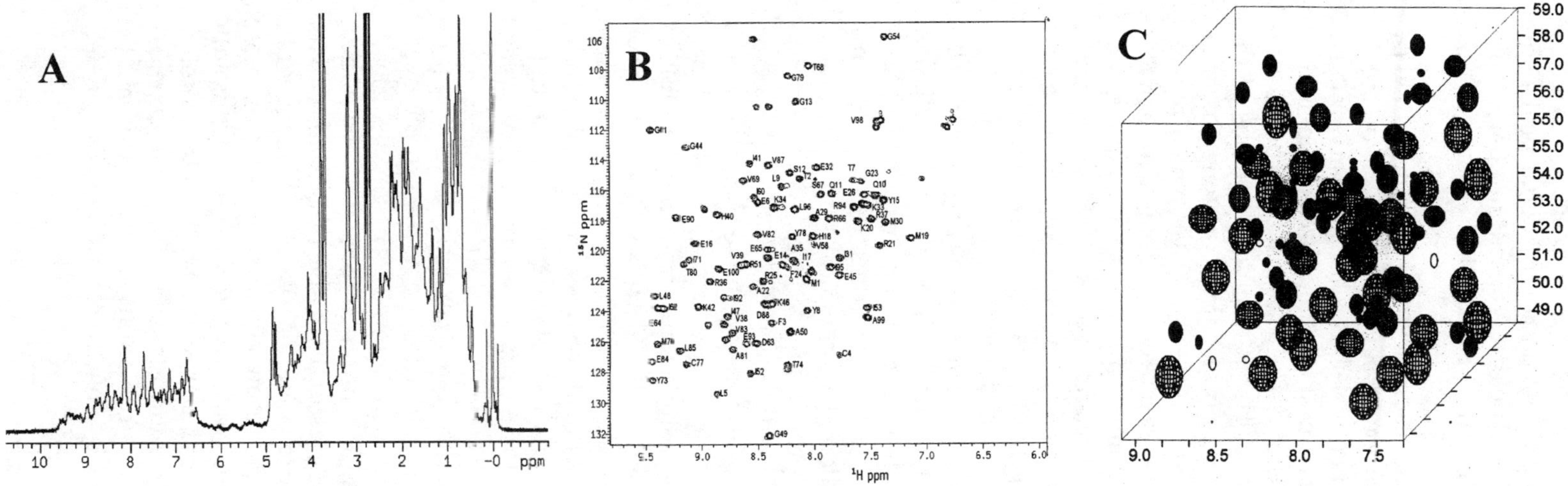

Figure 1 An illustration of how 1-D, 2-D, and 3-D NMR spectra are normally viewed. The amount of data collected and the difficulty of the experiment increase as the number of dimensions is increased.

is characterized by a series of sharp Lorentzian peaks spread over a range of frequencies. These peaks represent resonances or RF absorption frequencies for individual or individual groups of atoms or, more correctly, nuclei.

The position of these resonance peaks is determined by the chemical structure of the compound being studied. Resonance positions or frequencies are called chemical shifts. The intensity of these resonance peaks corresponds to the number of atoms or nuclei in the compound with identical chemical shifts. The width of these peaks corresponds to the speed of the local and global tumbling motions of the molecule, with faster moving atoms or molecules having narrower peaks. Many peaks also have a fine structure (splitting), which represents a coupling or direct linkage between two or more NMR-active nuclei. This splitting is called spin-spin coupling, scalar coupling, or J-coupling. There are other, less obvious, features to NMR spectra that can also be measured or observed such as dipolar couplings, relaxation times, nuclear Overhauser effects, (NOEs), cross-correlated relaxation rates, chemical exchange rates, and chemical shift anisotropy (CSA).

These additional NMR parameters may be measured by analyzing standard one-dimensional (1-D) NMR spectra (Figure 1A) or, more frequently, they are measured using two-dimensional (2-D) or three-dimensional (3-D) NMR methods (Figures 1B and 1C). In 2-D NMR, resonances are spread along two chemical shift axes, much as 2-D gel electrophoresis spreads protein peaks over two axes (molecule weight and isoelectric point). Two-dimensional NMR allows hidden peaks or resonances to be revealed by correlating the chemical shift of one atom with the chemical shift of a nearby atom (Wuthrich, 1986, 1995). The same principle of chemical shift correlation along multiple axes applies to 3-D and even four-dimensional (4-D) NMR. In the case of 3D NMR, the peaks are not displayed in a plane, but in a cube (Figure 1C). By spreading peaks over two or three axes, it is possible to individually identify and enumerate hundreds of resonances at a time. In this way, it is possible to use NMR to analyze very large and complex molecules such as proteins.

In a multidimensional NMR experiment, several (sometimes dozens of) short RF pulses are directed at the sample over the course of a few dozen milliseconds (Figure 2). These pulses lead to the exchange of energy or magnetization between coupled or nearby nuclei (Edison et al., 1994). Interestingly, if the pulses are delivered in the right way, this energy exchange, or *coherence transfer*, will lead to an NMR spectrum where the resonance intensities seem to oscillate, depending on how long one waits during the so-called evolution period. By acquiring a series of 1-D spectra with incrementally increased delay times during the evolution period, these frequencies can be determined (via Fourier transformation) and plotted in a 2-D spectrum. Different types of 2-D or 3-D NMR experiments can be designed and

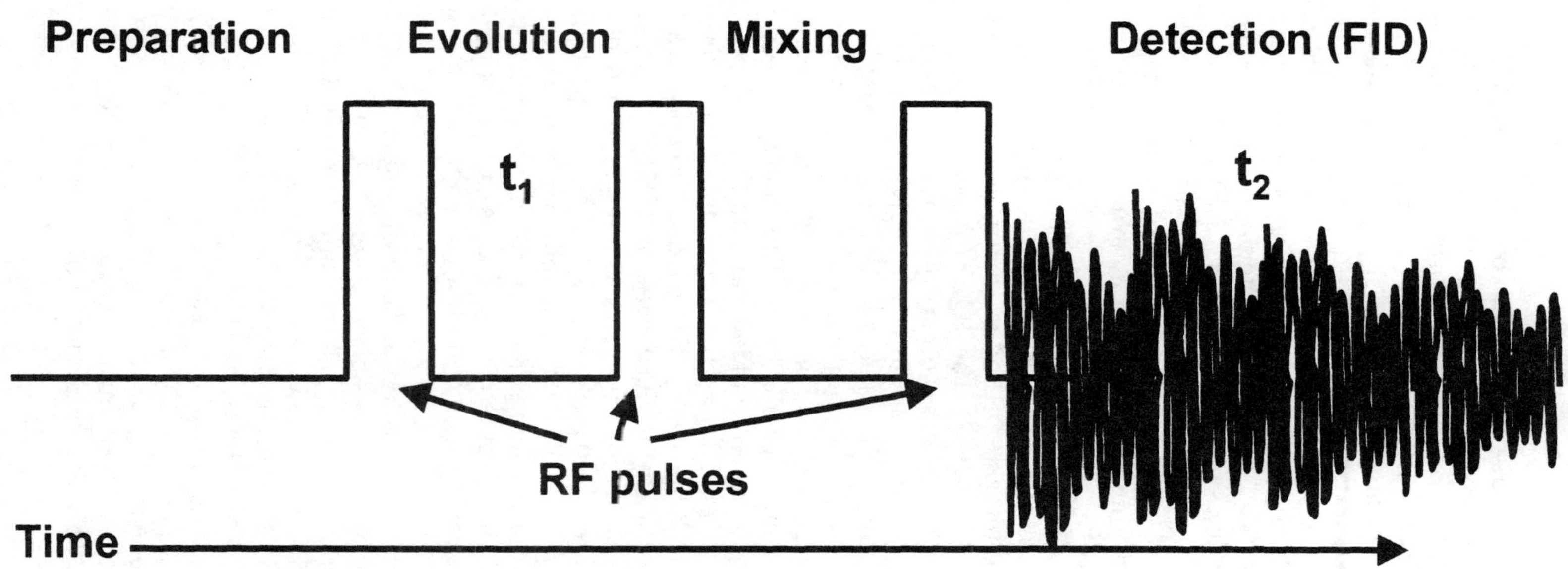

Figure 2 A schematic illustration of the multiple RF pulses and delays used in a 2-D (NOESY) experiment. Typically, the preparation period requires 2 seconds, the RF pulses are about 7 microseconds, the increments during the t_1 period are 100 microseconds (usually 512 increments are collected), the evolution period is about 75 milliseconds, and the detection period is about 300 milliseconds. Depending on the number of scans required to get sufficient signal to noise, the entire NOESY experiment may take from twelve to thirty-six hours.

implemented to detect or display specific combinations of chemical shifts, coupling constants, relaxation rates, or NOEs (Edison et al., 1994; Cavanagh et al., 1996). The COSY (correlated spectroscopy) is one of the original 2-D NMR experiments (Marion and Wuthrich, 1983). In the COSY experiment, chemical shifts between nuclei that are J-coupled (i.e., separated by two to three covalent bonds) are plotted. This results in every peak having two chemical shift values (an *X* and a *Y* value). Those that have identical *X* and *Y* values are called diagonal peaks. Those that have different *X* and *Y* values are called off-diagonal or cross peaks. Identifying COSY cross peaks (and the two atoms that gave rise to *X* shift and the *Y* shift, respectively) allows any two coupled spins to be identified by simple inspection. This can be very useful in determining the covalent connectivity of atoms within a molecule.

Another experiment called the TOCSY (total correlation spectroscopy) allows chemical shifts between nuclei that are J-coupled over a much greater number of bonds (five to six) to be plotted (Braunschweiler and Ernst, 1983). The TOCSY experiment is particularly useful in protein NMR, as it is often able to reveal the complete set of characteristic ^{1}H chemical shifts for individual amino acid residues. In other words, TOCSY spectra allow rapid identification of amino acid residue types. Another important type of 2-D NMR experiment is called the NOESY, or nuclear Overhauser enhancement spectroscopy (Kumar et al., 1980). The NOESY experiment yields a spectrum that looks almost identical to the TOCSY spectrum. However, in the NOESY experiment, there are extra peaks (often hundreds, even thousands for proteins). These extra "pseudo" peaks, or cross peaks, which show up between pairs of TOCSY peaks, correspond to interactions where nearby protons are within 5 Å of each other. The stronger the cross peak, the closer the two protons (or amino acids) are. Using NOESY data in combination with the TOCSY data, it is possible to fully assign smaller peptides and proteins (less than 10 kDa). In NMR, the term *assignment* means each set of peaks in the TOCSY spectrum can be assigned an atom name and a residue number.

When a sample is isotopically labeled with other NMR active nuclei (usually ^{13}C/^{15}N), it is possible to extend 2-D NMR to three and four dimensions by plotting ^{1}H, ^{13}C, and ^{15}N chemical shifts along three or four different axes. These heteronuclear 3-D and 4-D NMR experiments are typically named for the types of nuclei they measure (Ikura et al., 1990). Hence, an HNCA experiment (Kay et al., 1990) measures the ^{1}H amide shift of a residue on one axis ("H"), the ^{15}N amide shift on another axis ("N"), and the ^{13}Cα shift on the third axis ("CA") as seen in Figure 3. The HNCACB experiment (Wittekind and Mueller, 1993) measures the ^{1}H amide shift of a residue on one axis ("H"), the ^{15}N amide shift on another axis ("N"),

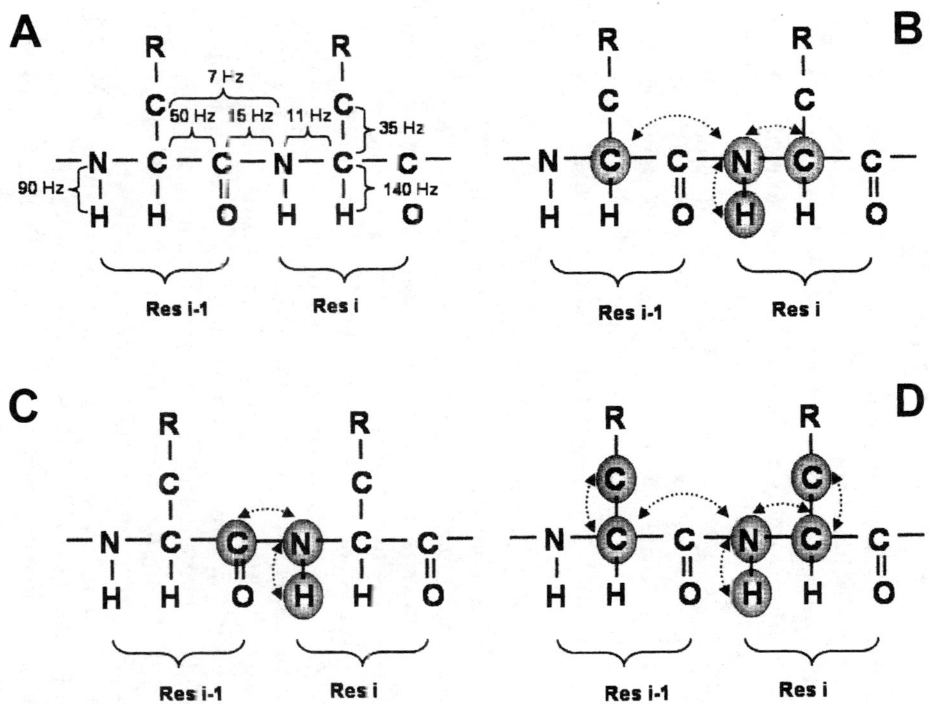

Figure 3 Heteronuclear 3-D NMR depends on the existence of modestly large one-bond, two-bond, and three-bond couplings between ^{1}H, ^{13}C, and ^{15}N nuclei. Shown in (A) is the magnitude of couplings typically seen in a dipeptide unit. In (B), the magnetization transfer route and the types of nuclei that can be detected from an HNCA experiment are shown. In part (C), the magnetization transfer route and the types of nuclei that can be detected from an HNCO experiment are shown. Finally, in part (D), the magnetization transfer route and the types of nuclei that can be detected from an HNCACB experiment are shown.

and the $^{13}C\alpha$ as well as the $^{13}C\beta$ shift on the third axis ("CACB"). The HNCACB spectrum also yields the $^{13}C\alpha$ and $^{13}C\beta$ shifts for the preceding residue (Figure 3). In the HNCO experiment, the amide ^{1}H shift and the amide ^{15}N shift for one residue are correlated (or plotted) with the ^{13}CO shift of the preceding residue (Figure 3). By identifying the linkages between amide shifts and the preceding ^{13}CO, $^{13}C\alpha$, and $^{13}C\beta$ shifts (from HNCO and HNCACB data), one can sequentially assign most of the ^{1}H, ^{13}C, and ^{15}N resonances in a protein (Ikura et al., 1990; Moseley et al., 2001).

NMR Parameters for Proteins

NMR is particularly rich in the number and type of spectroscopic observables it can offer. Chemical shifts, J-couplings, NOEs, dipolar couplings, line widths, exchange rates, cross-correlated relaxation rates, and chemical shift anisotropy (CSA) are but a few of the parameters that can be measured via NMR. However, for these NMR parameters to have some utility, they must be able to tell us something about the structure, dynamics, or character of the molecule we are studying. In this section, we will describe how many of these NMR parameters, if appropriately measured and analyzed, can provide quantitative structural or dynamic information about peptides and proteins.

Chemical Shifts

Chemical shifts are the mileposts to NMR spectroscopy. Just as roadside mileposts provide positional information about towns or villages within a country, chemical shifts provide positional information about individual NMR peaks, which in turn can be ascribed or assigned to individual atoms within a chemical structure. Unlike most other forms of spectroscopy, NMR reports its major spectroscopic values (chemical shifts) using a relative measure rather than an absolute measure. This is because chemical shifts are field dependent. As a result, all chemical shifts are measured in parts per million (ppm) as opposed to absolute frequencies (Hz). This convention provides a universal means by which NMR spectroscopists can report or share information about NMR spectra regardless of the magnetic field strength with which they are working. In biomolecular NMR, chemical shifts are normally reported to a chemical shift standard called DSS (2,2-dimethyl-2-silapentane-5-sulfonic acid), which defines the 0 ppm value for ^{1}H, ^{13}C, and, indirectly, ^{15}N nuclei (Wishart et al., 1995).

Chemical shifts are fundamentally an electronic phenomenon. They arise primarily due to the geometry or distribution of electrons surrounding the nuclei that are being observed by NMR. For a spherically symmetric atom, the chemical shift (δ) is given by Lamb's formula:

$$\delta = B_o 4\pi e^2/3mc^2 \int r\rho(r)dr \quad (1)$$

where $\rho(r)$ is the electron density at a distance r from the nucleus, B_o is the strength of the NMR's magnetic field, c is the speed of light, m is the mass of the electron, and e is its charge in cgs units. From this formula, we can see that when electrons are close to the nucleus (i.e., the nucleus is shielded), the chemical shift is large. When electrons are pushed away from the nucleus (i.e., the nucleus is deshielded), the chemical shift is small. In NMR, these changes are normally plotted in reverse so that shielded nuclei have low or upfield chemical shifts (around 1 ppm for ^1H), and deshielded nuclei have high or downfield chemical shifts (around 8 ppm for ^1H). Chemical shifts vary according to the characteristic electronic and nuclear structure of each element. Typically, ^1H chemical shifts in proteins range from –1 to 10 ppm, ^{15}N shifts range from 90 to 135 ppm, and ^{13}C shifts range from 15 to 185 ppm (Wishart and Case, 2001).

Chemical shifts provide exquisitely detailed information about the covalent structure of atoms and molecules. For instance, each of the twenty amino acids have very distinct patterns of ^1H and ^{13}C chemical shifts corresponding to their distinct covalent structures (Table 1). These patterns or so-called random coil chemical shift values for amino acid residues are routinely used in assigning protein NMR spectra. Interestingly, in proteins and other large biopolymers, chemical shifts can also provide very detailed information about noncovalent structure. Indeed, when properly analyzed, chemical shifts can be used to infer secondary structure, tertiary structure, dihedral angles, side chain orientation, solvent interactions, hydrogen bonding interactions, aromatic ring orientations, and electrostatic or ionic interactions (Case, 2000; Wishart and Case, 2001). However, to quantify or interpret these structural features, it is necessary to measure the difference between the observed chemical shifts (δ_{obs}) and their characteristic random coil chemical shifts (δ_{rc}). These secondary or so-called $\Delta\delta$ shifts are much smaller in magnitude (about 10 times less) than the intrinsic or random coil chemical shifts seen in Table 1. Nevertheless, they can easily and accurately be measured.

The sensitivity of secondary shifts to noncovalent structure is reflected in the many noncovalent factors that have been found to contribute to them. These contributions have been determined through detailed experiments, empirical studies, and quantum mechanical calculations (Wishart and Case, 2001). As a result, secondary shifts are typically written as a sum of multiple interactions:

$$\Delta\delta = \delta_{obs} - \delta_{rc} = \delta_{tor} + \delta_{ring} + \delta_{HB} + \delta_{e} + \delta_{side} + \delta_{misc} \quad (2)$$

where δ_{rc} is the random coil shift, δ_{tor} is the backbone torsional contribution, δ_{ring} is the ring current contribution, δ_{HB} is the contribution arising from hydrogen bonds, δ_e is the electric or ionic field effect, δ_{side} is the side-chain torsional contribution, and δ_{misc} includes all other contributions such as solvent, temperature, and dynamic averaging.

Amino Acid	^{1}HN	^{15}N	^{1}Hα	^{13}Cα	^{1}Hβ	^{13}Cβ	^{13}CO
Ala	8.24	123.8	4.32	52.5	1.39	19.1	177.8
Cys (red)	8.32	118.8	4.55	58.2	2.93/2.93	28.0	174.6
Cys (oxid)	8.43	118.6	4.71	55.4	3.25/2.99	41.1	174.6
Asp	8.34	120.4	4.64	54.2	2.72/2.65	41.1	176.3
Glu	8.42	120.2	4.35	56.6	2.06/1.96	29.9	176.6
Phe	8.30	120.3	4.62	57.7	3.14/3.04	39.6	175.8
Gly	8.33	108.8	3.96	45.1	—	—	174.9
His	8.42	118.2	4.73	55.0	3.29/3.16	29.0	174.1
Ile	8.00	119.9	4.17	61.1	1.87	38.8	176.4
Lys	8.29	120.4	4.32	56.2	1.84/1.75	33.1	176.6
Leu	8.16	121.8	4.34	55.1	1.62/1.62	42.4	177.6
Met	8.28	119.6	4.48	55.4	2.11/2.01	32.9	176.3
Asn	8.40	118.7	4.74	53.1	2.83/2.75	38.9	175.2
Pro	—	—	4.42	63.3	2.29/2.04	32.1	177.3
Gln	8.32	119.8	4.34	55.7	2.12/1.99	29.4	176.0
Arg	8.23	120.5	4.34	56.0	1.86/1.76	30.9	176.3
Ser	8.31	115.7	4.47	58.3	3.89/3.87	63.8	174.6
Thr	8.15	113.6	4.35	61.8	4.24	69.8	174.7
Val	8.03	119.2	4.12	62.2	2.08	32.9	176.3
Trp	8.25	121.3	4.66	57.5	3.29/3.27	29.6	176.1
Tyr	8.12	120.3	4.55	57.9	3.03/2.98	37.8	175.9

[a]Measured at 25°C, pH 5, in H_2O using DSS as a reference (direct and indirect).

Table 1 "Random Coil" Chemical Shifts for the Common Amino Acids (in ppm)[a]

Interestingly, torsion angle effects seem to dominate secondary shifts in proteins (Xu and Case, 2002; Neal et al., 2003). This has made $\Delta\delta$ particularly useful in analyzing protein structure.

Among the simplest and most useful applications of chemical shifts in protein NMR is in protein secondary structure determination (Wishart et al., 1991a; Wishart and Sykes, 1994a). These methods exploit the well-known upfield or downfield biases that certain secondary structures exert on secondary shifts. Using these chemical shift trends, it has been found that secondary structure content (%helix, %beta, %coil) can rapidly be quantified independently of any NMR assignment by integrating certain regions of 1-D ^{1}H or ^{13}C spectra or by counting peaks in certain kinds of 2-D spectra (Wishart et al., 1991b; Wishart and Sykes, 1994a). If a protein is already assigned, it is possible to determine the exact extent and location of secondary structures using techniques such as the chemical shift index (CSI) or other $\Delta\delta$ plots (Wishart et al., 1992; Wishart and Sykes, 1994b). In these methods, the secondary shifts are plotted along the sequence and then filtered to reveal patterns of upfield or downfield shifts that indicate the start and end of secondary structures. These simple methods have proved to be very accurate and are used widely in the biomolecular NMR community.

More recently, chemical shifts have been used to determine approximate backbone dihedral angles by comparing observed chemical shift values to databases of known protein chemical shifts (and measured dihedral angles) of solved protein structures. Programs such as TALOS and SHIFTOR are two examples of programs that allow many dihedral angle restraints to be determined within $\pm 15^\circ$ (Cornelescu et al., 1999; Wishart and Case, 2001). In addition to using chemical shifts to determine initial structural restraints, a number of NMR spectroscopists have used chemical shifts to refine or improve the quality of their final protein structures (Kuszewski et al., 1995). However, the practice of chemical shift refinement has not yet taken off. The recent development of several substantially improved protein chemical shift calculators (Xu and Case 2002; Neal et al., 2003) will likely open the door to much better chemical shift refinement schemes—and better quality protein structures.

The abundance of chemical shift data for proteins has led to the establishment of the BioMagResBank, or BMRB (Seavey et al., 1991). This is the central archive for all biomolecular chemical shift data. The BMRB currently contains more than 550,000 shifts from nearly 3000 peptide and protein entries. The BMRB has led the way in standardizing the reporting of chemical shifts and in developing approaches to check, verify, and validate chemical shifts in proteins. This resource has also allowed a number of groups to investigate protein chemical shifts in more detail and extract very useful methods or schemes for interpreting chemical shifts in proteins. These include approaches to identify the oxidation state of cysteines by Cβ chemical shifts (Sharma and Rajarathnam, 2000), methods to identify beta turn types and helix caps by chemical shifts (Gronenborn and Clore, 1994), methods to assist in automated or semiautomated chemical shift assignment (Bartels, 1997; Gronwald et al., 1998), methods to predict protein chemical shifts from sequence data (Wishart et al., 1997), and methods to automatically correct, re-reference, or validate chemical shift assignments using known 3-D structures (Zhang et al., 2003).

The utility of chemical shifts for interpreting, assigning, constraining, or refining protein structures is somewhat limited by the accuracy of chemical shift data. It is quite ironic that chemical shifts are far and away the most precisely measurable quantity in NMR but may be among the least accurately measured. Inconsistencies in chemical shift referencing and chemical shift reporting have recently been revealed, and they indicate that the problem is widespread with ~25% of proteins being mis-referenced (Zhang et al., 2003). Inconsistently or incorrectly referenced shifts can easily lead to serious misinterpretations of chemical shift data, so it is always important to ensure that any NMR data one may wish to use has properly been referenced using IUPAC conventions (Wishart et al., 1995). This can be done by using programs such as SHIFTCOR (Zhang et al., 2003) or by checking the chemical shift referencing information reported in the BMRB source file.

Scalar Coupling Constants

Many NMR resonances also display a fine structure characterized by split peaks (doublets or triplets) arising from magnetic coupling between adjacent, or covalently bonded nuclei (Figure 1A). This peak-splitting phenomenon is called scalar coupling (or J-coupling), and it too can be used to infer knowledge about a molecule's covalent geometry (Case et al., 1994; Case, 2000). Unlike chemical shifts, J-couplings are field independent. This means that they do not vary from one NMR instrument to another or from one day to the next. As a consequence, J-couplings are reported in absolute terms (in Hz) instead of ppm. Figure 4 illustrates a number of different coupling constants that are useful for protein NMR.

There are three major types of J-couplings: (1) single bond couplings (^{1}J), (2) two-bond, or geminal couplings (^{2}J), and (3) three-bond couplings, or vicinal couplings (^{3}J). Occasionally, four- and five-bond couplings can also be observed. Single bond, or ^{1}J-couplings occur between two NMR-active atoms separated by a single bond. Their magnitude is determined primarily by the gyromagnetic ratio (a constant dependent on the nuclear structure) of the two interacting nuclei, with values between 20 and 200 Hz being typical. (See Figure 3 for an illustration of important ^{1}J-couplings in amino acids.) Geminal, or ^{2}J-couplings occur between two NMR-active atoms separated by two covalent bonds. Their magnitude (10–30 Hz) is determined by the difference in

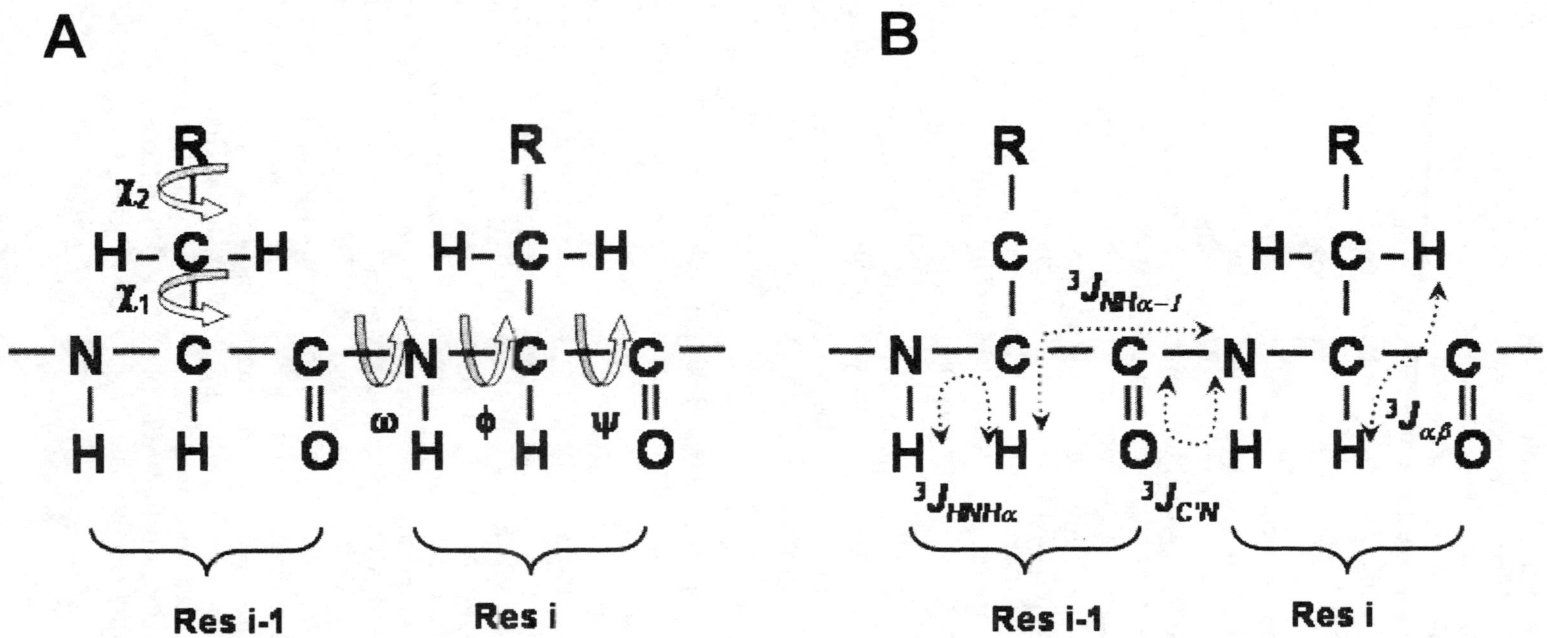

Figure 4 An illustration of (A) the torsion angles in polypeptides that can be measured using NMR and (B) the types of vicinal couplings (used to determine these torsion angles) that can be detected using modern NMR methods.

electronegativity between the two coupled (geminal) atoms. The most useful coupling constant in protein NMR is actually the vicinal coupling constant. Vicinal (^{3}J) coupling occurs between two atoms separated by three covalent bonds. Their magnitude (1–15 Hz) is determined by the dihedral or torsional angle between the two coupled atoms. The relationship between ^{3}J-couplings and torsional angles was first described by Karplus (1959). It is given by the following periodic function:

$$^3J = A + B \cos \theta + C \cos^2 \theta \tag{3}$$

where A, B, and C are coefficients that depend on atomic electronegativity, and θ corresponds to the dihedral angle.

The most frequently measured vicinal coupling constant in protein NMR is the $^3J_{HNH\alpha}$ coupling constant (see Figure 4). The $^3J_{HNH\alpha}$ coupling constant corresponds to the coupling between the amide proton (HN) and the alpha proton (Hα) within a given amino acid residue. Measuring $^3J_{HNH\alpha}$ allows one to determine the backbone ϕ angle. Typically, the magnitude of the $^3J_{HNH\alpha}$ couplings are between 3 and 10 Hz. Given that most protein linewidths are greater than any given $^3J_{HNH\alpha}$ coupling constant and given that most protein amide peaks are not resolvable in a 1-D NMR spectrum (too many peaks overlap), it was almost impossible to accurately measure protein $^3J_{HNH\alpha}$ couplings until the advent of 2-D and 3-D NMR. The introduction of phase-sensitive double-quantum filtered correlated spectroscopy (Rance et al., 1983), or DQ-COSY, opened the way to directly measuring $^3J_{HNH\alpha}$ for proteins. The DQ-COSY is a special kind of 2-D experiment that allows small couplings to be resolved as two antiphase peaks (one peak points up, the other points down). This antiphase display permits peak-to-peak measurements (i.e., J-couplings) to be directly measured. More recently, the development of E-COSY methods (Sorensen et al., 1994) have allowed couplings to be measured even more precisely. In E-COSY spectra, the peaks are separated by both the usual (active or COSY-like) J-coupling and by J-coupling to a third "passive" spin. This allows essentially unresolvable couplings to be measured by virtue of slight chemical shifts induced by the passive coupling to a third (usually a ^{15}N or ^{13}C) nucleus.

Additionally, much simpler approaches have emerged that only require the measurement of amide peak line widths from either TOCSY or ^{15}N-^{1}H HMQC spectra to extract accurate $^3J_{HNH\alpha}$ couplings (Wang et al., 1997; Wishart and Wang, 1998). In addition to their simplicity, these TOCSY or HMQC methods do not require the collection and assignment of extra spectra, since TOCSYs and HMQCs are normally required for most protein assignment protocols. Another kind of 2-D NMR experiment, called the HNHA, allows coupling constants to be measured not on the basis of peak-to-peak separation but rather on the basis of differential peak intensities (Bax et al., 1994). Because peak intensity is easier to quantify, these new quantitative

J-correlation methods can often be more precise than peak-to-peak measurements. Under the best of circumstances, all three methods (direct measurement, E-COSY, or quantitative J-correlation) typically have a minimum $^3J_{HNH\alpha}$ coupling error of ±0.6 Hz. The abundance of $^3J_{HNH\alpha}$ data, especially that from quantitative J-correlation, has allowed a "universal" Karplus equation to be derived:

$$^3J_{HNH\alpha} = 6.64 \cos^2(\phi - 60^\circ) - 1.43 \cos(\phi - 60^\circ) + 1.86 \quad (4)$$

which allows ϕ angles to be calculated to within $\pm 15^\circ$ (Wang and Bax, 1996).

Other types of scalar couplings have recently been observed that allow even more geometrical or constraint information to be derived. These include the measurement of internuclear $^3J_{NC'}$ couplings arising between hydrogen bonds (Cordier and Grzesiek, 1999), which allows identification of hydrogen bond donor-acceptor pairs), the measurement of $^3J_{H\alpha\text{-}1N}$ couplings, which provide information on protein ψ angles (Permi et al., 2000), and the measurement of $^3J_{C'H\alpha}$ couplings, via HCANC′ experiments, which provide information on backbone ϕ angles (Permi, 2003). The HCANC′ experiment transfers magnetization from the Hα nucleus of one residue through to the C=O (carbonyl) nucleus of the preceding nucleus, with the resulting signal being split by the $^3J_{C'H\alpha}$ coupling constant. Many other J-coupling constants, which provide main chain ϕ/ψ and side-chain χ_1 angle information, are also measurable using modified 2-D and 3-D heteronuclear experiments (Bax et al., 1994; Wang and Bax, 1996).

Residual Dipolar Couplings

Among the most exciting "new" parameters to emerge in protein NMR have been residual dipolar couplings, or RDCs (Tjandra and Bax, 1997; Bax, 2003). These are different from spin-spin or J-couplings in that the peak splitting arises when two nearby nuclear dipoles are differentially oriented relative to the external magnetic field. In solution, these dipolar couplings are normally averaged to zero, so we observe only scalar or J-couplings. In solids, these dipolar couplings are exceedingly large (20 kHz) and actually dominate the spectrum. If solution conditions can be found that give proteins a partial or residual orientation, much smaller dipolar couplings (–20 to +20 Hz) can be observed. By making use of dilute solutions of bicelles (Tjandra and Bax, 1997), filamentous phage particles (Hansen et al., 2000), or other large polymers that weakly bind to proteins, NMR spectroscopists have found that proteins can be weakly oriented to yield measurable residual dipolar couplings. Residual dipolar couplings D are given by the following equation:

$$D = D_a[(3 \cos^2 \theta - 1) + 3/2R \sin^2 \theta \cos 2\phi] \quad (5)$$

where $D_a = -\mu(h/2\pi)\gamma_A\gamma_B/(8\pi^2 r^3)$ is a constant that depends on the physical properties of the two nuclei (A and B), R is the rhombicity of the dipolar tensor (a value between 0 and 2/3), θ is the angle between the internuclear A-B vector and the magnetic field, and ϕ is the other polar angle in the principle axis system. Because these couplings depend on the orientation of the protein relative to the external magnetic field as well as the orientation of the dipole pair within the protein, RDCs can be used to extract so-called long-range information. In other words, RDCs can provide information about the orientation of two helices relative to each another or even about the orientation of two domains relative to each another (Fischer et al., 1999). The use of RDCs has become increasingly prevalent in protein structure determination and refinement, as these longer range constraints greatly improve the quality and precision of protein structures (Bax, 2003). More recently, protein structures have actually been generated using only RDC data (Rohl and Baker, 2002), which opens up the exciting possibility of doing NMR without NOEs.

NOEs

In NMR, nuclei can transfer energy or magnetization to each other, both through bonds and through space (for short distances). The through-space transfer process is called the nuclear Overhauser effect, or NOE (Roques et al., 1980). In 1-D NMR, the nuclear Overhauser effect can be seen as a change in intensity of an NMR resonance when the transitions of another nearby resonance are perturbed or when the resonance is irradiated with a sustained RF pulse. In 2-D NMR (NOESY), the NOE phenomenon is manifested as the appearance of new crosspeaks from spatially close nuclei. Hence, if 8.34 ppm and 9.21 ppm are the chemical shifts of two amide protons, then a cross peak at 8.34,9.21 in the NOESY spectrum would indicate that these two amides are within a few angstroms of each other. These NOESY cross peaks will build with intensity as the length of the NOESY mixing time (50–400 ms) is increased (Figure 2).

Fundamentally, NOEs arise from dipolar coupling interactions (as opposed to J-coupling). These dipole-dipole interactions affect how energy or magnetism is dissipated to other nuclei within the molecule (a phenomenon called relaxation). A complete description of NOEs is beyond the scope of this chapter, but the interested reader is advised to consult the text by Wuthrich (1986) for a more comprehensive introduction to protein NOEs. As seen below, the intensity of a NOE resonance is inversely proportional to the sixth power of the distance (r) between interacting nuclei:

$$\text{NOE} \sim f(\tau_c)/r^6 \tag{6}$$

where $f(\tau_c)$ indicates that the NOE is also a function of the Brownian motion or correlation time (τ_c) between the two nuclei. This r^{-6} dependence means that NOEs fall to very small values within relatively short distances (few are observed with r greater than 5 Å) compared to the dimension of most proteins (30 Å).

In protein NMR, the large number of nearby ^{1}Hs or other "dipoles" often causes a phenomenon known as spin diffusion (Roques et al., 1980). Spin diffusion leads to interference or disruption in the usual pairwise NOE interaction pathways and so the r^{-6} assumption about NOE intensity usually becomes incorrect, especially for longer mixing times. To get around the problem of spin diffusion, most protein NMR spectroscopists collect NOESY spectra over a range of mixing times. For proteins, it has been shown that the initial NOE buildup rate, plotted over shorter mixing times, is directly related to the distance between nuclei. At short mixing times (less than 100 milliseconds), the NOE buildup is usually linear. At longer times, this buildup curve will either level off or even fall. The key in working with protein NOEs is to collect the NOE intensity data or slope data in the linear portion of the curve (Wuthrich, 1986). From these NOE intensities, it is possible to calibrate to a well-known or fixed distance (the distance between two aromatic protons) and to classify NOEs as strong, medium, or weak (1.8–2.8, 1.8–4.0, 1.8–5.0 Å). Figure 5 illustrates the types of NOEs that are typically measurable in a polypeptide chain.

Interestingly, for protein NMR work it is not necessary to have accurately measured NOEs. In fact, it is far better to have a large number of approximate NOEs than a small number of exact NOEs (Guntert, 1998). An abundance of NOE information is necessary for most geometry optimization programs (distance geometry, simulated annealing) to converge to a consistent solution. In analogy to linear algebra, the problem must be overdetermined such that the number of knowns (i.e., NOEs) should exceed the number of unknowns (i.e., atom positions).

Without NOEs, protein structure determination via NMR would be almost impossible. This is because NOEs provide the vital through-space distance information that is necessary to orient, align, or bring together distal parts of the protein chain. However, there are limits to the utility of NOEs. For smaller peptides or proteins (500–2000 Da), NOEs are typically nulled due to the near match of the molecule's characteristic τ_c and the characteristic spectrometer frequency. This makes structure determination for smaller peptides difficult. For larger proteins, spin diffusion and spectral overlap make it exceedingly difficult to identify and accurately measure NOEs. The use of heteronuclear 3-D NOESY methods appears to have helped solve some of these problems. Nevertheless, one of the continuing bottlenecks in protein structure determination is the assignment of NOEs. Frequently, it takes longer to assign NOEs than it does to assign proteins.

For larger proteins (greater than 15 kDa), it turns out that most NOESY cross peaks are especially hard to assign because they are ambiguous. Ambiguous NOEs arise from protons having the same or

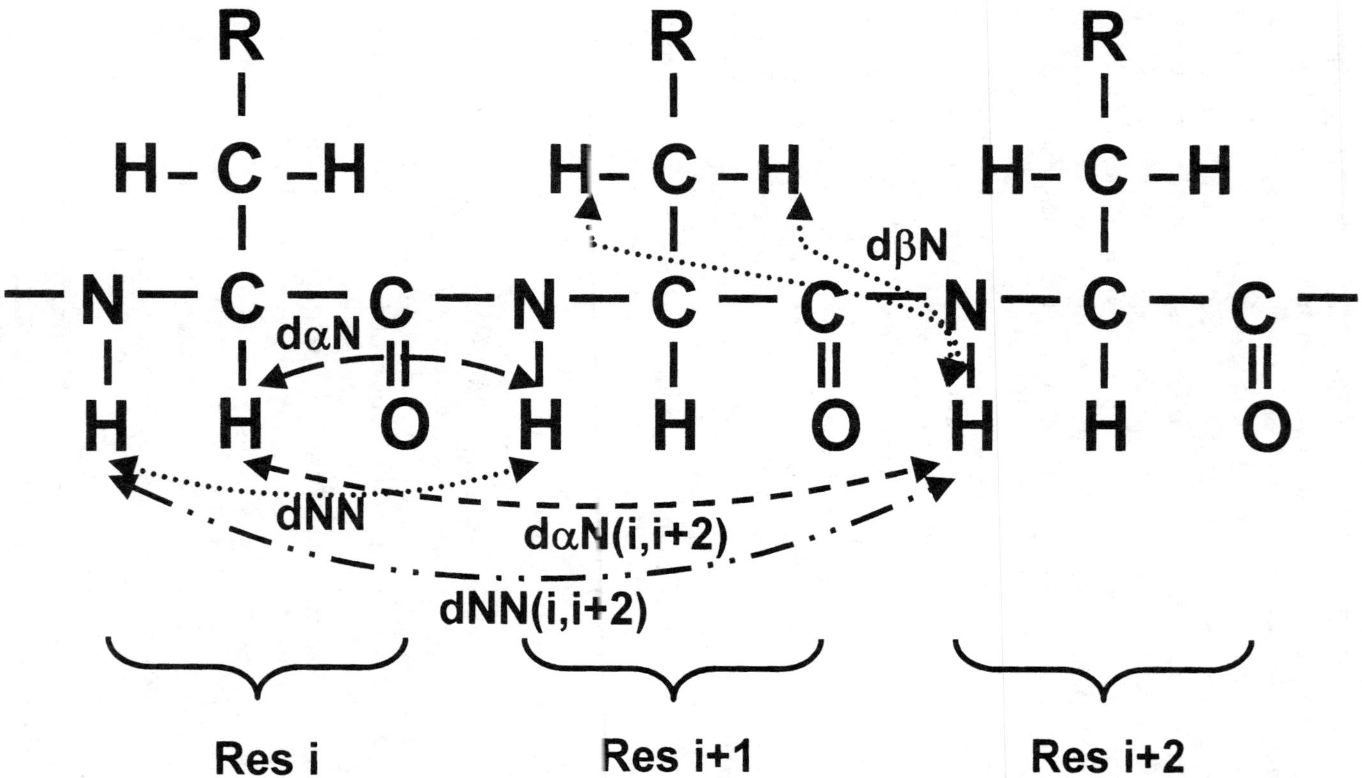

Figure 5 An illustration of the types of inter- and intra-residue ^{1}H NOEs (and distances) that can be detected in polypeptides. Not shown are the longer range NOEs commonly seen for helices such as $d_{\alpha N(i,i+3)}$ and $d_{\alpha N(i,i+4)}$

nearly identical ^{1}H chemical shifts. A program called ARIA (ambiguous restraints for iterative assignment) has recently been developed, which deals with ambiguous NOEs and automatically generates protein structures using only a chemical shift assignment list and a list of unassigned or partially assigned NOE cross peaks (Linge et al., 2001). ARIA uses an iterative, self-correcting approach to deal with wrong NOE assignments or overlapping NOEs. The success of ARIA has led to the development of other NOE-based automated structure generation programs such as ATNOS (Herrmann et al., 2002) and NOAH (Oezguen et al., 2002). Given suitably good quality spectra (or peak lists), these methods can automatically generate good quality 3-D structures in as little as 10 minutes.

Protein Structure Determination by NMR

As a general rule, protein structure determination via NMR can be divided into three steps: (1) sample preparation; (2) data collection and sequential assignment; and (3) structure generation. Here we shall review each of these methods, with a special emphasis on describing recent developments in these areas that are affecting research on proteins of pharmaceutical interest.

Sample Preparation

In NMR spectroscopy, one generally tries to obtain a protein sample that is pure (greater than 95%), mostly folded, monomeric, soluble for extended periods of time (days), and stable to concentrations of ~1 mM. Protein samples may be either unlabeled (^{1}H only) or labeled with certain NMR sensitive isotopes (i.e., ^{13}C and ^{15}N). Because ^{1}H is a naturally occurring and highly sensitive NMR nucleus, many smaller peptides and proteins need not be isotopically labeled. For larger proteins (greater than 10 kDa) or in the pursuit of higher quality structures, most NMR spectroscopists like to work with isotopically labeled proteins (Wuthrich, 1995). Isotopically labeled proteins are prepared by growing cells and expressing the protein in defined minimal media that has been enriched with ^{13}C and/or ^{15}N, and/or ^{2}H labeled substrates. The requirements for both quantity (tens of milligrams) and isotopic labeling usually mean that protein samples prepared for NMR must be cloned into high-expression microbial hosts.

A number of excellent microbial expression systems have been developed and have been shown to provide consistently excellent yields of isotopically labeled protein. Among bacterial hosts, there are several protease depleted (Lon^{-}, OmpT^{-}) *Escherichia coli* strains (such as BL21[DE3]) that allow a range of proteins to be expressed inside the cell without fear of proteolytic degradation. There are also a number of *E. coli* strains that have been specifically developed to permit residue-specific isotopic labeling (LeMaster and Richards, 1985; Waugh, 1996).

These auxotrophic strains have had certain key genes in their amino acid synthetic pathways knocked out, so they require specific amino acid supplements in their growth media. Residue-specific labeling is a very powerful approach to assigning or analyzing very large proteins by NMR, as it greatly simplifies their spectra and allows individual amino acid types to be unambiguously identified (Waugh, 1996; Goto et al., 1999). More recently, yeast, mammalian, and baculovirus systems have been developed for expressing labeled proteins for NMR (Hansen et al., 1992; Wood and Komives, 1999). These eukaryotic systems require a fairly sophisticated and expensive growth medium, but for proteins that cannot be expressed any other way, these eukaryotic systems are sometimes the only option.

To facilitate the search for an appropriate host (different strains of bacteria, yeast, baculovirus, etc.) the development of ligation independent cloning methods (LICs) has made the process much easier and much faster. LIC allows the simultaneous and facile preparation of expression vectors for a variety of different hosts, with a variety of different affinity tags (Stols et al., 2002). Affinity tags, which are cleavable peptide or protein tags that are genetically engineered into the protein of interest, represent another major innovation in protein expression and purification (Ford et al., 1991). With affinity tags, purification is often just a single step, and yields can approach 90%. The most popular affinity tag is the polyhistidine, or His_6 tag, which may be attached either to the N or C terminus of the protein of interest. His_6-tagged proteins can be purified quickly and easily using a chelating sepharose (sepharose-conjugated iminodiacetic acid) column.

Even after a protein has been purified, appropriately labeled, and placed in an NMR tube, many challenges still exist. Frequently, under seemingly benign conditions, proteins will precipitate or form large multimers after a short period of time. This obviously prevents good quality NMR spectra from being collected. Usually, NMR spectroscopists must spend a good deal of time finding optimal solution conditions for their protein to collect the necessary spectra. This is called sample conditioning (Bagby et al., 2001). Sample conditioning is not unlike the sampling process that x-ray crystallographers use to find crystals. In sample conditioning, a large number of pH, temperature, and salt conditions are iteratively tested to find the optimal solution conditions for collecting NMR spectra (Lepre and Moore, 1998). The so-called button test (Bagby et al., 1997) has been developed to facilitate this process. The button test uses a microtiter plate and dialysis film to allow large numbers of sample conditions to be evaluated and for precipitation events to be detected, using a minimal amount of protein. In addition to changing pH or salt conditions, occasionally the addition of other cosolvents (trifluoroethanol, dimethylsulfoxide), low viscosity additives, or detergents can help (Wand et al., 1998). Likewise, limited proteolytic digestion of loose, hydrophobic tails can sometimes lead to a more stable protein that behaves much better in the NMR tube (Bagby et al., 2001).

Data Collection and Assignment

Data collection in protein NMR is complicated not only by the number of signals one wants to measure (thousands), but also by the fact that all (or almost all) proteins must be dissolved in water. Water (H_2O) is a strongly protic solvent, meaning that it contains protons (^1H). In fact, the concentration of protons in water is ~110 M. Given that most proteins are dissolved in water at sub-millimolar concentrations, this means that the most intense ^1H signal arising from a protein is typically 100,000 times less than the ^1H signal arising from water. This signal intensity problem is compounded by the fact that the H_2O resonance typically falls near the center of the protein spectrum (around 4.8 ppm), where many of the most interesting and important protein resonances lie. Getting rid of the water signal or reducing it is a major challenge to protein NMR spectroscopists (Sklenar, 1990). Three approaches are possible. One is to use D_2O instead of H_2O as a solvent. When proteins are dissolved in D_2O, they behave essentially as they do in H_2O. However, the labile hydrogens, especially amide protons, tend to exchange with the D_2O solvent. This hydrogen exchange leads to a loss of amide proton signals as the amide protons are rapidly replaced with deuterons (which do not have a detectable ^1H NMR signal). Without the amide ^1H signals, it is frequently impossible to make the required assignments or measure the necessary NOEs to determine the protein structure. Despite these disadvantages, D_2O solvent exchange methods are frequently used to identify hydrogen bonds in proteins or to simplify certain kinds of 2-D NMR spectra (Wuthrich, 1986).

A second approach to water suppression is a technique called presaturation. In presaturation, a low-power continuous RF irradiation is directed at the sample. If the frequency of the RF irradiation matches the frequency of the water signal, then the water signal is "bleached" out. Typically, this bleaching leads to a loss or reduction of signal for not only the water but also the protein resonances 0.2–0.3 ppm around the water signal. Despite the fact that it can lead to a variety of spectral artifacts, presaturation is still widely used in protein NMR because it is so easy to implement. A third approach to water suppression takes advantage of pulsed field gradient (PFG) technology, which is widely available for most modern NMR spectrometers (Keeler et al., 1994). Using PFG methods, it is possible to suppress signals not only on the basis of their frequency but on the basis of their diffusion coefficients. Obviously, water molecules (18 Da) have very different diffusion coefficients than proteins (~18,000 Da). Using PFG schemes such as WATERGATE (Piotto et al., 1992), it is possible to obtain spectra that are essentially free of water suppression artifacts and that display most of the resonances that would normally have been bleached out with presaturation methods.

Once a protein sample is ready for NMR and suitable solution conditions (and water suppression methods) found, it is then subjected to a series of 2-D or 3-D NMR experiments. The intent of these experiments is to collect sufficient spectral data so that every NMR peak or resonance can be assigned to every NMR detectable atom in the protein (Wuthrich, 1986, 1995). This process is called sequential assignment. For unlabeled peptides and proteins, the assignment process is relatively straightforward. Two types of 2-D NMR experiments are collected. A TOCSY experiment is first collected, which allows individual amino acids to be identified by the distinct patterns of their chemical shifts. A NOESY experiment is then collected, which yields a spectrum that looks almost identical to the TOCSY spectrum. Both kinds of 2-D spectra may be processed and displayed using a variety of software packages such as NMRVIEW (Johnson and Blevins, 1994), NMRPIPE (Delaglio et al., 1995), ANSIG (Helgstrand et al., 2000), and XEASY (Eccles et al., 1991). These NMR-specific programs allow the spectra to be filtered and the signals to be enhanced using a variety of methods such as linear prediction, zero-filling, or shifted sine-bell filtering (Wuthrich, 1986; Cavanagh et al., 1996). A good deal of practice, patience, and skill is needed to produce acceptable looking 2-D spectra. Fortunately, with today's fast computers and advanced graphical interfaces, many of these spectral modifications can be done and viewed interactively.

Once processed, the NOESY data in combination with the TOCSY data can be used to determine which amino acids are sequentially proximal to each other (Basus, 1989). In this way, each set of peaks in the TOCSY spectrum can be assigned an atom name and a residue number. Once all the TOCSY peaks have been assigned an atom name and a residue number, the sequential assignment process is essentially complete. Although conceptually simple in theory, in practice homonuclear NMR is not particularly easy to do. Because of the complexity or resonance overcrowding innate to most homonuclear NMR spectra, it often takes months of painstaking, manually intensive effort to process, analyze, and assign these types of spectra. The intrinsic shortcomings of homonuclear NMR have recently been overcome with the development of more advanced heteronuclear NMR methods designed for isotopically (usually $^{13}C/^{15}N$ labeled) labeled proteins. Because ^{15}N and ^{13}C chemical shifts display much greater dispersion than ^{1}H chemical shifts, heteronuclear NMR spectra are much less crowded than homonuclear spectra.

When working with isotopically labeled proteins, the sequential assignment process does not actually require any NOESY data at all. Rather, it is possible to use a series of heteronuclear 3-D NMR experiments, such as the HNCA, HNCACB, and HNCO experiments, to identify individual amino acid types and to identify which pairs of amino acids are sequentially proximal to each other. Under favorable circumstances, it is actually possible to assign most of a protein's backbone using just a single 3-D NMR experiment (the HNCACB). More often, however, at least four NMR experiments (HNCA, HNCACB, HNCO, HCCH-TOCSY) are needed to assign all of a protein's main chain and side-chain resonances (Moseley et al., 2001).

Three-dimensional and 4-D NMR experiments are relatively easy to set up but typically require three- to fourfold more time than 2-D experiments. Indeed, a single 3-D or 4-D NMR experiment can sometimes take as long as five to six days. Given that as many as six or seven 3-D data sets may have to be acquired to fully assign a protein, this represents an enormous amount of time devoted exclusively to data collection. The development of projection reconstruction, GFT-NMR, and Hadamard spectroscopy along with other data processing approaches such as filter diagonalization has potentially shortened the data collection period by a factor of 10 or more (Chen et al., 2003; Freeman and Kupce, 2003). Now it is possible to collect many kinds of 3-D and 4-D NMR spectra in less than a few hours.

Three-dimensional and 4-D NMR spectra can be processed, displayed, and analyzed using many of the same software packages and processing techniques employed for 2-D spectra. However, with chemical shift data displayed on three or more axes, there is obviously much more data that must be seen or analyzed. Although it is possible to analyze 3-D NMR spectra manually, there is a growing trend to use automated and semiautomated approaches. Several automated or semiautomated NMR assignment programs have recently been described including AUTOASSIGN (Moseley et al., 2001), GARANT (Bartels et al., 1997), Smartnotebook (Slupsky et al., 2003), and CAMRA (Gronwald et al., 1998). All that is required to make these programs work is a list of peaks or a defined set of appropriately processed 2-D or 3-D spectra. Under ideal circumstances, it is possible to go from a set of processed spectra to a complete set of assignments in as little as ten seconds. However, it is almost impossible to obtain "ideal" spectra.

The assignment process, whether automated or manual, can be greatly aided if one has some prior knowledge about what the chemical shifts might be. In this regard, several approaches have been developed that allow researchers to use either sequence homology (to previously assigned proteins) or known coordinates (of identical or homologous x-ray structures) to predict chemical shifts. These methods are quite accurate and, if the sequence identity is high enough (greater than 35%), greatly improve the performance of any assignment strategy (Wishart and Case, 2001).

Structure Determination

Once the sequential assignment process is completed, it is possible to begin the structure determination process. This process is outlined in Figure 6. Both homonuclear and heteronuclear NMR

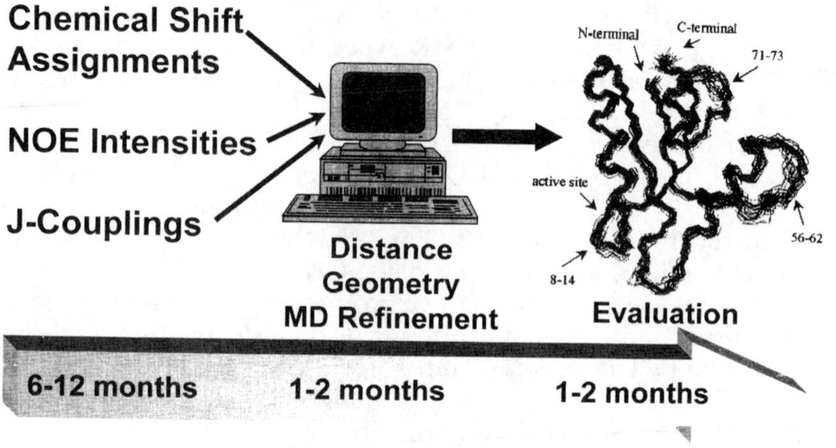

Figure 6 A schematic summary of the 3-D structure determination process. Experimental parameters are fed into a computer and run through simulated annealing or distance geometry programs to generate an ensemble of structures that satisfy the experimental constraints. Shown on the arrow is the typical amount of time required to perform each task using manual approaches. Recent advances are shortening these steps by a factor of 100 or more.

approaches depend on collecting multiple NOESY spectra. By using the sequential assignments already obtained, NOESY cross peaks are identified and quantified. As already mentioned, several NOESY spectra must be collected with different mixing times to determine more precisely the strength of the NOE interaction and to better calibrate the distances between each NOE-active pair of protons. Additional information about secondary structure (derived from chemical shifts or NOE patterns), coupling constants, hydrogen bonding (derived from hydrogen exchange data), and disulfide bonds is also added to a long list of approximate constraints (Guntert, 1998). Typically for a protein of 100 residues, one needs to obtain 1500 to 2000 constraints (15–20 constraints per residue) to generate a high-quality structure. The experimentally derived constraint file must also be supplemented with "prior knowledge" about the covalent geometry of each amino acid (characteristic bond lengths, bond angles, amino acid planarity and chirality information). The combined constraint file is then submitted to a distance geometry program (Braun, 1987), a simulated annealing program or energy minimization "engine" such as CNS/XPLOR (Brunger et al., 1998), AMBER (Wang and Kollman, 2001), CHARMM (Brooks et al., 1983), or DYANA (Guntert et al., 1997) for structure generation and refinement. These very sophisticated geometry optimization programs attempt to generate a protein structure or set of protein structures that are consistent with the experimental constraints. What normally results from these calculations is a set of possible protein structures (twenty-five to fifty), which must be individually evaluated for their fitness and agreement to the experimental data. Typically, multiple iterations of the geometry optimization process are necessary before a final set of structures is obtained.

Recent assessments of protein structures generated by NMR have pointed to a number of problems with their quality, stereochemistry, or assumed resolution—particularly for structures generated prior to 1995 (Clore et al., 1993; Laskowski et al., 1996). This has led to the search for novel kinds of structural constraints in an attempt to improve the accuracy and precision of NMR structures. Some of these new constraints include RDCs, new kinds of J-couplings, chemical shifts, and paramagnetic relaxation data. The inclusion of these kinds of experimental constraints, along with the incorporation of empirical database potentials or more theoretical constraints, has been a major focus in protein structure generation over the past five years (Clore and Schwieters, 2002). It has already been shown that many of these new constraints can significantly improve the quality of NMR-generated protein structures (Clore and Schweiters, 2002; Bax, 2003). Using constraints such as RDCs in structure generation, at least from the perspective of a user, is relatively painless. This is because nearly all NMR-based protein structure generation programs are being continually updated by teams of programmers to accommodate the latest set of measurable NMR parameters.

In addition to using new parameters to improve conventional NOE-based methods, a number of researchers are looking for completely new approaches to protein structure generation by NMR. Since the early 1990s, several reports have been published that demonstrate the possibility of going straight from raw spectral data to a solved 3-D structure, without the need for spectral assignments. The first effort was described by Kraulis (1994), who demonstrated the feasibility of generating 3-D structures of two small proteins (BPTI and GAL4) using $^{13}C/^{15}N$ heteronuclear NOESY data. More recently, Grishaev and Llinas (2002) have used a slightly different approach to generate 3-D structures of two small proteins (about seventy residues) using only homonuclear NOESY data. The basic concept involves using molecular dynamics and simulated annealing to condense a "gas" of unassigned, unconnected ^1H atoms into a structured proton distribution or cloud using only internuclear distances derived from NOE intensities and van der Waals repulsive terms as restraints. The protein sequence is then mapped to the cloud of protons to generate a complete, connected polypeptide structure.

In addition to these assignment-independent methods for structure generation, there is a growing trend to accelerate the structure generation process by integrating NMR data into powerful structure prediction or fold recognition methods. This integration is leading to

3-D structure generation methods that use little or no NOE information. For example, several groups have employed the idea of using chemical shifts, which provide residue-specific secondary structure information as experimental constraints for protein threading algorithms (Ayers et al., 1999; Xu et al., 2000). The inclusion of experimental data appears to improve the reliability of the threading alignment and hence the identification of the protein fold. More recently, the concept of chemical shift threading has been extended to the point where chemical shifts are iteratively calculated and used directly in the threading potential. A new program, called THRIFTY (Wishart and Case, 2001), which uses chemical shift threading, appears to be capable of generating high-quality 3-D structures without any NOE information.

Extending the Limits

The first NMR protein structures were determined in the mid 1980s. These early NMR structures were very crude and were limited to very small (less than seventy residues), well behaved proteins (Wuthrich, 1986). Since then, enormous strides in both NMR technologies and techniques have been made. Stronger magnets, improved probe sensitivity, more sophisticated hardware, faster computers, better isotopic labeling schemes, higher yielding expression systems, newly measurable spectral parameters, and greatly improved pulse sequences are now allowing larger proteins to be characterized much more quickly and with far greater precision (Dotsch an Wagner, 1998; Wider and Wuthrich, 1999). Proteins as large as 723 residues have been fully assigned (Tugarinov et al., 2002), and protein complexes as large as 900,000 Da have been analyzed (Fiaux et al., 2002). Indeed, a major focus of most protein NMR research today is to extend the size limits to which NMR methods can be applied. Given that many pharmaceutically interesting proteins are not particularly small, this trend toward extending the size limits will clearly have a positive effect on the pharmaceutical industry.

Outside of the development of 3-D and 4-D heteronuclear NMR, which occurred in the early 1990s, the most important innovation for NMR in terms of extending its size limits happened in 1997 (Pervushin et al., 1997). This innovation is known as transverse relaxation optimized spectroscopy, or TROSY. TROSY exploits the differential relaxation effects arising from chemical shift anisotropy (CSA) and dipole-dipole (DD) coupling. At higher magnetic fields (greater than 500 MHz), these differences are quite noticeable. The TROSY pulse sequence selects certain magnetization transfer pathways (coherences) so that only the narrowest peaks are selected and displayed. This coherence selection results in an NMR spectrum that can have at least a twofold improvement in resolution. TROSY spectroscopy has become widely adopted by many protein NMR labs because the improvements in resolution, especially with partially deuterated proteins, can be quite spectacular (Wider and Wuthrich, 1999). Interestingly, the

TROSY pulse sequence can be inserted into most standard 2-D, 3-D, and 4-D NMR experiments, meaning that TROSY experiments can be performed and the benefits realized quite easily. TROSY methods are now routinely employed in almost all NMR experiments involving proteins or protein complexes greater than 30 kDa. In fact, TROSY experiments were instrumental in characterizing the largest protein (80 kDa) ever assigned by NMR—malate synthase G (Tugarinov et al., 2002).

Applications to Pharmaceutically Important Proteins

Since 1980, the U.S. Food and Drug Administration (FDA) has approved more than 100 protein and peptide pharmaceuticals including a variety of growth factors, hormones, antibodies, and chemokines. Hundreds more are expected to be approved in the next decade. At the same time, thousands of small-molecule drugs have also been approved by the FDA, with more than 95% being targeted to pharmaceutically important proteins. NMR has played and continues to play a major role in the characterization of many of these protein drugs and protein drug targets. Not only does NMR reveal important structural information about pharmaceutically interesting proteins, it allows scientists to probe the dynamics, interactions, and binding events that are vital to understanding drug action or discovering new, small-molecule drug leads. Because of its versatility, NMR has found a number of useful niches in structure-aided drug discovery and development, particularly in areas where x-ray crystallography has proven inadequate. These include the characterization of (1) smaller, flexible hormones or peptides, (2) smaller, more disordered or more flexible proteins, (3) protein domains or modules, (4) protein formulations, and (5) protein-ligand interactions (Zuiderweg, 2002; Staunton et al., 2003).

NMR Applications in Structure Determination

A partial list of eighty pharmaceutically important peptides and proteins that have been solved by NMR is provided in Table 2. This list includes the protein/peptide name and Protein Data Bank (PDB) coordinates (x-ray or NMR, whichever is more recent) of each protein/peptide. Inspection of this table reveals that there are six general classes of proteins where NMR has been most frequently used in structure determination or structural characterization. These include the chemokines or cytokines (interleukins and interferons), growth factors or hormones (TGF, EGF, BFGF, HGH, glucagons, atrial natriuretic peptide), blood coagulation proteins (TPA, factor domains), antimicrobial peptides (hepcidins, bacteriocins, tachystatins), proteins involved in immune responses (FK506, cyclophilin, defensins), and viral targets (HIV and heptatitis viruses). These reflect the particular strengths of NMR in determining the structures of smaller, more flexible proteins and peptides. With recent advances in NMR technology,

Protein Name	PDB ID	Protein Name	PDB ID
Transforming growth factor alpha	2TGF	HIV Tat protein	1JFW
Interleukin 8	3IL8	Hepatitis C virus Ns3 protease	1BT7
Interleukin 1-beta	6ILB	HIV protease	1BVE
Epidermal growth factor (EGF)	1JL9	Human angioigenin	1AWZ
Cyclophilin	1FGLA	Plasminogen activator kringle	1HPK
FK506 binding protein	2FKE	Coagulation factor	1MGX
Interferon alpha	1ITF	EGF-like domain from factor VII	1BF9
Basic fibroblast growth factor (FGF)	1BFG	EGF-like domain from factor IX	1IXA
Interleukin 4	1BBN	Fibrin binding finger domain	1TPM
HIV RNase H (DNA binding domain)	1O1W	F1-G module of TPA	1TPG
Monocyte chemoattractant protein 3	1BO0	Atrial natriuetic peptide	1ANP
Ragweed allergen protein	2BBG	Human insulin (hexamer)	1AIY
Dihydrofolate reductase	3DFR	Human erythropoietin	1BUY
Insulin-like growth factor (IGF)	1GZZB	Human parathyroid hormone	1BWX
Human prion protein	1I4MA	Human chorionic gonadotropin	1DZ7
Birch allergen protein	1B6F	Neuromedin	1C(a
HIV integrase (DNA binding domain)	1IHV	Cortiocotropin releasing hormone	1GO9
Human defensin	1FD3D	Human betacellulin-2	1IOX
Human growth hormone (HGH)	1HWGA	Human epiregulin	1K36

Hepatitis A virus (HAV) 3C protease	1HAVA	Melanoma inhibitory activity peptide	1KOX
Hepatitis C virus (HCV) helicase	1JR5	Human glucagon	1KX9
p53 tumor suppressor protein	1HS5B	Leukemia inhibitory factor	1A7M
HIV nef protein	1QA5A	Human MIP-1a	1B50
Interleukin 13	1GA3	Eotaxin-2	1EIG
Lymphotactin chemokine	1J9OA	Eotaxin-3	1G2S
Alzheimer peptide	1BA4	Rantes	1HRJ
Macrophage inflammatory peptide	1HUMA	Human macrophage inhibitory protein 1	1HUN
Vascular endothelial growth factor	1KMX	Human macrophage inhibitory protein 2	1MI2
Tachystatin antimicrobial peptide	1CIXA	Interleukin 16	1I16
HIV matrix protein	1TAM	Interleukin 6	2IL6
Interleukin 5	1HULA	Interleukin 2	1IRL
Tyrosine phosphatase 1b	1G7FA	IL-1 receptor antagonist protein	1IRP
Cherry tree allergen protein	1H2O	Interleukin 18	1J0S
Hepcidin antimicrobial peptide	1M4EA	Granulin-1	1I8Y
Neuropeptide Y	1FVN	Stromal cell derived factor	2SDF
Human prolactin	1N9DA	Human melanoma growth peptide	1MGS

[a]Names: Protein Data Bank (PDB) accession IDs.

Table 2 A Partial List of Pharmaceutically Important Proteins or Peptides That Have Been Characterized by NMR[a]

larger proteins are being tackled, and it is likely that more enzymes (kinases, phosphatases, proteases, metabolic enzymes) will appear in the near future. Indeed, many of the proteins being studied by the pharmaceutical industry today fall into this category. Unfortunately, confidentiality issues often prevent information about these targets from being publicly released.

Structural studies by NMR of many of the chemokines and growth factors have revealed that their active sites or binding loops tend to be quite flexible, with relatively little defined structure (McInnes and Sykes, 1997; Rajarathnam, 2002). This is in contrast to the rigid, well-defined structure that is often implied through x-ray crystallography of the same or similar proteins. Interestingly, the rigid structures observed via x-ray now appear to be artifacts, largely due to crystal packing effects. In this regard, NMR has revealed a key insight into chemokine/growth factor function and activity that could not have been obtained via x-ray crystallography. Evidently, this active-site flexibility appears to be vital to facilitating or accommodating protein-protein or protein-receptor interactions (McInnes and Sykes, 1997; Rajarathnam, 2002). Studies of the active site structure of growth factors and chemokines have also led to the development of peptide/protein analogues and a number of potentially interesting small-molecule or minimal peptide drug leads (McInnes et al., 1998; Rajarathnam, 2002). In this regard, the structural characterization of prospective protein drugs is motivating the development of prospective small-molecule drug analogues (Booth et al., 2003).

As a general rule, comparative studies of many pharmaceutically interesting proteins solved by NMR have found both similarities and differences between their corresponding x-ray counterparts (Laskowski et al., 1996). The overall folding and topology of most proteins is invariably well conserved, but the differences often lie in the active sites or in more flexible loops. Some of these differences are likely significant, but many simply reflect the flexible nature of most protein structures. Particular care must be taken in interpreting protein-protein interactions in both solution and in crystals, as the high concentrations required to generate crystals or solve proteins by NMR often can lead to spurious or biologically irrelevant interactions (Rajarathnam et al., 1997). Although x-ray structures are generally more precise than NMR structures, it is not necessarily correct to assume that x-ray structures are more correct or accurate. Overall, it is better to have structural data from both x-ray and NMR as both methods can reveal important and complementary insights into the structure, function, and activity of proteins.

NMR Applications in Protein Formulations

Understanding how proteins behave when stored as lyophilized powders or as solutions in vials is important to understanding how to extend the shelf lives of many protein drugs. When stored in vials, proteins are subject to at least four noncovalent processes that lead to loss of activity. These include precipitation, adsorption, unfolding/denaturation, and aggregation. This means that protein drugs are far more likely than small-molecule drugs to lose their potency during the packaging, distribution, and dispensing process. NMR offers the opportunity to obtain atomic-level detail about the events leading to loss of activity. It can also provide detailed information about the atomic structures of proteins in liquid formulations or lyophilized powders.

For example, solid-state NMR has been applied to a number of lyophilized proteins, including gamma globulin, beta-galactosidase, and lysozyme formulated with dextran and/or a variety of sugars as excipients (Lam et al., 2002; Yoshioka et al., 2002). These studies investigated the effect of temperature, humidity, and sugar/dextran composition on relaxation times (T_1, $T_{1\rho}$, and T_2 measurements) of both the proteins and excipients. This work demonstrated that solid-state NMR data could be used to detect differences between excipient and protein relaxation times and to correlate these changes with protein aggregation, stability, and activity (Lam et al., 2002). These results may open the way to predicting optimal formulation and storage protocols for other pharmaceutically useful proteins.

NMR has long been used to study solution-state protein denaturation and aggregation (Fersht and Daggett, 2002). In fact, NMR has provided some of the most detailed insights into the structures of molten globules that seem to represent later stages of folding or early stages of unfolding and that may play a role in initiating protein aggregation or adsorption events. Studies on apomyoglobin and lactalbumin have revealed an unusual mix of native and non-native structure characterized by a condensed but flexible tertiary fold (Kuwajima, 1996; Cavagnero et al., 2001). The flexible nature of these structures essentially precludes their analysis by crystallization or x-ray crystallography. At the other end of the protein-folding spectrum, the use of chemical shift and NOE analyses has allowed residual structures or hydrophobic clusters to be detected in seemingly unstructured (by CD or fluorescence) or denatured proteins (Lietzow et al., 2002; Crowhurst and Forman-Kay, 2003). In fact, a number of NMR experiments have recently been developed to detect site-specific contributions to the pH dependence of protein stability, to detect weakly populated (less than 1%) folding intermediates, or to monitor detailed protein motions in the denatured state (Tollinger et al., 2003).

Our understanding of protein aggregation has grown considerably as a consequence of numerous NMR studies of prion proteins (Riesner, 2003). These efforts, which use hydrogen exchange, relaxation measurements, and chemical shift analyses, have identified potential destabilizing hot-spots in the native structure that seem to trigger aggregation. These same techniques are being applied to understand other protein aggregates such as the R6 human insulin hexamer (Jacoby et al., 1996).

This particular study explored the differences between the structures of insulin hexamers, monomers, and dimers, which were manifested by dramatic changes in chemical shifts and long-range NOEs. The role of phenol in stabilizing insulin hexamers was also studied and it was shown that certain flexible "gatekeeper" residues create a phenol binding portal. The use of chemical shift mapping allowed the location of the phenol binding site to be identified and the role of phenol as a stabilizing ligand to be better rationalized.

NMR Applications in Protein-Ligand Binding

No other form of spectroscopy and certainly no variant of x-ray crystallography can match the intrinsic ability of NMR to monitor and detect protein-ligand interactions. NMR is exquisitely sensitive, universally applicable, quantitative, amenable to analyzing complex mixtures, and uniquely able to provide atomic resolution details of ligand-protein interactions in solution (Dierks et al., 2001; Zuiderweg, 2002). Because of the enormous range of concentrations at which NMR signals can be detected and reliably measured (10^{-2} M to 10^{-6} M), NMR is able to detect the weakest protein-ligand interactions of any method so far identified. Furthermore, because NMR detects a universally abundant compound (^1H atoms), it is possible to design NMR-based screening protocols without any prior knowledge of a protein's function, substrates, or substrate analogues. In other words, NMR eliminates the need to develop bioassays that need specially designed fluorophores, chromophores, or substrate analogues.

In addition to its near-universality, NMR offers superb approaches to quantifying ligand binding. The ability to resolve individual resonances during an NMR titration allows the precise determination of binding constants, off-rates, on-rates, allosteric interactions, and many other kinetically (and thermodynamically) important parameters. Furthermore, the level of spectral resolution offered by NMR means that dozens of compounds can be monitored simultaneously during any screening or titration processes. This obviously facilitates high-throughput screening of ligand libraries. Finally, outside of x-ray crystallography, NMR is the only spectroscopic method that can provide precise 3-D structural details about how and where a ligand binds to a protein.

During the past ten years, NMR researchers, both inside and outside the pharmaceutical industry, have focused a great deal of effort on refining the technology to detect and deconvolve protein-ligand interactions. Essentially, two basic approaches have emerged (Diercks et al., 2001). One is based on using NMR to observe changes to the small-molecule ligand (or ligands) when it binds to the protein. This is called ligand-observed screening. The other approach is based on using NMR to observe changes to the target protein when it binds a ligand (or ligands). This is called target or protein-observed screening.

In ligand-observed screening, it is not necessary to have previously assigned the protein target, nor is it necessary to have a protein that is particularly amenable (small, soluble, stable) to NMR studies. This is because the resonances that are detected or measured are those of the ligand—not the protein. In ligand-observed screening, one relies on detecting NMR phenomena that are exacerbated by the size differences between the large protein target and the small molecule ligand (Figure 7). Typically, for ligand-based studies to work, the dissociation constant (K_D) should be between 10^{-7} and 10^{-3} M, and the mass difference between the target protein and prospective ligand should be greater than 50-fold. If these conditions are met, a variety of specially designed pulse sequences may be used to "filter" or select for the resonances of ligands that exhibit different affinities. In this way, researchers can work with mixtures of ligands yet only see the ligand of interest in the resulting NMR spectrum (Chen and Shapiro, 1999; Diercks et al., 2001).

Three things happen when small ligands transiently bind to a protein. Their diffusion rate slows, their linewidths broaden (fast relaxation), and their NOEs shift from positive to negative (Figure 5). A number of elegant approaches have been developed during the past five years that exploit the measurement of decreased diffusion coefficients via NMR. Techniques such as DOSY (diffusion-ordered spectroscopy), diffusion-modulated gradient COSY, or diffusion-encoded TOCSY experiments actually allow molecules (and spins) to be separated by their diffusion coefficients (Chen and Shapiro, 1999; Diercks et al., 2001). The main limitation to ligand-observed screening is that relatively little structural information can be obtained regarding how or where the ligand is interacting with the protein. Although some information about the ligand structure (or structural changes) is obtainable, the primary role for ligand-observed screening is to identify rapidly compounds that bind to a given protein target. Ultimately, it is the role of target-observed screening to figure out how and where these compounds bind.

In target-observed screening (or ligand mapping), one is focused on measuring and mapping the perturbations to the protein target resonances when a ligand binds (Diercks et al., 2001). Therefore, it is essential to have the resonances of the protein of interest fully assigned and, ideally, a structure determined. Unlike ligand-observed screening methods, there is no limitation on the affinity, binding constant, or on/off rates for target-observed screening. Furthermore, target-observed mapping allows one to obtain detailed information about the orientation and placement of the ligand relative to the protein. It also allows one to determine how that ligand affects the target protein structure. This kind of information is crucial to doing structure-aided drug design. However, the major limitation to target-observed mapping is that one is restricted to working with proteins that are sufficiently

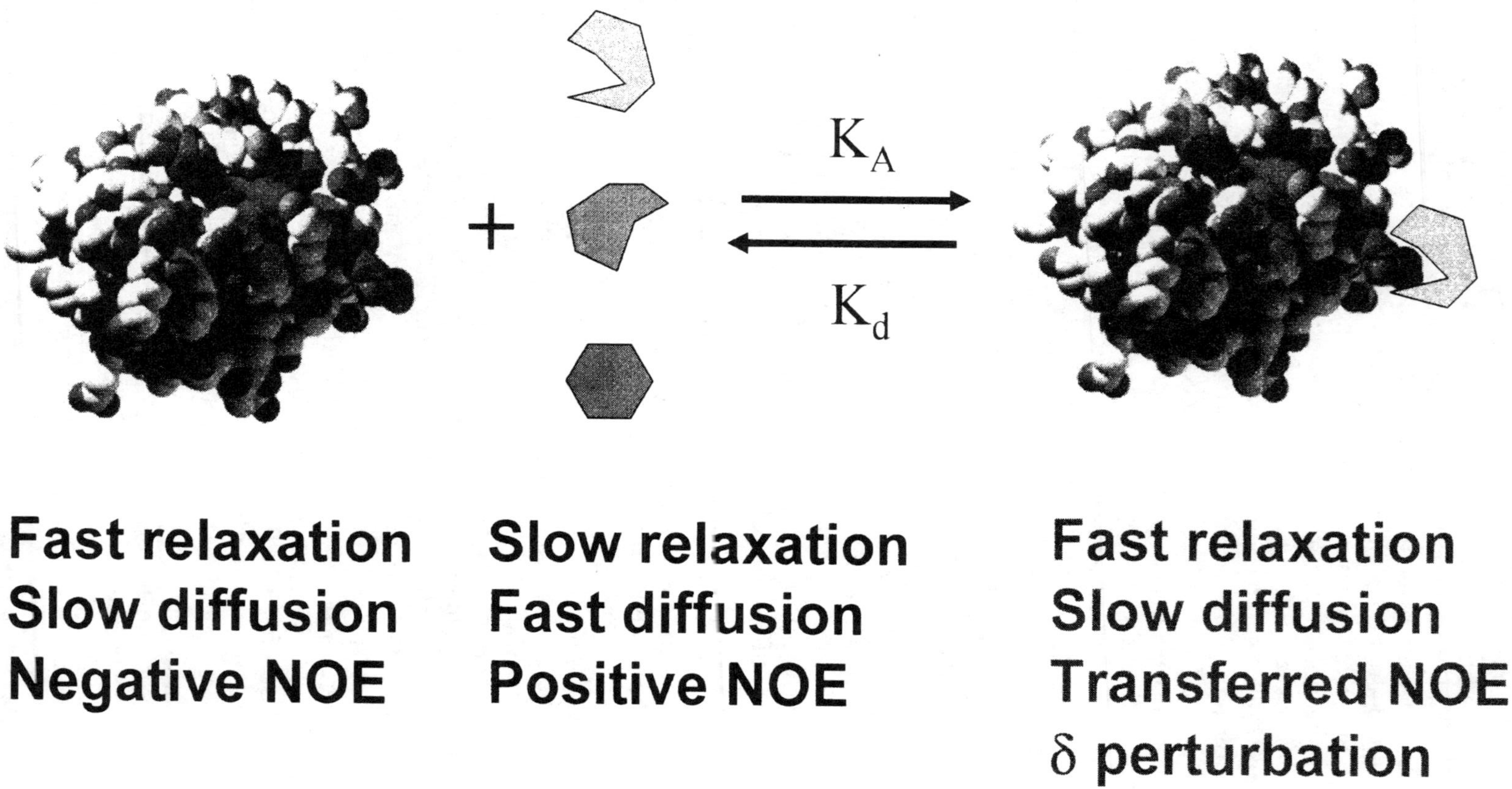

Figure 7 A simplified illustration of how many key NMR observables are changed when a small ligand binds a much larger protein.

small and well behaved so that their resonances may be fully observed and assigned.

Five major approaches to target-observed mapping exist. They include (1) mapping via intermolecular NOEs, (2) chemical shift mapping, (3) mapping with dynamics, (4) mapping with amide exchange, and (5) mapping with paramagnetically labeled ligands. We shall only review the first two approaches here. NOE mapping is essentially an extension of standard 3-D structure determination, except that two molecules are involved (the ligand and the protein) rather than just one. Target-observed screening via NOEs requires relatively tight binding (K_D less than 10 μM). To facilitate studies with larger molecules, many NOE mapping approaches use isotope-edited NOESY techniques (Fesik et al., 1988). This requires that the target protein be isotopically labeled (^{1}H, ^{13}N, ^{15}N) and the ligand unlabeled (or *vice versa*). Isotope editing allows one to unambiguously discriminate between NOEs within the protein target and between the two molecules. NOE-based approaches are particularly useful if one is trying to generate a detailed 3-D structure of a protein-ligand complex or in situations where the ligand binding induces substantial changes to the protein structure. The NOEs measured through these approaches can be used in standard distance geometry or simulated annealing programs to generate 3-D coordinates.

Chemical shift mapping has become the most popular approach to mapping ligand interactions in NMR. This is because it is both quick and easy to do. In essence, the ligand of interest is titrated with an isotopically labeled (usually ^{15}N) version of the protein of interest. With each addition of the ligand, an HSQC (heteronuclear single quantum coherence spectroscopy) spectrum of the protein is collected, and the chemical shift changes are recorded. Chemical shift mapping is best used when the perturbations to the protein upon ligand binding are relatively minor (which is generally true). Chemical shifts are extremely sensitive to changes in geometry, local charge effects, and hydrogen bonding. Consequently, most ligand binding events will lead to measurable protein chemical shift changes where the ligand is binding. Plotting the chemical shift changes through the course of a titration allows one to not only identify where the ligand is binding but also to determine the affinity, stoichiometry, specificity, and even the kinetics of binding (Diercks et al., 2001; Zuiderweg, 2002). Although chemical shift mapping cannot yet provide sufficient detail to generate 3-D structures of bound ligands, chemical shift perturbations can be used as constraints in molecular docking programs to generate approximate models showing where and how ligands have bound to their target protein (McKenna et al., 2003). Chemical shift mapping methods actually form the basis to a technique called SAR-by-NMR (structure activity relationships by NMR) that was developed and patented by Fesik and co-workers (Shuker et al., 1996). In

SAR-by-NMR, a series of weak-binding ligands are identified and their binding positions or orientations mapped on the target protein. By chemically linking adjacently mapped ligands, it is possible to enhance the binding and develop much higher affinity ligands. SAR-by-NMR has already been used to generate and refine a number of useful inhibitors for pharmaceutically important proteins (Hajduk et al., 1999).

Conclusions

NMR spectroscopy has evolved quickly to become a central pillar in structural biology. It is now being used in more than 500 structural biology labs around the world including a growing number of high-throughput structural genomics labs (Staunton et al., 2003). An average of one new protein NMR structure is being deposited into the Protein Data Bank every single day. However it is important to remember that NMR can be more than just a protein structure elucidation tool. NMR methods can reveal much about protein dynamics, protein interactions, protein kinetics, and protein function. Indeed, determining the structure of a protein by NMR is simply the first step in a process that can ultimately lead to the discovery and development of many pharmaceutically important products or ideas. In this chapter, we have seen how NMR is continuing to evolve and how technical or conceptual innovations are making NMR much more powerful and more useful over a much larger size range of molecular systems. Certainly, it is not unreasonable to expect that within three to five years, proteins as large as 100 kDa will have been fully assigned and characterized by NMR methods. It is also likely that NMR will have played a central role in the identification or design of at least a few blockbuster drugs. We may also expect to see the development of computational tools that will allow smaller peptides and proteins to be assigned and their structures automatically determined within minutes or even seconds of completing data collection. These trends toward simplifying and automating NMR appear to be a natural progression for many kinds of spectroscopy. In the 1930s, you needed a Ph.D. in spectroscopy to operate a UV spectrophotometer; today you don't. Currently, the view of most NMR spectroscopists is that you need a Ph.D. to operate an NMR spectrometer. Perhaps in five years you won't. Regardless of one's current or future qualifications, it is clear that NMR will be playing a vital role in just about every facet of drug discovery and development for a long time to come.

Abbreviations

1-D	One-dimensional
2-D	Two-dimensional

3-D	Three-dimensional
4-D	Four-dimensional
BMRB	BioMagResBank
COSY	Correlated spectroscopy
CSA	Chemical shift anisotropy
CSI	Chemical shift index
DOSY	Diffusion-ordered spectroscopy
DSS	2,2-Dimethyl-2-silapentane-5-sulfonic acid
DQ-COSY	Double quantum filtered COSY
E-COSY	Easy or Exclusive COSY
FID	Free induction decay
FT	Fourier transform
HM(S)QC	Heteronuclear multiple (single) quantum coherence spectroscopy
HCANC′	A 3-D NMR experiment for measuring coupling constants between Hα and ^{13}C′
HNCA	A 3-D NMR experiment for displaying H, ^{15}N, and ^{13}Ca chemical shifts
HNCACB	A 3-D NMR experiment for displaying H, ^{15}N, ^{13}Cα, and ^{13}Cβ chemical shifts
HNCO	A 3-D NMR experiment for displaying H, ^{15}N, and ^{13}C′ chemical shifts
Hz	Hertz
$^{n}J_{\alpha\beta}$	Coupling constant between nucleus "α" and "β" over "n" bonds
kDa	Kilodalton
LIC	Ligation independent cloning
NMR	Nuclear magnetic resonance
NOE	Nuclear Overhauser effect
NOESY	Nuclear Overhauser effect spectroscopy
PDB	Protein Data Bank
PFG	Pulsed field gradient
ppm	Parts per million
RDC	Residual dipolar coupling
RF	Radio frequency
SAR	Structure activity relationship
TOCSY	Total correlation spectroscopy
TROSY	Transverse relaxation spectroscopy

Acknowledgments

The author wishes to thank G. Amegbey for his assistance in preparing several figures and D. Waldman for her assistance in editing. Financial support from the Natural Sciences and Engineering Research Council (NSERC) and the Protein Engineering Network of Centres of Excellence (PENCE) is gratefully acknowledged.

References

1. Ayers DJ, Gooley PR, Widmer-Cooper A, and Torda AE. Enhanced Protein Fold Recognition Using Secondary Structure Information from NMR. *Protein Sci* 1999; 8:1127–1133
2. Bagby S, Tong KI, Liu D, Alattia JR, and Ikura M. The Button Test: A Small Scale Method Using Microdialysis Cells for Assessing Protein Solubility at Concentrations Suitable for NMR. *J Biomol NMR* 1997; 10:279–282
3. Bagby S, Tong KI, and Ikura M. Optimization of Protein Solubility and Stability for Protein Nuclear Magnetic Resonance. *Methods Enzymol* 2001; 339:20–41
4. Bartels C, Guntert P, Billeter M, and Wuthrich K. GARANT—A General Algorithm for Resonance Assignment of Multidimensional Nuclear Magnetic Resonance Spectra. *J Comput Chem* 1997; 18:139–149
5. Basus VJ. Proton Nuclear Magnetic Resonance Assignments. *Methods Enzymol* 1989; 177:132–149
6. Bax A, Vuister GW, Grzesiek S, Delaglio F, Wang AC, Tschudin R, and Zhu G. Measurement of Homo- and Heteronuclear J Couplings from Quantitative J Correlation. *Methods Enzymol* 1994; 239:79–105
7. Bax A. Weak Alignment Offers New NMR Opportunities to Study Protein Structure and Dynamics. *Protein Sci* 2003; 12:1–16
8. Booth V, Slupsky CM, Clark-Lewis I, and Sykes BD. Unmasking Ligand Binding Motifs: Identification of a Chemokine Receptor Motif by NMR Studies of Antagonist Peptides. *J Mol Biol* 2003; 327:329–334
9. Braun W. Distance Geometry and Related Methods for Protein Structure Determination from NMR Data. *Q Rev Biophys* 1987; 19:115–157
10. Braunschweiler L, and Ernst RR. Coherence Transfer by Isotropic Mixing. *J Magn Reson* 1983; 53:521–526
11. Brooks BR, Bruccoleri RE, Olafson BD, States DJ, Swaminathan S, and Karplus M. CHARMM: A Program for Macromolecular Energy Minimization and Dynamics Calculations. *J Comput Chem* 1983; 4:187–217

12. Brunger AT, Adams PD, Clore GM, DeLano WL, Gros P, Grosse-Kunstleve RW, Jiang JS, Kuszewski J, Nilges M, Pannu NS, Read RJ, Rice LM, Simonson T, and Warren GL. Crystallography & NMR System: A New Software Suite for Macromolecular Structure Determination. *Acta Crystallogr D Biol Crystallogr* 1998; 54:905–921

13. Case DA. Interpretation of Chemical Shifts and Coupling Constants In Macromolecules. *Curr Opin Struct Biol* 2000; 10:197–203

14. Case DA, Dyson HJ, and Wright PE. Use of Chemical Shifts and Coupling Constants in Nuclear Magnetic Resonance Structural Studies on Peptides and Proteins. *Methods Enzymol* 1994; 239:392–416

15. Cavagnero S, Nishimura C, Schwarzinger S, Dyson HJ, Wright PE. Conformational and Dynamic Characterization of the Molten Globule State of an Apomyoglobin Mutant with an Altered Folding Pathway. *Biochemistry* 2001; 40:14459–14467

16. Cavanagh J, Fairbrother WJ, Palmer AG, and Skelton NJ. *Protein NMR Spectroscopy: Principles and Practice*. New York: Academic Press; 1996

17. Chen A, and Shapiro MJ. Affinity NMR. *Anal Chem* 1999; 71:669A–675A

18. Chen J, De Angelis AA, Mandelshtam VA, and Shaka AJ. Progress on the Two-Dimensional Filter Diagonalization Method. An Efficient Doubling Scheme for Two-Dimensional Constant-Time NMR. *J Magn Reson* 2003; 162:74–89

19. Clore GM, and Schwieters CD. Theoretical and computational Advances in Biomolecular NMR Spectroscopy. *Curr Opin Struct Biol* 2002; 12:146–153

20. Clore GM, Robien MA, and Gronenborn AM. Exploring the Limits of Precision and Accuracy of Protein Structures Determined by Nuclear Magnetic Resonance Spectroscopy. *J Mol Biol* 1993; 231:82–102

21. Cordier F, and Grzesiek S. Direct Observation of Hydrogen Bonds in Proteins by Interresidue $^{3h}J_{NC'}$ Scalar Couplings. *J Am Chem Soc* 1999; 121:1601–1602

22. Cornilescu G, Delaglio F, and Bax A. Protein Backbone Angle Restraints from Searching a Database for Chemical Shift and Sequence Homology. *J Biomol NMR* 1999; 13:289–302

23. Crowhurst KA, and Forman-Kay JD. Aromatic and Methyl NOEs Highlight Hydrophobic Clustering in the Unfolded State of an SH3 Domain. *Biochemistry* 2003; 42:8687–8695

24. Delaglio F, Grzesiek S, Vuister GW, Zhu G, Pfeifer J, and Bax A. NMRPipe: A Multidimensional Spectral Processing System Based on UNIX Pipes. *J Biomol NMR* 1995; 6:277–293

25. Diercks T, Coles M, and Kessler H. Applications of NMR in Drug Discovery. *Curr Opin Chem Biol* 2001; 5:285–291

26. Dotsch V, and Wagner G. New Approaches to Structure Determination by NMR Spectroscopy. *Curr Opin Struct Biol* 1998; 8:619–623

27. Eccles C, Guntert P, Billeter M, and Wuthrich K. Efficient Analysis of Protein 2D NMR Spectra Using the Software Package EASY. *J Biomol NMR* 1991; 1:111–130

28. Edison AS, Abildgaard F, Mooberry ES, and Markley JL. Practical Introduction to Theory and Implementation of Multinuclear, Multidimensional Nuclear Magnetic Resonance Experiments. *Meth Enzymol* 1994; 239:3–78

29. Ernst RR. Nobel Lecture. Nuclear Magnetic Resonance Fourier Transform Spectroscopy. *Biosci Rep* 1992; 12:143–187

30. Fersht AR, and Daggett V. Protein Folding and Unfolding at Atomic Resolution. *Cell* 2002; 108:573–582

31. Fesik SW, Luly JR, Erickson JW, and Abad-Zapatero C. Isotope-Edited Proton NMR Study on the Structure of a Pepsin/Inhibitor Complex. *Biochemistry* 1988; 27:8279–8301

32. Fiaux J, Bertelsen EB, Horwich AL, and Wuthrich K. NMR Analysis of a 900K GroEL GroES Complex. *Nature* 2002; 418:207–211

33. Fischer MW, Losonczi JA, Weaver JL, and Prestegard JH. Domain Orientation and Dynamics in Multidomain Proteins from Residual Dipolar Couplings. *Biochemistry* 1999; 38:9013–9022

34. Ford CF, Suominen I, and Glatz CE. Fusion Tails for the Recovery and Purification of Recombinant Proteins. *Protein Expr Purif* 1991; 2:95–107

35. Freeman R, and Kupce E. New Methods for Fast Multidimensional NMR. *J Biomol NMR* 2003; 27:101–113

36. Goto NK, Gardner KH, Mueller GA, Willis RC, and Kay LE. A Robust and Cost-Effective Method for the Production of Val, Leu, Ile (Delta 1) Methyl-Protonated 15N-, 13C-, 2H-Labeled Proteins. *J Biomol NMR* 1999; 13:369–374

37. Grishaev A, and Llinas M. CLOUDS, a Protocol for Deriving a Molecular Proton Density via NMR. *Proc Natl Acad Sci U S A* 2002; 99:6707–6712

38. Gronenborn AM, and Clore GM. Identification of N-Terminal Helix Capping Boxes by Means of 13C Chemical Shifts. *J Biomol NMR* 1994; 4:455–458

39. Gronwald W, Sonnichsen FD, Wishart DS, and Sykes BD. CAMRA: Chemical Shift Based Computer Aided Protein NMR Assignments. *J Biomol NMR* 1998; 10:165–179

40. Guntert P, Mumenthaler C, and Wuthrich K. Torsion Angle Dynamics for NMR Structure Calculation with the New Program DYANA. *J Mol Biol* 1997; 273:283–298

41. Guntert P. Structure Calculation of Biological Macromolecules from NMR Data. *Q Rev Biophys* 1998; 31:145–237

42. Hajduk PJ, Meadows RP, and Fesik SW. NMR-Based Screening in Drug Discovery. *Q Rev Biophys* 1999; 32:211–240

43. Hansen AP, Petros AM, Mazar AP, Pederson TM, Rueter A, and Fesik SW. A Practical Method for Uniform Isotopic Labeling of Recombinant Proteins in Mammalian Cells. *Biochemistry* 1992; 31:12713–12718

44. Hansen MR, Hanson P, and Pardi A. Filamentous Bacteriophage for Aligning RNA, DNA, and Proteins for Measurement of Nuclear Magnetic Resonance Dipolar Coupling Interactions. *Methods Enzymol* 2000; 317:220–240

45. Helgstrand M, Kraulis P, Allard P, and Hard T. Ansig for Windows: An Interactive Computer Program for Semiautomatic Assignment of Protein NMR Spectra. *J Biomol NMR* 2000; 18:329–336

46. Herrmann T, Guntert P, and Wuthrich K. Protein NMR Structure Determination with Automated NOE-Identification in the NOESY Spectra Using the New Software ATNOS. *J Biomol NMR* 2002; 24:171–189

47. Ikura M, Kay LE, and Bax A. A Novel Approach for Sequential Assignment of 1H, 13C, and 15N Spectra of Proteins: Heteronuclear Triple-Resonance Three-Dimensional NMR Spectroscopy. Application to Calmodulin. *Biochemistry* 1990; 29:4659–4667

48. Ishima R, and Torchia DA. Protein Dynamics from NMR. *Nat Struct Biol* 2000; 7:740–743.

49. Jacoby E, Hua QX, Stern AS, Frank BH, and Weiss MA. Structure and Dynamics of a Protein Assembly. 1H NMR Studies of the 36kDa R6 Insulin Hexamer. *J Mol Biol* 1996; 258:136–157

50. Johnson BA, and Blevins J. NMRView: A Computer Program for the Visualization and Analysis of NMR Data. *J Biomol NMR* 1994; 4:603–614

51. Karplus, M. Contact Electron-Spin Coupling of Nuclear Magnetic Moments. *J Phys Chem* 1959; 30:11–15

52. Kay LE, Ikura M, Tschudin R, and Bax A. 3-Dimensional Triple-Resonance NMR-Spectroscopy of Isotopically Enriched Proteins. *J Magn Reson* 1990; 89:496–501

53. Keeler J, Clowes RT, Davis AL, and Laue ED. Pulsed-Field Gradients: Theory and Practice. *Methods Enzymol* 1994; 239:145–207

54. Kraulis P. Protein Three-Dimensional Structure Determination and Sequence-Specific Assignment of 13C and 15N-Separated NOE Data. A Novel Real-Space ab initio. *J Mol Biol* 1994; 243:696–718

55. Kumar A, Ernst RR, and Wuthrich K. A Two-Dimensional Nuclear Overhauser Enhancement (2D NOE) Experiment for the Elucidation

of Complete Proton-Proton Cross-Relaxation Networks in Biological Macromolecules. *Biochem Biophys Res Commun* 1980; 95:1–6

56. Kuszewski J, Qin J, Gronenborn AM, and Clore GM. The Impact of Direct Refinement Against 13C Alpha and 13C Beta Chemical Shifts on Protein Structure Determination by NMR. *J Magn Reson B* 1995; 106:92–96

57. Kuwajima K. The Molten Globule State of Alpha-Lactalbumin. *FASEB J* 1996; 10:102–109

58. Lam YH, Bustami R, Phan T, Chan HK, and Separovic F. A Solid-State NMR Study of Protein Mobility in Lyophilized Protein-Sugar Powders. *J Pharm Sci* 2002; 91:943–951

59. Laskowski RA, Rullmann JA, MacArthur MW, Kaptein R, and Thornton JM. AQUA and PROCHECK-NMR: Programs for Checking the Quality of Protein Structures Solved by NMR. *J Biomol NMR* 1996; 8:477–486

60. LeMaster DM, and Richards FM. 1H-15N Heteronuclear NMR Studies of Escherichia coli Thioredoxin in Samples Isotopically Labeled by Residue Type. *Biochemistry* 1985; 24:7263–7268

61. Lepre CA, and Moore JM. Microdrop Screening: A Rapid Method to Optimize Solvent Conditions for NMR Spectroscopy of Proteins. *J Biomol NMR* 1998; 12:493–499

62. Lietzow MA, Jamin M, Jane Dyson HJ, and Wright PE. Mapping Long-Range Contacts in a Highly Unfolded Protein. *J Mol Biol* 2002; 322:655–662

63. Linge JP, O'Donoghue SI, and Nilges M. Automated Assignment of Ambiguous Nuclear Overhauser Effects with ARIA. *Methods Enzymol* 2001; 339:71–108

64. Marion D, and Wuthrich K. Application of Phase Sensitive Two-Dimensional Correlated Spectroscopy (COSY) for Measurements of 1H-1H Spin-Spin Coupling Constants in Proteins. *Biochem Biophys Res Commun* 1983;113:967–974

65. McInnes C, and Sykes BD. Growth Factor Receptors: Structure, Mechanism, and Drug Discovery. *Biopolymers* 1997; 43:339–366

66. McInnes C, Wang J, Al Moustafa AE, Yansouni C, O'Connor-McCourt M, and Sykes BD. Structure-Based Minimization of Transforming Growth Factor-alpha (TGF-alpha) Through NMR Analysis of the Receptor-Bound Ligand. Design, Solution Structure, and Activity of TGF-alpha 8-50. *J Biol Chem* 1998; 273:27357–27363

67. McKenna S, Moraes T, Pastushok L, Ptak C, Xiao W, Spyracopoulos L, and Ellison MJ. An NMR-Based Model of the Ubiquitin-Bound Human Ubiquitin Conjugation Complex Mms2.Ubc13. The Structural Basis for Lysine 63 Chain Catalysis. *J Biol Chem* 2003; 278: 13151–13158

68. Moseley HN, Monleon D, and Montelione GT. Automatic Determination of Protein Backbone Resonance Assignments from Triple Resonance Nuclear Magnetic Resonance Data. *Methods Enzymol* 2001; 339:91–108

69. Neal S, Zhang H, Nip AM, and Wishart DS. Rapid and Accurate Calculation of Protein 1H, 13C and 15N Chemical Shifts. *J Biomol NMR* 2003; 26:215–240

70. Oezguen N, Adamian L, Xu Y, Rajarathnam K, and Braun W. Automated Assignment and 3D Structure Calculations Using Combinations of 2D Homonuclear and 3D Heteronuclear NOESY Spectra. *J Biomol NMR* 2002; 22:249–263

71. Palmer AG 3rd. NMR Probes of Molecular Dynamics: Overview and Comparison with Other Techniques. *Annu Rev Biophys Biomol Struct* 2001; 30:129–155

72. Permi P. Determination of Three-Bond Scalar Coupling Between 13C′ and 1H(alpha) in Proteins Using an iHN(CA),CO(alpha/beta-J-COHA) Experiment. *J Magn Reson* 2003; 163:114–120

73. Permi P, Kilpelainen I, and Annila A. Determination of Backbone Angle psi in Proteins Using a TROSY-Based alpha/beta-HN(CO)CA-J Experiment. *J Magn Reson* 2000; 146:255–259

74. Pervushin K, Riek R, Wider G, and Wuthrich K. Attenuated T2 Relaxation by Mutual Cancellation of Dipole-Dipole Coupling and Chemical Shift Anisotropy Indicates an Avenue to NMR Structures of Very Large Biological Macromolecules in Solution. *Proc Natl Acad Sci U S A* 1997; 94:12366–12371

75. Piotto M, Saudek V, and Sklenar V. Gradient-Tailored Excitation for Single-Quantum NMR Spectroscopy of Aqueous Solutions. *J Biomol NMR* 1992; 2:661–665

76. Rajarathnam K. Designing Decoys for Chemokine-Chemokine Receptor Interaction. *Curr Pharm Des* 2002; 8:2159–2169

77. Rajarathnam K, Kay CM, Clark-Lewis I, and Sykes BD. Characterization of Quaternary Structure of Interleukin-8 and Functional Implications. *Methods Enzymol* 1997; 287:89–105

78. Rance M, Sorensen OW, Bodenhausen G, Wagner G, Ernst RR, and Wuthrich K. Improved Spectral Resolution in Cosy 1H NMR Spectra of Proteins Via Double Quantum Filtering. *Biochem Biophys Res Commun* 1983; 117:479–485

79. Riesner D. Biochemistry and Structure of PrP(C) and PrP(Sc). *Br Med Bull* 2003; 66:21–33

80. Rohl CA, and Baker D. De novo Determination of Protein Backbone Structure from Residual Dipolar Couplings Using Rosetta. *J Am Chem Soc* 2002; 124:2723–2729

81. Roques BP, Rao R, and Marion D. Use of Nuclear Overhauser Effect in the Study of Peptides and Proteins. *Biochimie* 1980; 62:753–773
82. Seavey BR, Farr EA, Westler WM, and Markley JL. A Relational Database for Sequence-Specific Protein NMR Data. *J Biomol NMR* 1991; 1:217–226
83. Sharma D, and Rajarathnam K. 13C NMR Chemical Shifts Can Predict Disulfide Bond Formation. *J Biomol NMR* 2000; 18:165–171
84. Shuker SB, Hajduk PJ, Meadows RP, and Fesik SW. Discovering High-Affinity Ligands for Proteins: SAR by NMR. *Science* 1996; 274:1531–1534
85. Sklenar V. Selective Excitation Techniques for Water Suppression in One- and Two-Dimensional NMR Spectroscopy. *Basic Life Sci* 1990; 56:63–84
86. Slupsky CM, Boyko RF, Booth VK, and Sykes BD. Smartnotebook: A Semi-Automated Approach to Protein Sequential NMR Resonance Assignments. *J Biomol NMR* 2003; 27:313–321
87. Sorensen MD, Led JJ, and Sorensen OW. A New 2D NMR Method for Measurement of JHH Coupling Constants. *J Biomol NMR* 1994; 4:135–141
88. Staunton D, Owen J, and Campbell ID. NMR and Structural Genomics. *Acc Chem Res* 2003; 36:207–214
89. Stols L, Gu M, Dieckman L, Raffen R, Collart FR, and Donnelly MI. A New Vector for High-Throughput, Ligation-Independent Cloning Encoding a Tobacco Etch Virus Protease Cleavage Site. *Protein Expr Purif* 2002; 25:8–15
90. Tjandra N, and Bax A. Direct Measurement of Distances and Angles in Biomolecules by NMR in a Dilute Liquid Crystalline Medium. *Science* 1997; 278:1111–1114
91. Tollinger M, Crowhurst KA, Kay LE, and Forman-Kay JD. Site-Specific Contributions to the pH Dependence of Protein Stability. *Proc Natl Acad Sci U S A* 2003; 100:4545–4550
92. Tugarinov V, Muhandiram R, Ayed A, and Kay LE. Four-Dimensional NMR Spectroscopy of a 723-Residue Protein: Chemical Shift Assignments and Secondary Structure of Malate Synthase g. *J Am Chem Soc* 2002; 124:10025–10035
93. Wand AJ, Ehrhardt MR, and Flunn PF. High-Resolution NMR of Encapsulated Proteins Dissolved in Low-Viscosity Fluids. *Proc Natl Acad Sci U S A* 1998; 95:15299–15302
94. Wang AC, and Bax A. Determination of the Backbone Dihedral Angles ϕ in Human Ubiquitin from Reparametrized Empirical Karplus Equations. *J Am Chem Soc* 1996; 118:2483–2494
95. Wang J, and Kollman PA. Automatic Parameterization of Force Field by Systematic Search and Genetic Algorithms. *J Comput Chem* 2001; 22:1219–1228

96. Wang Y, Nip AM, and Wishart DS. A Simple Method to Quantitatively Measure Polypeptide JHNH Alpha Coupling Constants from TOCSY or NOESY Spectra. *J Biomol NMR* 1997; 10:373–382

97. Waugh DS. Genetic Tools for Selective Labeling of Proteins with alpha-15N-Amino Acids. *J Biomol NMR* 1996; 8:184–192

98. Wider G, and Wuthrich K. NMR Spectroscopy of Large Molecules and Multimolecular Assemblies in Solution. *Curr Opin Struct Biol* 1999; 9:94–601

99. Wishart DS, and Case DA. Use of Chemical Shifts in Macromolecular Structure Determination. *Methods Enzymol* 2001; 338:3–34

100. Wishart DS, and Sykes BD. Chemical Shifts as a Tool for Structure Determination. *Methods Enzymol* 1994a; 239:363–392.

101. Wishart DS, and Sykes BD. The 13C Chemical-Shift Index: A Simple Method for the Identification of Protein Secondary Structure Using 13C Chemical-Shift Data. *J Biomol NMR* 1994b; 4:171–180

102. Wishart DS, and Wang Y. Facile Measurement of Polypeptide JHNH Alpha Coupling Constants from HMQC-J Spectra. *J Biomol NMR* 1998; 11:329–336

103. Wishart DS, Sykes BD, and Richards FM. Relationship Between Nuclear Magnetic Resonance Chemical Shift and Protein Secondary Structure. *J Mol Biol* 1991a; 222:311–333

104. Wishart DS, Sykes BD, and Richards FM. Simple Techniques for the Quantification of Protein Secondary Structure by 1H NMR Spectroscopy. *FEBS Lett* 1991b; 293:72–80

105. Wishart DS, Sykes BD, and Richards FM. The Chemical Shift Index: A Fast and Simple Method for the Assignment of Protein Secondary Structure Through NMR Spectroscopy. *Biochemistry* 1992; 31:1647–1651

106. Wishart DS, Bigam CG, Yao J, Abildgaard F, Dyson HJ, Oldfield E, Markley JL, and Sykes BD. 1H, 13C and 15N Chemical Shift Referencing in Biomolecular NMR. *J Biomol NMR* 1995; 6:135–140

107. Wishart DS, Watson MS, Boyko RF, and Sykes BD. Automated 1H and 13C Chemical Shift Prediction Using the BioMagResBank. *J Biomol NMR* 1997; 10:329–336

108. Wittekind M, and Mueller L. HNCACB, a High Sensitivity 3D NMR Experiment to Correlate Amide-Proton and Nitrogen Resonances with the alpha Carbon and beta-Carbon Resonances in Proteins. *J Magn Reson Ser B* 1993; 101:206–210

109. Wood MJ, and Komives EA. Production of Large Quantities of Isotopically Labeled Protein in Pichia pastoris by Fermentation. *J Biomol NMR* 1999; 13:149–159

110. Wuthrich K. *NMR of Proteins and Nucleic Acids*. New York: John Wiley & Sons; 1986

111. Wuthrich K. NMR—This Other Method for Protein and Nucleic Acid Structure Determination. *Acta Crystallogr D* 1995; 51:249–270

112. Xu XP, and Case DA. Probing Multiple Effects on 15N, 13C alpha, 13C beta, and 13C′ Chemical Shifts in Peptides Using Density Functional Theory. *Biopolymers* 2002; 65:408–423

113. Xu Y, Xu D, Crawford OH and Einstein JR. A Computational Method for NMR-Constrained Protein Threading. *J Comput Biol* 2000; 7: 449–467

114. Yoshioka S, Aso Y, and Kojima S. Different Molecular Motions in Lyophilized Protein Formulations as Determined by Laboratory and Rotating Frame Spin-Lattice Relaxation Times. *J Pharm Sci* 2002; 91:2203–2210

115. Zhang H, Neal S, and Wishart DS. RefDB: A Database of Uniformly Referenced Protein Chemical Shifts. *J Biomol NMR* 2003; 25:173–195

116. Zuiderweg ER. Mapping Protein-Protein Interactions in Solution by NMR Spectroscopy. *Biochemistry* 2002; 41:1–7

7

Light Scattering

JOSEPH DEMEESTER, STEFAAN S. DE SMEDT, NIEK N. SANDERS, and JURGEN HAUSTRAETE

Laboratory of General Biochemistry and Physical Pharmacy, Department of Pharmaceutical Sciences, Ghent University, Gent, Belgium

Introduction

Scattering of radiation is an important phenomenon of electromagnetic waves. The theory of light scattering (concerning wavelengths between 400 and 700 nm) was developed by some of the greatest scientists of the twentieth century; among them, Einstein (1910), Lord Rayleigh (1910), Raman (1927), and Debye (1944). Rayleigh (1910) explained how diluted gases and air scatter light. He found that the intensity of the scattered light is predominantly in the blue end of the spectrum. Indeed, he showed that it is proportional to the inverse of the fourth power of the wavelength of the incident light, which explains many well-known phenomena like the blue color of the sky. The wavelength of the scattered light is identical to the wavelength of the incident light, as the photons are scattered elastically. However, a very small number of nonelastically scattered photons appear at higher or lower wavelengths (Raman scattering; see Chapter 5 of this book). Although the theory that explains light scattering was developed a long time ago, routine use did not begin until the commercial availability of stable lasers as light sources instead of mercury arcs.

The first part of this chapter explains the theory of static and dynamic light scattering (LS) by solutions. If light traverses a solution, attenuation of the light beam may occur both by absorption and scattering. Scattering only occurs when nonhomogeneous regions are present in the solution. In static light scattering (SLS), the average value of the fluctuating intensity of the scattered light is considered; we will explain how it can provide us with information on the size and molecular weight of the solute. The solute can be, among others, a protein or an aggregate. Due to the Brownian motion of the solute, the intensity of the scattered light is fluctuating. In dynamic light scattering (DLS), sometimes called quasi-elastic light scattering (QELS) or photon correlation spectroscopy (PCS), information on the dynamic (diffusion) properties of the solute is obtained from the microsecond fluctuations of the intensity.

The second part of this chapter deals with LS in the field of proteins. Although the "noninvasive" character of the method is a clear advantage in general, the need to remove small (dust) particles from solvents and solutions is often an experimental problem, especially for the characterization of aqueous (protein) solutions. Also, although commercial instruments make LS feasible in routine practice, researchers with little experience of LS should beware of misinterpretations of LS data.

Static Light Scattering by Small Molecules

Scattering Theory

In the nineteenth century, Lord Rayleigh (1881) developed the theory describing the scattering of light by particles/molecules with a small size compared to the wavelength of the incident light. Tanford (1961) described a very comprehensive outlay of this scattering theory. In particles/molecules smaller than about 5% of the wavelength of the light, all regions in the particle/molecule are subjected to a similar electric field, and the scattered waves are in phase. As most pharmaceutical proteins are smaller than about 30 nm, the information below describes the scattering properties of (nonaggregated) proteins in solution.

Importantly, the early theories on the scattering of light by small solutes already showed that the intensity of the scattered light is proportional to the sixth (!) power of the particle radius and to the reciprocal value of the fourth power of the wavelength. The basic equations that describe the scattering of light by infinite dilute dispersions or solutions of uniform small particles/macromolecules are summarized below. The normalized intensity of the scattered light (R_θ), also called the *Rayleigh ratio*, is defined as:

$$R_\theta = \frac{I_\theta}{I_o} r^2 = KcM \tag{1}$$

where I_o is the intensity of the incident light, I_θ is the intensity of the scattered light (per unit scattering volume) measured by the detector at a distance r from the scattering volume, c is the concentration (in g/mL), and M is the molecular weight of the solute. K, the optical constant, is explained below. To obtain I_θ, the scattering intensity of the solvent has to be measured and subtracted from the scattering intensity of the solution to know the contribution of the particles/macromolecules. R_θ is expressed in cm^{-1} and can be interpreted as the fraction of the intensity of the incident light, which is scattered per unit path length.

The equation used to calculate the optical constant K depends on the type of incident light. Using nonpolarized incident light, K is given by Equation 2; in the case of vertically polarized light, K is calculated from Equation 3:

$$K = \frac{2\pi^2 n_o^2 (\partial n/\partial c)^2 (1 + \cos^2\theta)}{\lambda_o^4 N_A} \tag{2}$$

$$K = \frac{4\pi^2 n_o^2 (\partial n/\partial c)^2}{\lambda_o^4 N_A} \tag{3}$$

where n_o is the refractive index of the solvent, $\partial n/\partial c$ is the refractive index increment, N_A is Avogadro's number, λ_o is the wavelength of the light *in vacuo*, and θ is the *scattering angle*, being the angle between the incident beam and the position of the detector. Nowadays, light-scattering instruments (see Figure 1) mostly use vertically polarized laser light, which leads to the important conclusion that I_θ is independent of the scattering angle. Consequently, measuring the intensity of the scattered light at only one angle (usually 90°) is sufficient to determine the molecular weight of the solute. K, and consequently I_θ (see Equation 1), depends on $\partial n/\partial c$. In other words, the particles/solute molecules must have a good contrast (i.e., the refractive index n of the solution has to differ significantly from n_o) to sufficiently produce scattered light at a reasonable power of the incident

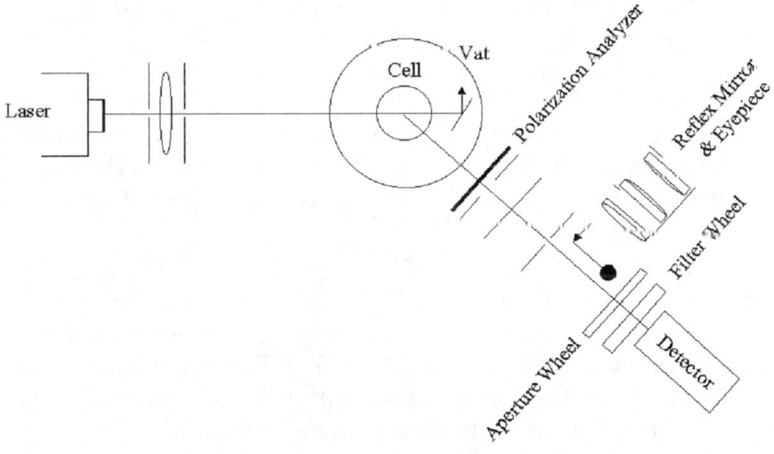

Figure 1 Optical system of a BI-200SM goniometer LS instrument (from Brookhaven Instruments, Long Island, New York, with permission). The cell is surrounded by an index matching Vat, and via lenses, filters, and apertures, the scattered light is measured by a detector at angle θ with respect to the outgoing laser beam.

light. Note that K depends on $\partial n/\partial c$ squared, which necessitates a very accurate determination of $\partial n/\partial c$ by a differential refractometer (see below).

The equations above are only valid for *diluted, monodisperse* solutions. For *nondiluted, polydisperse* solutions (with only one type of solute, i.e., a constant $\partial n/\partial c$ value), the scattering of the light is described by Equation 4, where M_w is the weight average molecular weight of the solute, and A_2 and A_3 are the second and third virial coefficients, respectively (note that in many textbooks, B and C are used, instead of A_2 and A_3):

$$\frac{K_c}{R_\theta} = \frac{1}{M_w} + 2A_2c + 3A_3c^2 + \dots \tag{4}$$

Plotting Kc/R_θ in function of the solute concentration shows a linear or parabolic relationship, and by regression analysis the weight average molecular weight M_w (i.e., the reciprocal value of the intercept), A_2 (which equals the slope), and A_3 can be obtained. M_w obtained in this way differs from (i) the (number) average molecular weight (M_n) obtained from colligative property measurements (such as osmotic pressure and freezing point depression), (ii) the z-average molecular weight obtained from sedimentation measurements, and (iii) the viscosimetric average molecular weight obtained from intrinsic viscosity measurements (M_v). Note that M_n is less than or equal to M_w, which is less than or equal to M_z. The molecular weight ratio M_w/M_n is often used as an indicator for the polydispersity of the molecular weight of solutes (macromolecules) in the solution. For proteins, M_w/M_n equals one; however, polysaccharides and synthetic polymers mostly have broad molecular weight distributions, and M_w/M_n values are often higher than two. Also notice that the obtained virial coefficients are average values; the second virial coefficient obtained by LS is a z-average value, the one obtained from osmotic pressure measurements is a weight average value (Tanford, 1961).

The Specific Refractive Index Increment

As explained above, good knowledge of $\partial n/\partial c$ (which is a solute-solvent dependent variable) is a crucial requirement to be able to calculate the optical constant K (by Equation 3) and thus the weight average molecular weight of the solute (by Equation 4). If experimentally possible, the $\partial n/\partial c$ value should be determined with a differential refractometer by measuring the difference in refractive index between the solution (n) and the solvent (n_o) as a function of the solute concentration:

$$\frac{\Delta n}{c} = \frac{n - n_o}{c} = \alpha_1 + \alpha_2 c \tag{5}$$

The intercept α_1 corresponds to $\partial n/\partial c$, which is wavelength, solvent, and temperature dependent. Huglin (1972, 1989) compiled $\partial n/\partial c$ values of many solutes. Note that $\partial n/\partial c$ depends on the wavelength of the light used. However, the wavelength (λ_o) of the beam in modern laser LS instruments (e.g., 632.8 nm in the case of a He-Ne laser) differs from the wavelength of the light available in differential refractometers (which is mostly 436 or 546 nm, as often mercury lamps are used). One can calculate the $\partial n/\partial c$ value at the wavelength of the laser beam in the LS instrument when the $\partial n/\partial c$ is known for at least two different wavelengths using the Cauchy plot, which describes a linear relationship between $\partial n/\partial c$ and $1/\lambda_o^2$ (Huglin, 1972).

For protein solutions a $\partial n/\partial c$ value of 0.189 mL/g is a sufficient approximation in many applications (Graupner et al., 1999). As $\partial n/\partial c$ depends on the chemical composition of the solute (higher order structures of the solute have almost no influence), $(\partial n/\partial c)_p$ for proteins is a function of the particular amino acid composition (McMeekin et al., 1962; Kendrick et al., 2001). Hence, for accurate measurements, $(\partial n/\partial c)_p$ must be determined experimentally. Proteins are often conjugated (to, e.g., polysaccharides), and the $(\partial n/\partial c)_{cp}$ value for the conjugated protein can be calculated from $(\partial n/\partial c)_p$ before conjugation and the $(\partial n/\partial c)_c$ value of the conjugating species (Kendrick et al., 2001).

The $\partial n/\partial c$ values are strongly affected by the interaction between solvent and polymer.

The approximate relationship between $\partial n/\partial c$ and solvent refractive index is given by

$$\frac{\partial n}{\partial c} = \left(\frac{\partial n}{\partial c}\right)_{\text{ref}} - \frac{dV}{dm}(n - n_{\text{ref}}) \tag{6}$$

where ref refers to a reference solution in which the refractive index increment and the partial specific volume (dV/dm) are known. For proteins, the partial specific volume is about 0.73 mL/g. For heteropolymers, the refractive index increment can be expressed as:

$$\frac{\partial n}{\partial c} = \Sigma w_i \left(\frac{\partial n}{\partial c}\right)_i \tag{7}$$

where w_i and $(\partial n/\partial c)_i$ are the weight fraction and the refractive index increment of the monomer, respectively (Liu and Chu, 2002).

Instrumentation

As explained in the previous section, LS measurements on small solutes only require a simple instrument, because a measurement of the intensity of the scattered light at only one (fixed) scattering angle (e.g., 90°) is sufficient to determine the Rayleigh ratio (Equations 2 and 3). Many LS instruments allow measuring at three or more scattering angles and are thus perfectly suited for the characterization of small solutes. A stable light source is crucial. Many older instruments used mercury lamps with filters, which explains why most LS data in literature were obtained using 436 nm and 546 nm light (Brice et al. 1950; Huglin, 1972). More recent LS apparatus use laser light sources. Most instruments are equipped with a He-Ne or a Ga-As laser, which produce fairly long waves of 633 nm and 690 nm, respectively, although the intensity of the scattered light decreases when an incident laser beam of higher wavelength is used (see Equations 2 and 3).

Instruments can be calibrated on an absolute basis using information on the optics and geometry of the instrument as described by Huglin (1972). To do so, after attenuation with filters of known transmittance, one measures the intensity of the incident beam (I_o) using the same detector that measures the intensity of the scattered light. Therefore, to reduce reflections, the LS cell is immersed in a liquid with the same refraction index as the wall of the LS cell. Taking into account attenuation factors, refractive indices of cell and solvent, and the geometry of the instrument, the Rayleigh ratio can be calculated on an absolute basis. However, a more common technique to calibrate the instrument makes use of a pure organic solvent with a known Rayleigh ratio, for example, toluene (Wyatt, 1993; Santos et al., 1996). For toluene, R_θ equals 14.02×10^{-6} cm^{-1} when measured at 90° using 633 nm laser light (Kaye and McDaniel, 1974):

$$R_\theta = R_{\theta,\text{toluene}} \cdot \frac{n_o^2 I_\theta}{n_{\text{toluene}}^2 I_{\theta,\text{toluene}}} \tag{8}$$

Alternatively, a standard with known M_w can be used (see Equation 4).

Most differential refractometers, used to determine $\partial n/\partial c$, measure the deflection of a light beam after passing consecutively through a cell consisting of two chambers separated by a flat glass or quartz window, inclined with respect to the incident beam. One chamber contains the solution; the other chamber contains the (pure) solvent. The deflection of the light beam is proportional to the difference in refractive index between solvent and solution (Brice and Halwer, 1951). A good thermal stabilization of the refractometer assures almost the same temperature in the solvent and solution compartments, which is necessary as temperature variations are the major source of variations in refractive index measurements. Note also that the differential refractometers used as detectors in chromatography instruments are perfectly suited to measure $\partial n/\partial c$, given they operate at the same wavelength as the LS instrument. Interferometric refractometers are also suitable for sensitive measurements of $\partial n/\partial c$ (Wyatt, 1993).

Potentials and Limitations of Single Angle Light Scattering in Protein Research

One main advantage of LS is that it allows an accurate determination of the (weight average) molecular weight (error less than 5%) of a polymer or protein in solution in an absolute, nondestructive way. Following the specifications reported for today's instruments, the lowest molecular weight that can be measured accurately is estimated to be about 2000 g/mol. Small variations in molecular weight (e.g., in cases of protein deamidation) cannot be detected; mass spectrometry is by far better suited for this purpose.

When LS measurements are done at sufficiently high protein concentrations, the second virial coefficient (which characterizes the protein-solvent interactions, being positive in cases of repulsive forces, being negative in cases of attractive forces, being zero when good and poor solvent are balanced and the solvent is a theta solvent) and even the third virial coefficient can be obtained. For solutes with narrow molecular weight distribution, the virial coefficients are identical to those obtained from osmotic pressure measurements. LS has an advantage in that semipermeable membranes are not necessary to do the measurements (Tombs and Peacocke, 1974). It is therefore the recommended technique to determine virial coefficients (Narayanan and Liu, 2003). In x-ray crystallography of proteins, the purification and analysis of the x-ray scattering data is now relatively straightforward. The growth of high-quality crystals is, however, often a stumbling block (McPherson, 1990; Wilson, 2003). Second virial coefficients are used to predict if the crystallization conditions will allow crystal growth without waiting for too long a time (Farnum and Zukoski, 1999; Wilson, 2003). A_2 has to be in the *crystallization window*: -8×10^{-4} mL mol/g^2 less than A_2 less than -2×10^{-4} mL mol/g^2 (George and Wilson, 1994; Narayan and Liu, 2003).

Removal of particles of extraneous matter from the solvents and solute solutions is a serious practical problem in all LS experiments. In protein and polymer solutions, microgel particles can exist, which enormously contribute to the scattered light (Huglin, 1972). Also, the high refractive index of dust particles (about 1.5 compared to 1.34 for water) and their large dimensions make scattering measurements impossible (remember that I_θ is proportional to the sixth power of the particle diameter). Filters with pore sizes down to 20 nm are used to remove particles, but when small volumes (less than about 0.3 mL) are used, syringe filters cannot be used, but special filter holders with small dead volume are appropriate. Alternatively, centrifugation can be used.

Also, the cell walls and the index matching liquid used should be free of dust particles. Standard operating procedures in the pharmaceutical industry should contain a rigorous description of "good scattering practice" including a rigorous description of particle removal.

Static Light Scattering by Large Molecules

Scattering Theory

The LS discussion in the preceding paragraphs was limited to particles/solutes that are much smaller than the wavelength of the incident light, typically smaller than about 30 nm (Tanford, 1961). In larger particles, such as protein aggregates, destructive interference of the scattered light occurs. At very low scattering angles, the scattered beams are almost in phase and no destructive interference occurs. However, the larger the scattering angle, the more the scattered beams are out of phase, the stronger destructive interference occurs. Clearly, contrary to small solutes, the intensity of the scattered light depends on the scattering angle. The following $P(\theta)$ function can be defined:

$$P(\theta) = \frac{\text{scattered intensity for a large particle}}{\text{scattered intensity without interference}} \tag{9}$$

It follows that Equation 4 only remains valid when the intensity of the scattered light is measured at very low scattering angles. Alternatively, one can measure the intensity of the scattered light at larger scattering angles and extrapolate the obtained values for scattering angle zero.

At first sight, the destructive interference seems a nuisance. However, as discussed below, it will give us very useful supplementary information on the molecular configuration of the solute (polymer). The form of $P(\theta)$ depends on the geometric shape of the scattering particle: as Equation 10 shows, $P(\theta)$ is determined by the radius of gyration of the scattering particle (Tanford, 1961).

$$P(\theta) = 1 - \frac{16\pi^2}{3\lambda^2} R_G^2 \sin^2 \frac{\theta}{2} \tag{10}$$

As the equation for $P(\theta)$ is only valid when particles are independent of each other, it is necessary to extrapolate scattering data at each angle to zero concentration before we can use Equation 10.

For solutions of large, nonuniform solutes, the dependence of the intensity of the scattered light on the solute concentration and scattering angle is given by the very important Equation 11, in which $\langle R_G^2 \rangle$ is the square of the z-average radius of gyration:

$$\frac{K_c}{R_\theta} = \frac{1}{M_w}\left(1 + \frac{16\pi^2}{3\lambda^2}\langle R_G^2\rangle_z \sin^2\frac{\theta}{2} + \ldots\right) + 2A_2c + 3A_3c^2 + \ldots \tag{11}$$

$$\langle R_G^2\rangle_z = \frac{\Sigma M_i c_i (R_G^2)_i}{\Sigma M_i c_i} \tag{12}$$

The angle brackets indicate that, if the solute molecules are flexible (e.g., certain polymers), the obtained R_G^2 is an average value of all conformations. The radius of gyration is defined by:

$$R_G = \sqrt{\frac{\Sigma m_i R_i^2}{\Sigma m_i}} \tag{13}$$

where m_i and R_i are, respectively, the mass of mass point i of the particle and the distance of this mass point from the center of mass of the particle. Despite the somewhat complicated definition, the radius of gyration is very unique. The radius of gyration can be obtained even without knowledge of refractive index, refraction index increment, or molecular weight (cf. Equation 10).

Indeed, its definition does not depend on the shape of the particle; consequently, one need not make assumptions on the shape of the solute when calculating R_G from LS experiments. In the calculation of other size parameters, such as the hydrodynamic radius from dynamic LS, one assumes the solute to have a certain shape (e.g., spherical).

The Zimm Plot

Considering Equation 11, it is clear that to obtain, for example, the molecular weight of larger solutes from LS, one should measure the intensity of the scattered light at different solute concentrations and different scattering angles. Zimm (1948a,b) found a simple method to plot the complex function of scattering on a planar surface. By plotting Kc/R_θ in function of $\sin^2(\theta/2) + Sc$, where S is an arbitrary constant, the data can be spread. The method only holds for solutes larger than 5% and smaller than about half of the wavelength of the incident light. If the solute molecules are larger, $P(\theta)$ becomes too small to make accurate measurements. Indeed, technically it is impossible to measure the scattered light at scattering angles very close to zero due to interferences with the incident beam. The spreading factor S can be chosen arbitrarily and gives approximately equal weight to the concentration and angular contributions (Wyatt, 1993).

As can be seen in Figure 2, the data are equally spread using four concentrations and nine scattering angles, giving thirty-six data points that are extrapolated by parabolic regression analysis (or linear regression if no curvature is observed). Thus, we obtain two limiting plots.

Extrapolating to $\theta = 0°$, we obtain $P(\theta) = 1$ and can fit the data according to the equation:

$$\lim_{\theta \to 0} \frac{K_c}{R_\theta} = \frac{1}{M_w} + 2A_2c + 3A_3c^2 \tag{14}$$

This is the "$\theta = 0$" curve from which we obtain $M_w = 1{,}285{,}000$ g/mol, $A_2 = 1.4 \times 10^{-4}$ mol mL g^{-2}, and $A_3 = 0.057$ mol mL^2 g^{-3}. Extrapolation

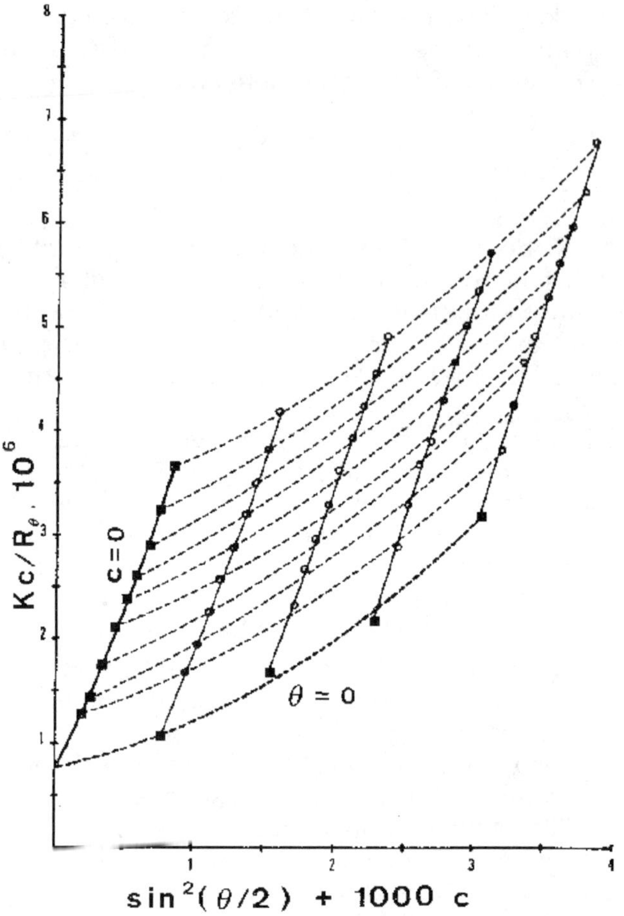

Figure 2 Example of a Zimm plot. (Reprinted with permission from Advances in Chemistry Series 181, p. 98. Copyright 1979 American Chemical Society.) The data are for hydroxyethylcellulose in water (Demeester et al., 1979).

to $c = 0$ causes the disappearance of the virial terms:

$$\lim_{c \to 0} \frac{K_c}{R_\theta} = \frac{1}{M_w}\left(1 + \frac{16\pi^2}{3\lambda^2}\langle R_G^2\rangle_z \sin^2\frac{\theta}{2} + \ldots\right) \tag{15}$$

This is the "$c = 0$" curve from which we obtain $M_w = 1{,}258{,}000$, and the z-average radius of gyration = 81 nm. Both extrapolated curves cross the Kc/R_θ axis at $1/M_w$.

Zimm plots are generated automatically by the software accompanying the LS instrument. However, it is recommended not to use this facility as a "black box." When multiple species are present, the data analysis becomes more complicated. A typical example of a situation where Zimm analysis becomes tricky is when nonprotein aggregates are present, coexisting with proteins (Liu and Chu, 2002).

From the theoretical considerations above, it is clear that the $\partial n/\partial c$ has to be known to determine M_w, A_2, and A_3. The calculated radius of gyration, however, is not dependent on $\partial n/\partial c$.

Instrumentation

To characterize large molecules by LS, one needs to measure at different scattering angles to evaluate $P(\theta)$. Zimm (1948b) had already described an instrument able to measure between 26° and 143°. Brice et al. (1950) conceived the first commercial instrument used during the second part of the twentieth century in many research laboratories. His design is still used in many commercial scanning goniometers of today: the sample cell is surrounded by a refractive index matching liquid while a stepping motor controls the motion of the detector, being a photomultiplier or an avalanche photodiode. The angular range is usually between about 15° and 150°. With instruments where only 2 or 3 fixed angles (usually 45°, 90°, and 185°) are available, it is difficult to estimate $P(\theta)$.

A good alternative to scanning goniometers is the use of multiangle laser light scattering instruments (Figure 3), produced commercially by different manufacturers. Such instruments have a number of detectors (three to about eighteen) placed at fixed scattering angles, giving the possibility of measuring simultaneously at the different angles. This is an advantage in chromatographic detection or for following time-dependent scattering (Wyatt, 1993).

Dynamic Light Scattering

Scattering Theory

With the introduction of lasers with a collimated beam and coherence, it became possible to record the temporal variations of the scattered light. The fluctuations in the scattered light are due to Brownian motion, caused by the bombardment by surrounding molecules. Indeed, small solutes move quickly, and the intensity variations are faster than for large particles, as can be seen in Figure 4.

The dynamic properties of the scattering solute molecules are derived from the (auto)correlation function $[G(\tau)]$ calculated from the fluctuations of the scattered light. $G(\tau)$ of the scattered intensity $I(t)$ is the average of the product of the scattered intensity $I(t)$ at time t multiplied by the scattered intensity measured a time interval τ later [i.e., $I(t + \tau)$]:

$$G(\tau) = \langle I(t) \times I(t + \tau) \rangle \tag{16}$$

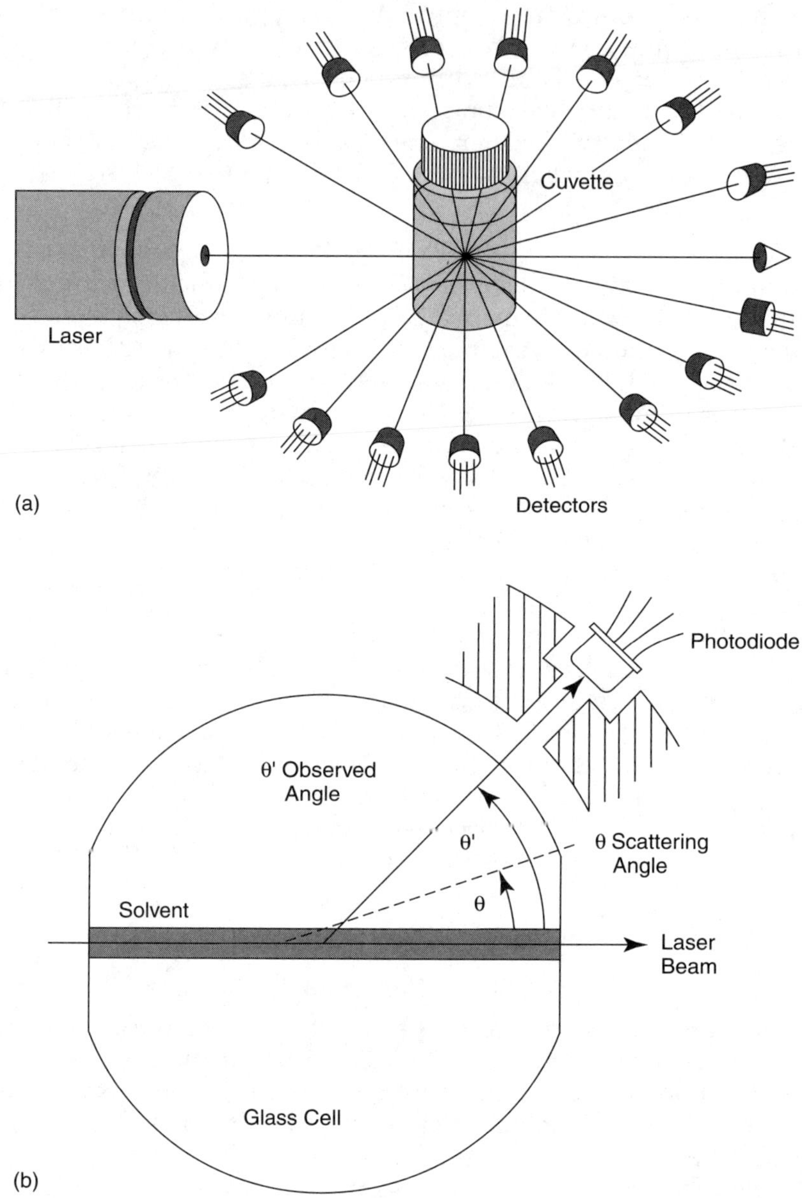

Figure 3 (a) Dawn detector array. (b) The flow cell includes a glass cell, with a 2-mm-diameter throughbore. The ultrafine illuminating laser beam passes through the entire cell. Refractions: θ, actual scattering angle; θ', apparent (and instrument) scattering angle (from Wyatt Technology Corp., Santa Barbara, CA, with permission).

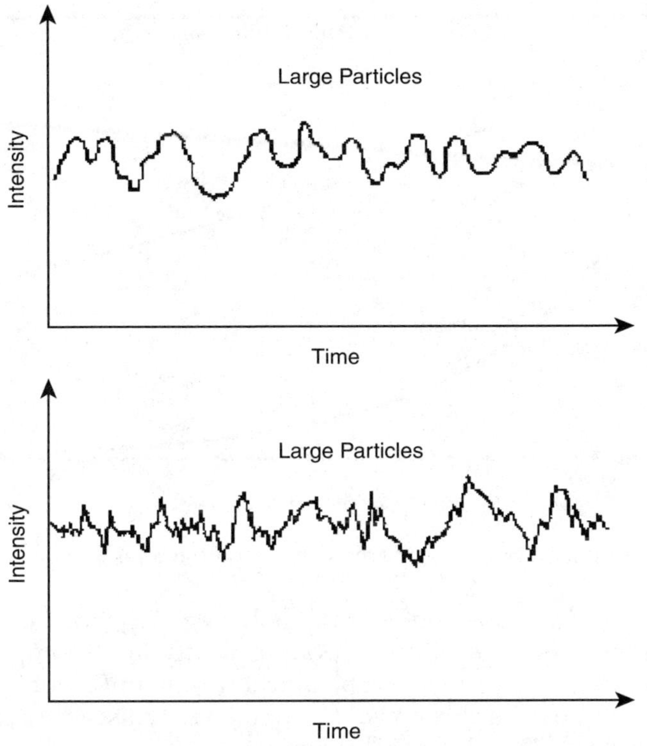

Figure 4 Typical intensity fluctuations for large and small particles.

At low τ values, the positions of the solute molecules and thus the intensities of the scattered light are highly correlated: $G(\tau)$ is large (see Figure 5). Later on (i.e., at higher τ values), the positions of the solute molecules are no longer correlated, which explains the decrease in $G(\tau)$. Indeed, it can be shown that when a large number of monodisperse solute molecules/particles are moving randomly, $G(\tau)$ decays exponentially in function of the time delay τ:

$$G(\tau) = A + B \exp(-2\Gamma\tau) \tag{17}$$

A and B are instrumental factors, the decay rate (Γ) is related to the translational diffusion coefficient (D_τ) of the particles, and q is the scattering vector:

$$\Gamma = D_\tau q^2 \tag{18}$$

$$q = \frac{4\pi n}{\lambda_o} \sin(\theta/2) \tag{19}$$

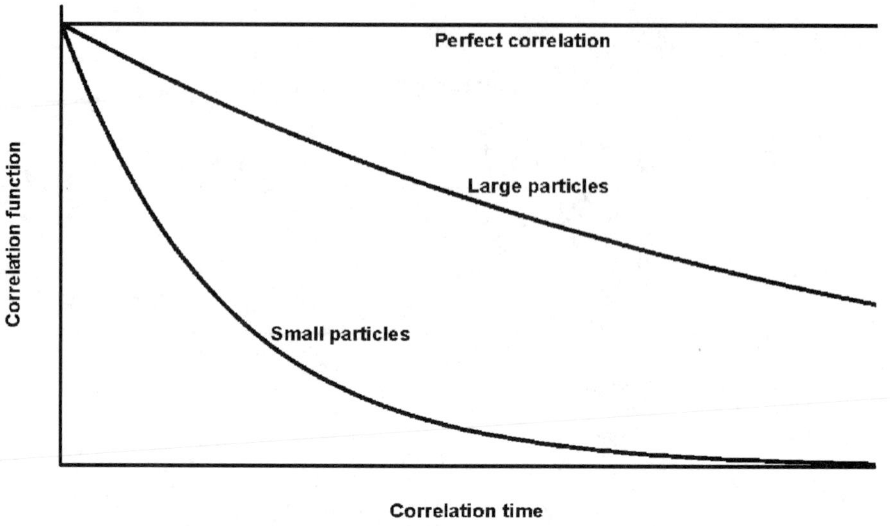

Figure 5 Correlation function for small and large particles.

The decay rate (Γ) is affected by the "polydispersity" of the sample: polydispersity in size distribution, polydispersity in the properties of the material, polydispersity in shape all contribute to Γ. Note that it is hard to get complete distribution data from the polydispersity analysis: for example, many different size distributions give very similar correlation functions.

In pharmaceutical applications, knowledge on both the diffusion coefficient as well as on solute/particle size is required. There is no universal relation between diffusion coefficient and size, but for non-interacting monodisperse rigid spheres, the Stokes-Einstein equation can be applied:

$$R_H = \frac{k_B T}{6\pi\eta D_\tau} \tag{20}$$

where k_B is Boltzmann's constant (1.3807×10^{-23} J K^{-1}), T is the absolute temperature (K), η is the solvent viscosity, and R_H is the hydrodynamic radius.

The description of the different methods that allow fitting of the correlation function of multimodal or polydisperse samples (where Equation 17 is not valid anymore) is outside the scope of this chapter, and the interested reader is referred to literature (e.g., Chu, 1991; Finsy, 1994). Also, a basic assumption in dynamic LS is that multiple scattering does not occur. In concentrated samples, especially for large solute molecules/particles, multiple scattering occurs, which strongly contributes to $G(\tau)$. Special methods, which are also outside the scope of this contribution, are needed to analyze the fluctuations of the scattered light.

Instrumentation

Chu (1991) has presented an excellent overview of both instrumental principles and data acquisition in dynamic LS. Dynamic LS invariably makes use of laser light sources, as coherence is required. He-Ne lasers (633 nm) are most often used while Ar lasers (488 nm) are also popular, as the intensity of scattered light at 488 nm is about 3 times higher than at 633 nm (see Equation 2). Most static LS instruments also allow dynamic scattering experiments. Small collimating pinholes (about 200 μm), or an optical fiber, are necessary so that one coherence area ("speckle") falls on the detector face. As discussed above, for large particles, the intensity of the scattered light depends on the scattering angle. Therefore, when a multi-angle goniometer is available on the instrument, to enhance the signal to noise ratio, it is recommended to perform the dynamic LS measurements at the angle resulting in the highest intensity of scattered light.

To calculate $G(\tau)$, digital correlators are used that operate preferentially in real-time, so that every incoming pulse (generated by a photon) can be processed. Sampling times are generally of the order about 100 ns to thousands of seconds to cover a wide range of scattering particles (in the range of less than 1 nm up to more than 1 μm).

Potentials and Limitations

Generally, dynamic LS is a very simple, noninvasive technique for the determination of diffusion coefficients and hydrodynamic radii. However, the size resolution is rather low and, despite the use of the best statistical analysis methods, the obtained size distribution is rather inaccurate. Indeed, particles must differ in radius by about twofold (or mass about eightfold) to be resolved. Typically, analysis of dynamic LS measurements shows intensity of scattered light in function of the size of the particles. Converting the intensity distribution to a weight or volume distribution requires the division of each point by the third power of the radius. This makes the weight distribution much less accurate than the intensity distribution. The intensity of scattered light being proportional to the sixth power of the particle radius, it is easy to understand that few large particles produce highly intense scattered light. This makes it very difficult, if not impossible, to determine a number distribution, that is, the number of particles in function of size: in presence of large particles, small ones are very difficult to detect. Finally, due to sedimentation, diffusion coefficient and size of particles larger than about 5 μm cannot be determined by dynamic LS.

Whenever the refractive index of the particles is sufficiently different from the solvent, as to be detectable, we need make no assumption about the optical properties to determine the hydrodynamic radius.

Zeta Potential Measurement

The zeta potential of charged particles such as protein complexes is often determined by electrophoretic dynamic light scattering. Dynamic LS instruments that allow both size and zeta potential measurements are widely available. When an electrical field is applied to a charged particle, its motion vector is the vector sum of its random Brownian motion and its charge dependent motion in the direction of the oppositely charged electrode. Using laser Doppler velocimetry to measure the particle velocity, we can thereby calculate the zeta potential (Miller et al., 1991).

According to the classical DLVO (Derjaguin, Landau, Vervey, and Overbeek) theory (Hunter, 1993), the physical stability of a dispersed system is determined by the charge of the particles causing repulsive electrostatic forces. When a particle moves, ions and solvent in a boundary move together with the particle. The potential at this boundary, which is at a short but undefined distance from the real particle surface, is the zeta potential, having a lower absolute value than the surface potential. When the zeta potential is low—absolute value typically below about 30 mV—particle aggregation may occur. For proteins, the zeta potential decreases with increasing pH, and the isoelectric point is the pH at which the zeta potential equals zero.

Light Scattering Detection in Separation Techniques

From the previous chapters, it is clear that LS is highly recommended to measure properties like average molecular weights or radii, although it is less suited to obtain information on molecular weight distributions or size distributions. However, combined with separation techniques, LS does allow constructing the molecular weight or size distribution upon analysis of the monodisperse fractions. LS is often combined with size exclusion (gel permeation) chromatography (Brown, 1996) and recently with field flow fractionation (Thielking et al., 1995; Thielking and Kulicke, 1996; Korgel et al., 1998), which are described in other chapters of this book.

In size exclusion experiments where the relationship between the elution time and the molecular weight or size of the solute cannot be obtained from a proper calibration of the column, LS detection offers all the information one needs; for each fraction that elutes from the column, the concentration c_i, the molecular weight M_i, the radius of gyration, and even the hydrodynamic radius can be obtained. Furthermore, not only M_n, M_w, M_z . . . of the sample, but also the number, weight, and z-average radii of gyration can be calculated (Wyatt, 1993).

Large particulates (such as dust) are easily separated and do not interfere with the species of interest.

When used in combination with a gel permeation chromatography setup, a small flow-through cell is mounted in the LS instrument described above which is further connected to a concentration-sensitive detector. Low-angle or multi-angle LS instruments connected to a refractive index detector are most suited (Wyatt, 1993; Brown, 1996). For "difficult samples," such as protein conjugates (see below) where the refractive index increment is not constant, a third (often absorbance) detector is necessary (Kendrick et al., 2001). Wen et al. (1996) have published an excellent review on the advantages of adding an absorbance detector. Adding a viscosimetric detector even allows measuring the intrinsic viscosity of each fraction and to determine the parameters of the Mark Houwink equation and M_v.

Evaporative LS detectors are universal concentration detectors for most separation techniques (liquid chromatography, capillary electrophoresis, etc.). The detection is based on nebulization and evaporation of the solvent. The remaining particles enter the detector cell, causing light scattering in the detector chamber with intensity being exponentially related to the mass of the particles (Stolyhwo et al., 1983).

Applications of Light Scattering for Characterizing Proteins

Light Scattering to Study the "Pharmaceutical Quality" of Therapeutic Proteins

Characterization of Monomeric and Oligomeric Proteins by Light Scattering

Measuring the hydrodynamic radius by dynamic LS with cumulants analysis is usually sufficient to distinguish between monomeric protein, protein monomer accompanied by oligomers (dimers, trimers, etc.), and protein monomer with some large protein aggregates (Wilson, 2003). As an example, Baldini et al. (1999) described the use of dynamic LS to study the equilibrium between monomers and dimers of β-lactoglobulin; they also described how the dissociation constant of the dimer can be calculated from dynamic LS measurements. Based on dynamic LS measurements, Schibli et al. (2002) detected that human β-defensin dimers are asymmetric (ellipsoidal). In this study, they used the relationship (Equation 21) between molecular weight M_{app} and hydrodynamic radius, which is valid for globular proteins and thus allows an estimation of the protein molecular weight from dynamic LS measurements (Creighton, 1992):

$$M_{\text{app}} = 736\ R_H^{2.34} \tag{21}$$

In this context, we would like to emphasize that, especially at low ionic strength, the hydrodynamic radii of proteins as obtained by dynamic LS may be overestimated (Schurr, 1980; Hernandez-Contreras et al., 1997), as they show a typical polyelectrolyte behavior.

Although the current compendia such as the European, Japanese, or United States pharmacopoeias do not describe LS tests in their monographs, in the pharmaceutical industry LS is used for in-house quality control. However, literature on LS studies dealing with the stability of therapeutic proteins is rather scarce, possibly because pharmaceutical companies prefer not to publish on difficulties they encounter in formulating their proteins. We would like to refer to LS work on lysozyme in different media, probably the most intensively studied protein in pharmaceutical and structural research. Literature on lysozyme gives a good idea indeed on how LS combined with other tools gives information on protein behavior in solution (Poznanski et al., 2003; Zhang and Liu, 2003).

Characterization of Conjugated Proteins by Light Scattering

Many therapeutic proteins are glycosylated or PEGylated. The European Agency for the Evaluation of Medicinal products (EMEA, 2003) states that "The polymer reagent used in the manufacture should be well characterized, including determination of the polydispersity of mass. . . . Methods for routine verification of this higher order structure, including the corresponding total molecular size, should be selected and validated. Furthermore, methods should be in place for quantification of any free, unconjugated protein."

In contrast to nonconjugated proteins, there is no general relationship anymore between molecular weight and size of the conjugated protein. Typically, conjugated proteins are analyzed by size exclusion chromatography (Wen et al., 1996). As an example, Kendrick et al. (2001) give detailed information on the analysis of conjugated proteins by a size exclusion instrument equipped with a "triple detection system" (static LS, UV, and IR) to determine the degree of conjugation, the molecular weight of the conjugate, and the polypeptide mass.

Characterization of Protein Aggregation by Light Scattering

In a recent paper, Liu and Chu (2002) excellently overviewed the influence of different ions, pH conditions, and heat on the aggregation and conformation of proteins and how this can be studied by LS. In the preparation of protein pharmaceuticals, the proteins are often purified from fermentation broths. In many of the purification schemes, microfiltration or ultrafiltration is necessary, and filters in the size range of 0.01 to 1 μm are often used (e.g., for sterilization purposes). By changing pH, ion composition, or temperature, optimum conditions for filtration without loss of protein can be found by size control based on LS measurements. In other processes, proteins are precipitated, followed by filtration or centrifugation. In this case, aggregation must be sufficient. Knowledge of size and zeta potential are useful for the optimization of ultrafiltration, a process governed by electrostatic and hydrodynamic resistance effects (Bowen and Jenner, 1995; Bowen and Williams, 1996; Bowen et al., 1996). LS analysis before and after filtration of protein pharmaceuticals gives useful information about eventual damage caused by the filters in the manufacturing process (e.g., viral removal). High filtration rates with fine filters cause high shear forces, may damage proteins, and introduce new aggregation.

In protein formulations, aggregates have to be avoided, as they may lead to an immunogenic response and clogging upon parenteral delivery. Drastic protein aggregation can be seen by the naked eye. However, in pharmaceutical development, one should be able to detect very early changes of only a few percent. When protein solutions degrade, they often form a limited amount of white, often amorphous visible particles after months of storage. In the *United States Pharmacopoeia* guidelines for submitting a request for revision (USP, 2003), there is now a recommendation to detect aggregates: "Typical procedures for detection and quantification of aggregates include size exclusion chromatography, often with on-line LS, PAGE under native conditions, or capillary electrophoresis."

Dynamic LS is particularly suited to predict formation of aggregates below 0.01%. This is illustrated in Figure 6. As an example, using dynamic (and static) LS, Weijers et al. (2002) studied the effect of ionic strength on the aggregation of ovalbumin; they found linear chains with little branching until close to the gel point at low ionic strength and branched aggregates at higher ionic strengths. Also, aggregation of bovine serum albumin (Hagiwara et al., 1996) and β-lactoglobulin (Gimel et al., 1994; Le Bon et al., 1999) were studied by dynamic and static LS.

Size exclusion chromatography (SEC) suffers from being unable to resolve very large aggregates, and dilution (when eluting) or change in solvent conditions can change aggregate structures. Static LS detection helps the interpretation of SEC chromatograms by detecting the molecular mass independent of conformation. Often, aggregates elute near the exclusion limit, and the M_w value can be calculated—even without complete resolution. An example is given in Figure 7.

Techniques such as circular dichroism, calorimetry, or Fourier transform infrared (FTIR) spectroscopy, suited to study the unfolding of secondary protein structures, are not suited to predict protein aggregation, as most proteins conserve the majority of their secondary structure after aggregation (Fish et al., 1985; Hsu and Arakawa, 1985). Indeed, rather than changes in secondary structure, interchain disulfide bond formation or association of different hydrophobic surfaces are major factors in protein aggregate formation (Arakawa et al., 2001). Clearly, LS cannot tell us the complete story of protein aggregation. It should be combined with other nondenaturating analysis methods like analytical ultracentrifugation, capillary or native gel electrophoresis, and chromatography (see related chapters in this book) and with denaturating methods such as disulfide bond reduction or SDS-PAGE

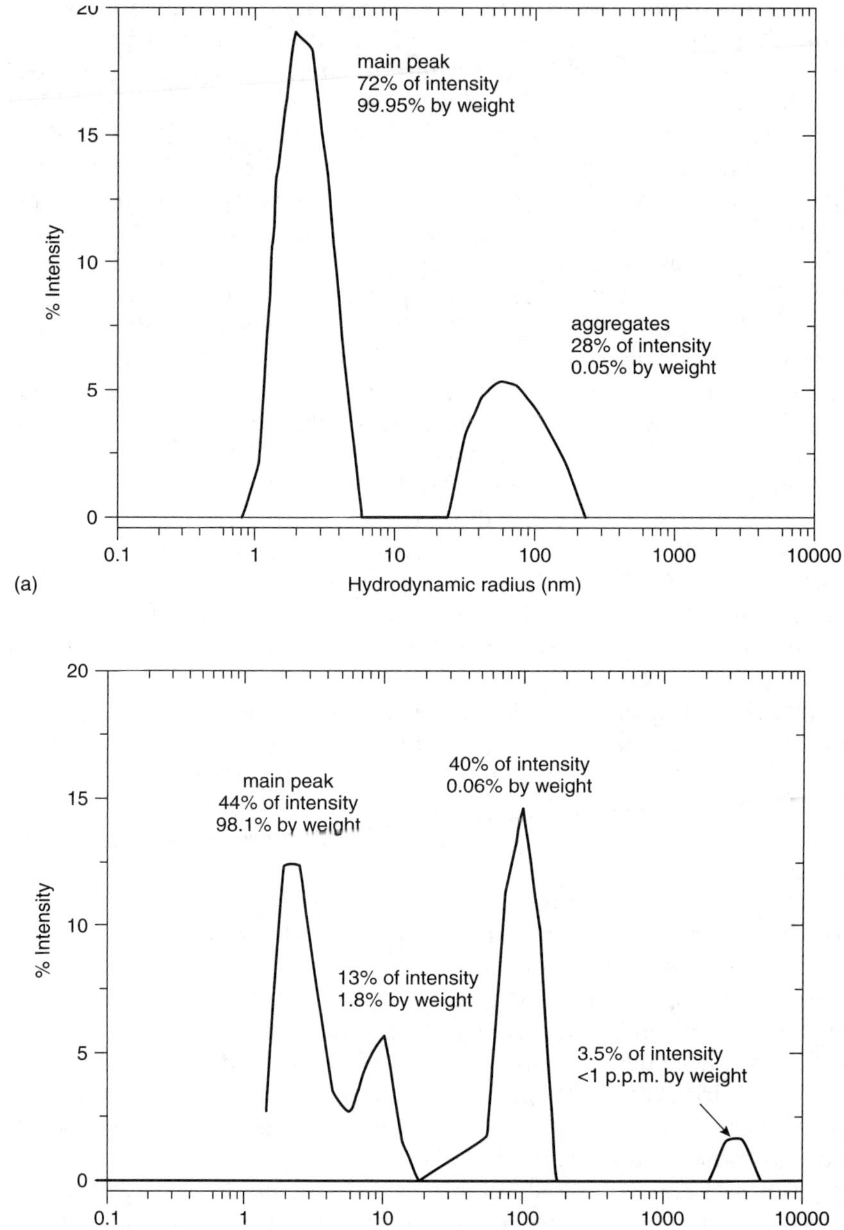

Figure 6 Dynamic LS data for two different lots of a protein. In (a), the main peak represents 72% of the scattering intensity and an estimated 99.95% fraction by weight. The aggregate peak is 28% of the intensity but only 0.05% by weight. The other lot (b) contains much more aggregate, and the aggregate resolves into 3 peaks, but this lot is still 98.1% main peak. The aggregate peak at ~3000 nm (the upper size limit for this instrument) is less than 1 part per million by weight [data from John Philo (2004)].

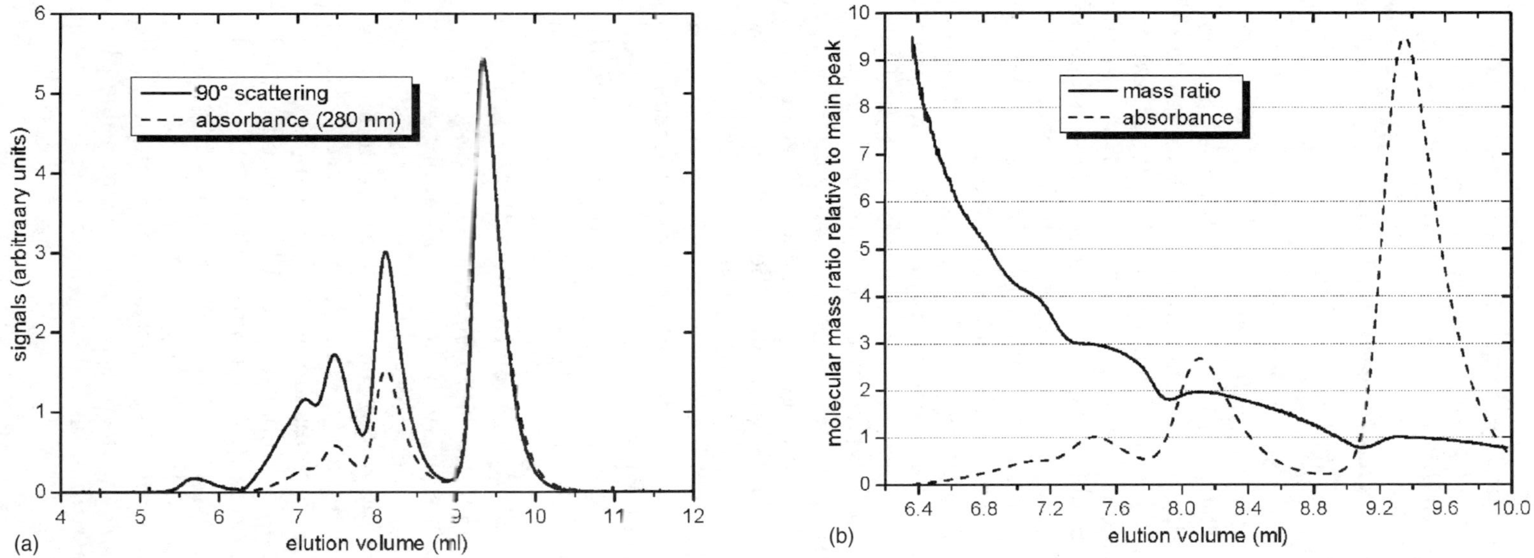

Figure 7 (a) Size exclusion chromatography data for an Fc-fusion protein with a strong tendency to form noncovalent oligomers. Those oligomers were metastable—upon dilution, they took hours to dissociate and thus were easily separated by SEC. The solid curve shows the 90° static LS chromatogram and the dashed curve the absorbance chromatogram. The curves have been scaled to match the peak heights for the main peak (monomer). This clearly demonstrates that the oligomers give much larger LS signals relative to the UV data. Also note the peak at ~5.7 mL (approximately the exclusion volume for this column) that is present in the LS data but absent in the UV data. That peak represents trace amounts of very high molecular mass aggregates (amounts too low to produce a UV peak, but nonetheless producing strong scattering signals). (b) This graph shows the molecular mass data calculated from the same chromatograms. The vertical axis is scaled so that the mass of the main monomer peak is defined to be one unit and directly counts the number of monomers in each species. The mass ratio plateaus have values of 1, 2, 3, and 4, clearly identifying the peaks at ~7.0, 7.25, 8.1, and 9.3 mL as tetramer, trimer, dimer, and monomer [data from John Philo (2004)].

(sodium dodecyl sulfate-polyacrylamide gel electrophoresis). Size exclusion chromatography combined with LS detection is an important tool in obtaining a good insight into the native, partially unfolded and unfolded conformations of proteins in solution and to understand how the hydrophobic surface area increases, leading often to protein aggregation (Remmele et al., 1999; Matsuura et al., 2001).

Light Scattering to Study Protein Behavior Preceding Crystallization

A combination of static and dynamic LS is probably the most revealing method for optimizing the crystallization conditions in protein solutions (Muschol and Rosenberger, 1995). The crystallization times can be drastically enhanced from months to days (Habel et al., 2001; Huang et al., 2002). LS allows us to study the different steps in the conversion of individual protein molecules into a crystal with good optical quality for x-ray analysis of the three-dimensional structure. Wilson (2003) gives an excellent overview of the diagnostic value of LS in this field and of the numerous publications on nucleation. Besides sequencing, revealing the three-dimensional structure of proteins by x-ray crystallography is crucial for understanding the functions of proteins that will be discovered by proteomics technologies.

As previously discussed in the paragraphs on static LS, the knowledge of the second virial coefficient tells us if attractive or repulsive forces are present between the molecules. Besides static LS, dynamic LS is also useful for optimizing protein crystal growth experiments. In contrast to dynamic LS, microscopy is much less sensitive; when crystals become visible already too many nuclei have formed. Saridakis et al. (2002) explained that the optimal time to decouple the nucleation growth corresponds to the moment where dynamic LS shows a significant change in size distribution.

Light Scattering Study of Protein Interactions

Light Scattering Study of Protein Assembly

The use of dynamic LS in the study of protein assembly, an organized protein aggregation phenomenon, will not be discussed in detail in this chapter, and the interested reader is referred to a review by Lomakin et al. (1999). The review gives a detailed survey of fibrillogenesis. Protein assembly reactions are of biochemical (e.g., actine polymerization, formation of ion channels) or of pathological relevance (Alzheimer disease, prion diseases, amyloidosis, etc.), and LS can help to explain the underlying mechanisms and in the future help in developing fibrillogenesis-inhibiting drugs.

Rapid aggregation phenomena such as fibrin formation require special adaptation of LS experiments. Static LS with Zimm-like plots can be used when the LS instrument is linked to a stopped flow mixer

(Bernocco et al., 2000). This is a good example of the use of a multi-angle instrument to obtain the time-resolved values of M_w and radius of gyration.

Dynamic LS is not suited for rapid aggregation phenomena, as the growth of the aggregates interferes with the correlation measurement. During protein refolding, aggregation is common. Suppression of aggregation is caused by a shift from attractive to repulsive intermolecular interactions. This is reflected in a shift from negative to positive second virial coefficient (Ho et al., 2003).

Many studies use only dynamic LS, but the study of structure switching by pH-induced conformational changes of HIV-1 capsid protein (CA) is the subject of an instructive paper by Ehrlich et al. (2001), showing how the combined size information from both techniques (described in detail in this chapter) leads to an understanding of the switch from spherical to rod-like protein assemblies. At pH 6.8, the ratio of the radius of gyration (R_G) to the hydrodynamic radius (R_H) was found to be 0.75. This value is very close to the theoretical value $\sqrt{3/5}$ for solid spheres (Tanford, 1961). On the other hand, at pH 7.0, R_G/R_H increased to 1.83, consistent with a rod-like morphology (Tanford, 1961).

Characterization of Proteins in Delivery Systems by Light Scattering

In nonviral gene therapy, DNA is condensed to cationic carriers such as cationic liposomes, cationic polymers (De Smedt et al., 2000), and positively charged proteins such as poly-lysine. The characterization of such DNA complexes involves the determination of zeta potential, hydrodynamic radius by dynamic LS, and molecular weight and radius of gyration by static LS. Also, block copolymers with a protein segment [e.g., poly(ethylene glycol)-poly(aspartic acid)] are studied to develop micelles with a *core shell structure* for drug delivery. Harada and Kataoka (1998, 1999) successfully applied dynamic LS and Zimm analysis to elucidate the structures of such core shell micellar structures.

Characterization of the Enzymatic Activity of Proteins by Light Scattering

In cases where a protein degrades water-soluble macromolecular substrates, its hydrolytic (enzymatic) activity may be studied by LS. Static LS analysis can elucidate the kinetics of degradation by following the M_w decrease of the polymeric substrate in function of the degradation time. LS has been successfully applied to study the activity of therapeutic enzymes like hyaluronidase (Reed et al., 1989), endocellulases (Demeester et al., 1979, 1988; Vlasenko et al., 1998; Richardson et al., 2002), and alpha amylases (Yilmaz et al., 2002).

Characterization of the Stoichiometry of Protein Complexes by Light Scattering

When the molecular weights of two proteins A and B are known, LS is a unique tool to determine the stoichiometry of the A_mB_n complex the two proteins form in solution, by determining the molecular weight of A_mB_n. This stoichiometry holds for A-B complexes in solution and can be different from the one obtained by x-ray crystallography.

Conclusions

LS is a powerful, nondestructive technique to elucidate the behavior of particles and macromolecules in their natural liquid environment. We explained how static LS provides us with information on the weight average molecular weight (M_w), the second and third virial coefficients (A_2 and A_3), and, for large particles, the radius of gyration (R_G). Dynamic LS gives the translational diffusion coefficient (D_τ), the hydrodynamic radius (R_H), and an idea about the polydispersity. Combining dynamic and static LS allows a determination of the shape of the molecules/particles. By dynamic electrophoretic LS, we obtain the zeta potential. In combination with size exclusion chromatography, molecular weight and size distributions are obtained, in addition to the number average (M_n), weight average, z average (M_z), and different average hydrodynamic and gyration radii.

Tools for characterizing primary or three-dimensional crystal structures of proteins give no elusive information on the correct folded protein structure in solution. Clearly, this information on proteins in solution can be extracted from LS experiments. However, LS does not analyze the particles/solute molecules, but rather the light scattered from the particles and hence gives no sharp resolution and little information on shape.

The information on solute behavior of proteins obtained by LS is never complete and should be supplemented with information obtained by techniques such as analytical ultracentrifugation or SDS-PAGE.

A clear potency of LS is that it is the only method that allows fast molecular weight determinations from a few thousand to one billion on small samples (only about 10 to 100 µg of material is required). On one hand, the extreme sensitivity to small amounts of large particles is a major drawback, as dust-free samples have to be prepared. On the other hand, LS also provides the power to detect sensitively aggregation in protein solutions long before particles are visible to the naked eye.

Although this chapter focuses on LS applications in the therapeutic protein field, the methods described can equally be applied to, for example, pharmaceutical polysaccharides, synthetic polymers, and pharmaceutical micro- and nanoparticles.

Abbreviations

A_2	Second virial coefficient
A_3	Third virial coefficient
D_τ	Translational diffusion coefficient
$G(\tau)$	Correlation function
I_θ	Scattered intensity
I_o	Incident intensity
LS	Light scattering
M_n	Number average molecular weight
M_v	Viscosimetric average molecular weight
M_w	Weight average molecular weight
M_z	z average molecular weight
$P(\theta)$	Ratio of scattered intensity for large particle over scattered intensity without interference
q	Scattering vector
R_G	Radius of gyration
R_H	Hydrodynamic radius
R_θ	Rayleigh ratio for scattering at an angle θ
r	Distance between scattering volume and detector
Γ	Decay rate
$\partial n/\partial c$	Specific refractive index increment
θ	Scattering angle
τ	Time interval

Acknowledgment

The authors thank Dr. John Philo of Alliance Laboratories for illustrative examples and for his critical reading of this manuscript.

References

1. Arakawa T, Philo JS, and Kita Y. Kinetic and Thermodynamic Analysis of Thermal Unfolding of Recombinant Erythropoietin. *Biosci Biotechnol Biochem* 2001; 65:1321–1327
2. Baldini G, Beretta S, Chirico G, Franz H, Maccioni E, Mariani P, and Spinozzi F. Salt-Induced Association of Beta-Lactoglobulin by Light and X-ray Scattering. *Macromolecules* 1999; 32(19):6128–6138
3. Bernocco S, Ferri F, Profumo A, Cuniberti C, and Rocco M. Polymerization of Rod-Like Macromolecular Monomers Studied by

Stopped-Flow, Multiangle Light Scattering: Set-up, Data Processing, and Application to Fibrin Formation. *Biophys J* 2000; 79(1):561–583

4. Bowen WR, and Jenner F. Dynamic Ultrafiltration Model for Charged Colloidal Dispersions: A Wigner-Seitz Cell Approach. *Chem Eng Sci* 1995; 50:1707–1736
5. Bowen WR, and Williams PL. Dynamic Ultrafiltration Model for Proteins—A Colloidal Interaction Approach. *Biotechnol Bioeng* 1996; 50:125–135
6. Bowen WR, Mongruel A, and Williams PM. Prediction of the Rate of Crossflow Membrane Ultrafiltration: A Colloidal Interaction Approach. *Chem Eng Sci* 1996; 51:4321–4333
7. Brice B, and Halwer M. A Differential Refractometer. *J Opt Soc Am* 1951; 41:1033–1037
8. Brice BA, Halwer M, and Speiser R. Photoelectric Light-Scattering Photometer for Determining High Molecular Weights. *J Opt Soc Am* 1950; 40:768–778
9. Brown W. *Light Scattering-Principles and Development*. Oxford: Clarendon Press; 1996. 528p.
10. Chu B. *Laser Light Scattering-Basic Principles and Practice*. Boston: Academic Press; 1991. 343p.
11. Creighton TE. *Proteins: Structure and Molecular Properties*. New York: WH Freeman; 1992. 512p.
12. Debye P. Light Scattering in Solutions. *J Appl Phys* 1944; 15:338
13. Demeester J, Bracke M, and Lauwers A. Absolute Viscosimetric Method for the Determination of the Endocellulase (C_x) Activities Based upon Light-Scattering Interpretations of Gel Chromatographic Fractionation Data. In: Brown RD Jr, and Jurasek L. *Hydrolysis of Cellulose: Mechanisms of Enzymatic and Acid Catalysis*. Advances in Chemistry Series 181. Washington DC: American Chemical Society; 1979. 399p.
14. Demeester J, Eigner WD, Huber A, and Glatter O. Light-Scattering Experiments to Follow the Degradation Process of Hydroxethyl-Cellulose Due to Trichoderma-Reesei Cellulase. *J Wood Chem Technol* 1988; 8:135–153
15. De Smedt SC, Demeester J, and Hennink WE. Cationic Polymer Based Gene Delivery Systems. *Pharmaceut Res* 2000; 17:113–126
16. Ehrlich LS, Liu T, Scarlata S, Chu B, and Carter CA. HIV-1 Capsid Protein Forms Spherical (Immature-Like) and Tubular (Mature-Like) Particles in Vitro: Structure Switching by pH-induced Conformational Changes. *Biophys J* 2001; 81:586–594
17. Einstein A. Theorie der Opaleszens von homogenen Flüssigkeiten und Flüssigkeitsgemischen in der Nähe des kritischen Zustandes. *Ann Phys* 1910; 33:1275–1298

18. EMEA. *Guidance on the Description of Composition of Pegylated (Conjugated) Proteins in the SPC*. London: The European Agency for the Evaluation of Medicinal Products; 2003

19. Farnum M, and Zukoski C. Effect of Glycerol on the Interactions and Solubility of Bovine Pancreatic Trypsin Inhibitor. *Biophys J* 1999; 76:2716–2726

20. Finsy R. Particle Sizing by Quasi-Elastic Light Scattering. *Adv Colloid Interfac* 1994; 52:79–143

21. Fish WW, Danielsson A , Nordling K, Miller SH, Lam CF, and Bjork I. Denaturation Behavior of Antithrombin in Guanidinium Chloride—Irreversibility of Unfolding Caused by Aggregation. *Biochemistry* 1985; 24:1510–1517

22. George A, and Wilson WW. Predicting Protein Crystallization from a Dilute Solution Property. *Acta Crystallogr D* 1994; 50:361–365

23. Gimel JC, Durand D, and Nicolai T. Structure and Distribution of Aggregates Formed after Heat-Induced Denaturation of Globular-Proteins. *Macromolecules* 1994; 27(2): 583–589

24. Graupner M, Haalck L, Spener F, Lindner H, Glatter O, Paltauf F, and Hermetter A. Molecular Dynamics of Microbial Lipases as Determined from Their Intrinsic Tryptophan Fluorescence. *Biophys J* 1999; 77:493–504

25. Habel JE, Ohren JF, and Borgstahl GEO. Dynamic Light-Scattering Analysis of Full-Length Human RPA14/32 Dimer: Purification, Crystallization and Self-Association. *Acta Crystallogr D* 2001; 57:254–259

26. Hagiwara T, Kumagai H, and Nakamura K. Fractal analysis of aggregates formed by heating dilute BSA solutions using light scattering methods. *Biosci Biotechnol Biochem* 1996; 60:1757–1763

27. Harada A, and Kataoka K. Novel Polyion Complex Micelles Entrapping Enzyme Molecules in the Core: Preparation of Narrowly-Distributed Micelles from Lysozome and Poly(ethylene glycol)-Poly(aspartic acid) Block Copolymer in Aqueous Medium. *Macromolecules* 1998; 31:288–294

28. Harada A, and Kataoka K. On-Off Control of Enzymatic Activity Synchronizing with Reversible Formation of Supramolecular Assembly from Enzyme and Charged Block Copolymers. *J Am Chem Soc* 1999; 121:9241–9242

29. Hernandez-Contreras M, Alarcon-Waess O, and Medina-Noyola M. Rotational-Translational Electrolyte Friction on Nonspherical Particles. *J Chem Phys* 1997; 106:2492–2501

30. Ho JGS, Middelberg APJ, Ramage P, and Kocher HP. The Likelihood of Aggregation During Protein Renaturation Can Be Assessed Using the Second Virial Coefficient. *Protein Sci* 2003; 12:708–716

31. Hsu YR, and Arakawa T. Structural Studies on Acid Unfolding and Refolding of Recombinant Human Interferon-Gamma. *Biochemistry US* 1985; 24:7959–7963

32. Huang MD, Weissman JT, Wang CQ, Balch WE, and Wilson IA. Protein Engineering for Crystallization of the GTPase Sar1 that Regulates ER Vesicle Budding. *Acta Crystallogr D* 2002; 58:700–703

33. Huglin M. Specific Refractive Index Increments. In: Huglin M. *Light Scattering from Polymer Solutions*. London: Academic Press; 1972:165–331

34. Huglin M. Specific Refractive Index Increments of Polymers in Dilute Solutions. In: Brandup J, Immergut E. *Polymer Handbook*. New York: John Wiley & Sons; 1989:VII/409–484

35. Hunter RJ. Transport Properties of Suspensions. In: *Foundations of Colloid Science,* Vol. 1. New York: Oxford University Press; 1993:494–563

36. Kaye W, and McDaniel JB. Low-Angle Laser Light Scattering: Rayleigh Factors and Depolarization Ratios. *Appl Opt* 1974; 13:1934–1937

37. Kendrick BS, Kerwin BA, Chang BS, and Philo JS. Online Size-Exclusion High-Performance Liquid Chromatography Light Scattering and Differential Refractometry Methods to Determine Degree of Polymer Conjugation to Proteins and Protein-Protein or Protein-Ligand Association States. *Anal Biochem* 2001; 299:136–146

38. Korgel BA, van Zanten JH, and Monbouquette HG. Vesicle Size Distributions Measured by Flow Field-Flow Fractionation Coupled with Multiangle Light Scattering. *Biophys J* 1998; 74:3264–3272

39. Le Bon C, Nicolai T, Kuil ME, and Hollander JG. Self-Diffusion and Cooperative Diffusion of Globular Proteins in Solution. *J Phys Chem B* 1999; 103:10294–10299

40. Liu T, and Chu B. Light-Scattering by Proteins. In: Somasundaran P. *Encyclopedia of Surface and Colloid Science*. New York: Marcel Dekker; 2002:3023–3043

41. Lomakin A, Benedek GB, and Teplow DB. Monitoring Protein Assembly Using Quasielastic Light Scattering Spectroscopy. In: Wetzel R. *Methods in Enzymology Volume 309—Amyloid, Prions and Other Protein Aggregates*. San Diego: Academic Press; 1999:429–460

42. Matsuura JE, Morris AE, Ketchem RR, Braswell EH, Klinke R, Gombotz WR, and Remmele RL Jr. Biophysical Characterization of a Soluble CD40 Ligand (CD154) Coiled-Coil Trimer: Evidence of a Reversible Acid-Denaturated Molten Globule. *Arch Biochem Biophys* 2001; 392:208–218

43. McMeekin T, Wilensky M, and Groves M. Refractive Indices of Proteins in Relation to Amino Acid Composition and Specific Volume. *Biochem Biophys Res Commun* 1962; 7:151–156

44. McPherson A. Current Approaches to Macromolecular Crystallization. *Eur J Biochem* 1990; 189:1–23

45. Miller JF, Schätzel K, and Vincent B. The Determination of Very Small Electrophoretic Mobilities in Polar and Nonpolar Colloidal Dispersions Using Phases Analysis Light Scattering. *J Colloid Interface Sci* 1991; 143:532–554

46. Muschol M, and Rosenberger F. Interactions in Undersaturated and Supersaturated Lysozyme Solutions: Static and Dynamic Light Scattering Results. *J. Chem Phys* 1995; 103:10424–10432

47. Narayanan J, and Liu XJ. Protein Interactions in Undersaturated and Supersaturated Solutions: A Study Using Light and X-Ray Scattering. *Biophys J* 2003; 84:523–532

48. Poznanski J, Georgalis Y, Wehr L, Saenger W, and Zielenkiewicz P. Comparison of Two Different Lysozyme Types Under Native and Crystallization Conditions Using Two-Dimensional NMR and Dynamic Light Scattering. *Biophys Chem* 2003; 104:605–616

49. Raman C. Relation of Tyndall Effect to Osmotic Pressure in Colloidal Solutions. *Indian J Phys* 1927; 2:1–6

50. Rayleigh JWS. The Incidence of Light upon a Transparent Sphere of Dimensions Comparable with the Wave-Length. *Proc Roy Soc A* 1910; 84:25–46

51. Rayleigh JWS. On the Electromagnetic Theory of Light. *Philos Mag* 1881; 12:81–101

52. Reed WF, Reed CE, and Byers LD. Random Coil Scission Rates Determined by Time-Dependent Total Intensity Light-Scattering-Hyaluronate Depolymerization by Hyaluronidase. *Biopolymers* 1990; 30:1073–1082

53. Remmele RL Jr, Bhat SD, Phan DH, and Gombotz WR. Minimization of Recombinant Human Fit3 Lignad Aggregation at the Tm Plateau: A Matter of Thermal Reversibility. *Biochemistry* 1999; 38:5241–5247

54. Richardson S, Lundqvist J, Wittgren B, Tjerneld B, Tjerneld F, and Gorton L. Initial Characterization of Ethyl(hydroxyethyl) Cellulose Using Enzymic Degradation and Chromatographic Methods. *Biomacromolecules* 2002; 3:1359–1363

55. Santos NC, and Castanho MA. Teaching Light Scattering Spectroscopy: The Dimensions and Shape of Tobacco Mosaic Virus. *Biophys J* 1996; 71:1641–1650

56. Saridakis E, Dierks K, Moreno A, Dieckmann MWM, and Chayen NE. Separation Nucleation and Growth in Protein Crystallization using Dynamic Light Scattering. *Acta Crystallogr D* 2002; 58:1597–1600

57. Schibli DJ, Hunter HN, Aseyev V, Starner TD, Wiencek JM, McCray PB, Tack BF, and Vogel HJ. The Solution Structures of the Human Beta-Defensins Lead to a Better Understanding of the Potent Bactericidal

Activity of HBD3 against Staphylococcus aureus. *J Biol Chem* 2002; 277:8279–8289

58. Schurr JM. A Theory of Electrolyte Friction on Translating Polyelectrolytes. *Chem Phys* 1980; 45:119–132
59. Stolyhwo A, Colin H, and Guiochon G. Use of Light Scattering as a Detector Principle in Liquid Chromatography. *J Chromatogr* 1983; 265:1–18
60. Tanford C. *Physical Chemistry of Macromolecules.* New York: John Wiley & Sons; 1961. 710p.
61. Thielking H, and Kulicke W-M. On-Line Coupling of Flow Field-Flow Fractionation and Multiangle Laser Light Scattering for the Characterization of Macromolecules in Aqueous Solution as Illustrated by Sulfonated Polystyrene Samples. *Anal Chem* 1996; 68:1169–1173
62. Thielking H, Roessner D, and Kulicke W-M. On-Line Coupling of Flow Field-Flow Fractionation and Multiangle Laser Light Scattering for the Characterization of Polystyrene Particles. *Anal Chem* 1995; 67:3229–3233
63. Tombs MP, and Peacocke AR. *The Osmotic Pressure of Biological Macromolecules.* Oxford: Clarendon Press; 1974. 143p.
64. USP. USP Guideline for Submitting Requests for Revision to the USP/NF. New York: The USP Convention; 2003. 111p.
65. Vlasenko EY, Ryan AI, Shoemaker CF, and Shoemaker SP. The Use of Capillary Viscometry, Reducing End-Group Analysis, and Size Exclusion Chromatography Combined with Multi-Angle Laser Light Scattering to Characterize Endo-1, 4-β-D-Glucanases on Carboxymethylcellulose: A Comparative Evaluation of the Three Methods. *Enzyme Microb Tech* 1998; 23:350–359
66. Weijers M, Visschers RW, and Nicolai T. Light Scattering Study of Heat-Induced Aggregation and Gelation of Ovalbumin. *Macromolecules* 2002; 35(12):4753–4762
67. Wen J, Arakawa T, and Philo JS. Size-Exclusion Chromatography with On-Line Light-Scattering, Absorbance, and Refractive Index Detectors for Studying Proteins and Their Interactions. *Anal Biochem* 1996; 240:155–166
68. Wilson WW. Light Scattering as a Diagnostic for Protein Crystal Growth—A Practical Approach. *J Struct Biol* 2003; 142:56–65
69. Wyatt PJ. Light Scattering and the Absolute Characterization of Macromolecules. *Anal Chim Acta* 1993; 272:1–40
70. Yilmaz G, Öngen G, Jongboom ROJ, Feil H, van Dijk C, and Hennink WE. Modulated Release of a Volatile Compound from Starch Matrixes

via Enzymatically Controlled Degradation. *Biomacromolecules* 2002; 3:305–311

71. Zhang J, and Liu XY. Effect of Protein-Protein Interactions on Protein Aggregation Kinetics. *J Chem Phys* 2003; 119:10972–10976
72. Zimm BH. The Scattering of Light and the Radial Distribution Function of High Polymer Solutions. *J Chem Phys* 1948a; 16:1093–1099
73. Zimm BH. Apparatus and Methods for Measurement and Interpretation of the Angular Variation of Light Scattering; Preliminary Results on Polystyrene Solutions. *J Chem Phys* 1948b; 16:1099–1116

8

Liquid Chromatography

RAINER BISCHOFF and BEGONA BARROSO

University of Groningen, Center of Pharmacy, Groningen, The Netherlands

Introduction

The chromatographic principle has been used for many years to separate molecules. It is next to electrophoresis the most widely applied separation method. Chromatography is based on an exchange of molecules between a mobile and stationary phase. As the name implies, molecules that are bound to the stationary phase do not move, whereas those present in the mobile phase travel at a constant speed down the chromatographic column. The overall speed with which molecules travel down a chromatographic column depends thus on their affinity for the stationary phase under the conditions determined by the mobile phase. Macromolecules have a tendency to bind either very strongly to a given stationary phase or not at all depending on the mobile phase conditions. This renders isocratic elution of macromolecules such as proteins, peptides, or nucleic acids essentially impossible or restricted to a very narrow and defined mobile phase composition. It is thus more practical to change the mobile phase composition over time in a so-called gradient so that proteins desorb at critical mobile phase compositions and are thus eluted from the column. This is exemplified in Figure 1, where the amount of protein adsorbed to the stationary phase is plotted against its concentration in the mobile phase. Under mobile conditions where the affinity to the stationary phase is strong (high b value), the isotherm shows a very steep rise. As the mobile phase is changing toward eluting conditions, b decreases, reaching a point where elution occurs. This point is relatively sharply defined for biological macromolecules such as proteins. The nature of the gradient and the mobile phase is governed by the nature of the interaction between the protein and the stationary phase. Various interaction modes will be described in the section "Different Modes of Chromatography," below.

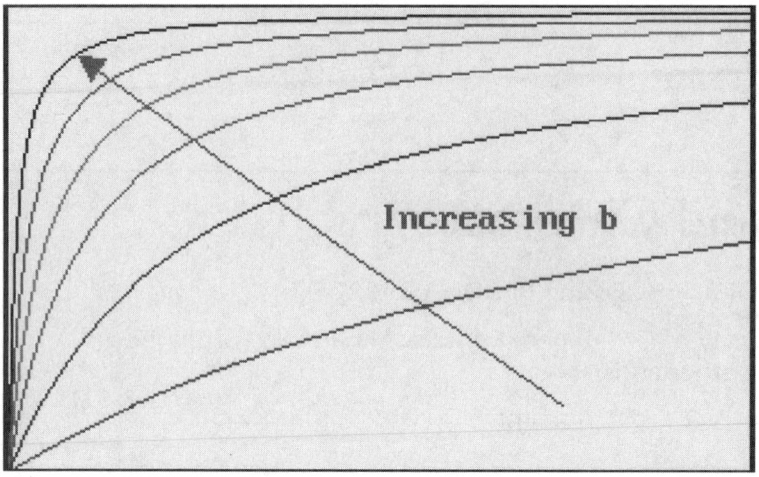

Figure 1 Protein adsorption to a stationary phase as exemplified by the Langmuir isotherm. The y-axis shows the amount of protein that is adsorbed at a given concentration of protein in the mobile phase (x-axis) at a fixed temperature. The parameter b is a measure of the affinity of the protein for the stationary phase.

The chromatographic principle of exchange between molecules in the mobile and in the stationary phase necessitates that they travel between these two phases. This implies that kinetic principles play an important role in chromatography next to the thermodynamically governed adsorption behavior. Understanding the kinetics of mass transfer between the stationary and the mobile phases is at the center of improving the efficiency of chromatographic separations. Especially for proteins, it is pivotal that the pore size of the stationary phase is sufficiently large to allow easy access of the proteins to the large internal surface area. Reduced mass transfer kinetics is the main limiting factor for obtaining highly efficient and thus rapid separations. This *mass transfer* term is determined by the length of the diffusion path between molecules in the mobile phase and the stationary phase as well as by the accessibility of the stationary phase. The quest for increasing efficiency has thus led to a reduction in particle diameter down to a practical limit of about 3 μm. This limit is determined by the increase in pressure drop across the column, which becomes unpractical for very small particles.

Novel developments in high-performance liquid chromatography (HPLC) are continuing to advance the boundaries of separation efficiency. With the advent of monolithic stationary phases that no longer consist of particles but a continuous rod throughout the column, it is feasible to produce columns with the efficiency of a 3-μm particle at a pressure drop corresponding to a 20-μm particle packed column (Leinweber et al., 2002).

Surface area of the stationary phase is of great importance, as it determines the binding capacity. Although nonporous particles deliver the fastest mass transfer rates due to the fact that there is no pore diffusion, they have rather low surface areas and consequently cannot adsorb large amounts of protein. It depends on the analytical application whether this is acceptable in favor of rapid, highly efficient separations. When detection of minor constituents in a protein preparation is the goal, binding capacity is relevant, as the strong binding major components may easily displace the minor contaminant or by-product. This makes the use of porous media a necessity. For analytical purposes, silica is still the most widely used base material for high-performance stationary phases, but polymers such as styrene-divinylbenzene copolymers of high quality are available as well. Polymeric stationary phases have the advantage of being more resistant toward harsh cleaning procedures to remove strongly bound proteins or other components from the sample.

Size exclusion chromatography is not based on protein adsorption to a stationary phase but rather on the distribution between the inner pore volume containing a more or less stagnant mobile phase and the flowing mobile phase. Inclusion into the inner pore volume depends on the size and shape (hydrodynamic volume) of the molecules, with smaller molecules having easier access to the internal pore volume and thus being more retained on a size exclusion column. Larger molecules such as proteins have only partial access to this inner volume and are thus eluted earlier.

Different Modes of Chromatography

Chromatographic modes are primarily defined by the stationary phase. The stationary phase furthermore determines the interactions that lead to adsorption and consequently the mobile phase compositions that are effective in desorbing the bound proteins. In saying this, it should be clear that there is no "pure" interaction mode in chromatography, and that proteins bind to stationary phases through multiple interactions. In selected cases, it is even possible to make use of multiple interactions in a deliberate way to obtain selectivities not obtainable through a single type of interaction. These so-called mixed-mode chromatographic systems have not only been used for proteins but also for other biological macromolecules such as nucleic acids (Bischoff et al., 1985). For the sake of simplicity, this section is organized according to the main chromatographic methods used for protein separation.

Chromatography Based on Hydrophobic Interactions

Proteins maintain their three-dimensional structure through different types of interactions. Notably, hydrophobic interactions are important for the overall stability of proteins, as they stabilize the so-called hydrophobic core of many proteins. For molecules that naturally exist in an aqueous environment like most proteins, it is thermodynamically favorable to avoid contact between the side chains of hydrophobic amino acids such as valine, isoleucine, leucine, or phenylalanine with water. A large group of proteins that forms an exception to this rule are transmembrane proteins, which reside partially within the hydrophobic part of a lipid bilayer. These proteins are as such not soluble or stable in an aqueous milieu but rather require surfactants to render the hydrophobic transmembrane regions more hydrophilic. We will not deal specifically with membrane proteins in this chapter, as they are rarely used as biopharmaceuticals. They form, however, the largest class of proteins that are currently targeted by low-molecular-weight therapeutics and thus of utmost importance in drug discovery.

Hydrophobic Interaction Chromatography

Hydrophobic interaction chromatography (HIC) is an important chromatographic mode of separation, especially for the isolation of biologically active proteins. HIC requires a fairly high ionic strength in solution and is thus not preferred as an analytical tool in cases where mass spectrometry is needed for the identification of contaminants or by-products. To induce hydrophobic interactions between a mildly hydrophobic stationary phase and proteins, it is necessary to add a so-called anti-chaotropic salt to the mobile phase. Most widely used are sulfates of sodium or ammonium at concentrations above 1 M. Sulfate as a doubly charged anion stabilizes the structure of water and promotes hydrophobic interactions with the stationary phase. There are other anions and cations that induce hydrophobic interactions, and their relative strength to do so is classified in the Hofmeister series, which ranks ions according to their capacity to induce "water-structuring" effects (Table 1). Proteins are eluted by decreasing the salt concentration in the mobile phase, thereby reducing the strength of hydrophobic interactions. HIC can be a powerful method to detect impurities in biopharmaceutical products, as they often affect the hydrophobicity of the protein surface. Monoclonal antibodies, a growing family of biopharmaceuticals, have been analyzed for impurities by HIC pointing to the presence of deamidated or isomerized forms (Figure 2) (Harris et al., 2001). Forced stability studies at elevated temperature revealed the sites of deamidation or isomerization to reside in Asn

Anions	PO_4^{3-}	SO_4^{2-}	Cl^-	Br^-	NO_3^-	ClO_4^-
Cations	Al^{3+}	Mg^{2+}	Ca^{2+}	Na^+	K^+	NH^{4+}

Table 1 Hofmeister Series of Anions and Cations Arranged According to Their Decreasing Capacity to Structure Water (Left to Right).

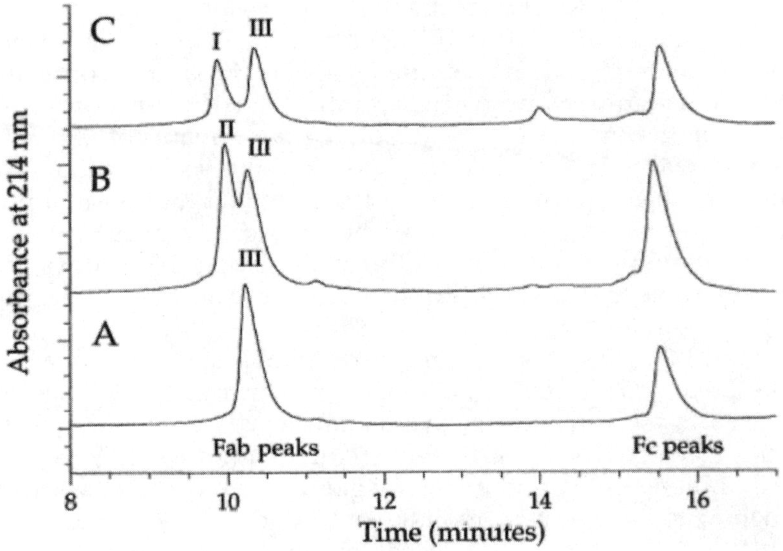

Figure 2 Hydrophobic interaction chromatography of a monoclonal antibody against the *her2/neu* gene product following digestion with papain. (A) Fab and Fc fragments of the main product (no deamidation or isomerization); (B) Fab and Fc fragments of an impurity related to deamidation of Asn-30 of one light chain and isomerization of Asp-102 to iso-Asp of one heavy chain. (C) Fab and Fc fragments of an impurity related to deamidation of Asn-30. Reprinted with permission from Harris et al. (2003).

residue 30 or Asp residue 102. Asn-55 was also deamidated albeit at a much-reduced rate. It is likely that deamidation affects the overall hydrophobicity of the antibody by introducing an additional negative charge on the protein's surface. Isomerization from Asp to isoAsp is likely to lead to a considerable conformational change, which affects the local structure of the protein. Obviously, these structural changes happen within or close to the interaction interface with the stationary phase, leading to a change in the strength of interactions. Figure 2 shows that all of the deamidated forms of the antibody elute earlier than the main product. This is expected, as Asp (relative hydrophobicity of –0.20) is less hydrophobic than Asn (relative hydrophobicity of +0.25) at neutral pH.

HIC is usually performed on modestly hydrophobic stationary phases containing propyl, butyl, or phenyl groups. There are a number of manufacturers that sell HIC materials but mainly for preparative purposes. The use of HIC in protein analysis is not widespread but probably of greater potential than widely appreciated. The fact that it requires high salt concentrations puts, however, some restrictions on the analytical HPLC equipment, which makes, for example, capillary HPLC unsuitable.

Reversed-Phase Chromatography

Reversed-phase HPLC (RP-HPLC) is the most widely used chromatographic method for peptide analysis. It is the cornerstone method for peptide mapping to ascertain the correct primary structure of proteins and to detect and isolate contaminants or impurities. RP-HPLC was initially developed for low-molecular-weight compounds and has inherited its name from the fact that normal phase chromatography, for example on silica stationary phases, was the standard method twenty-five to thirty years ago. Nowadays, even for small molecules, RP-HPLC has replaced normal phase chromatography in most cases.

RP-HPLC is based on a highly hydrophobic stationary phase and a more polar mobile phase. Stationary phases for RP-HPLC are to a large extent based on alkyl chains that are covalently bound to spherical silica beads. Numerous manufacturers produce RP stationary phases of high quality and generally good reproducibility, which has also helped to make this method the most widely used HPLC separation method. Although silica-based stationary phases still dominate the market, more recently there has been a range of polymeric materials mostly based on copolymers of styrene and divinylbenzene that allow high-efficiency separations. These polymeric materials have the advantage of being chemically more resistant, especially to very high pH values that are sometimes needed for cleaning of the stationary phases. Stationary phases based on zirconium oxide are also advancing with the added advantage of high pH stability. Moreover, recent developments in rendering silica-based RP stationary phases more pH stable have extended the useful range into the basic pH region, sometimes even up to pH 10. Naturally, there is no ideal RP stationary phase for all applications, but it can be said that most peptide mapping studies have been performed on silica-based materials with alkyl chains bound to them.

Separation of complete proteins is also possible using RP-HPLC, although the harsh phase transition between a water-rich polar mobile phase and the very hydrophobic stationary phase may lead to irreversible conformational changes in proteins (denaturation), which render them biologically inactive. RP-HPLC has therefore only a limited applicability for purifying native proteins depending on their stability. Moreover, elution for RP-HPLC necessitates water-miscible organic solvents such as methanol or acetonitrile, which again favor denaturation of proteins at higher concentrations. To make matters even worse, most mobile phases for RP-HPLC use low pH (2–3) brought about by the addition of acids such as trifluoroacetic acid (TFA), acetic acid, or formic acid. This is an additional factor favoring protein denaturation. Nevertheless, there are some smaller proteins stabilized by disulfide bonds that can withstand the harsh conditions of RP-HPLC. Examples of biopharmaceutical products purified by RP-HPLC are recombinant human insulin produced in *Escherichia coli* or *Saccharomyces cerevisiae* and recombinant hirudin produced in *S. cerevisiae* (Bischoff et al., 1989).

Reversed-phase chromatographic conditions pose little problems to peptides, which contain generally no three-dimensional structural elements and are thus not prone to denaturation. The digestion of larger proteins to peptides using proteases that cleave before or after defined amino acids is an important method for the analysis of their primary structure. With the increasing importance of biopharmaceuticals based on recombinant proteins, identification of impurities is a crucial part of process development and quality control. Figure 3 gives a detailed view of a peptide map of the monoclonal antibody herceptin directed against the human *her/neu2* gene product. Different forms of this therapeutic antibody were obtained during forced stability studies, which could be separated by HIC after papain cleavage as shown in Figure 2. Digestion of the various forms with trypsin,

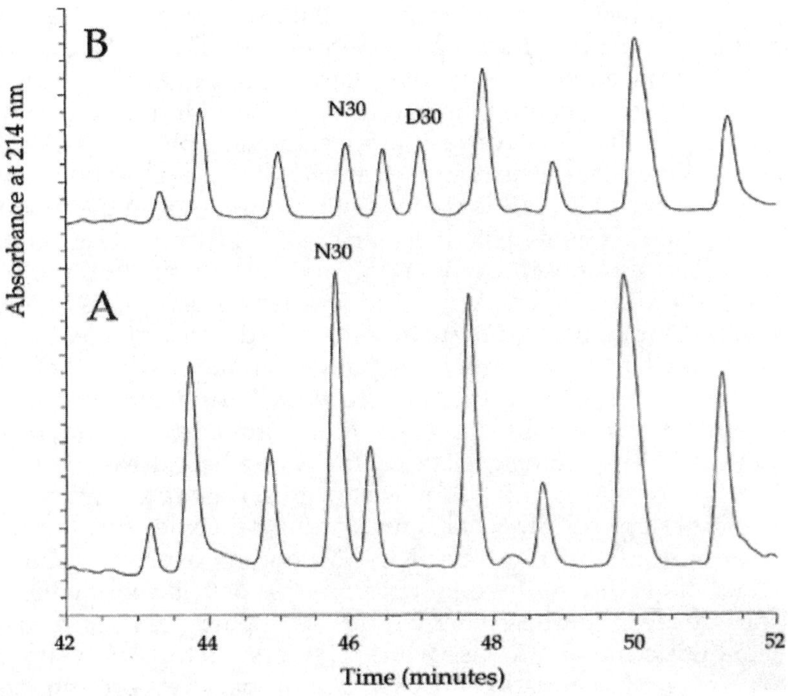

Figure 3 Tryptic peptide maps of forms (A) IEX-3 and (B) IEX-1 of the monoclonal antibody herceptin obtained by RP-HPLC. In this view of a region of the complete chromatogram, an additional peak in form 3 is clearly visible, pointing toward a modification in this peptide. Analysis of this peptide by mass spectrometry and protein sequencing using Edman degradation showed that asparagine 30 (N30) has been deamidated to aspartic acid (D). Reprinted with permission from Harris et al. (2001).

which cleaves after lysine (K) or arginine (R) residues [except when followed by proline (P)], shows that even small differences such as the deamidation of asparagine (N) residue 30 to aspartic acid (D) lead to a chromatographic shift that can easily be detected by RP-HPLC. Deamidation is a fairly common source of heterogeneity in proteins both natural and recombinant and often does not jeopardize the use of the protein as pharmaceutical product if it can be controlled within given limits. A potential risk is that modified proteins may lead to immunological reactions directed at the newly generated epitope, but this has in most cases been shown not to be a source of side reactions in patients. The occurrence of iso-aspartic acid (see Figure 2, trace B), however, leads to a significant structural change in the protein, which may affect both its biological activity and its immunogenicity. Such modifications are generally undesirable in a pharmaceutical product.

Recently, novel types of materials for HPLC have been developed based on a continuous polymer (silica or organic) within the column. These materials, which are generally termed *monoliths*, hold promise to push the boundaries of efficiency in HPLC a step further. A few monolithic stationary phases have become commercially available and have found acceptance in the pharmaceutical industry for the rapid analysis of drug compounds or metabolites. Monoliths based on silica polymers can be derivatized to RP stationary phases in a similar manner to silica beads. The main advantage of these monoliths is that chromatographic efficiency is somewhat uncoupled from the pressure drop across the column. This has been achieved through a bimodal pore size distribution within these polymers with large flow through pores on the order of micrometers and smaller diffusive mesopores of diameters comparable to regular silica beads (nanometer range) (Figure 4). This leads to stationary phases having a separation efficiency comparable to a column packed with 3-μm beads but with a pressure drop comparable to a column containing 20-μm beads. Although extremely high separation efficiencies had already been obtained with nonporous 1.5-μm particles, this was at the expense of very high column pressures and low binding capacities due to the reduced surface area of these particles. Monoliths overcome some of these limitations due to the bimodal pore size distribution, which allows surface areas comparable to porous silica particles (Figure 5).

Monolithic materials based on polymers such as polystyrene-divinylbenzene have also been used for preparative protein isolation either in column or disk format as well as for protein and peptide analysis (Xie et al., 1999; Branovic et al., 2000). The development of various kinds of monoliths is likely to continue both for protein analysis and protein production.

Miniaturization of HPLC systems from regular columns (2–4 mm diameters) to capillary (100–500 μm diameters) or "nano" columns (25–100 μm) allows the analysis of smaller amounts of proteins and

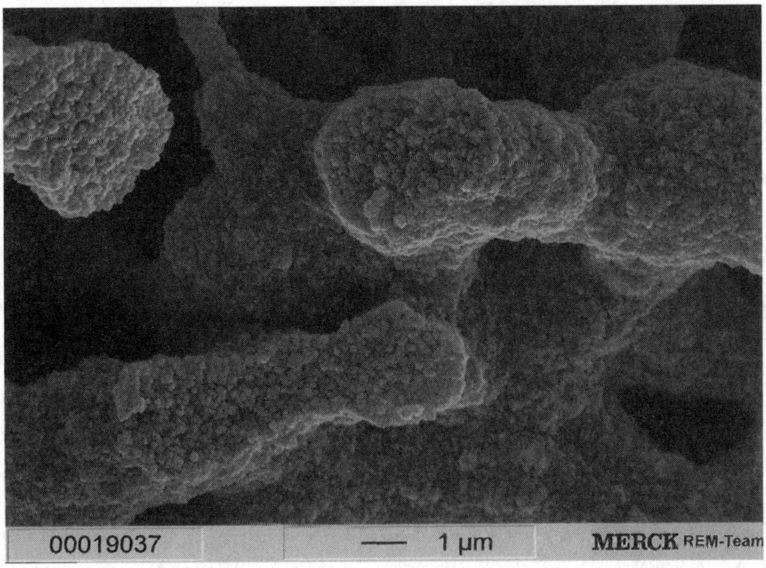

Figure 4 Microscopic structure of a silica monolith showing the bimodal pore size distribution. Reproduced with permission from Barroso et al. (2003).

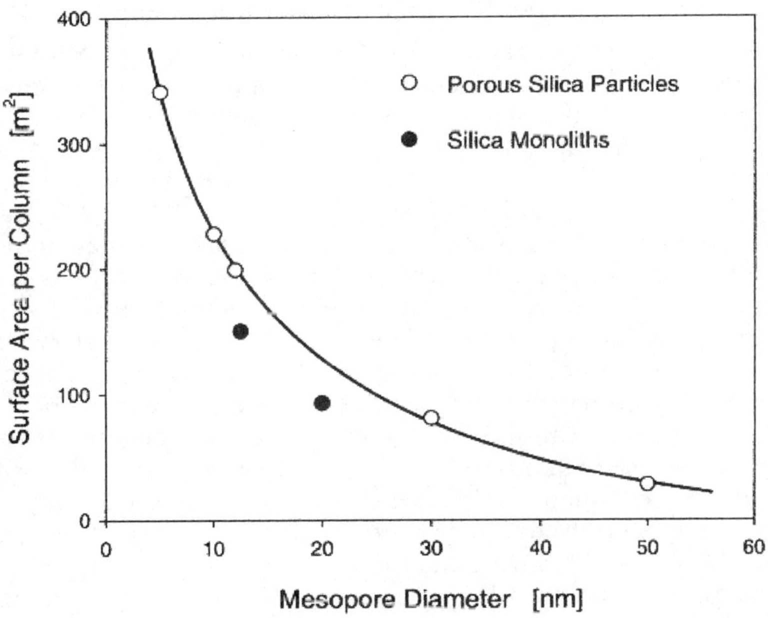

Figure 5 Surface area of a 4.6 mm × 10 cm HPLC column packed with porous silica particles or filled with a silica monolith depending on the mesopore diameter. Reproduced with permission from Leinweber et al. (2002).

peptides. This trend has been greatly accelerated by the widespread use of on-line HPLC electrospray ionization (ESI) mass spectrometry (LC-MS), where ionization efficiency and ion transmission are significantly improved with decreasing flow rates resulting in overall enhanced sensitivity. This is less of an issue for producers of recombinant proteins such as biopharmaceuticals, as the amount of material is generally not a limiting factor. For discovery research, however, where it is often difficult to obtain enough protein for unambiguous identification, the issue of sensitivity is pivotal.

In proteomics research, complex protein mixtures are often separated by two-dimensional gel electrophoresis prior to identification. To this end, proteins are first digested with trypsin in the gel piece and the generated peptides analyzed by LC-MS. The availability of low pmol to fmol amounts makes high-sensitivity techniques mandatory. Particle-packed capillary HPLC columns have been in use for a number of years, but capillary monoliths are an important addition to the toolbox of analytical chemists. Figure 6 demonstrates the capability of RP monoliths in capillary format (100-μm inner diameter) to produce peptide maps from in-gel digested proteins. Coupling of capillary RP-HPLC to ESI mass spectrometry then allows identification of the respective protein based on the molecular mass of the different peptides and limited sequence information through collision-induced dissociation.

Chromatography Based on Ionic Interactions

Proteins and peptides are amphiphiles containing both positive and negative charges. Ion-exchange chromatography is based on electrostatic interactions between positively or negatively charged amino acids in proteins and the oppositely charged stationary phase. A number of functional groups have been widely used to create stationary phases with either cation- or anion-exchange properties (Table 2). So-called strong ion-exchangers maintain their charge state and capacity over a wide pH range. In practice, these stationary phases can be used for protein separation over the complete pH range that is useful for most proteins. In the case of cation-exchangers, such stationary phases contain most often immobilized sulfonic acid groups, whereas for anion-exchangers quaternary amine groups are used. Interactions between proteins and strong cation- or anion-exchangers can be modulated by the mobile phase pH and ionic strength. As proteins contain weakly basic and acidic functional groups (e.g., primary amines, carboxylic acids), their overall charge changes with pH. This is most easily demonstrated by measuring the isoelectric point (pI) of a protein by gel electrophoresis in a pH gradient. As the protein reaches its pI, the sum of all of its negative charges equals the sum of all the positive charges. The protein therefore does no longer experience a force in the electric field and will remain focused at a well-defined position in the pH gradient. When similar experiments are performed in

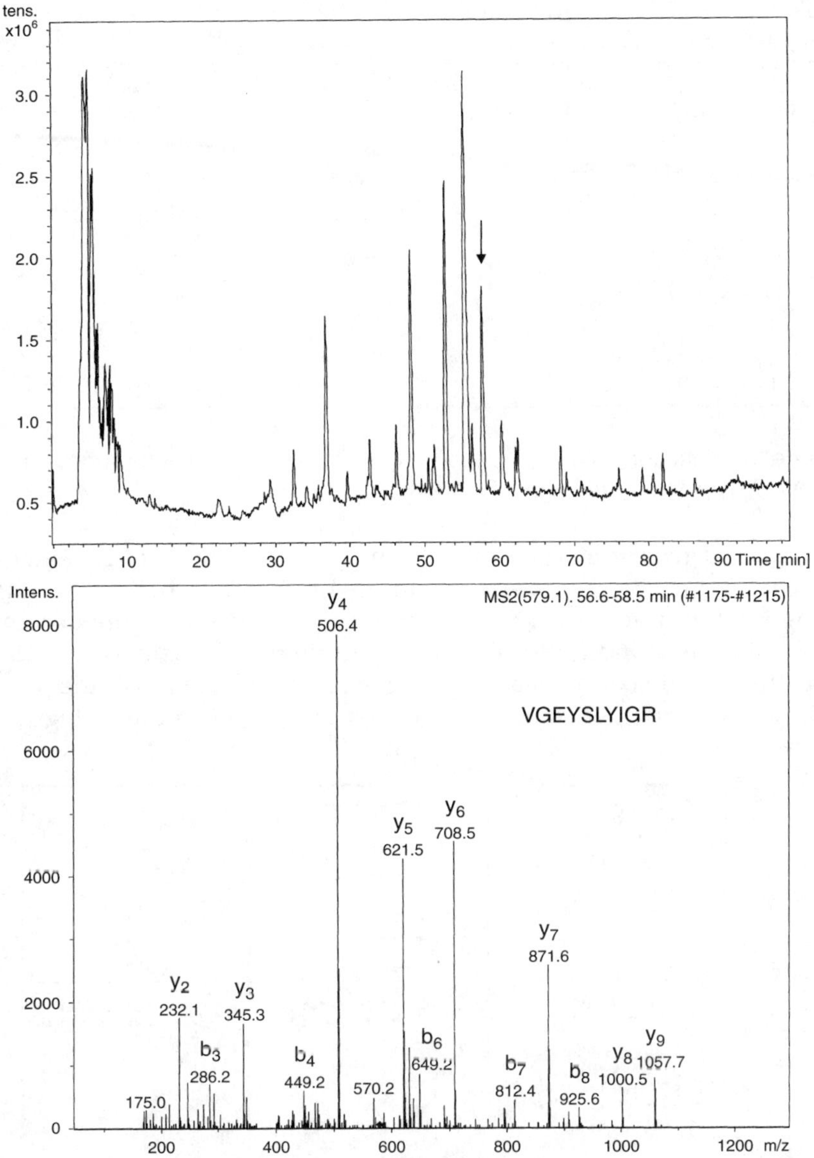

Figure 6 Peptide map of a protein separated by gel electrophoresis and digested with trypsin. The peptides were separated by RP-HPLC on a monolithic capillary (100 μm × 50 cm) and analyzed on-line by ESI mass spectrometry using an ion trap. Upper trace: Total ion current (TIC). Lower trace: CID spectrum of the precursor ion at $m/z = 579.1$ corresponding to one peptide in the map (see arrow in upper trace) elucidating its sequence. Reproduced with permission from Barroso et al. (2003).

Type of Exchanger	Functional Group	Type	Range
Strong anion	$-N^{+}-CH_3$ (with two further bonds on N)	Quaternary amine	1–11
Weak anion	$-NH_2$	Primary amine	1–8
	$-NH$	Secondary amine	1–7
	$-N$ (with two further bonds on N)	Tertiary amine	1–6
Strong cation	$-SO_3^-$	Sulfonic acid	1–14
Weak cation	$-COO^-$	Carboxylic acid	6–14

Table 2 Functional Groups Used in Ion-Exchange Chromatography with Their Respective Working pH Ranges.

ion-exchange chromatography, a more complex picture emerges. When binding proteins to strong anion- or cation-exchangers and running a pH gradient over the column, proteins do not elute exactly at the pH corresponding to their pI value (Figure 7). This is due to the fact that positive and negative charges are distributed over the protein surface in a manner that is characteristic for each protein (Figure 8).

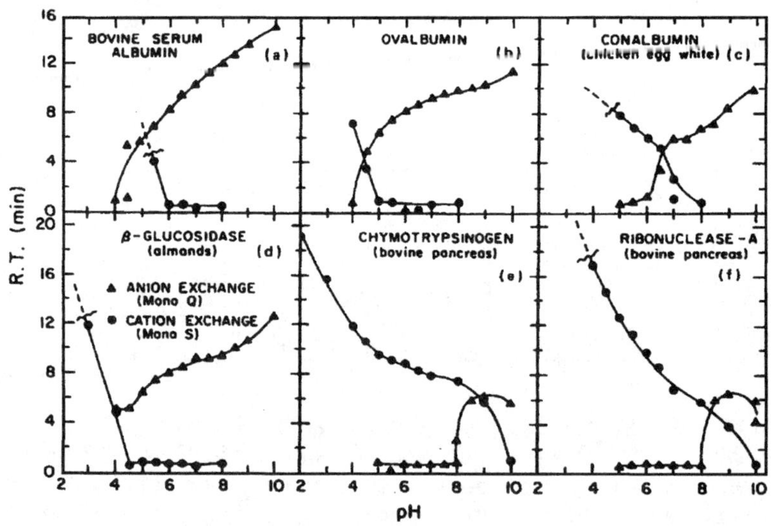

Figure 7 Retention of different proteins on both strong cation- and anion-exchangers over a given pH range. The isoelectric points of these proteins were BSA 5.07; OVA 4.7; CON 6.3; β-GLU 7.3; CHY 9.2; and RIB 8.7. Reproduced with permission from *HPLC of Biological Macromolecules*, 2nd ed. (Karen M. Gooding and Fred E. Regnier, eds.), Marcel Dekker, New York, 2002.

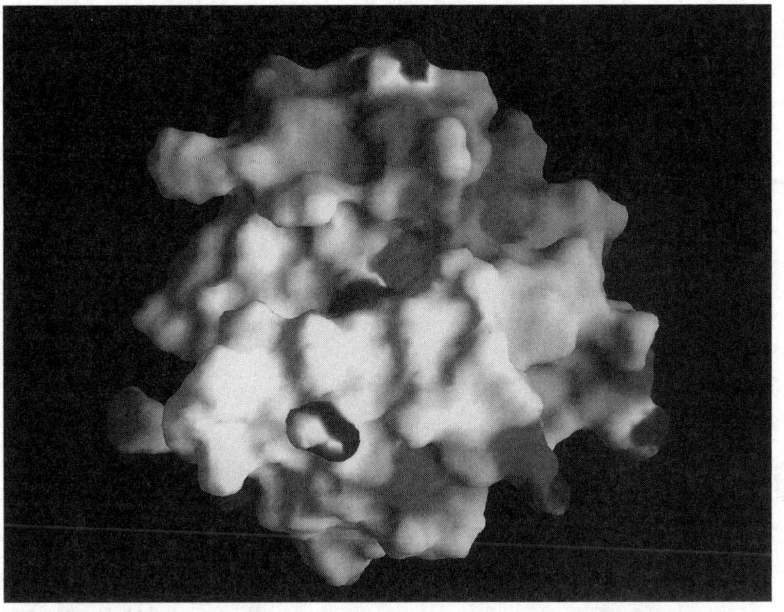

Figure 8 Distribution of positive (lighter areas) and negative charges (dark areas) on the surface of matrix metalloprotease 9 (MMP-9). Negative charges are clustered in a few regions, making these the predominant interacting sites with anion-exchangers, whereas positive charges are more scattered over the surface.

Thus, proteins will orient themselves with their oppositely charged surface toward the stationary phase, which often means that they bind to both strong anion- and cation-exchangers at their pI, as illustrated for some examples in Figure 7.

The most common way of breaking electrostatic interactions is by adding salt to the mobile phase. Again, because proteins have very steep adsorption isotherms due to multi-point interactions with the stationary phase, it is necessary to increase the salt concentration in the mobile phase over time from low to high in a gradient. Salts that are most commonly used are sodium or potassium chloride, but there is little limit to the imagination of the scientist in designing mobile phases. As often in chromatography, there is no pure interaction mechanism. Thus, most ion-exchangers exhibit some sites for hydrophobic interactions as well. This may be used in an advantageous way to achieve separations based on so-called mixed-mode interactions, but optimization of the conditions is often more complex. Hydrophobic interactions may be suppressed to a certain extent by adding an organic solvent such as acetonitrile to the mobile phase. Practical limits to this are the solubility of the proteins (risk of denaturation and precipitation) and the possible need to remove the solvent afterwards. Figure 9 shows the effect of increasing acetonitrile concentrations in the mobile phase on the retention

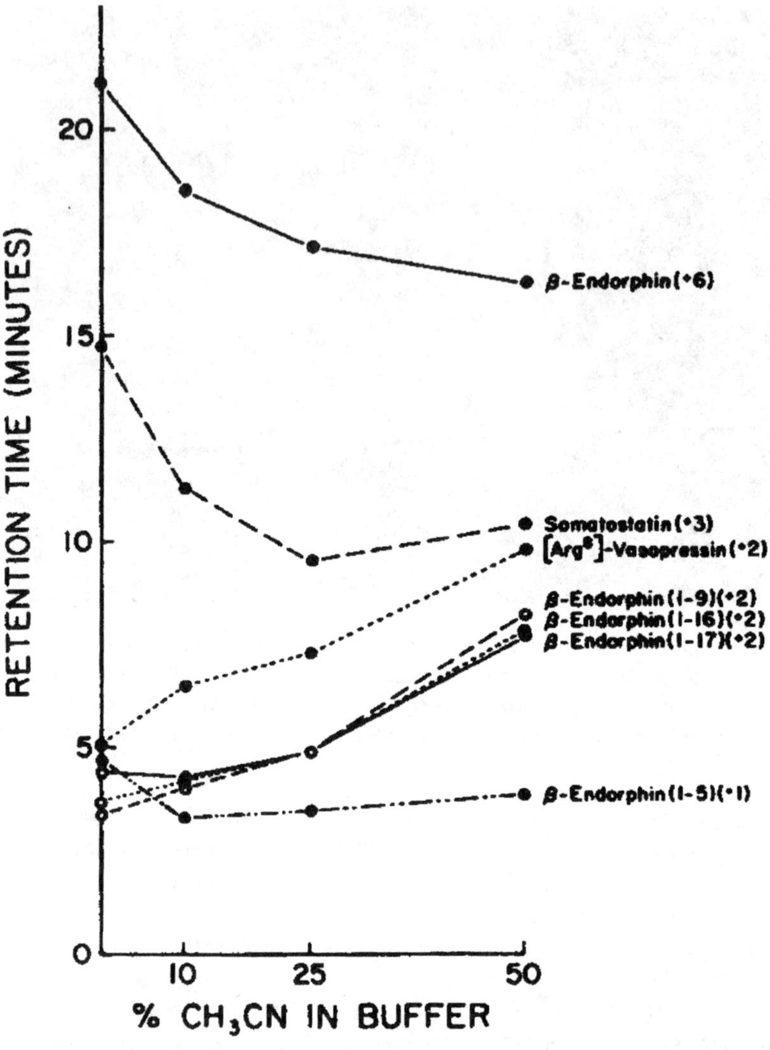

Figure 9 Influence of acetonitrile on the retention time of peptides on the strong cation-exchanger poly(2-sulfoethyl aspartamide)-silica. Note that some peptides show an increased retention on the column upon an increase in acetonitrile concentration. Reprinted with permission from Alpert and Andrews (1988).

time of some peptides on a poly (2-sulfoethyl aspartamide)-silica strong cation-exchange stationary phase (Alpert and Andrews, 1988). Figure 9 shows both a decrease in retention (as expected when taking only hydrophobic interactions into account) as well as an increase in retention with increasing acetonitrile concentrations. This "contradictory" behavior is not uncommon for proteins and peptides, as changes in the three-dimensional structure or in local secondary structural elements may render binding sites more or less accessible to the stationary phase. These effects, which are protein dependent, overlap with the overall tendency of decreasing hydrophobic interactions when adding an organic solvent. Furthermore, electrostatic interactions are strengthened when decreasing the dielectric constant of the solvent, as is the case when adding organic solvents to the aqueous mobile phase.

Ion-exchange chromatography is often used early on in a purification scheme, as it is rather robust with respect to other components in the sample. The only requirement is that the salt concentration should be low enough for the proteins of interest to bind. Ion-exchange chromatography is highly complementary to reversed-phase HPLC, as separation is primarily based on ionic as compared to hydrophobic interactions. Thus, the combination of ion-exchange and reversed-phase HPLC is very popular in so-called two-dimensional HPLC systems. Giddings realized that if one could design a separation system based on two "orthogonal" principles, where components co-eluting in the first dimension would be separated in the ensuing second dimension and at the same time components separated in the first dimension would not be mixed again in the second dimension, that the maximum number of components that could be separated would be (Davis and Giddings, 1983):

$$\text{(the number of separated components in the first dimension)} \times \text{(the number of components separated in the second dimension)}.$$

This means, for example, that if the ion-exchange dimension can separate 50 and the reversed-phase dimension 100 components, an overall capacity of 5000 components could be reached. In reality, no two separation dimensions are ever fully orthogonal, but practical systems have come quite close to this ideal situation. An early example is the separation of proteins in an *E. coli* lysate by a combination of cation-exchange and reversed-phase HPLC (Opiteck et al., 1997). By representation of the chromatographic data in a two-dimensional plot resembling two-dimensional gel electrophoresis, the partial orthogonality of the separation can be appreciated; however, overall peak capacity (in this case 512) is still well below that of two-dimensional gels. Further improvements in both the setup of the chromatographic system as well as in stationary phase chemistry (see, for example, the above-mentioned monolithic stationary phases) have led to ever increasing peak capacities, especially for smaller proteins and peptides. Integration of a sample preparation step with two-dimensional HPLC has been shown to be efficient for the detailed analysis of complex samples such as human hemofiltrate (Wagner et al., 2002). Using restricted access media (RAM) chromatography combined with strong cation-exchange in the first dimension and 4 parallel functioning fast reversed-phase columns in the second dimension has allowed reaching peak capacities on the order of 5000. An example of some of the second-dimension rapid separations is shown in Figure 10. This is comparable with the separation capacity of the most advanced two-dimensional gel electrophoresis methods albeit that gel electrophoresis performs better for higher molecular weight complete proteins whereas HPLC shows its advantages for smaller proteins and peptides. Figure 11 shows that analysis of selected peaks from these chromatograms by matrix assisted laser desorption ionization-time of flight mass spectrometry (MALDI-TOF-MS) leads to spectra that can easily be interpreted. Thus, two-dimensional gels and multidimensional HPLC are highly complementary.

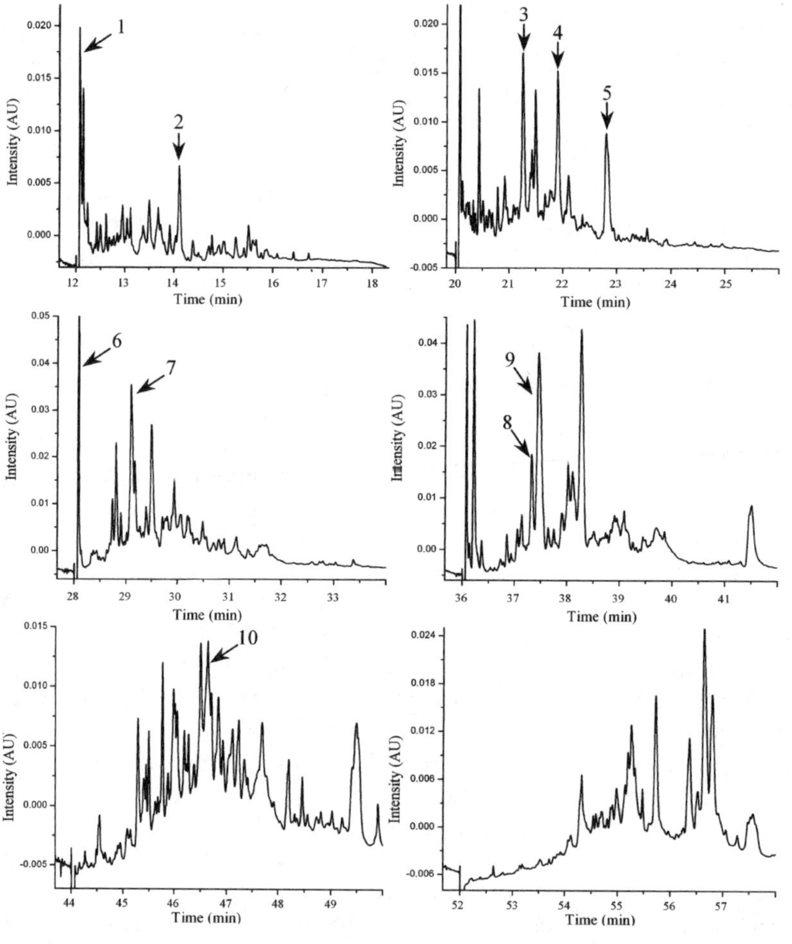

Figure 10 Reversed-phase chromatograms of an automated two-dimensional separation of human hemofiltrate based on strong cation-exchange HPLC in the first dimension. Selected fractions from the cation-exchange run are shown separated by fast RP-HPLC on non-porous C-18 silica particles. Reprinted with permission from Machtejevas et al. (2004).

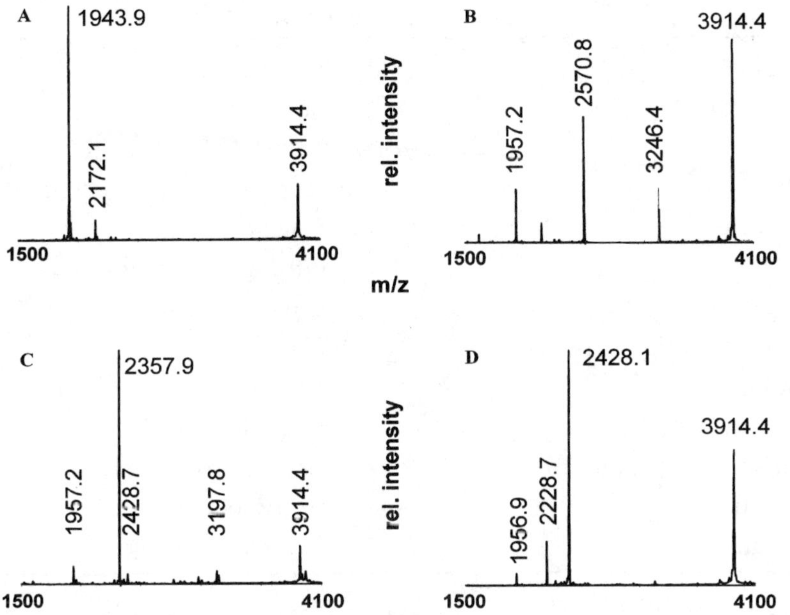

Figure 11 MALDI-TOF mass spectra of RP-HPLC fractions (A) 4, (B) 7, (C) 8, and (D) 9 corresponding to numbered peaks of Figure 11. *m/z* 3914.4 (doubly charged: 1957.2) is an internal peptidic standard for mass calibration. Measurement was performed in the linear positive operation mode using a matrix of α-cyano-4-hydroxycinnamic acid mixed with fucose. Reprinted with permission from Machtejevas et al. (2004).

Integrated multidimensional HPLC has not been used extensively for the analysis of biopharmaceuticals. This is mainly due to the fact that these proteins are obtained in fairly large quantities, making fraction collection and reinjection possible. Moreover, the number of components in one sample is often low. However, it is worth considering if such an approach would not speed up the isolation and identification of minor impurities in addition to being more reproducible than manual sample handling. The future will show whether multidimensional HPLC-MS will be more frequently applied in biopharmaceutical research and quality control.

Affinity Chromatography

Affinity chromatography (AC) can be considered a special kind of chromatography, as it relies on highly specific interactions between the stationary phase and the target proteins. AC is thus less dependent on the efficiency of the stationary phase in terms of plate height to achieve the desired goal of enriching a given protein. On the other hand, AC is crucially dependent on the selectivity of the immobilized affinity ligand. Over the years, many ligands have

Ligand	Binding partners
Avidin	Biotin derivatives
Aprotinin	Serine proteases
Biotin	Avidin, streptavidin
Concanavalin A	Glycoproteins, oligosaccharides
Gelatin	Gelatinases
Glutathione	Glutathione transferases
Heparin	Coagulation factors
Iminodiacetic acid	Interferon
Lysine	Plasminogen activator
Protein A	Immunoglobulin G
Phosphorylethanolamine	C-reactive protein
Protein G	Immunoglobulin G
Protamine	Immunoglobulin M

[a]This list gives only a glimpse of possible affinity ligand-target pairs.

Table 3 Examples of Affinity Ligands and Their Corresponding Target Molecules.[a]

been developed with more or less high selectivity (Table 3). Some of them, like lysine, are also used on a preparative scale to purify biopharmaceutical products, but this is more the exception. Most affinity ligands are relatively expensive, which makes their use on a preparative scale costly. Leakage of affinity ligands from the stationary phase can also pose a problem, as they may induce adverse side effects, such as an immune response against proteinaceous affinity ligands.

The most versatile and widely applied ligands are those that bind to a range of proteins that have a common feature. Protein A and protein G fall into this category, as they bind to structural determinants in certain subclasses of immunoglobulins. This can be used in the rapid and efficient enrichment of antibodies from cell culture supernatants. Widely used purification strategies in the case of recombinant proteins are based on fusions with proteins or peptides that bind to specific affinity ligands. These comprise glutathione-*S*-transferase (GST) and oligo-histidine tags. The former bind to immobilized glutathione, whereas the latter bind to immobilized Ni^{2+} ions. Again, these methods are primarily of importance in research, as the fusion part of the protein needs to be fully and specifically removed for clinical applications, which is far from easy. Many other fusions have been tried and some of them are being used successfully mainly on an

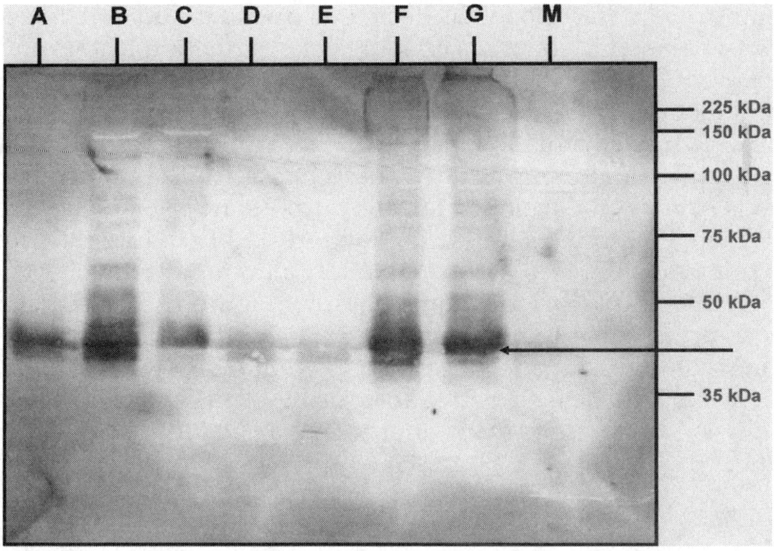

Figure 12 Analysis of the expression and partial purification on a His-trap column of the dopamine D2S receptor containing an internal hexa-hisitidine tag (SDS-PAGE and Western blotting). A = membrane preparation; B = supernatant after solubilization (binding: 0.045 pmol/mg protein); C–E = flow-through of His-trap column; F and G = elution from His-trap column with 0.5 M imidazole (binding: 11.0 pmol/mg protein), Reprinted with permission from de Jong et al. (2004).

analytical scale, for example for so-called pull-down experiments (Nilsson et al., 1997). Even in the case of integral membrane proteins, fusion with affinity tags can be useful to enrich the proteins under mild conditions, where activity is maintained. This is illustrated on the example of the dopamine D2S receptor expressed in *Pichia pastoris*, which was enriched 244-fold in a single step (Figure 12) (de Jong et al., 2004).

Antibodies form not only a large group of new biopharmaceutical products but are also versatile affinity ligands. Notably, the capability to generate a large variety of binding sites in an efficient manner has led to a more widespread use of antibodies as affinity ligands. Minimal size antibodies such as those derived by phage display are attractive, as they can be produced in fast-growing microorganisms such as *E. coli* and exhibit a rather good stability (Hoogenboom and Chames, 2000). Display technologies have also opened ways to engineer specific affinity ligands based on other protein scaffolds as well as enzymes with tailored activities (Wittrup, 2001).

Another example of the use of affinity chromatography is the removal of contaminating proteins. This is often a problem when isolating proteins from natural sources such as blood. Although production of proteins from human blood has decreased due to the advent of recombinant proteins, it is still a major source of coagulation factors and albumin. Problems with blood-derived products relate to potential contamination with viruses and the isolation of low-abundance labile proteins such as coagulation factor VIII from amidst highly abundant proteins such as albumin. Although highly efficient immunoaffinity methods have been developed for this purpose, removal of albumin is an additional option to simplify the mixture and to reduce the protein load prior to AC. As an example of albumin removal by immunoaffinity or dye ligand chromatography, Figure 13 shows that both methods are capable of removing the major portion of albumin from human serum without affecting other serum proteins (Govorukhina et al., 2003). Nevertheless, sensitive analysis by LC-MS reveals that some albumin is still present even after removal albeit at the 1% level (Figure 14).

Size Exclusion Chromatography

Size exclusion chromatography (SEC), also often referred to as gel filtration, is probably the easiest chromatographic method to understand from a theoretical point of view. Separation in SEC relies on the different sizes or better hydrodynamic volumes of proteins relative to a given pore diameter of the stationary phase. Ideally, there should be no interaction between the proteins and the stationary phase, a criterion

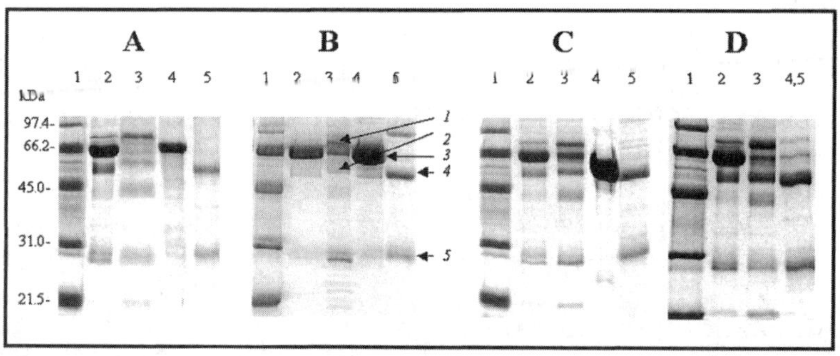

Figure 13 Removal of albumin from human serum. (A) Immunoaffinity chromatography; (B) dye ligand affinity chromatography (HiTrap Blue); (C) dye ligand affinity chromatography (albumin removal column, Merck KGaA); (D) dye ligand affinity chromatography (Aurum Serum Protein column, BioRad).
Lanes: 1, standards; 2, crude serum; 3, depleted serum; 4, bound protein eluted from albumin depleting columns with either 12 mM HCl (A, D), or with 1.5 M KCl in 20 mM KH_2PO_4 pH 7.0 (B), or with 25 mM Na_2HPO_4, 2.5 M NaCl, pH 8.0 (C). Indicated proteins were identified by in-gel tryptic digestion and peptide mass mapping by MALDI-TOF-MS: 1, serotransferrin; 2, α_1-antitrypsin; 3, albumin; 4,5, γ-globulins heavy and light chain, respectively. Lane 5 shows the removal of IgGs using either protein A or G columns. Reprinted with permission from Govorukhina et al. (2003).

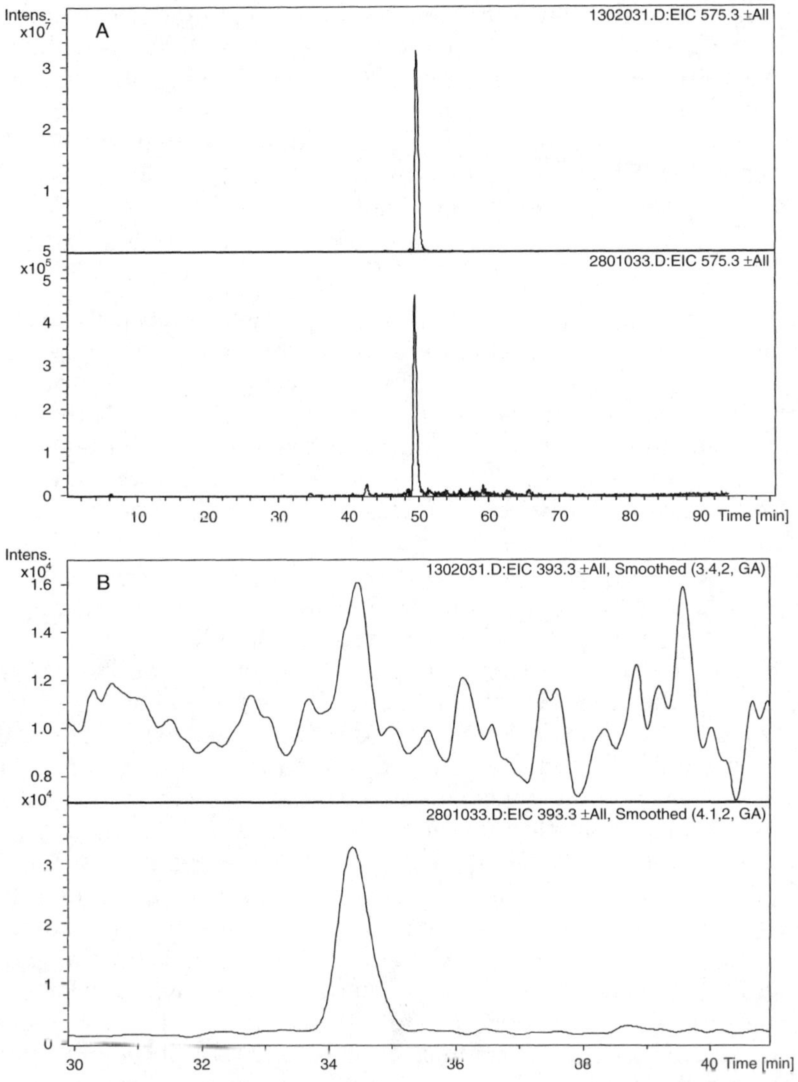

Figure 14 (A) Efficiency and (B) selectivity of albumin removal from human serum using an anti-albumin immunoaffinity column assessed by LC-MS. (A) Extracted ion chromatogram of $m/z = 575.3$ (doubly charged molecular ion of peptide LVNEVTEFAK; positions 41-50 in human serum albumin) of tryptic digests of human serum (upper trace; peak height 3.2×10^7) or of human serum after depletion with an anti-albumin immunoaffinity column (lower trace; peak height 4.6×10^5). (B) Extracted ion chromatogram of $m/z = 393.3$ (doubly charged molecular ion of peptide IVDLVK; positions 193-198 in human α_1-antitrypsin) of tryptic digests of human serum (upper trace; peak height 16,052) or of human serum after depletion with an anti-albumin immunoaffinity column (lower trace; peak height 32,607). Note the much cleaner detection of this peptide fragment after depletion and the increased overall peak height. Reprinted with permission from Bischoff and Luider (2004).

that is difficult to fulfill in practice. However, most modern stationary phases for SEC have been surface modified in a way that adsorption is minimized. In the ideal situation of perfect SEC, the retention volume (V_r) equals the so-called exclusion volume (V_o) plus part of the internal volume (V_i) of the stationary phase. The portion of V_i that is accessible to the respective protein is defined as the so-called distribution coefficient K_D, resulting in the simple equation:

$$V_r = V_o + K_D V_i \tag{1}$$

Equation 1 signifies that a protein can only elute between the exclusion volume and the internal volume, as the two limits of the separation space in SEC. Consequently, SEC is not really a high-resolution technique but has its unique advantages when it comes to separating proteins from low-molecular-weight constituents (e.g., buffer salts). Furthermore, SEC is generally a very gentle separation method due to the absence of strong adsorptive forces and can thus be used to separate multiprotein complexes or organelles. During the production of biopharmaceuticals, SEC is often used at the end of a purification scheme to "polish" the final product and to formulate it in the appropriate buffer including additives for enhanced stability.

SEC also has an important place in quality control of biopharmaceuticals, notably the detection of aggregates. Protein aggregates are often detrimental to the biological activity of a protein but more importantly may be recognized by the immune and inflammatory system of the organism as "foreign" material eliciting a response. Moreover, protein aggregates indicate that the formulation of a given product is unstable and thus prone to further aggregation during storage (see Figure 16).

One limitation of SEC is that proteins cannot be concentrated due to the fact that adsorption is avoided. This means that the sample volume should not exceed approximately 2% of the column volume not to affect the separation too much. Furthermore, there will always be a dilution of the applied sample due to diffusion during the run. As the diffusion coefficient is proportional to the square root of the molecular weight, diffusional band broadening is less severe for bigger proteins than for low-molecular-weight components. It is important to note that the elution properties of proteins on SEC can also be dependent on the mobile phase composition, especially pH and ionic strength, as these may influence the hydrodynamic radius or "shape" of the molecule. For example, a spherical molecule of 500 kDa would have a radius of gyration (R_G) of about 4.5 nm, whereas a rod-shaped molecule of the same molecular weight would have an R_G of somewhere between 40 and 100 nm depending on the diameter of the rod. Last but not least, the flowrate of the mobile phase has a considerable impact on resolution in SEC. SEC is best performed at rather low linear velocities of the mobile phase, when compared to adsorptive chromatography methods, to allow the proteins to diffuse between the stagnant inner pore volume and the advancing mobile phase. At very low linear flow, this is counterbalanced by the effect of longitudinal diffusion, leading to band spreading. As in all other modes of chromatography, a smaller particle diameter allows for higher flow rates without compromising separation efficiency. In practice, one needs to find a suitable compromise between the time per run and the required resolution.

Applications

The following paragraphs describe applications from the production of recombinant proteins for pharmaceutical purposes. The chapters are organized according to the increasing complexity of the host organism starting with bacteria and ending with living animals. The problem of characterizing glycoproteins is particularly highlighted, as most of the present-day biopharmaceuticals are glycoproteins.

α_1-Antitrypsin Produced in *E. coli*

The expression of human proteins in microorganisms is often accompanied by inclusion body formation. This is due to the generally very high expression level leading to aggregation and precipitation and the absence of chaperones that prevent aggregation and facilitate correct protein folding. Although most of the times inclusion body formation is a disadvantage, as active protein is difficult to recover after refolding, it can be turned to an advantage in case refolding conditions do allow the recovery of functional protein (Middelberg, 2002). Another option is to try to reduce the formation of inclusion bodies through engineering the expression host and/or the protein itself. The latter was evaluated for the protease inhibitor α_1-antitrypsin (AAT), which is deficient in patients developing hereditary emphysema (Coakley et al., 2001). AAT is a glycoprotein of 52 kDa normally produced by the liver. AAT circulates in the blood at mg/mL levels but is severely reduced in patients with mutations in the coding sequence, leading to intracellular aggregation and reduced blood levels. Supplementation of AAT to these patients can normalize blood levels and possibly prevent the development of pulmonary emphysema. Clinical grade AAT is currently obtained as a by-product of blood fractionation and thus prone to risks of viral contamination. Furthermore, it is not cost-effective to obtain large amounts of this protein from blood. A number of expression systems have therefore been tried to obtain high levels of recombinant AAT among which the expression in transgenic animals is particularly promising (Bischoff et al., 1992; Tebbutt, 2000).

As an example on how protein engineering can affect inclusion body formation, the production of soluble AAT within the cytoplasm of *E. coli* has been influenced by selected mutations within the extended

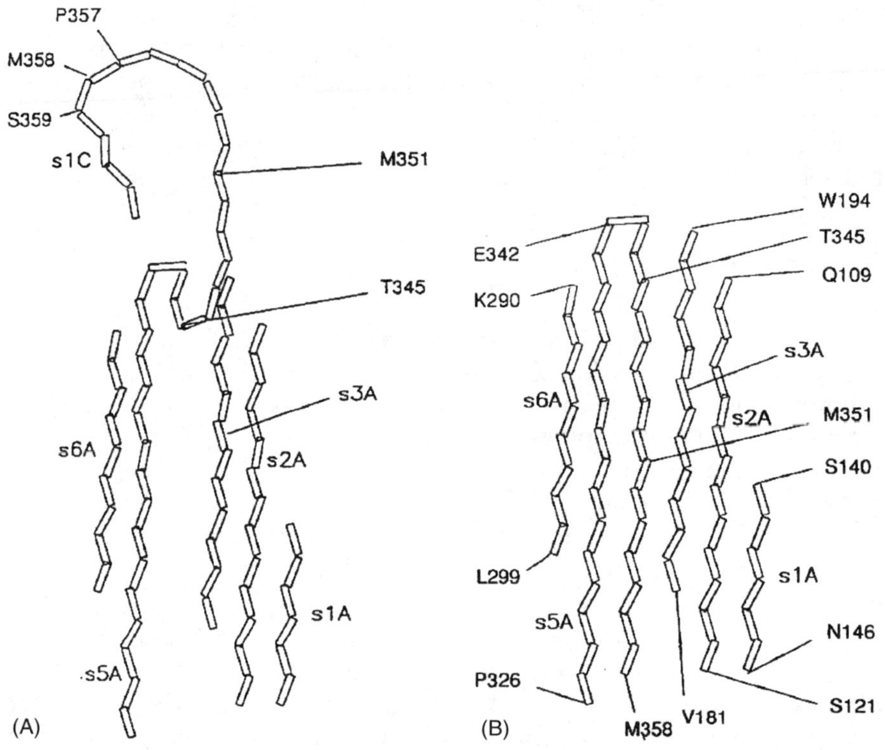

Figure 15 Schematic representation of the polypeptide fold of β-sheet A and the active site loop of (A) native and (B) proteolytically modified α_1-antitrypsin. Insertion of β-sheet A from the same molecule (intramolecular) or an adjacent one (intermolecular) leads to a stable form. In the case of intermolecular insertion, this can lead to aggregate formation. Aggregate formation can occur even in the absence of proteolytic cleavage. Reproduced with permission from Schulze et al. (1994).

loop region between residues 345 and 359 (Figure 15) (Schulze et al., 1994). This region is prone to insertion into another AAT molecule, leading to extensive aggregate formation. For example, replacement of Met-351 with Glu and Met-358 with Arg leads to an increase in the ratio of soluble AAT to total protein from 11.4 mg/g to 28.5 mg/g by rendering loop insertion more difficult. Purification of these AAT variants is thus greatly facilitated and yields are improved. AAT variants have been purified from the cytoplasmic fraction of *E. coli* by a sequence of chromatographic steps after an initial clarification step by tangential flow micro- and ultrafiltration. A number of analytical methods were used to ensure the absence of contaminants, to assess the biological activity, and to determine purity and identity. Table 4 shows the overall purification in terms of yield and endotoxin content. Endotoxin also known as LPS (lipopolysaccharide) is a potent inflammatory stimulator present in the cellular membrane of bacteria that needs to be removed to

Purification Step	α_1-Antitrypsin (mg)	Total Protein (mg)	Purification Factor	Yield (%)	Endotoxin (units/mg)
Starting material	533	77,694	1	100	1.2×10^7
Retentate 10 kDa ultrafiltration	508	32,098	2.3	95	1.9×10^5
Anion exchange	465	10,810	6.3	87	1.6×10^4
IMAC	348	590	86	65	Less than 0.22
Final material	294	294	146	55	Less than 0.22

[a]Reproduced with permission from Bischoff et al. (1991).

Table 4 Purification Chart of α_1-Antitrypsin Produced in *Escherichia coli*.[a]

very low levels from any biopharmaceutical product. The purification table shows that immobilized metal affinity chromatography (IMAC) on Zn^{2+} ions proved to be highly efficient for this purpose.

As examples of analyses that were done to characterize the final product, size exclusion chromatography was used to ascertain the absence of aggregates (Figure 16). This is particularly important in the case of AAT, as the protein is prone to aggregation as discussed before. The molecular mass of the complete protein was furthermore measured by ESI-MS. The measured mass of 43,965 ± 10 Da was in good agreement with the calculated value of 43,928 Da. However, due to the spread in the MS measurements, it was not possible to ascertain that the recombinant AAT was identical to the AAT as predicted from the sequence, as deamidations (+1 Da) or oxidations (+16 Da) would escape this analysis. Indeed, comparative analyses of different production lots by isoelectric focusing gel electrophoresis in a pH gradient between pH 5 and 6 revealed some microheterogeneity (Figure 17). This is not unusual for larger proteins, which are rarely found as a single form. Important is that this heterogeneity in overall surface charge is reproducible from batch to batch.

Structural analyses as described above need to be combined with functional assays of biological activity. This is easily done in the case of enzymes or enzyme inhibitors, as a simple biochemical assay can be developed. As the major target protease for the AAT variant with Ala in position 357 and Arg in 358 is plasma kallikrein, an assay for inhibition of this enzyme was developed giving a second-order rate constant of 1.8×10^5 M^{-1} s^{-1} (Bischoff et al., 1991). Although there is

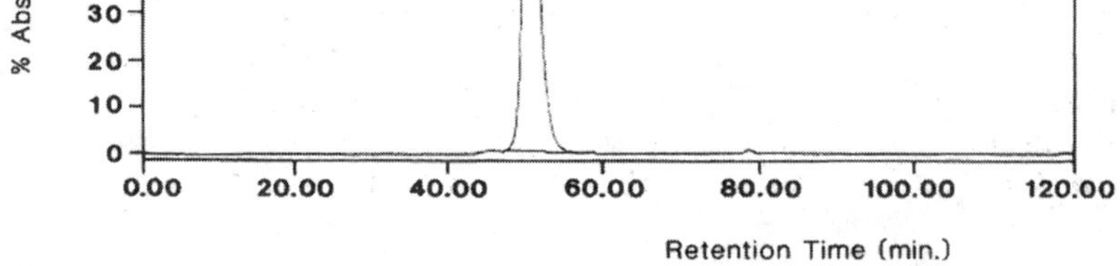

Figure 16 Size exclusion chromatography of recombinant α_1-antitrypsin variant Ala^{357}-Arg^{358} produced in *E. coli.* Analysis was performed on a set of tandem TSK-G 2000 SW columns (60 × 0.75 cm each, 10-μm particle diameter) in 10 mM Tris-HCl, pH 7.5, 150 mM NaCl at 0.5 mL/min. Reproduced with permission from Bischoff et al. (1991).

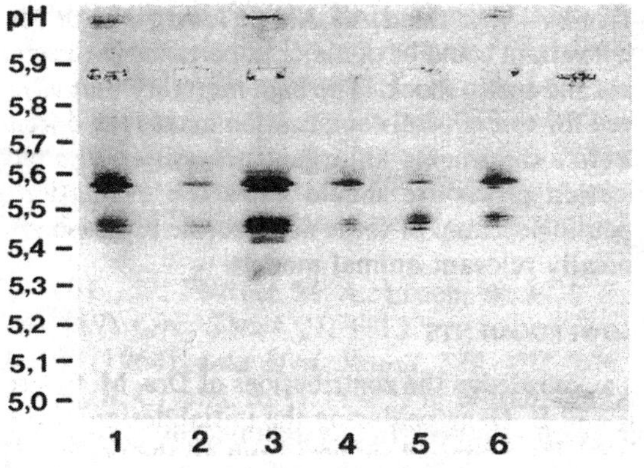

Figure 17 Isoelectric focusing of α_1-antitrypsin Ala-375, Arg-358 obtained from different production batches in an immobilized pH gradient between 5 and 6. β-Lactoglobulin A (pI = 5.2) and bovine carbonic anhydrase (pI = 5.85) were used as isoelectric point markers (not shown). Gels were run at 500 V (5 mA) overnight and stained with Coomassie Brilliant Blue R-250. Reproduced with permission from Bischoff et al. (1991).

currently no internationally accredited standard available for this inhibitor, the value is in close agreement with previously published data (Patston et al., 1990).

This example of expression of a human serum protein in a prokaryotic microorganism shows that there can be numerous hurdles that need to be taken before a biologically active protein is obtained at fairly high levels. Quality control of this protein requires a range of techniques, of which some were outlined in this paragraph. Among them, chromatography plays a crucial role.

Hirudin Produced in *Saccharomyces cerevisiae*

Hirudin is a naturally occurring, very potent inhibitor of thrombin, one of the central enzymes in blood coagulation. Hirudin is produced by the blood leech *Hirudo medicinalis* and serves to prevent coagulation during feeding. With an inhibition constant in the fM range, hirudin is to date the most potent inhibitor of thrombin. Because the natural host produces only minute quantities of hirudin, it is not possible to use it for making large amounts of a homogenous product. Moreover, hirudin in the blood leech is produced in a number of variants, each of which may be partially degraded at its C-terminus due to proteolytic activity. In order to evaluate the potential of hirudin as an anticoagulant, it is thus necessary to produce it in a homogenous fashion as a recombinant protein in heterologous hosts. Initial trials to produce hirudin in *E. coli* were not highly successful, as the production yield was low and the protein existed in a number of modified forms. *S. cerevisiae* proved to be the most suitable host organism to express hirudin at high levels and to secrete it into the culture supernatant. Considerable optimization of the host strain was however required, notably the deletion of a number of proteases located in the secretory pathway of *S. cerevisiae.* This prevented C-terminal cleavage of recombinant hirudin and led finally to a production level of close to 1 g/L culture supernatant with little degradation. Hirudin is today a pharmaceutical product with its main application in deep venous thrombosis.

Purification from culture supernatant of hirudin starts in fact with the engineering of the host organism, as it is much easier to use culture supernatant as the starting material than cellular extracts. The main challenge initially is to reduce the volume of the supernatant without losing significant amounts of the product. This can be done by tangential flow micro- and ultrafiltration to first remove cellular debris and other insoluble material and subsequently to concentrate the product. Purification is often less demanding, as the overall protein load is rather low and the product constitutes the major protein. Nevertheless, even highly optimized production systems cannot assure that the final quality is sufficient for clinical use. Furthermore, there may be issues with stability of the final product. All this requires careful analytical monitoring of product quality.

Next to C-terminal degradation, which is mainly a problem during protein expression and secretion itself, deamidation of Asn residues turned out to be an issue for protein stability. Deamidation of Asn and to a lesser extent also Gln residues is a well-known reaction that proceeds more or less rapidly depending on the structural and sequence context (Bischoff and Kolbe, 1994). Next to the undoubtedly important role of deamidation in controlling protein stability *in vivo*, it is an important mechanism of introducing modifications into proteins during or after purification. Hirudin is just a case example of the many proteins that have been shown to undergo spontaneous deamidation. Deamidation is catalyzed both at basic and acidic pH and the stability optimum of a given Asn or Gln residue depends in part on sequence and structure. Deamidation is most often detected by analytical methods that respond to changes in surface charge. Prime examples are isoelectric focusing and ion-exchange HPLC. Charge heterogeneity in a recombinant therapeutic monoclonal antibody HER2 was discerned by both cation-exchange HPLC and isoelectric focusing gel electrophoresis (IEF) (Figure 18). It is evident that IEF has a much higher charge-dependent resolution than HPLC, revealing further microheterogeneity in the isolated HPLC fractions.

The mechanism of deamidation has been studied in detail, and some of the intermediates may have sufficient stability to appear as impurities in the final product. This was the case for rec. hirudin, where reversed-phase HPLC analysis revealed a minor impurity with slightly increased hydrophobicity (Bischoff et al., 1993). Detailed analyses by peptide mapping and mass spectrometry proved this impurity to be

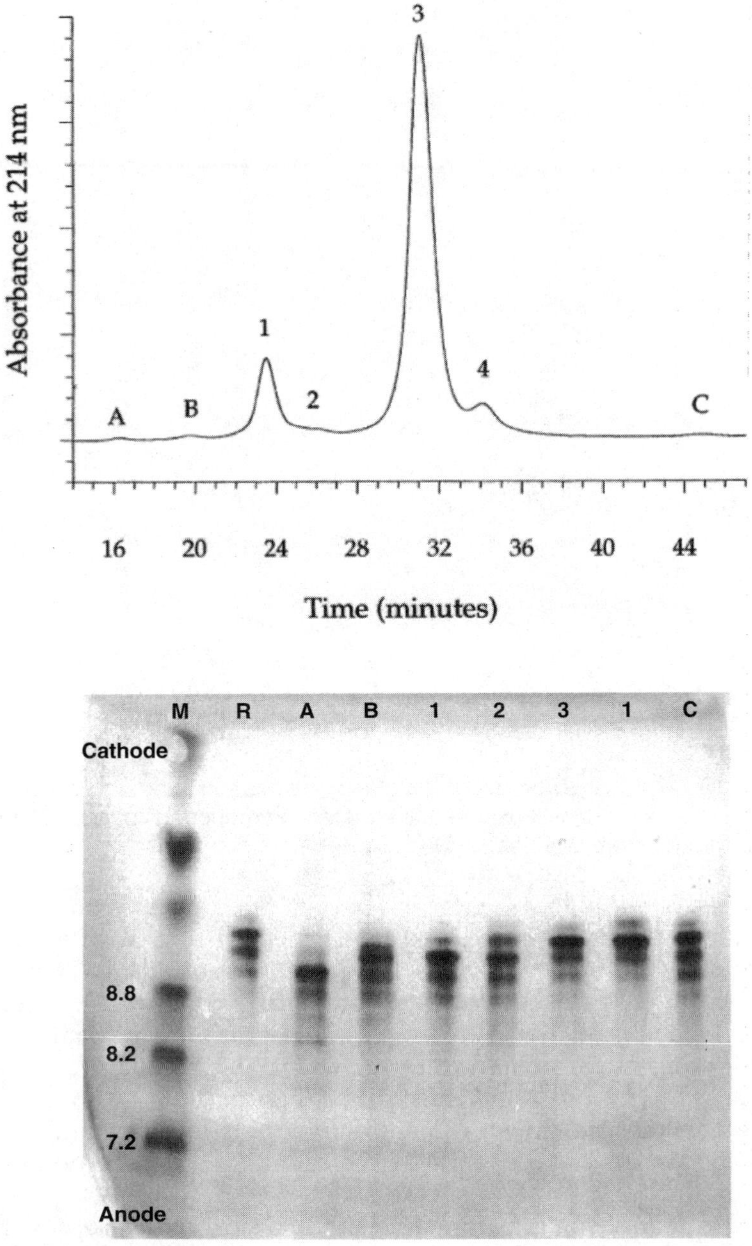

Figure 18 Analysis of the therapeutic monoclonal antibody HER2 for surface charge heterogeneity by (A) cation-exchange HPLC and (B) isoelectric focusing (IEF). The fractions shown in (A) were analyzed by IEF showing that protein forms eluting earlier from the cation-exchange column have a more acidic isoelectric point as expected. It is noteworthy that IEF shows further microheterogeneity not discriminated by HPLC. Reproduced with permission from Harris et al. (2003).

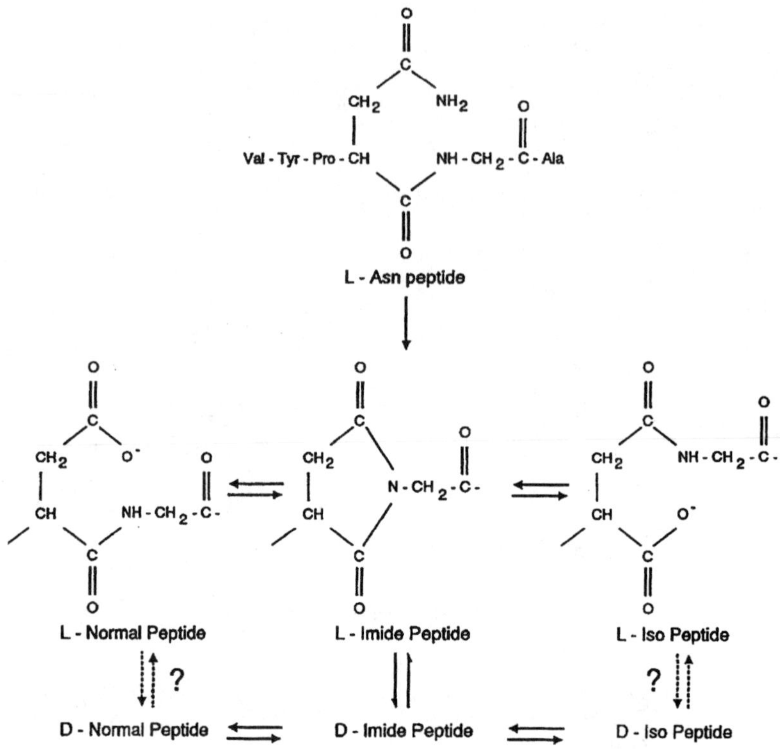

Figure 19 Pathways for deamidation, isomerization, and racemization for aspartyl- and asparaginyl hexapeptides related to adrenocorticotropin. Reprinted with permission from Bischoff and Kolbe (1994).

due to stable succinimide intermediates of the deamidation at Asn-33 and Asn-53. These intermediates are formed by ring closure between the amide group of the Asn side chain and the peptidic α-NH group of the following Gly (Figure 19). It is noteworthy that both Asn residues are located in highly flexible surface loops of hirudin and that both are being followed by the small amino acid Gly, which allows for ring closure without major steric constraints (Figure 20). As an example, Figure 21 shows the power of peptide mapping to detect and localize such modifications and of LC-MS to pinpoint to the underlying structural modification.

Deamidations at Asn can also lead to a shift of the peptide backbone from the customary α position to generate isoaspartic acid (Iso-Asp) (see Figure 19). Such modifications can be detected and localized through the incorporation of radiolabeled methyl groups from the substrate *S*-adenosyl-L-[methyl-^{3}H] methionine using the enzyme protein isoaspartyl methyltransferase (PIMT; EC 2.1.1.77).

Taken together, this gives the analytical chemist a wide range of methods to detect and localize deamidation events in proteins as well as

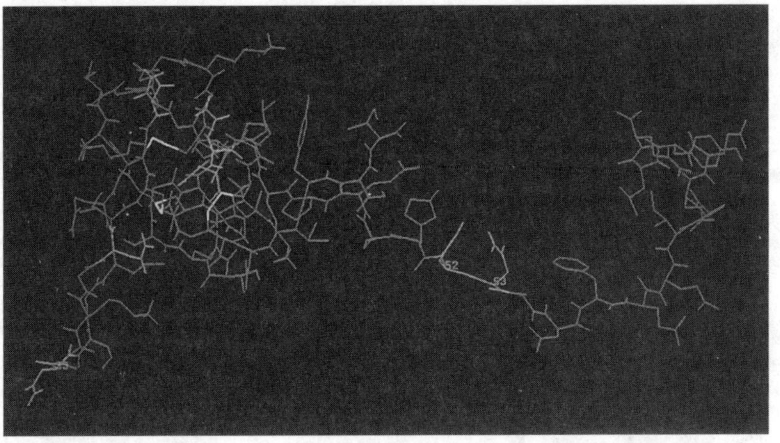

Figure 20 Representation of the three-dimensional structure of the hirudin (rHV2-Lys^{47}) α-thrombin complex. Asn^{52} and Asn^{53} (in the linker region right) and Asn^{33} (in the flexible loop, lower left corner) are marked with numbers. It is noteworthy that mainly these three Asn residues, which are located in solvent accessible, flexible regions of the proteins, are prone to deamidation. Reprinted with permission from Bischoff and Kolbe (1994).

their different intermediates and end products. It is without doubt that deamidation is one of the most frequent reasons for protein instability. Whether deamidation affects the biological function of the pharmaceutical product, its half-life, or immunogenicity needs to be determined on a case-by-case basis. It can be said that deamidation is a "natural" process that will likely also occur once a protein has been administered to the patient. In the case of rec. hirudin, there was no effect of succinimide formation at Asn-33 or 53 on thrombin inhibition as determined by measuring the inhibition constant.

Analysis of Glycoproteins: Recombinant C1-Inhibitor from Rabbit Milk

Glycoprotein Analysis

Glycosylation is one of the most common post-translational modifications encountered in proteins. The majority of the cell surface and secreted proteins as well as many biopharmaceuticals in the market today and in different phases of clinical development are glycoproteins (Dembowsky and Stadler, 2001).

Glycoproteins are characterized by the presence of different types of oligosaccharides attached to the peptide chain. These oligosaccharides are short, usually branched chains of specific sugars and nitrogen-containing amino sugar units bonded covalently to specific amino acid residues. Many examples of glycosylated therapeutic proteins can be found in the literature such as factor VIII, most of the

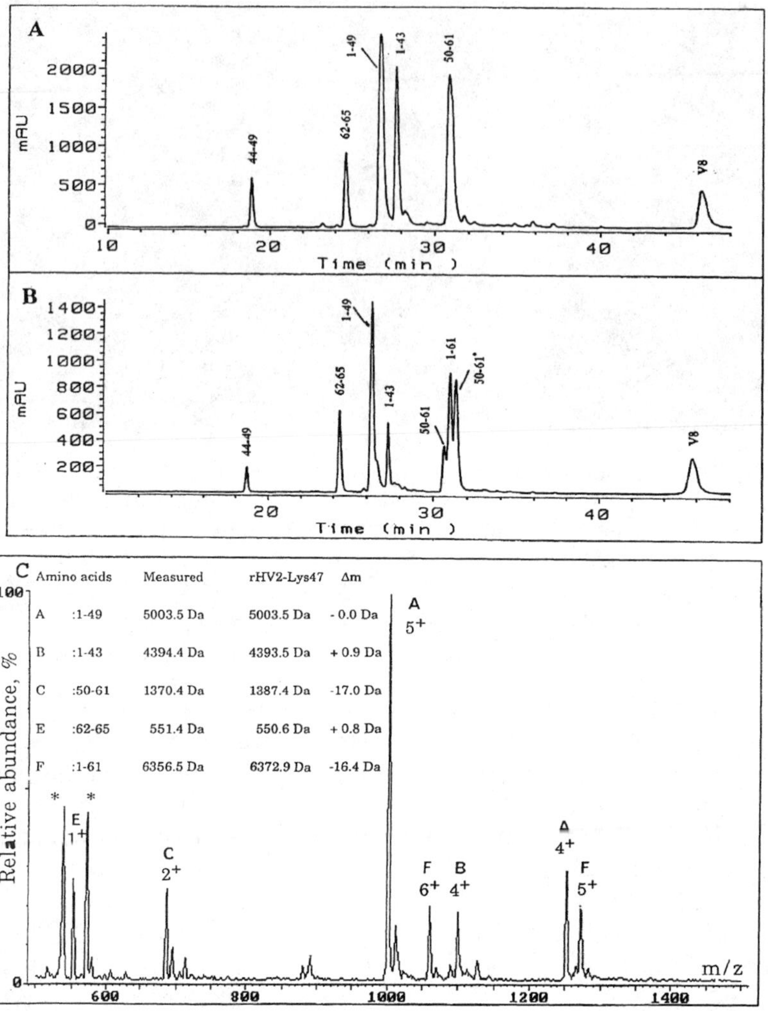

Figure 21 Mapping of the peptides from rHV2-Lys^{47} (hirudin) by LC-MS after digestion with Glu-C protease. (A) Unmodified hirudin. (B) Impurity containing a succinimide between Asn^{53} and Gly^{54}. (C) Mass spectrometric analysis of the proteolytic digest by ESI-MS showing that a 17 Da mass reduction is observed in peptide fragment 50-61 in agreement with the loss of an NH_3 moiety upon succinimide formation (see Figure 19). Reprinted with permission from Bischoff et al. (1993).

monoclonal antibody drugs, and cytokines such as interferon α and β and IL-2. (Mountain et al., 1998).

The presence of glycans is absolutely necessary for full functionality and optimal efficacy of the glycoprotein, as the carbohydrate components play crucial roles in the modulation of the physicochemical properties of the protein. They influence protein folding, oligomer assembly, and secretion processes, as well as the clearance of glycoproteins from the bloodstream.

Successful production of recombinant glycoproteins to be used as therapeutic agents requires the use of host systems able to mimic the structures synthesized by humans. Different organisms perform different types of glycosylation; for example, the N-glycans of recombinant glycoproteins expressed in insect cells contain mainly high mannose or tri-mannose structures, which are truncated forms of the sialylated N-glycans found in mammalian cells (Chang et al., 2003). If the expression system produces incomplete glycosylation, it results in shorter half-life of the protein in the blood circulation and decreased clinical effectiveness.

Certain carbohydrate structures have also been found to be antigenic, and regulatory authorities such as the U.S. Food and Drug Administration are demanding increasingly sophisticated carbohydrate analysis as part of the product or process validation (Jenkis et al., 1996).

Modifications in the carbohydrate chain of a glycoprotein are a reflection of its movement through the secretory pathway because sugar-chain processing takes place in a stepwise and unidirectional manner. Therefore, study of glycans provides significant information about the working of the glycosylation machinery of the cell and constitutes a key aspect of glycoprotein characterization.

There are two main types of glycosidic linkage in glycoproteins (Figure 22):

- N-glycosidic link via the amide nitrogen of an asparagine residue
- O-glycosidic link via the hydroxyl (-OH) group of serine, threonine, tyrosine, hydroxylysine, or hydroxyproline. Glycan units may also be O-linked to serine and threonine residues via a phosphate group, specifically known as phosphodiester linkage.

N-linked glycans almost always have an *N*-acetylglucosamine residue at the reducing terminus and are beta linked to the amide nitrogen of

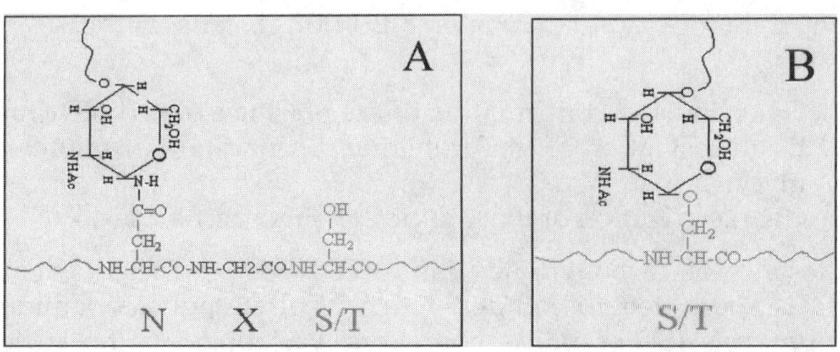

Figure 22 Scheme corresponding to (A) N-linked oligosaccharides and (B) O-linked oligosaccharides.

an asparagine residue in the consensus sequence Asn-Xaa-Ser/Thr/Cys where Xaa is any amino acid except proline.

There are three main classes of N-linked oligosaccharides: those containing only mannose residues (high-mannose type), those in which the 1-6 arm contains only mannose residues while the 1-3 arm is further processed (hybrid type), and those in which both arms are processed (complex type).

Contrasting with N-glycosylation, there is no single motif for prediction of O-glycosylation most probably because there is a bigger number of monosaccharides and amino acids involved in *O*-glycan linkages. O-linked oligosaccharides are characterized by the presence of fucose, galactose, *N*-acetylglucosamine, and *N*-acetylgalactosamine and may have acidic sugar residues such as *N*-acetylneuraminic acid (sialic acid), *N*-glycolylneuraminic acid, 2-keto-3-deoxynonulosonic acid, and glucuronic acid (Table 5). They can also contain sulfate and phosphate residues.

Glycans from glycoproteins are often complex mixtures of neutral and charged (sialylated) oligosaccharides. Most glycoproteins possess several glycosylation sites, with each site having various different glycan structures within the protein population. Hence, a single glycoprotein can exist in many different glycoforms. Separation of glycans released from a glycoprotein yields a fingerprint, also referred to as oligosaccharide profile or map.

In the pharmaceutical and biotechnology industries, mapping techniques for glycoprotein carbohydrate structures are used:

- during initial characterization for comparison of the native and recombinant protein glycosylation,
- to separate and identify the oligosaccharide structures present,
- to monitor consistency of glycosylation and identify changes that may have resulted from alterations in cell culture conditions or during the manufacturing process, and
- to monitor changes in glycosylation that occur as a result of expression in different cell lines.

The general strategy for characterization of glycans comprises three main steps:

1. Detection and confirmation of the presence of carbohydrates.
2. Deglycosylation of the protein and separation of the released glycans.
3. Characterization of the individual oligosaccharides.

Presence of carbohydrates can already be suspected when the protein is submitted to sodium dodecyl sulfate-polyacrylamide gel electrophoresis (SDS-PAGE) separation. Variations in the structure and degree of glycosylation contribute to overall mass heterogeneity. Moreover, it is common that these glycans carry *N*-acetylneuraminic

Monosaccharide	Abbreviation	Composition	Monoisotopic Mass	Average Mass
Pentose[a]	Pen	$C_5O_5H_{10}$	132.12	132.04
Deoxypentose	DeoxyPen	$C_5O_4H_{10}$	116.12	116.05
Hexose[b]	Hex	$C_6O_6H_{12}$	162.14	162.05
Deoxyhexose	DeoxyHex	$C_6O_5H_{12}$	146.14	146.06
Hexosamine	HexN	$C_6O_5NH_{13}$	161.16	161.07
N-Acetylhexosamine[c]	HexNAc	$C_8O_6NH_{15}$	203.19	203.08
Hexuronic acid	HexA	$C_6O_7H_{10}$	176.13	176.03
N-Acetylneuraminic acid	NeuAc	$C_{11}O_9NH_{19}$	291.26	291.09
N-Glycolylneuraminic acid	NeuGc	$C_{11}O_{10}NH_{19}$	307.26	307.09

[a]For example: rhamnose (Rha), fucose (Fuc).
[b]For example: glucose (Glc), mannose (Man), galactose (Gal).
[c]For example: glucosamine (GlcNAc), galactosamine (GalNAc).

Table 5 Monosaccharide Residues Commonly Found in Glycoconjugates.

acid (sialic acid), sulfate, or phosphate groups. Changes in the degree of these modifications affect both the mass and charge of a glycoprotein. All this heterogeneity results in diffuse bands on the gels.

Glycoprotein bands on PAGE gels and/or Western blots can be visualized by colorimetric reactions such as the periodic acid-Schiff method or fluorescent labeling with different reagents (Steinberg et al., 2001).

Purification processes are generally required before the produced glycoprotein can be further characterized. A very common and selective strategy for the isolation/purification of a glycoprotein is to use their glycan component as a target for affinity chromatography. The affinity matrices of choice are lectins. Lectins comprise a big group of proteins of nonimmune origin that bind in a highly specific manner to sugars, being able to discriminate subtle differences in glycan structures. They can be found in cells and biological fluids. A large number of lectins has been identified from different sources such as plant seeds, bulbs, as well as animal cells. Lectins are involved in many biological processes such as defense against invading organisms, recognition of host cells by viruses, cellular adhesion and recognition, and cell growth and differentiation.

There are many available lectins. Some of the most commonly used are

- Concanavalin A (ConA): binds specifically to glucosyl and mannosyl residues of N–linked oligosaccharides.
- Wheat germ lectin (WGL): specific for the chitobiose core (di-*N*-acetylglucosamine) and *N*-acetylneuraminic acid residues.
- Peanut agglutinine (PNA): shows specificity for the T antigen (Galβ1-3GalNAc) found in *O*-glycans of mucin-type proteins.
- Aleuria aurantia (AAL): specific for L-fucose–containing oligosaccharides.
- Galectines: specific for *N*-acetyllactoseamine (LacNAc)-containing glycans found in both *N*- and *O*-glycans.

The list can easily be enlarged. For further information about lectins, there are reviews available (Helmholz et al., 2003).

Lectin affinity chromatography is rather similar to other types of affinity chromatography. The stationary phase consists of a specific lectin immobilized (covalently attached) to the solid matrix (usually agarose beads or sepharose). When a mixture of heterogeneous proteins is loaded, the glycoprotein with a specific sugar sequence gets adsorbed onto the matrix due to its interaction with the immobilized lectin, whereas the other proteins and glycoproteins in the mixture just pass through the column without retention. After loading, a washing step is applied using the loading buffer to wash off any remaining nonspecifically adsorbed material. Finally, the adsorbed glycoprotein is eluted from the column using a sugar with a complementary structure.

Affinity of sugars for lectins is much lower than for the corresponding carbohydrate specific antibodies (10^2–10^6 M^{-1} *vs.* 10^4–10^8 M^{-1}). Because of this, lectins are preferred as affinity ligands when the biological activity of the glycoprotein has to be maintained after purification. Carbohydrate-specific antibody columns require harsh and often non-physiological conditions for elution.

Analysis of the oligosaccharide structure of glycoproteins usually requires enzymatic or chemical methods of deglycosylation. The method of choice depends on the nature of the glycan to be released as well as on the information to be extracted from the analysis. Chemical deglycosylation methods are simple, very effective, and require no expensive reagents, but they employ harsh conditions resulting sometimes in destruction of the protein or of the glycans. The most often used chemical methods are described below.

Hydrazinolysis results in the complete release of unreduced O- and N-linked oligosaccharides. Selective and sequential release of oligosaccharides can be accomplished by initial mild hydrazinolysis of the O-linked oligosaccharides at 60 °C followed by release of N-linked oligosaccharides at 95 °C (Patel et al., 1993). Hydrazine hydrolysis destroys the protein component and sometimes also degrades and modifies the released sugars (Egge et al., 1983).

Alkaline-β-elimination uses alkali plus a reducing agent (sodium borohydride) to release O-linked oligosaccharides. (Piller and Piller, 1993). The method is very efficient, but it yields the glycans in the form of reduced sugar alcohols (alditols) therefore without reactivity for further derivatizations (e.g., introduction of fluorescent labels). On the other hand, after β-elimination, the formerly O-glycosylated Ser and Thr residues are transformed into unique amino acid residues that can be labeled for further studies (Wells et al., 2002). This allows detection of the original glycosylation site.

Acid hydrolysis or methanolysis releases the monosaccharides as reducing sugars or methyl glycosides, respectively, which can be further separated by GC or LC. This is valuable to determine the overall carbohydrate composition of the glycoprotein. Compositional analysis provides information on the number and type of monosaccharide residues present in a glycoconjugate. This is analogous to amino acid composition analysis in proteins.

Trifluoromethanesulfonic acid (TFMS) hydrolysis leads to total deglycosylation (Burgess and Norman, 1988) by solvolytic cleavage leaving the protein core intact. The reaction must be carried out without a solvent and in the presence of a scavenger to protect the amino acids. This procedure has the advantage that all glycans can be removed regardless of composition and structure in one step, however, they are destroyed (Edge, 2003).

Enzymatic deglycosylation methods are generally preferred due to the use of mild conditions allowing further structural analysis of both the glycan and the protein part. Several endoglycosidases have been described for this purpose. Peptide-*N*-glycosidase F (PNGase F, *N*-glycanase, N-Gly F) is the most effective enzyme to remove almost all N-linked oligosaccharides from glycoproteins. N-Gly F is an amidase that releases the N-linked oligosaccharide from the peptide chain as an N-glycoside, which spontaneously hydrolyzes to yield the free, intact oligosaccharide in a suitable way for further analysis. The asparagine residue in the protein from which the glycan was removed is deamidated to aspartic acid.

Not all carbohydrate chains on a glycoprotein are equally sensitive to N-Gly F. In some cases, denaturation of the protein using detergents and/or reduction of the disulfide bridges prior to the enzymatic treatment is necessary. Another limitation for this enzyme is the presence of α(1–3) bound core fucose encountered in some plant and invertebrate glycoproteins. In that case, PNGase A must be used. However, this enzyme is ineffective when sialic acid is present on the N-linked oligosaccharide.

Other useful endoglycosidases are the endo-beta-*N*-acetylglucosidases. These enzymes cleave between the two GlcNAc residues in the N-linked oligosaccharide core. The released oligosaccharide is thus shortened by one GlcNAc residue, whereas the other GlcNAc remains linked to the Asn residue in the glycosylation site. Endoglycosidase H and F1 (acts only on high mannose and hybrid type structures) and endoglycosidase F2 (prefers biantennary complex type chains) belong to this type. They are less extensively used for general deglycosylation due to their limited specificity and because they leave one *N*-acetylglucosamine residue attached to the asparagine residue.

For the removal of intact O-linked oligosaccharides, there is no enzyme comparable to N-Gly F. The strategy consists of the sequential hydrolysis of monosaccharides using a series of exoglycosidases until only the Gal-β(1-3)-GalNAc core is left. After that, *O*-glycosidase is able to remove the core structure intact, leaving the serine or threonine residue unchanged. Modifications of the core structure (such as different degrees of sialylation that are commonly found) can hinder the action of this enzyme. In that case, α2-(3,6,8.9)-neuraminidase can be used to cleave the NeuAc-α(2-8)-NeuNAc bond.

It is also common to find O-linked oligosaccharides containing structures composed of β(1-4)-linked galactose, β(1-6)-linked *N*-acetylglucosamine, and sialic acid. In that case, apart from neuraminidase, also a β(1-4)-specific galactosidase and an *N*-acetylglucosaminidase are required. There are more exoglycosidases available for other types of modifications, and they can be used in a sequential fashion depending on the desired removal step. Only when fucose and mannose residues are directly O-linked to proteins they cannot be enzymatically removed.

One of the drawbacks to using enzymes to release oligosaccharides is that peptide structure can often influence the effectiveness of the enzymatic release of the glycan. It is often required to optimize the protocol for each glycopeptide. Moreover, some peptides are refractory to one or more of these enzymes. For example, N-Gly F cannot cleave oligosaccharides from amino- or carboxy-terminal Asn residues. In the case of O-linked oligosaccharides, there is no good "O-glycanase" that can be universally applied to O-deglycosylation.

After release from the glycoprotein, the oligosaccharides can be separated by chromatographic or electrophoretic methods producing a profile that is called a glycan map. Capillary electrophoresis (CE) and HPLC, both in normal and reversed-phase, are broadly applied to the separation of oligosaccharides. Capillary electrophoresis offers the potential of highly efficient separations, high mass sensitivity, and rapid analysis times, whereas HPLC allows much larger injection volumes. Due to the lack of a specific chromophore or fluorophore, most of these methods require the use of chemical derivatization (Lamari et al., 2003).

Many different derivatization reagents are available, and the selection is dictated by the desired modification in the properties of the glycan: for example, degree of hydrophobicity (reversed-phase HPLC), attachment of basic groups (more efficient MS), or fluorophore properties.

Reductive amination in which a saccharide is converted to glycamine in the presence of a reducing agent is one of the most used derivatizations for HPLC and CE. There are many different fluorogenic compounds that can be attached to oligosaccharides following this scheme: 2-aminobenzoic acid (Anumula and Du, 1999), 8-aminonaphtalene-1,3,6-trisulfonic acid (Klockow et al., 1995), 2-aminoacridone (Okafo et al., 1996), and 2-aminopyridine (2AP).

Derivatization with 2-aminopyridine was first reported by Hase in 1978 (Hase et al., 1978) and is one of the most popular methods. With this derivatization reagent, detection can be accomplished photometrically (λ = 240 nm) or fluorimetrically (λ_{em} = 375 nm). Interesting mapping methodologies have been developed for 2AP-labeled oligosaccharides: for example, the two-dimensional or three-dimensional mapping designed by Nakagawa et al. (1995) in which the structures of AP-labeled oligosaccharides from glycoproteins were easily confirmed by plotting the elution data collected on a reversed-phase octadecylsilyl (ODS) column, a normal-phase amide column, and an ion-exchange diethylaminoethyl (DEAE) column, respectively. Furthermore, 2AP oligosaccharides are good candidates for structural characterization by means such as mass spectrometry due to their easy protonation. However, the stability of some substituents (e.g., sialic acids) to the 2AP derivatization is still uncertain.

One of the most popular methodologies currently is the separation of the different glycans in their native state using high pH anion exchange chromatography (HPAEC). Detection is accomplished with pulsed amperometric detection (PAD), taking advantage of the presence of multiple hydroxyl groups in the molecules. HPAEC is selective for not only composition, size, and anomerity, but also ring substitutions and branching, making this ideal for the isomeric nature of oligosaccharides (Townsend and Hardy, 1991).

Separation is achieved using a sodium acetate gradient developed in the presence of sodium hydroxide, using a CarboPac PA-1 or PA-100 column (DionexCorporation, Sunnyvale, CA, USA) that is calibrated with pure standards. Typically, 5% of the oligosaccharides released from 100 µg of the glycoprotein fetuin are sufficient for one profile. The technique is fairly simple and sensitive (in the pmol range) and is applied for routine fingerprint assays of protein oligosaccharide content in many laboratories. The molar sensitivity is enhanced by increasing the molecular weight because the efficiency of oxidation depends on the number of hydroxyl groups per oligosaccharide. However, the electrochemical response factor differs for the different sugars, therefore quantification and identification are done based on comparison with known standard oligosaccharides.

The use of an alkaline mobile phase may cause some extent of base-catalyzed modifications such as "peeling" or epimerization of the oligosaccharides. Nevertheless, it is possible to isolate preparatively isomerically pure oligosaccharides from HPAEC/PAD for thorough later structural characterization. HPAEC/PAD is also one of the best approaches to separate anionic oligosaccharides, such as those containing sialic acid, sulfate, or phosphate. In Figure 23, the HPAEC-PAD profile corresponding to the oligosaccharides released from fetuin is presented.

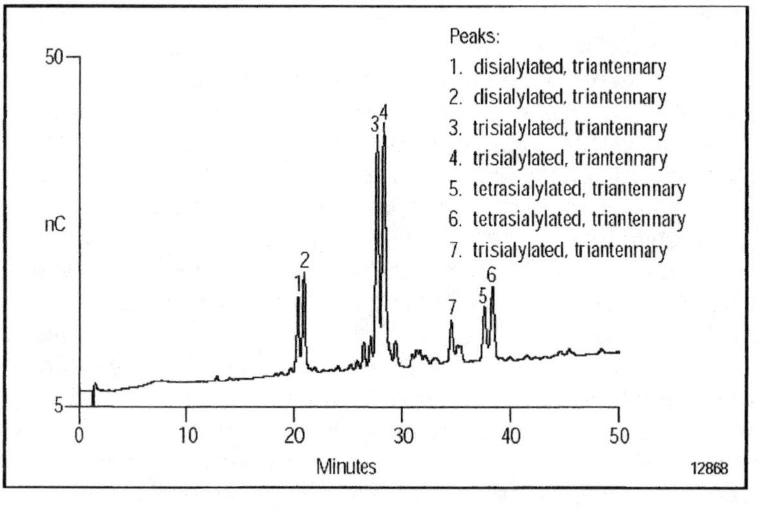

Figure 23 HPAEC-PAD of fetuin N-linked oligosaccharide alditols. Adapted from "Glycoprotein Oligosaccharide Analysis Using High-Performance Anion-Exchange Chromatography," Dionex, application number TN42.

Characterization of individual oligosaccharides involves not only the determination of the identity of the monosaccharide components, but also the linkage type and anomeric configuration for each intermonomer glycosidic bond, as well as the sequential arrangement of sugar residues. Taking this into account, it is obvious that no technique can provide all this information. The strategy consists of combining different methods to extract as much information as possible.

NMR spectroscopy has been applied to the structural and conformational analysis of oligosaccharides (Lommerse et al., 2002). Chemical shifts and spin-spin coupling constants are helpful in determining the composition of sugar residues and the positions and configurations of the glycosidic linkage. Recent developments in stable isotope labeling (e.g., ^{13}C-labeling) techniques have shown to be useful in conformational analysis. However, the limitation in sensitivity and laborious sample preparation have directed most researchers toward mass spectrometry.

Various mass spectrometric methods are applicable to glycoconjugate analysis. To date, MALDI-TOF has been the most popular combination to analyze protein oligosaccharides (Zeng and Biemann, 1999; Harvey et al. 2000, 2003), although ESI coupled to different types of detectors such as qTOF is becoming increasingly used (Charlwood et al., 1999). This is mainly due to the compatibility of ESI with on-line LC-MS or CE-MS providing extra information from a chromatographic or electrophoretic "fingerprint" of a tryptic digest or a pool of released oligosaccharides. The possibility of performing tandem MS adds structural and compositional information to the molecular weight through the fragmentation of the oligosaccharide.

Interpretation of mass spectrometric data can be laborious but with the appearance of new computational tools is becoming easier. One example is GlycoMod, which is included in the proteomics suite of tools available on Expasy (http://www.expasy.org/tools/glycomod). Glycomod is linked to the information available in the SWISS-PROT database and enables the prediction of N- and O-linked oligosaccharides on glycopeptides using experimentally obtained mass data. It also allows a range of options of oligosaccharide release and derivatization strategies to be included in the calculation (Cooper et al., 2001).

Although mass spectrometric methods can provide accurate masses of molecular or fragment ions, they are unable to distinguish between isomeric monosaccharide residues (e.g., all neutral hexoses have the same mass of 162 Da). In that case NMR, chemical, or exoglycosidase sequential degradation is of great utility. Exoglycosidases have a high specificity. The beta-galactosidase from jack bean is able to cleave a nonreducing terminal galactose (Gal) residue when it is beta-linked to the 6 or 4 position of the penultimate sugar residue but is almost inactive toward a Gal (beta,1-3) linkage. Sequential exoglycosidase digestion, by means of an array of exoglycosidases of known specificity, can be combined with the chromatographic (or electrophoretic) mapping techniques and mass spectrometry to sequence an oligosaccharide.

The strategy for characterization of the attachment site described till now (glycoprotein deglycosylation + glycan map) presents an important drawback: destruction of the information about the sites of attachment. There are different ways to get this important information. One of them is to perform a digestion of the whole glycoprotein followed by a peptide map. Location of the glycopeptides in the whole mixture can be done after treatment of the digest with an appropriate deglycosylating enzyme. After deglycosylation, a shift of the retention times of the peaks corresponding to glycopeptides will be observed. In case of reversed-phase separations, slightly later elution will take place, as the presence of many OH groups in the sugar part gives glycopeptides a more hydrophilic character.

Another alternative is the use of lectins. Glycoproteins can be trapped by an affinity adsorbent bearing a certain type of lectin. Adsorbed glycoproteins are eluted and subjected to extensive proteolysis. The resultant peptides are applied to the same lectin column to select target glycopeptides that can be further separated and sequenced (Hirabayashi and Kasai, 2003).

An easier alternative, which does not require deglycosylation steps, is the use of mass spectrometric methods (Huddleston et al., 1993). Collision-induced fragment ions of glycopeptides can be exploited for their selective detection in complex digests. Neutral loss scanning and parent or precursor ion scanning are useful approaches. Monitoring of oxonium ions Hex+ (m/z 163), HexNAc+ (m/z 204), NeuAc+ (m/z 292), and Hex-HexNAc+ (m/z 366) in an LC-MS chromatogram gives a clear indication of the position of the glycopeptide. In Table 5, the monosaccharide residues commonly found in glycoconjugates together with their monoisotopic mass are presented.

Recombinant Human C1-Inhibitor Produced in Milk of Transgenic Rabbits

C1-Inhibitor (C1 INH) is a plasma protease inhibitor essential for regulation of activation of several different proteolytic systems in plasma: the coagulation, the complement, and the kinin generating system. This protein is a member of the serpin family of serine protease inhibitors. Other members of this family are α_1-antitrypsine, antithrombin III, and α_2-antiplasmin. (Davis, 1989).

A deficiency (hereditary or acquired) of C1 INH causes angioedema, a disease characterized by attacks of swelling of soft tissue (lips, other parts of the mouth, throat, extremities, abdomen, etc.). Swelling of the airway may produce difficulties in breathing and can be life-threatening. Current treatment of C1 INH deficiency consists of prophylactic/preventive therapy (androgenic steroids) and intravenous administration of concentrated C1 INH (purified from human blood) in case of acute attacks. However, the use of plasma-derived products presents some drawbacks in terms of safety, acceptance, and product supply. An alternative is the use of recombinant human C1 INH (rh-C1 INH) produced in the milk of transgenic animals.

C1 INH is heavily glycosylated. Approximately 50% of its mass is carbohydrate. Characterization of the protein after production requires not only the development of a peptide map to check the amino acid sequence but also a glycan map. For this last task, first the oligosaccharides are released from the core protein: N-linked glycans enzymatically with *N*-glycosidase F and O-linked glycans chemically with β-elimination (Barroso et al., 2002).

Release of the N-linked oligosaccharides can be performed by boiling the glycoprotein in *N*-octylglucoside at 100 °C for 2 minutes and then adding N-Gly F in 20 mM sodium phosphate and 50 mM EDTA. After incubation at 37 °C for 48 hours, the digest can be directly injected in the HPAEC-PAD system. The differences in acidity of the oligosaccharides results in their differential retention on pellicular quaternary amine-bonded resins. Strongly retained glycans are eluted with a concentration gradient of sodium acetate in the eluent. In Figure 24, one of these profiles is presented. The retention time in the HPAEC-PAD chromatogram gives already an indication of the type of glycan. However, for better structural characterization, the fractions can be collected and submitted to MS analysis.

A limitation when dealing with these anion exchange fractions is the high salt concentration present in the samples, which makes MS analysis impossible. A good way of desalting the fractions is the use of HPLC employing a porous graphitized carbon (PGC) column (HypercarbThermo Electron Corporation, Cheshire, UK) coupled

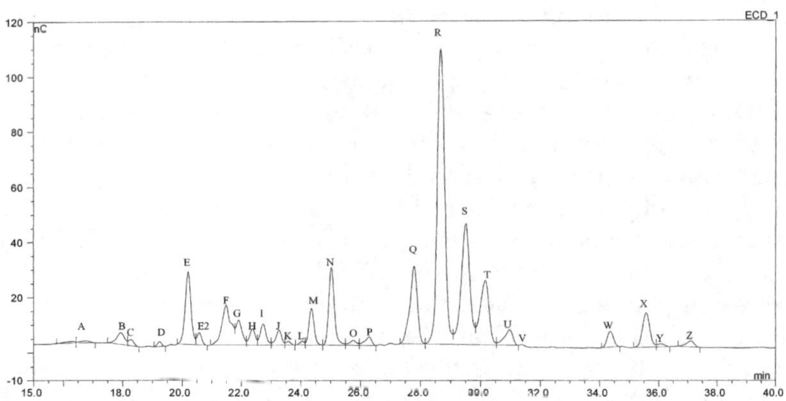

Figure 24 HPAEC-PAD chromatogram of N-linked oligosaccharides released from recombinant human C1-inhibitor. Each of the labeled fractions was isolated and submitted to HPLC-MS.

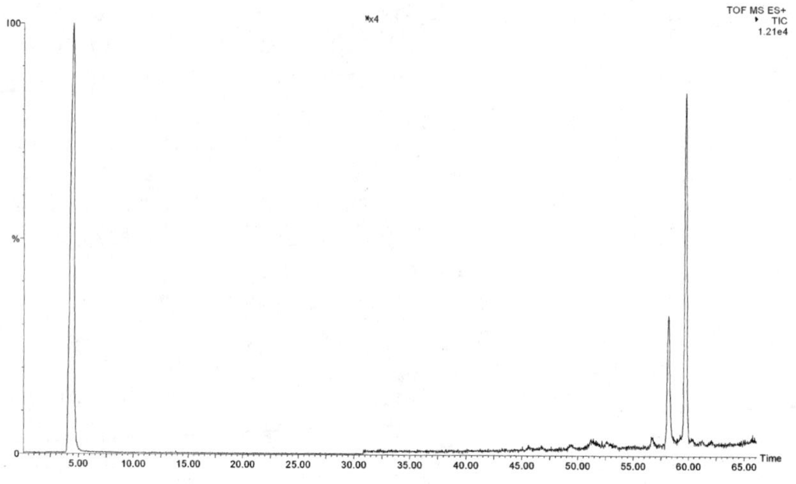

Figure 25 HPLC-MS chromatogram of one of the HPAEC-PAD fraction (R) containing two N-linked oligosaccharides.

on-line to the mass spectrometer. The use of an acetonitrile gradient gives a sort of reversed-phase separation.

In Figure 25, the total ion current (TIC) chromatogram obtained for one of the fractions containing two oligosaccharides can be observed, and in Figure 26 the corresponding mass spectra are shown. The fragment ions are represented according to the nomenclature proposed by Domon and Costello (1988). The major peak in the chromatogram is a monosialylated, galactosylated biantennary complex N-linked oligosaccharide with a molecular weight of 1932.15 Da (Figure 26A). The doubly charged ion (m/z 967.13) is also present. Fragmentation can be directly related to the glycan structure, as mainly cleavage of the glycosidic bonds occur. The dominant ion at m/z 657.4 is typical of the sialylated antennae (protonated NeuAc-Hex-HexNAc). Another characteristic ion is m/z 366.28 corresponding to protonated Hex-HexNAc.

The minor chromatographic peak is a sialylated monoantennary oligosaccharide (Figure 26B) with a molecular weight of 1729.06 Da. Fragmentation is quite similar to the previous case.

O-linked oligosaccharides can be released from C1 INH by reductive alkaline β-elimination suspending the protein in 50 mM NaOH/1 M $NaBH_4$ and incubating overnight at 37 °C. After stopping the reaction and lowering the pH with acetic acid, the mixture is dried and washed with MeOH/1% acetic acid and submitted to cation exchange (using AG50WX12). After elution of the alditols with water, they can be directly loaded in the HPAEC-PAD system and collected for MS analysis after chromatographic separation with the same PGC column.

In Figure 27, the TIC chromatogram of one of these alditols is presented. The spectrum (Figure 28) is less complicated than in the

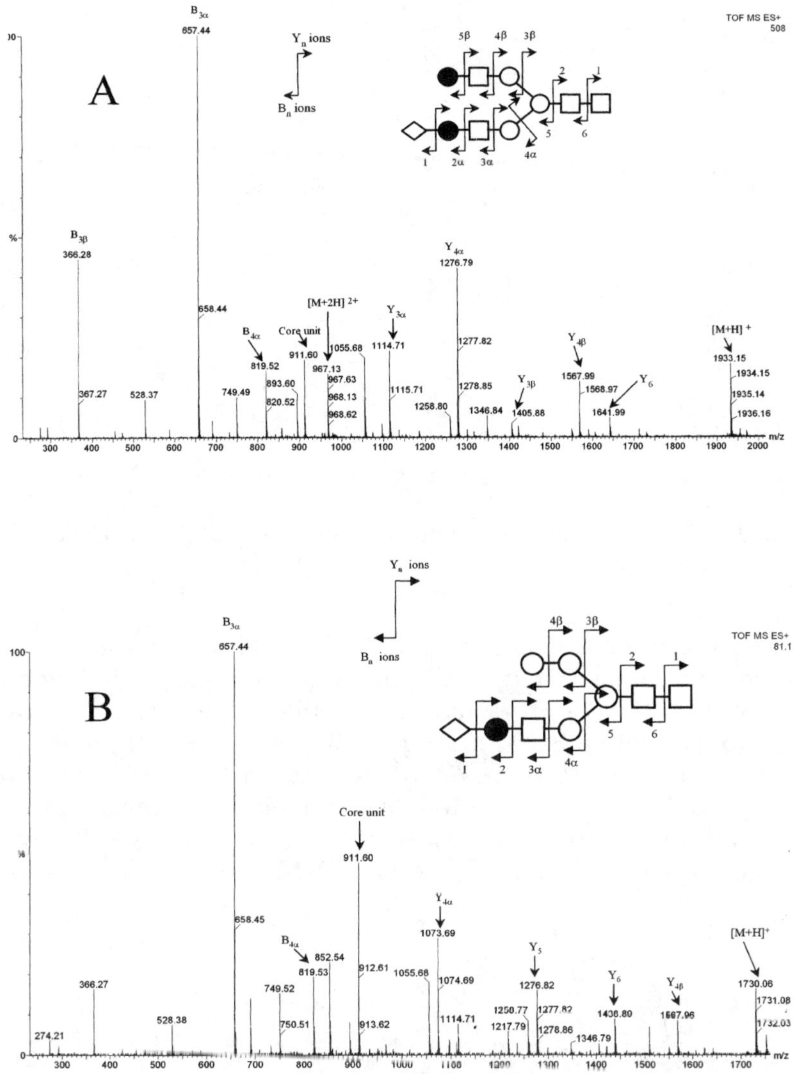

Figure 26 Spectrum corresponding to the (A) major and (B) minor peaks observed in the chromatogram from Figure 2. Symbols: □, *N*-acetylhexosamine; ○, hexose (mannose); ●, hexose (galactose); ◇, *N*-acetylneuraminic acid; ▽, fucose.

previous case, due to the smaller size of the O-linked oligosaccharides. A molecular mass of 676.47 Da can easily be deduced. Other important diagnostic ions detected are 386.3 (loss of *N*-acetylneuraminic acid from the molecular ion) and its counterpart 292.21 (protonated NeuAc).

Another alternative to the use of HPAEC-PAD is the direct analysis of the released pool of oligosaccharides by LC-MS with a PGC

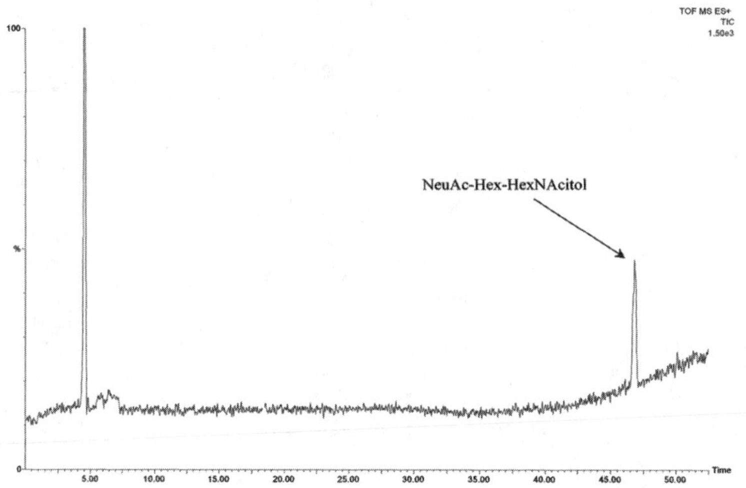

Figure 27 TIC chromatogram obtained after injection of an HPAEC-PAD fraction containing one O-linked oligosaccharide alditol.

column. In this case, it is convenient to run N-linked and O-linked oligosaccharides separately. In Figure 29, a complete *N*-glycosidase F digest of rhC1 INH is presented. The mobile phase was a mixture of ammonium acetate 100 mM pH 8.0 (A) and acetonitrile (B) with a gradient of 1% increase of B/min. However, it must be pointed out that this methodology is less sensitive than the electrochemical detection and higher amounts of sample are necessary. For higher sensitivity, the system can be downscaled to nano-HPLC/MS.

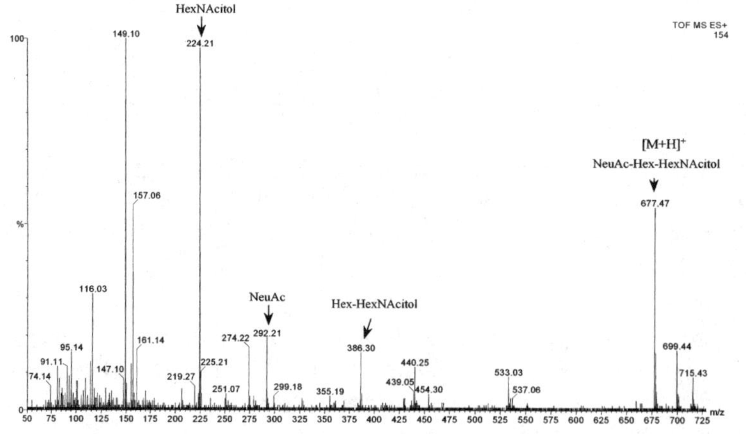

Figure 28 Spectrum corresponding to the O-linked oligosaccharide alditol NeuAc-Hex-HexNAcitol in Figure 28.

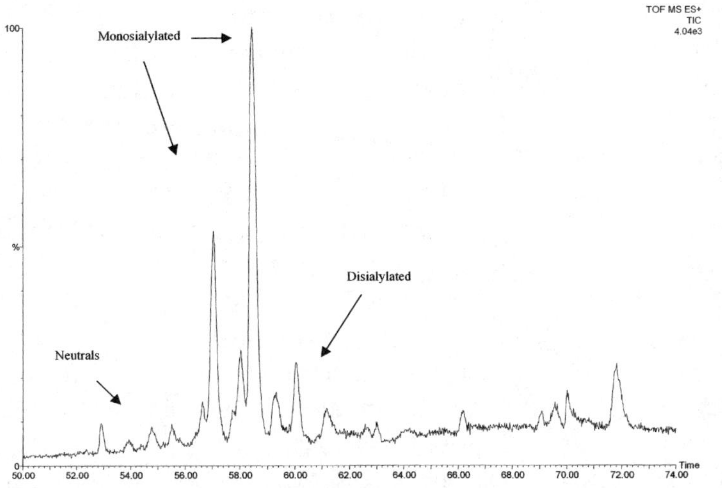

Figure 29 HPLC/MS chromatogram obtained after injection of a whole *N*-glycosidase F digest of rh-C1-inhibitor.

Conclusions

In this chapter, we have outlined the different modes of chromatography with special emphasis on protein separations. No doubt there is much more to say about the widespread use of chromatography in biotechnology and in particular the production and quality control of biopharmaceuticals, but we hope to have given the reader a starting point for further, more detailed studies. The applications were chosen according to increasing complexity of the production organism, ending with a more thorough description of the analysis of glycoproteins. Glycoproteins play a particularly important role as biopharmaceuticals, as most of the currently marketed drugs belong to this category. Analysis of biopharmaceuticals is based on the combination of many techniques, as nicely laid out in this book. Our chapter relates in particular to the chapter on mass spectrometry, which we invite the reader to consult for further details on this technique.

Abbreviations

2AP	2-Aminopyridine
AAT	α1-Antitrypsin
AC	Affinity chromatography
CE	Capillary electrophoresis
ConA	Concanavalin A
DEAE	Diethylaminoethyl

ESI	Electrospray ionization
GST	Glutathione-*S*-transferase
HIC	Hydrophobic interaction chromatography
HPAEC-PAD	High-performance anion-exchange chromatography–pulsed amperometric detection
HPLC	High-performance liquid chromatography
IEF	Isoelectric focusing
IL-2	Interleukin-2
IMAC	Immobilized metal affinity chromatography
LC-MS	Liquid chromatography-mass spectrometry
LPS	Lipopolysaccharide
MALDI-TOF	Matrix assisted laser desorption ionization-time of flight
MMP	Matrix metalloprotease
N-Gly F	*N*-glycosidase F
NMR	Nuclear magnetic resonance
ODS	Octadecylsilyl
PAGE	Polyacrylamide gel electrophoresis
pI	Isoelectric point
qTOF	Quadrupole time of flight
R_G	Radius of gyration
RAM	Restricted access matrix
rHV2-Lys^{47}	Recombinant hirudin variant with lysine in position 47
RP	Reversed-phase
SDS	Sodium dodecyl sulfate
SEC	Size exclusion chromatography
TFA	Trifluoroacetic acid
TFMS	Trifluoromethylsulfonic acid
TIC	Total ion current

Acknowledgments

The authors thank the many collaborators that we worked with throughout our industrial and academic careers. At the same time our apologies go to the many excellent pieces of work that had to be left out. The world of biopharmaceuticals is certainly alive and commercially successful, which is due to the many pioneers that continue to overcome obstacles be they scientific or other.

A special thank you to our secretary, Mrs. Jolanda Meindertsma, for helping us with the formatting of this chapter and with many other details.

References

1. Alpert AJ, and Andrews PC. Cation-Exchange Chromatography of Peptides on Poly(2-Sulfoethyl Aspartamide)-Silica. *J Chromatogr* 1988; 443:85–96
2. Anumula KR, and Du P. Characterization of Carbohydrates Using Highly Fluorescent 2-Aminobenzoic Acid Tag Following Gel Electrophoresis of Glycoproteins. *Anal Biochem* 1999; 275:236–242
3. Barroso B, Dijkstra R, Geerts M, Lagerwerf F, van Veelen P, and de Ru A. On-line High-Performance Liquid Chromatography/Mass Spectrometric Characterization of Native Oligosaccharides from Glycoproteins. *Rapid Commun Mass Spectrom* 2002; 16:1320–1329
4. Barroso B, Lubda D, Bischoff R. Applications of Monolithic Silica Capillary Columns in Proteomics. *J Proteome Res* 2003; 2:633–642
5. Bischoff R, and Kolbe HV. Deamidation of Asparagine and Glutamine Residues in Proteins and Peptides: Structural Determinants and Analytical Methodology. *J Chromatogr B Biomed Appl* 1994; 662:261–278
6. Bischoff R, and Luider TM. Methodological Advances in the Discovery of Protein and Peptide Disease Markers. *J Chromatogr B* 2004; 803:27–40
7. Bischoff R, and McLaughlin LW. Isolation of Specific tRNAs Using an Ionic-Hydrophobic Mixed-Mode Chromatographic Matrix. *Anal Biochem* 1985; 151:526–533
8. Bischoff R, Clesse D, Whitechurch O, Lepage P, and Roitsch C. Isolation of Recombinant Hirudin by Preparative High-Performance Liquid Chromatography. *J Chromatogr* 1989; 476:245–255
9. Bischoff R, Speck D, Lepage P, Delatre L, Ledoux C, Brown SW, and Roitsch C. Purification and Biochemical Characterization of Recombinant Alpha 1-Antitrypsin Variants Expressed in Escherichia Coli. *Biochemistry* 1991; 30:3464–3472
10. Bischoff R, Degryse E, Perraud F, Dalemans W, Ali-Hadji D, Thepot D, Devinoy E, Houdebine LM, and Pavirani A. A 17.6 kbp Region Located Upstream of the Rabbit WAP Gene Directs High Level Expression of a Functional Human Protein Variant in Transgenic Mouse Milk. *FEBS Lett* 1992; 305:265–268
11. Bischoff R, Lepage P, Jaquinod M, Cauet G, Acker-Klein M, Clesse D, Laporte M, Bayol A, Van Dorsselaer A, and Roitsch C. Sequence-Specific Deamidation: Isolation and Biochemical Characterization of

Succinimide Intermediates of Recombinant Hirudin. *Biochemistry* 1993; 32:725–734

12. Branovic K, Buchacher A, Barut M, Strancar A, and Josic D. Application of Monoliths for Downstream Processing of Clotting Factor IX. *J Chromatogr A* 2000; 903:21–32
13. Burgess AJ, and Norman RI. The Large Glycoprotein Subunit of the Skeletal Muscle Voltage-Sensitive Calcium Channel. Deglycosylation and Development. *Eur J Biochem* 1988; 178:527–533
14. Chang GD, Chen CJ, Lin CY, Chen HC, and Chen H. Improvement of Glycosylation in Insect Cells with Mammalian Glycosyltransferases. *J Biotechnol* 2003; 102:61–71
15. Charlwood J, Langridge J, and Camilleri P. Structural Characterisation of N-Linked Glycan Mixtures by Precursor Ion Scanning and Tandem Mass Spectrometric Analysis. *Rapid Commun Mass Spectrom* 1999; 13:1522–1530
16. Coakley RJ, Taggart C, O'Neill S, and McElvaney NG. Alpha1-Antitrypsin Deficiency: Biological Answers to Clinical Questions. *Am J Med Sci* 2001; 321:33–41
17. Cooper CA, Gasteiger E, and Packer NH. GlycoMod—A software Tool for Determining Glycosylation Compositions from Mass Spectrometric Data. *Proteomics* 2001; 1:340–349
18. Davis III AE. C1-Inhibitor (C1-INH): Genes, Biosynthesis and Biology. *Behring Inst Mitt* 1989; 84:142–150
19. Davis JM, and Giddings JC. Statistical Theory of Component Overlap in Multicomponent Chromatograms. *Anal Chem* 1983; 55:418–424
20. de Jong LAA, Grünewald S, Franke JP, Uges DRA, and Bischoff R. Purification and Characterization of the Recombinant Dopamine D2S Receptor from Pichia pastoris. *Protein Expr Purif* 2004; 33:176–184
21. Dembowsky K, and Stadler P. *Novel Therapeutic Proteins. Selected Case Studies*. Wiley VCH; 2001. 358p.
22. Domon B, and Costello CE. A Systematic Nomenclature for Carbohydrate Fragmentation in FAB-MS/MS Spectra of Glycoconjugates. *Glycoconjugate J* 1988; 5:394–409
23. Edge AS. Deglycosylation of Glycoproteins with Trifluoromethane-sulfonic Acid: Elucidation of Molecular Structure and Function. *Biochem J* 2003; 376:339–350
24. Egge H, Peter-Katalinic J, Paz-Parente J, Strecker G, Montreuil J, and Fournet B. Carbohydrate Structures of Hen Ovomucoid. A Mass Spectrometric Analysis. *FEBS Lett* 1983; 156:357–362
25. Gooding KM, and Regnier FE. *HPLC of Biological Macromolecules*, 2nd ed. New York: Marcel Dekker; 2002

26. Govorukhina NI, Keizer-Gunnink A, van der Zee AGJ, de Jong S, de Bruijn HWA, and Bischoff R. Sample Preparation of Human Serum for the Analysis of Tumor Markers: Comparison of Different Approaches for Albumin and [Gamma]-Globulin Depletion. *J Chromatogr A* 2003; 1009:171–178

27. Harris RJ, Kabakoff B, Macchi FD, Shen FJ, Kwong M, Andya JD, Shire SJ, Bjork N, Totpal K, and Chen AB. Identification of Multiple Sources of Charge Heterogeneity in a Recombinant Antibody. *J Chromatogr B Biomed Sci Appl* 2001; 752:233–245

28. Harvey DJ. Matrix-assisted Laser Desorption/Ionization Mass Spectrometry of Carbohydrates and Glycoconjugates. *Int J Mass Spectrom* 2003; 226:1–35

29. Harvey DJ, Wing DR, Kuster B, and Wilson IBH. Composition of N-linked Carbohydrates from Ovoalbumin and Co-Purified Glycoproteins. *J Am Soc Mass Spectrom* 2000; 11:564–571

30. Hase S, Ikenaka T, and Matsushima Y. Structure Analysis of Oligosaccharides by Tagging of the Reducing End Sugars with a Fluorescent Compounds. *Biochem Biophys Res Commun* 1978; 85:257–263

31. Helmholz H, Cartellieri S, He L, Thiesen P, and Niemeyer B. Process Development in Affinity Separation of Glycoconjugates with Lectins as Ligands. *J Chromatogr A* 2003; 1006:127–135

32. Hirabayashi J, and Kasai K. Separation Technologies for Glycomics. *J Chromatogr B* 2002; 771:67–87

33. Hoogenboom HR, and Chames P. Natural and Designer Binding Sites Made by Phage Display Technology. *Immunology Today* 2000; 21:371–378

34. Huddleston MJ, Bean MF, and Carr SA. Collisional Fragmentation of Glycopeptides by Electrospray Ionization LC/MS and LC/MS/MS: Methods for Selective Detection of Glycopeptides in Protein Digest. *Anal Chem* 1993; 65:877–884

35. Jenkis N, Parekh RB, and James DC. Getting the Glycosylation Right: Implications for the Biotechnology Industry. *Nat Biotechnol* 1996; 14:975–981

36. Klockow A, Amado R, Widmer H, and Paulus A. Separation of 8-Aminonaphtalene-1,3,6-Trisulfonic Acid Labelled Neutral And Sialylated N-Linked Complex Oligosaccharides by Capillary Electrophoresis. *J Chromatogr A* 1995; 716:241–247

37. Lamari FN, Kuhn R, and Karamanos NK. Derivatization of Carbohydrates for Chromatographic, Electrophoretic and Mass Spectrometric Analysis. *J Chromatogr B* 2003; 793:15–36

38. Leinweber FC, Lubda D, Cabrera K, and Tallarek U. Characterization of Silica-Based Monoliths with Bimodal Pore Size Distribution. *Anal Chem* 2002; 74:2470–2477

39. Lommerse JPM, van Rooijen JJM, Kroon-Batenburg LMJ, Kamerling JP, and Vliegenthart JFG. Conformational Analysis of Two Xylose-Containing N-Glycans in Aqueous Solution by Using 1H NMR ROESY and NOESY Spectroscopy in Combination with MD Simulations. *Carbohydr Res* 2002; 337:2279–2299

40. Machtejevas E, John H, Wagner K, Ständker L, Marko-Varga G, Forssmann WG, Bischoff R, and Unger KK. Automated Multidimensional HPLC: Sample Preparation and Identification of Peptides from Human Blood Filtrate. *J Chromatogr B* 2004; 803, 121–130

41. Middelberg APJ. Preparative Protein Refolding. *Trends Biotechnol* 2002; 20:437–443

42. Mountain A, Ney U, and Schomburg D. Biotechnology, Volume 5A, Recombinant Proteins, Monoclonal Antibodies, and Therapeutic Genes. Wiley VCH, 1998. 562p.

43. Nakagawa H, Kawamura Y, Kato K, Shimada I, Arata Y, and Takahashi N. Identification of Neutral and Sialyl N-Linked Oligosaccharide Structures from Human Serum Glycoproteins Using Three Kinds of High-Performance Liquid Chromatography. *Anal Biochem* 1995; 226:130–138

44. Nilsson J, Stahl S, Lundeberg J, Uhlen M, and Nygren PA. Affinity Fusion Strategies for Detection, Purification, and Immobilization of Recombinant Proteins. *Protein Expr Purif* 1997; 11:1–16

45. Okafo G, Burrow L, Carr SA, Roberts GD, Johnson W, and Camilleri PA Coordinated High Performance Liquid Chromatographic, Capillary Electrophoretic and Mass Spectrometric Approach for the Analysis of Oligosaccharide Mixtures Derivatized with 2-Aminoacridone. *Anal Chem* 1996; 68:4424–4430

46. Opiteck GJ, Lewis KC, Jorgenson JW, and Anderegg RJ. Comprehensive On-Line LC/LC/MS of Proteins. *Anal Chem* 1997; 69:1518–1524

47. Patel T, Bruce J, Merry A, Bigge C, Wormald M, Jaques A, and Parekh R. Use of Hydrazine to Release in Intact and Unreduced Form Both N- and O-Linked Oligosaccharides from Glycoproteins. *Biochemistry* 1993; 32:679–693

48. Patston PA, Roodi N, Schifferli JA, Bischoff R, Courtney M, and Schapira M. Reactivity of Alpha 1-Antitrypsin Mutants Against Proteolytic Enzymes of the Kallikrein-Kinin, Complement, and Fibrinolytic Systems. *J. Biol Chem* 1990; 265:10786–10791

49. Piller F, and Piller V. In: Fukuda, and Kobata A. *Glycobiology: A Practical Approach*. IRL/Oxford University Press; 1993:291–328

50. Schulze AJ, Degryse E, Speck D, Huber R, and Bischoff, R. Expression of Alpha 1-Proteinase Inhibitor in Escherichia coli: Effects of Single

Amino Acid Substitutions in the Active Site Loop on Aggregate Formation. *J Biotechnol* 1994; 32:231–238

51. Steinberg TH, On Top KP, Berggren KN, Kemper C, Jones L, Diwu Z, Haugland RP, and Patton WF. Rapid and Simple Single Nanogram Detection of Glycoproteins in Polyacrylamide Gels and on Electroblots. *Proteomics* 2001; 1:841–855
52. Tebbutt SJ. Technology Evaluation: Transgenic Alpha-1-Antitrypsin (AAT), PPL Therapeutics. *Curr Opin Mol Ther* 2000; 2:199–204
53. Townsend RR, and Hardy MR. Analysis of Glycoprotein Oligosaccharides Using High-pH Anion Exchange Chromatography. *Glycobiology* 1991; 1:139–147
54. Wagner K, Miliotis T, Marko-Varga G, Bischoff R, and Unger KK. An Automated On-Line Multidimensional HPLC System for Protein and Peptide Mapping with Integrated Sample Preparation. *Anal Chem* 2002; 74:809–820
55. Wells L, Vosseller K, Cole RN, Cronshaw JM, Matunis MJ, and Hart GW. Mapping sites of O-GlcNAc Modification Using Affinity Tags for Serine and Threonine Post-Translational Modifications. *Mol Cell Proteomics* 2002; 10:791–804
56. Wittrup KD. Protein Engineering by Cell-Surface Display. *Curr Opin Biotechnol* 2001; 12:395–399
57. Xie S, Allington RW, Svec F, and Frechet JMJ. Rapid Reversed-Phase Separation of Proteins and Peptides Using Optimized 'Moulded' Monolithic Poly(Styrene-Co-Divinylbenzene) Columns. *J Chromatogr A* 1999; 865:169–174
58. Zeng C, and Biemann K. Determination of N-linked Glycosilation of Yeast External Invertase by Matrix Assisted Laser Desorption/ Ionization Time of Flight Mass Spectrometry. *J Mass Spectrom* 1999; 34:311–329

9

Capillary Electrophoresis

JONATAN R. CATAI, GERHARDUS J. DE JONG,
and GOVERT W. SOMSEN

Department of Biomedical Analysis, Utrecht University, Utrecht,
The Netherlands

Introduction

Capillary electrophoresis (CE) is a modern analytical separation technique that can be used for a variety of compounds such as drugs (and their enantiomers), amino acids, peptides, DNA, and numerous other ionic species. In recent years, CE has played a major role in the Human Genome Project where it provided a huge and decisive boost to the project's completion. The vast majority of the sequence of the human genome was determined with CE-based high-throughput DNA sequencers. CE is also very well suited for the separation of proteins (Jenkins and Guerin, 1997; Hu and Dovichi, 2002). The CE separation mechanism is governed by electrophoretic mobility, which is a function of charge and size of the studied ionic species. Differences between a (therapeutic) protein and, for example, its degradation products are often small relative to the molecular size of the protein (10–200 kDa) but usually involve changes in net protein charge and/or shape. These modifications frequently are sufficient for separation by CE. Even glycoforms of one protein might be separable by CE.

During the past decade, CE has emerged as a powerful tool for the characterization of pharmaceutical and other biotechnologically derived proteins (Lagu, 1999; Chen and Canova-Davis, 2001; Ma and Nashabeh, 2001; Patrick and Lagu, 2001) and established a position next to more conventional protein separation techniques like polyacrylamide gel electrophoresis (PAGE) and high-performance liquid chromatography (HPLC). CE is a rapid analytical technique, requiring small sample volumes and limited quantities of reagents, and is applicable to a wide range of analytes. In addition, CE is a highly efficient separation technique capable of yielding much higher plate numbers than PAGE and HPLC. The high electrical resistance of a buffer-filled capillary and its favorable surface area-to-volume ratio (good heat dissipation) enables the application of very high electrical fields (100–500 V/cm) without causing detrimental heat generation. CE is carried out in "free" buffer solutions, so sources of band-broadening related to the stationary phase (as are common in HPLC) are absent. Furthermore, whereas in HPLC the pressure-driven flow profile is parabolic (laminar flow), in CE the electrically driven flow has basically a flat profile. Overall, this induces the analytes to migrate as very narrow zones and ideally undergo dispersion only from axial diffusion. The mild separation conditions that can be chosen in CE present the possibility to separate and study proteins in their natural (nondenatured) state, which can be important when characteristics of pharmaceutical proteins have to be assessed. Though it was originally thought that CE would compete with HPLC as a separation technique, the two methods are now increasingly deemed complementary. In this respect, CE and reversed-phase HPLC often are referred to as "orthogonal" separation techniques providing noncorrelated information on separated species.

In this chapter, we will highlight the (potential) role of CE in the analysis of biopharmaceuticals. CE offers various modes of separation, which can be useful in protein analysis. These include capillary zone electrophoresis (CZE; charge-size ratio–based separation), capillary gel electrophoresis (CGE; molecular-weight–based separation), and capillary isoelectric focusing (CIEF; isoelectric-point–based separation). After first describing the general basics and properties of CE, the principles and applications of each of these modes are outlined. Selected examples are given to illustrate the usefulness of the CE systems, and comprehensive tables provide an overview of applications of CE for the characterization of biotechnologically produced proteins. As mass spectrometric (MS) detection is of increasing importance in protein analysis, specific attention is devoted to the on-line coupling of CE and MS for the characterization of intact proteins. Also, affinity capillary electrophoresis (ACE), which permits direct determination of the binding stoichiometry and the thermodynamic equilibrium constants of protein-ligand complexes, is briefly treated.

Principles of Capillary Electrophoresis

Setup, Theory, and Practical Aspects

In CE, separation of analyte ions is carried out in a buffer or background electrolyte (BGE) present in a narrow fused-silica capillary (25–100 µm i.d.). The ends of the capillary are inserted into vials filled with BGE, which also contain electrodes connected to a high-voltage power supply to complete an electrical circuit (Figure 1). The sample solution is introduced as a small plug at the inlet side of the capillary. With the application of high voltage (5–30 kV) across the capillary, different zones of analyte ions are developed and start to migrate with

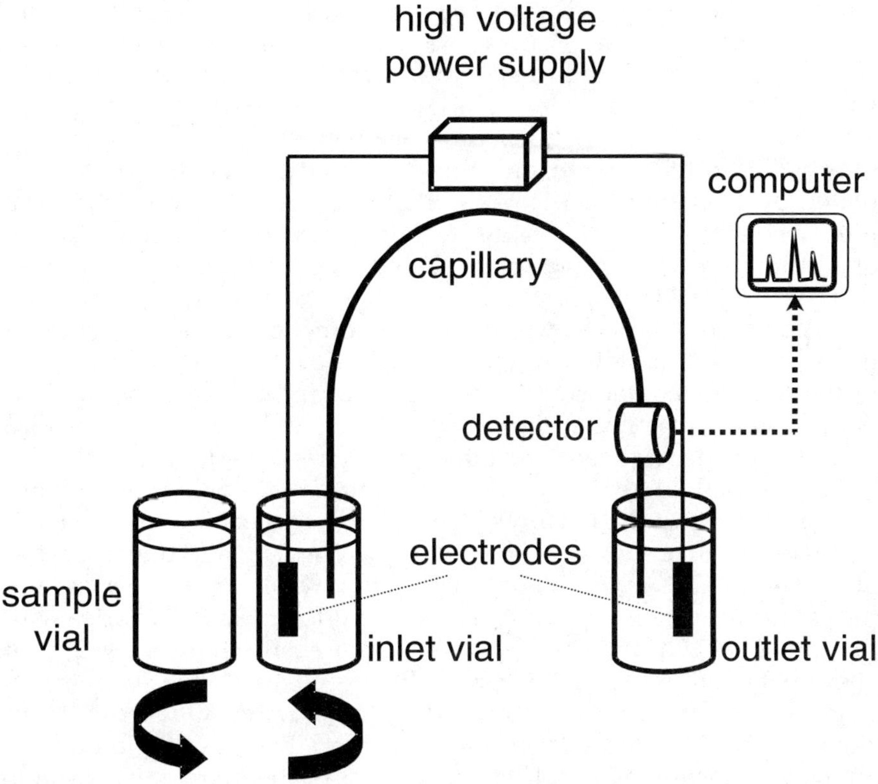

Figure 1 Basic configuration of a CE system. To start an analysis, the capillary is first filled with BGE. Subsequently, a small volume of sample is loaded onto the capillary by replacing the inlet vial filled with BGE for a vial filled with sample and applying either a voltage or an external pressure for typically 1–10 seconds. After replacing the inlet buffer vial, high voltage is applied and the separation performed.

different velocities toward the outlet side of the capillary. Before reaching the end of the capillary, most commonly, the separated analyte bands (peaks) are registered by absorbance detection directly through the capillary wall yielding an electropherogram.

Separation by electrophoresis is based on differences in electrophoretic mobility (μ) of ionic species. The mobility is the result of the electrostatic force exerted on an ion by the electric field and the opposing viscous force (frictional drag) that the ion meets when it moves in solution. For a spherical ion, the mobility μ can be written as:

$$\mu = \frac{q}{6\pi\eta r} \tag{1}$$

where q is the charge of the ion, η is the viscosity of the solution, and r is the hydrodynamic radius of the ion. Different ions can be separated when their charge-to-size ratio differs. As proteins are multivalent ionic species covering a wide range of sizes (molecular weights), they—in principle—are very well suited for CE analysis. Moreover, protein charge, and thus selectivity, can easily be manipulated by adjusting the pH of the BGE. It should be noted that also purely size-based or pI-based selectivities can be achieved with CE by adding a sieving polymer gel or pH gradient forming ampholytes to the BGE.

A fundamental electrophoretic phenomenon occurring in CE is the electroosmotic flow (EOF). This bulk flow of BGE is usually directed toward the detector. Upon filling a capillary with an aqueous buffer solution, silanol groups at the inside surface of the fused-silica will (partly) deprotonate, forming an electric double layer. When a voltage is applied, the net positively charged solution in the capillary will move toward the negative electrode. The velocity of the EOF depends on the magnitude of the applied electric field and the properties of the used BGE. As the silanols are mainly weak acids, the charge of the capillary surface, and thus the magnitude of the EOF, varies with the pH. The EOF increases with raising pH, reaching a plateau at about pH 8, while at pH 2 it essentially ceases due to the silanols getting fully protonated. At neutral and higher pH, the EOF often is sufficiently strong to move even anions toward the negative electrode, allowing detection of all solutes (regardless of their charge) in a single CE run. So, the observed (or apparent) mobility of an ion (as determined from the electropherogram) usually is the sum of two mobilities: the electroosmotic mobility and the ionic (or effective) mobility. Obviously, separation only takes place when ionic species have different effective mobilities.

As for any other separation technique, the widths of the separated analyte zones (peaks) preferably should be as narrow as possible in order to enhance resolution. Assuming Gaussian zone profiles in CE, band widths are usually given as the standard deviation (σ) or variance (σ^2) of the peak. Like in chromatography, in CE the ability of a separation system to produce narrow peaks (i.e., the efficiency) is expressed by the plate number (N):

$$N = \left(\frac{t_m}{\sigma}\right)^2 \tag{2}$$

where t_m is the migration time of the analyte. In CE, plate numbers of several hundreds of thousands to even a few millions are very well possible, whereas in HPLC plate numbers rarely exceed 10,000. These high efficiencies can be achieved because several potential sources of band broadening are absent or can be effectively minimized. Ideally, axial diffusion is the only source of band broadening and the ultimate plate number is then given by:

$$N = \frac{\mu V}{2D} \tag{3}$$

where V is the applied voltage and D is the diffusion coefficient of the analyte. Notably, the plate number does not depend on the length of the capillary. Larger molecules have smaller diffusion coefficients, and indeed highest efficiencies in CE have been obtained for macromolecules such as DNA and proteins. In practice, however, for proteins these high plate numbers are not always easy to achieve due to adsorption of proteins to the capillary wall, which can adversely affect the peak widths. Methods to decrease protein adsorption are discussed below.

Equation 3 also indicates that the applied voltage in principle should be maximized to attain high efficiencies. Using commercial CE instruments, commonly voltages of up to 30 kV can be applied (cf. in PAGE, applied voltages typically are 100–150 V). In fact, the favorable plate numbers in CE are directly related to the high voltages that can be used. An inherent effect of applying voltage is the production of heat in the BGE in the capillary resulting from the electric current passing through. Advanced CE instruments use capillary thermostatting to provide an effective way of dissipating this so-called Joule heat and a more stable performance. Joule heating can cause band broadening when the current gets too high (e.g., as result of the high conductivity of the used BGE). To avoid this, the voltage has to be lowered or the concentration of the BGE has to be reduced. Another option is to select organic and/or zwitterionic buffers, such as TAPS, CHES, and HEPES, as BGE. These buffers add to the ionic strength, but less or not to the conductivity, and thus allow high voltages to be applied at acceptable currents.

Measurement of the absorption of UV or visible light by the separated analytes is the predominant detection mode in CE. For absorbance (and also fluorescence) detection, a certain length of the capillary near the outlet is used as detector "cell." It is created by removing a piece of the polyimide coating from the outside of the fused-silica capillary which in itself is light-transparent down to 190 nm. Detection apertures with a typical length of 0.2–0.8 mm are used to create a well-defined detection volume. The size of the detection window has to be limited to prevent significant contribution of the detection to band broadening and, thus, to allow proper detection of the sharp analyte zones.

In CE, the sample is introduced as a small zone at the inlet of the capillary. In order to minimize the effect of the injection on band broadening, as a rule of thumb, the length of the sample plug should be within 1% of the total length of the capillary. This corresponds to maximum injection volumes of about 2–50 nL only, depending on the length and diameter of the used capillary. For a 100 μg/mL sample of a 20-kDa protein, for example, this implies that the analyzed protein amounts are in the femtomol range. An important aspect to consider in CE is the concentration of salts in the sample. To ensure high efficiency, the ionic strength of the sample should not exceed the ionic strength of the BGE. On the other hand, when the ionic strength of the sample is lower than the BGE, sharpening of analyte zones (so-called stacking) occurs. This often allows larger injection volumes to be loaded into the capillary and, thus, lower detection limits to be achieved.

Although mass sensitivities can be impressive in CE, the attainable concentration sensitivity can be considerably less favorable. This is in part caused by the restricted sample loadability, but most of all (bearing Beer's law in mind) by the limited optical pathlength for detection, which is not larger than the capillary diameter (25–100 μm). The effect of the small pathlength is to some extent counteracted by the relatively high peak concentrations (resulting from the high efficiency) in CE. Overall, however, with UV absorbance detection in CE, concentration detection limits generally are about two orders of magnitude higher than in HPLC. Sometimes this loss can partly be compensated for by selecting a more favorable detection wavelength. For instance, the molar absorptivity of proteins at 190–200 nm is much higher than at 280 nm, the commonly used wavelength in protein HPLC. Detection in the far UV region in HPLC can be problematic, but often is very well possible in CE (due to the short optical pathlength and/or high separation selectivity), even when a moderately UV-absorbing BGE is used.

In order to improve detection limits in CE, proteins may be derivatized with fluorescent reagents and detected with highly sensitive laser-induced fluorescence (LIF) detection. However, as one protein normally contains several reactive moieties (e.g., lysine residues), upon derivatization, usually multiple products are formed. These give rise to a wide peak pattern when analyzed by CZE or CIEF, which obviously is undesirable if, for instance, quality control of a biopharmaceutical sample is pursued. In CGE, which has a purely size-based separation mechanism, this problem does not occur, and LIF can be used effectively to obtain better sensitivity for proteins.

Capillary Coating

Proteins tend to adsorb to surfaces, which in CE can lead to band broadening and peak tailing, or even loss of analyte(s) when the binding is irreversible. Moreover, changes in the surface of the capillary may cause nonuniform liquid flow profiles and an uncontrollable EOF leading to poor migration time reproducibilities. Adsorption is caused by electrostatic and/or hydrophobic interactions of proteins with the capillary inner wall. By using BGEs with extreme pH values, protein adsorption phenomena sometimes can be reduced. At a pH lower than 2, the silanol groups of the fused-silica are fully protonated and the surface charge will approach zero. This means that electrostatic interaction of the proteins (which are positively charged at low pH) with the capillary wall will be minimized. At pH values higher than 10, the capillary surface will be highly negatively charged, while most proteins also will bear a net negative charge. So in this situation, electrostatic repulsion will reduce the chance of adsorption of proteins. Contrary to HPLC, in CE there are no technical constraints to perform separations at very low or high pH. Nevertheless, from the viewpoint of protein stability and integrity, the use of BGEs with extreme pH might not be desirable. Moreover, for separation optimization, the possibility to vary the pH of the BGE might be required.

In CE of proteins, coated capillaries are often used in order to minimize protein interaction/adsorption and stabilize or cancel out the EOF (Horvath and Dolnik, 2001). Before use, the fused-silica capillary wall surface is then coated with a suitable layer that masks the silanol groups. Several coating approaches can be discerned. The first applied, and still quite popular method of producing a capillary coating involves the use of polymers (*in situ* polymerized or preformed) that are covalently bound to the capillary internal wall (Figeys and Aebersold, 1997; Belder et al., 2001). These coating materials, like for example polyacrylamide or poly(vinyl alcohol), cover the capillary surface and are less susceptible to protein adsorption. Chemically coated capillaries are widely used to improve efficiencies and, especially in CIEF, to eliminate the EOF. Disadvantage of this type of coating is that the preparation is often based on a multi-step chemical reaction scheme that may cause a large variance between capillaries. Besides, covalent coatings are usually stable only in a restricted pH range, have limited lifetimes, and can be quite expensive when purchased commercially.

Another approach to prevent protein adsorption and control the EOF is the addition of compounds (such as amines or surfactants) to the BGE that interact electrostatically with the silanol groups. The preparation of these so-called dynamically coated capillaries is quite simple and fast, comprising a flushing procedure only (Corradini, 1997; Righetti et al., 2001). As an example, Figure 2 shows that adding triethylamine to the BGE can strongly improve the efficiency and, thus, resolution of a CE separation of a drug-protein conjugate sample. Removal of dynamic coatings can easily be accomplished by changing the pH. This also implies a drawback because analyses can only be carried out in a certain pH range. Furthermore, the presence of coating additives in the BGE may influence the CE selectivity and, depending on their character, additives may interfere with UV absorbance or MS detection.

A third coating development involves the use of noncovalently coated capillaries, which are prepared by flushing the capillary with a solution of an ionic polymer prior to CE analysis. Initially, capillaries

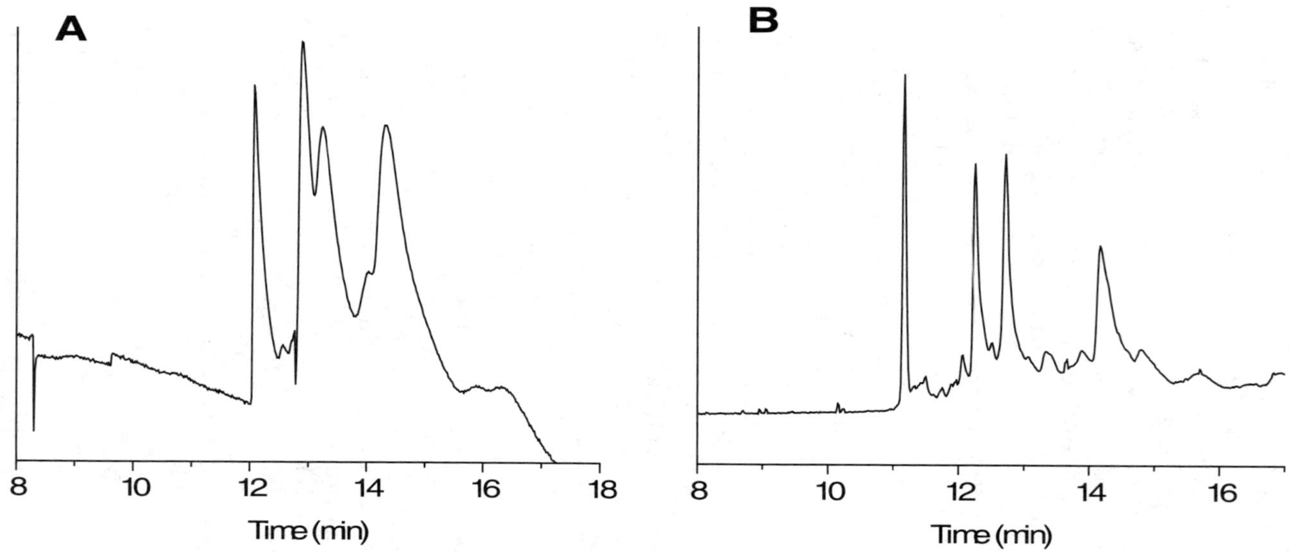

Figure 2 CZE of lysozyme conjugated with captopril (reaction ratio 1:2) using a BGE of 50 mM phosphate buffer, pH 2.5, (A) without and (B) with 40 mM triethylamine. As lysozyme contains multiple reaction sites (lysine groups), the reaction yields various differently conjugated protein molecules (Somsen et al., unpublished results).

were coated with a monolayer of positively charged polymers, such as polybrene and polyethylenimine (Erim et al., 1995; Córdova et al., 1997; Wang and Dubin, 1999). As the capillary wall in these systems is positively charged, the EOF is reversed and, therefore, the voltage polarity of the CE system also has to be switched so that the EOF will be directed toward the detector. In such a case, positively charged proteins will migrate slower than the EOF. Recently, it has been demonstrated that a second, anionic polymer layer can be electrostatically adsorbed on the first cationic layer, creating a doubly noncovalently coated capillary. These bilayer coatings show a high endurance and uniformity and seem to be very promising for the highly efficient and reproducible analysis of peptides and proteins (Katayama et al., 1998; Catai et al., 2004).

Summarizing, one can state that currently there is not one best coating system for protein analysis by CE. The success of the coating procedure often will rely on the characteristics of the analyzed protein and cannot easily be predicted beforehand. In this respect, it is most attractive to use dynamic or noncovalent/physically adsorbed coatings because they are easy to prepare and can be evaluated in a relatively short time.

Capillary Zone Electrophoresis

Separation Conditions

CZE is the most used CE mode for analysis of pharmaceutical proteins. Simple buffers can be used as separation medium, making the technique highly versatile and easy to use. Electrophoretic mobilities can be manipulated by changing the pH and/or ionic strength of the BGE, and separation optimization can be carried out quite straightforwardly and fast. BGE constituents such as additives for dynamic coating and also capillary temperature can be important parameters to consider when optimizing selectivity (Guzman et al., 1992).

Plain buffers such as phosphate and borate buffers are regularly used in CZE, but also organic buffers as frequently used in biochemical procedures (e.g., TRIS, HEPES, etc.) can be applied as BGE (Righetti et al., 2000; Janini and Issac, 2001). This means that CE offers the possibility to analyze the proteins under relatively mild conditions. In this respect, CZE can be referred to as being "biocompatible." This can be important when one wants to gain information on the therapeutic protein in its natural state and needs to be sure that the analytical conditions do not induce (conformational) changes of the protein.

Applications

In CZE, separations are merely based on analyte charge and size, which makes the technique quite useful for quality control. CZE is frequently used for the detection of impurities and/or degradation products, for instance during process control or batch release. In this respect, CZE is often chosen as an alternative or complementary technique to HPLC in order to increase the overall confidence level of impurity profiling. Another important application of CZE is the separation of glycoforms of proteins such as erythropoietin (r-EPO) and human deoxyribonuclease (r-DNase). In fact, a CZE-based method for the characterization of the glycoform pattern of pharmaceutical EPO has been included in the 2002 *European Pharmacopoeia*, substituting the conventional isoelectrofocusing test. Table 1 presents an overview of the applications of CZE for the analysis of pharmaceutical proteins. Some selected cases are discussed below in order to illustrate the potential of CE as a "stand-alone" as well as a complementary technique to HPLC.

Several reported applications of CZE deal with the analysis of insulin, a widely used and produced therapeutic protein. Mandrup described a CZE method for the separation of insulin and deamidation products (impurities) using untreated fused-silica capillaries and a BGE containing CHES, triethylamine, and 10% acetonitrile (Mandrup, 1992). Acidic and neutral desamido-insulin in formulated human insulin could be separated within 20 minutes with the relative standard deviation (RSD) of the migration times being below 1%. In reversed-phase HPLC, the neutral desamido-insulin coeluted with the insulin, whereas with CZE in comparison with ion-exchange chromatography (IEC) better resolution was obtained in a shorter analysis time. Klyushnichenko et al. showed how CZE can be incorporated in an analytical scheme for the monitoring of recombinant human insulin during the various steps of its production (Klyushnichenko et al., 1994; Sergeev et al., 2001). Simple borate or phosphate buffers (pH 9–11) were used with uncoated fused-silica capillaries. CZE was applied, for example, for the assessment of the isolation and purification of the fusion protein (used for expression), for the control of the isolation and purification of the intermediate di-Arg$^{B31\text{-}B32}$- insulin from the reaction mixture, and for control of the final purification of the insulin by IEC. The latter is illustrated in Figure 3, which shows the CZE analysis of three IEC fractions revealing the presence of several impurities from preparation (Arg-insulin) and degradation (unidentified desamido-insulins) in the isolated insulin. The degradation impurities were not detected with HPLC.

Guzman et al. used CZE to separate and quantify two recombinant cytokines (leukocyte A interferon and interleukin-1α) in actual pharmaceutical dosage forms (Guzman et al., 1991). Borate buffer (50 mM, pH 8.3) was used in combination with uncoated capillaries. Lithium chloride (25 mM) was added to the BGE to improve resolution, probably as a result of stabilized tertiary protein structures and reduced protein adsorption. Both cytokines were well separated

Protein[a]	Objective	BGE	Capillary Coating[b]	Detection	Ref.
r Acidic fibroblast growth factor	Protein stability assessment Impurity profiling	50 mM phosphate (pH 2.5) with 0.25% HPMC	Uncoated	UV (214 nm)	Roddy et al. (1997)
Alkaline phosphatase from human placenta	Quality control Impurity profiling	50 mM sodium borate (pH 8.5) with 2 mM diaminobutane	Uncoated	UV (200 nm)	Eriksson et al. (2001)
Alkaline phosphatase from human placenta	Determination of degradation products	50 mM sodium borate (pH 8.5) with 2 mM diaminobutane	Uncoated	UV (200 nm)	Eriksson et al. (2003)
r Bone morphogenic protein 2	Separation of glycoforms Protein characterization	100 mM sodium phosphate (pH 2.5)	Uncoated	UV (200 nm)	Yim et al. (1995)
r Deoxyribonuclease	Separation of glycoforms Protein characterization	200 mM 6-amino-*n*-caproic acid (pH 4.8) with 5 mM Ca^{2+}	PVA	UV (214 nm)	Felten et al. (1999)
r Erythropoietin	Separation of glycoforms Protein characterization	10 mM tricine, 10 mM sodium chloride, 10 mM sodium acetate, 7 M urea, and 2.5 mM diaminobutane (pH 5.5)	DB-1; DB-17	UV (214 nm)	Kinoshita et al. (2000)

Table 1 CZE of Pharmaceutical Proteins.

(continued on following page)

(continued from previous page)

Protein[a]	Objective	BGE	Capillary Coating[b]	Detection	Ref.
r Erythropoietin	Discrimination of recombinant and urinary origin	10 mM tricine, 10 mM sodium chloride, 10 mM sodium acetate, 7 M urea, and 3.9 mM diaminobutane (pH 5.5)	Uncoated	UV (214 nm)	de Frutos et al. (2003)
r Erythropoietin	Separation of glycoforms Protein characterization	10 mM tricine, 10 mM NaCl, 10 mM sodium acetate, 7 M urea, and 3.9 mM diaminobutane (pH 5.5)	Polyacrylamide	UV (214 nm)	Cifuentes et al. (1999)
r Erythropoietin	Separation/quantification in formulations containing HSA	200 mM sodium phosphate (pH 4.0) with 1 mM nickel chloride	eCAP amine (Beckman)	UV (200 nm)	Bietlot and Girard (1997)
r Erythropoietin	Separation of glycoforms	10 mM tricine, 10 mM NaCl, 10 mM sodium acetate, 7 M urea, and 3.9 mM diaminobutane (pH 5.5)	Uncoated	UV (214 nm)	Lopez-Soto-Yarritu et al. (2002b)
r Erythropoietin	Separation of glycoforms	400 mM ammonium acetate (pH 4.75)	Polybrene	UV (214 nm)	Sanz-Nebot et al. (2003)
r Follicle-stimulating hormone	Prediction of biological potency	100 mM sodium borate (pH 8.9) with 5 mM 1,3 diaminopropane	Uncoated	UV (200 nm)	Storring et al. (2002)

r Granulocyte-colony–stimulating factor	Separation of glycoforms Protein characterization	50 mM phosphate-50 mM borate (pH 8.0) with and without 2.5 mM diaminobutane	Uncoated	UV (214 nm)	Watson and Yao (1993)
r Insulin	Quantification	50 mM acetate, 850 mM CHES, 10% acetonitrile (v/v) (pH 7.8)	Uncoated	UV (214 nm)	Kunkel et al. (1997)
r Insulin	Quantification Determination of degradation products	50 mM acetate, 850 mM CHES, 10% acetonitrile (v/v) (pH 7.8)	Uncoated	UV (214 nm)	Mandrup (1992)
r Insulin	Characterization in formulation Quantitation	10 mM tricine, 5.8 mM morpholine, 20 M potassium chloride (pH 8.0)	Uncoated	UV (214 nm)	Lookabaugh et al. (1991)
r Insulin	Monitoring production process Determination of post-translational modifications and degradation products	50 mM sodium phosphate pH 11 and sodium phosphate (pH 9) with 10% MeOH	Uncoated	UV (214 nm)	Sergeev et al. (2001)
r Insulin	Quality control Impurity profiling	100 mM sodium borate (pH 9.3)	Uncoated	UV (214 nm)	Klyushnichenko et al. (1994)
r Interleukin-1 alpha	Separation in injectable dosage form containing HSA Determination of degradation products	50 mM sodium tetraborate (pH 8.3) with 2.5 mM lithium chloride	Uncoated	UV (214 nm)	Guzman et al. (1991)

(continued on following page)

(continued from previous page)

Protein[a]	Objective	BGE	Capillary Coating[b]	Detection	Ref.
Natural and *r* interleukin-2	Characterization	100 mM sodium phosphate (pH 2.5)	Uncoated	UV (200 nm)	Knüver-Hopf and Mohr (1995)
r Interleukin-6	Separation and identification of degradation products	20 mM ammonium acetate (pH 4.2)	Uncoated	UV (200 nm), ESI-MS	Gysler et al. (1999)
r Leukocyte A interferon	Separation in injectable dosage form containing HSA	50 mM sodium tetraborate (pH 8.3) with 2.5 mM lithium chloride	Uncoated	UV (214 nm)	Guzman et al. (1991)
r Monoclonal antibody anti-TAC	Separation/quantification in formulation	50 mM sodium tetraborate (pH 8.3) with 2.5 mM lithium chloride	Uncoated	UV (214 nm)	Guzman et al. (1992)
r Tissue plasminogen activator	Characterization Glycoform separation	200 mM ε-amino-*n* caproic acid with 50 mM acetic acid (pH 5.1)	PVA	UV (214 nm)	Thorne et al. (1996)

[a] *r*, recombinant.

[b] DB-1, dimethyl polysiloxane; DB-17, (50% phenyl)-methyl polysiloxane.

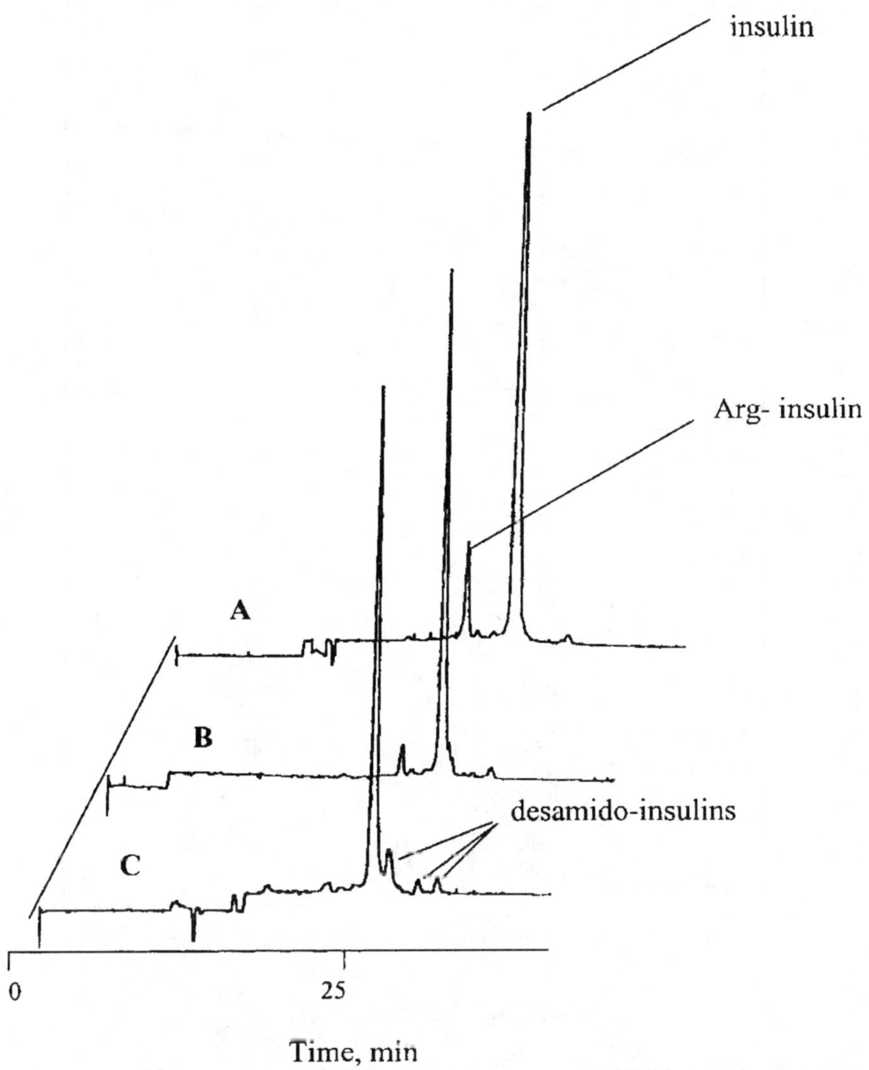

Figure 3 CZE of fractions obtained after ion-exchange chromatography of crude insulin sample. (A) Front, (B) central part, and (C) tail of IEC peak (Sergeev et al., 2001). Reproduced by permission of Elsevier.

from the excipients HSA and *N*-acteyltryptophan (Figure 4). The control of the temperature of the separation appeared to be essential to yield an acceptable reproducibility of the migration times during analysis of the dosage form. The RSD for migration time was within 0.5%, and the RSD for peak area (0.5 mg/mL protein injected) was below 3.6%.

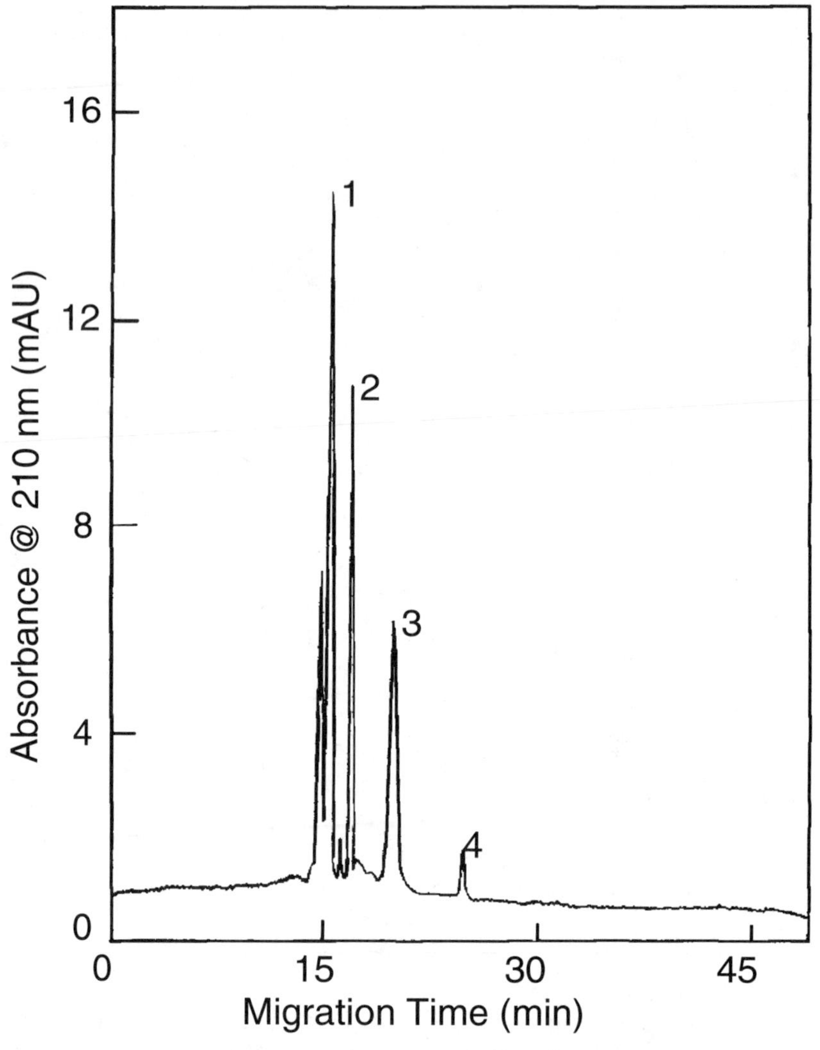

Figure 4 CZE of a pharmaceutical dosage form containing recombinant interleukin-1 alpha (peak 2), HSA (peak 3), *N*-acetyltryptophan (peak 4), and spiked with recombinant leukocyte A interferon (peak 1) (Guzman et al., 1991). Reproduced by permission of Elsevier.

Capillary Isoelectric Focusing

Principles and Separation Conditions

Capillary isoelectric focusing (CIEF) is a mode of CE that separates amphoteric compounds, such as proteins and peptides, according to their differences in isoelectric point (pI). CIEF offers the possibility to combine the high resolving power of IEF (as known from conventional gel-IEF) with the advantages of CE instrumentation (i.e., automation, on-line detection, and use of high electric field strengths). Proteins that differ 0.005 pI units or even less have successfully been separated with CIEF. In CIEF, a pH gradient is formed along the capillary using ampholytes (zwitterionic compounds), which have slightly different pI values spanning the desired pH range (e.g., pH 3–9). The ampholytes act as strong buffers at their pI. Upon analysis, the capillary is filled with a mixture of both the ampholytes and the sample. The capillary inlet and anode are then placed in an acidic solution (anolyte; typically phosphoric acid) and the capillary outlet and cathode in a basic solution (catholyte; typically sodium hydroxide). After application of the high voltage, a pH gradient will be formed by the ampholytes ranging from low (inlet) to high (outlet) pH. According to their electrophoretic mobility, the sample proteins will migrate through the gradient until they reach a pH region equal to their pI where they become net uncharged and stop migrating. The protein zones remain narrow (focused), as a protein that enters a zone of different pH will become charged and migrate back. The overall focusing process can be followed by monitoring the current, which approaches zero when the ion movement inside the capillary stops. After focusing, the zones are passed through the UV detector to produce an electropherogram. This is most commonly accomplished by adding a salt (e.g., sodium chloride) to the catholyte or by applying pressure to the capillary inlet.

In CIEF, the EOF needs to be eliminated because it can sweep the ampholytes from the capillary before focusing is complete. So, usually capillaries with neutral coatings are used, which are also helpful in reducing protein adsorption. An important advantage of CIEF is that, in contrast with most other CE techniques, relatively large sample volumes can be introduced into the capillary, as after injection the analytes are focused to sharp zones. Clearly, this concentration effect favors analyte detectability, although there also is a danger that the proteins aggregate and/or precipitate having minimum solubility at their pI. Most commonly, CIEF employs UV absorbance detection. Unfortunately, most ampholytes exhibit strong absorbance at wavelengths below 240 nm, so that proteins have to be detected at 280 nm and cannot be detected in the far UV where protein molar absorptivities are much higher. In other words, in CIEF the gain in analyte peak concentration is more or less counterbalanced by a loss in detection sensitivity.

Applications

Chemical changes in proteins often involve their basic or acidic moieties and, as a result, changes in charge and pI of the protein. Especially, deamidation is a major chemical degradation reaction, resulting in conversion of (neutral) amides to (negative) carboxyl groups. Also, differences in post-translational modifications (phosphorylation, glycosylation, etc.) can affect the pI of proteins, allowing, for example, separation of glycoforms. Consequently, in the analysis of protein pharmaceuticals, CIEF has been applied to a large extent to samples where protein (micro-) heterogeneity has to be revealed and/or closely related compounds have to be separated and, to a lesser extent, for overall quality analysis of biopharmaceutical samples. Table 2 gives an overview of analyses of pharmaceutical proteins by CIEF. Two examples are discussed in more detail below.

Recombinant human erythropoietin (r-EPO), a well-known pharmaceutical protein against renal anemia, is a glycoprotein carrying carbohydrate structures with varying sialic acid content. As each sialic acid contributes a negative charge to the protein, the relative proportion of glycoforms in principle can be studied by CIEF. Cifuentes et al. showed that by using polyacrylamide-coated capillaries and a mixture of broad and narrow pH-range ampholytes, and by adding urea to the r-EPO sample, resolution of r-EPO glycoforms can be achieved (Cifuentes et al., 1999). With an optimized method using hydrodynamic mobilization, the separation and quantification of seven glycoform bands with apparent pIs of 3.78–4.69 was accomplished in 12 minutes (Figure 5). The migration time reproducibility for r-EPO analyzed by CIEF was

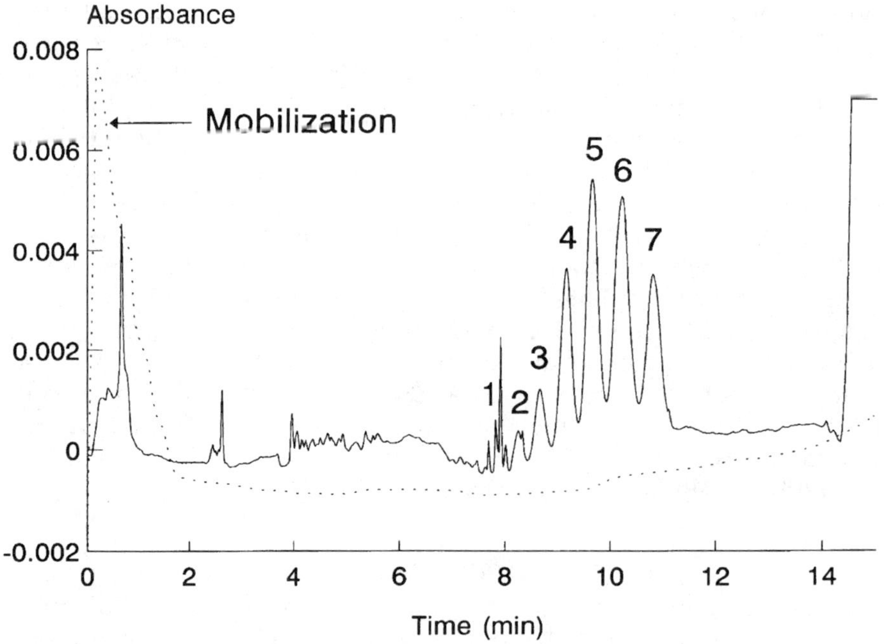

Figure 5 CIEF of r-EPO from a desalted sample. Dashed line corresponds to electric current profile. 1–7, EPO glycoforms (Cifuentes et al., 1999). Reproduced by permission of Elsevier.

Protein[a]	Objective	Separation Conditions[b]	Capillary Coating[c]	Detection	Ref.
r Erythropoietin	Separation of glycoforms Characterization	*Cat*: 20 mM sodium hydroxide titrated to pH 11.86 with phosphoric acid *Amp*: Sample mixed with ampholytes [pH range of 3–10 and 2–4 in the ratio 1:2 (v/v)] in CIEF gel (Beckman) in the ratio 1:5 (v/v), 7 M urea *Ano*: 91 mM phosphoric acid in CIEF gel (Beckman) *Foc*: 1 min at +25 kV *Mob*: 15 min at 0.5 psi	Polyacrylamide and eCAP neutral (Beckman)	UV (280 nm)	Lopez-Soto-Yarritu et al. (2002a)
r Erythropoietin	Separation of glycoforms Characterization	*Cat*: 40 mM sodium hydroxide *Amp*: Sample mixed with ampholytes [pH range of 3–10 and 2.5–5 in the ratio 1:2 (v/v)] in CIEF gel (Beckman) in the ratio 1:5 (v/v), 7 M urea *Ano*: 20 mM phosphoric acid *Foc*: 6 min at +25 kV *Mob*: 20 min at 0.5 psi	Polyacrylamide	UV (280 nm)	Cifuentes et al. (1999)

Table 2 CIEF of Pharmaceutical Proteins.

(continued on following page)

(continued from previous page)

Protein[a]	Objective	Separation Conditions[b]	Capillary Coating[c]	Detection	Ref.
r Erythropoietin	Separation of glycoforms	*Cat*: 20 mM sodium hydroxide *Amp*: Sample (390 μg/μL) mixed with ampholytes [Servalyt pH range of 3–10 and 2.5–5 in the ratio 1:1 (v/v)], 0.4% HPMC, and 4 M urea *Ano*: 20 mM phosphoric acid *Foc*: 16 min at +30 kV *Mob*: 50 mbar maintaining +30 kV	Uncoated	UV (280 nm)	Kubach and Grimm (1996)
r HIV envelope glycoprotein rgp-160	Characterization of two sub-populations of HIV virus	*Cat*: 20 mM sodium hydroxide *Amp*: 5% of Ampholines in the pH range of 3.5–5, 5–8, and 3.5–10 in the ratio 71:12:17 (v/v/v), 1% TEMED, 8.5 mM CAPS, 6% saccharose in 0.1% HPMC *Ano*: 100 mM phosphoric acid *Foc* and *mob*: –20 kV	PVA	UV (280 nm)	Tran et al. (2000)
r Immunoglobulin-1	Quality control Quantification of glycoforms Characterization	*Cat*: 20 mM NaOH in 0.4% methylcellulose *Amp*: 2% Pharmalyte pH 3–10 in 0.4% methylcellulose *Ano*: 120 mM phosphoric acid in 0.4% methylcellulose *Foc*: 20 min at +30 kV *Mob*: 20 min at 0.5 psi	DB-1	UV (280 nm)	Tang et al. (1999)

r Monoclonal antibody HER2	Quality control Separation of glycoforms Characterization	*Cat*: 40 mM sodium hydroxide *Amp*: a solution (2% w/v) consisting of a 8:1:1 ratio Pharmalyte pH 8–10.5, Bio-lyte pH 7–9, Bio-lyte pH 3–10 containing 0.5% (v/v) TEMED, and 0.2% (w/v) HPMC *Ano*: 20 mM phosphoric acid *Foc*: 17 min at +15 kV *Mob*: 30 min at 0.5 psi	Polyacrylamide	UV (280 nm)	Hunt et al. (1996)
Ribonuclease A (bovine pancreatic)	Characterization of refolding process	*Cat*: 20 mM sodium hydroxide *Amp*: 0.05% Pharmalyte pH 3–10 and 0.05% TEMED *Ano*: 20 mM phosphoric acid *Foc*: 10 min at +10 kV *Mob*: raising the inlet side to 10 cm above the level of the electrospray needle	Polyacrylamide	ESI-MS	Jensen et al. (1998)
r Tissue-type plasminogen activator	Separation of glycoforms Characterization	*Cat*: 20 mM sodium hydroxide *Amp*: Sample (125 µg/mL), 4 M urea, 0.1% (w/v) HPMC, 0.75% (v/v) TEMED, and 3% (w/v) ampholytes pH 3–10 (Pharmalyte, Ampholine, Bio-lyte and Servalyt) *Ano*: 10 mM phosphoric acid *Foc*: and *Mob*: 10 min at –13.5 kV	eCAP neutral (Beckman)	UV (280 nm)	Chen et al. (1996)

(continued on following page)

(continued from previous page)

Protein[a]	Objective	Separation Conditions[b]	Capillary Coating[c]	Detection	Ref.
r Tissue-type plasminogen activator	Separation of glycoforms	*Cat*: 20 mM sodium hydroxide *Amp*: Sample (390 μg/μL) mixed with ampholytes (Servalyt pH range of 3–10 and 2.5–5 in the ratio 1:1 (v/v)), 0.4% HPMC, and 4 M urea *Ano*: 20 mM phosphoric acid *Foc*: 16 min at −30 kV *Mob*: 50 mbar maintaining +30 kV	Uncoated	UV (280 nm)	Kubach and Grimm (1996)
r Tissue-type plasminogen activator	Separation of glycoforms Characterization	*Cat*: 10 mM sodium hydroxide *Amp*: Sample (320 μg/μL) mixed with 1.3% (v/v) ampholytes pH range of 3–10 (Ampholine, Pharmalyte and Servalyt), and 4.8 M urea *Ano*: 91 mM phosphoric acid *Foc*: 2 min at +13.5 kV *Mob*: 35 mbar	eCAP neutral (Beckman)	UV (280 nm)	Thorne et al. (1996)

[a] *r*, recombinant.
[b] *Cat*, catholyte; *Amp*, ampholyte; *Ano*, anolyte; *Foc*, focusing; *Mob*, mobilization.
[c] DB-1, dimethyl polysiloxane.

about 1% (RSD), and the RSD of the relative peak area of the most intense peak was below 1%. Compared with conventional gel-IEF, better resolution was obtained in a shorter time while allowing quantitative determination. It should be noted that also CZE can be used to produce characteristic profiles of EPO glycoforms, allowing distinction of EPO from different sources (Bietlot and Girard, 1997; Lopez-Soto-Yarritu et al., 2002b).

One of the strategies to develop an efficient anti–HIV-1 vaccine has been the production of recombinant HIV envelope glycoprotein (rgp 160). Tran et al. applied CIEF to study the charge heterogeneity of a soluble recombinant rgp-160 variant that comprises a large number of glycoforms (Tran et al., 2000). A blend of three different ampholyte mixtures in a 85 mM CAPS buffer containing 1% TEMED, 6% saccharose, and 0.1% HPMC in combination with a poly(vinyl alcohol)-coated capillary yielded optimal separation conditions. Although the CIEF method appeared not suitable for the quantitation of individual glycoforms, characteristic and reproducible profiles were obtained. The method was capable of differentiating the glycoform patterns of the rgp 160 of two subpopulations of HIV-1 originating from Europe (clade B) and from Asia (clade E) (Figure 6).

Capillary Gel Electrophoresis

Principles and Separation Conditions

Capillary gel electrophoresis (CGE) is a CE mode in which separation is based on differences in molecular size. Separation is achieved by electrophoresis of charged macromolecular analytes through a capillary filled with a "gel" that acts as a molecular sieve. During their migration through the gel, larger molecules are hindered more than smaller ones. Having a similar separation mechanism, CGE is readily comparable with traditional slab gel electrophoresis (PAGE). However, CGE offers the advantage of up to 100-fold higher electric fields, on-capillary detection, instrumental automation, and ease of use. The preparative and multilane capacity of PAGE is difficult to reproduce in the capillary format, although the much shorter analysis times of CGE and the recent introduction of multicapillary CE systems (partly) compensate for this. Analogous to PAGE, in CGE rigid, cross-linked polyacrylamide was used initially as separation matrix in the capillary, but these gels were not easy to make reproducibly and suffered from clogging and instability. An important breakthrough for CGE came in the 1990s with the advent of linear, non–cross-linked polymers, which have relatively low viscosity and can easily be replaced. These "liquid" polymers form entangled networks in solution and give excellent separations of DNA, RNA, and protein molecules of different size. Best results are obtained under EOF-suppressed conditions using permanently or dynamically coated capillaries.

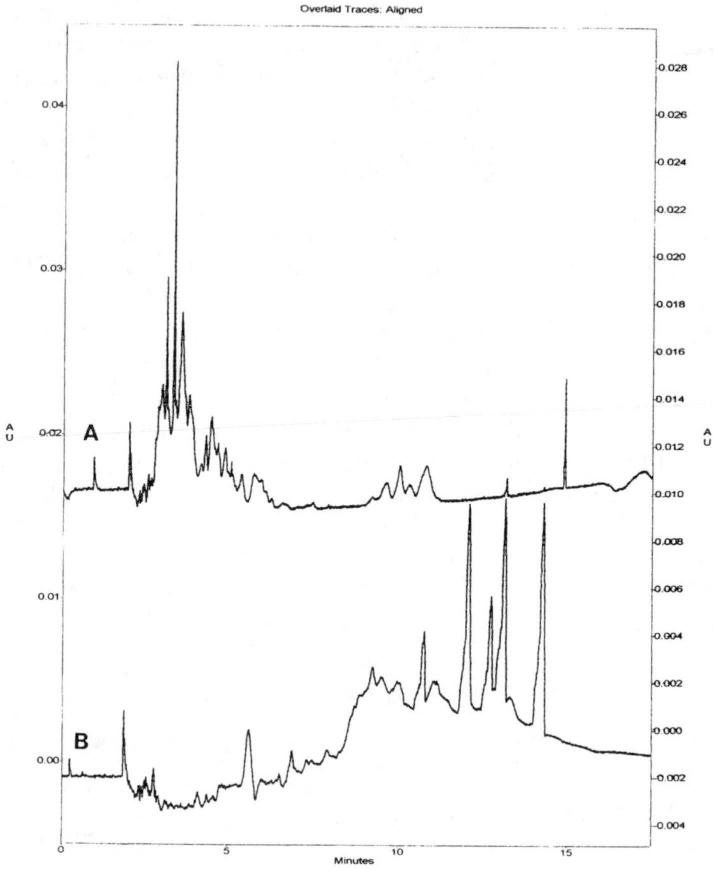

Figure 6 Comparison of the glycoform profile of rgp160 from two subpopulations of the HIV-1 virus: (A) clade B and (B) clade E (Tran et al., 2000). Reproduced by permission of Elsevier.

In order to achieve purely size-based separations of proteins in CGE, the proteins of interest are usually denaturated and complexed with sodium dodecyl sulfate (SDS) prior to analysis. Because the charge-to-size ratio of SDS-protein complexes is almost protein independent, the proteins will have virtually identical effective electrophoretic mobilities and, therefore, will only be discriminated on their size by the sieving polymer matrix (which also contains SDS). By running protein standards as well, good estimates of the molecular weight of unknown proteins can be achieved. It should be noted that this may not be true for glycoproteins, as the carbohydrate structures do not bind the negatively charged SDS molecules. An additional advantage of SDS-CGE is that fluorescent labeling of proteins does not affect the separation performance. As derivatization with the commonly small fluorescent labels hardly affects the overall size of the protein and complexation with SDS cancels out charge differences, in SDS-CGE one peak will be obtained for one protein, irrespective of the number of labels bound.

Applications

CGE monitors protein size and therefore is more useful for providing an overall picture of the composition of a protein sample and less suited for the analysis of subtle changes in a protein. Thus, in the analysis of pharmaceutical proteins, CGE is mainly used to check for the presence of other proteins, protein aggregates, or protein degradation fragments during characterization, purity, and/or stability studies. Table 3 summarizes the application of CGE in biopharmaceutical analysis; below some examples are outlined.

Farchaus et al. applied CGE to study the purity of the active compound, termed protective antigen (PA), of the human anthrax vaccine (Farchaus et al., 1998). PA is the B subunit (M_r 83,000) of the AB-type exotoxin of *Bacillus anthracis* and is produced recombinantly. Figure 7 shows the CGE analysis of the final PA product obtained after purification by anion-exchange chromatography (AEC). Before analysis, the PA sample was diluted in buffer containing SDS and 2-mercaptoethanol (for denaturation), and for separation a commercial sieving polymer was used with a 42-cm fused-silica capillary. Next to the main compound (3) and the internal standard (1), various proteinaceous impurities (2, 4) could be observed (Figure 7). Overall integration of all the peaks yielded a final PA purity of 90% for this lot. CGE analysis of the same PA after purification with hydrophobic interaction chromatography (HIC) revealed a final purity of 98% (not shown) and indicated that the HIC step was extremely effective in removing impurity 4.

The advantages of using CGE with LIF detection was well illustrated by Hunt and Nashabeh for the analysis of humanized recombinant IgG_1-monoclonal antibodies (r-MAb), a class of proteins that is increasingly applied for therapeutic purposes (Hunt and Nashabeh, 1999). To allow CGE-LIF, the r-MAb was first derivatized with a rhodamine succinimidyl ester (5-TAMRA.SE) and then incubated and heated in the presence of SDS. A commercial CGE buffer was used with a 24-cm fused silica capillary installed in a CE instrument equipped with a LIF detector employing 488 nm for excitation. Figure 8 shows the results of the CGE-LIF analysis of nonreduced and reduced preparations of a therapeutic r-MAb. The light and heavy chains (and their combinations) originating from the antibody were nicely separated, and the obtained profiles compared well with silver-stained PAGE, except that the light chain apparently was more reactive with the silver stain than the heavy chain (Figure 8). Repeated analysis showed that RSD values for the migration time were below 1%, and the RSD of the area of the main peak was lower than 0.6%. Compared with CGE-UV,

Protein[a]	Objective	Separation Conditions	Capillary Coating	Detection	Ref.
Alkaline phosphatase from human placenta	Characterization	CE-SDS protein kit (BioRad)	Uncoated	UV (200 nm)	Eriksson et al. (2001)
r Apoliprotein E3	Characterization Quantification	eCAP SDS 14-200 gel buffer (Beckman)	Uncoated	UV (214 nm)	Schlenck et al. (1999)
r Brain derived neurotrophic factor	Characterization	100 mM CHES (pH 8.5) with 0.1% SDS and 5% poly (ethylene oxide) (MW 100,000)	Uncoated	UV (214 nm)	Benedek and Thiede (1994)
r Ciliary neurotrophic factor	Characterization of reduced and nonreduced forms	eCAP SDS-200 protein kit (Beckman)	eCAP SDS-200 (Beckman)	UV (214 nm)	Guttman et al. (1993)
Immunoglobulin G (various sources)	Characterization Determination of degradation products	CE-SDS protein kit (BioRad)	Uncoated	UV (220 nm)	Lee (2000)
r Immunoglobulin 1-monoclonal antibody	Impurity profiling Separation of digested fragments	CE-SDS protein kit (BioRad)	Uncoated	UV (220 nm) and LIF (excitation 488 nm, detection 560 nm)	Hunt and Nashabeh (1999)

r Monoclonal antibody HER2	Protein stability assessment Quantification Impurity profiling	SDS-running buffer (BioRad)	Uncoated	UV (214 nm)	Hunt et al. (1996)
r NAD(+)-dependent formate dehydrogenase	Characterization of purification steps	eCAP SDS-200 protein kit (Beckman)	eCAP SDS-200 (Beckman)	UV (214 nm)	Klyushnichenko et al. (1997)
Protective antigen of lethal toxin and edema toxin secreted by *Bacillus anthracis*	Characterization	ProSort SDS-protein analysis reagent (Perkin Elmer Applied Biosystems)	Uncoated	UV (215 nm)	Farchaus et al. (1998)
r Soluble CD4-PE	Characterization Quantification	375 mM Tris (pH 8.8) with 0.1% SDS-ethylene glycol	γ-(acryloxypropyl) methyldichloro-silane coated	UV (214 nm)	Tsuji (1991)
r Tissue plasminogen activator	Characterization	eCAP SDS 14-200 gel buffer (Beckman)	eCAP neutral (Beckman)	UV (280 nm)	Thorne et al. (1996)

[a] *r*, recombinant.

Table 3 CGE of Pharmaceutical Proteins.

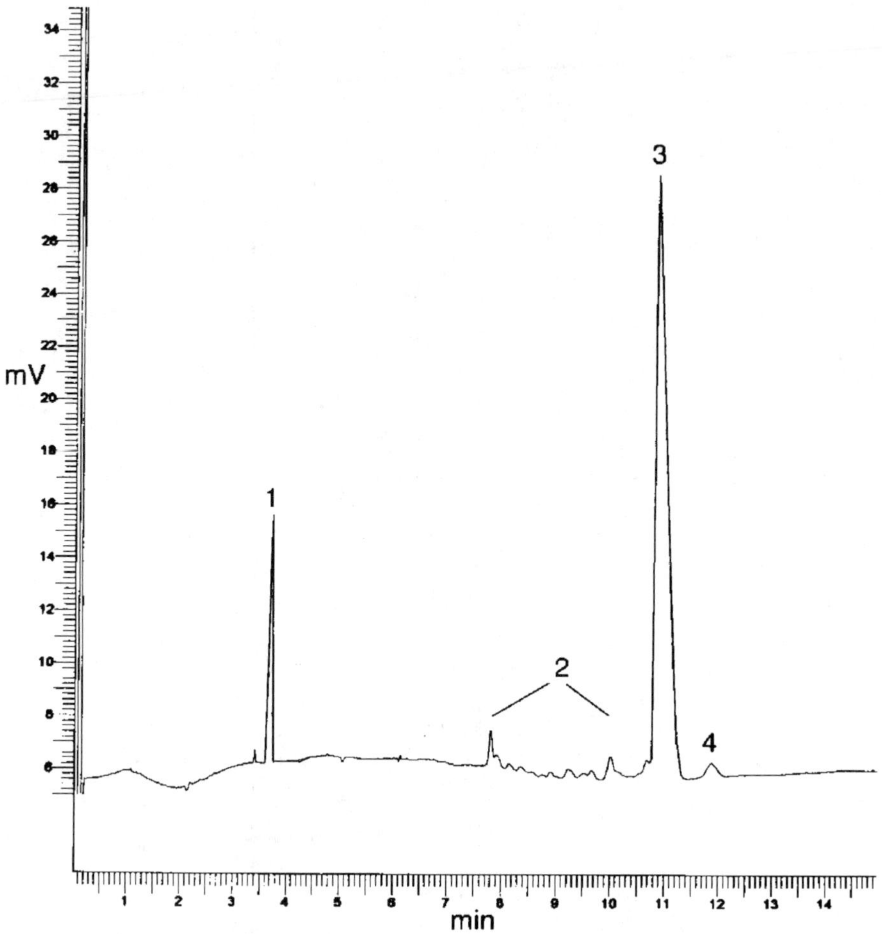

Figure 7 CGE of sample of protective antigen (PA) obtained after anion-exchange chromatography. 1, internal standard; 2 and 4, impurities; 3, PA (Farchaus et al., 1998). Reproduced by permission of the American Society for Microbiology.

a 140-fold increase of sensitivity was achieved with CGE-LIF, allowing detection of minor peaks, which were undetected with UV detection. It was further shown that also nonrelated proteinaceous impurities in r-MAb preparations could be detected at low levels and that the CGE-LIF system could readily resolve various common r-MAb proteolytic fragments. These applications demonstrate the usefulness of this technology for detecting manufacturing inconsistencies in recombinant protein production. The CGE-LIF method was validated according to ICH guidelines and is currently used at Genentec, Inc. (South San Francisco, CA) for release analysis of a r-MAb pharmaceutical.

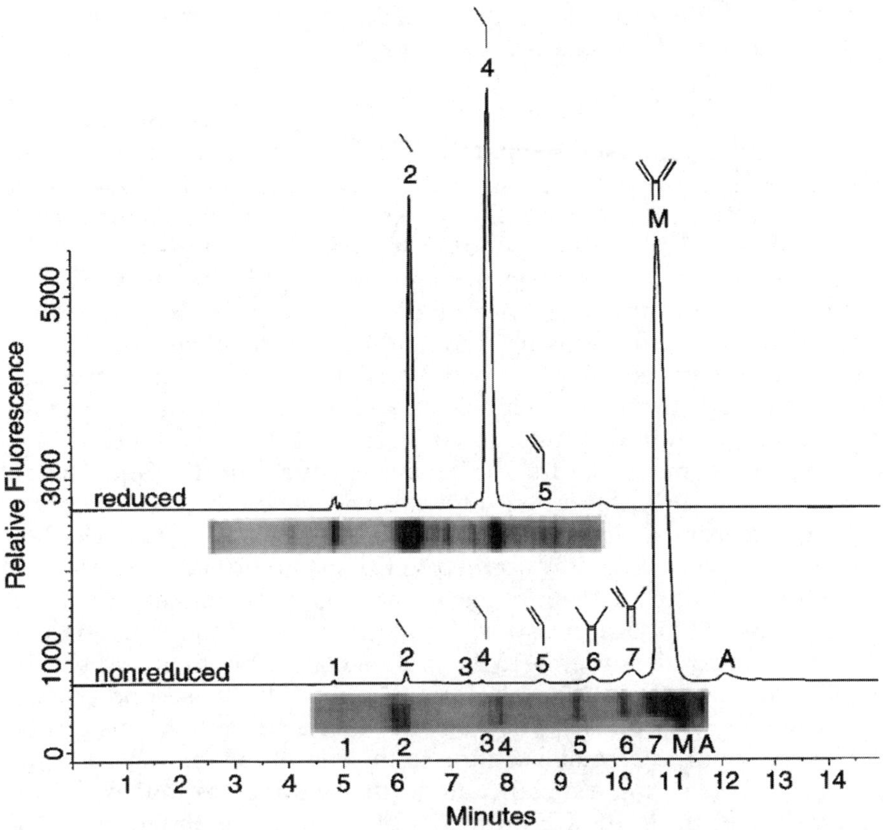

Figure 8 CGE-LIF of reduced and nonreduced preparations of a therapeutic r-MAb labeled with 5-TAMRA.SE. Insets show silver-stained SDS-PAGE traces of the same sample preparations. M, monomer; A, aggregate; 1–7, antibody chains (Hunt and Nashabeh, 1999). Reproduced by permission of the American Chemical Society.

Capillary Electrophoresis-Mass Spectrometry

Mass spectrometry (MS) is nowadays an important and powerful tool for the analysis and characterization of proteins (see Chapters 12 and 13 in this book). The multiple charging of proteins observed in electrospray ionization (ESI) allows their analysis on MS instruments with mass ranges far below the molecular masses of the proteins. MS detection can considerably enhance the utility of CE by providing information about the identity of the separated compounds. In purity and stability studies, the availability of MS data is highly desirable. Characterization of peaks by molecular mass strongly adds to the reliability of their assignment and is very helpful when conditions are changed and cross-correlations have to be made. Therefore, for quality control (e.g., when changes in proteins during storage or exposure are monitored), CE-MS is an attractive approach.

Coupling Modes and Separation Conditions

In CE-MS, the outlet of the separation capillary has to be connected to the ion source of the mass spectrometer, and obviously, no outlet buffer vial can be used. Still, CE requires a closed electrical circuit and, therefore, a CE-MS interface should provide a means to apply voltage to capillary outlet. Two approaches can roughly be discerned to achieve this (Schmitt-Kopplin and Frommberger, 2003). In the first one, an additional liquid is used to apply the voltage indirectly to the CE buffer. A very popular and widely used example of this approach is the so-called sheath liquid interface. In this interface, the separation capillary is surrounded by a second tube of larger diameter in a coaxial arrangement (Figure 9). The supportive "sheath" liquid (e.g., acetonitrile-water acidified with formic acid) is led through the outer tube and merges with the CE effluent at the capillary outlet. The terminating CE voltage is now applied to the sheath liquid. Most often, a third tube guiding a stream of gas surrounds this arrangement to aid nebulization into the electrospray (ES) ion source. As the sheath-liquid flow rates are of the order 3–15 μL/min, a conventional ES ion source as normally used for HPLC-MS can be applied. In the second CE-MS approach, the voltage is applied directly to the CE buffer. This can be achieved by putting a metal coating to the end of the separation capillary or by connecting a metal-coated, full metal, or conductive polymeric sprayer tip to the CE outlet. Another option is the insertion of microelectrode through the wall of the capillary end. The inherently low flow emerging from the CE capillary requires the use of special nano-electrospray ion sources. Although less robust when compared

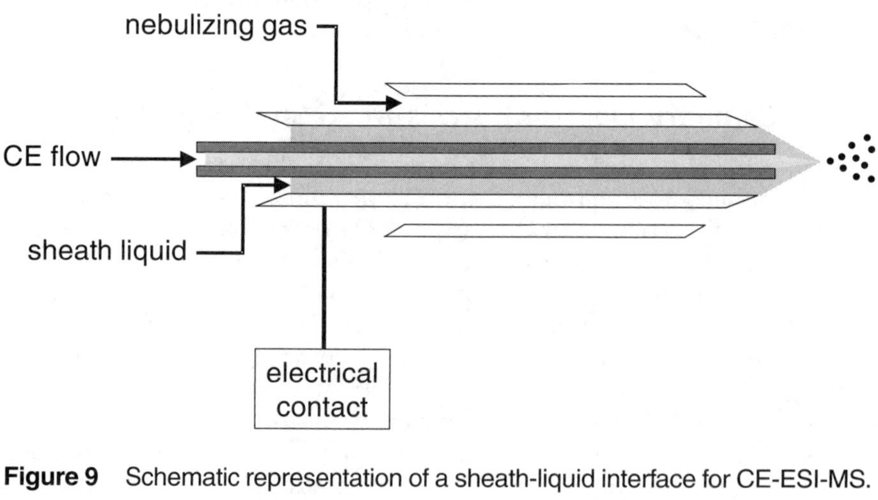

Figure 9 Schematic representation of a sheath-liquid interface for CE-ESI-MS.

to sheath-liquid interfaces, these sheathless setups offer a strongly improved sensitivity.

The presence of nonvolatile buffer constituents can be problematic when CE is coupled on-line to MS. Especially species like phosphate and SDS can give serious suppression of the intensity of MS signals of proteins when ESI is used. This phenomenon practically precludes the possibility of coupling, for example, CGE (which uses high SDS concentrations in the separation buffer) with ESI-MS. Thus, until now in CZE-MS, most often BGEs such as formic acid or ammonium acetate are applied, which are volatile and can be evaporated in the ESI interface. It should be noted that these type of buffers may not provide the conditions for optimal efficiency and/or selectivity in CE. Recently, it was demonstrated that also borate buffers might be used for on-line CE-MS of proteins (Eriksson et al., 2004).

Applications

During the production of biopharmaceuticals, analytical techniques that are able to identify compounds present in process samples and final formulations are of great value. Using a CE-MS approach for this purpose, ideally CE provides the efficient separation of sample constituents, while MS analysis allows the identification of the separated compounds. Several studies on the use of CE-MS for the analysis of pharmaceutical proteins have been reported. In one example, CE-MS is used for the identification of variants of human insulin-like growth factor 1 (IGF-1), a protein that mediates many of the anabolic effects of human growth hormone. Recombinant DNA technology is used to produce substantial quantities of IGF-1; however, during the cellular production of the protein, several variants are obtained. Nashabeh et al. designed a CZE method to resolve these closely related variants and combined the CE instrument with ESI-MS using a sheath-liquid interface (Nashabeh et al., 1994). Applying a polyacrylamide-coated capillary with a BGE containing aminocaproic acid, acetic acid, and DAPS, four out of the five variants present in the IGF 1 samples could be separated. Interestingly, IGF-1 and improperly folded IGF-1, which have identical molecular weights, were baseline resolved by CZE. The MS provided the additional selectivity to distinguish and identify all five IGF-1 variants, which were at the 100-femtomol level in the sample. The molecular weight calculated from the measured spectra nicely matched the ones expected for the respective variants (Table 4).

The use of CZE with UV and MS detection also has been reported for the identification of glycoforms of recombinant human bone morphogenic protein-2 (r-BMP-2), a high mannose-containing protein that has been found to induce bone growth (Yeung et al., 1997). A 50 mM beta-alanine buffer (pH 3.5) was used for separation in a capillary coated with polyacrylamide. A coaxial sheath-liquid interface was

Compound	Measured Molecular Mass (Da)	Expected Molecular Mass (Da)
des-Gly^{1}Pro^{2}Glu^3-IGF-1	7364.0 ± 0.9	7365
des-Gly^1-IGF-1	7591.2 ± 1.1	7591
Met-O-IGF-1	7662.9 ± 1.3	7664
Improperly folded IGF-1	7648.6 ± 0.9	7648
IGF-1	7646.9 ± 0.7	7648

Table 4 Molecular Weight Determination of IGF-1 Variants.

used for CE-MS coupling. After reduction and alkylation of the rh-BMP-2 sample, with CE-UV a nicely resolved glycoform pattern was obtained (Figure 10A). CE-MS of the same sample resulted in a less resolved profile (Figure 10B). The individual glycoforms could still be identified by taking advantage of the added selectivity provided by the MS detection. Figure 10C shows the MS spectra obtained by summing the spectra measured under each main peak (Figure 10B). Peaks I–III appeared to originate from the extended, mature, and pyroGlu forms of r-BMP-2, respectively. The charge-state distributions in the summed spectra revealed the various glycoforms present (Table 5). As the mass spectrum of peak IV showed masses corresponding to both the mature and pyroGlu forms, it was speculated that peak IV originated from the corresponding noncovalent aggregate, which broke apart during the ESI process.

Further examples of the application of CE-MS for the analysis of protein pharmaceuticals include a degradation study of recombinant human interleukin-6 (r-IL-6) sample, containing lysozyme and cytochrome c, incubated at 45 °C for 48 hours (Gysler et al., 1999). Separation and identification of the three proteins and several r-IL-6 degradation products was accomplished. CE-MS analysis was also used to monitor metal exchange in recombinant mouse liver metallothioneins Zn-7 complexes (Pawlak et al., 2002) and to characterize recombinant *Escherichia coli* proteins (Tang et al., 1997). The refolding process of RNAase, induced by disulfide formation, was also successfully monitored by on-line CIEF-MS analysis (Jensen et al., 1998). In the latter example, despite interferences of the ampholytes from the BGE with MS detection, still significant protein signals could be obtained.

Affinity Capillary Electrophoresis

In the previous sections of this chapter, CE in its various modes was applied as a mere separation tool, particularly used to achieve resolution of sample constituents of biopharmaceutical samples. However,

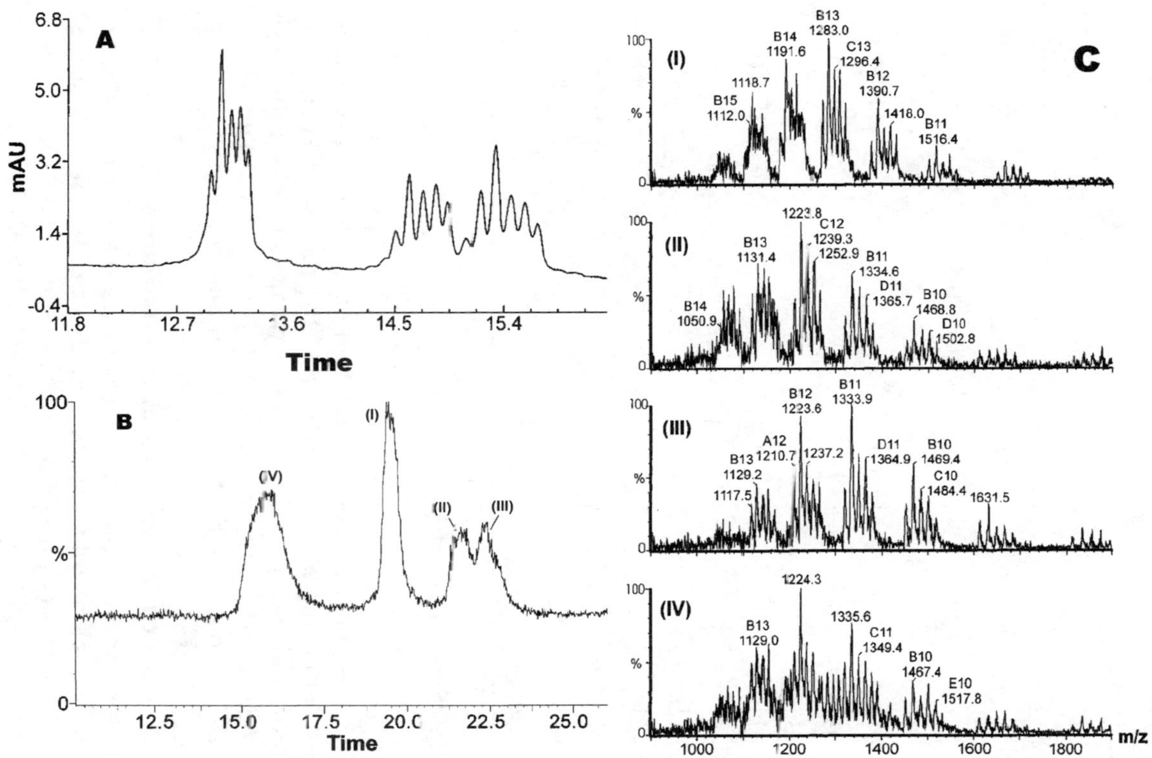

Figure 10 CE-UV and CE-MS of r-BMP-2 monomer obtained after reduction, alkylation, and desalting. (A) UV trace, (B) TIC trace, and (C) mass spectra from the peaks labeled in B (Yeung et al., 1997). Reproduced by permission of the American Chemical Society.

Carbohydrate Structure	Mass of Protein	
	Calculated	Observed
	Extended Form (Peak I)	
$(HexNAc)_2Hex_5$	16503	16506
$(HexNAc)_2Hex_6$	16665	16668
$(HexNAc)_2Hex_7$	16827	16826
$(HexNAc)_2Hex_8$	16989	16987
$(HexNAc)_2Hex_9$	17152	17155
	Mature Form (Peak II)	
$(HexNAc)_2Hex_5$	14521	14516
$(HexNAc)_2Hex_6$	14683	14684
$(HexNAc)_2Hex_7$	14845	14853
$(HexNAc)_2Hex_8$	15007	15007
$(HexNAc)_2Hex_9$	15170	15176
	PyroGlu Form (Peak III)	
$(HexNAc)_2Hex_5$	14504	14509
$(HexNAc)_2Hex_6$	14666	14668
$(HexNAc)_2Hex_7$	14828	14835
$(HexNAc)_2Hex_8$	14990	15000
$(HexNAc)_2Hex_9$	15153	15152

Table 5 CE-MS Results of r-BMP-2 Monomers.

as the electrophoretic mobility is a function of molecular size, shape, and charge, CE in principle can also be used to probe physicochemical parameters of analyzed proteins. In affinity capillary electrophoresis (ACE), the measured electrophoretic characteristics of a protein-ligand complex are used to assess (the extent of) protein binding.

Principles and Separation Conditions

ACE is a relatively new technique that can be used for the study of biomolecular noncovalent interactions and determination of binding/ dissociation constants of protein complexes (Shimura and Kasai, 1997; Heegaard et al., 1998; Heegaard and Kennedy, 1999). In ACE, changes in the electrophoretic mobility of a protein are correlated to the binding of the protein with a ligand. In a typical ACE experiment, the ligand (or protein) under study is added to the BGE at various concentrations. At each concentration, the potentially binding protein (or ligand) is injected into the capillary and its migration time is measured. During electrophoresis, protein-ligand interactions might occur that lead to changes in the electrophoretic mobility of the protein. Quantitative measurement of these mobility shifts allows determination of the protein-ligand binding constants. The advantage of ACE is that the analyzed sample does not have to be pure and that more proteins can be assessed in one run. Requirement for the described approach is that the protein-ligand complexation kinetics (the "on-off rate") is relatively fast, which often is the case. When the kinetics for establishing equilibrium conditions in protein-ligand complex formation are slow with respect to the time required for analysis, the protein and the ligand can be incubated until equilibrium is reached. Then, an aliquot of the mixture is injected, and CE is used to differentiate quantitatively between the free and bound forms of the protein.

Applications

The interaction of proteins with specific ligands such as drugs and toxins is among the important aspects of biological studies in drug discovery and pharmaceutical development processes. Until now, ACE has mainly been applied for binding studies of low-molecular-weight drugs with serum proteins. Some researchers, however, have recognized that ACE can also be useful for the investigation of the binding of protein pharmaceuticals (and related species) with compounds of interest. Table 6 lists several of these ACE applications and some examples are treated in more detail below.

One of these ACE studies involves the interaction of the chromoprotein kedarcidin with its chromophore (Liu et al., 1998). Kedarcidin has been identified as a highly potent and very active antitumor antibiotic. The antitumor activity of the chromoprotein is due mainly to the chromophore, and the apoprotein's function is to stabilize and deliver the chromophore. The molecular interaction between the chromophore and apoprotein was studied to provide useful information on the understanding of the antitumor mechanism. ACE binding constant measurements were carried out in buffer (citrate-MES, pH 6.0) with varying concentrations of kedarcidin chromophore (0–20 μM), and the migration time of the injected kedarcidin apoprotein was monitored (Figure 11). Using the effective mobilities of the apoprotein and Scatchard analysis, the binding constant K_d was calculated to be 5.6 μM. Addition of small, nondenaturating concentrations of acetonitrile to the BGE showed that the binding affinity is reduced as the organic solvent in the aqueous phase is increased.

Caldwell et al. studied the binding of several EPO mimetic, cyclic peptides (EMPs) to the ligand binding domain of the EPO receptor (EBP) that was expressed in *Esherichia coli* by ACE (Caldwell et al., 1999). The produced EBP comprises several isoforms. The binding characteristics for the EBP-EMP complexes were determined by adding the EMPs individually to the BGE (20 mM sodium citrate, pH 7.2) in

Protein[a]	Ligand[b]	Objective	Separation Conditions	Capillary Coating	Detection	Ref.
r Extracellular ligand binding domain of erythropoietin receptor	Mimetic peptides EMP1 and EMP37	Determination of binding constant Stoichiometry of complexation	20 mM sodium citrate (pH 7.2)	eCAP neutral (Beckman)	UV (214 nm)	Caldwell et al. (1999)
r Galectin-1	Lactose	Determination of binding constant	100 mM Tris acetate (pH 7.9) with 0.02% sodium azide	Uncoated	UV (214 nm)	Shimura et al. (2002)
r Galectin-1	β-galactoside-binding lectin	Determination of binding constant	100 mM Tris acetate (pH 7.9) with 0.02% sodium azide	Uncoated	UV (214 nm)	Shimura et al. (2001)
r Galectin-3	Lactose	Determination of binding constant	100 mM Tris acetate (pH 7.9) with 0.02% sodium azide	Uncoated	UV (214 nm)	Shimura et al. (2002)
r HIV gp41 binding protein P50	*h* Interferon-α and β	Determination of possible complexation	100 mM sodium borate (pH 8.4)	Uncoated	UV (280 nm)	Yu et al. (2000)

r Kedarcidin apoprotein	*r* Kedarcidin chromophore	Determination of binding constant	Citrate-MES (pH 6.0)	Sulfonic acid	UV (280 nm)	Liu et al. (1998)
h Monoclonal antibodies 8F5 and 3B10	*h* Rhinovirus 2	Stoichiometry of complexation	100 mM sodium borate (pH 8.3) with 10 mM SDS	Uncoated	UV (205 nm)	Okun et al. (2001a)
r Soluble very low density lipoprotein receptor fragment/maltose-binding protein complex 1–8	*h* Rhinovirus 2	Characterization of the virus-receptor complex Stoichiometry of complexation	100 mM sodium borate (pH 8.3) with 10 mM SDS	Uncoated	UV (205 nm)	Okun et al. (2001b)

[a] *r*, recombinant.
[b] *h*, human.

Table 6 ACE of Pharmaceutical Proteins.

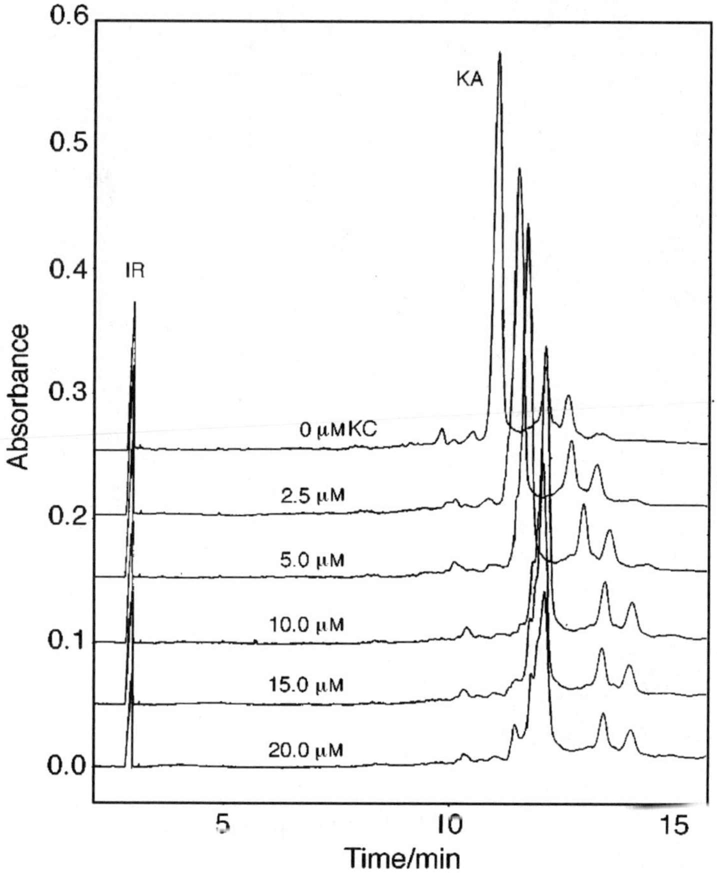

Figure 11 ACE of kedarcidin apoprotein (KA) as a function of the concentration of kedarcidin chromophore (KC) in a separation buffer of citrate-MES (pH 6.0). IR, internal reference (Liu et al., 1998). Reproduced by permission of The Royal Society of Chemistry.

concentrations ranging from 0.5 to 50 μM and monitoring the electrophoretic migration of injected EBP. A distinction between strongly and weakly binding EMPs could be made, indicating that ACE can be used to screen peptide-EBP interactions.

Conclusions

This chapter provides a general overview of the various ways CE can be used in the analysis of pharmaceutical proteins. In CE, diverse selectivities can be achieved with different modes of separation. The application of a particular mode depends on the analytical question posed. CIEF is especially useful when small changes or differences in variants of one protein have to be monitored. CZE can also be used to separate closely related protein species but also is well suited for the profiling of samples on impurities and degradation products. With CGE, biopharmaceuticals can be checked for the presence of other proteinaceous material. ACE approaches allow the determination of binding data of pharmaceutical proteins with potential ligands.

The quality control of biopharmaceuticals often presents a mixture analysis problem, and consequently separation techniques play (and will play) a pivotal role. As is demonstrated in this chapter, the application of CE for the characterization of protein pharmaceuticals has drastically increased during the past decade. Admittedly, the use of CE is still limited, although it should be realized that a substantial part of the analytical work on protein pharmaceuticals (including the accompanying method development) is carried out within the biopharmaceutical companies and often is not published in the scientific literature. In several of these industries, CE has already been recognized as a highly valuable tool that has enriched the arsenal of analytical methodologies that currently is applied to characterize biopharmaceuticals. Thus, CE provides a highly efficient and complementary technique for the evaluation of purity, stability, heterogeneity, and identity of pharmaceutical protein samples.

In biopharmaceutical analysis, until now, CE has mainly been explored as an efficient separation technique to distinguish proteins from other mixture constituents such as impurities or isoforms. However, CZE potentially could also provide information on the condition and structure of the protein itself. The electrophoretic mobility is governed by charge and size of the protein. This means that changes in protein charge by conjugation or changes in protein size by unfolding and/or aggregation will be reflected in the observed electrophoretic mobility. As in CZE separations, conditions can be chosen such that protein structure is not affected (nondenaturing circumstances), and the CE migration data truly reflect the native state of the protein (i.e., the state in which biopharmaceuticals are actually administered to patients). Moreover, next to protein identity, MS detection can also provide information on protein conformation. Because the charge states of a protein relate to its conformational state, mass spectra of proteins recorded using ESI in principle can be used to characterize the folding of the protein. So far, these aspects of CZE and CZE-MS have hardly been explored, but they most probably will be in the near future.

Abbreviations

ACE	Affinity capillary electrophoresis
BGE	Background electrolyte
BMP	Bone morphogenic protein
CAPS	3-Cyclohexylamino-1-propanesulfonic acid

CE	Capillary electrophoresis
CGE	Capillary gel electrophoresis
CHES	2-(Cyclohexylamino)ethanesulfonic acid
CIEF	Capillary isoelectric focusing
CZE	Capillary zone electrophoresis
DAPS	*N*-dodecyl-*N*,*N*-dimethyl-3-amino-1-propanesulfonate
DNase	Deoxyribonuclease
EBP	EPO binding protein
EOF	Electroosmotic flow
EPO	Erythropoietin
ES	Electrospray
ESI	Electrospray ionization
FDA	Food and Drug Administration
Gal	Galectin
h	Human
HEPES	4-(2-Hydroxyethyl)-1-piperazineethanesulfonic acid
HIC	Hydrophobic interaction chromatography
HIV	Human immunodeficiency virus
HPLC	High-performance liquid chromatography
HPMC	Hydroxypropylmethyl cellulose
HSA	Human serum albumin
I	Insulin
ICH	International Committee on Harmonization
i.d.	Internal diameter
IEC	Ion-exchange chromatography
IGF	Insulin-like growth factor
IgG	Immunoglobulin
IL	Interleukin
LIF	Laser-induced fluorescence
MAb	Monoclonal antibody
MES	4-Morpholinoethanesulfonic acid
M_r	Molecular weight
MS	Mass spectrometry
NAD(+)	Nicotinamide adenine dinucleotide
PA	Protective antigen
PAGE	Polyacrylamide gel electrophoresis
PVA	Poly(vinyl alcohol)
r	Recombinant

RSD	Relative standard deviation
SDS	Sodium dodecyl sulfate
TAPS	*N*-[Tris(hydroxymethyl)methyl]-3-aminopropanesulfonic acid
TEMED	1,2-Di(dimethylamino)ethane
TIC	Total ion current
TRIS	Tris(hydroxymethyl)aminomethane
UV	Ultraviolet
Vis	Visible

References

1. Belder D, Deege A, Husmann H, Kohler F, and Ludwig M. Cross-Linked Poly(Vinyl Alcohol) as Permanent Hydrophilic Column Coating for Capillary Electrophoresis. *Electrophoresis* 2001; 21:3813–3818
2. Benedek K, and Thiede S. High-Performance Capillary Electrophoresis of Proteins Using Sodium Dodecyl Sulfate-Poly(Ethylene Oxide). *J Chromatogr A* 1994; 676:209–217
3. Bietlot HP, and Girard M. Analysis of Recombinant Human Erythropoietin in Drug Formulations by High-Performance Capillary Electrophoresis. *J Chromatogr A* 1997; 759:177–184
4. Caldwell GW, McDonnell PA, Masucci JA, Johnson DL, and Jolliffe LK. Using Affinity Capillary Electrophoresis to Study the Interaction of the Extracellular Binding Domain of Erythropoietin Receptor with Peptides. *J Biochem Biophys Meth* 1999; 40:17–25
5. Catai JR, Somsen GW, and de Jong GJ. Efficient and Reproducible Analysis of Peptides by Capillary Electrophoresis Using Noncovalently Bilayer-Coated Capillaries. *Electrophoresis* 2004; 25:817–824
6. Chen AB, and Canova-Davis E. Capillary Electrophoresis in the Development of Recombinant Protein Biopharmaceuticals. *Chromatographia* 2001; 53:S7–S17
7. Chen AB, Rickel CA, Flanigan A, Hunt G, and Moorhouse KG. Comparison of Ampholytes Used for Slab Gel and Capillary Isoelectric Focusing of Recombinant Tissue-Type Plasminogen Activator Glycoforms. *J Chromatogr A* 1996; 744:279–284
8. Cifuentes A, Moreno-Arribas MV, de Frutos M, and Diez-Masa JC. Capillary Isoelectric Focusing of Erythropoietin Glycoforms and Its Comparison with Flat-Bed Isoelectric Focusing and Capillary Zone Electrophoresis. *J Chromatogr A* 1999; 830:453–463
9. Córdova E, Gao J, and Whitesides GM. Noncovalent Polycationic Coatings for Capillaries in Capillary Electrophoresis of Proteins. *Anal Chem* 1997; 69:1370–1379

10. Corradini D. Buffer Additives Other Than the Surfactant Sodium Dodecyl Sulfate for Protein Separations by Capillary Electrophoresis. *J Chromatogr B* 1997; 699:221–256
11. de Frutos M, Cifuentes A, and Diez-Masa JC. Differences in Capillary Electrophoresis Profiles of Urinary and Recombinant Erythropoietin. *Electrophoresis* 2003; 24:678–680
12. Eriksson HJC, Somsen GW, Hinrichs WLJ, Frijlink HW, and de Jong GJ. Characterization of Human Placental Alkaline Phosphatase by Activity and Protein Assays, Capillary Electrophoresis and Matrix-Assisted Laser Desorption/Ionization Time-of-Flight Mass Spectrometry. *J Chromatogr B* 2001; 755:311–319
13. Eriksson HJC, Wijngaard M, Hinrichs WLJ, Frijlink HW, Somsen GW, and de Jong GJ. Potential of Capillary Electrophoresis for the Monitoring of the Stability of Placental Alkaline Phosphatase. *J Pharm Biomed Anal* 2003; 31:351–357
14. Eriksson JHC, Mol R, Somsen GW, Hinrichs WLJ, Frijlink HW, and de Jong GJ. Feasibility of Nonvolatile Buffers in Capillary Electrophoresis-Electrospray Ionization-Mass Spectrometry of Proteins. *Electrophoresis* 2004; 25:43–49
15. Erim FB, Cifuentes A, Poppe H, and Kraak JC. Performance of a Physically Adsorbed High-Molecular-Mass Polyethyleneimine Layer as Coating for the Separation of Basic Proteins and Peptides by Capillary Electrophoresis. *J Chromatogr A* 1995; 708:356–361
16. Farchaus JW, Ribot WJ, Jendrek S, and Little SF. Fermentation, Purification, and Characterization of Protective Antigen from a Recombinant, Avirulent Strain of Bacillus anthracis. *Appl Environ Microbiol* 1998; 64:982–991
17. Felten C, Quan CP, Chen AB, Canova-Davis E, McNerney T, Goetzinger WK, and Karger BL. Use of Acidic and Basic pH and Calcium Ion Addition in the Capillary Zone Electrophoretic Characterization of Recombinant Human Deoxyribonuclease, A Complex Phosphoglycoprotein. *J Chromatogr A* 1999; 853:295–308
18. Figeys D, and Aebersold R. Capillary Electrophoresis of Peptides and Proteins at Neutral pH in Capillaries Covalently Coated with Polyethyleneimine. *J Chromatogr B* 1997; 695:163–168
19. Guttman A, Nolan JA, and Cooke N. Capillary Sodium Dodecyl-Sulfate Gel-Electrophoresis of Proteins. *J Chromatogr* 1993; 632:171–175
20. Guzman NA, Ali H, Moschera J, Iqbal K, and Malick AW. Assessment of Capillary Electrophoresis in Pharmaceutical Applications: Analysis and Quantification of a Recombinant Cytokine in an Injectable Dosage Form. *J Chromatogr* 1991; 559:307–315
21. Guzman NA, Moschera J, Iqbal K, and Malick AW. Effect of Buffer Constituents on the Determination of Therapeutic Proteins by Capillary Electrophoresis. *J Chromatogr* 1992; 608:197–204

22. Gysler J, Helk B, Dambacher S, Tjaden UR, and van der Greef J. Characterization of Recombinant Cytokine Fragments Using Isotachophoresis-Capillary Zone Electrophoresis, Reversed-Phase High Performance Liquid Chromatography, and Mass Spectrometry. *Pharm Res* 1999; 16:695–701

23. Heegaard NHH, and Kennedy RT. Identification, Quantitation, and Characterization of Biomolecules by Capillary Electrophoretic Analysis of Binding Interactions. *Electrophoresis* 1999; 20:3122–3133

24. Heegaard NHH, Nilsson S, and Guzman NA. Affinity Capillary Electrophoresis: Important Application Areas and Some Recent Developments. *J Chromatogr B* 1998; 715:29–54

25. Horvath J, and Dolnik V. Polymer Wall Coatings for Capillary Electrophoresis. *Electrophoresis* 2001; 22:644–655

26. Hu S, and Dovichi NJ. Capillary Electrophoresis for the Analysis of Biopolymers. *Anal Chem* 2002; 74:2833–2850

27. Hunt G, and Nashabeh W. Capillary Electrophoresis Sodium Dodecyl Sulfate Nongel Sieving Analysis of a Therapeutic Recombinant Monoclonal Antibody: A Biotechnology Perspective. *Anal Chem* 1999; 71:2390–2397

28. Hunt G, Moorhouse KG, and Chen AB. Capillary Isoelectric Focusing and Sodium Dodecyl Sulfate Capillary Gel Electrophoresis of Recombinant Humanized Monoclonal Antibody HER2. *J Chromatogr A* 1996; 744:295–301

29. Janini GM, and Issac HJ. Selection of Buffer in Capillary Zone Electrophoresis: Application to Peptide and Protein Analysis. *Chromatographia* 2001; 53:S18–S26

30. Jenkins MA, and Guerin MD. Capillary Electrophoresis Procedures for Serum Protein Analysis: Comparison with Established Techniques. *J Chromatogr B* 1997; 699:257–268

31. Jensen PK, Harrata AK, and Lee CS. Monitoring Protein Refolding Induced by Disulfide Formation Using Capillary Isoelectric Focusing Electrospray Ionization Mass Spectrometry. *Anal Chem* 1998; 70:2044–2049

32. Katayama H, Ishihama Y, and Asakawa N. Stable Capillary Coating with Successive Multiple Ionic Polymer Layers. *Anal Chem* 1998; 70:2254–2260

33. Kinoshita M, Murakami E, Oda Y, Funakubo T, Kawakami D, Kakehi K, Kawasaki N, Morimoto K, and Hayakawa T. Comparative Studies on the Analysis of Glycosylation Heterogeneity of Sialic Acid Containing Glycoproteins Using Capillary Electrophoresis. *J Chromatogr A* 2000; 866:261–271

34. Klyushnichenko VE, Koulich DM, Yakimov SA, Maltsev KV, Grishina GA, Nazimov IV, and Wulfson AN. Recombinant Human Insulin 3.

High-Performance Liquid-Chromatography and High-Performance Capillary Electrophoresis Control in the Analysis of Step-By-Step Production of Recombinant Human Insulin. *J Chromatogr A* 1994; 661:83–92

35. Klyushnichenko V, Tishkov V, and Kula MR. Rapid SDS-GEL Capillary Electrophoresis for the Analysis of Recombinant NADP(+)-Dependent Formate Dehydrogenase During Expression in Escherichia Coli Cells and Its Purification. *J Biotechnol* 1997; 58:187–195
36. Knüver-Hopf J, and Mohr H. Differences Between Natural and Recombinant Interleukin-2 Revealed by Gel-Electrophoresis and Capillary Electrophoresis. *J Chromatogr A* 1995; 717:71–74
37. Kubach J, and Grimm R. Non-native Capillary Isoelectric Focusing for the Analysis of the Microheterogeneity of Glycoproteins. *J Chromatogr A* 1996; 737:281–289
38. Kunkel A, Gunter S, Dette C, and Watzig H. Quantitation of Insulin by Capillary Electrophoresis and High-Performance Liquid Chromatography—Method Comparison and Validation. *J Chromatogr A* 1997; 781:445–455
39. Lagu AL. Applications of Capillary Electrophoresis in Biotechnology. *Electrophoresis* 1999; 20:3145–3155
40. Lee HG. High-Performance Sodium Dodecyl Sulfate-Capillary Gel Electrophoresis of Antibodies and Antibody Fragments. *J Immunol Methods* 2000; 234:71–81
41. Liu JP, Abid S, Hail ME, Lee MS, Hangeland J, and Zein N. Use of Affinity Capillary Electrophoresis for the Study of Protein and Drug Interactions. *Analyst* 1998; 123:1455–1459
42. Lookabaugh M, Biswas M, and Krull IS. Quantitation of Insulin Injection by High-Performance Liquid-Chromatography and High-Performance Capillary Electrophoresis. *J Chromatogr* 1991; 549:357–366
43. Lopez-Soto-Yarritu P, Diez-Masa JC, Cifuentes A, and de Frutos M. Improved Capillary Isoelectric Focusing Method for Recombinant Erythropoietin Analysis. *J Chromatogr A* 2002a; 968:221–228
44. Lopez-Soto-Yarritu PL, Diez-Masa JC, de Frutos M, and Cifuentes A. Comparison of Different Capillary Electrophoresis Methods for Analysis of Recombinant Erythropoietin Glycoforms. *J Sep Sci* 2002b; 25:1112–1118
45. Ma S, and Nashabeh W. Analysis of Protein Therapeutics by Capillary Electrophoresis. *Chromatographia* 2001; 53:S75–S89
46. Mandrup G. Rugged Method for the Determination of Deamidation Products in Insulin Solutions by Free Zone Capillary Electrophoresis Using an Untreated Fused-Silica Capillary. *J Chromatogr* 1992; 604:267–281

47. Nashabeh W, Greve KF, Kirby D, Foret F, Karger BL, Reifsnyder DH, and Builder SE. Incorporation of Hydrophobic Selectivity in Capillary Electrophoresis: Analysis of Recombinant Insulin-like Growth-Factor-I Variants. *Anal Chem* 1994; 66:2148–2154

48. Okun VM, Moser R, Blaas D, and Kenndler E. Complexes Between Monoclonal Antibodies and Receptor Fragments with a Common Cold Virus: Determination of Stoichiometry by Capillary Electrophoresis. *Anal Chem* 2001a; 73:3900–3906

49. Okun VM, Moser R, Ronacher B, Kenndler E, and Blaas D. VLDL Receptor Fragments of Different Lengths Bind to Human Rhinovirus HRV2 with Different Stoichiometry. An Analysis of Virus-Receptor Complexes by Capillary Electrophoresis. *J Biol Chem* 2001b; 276: 1057–1062

50. Patrick JS, and Lagu AL. Review Applications of Capillary Electrophoresis to the Analysis of Biotechnology-Derived Therapeutic Proteins. *Electrophoresis* 2001; 22:4179–4196

51. Pawlak KP, Palacios O, Capdevila M, Gonzalez-Duarte P, and Lobinski R. Monitoring of the Metal Displacement from the Recombinant Mouse Liver Metallothionein Zn-7-Complex by Capillary Zone Electrophoresis with Electrospray MS Detection. *Talanta* 2002; 57:1011–1017

52. Righetti PG, Gelfi C, Bossi A, Olivieri E, Castelletti L, Verzola B, and Stoyanov AV. Capillary Electrophoresis of Peptides and Proteins in Isoelectric Buffers: An Update. *Electrophoresis* 2000; 21:4046–4053

53. Righetti PG, Gelfi C, Verzola B, and Castelletti L. The State of the Art of Dynamic Coatings. *Electrophoresis* 2001; 22:603–611

54. Roddy TP, Molnar TE, McKean RE, and Foley JP. Method of Analysis of Recombinant Acidic Fibroblast Growth Factor by Capillary Electrophoresis. *J Chromatogr B* 1997; 695:49–58

55. Sanz-Nebot V, Benavente F, Vallverdú A, Guzman NA, and Barbosa J. Separation of Recombinant Human Erythropoietin Glycoforms by Capillary Electrophoresis Using Volatile Electrolytes. Assessment of Mass Spectrometry for the Characterization of Erythropoietin Glycoforms. *Anal Chem* 2003; 75:5220–5229

56. Schlenck A, Schiele F, Barbier A, Shuvaev VV, Visvikis S, and Siest G. Capillary Electrophoretic Analysis of Recombinant Human Apolipoprotein E—Calibration Mode of a Protein Reference Material. *J Chromatogr A* 1999; 853:237–241

57. Schmitt-Kopplin P, and Frommberger M. Capillary Electrophoresis-Mass Spectrometry: 15 Years of Developments and Applications. *Electrophoresis* 2003; 24:3837–3867

58. Sergeev NV, Gloukhova NS, Nazimov IV, Gulyaev VA, Shvets SV, Donetsky IA, and Miroshnikov AI. Monitoring of Recombinant Human Insulin Production by Narrow-Bore Reversed-Phase High-Performance

Liquid Chromatography, High-Performance Capillary Electrophoresis and Matrix-Assisted Laser Desorption Ionisation Time-of-Flight Mass Spectrometry. *J Chromatogr A* 2001; 907:131–144

59. Shimura K, and Kasai K. Affinity Capillary Electrophoresis: A Sensitive Tool for the Study of Molecular Interactions and Its Use in Microscale Analyses. *Anal Biochem* 1997; 251:1–16
60. Shimura K, Uchiyama N, and Kasai K. Prevention of Evaporation of Small-Volume Sample Solutions for Capillary Electrophoresis Using a Mineral-Oil Overlay. *Electrophoresis* 2001; 22:3471–3477
61. Shimura K, Arata Y, Uchiyama N, Hirabayashi J, and Kasai K-i. Determination of the Affinity Constants of Recombinant Human Galectin-1 and -3 for Simple Saccharides by Capillary Affinophoresis. *J Chromatogr B* 2002; 768:199–210
62. Storring PL, Das REG, Mulders JWM, and Halder M. Physicochemical Methods for Predicting the Biological Potency of Recombinant Follicle Stimulating Hormone: An International Collaborative Study of Isoelectric Focusing and Capillary Zone Electrophoresis. *Biologicals* 2002; 30:217–234
63. Tang Q, Harrata AK, and Lee CS. Two-Dimensional Analysis of Recombinant E-Coli Proteins Using Capillary Isoelectric Focusing Electrospray Ionization Mass Spectrometry. *Anal Chem* 1997; 69:3177–3182
64. Tang S, Nesta DP, Maneri LR, and Anumula KR. A Method for Routine Analysis of Recombinant Immunoglobulins (rIgGs) by Capillary Isoelectric Focusing (cIEF). *J Pharm Biomed Anal* 1999; 19:569–583
65. Thorne JM, Goetzinger WK, Chen AB, Moorhouse KG, and Karger BL. Examination of Capillary Zone Electrophoresis, Capillary Isoelectric Focusing and Sodium Dodecyl Sulfate Capillary Electrophoresis for the Analysis of Recombinant Tissue Plasminogen Activator. *J Chromatogr A* 1996; 744:155–165
66. Tran NT, Taverna M, Chevalier M, and Ferrier D. One-Step Capillary Isoelectric Focusing for the Separation of the Recombinant Human Immunodeficiency Virus Envelope Glycoprotein Glycoforms. *J Chromatogr A* 2000; 866:121–135
67. Tsuji K. High-Performance Capillary Electrophoresis of Proteins—Sodium Dodecyl-Sulfate Polyacrylamide Gel-Filled Capillary Column for the Determination of Recombinant Biotechnology-Derived Proteins. *J Chromatogr* 1991; 550:823–830
68. Wang Y, and Dubin PL. Capillary Modification by Noncovalent Polycation Adsorption: Effects of Polymer Molecular Weight and Adsorption Ionic Strength. *Anal Chem* 1999; 71:3463–3468

69. Watson E, and Yao F. Capillary Electrophoretic Separation of Recombinant Granulocyte-Colony-Stimulating Factor Glycoforms. *J Chromatogr* 1993; 630:442–446

70. Yeung B, Porter TJ, and Vath JE. Direct Isoform Analysis of High-Mannose Containing Glycoproteins by On-Line Capillary Electrophoresis Electrospray Mass Spectrometry. *Anal Chem* 1997; 69:2510–2516

71. Yim K, Abrams J, and Hsu A. Capillary Zone Electrophoretic Resolution of Recombinant Human Bone Morphogenetic Protein-2 Glycoforms. An Investigation into the Separation Mechanisms for an Exquisite Separation. *J Chromatogr A* 1995; 716:401–412

72. Yu T, Xiao Y, Bai Y, Ru Q, Luo G, Dierich MP, and Chen Y-H. Human Interferon-[beta] Inhibits Binding of HIV-1 gp41 to Lymphocyte and Monocyte Cells and Binds the Potential Receptor Protein P50 for HIV-1 gp41. *Immun Lett* 2000; 73:19–22

10

Analytical Ultracentrifugation

JOHN S. PHILO

Alliance Protein Laboratories, Thousand Oaks, CA

Introduction

Analytical ultracentrifugation (AUC) is one of the oldest methods for characterizing the solution structure and conformation of proteins. Indeed, the pioneering AUC studies by Thé Svedberg in the 1930s first established that proteins were unique molecular entities with well defined molecular masses and conformations, and not amorphous colloids, as was thought at that time. Despite this long history, however, this technique continues to see new innovations in instrumentation and particularly in data analysis. Those innovations have largely been driven by the need to characterize therapeutic proteins for the biotechnology industry, and today those applications are far more useful and powerful than they were just a few years ago.

An analytical centrifuge is simply a high-speed centrifuge to which an optical system is added. When used with specialized analytical rotors and cells, the optical system can measure the distribution of the macromolecule within the cell while the rotor is spinning. Those concentration distributions can be predicted based on rather simple physical principles, and ironically that simplicity is one important reason why AUC methods are powerful. This chapter will introduce the two primary AUC methods, sedimentation velocity and sedimentation equilibrium, including a brief discussion of their underlying theory. After some description of currently available equipment and data analysis approaches, several case studies will illustrate practical applications to protein therapeutics. The reader is referred to Lebowitz et al. (2002) for an excellent review and tutorial on modern AUC methods and to the March 2004 issue (vol. 108) of *Biophysical Chemistry*, which has many articles on state-of-the-art AUC applications.

Theory and Principles of Sedimentation Velocity

In a sedimentation velocity experiment, the centrifuge rotor is spun at such a high speed that all of the protein molecules will be driven completely to the outside of the rotor within a few hours. The rate at which this sedimentation occurs, which we characterize by a molecular parameter called the sedimentation coefficient, is measured via a series of optical scans at different times during the run. When multiple independent molecular species are present (e.g., a protein monomer plus irreversible aggregates), each species sediments at its own characteristic rate, so species are separated on the basis of their sedimentation coefficients. As we will see in more detail below, those sedimentation coefficients depend on molecular mass, on molecular shape (conformation), on the solvent density and viscosity, and (at high concentrations) also on the protein concentration. Thus, sedimentation velocity is fundamentally a separation method, and in many ways it is analogous to size-exclusion chromatography (SEC, also known as gel filtration). As will hopefully become clear, however, sedimentation velocity has many significant advantages over SEC.

The Buoyant Mass

One key concept for all AUC methods is that of the *buoyant mass*. Consider a molecule in solution placed in the spinning rotor. The centrifugal force acting on this molecule is strongly opposed by a buoyant force, which is proportional to the mass of solvent this molecule displaces (remember Archimedes!). Thus, the *effective* mass that produces a net sedimentation force is not the true molecular mass M, but rather it is the difference between M and the mass of the displaced solvent, a quantity we call the buoyant mass, M_b.

The size of the buoyancy effect is related to the density of the solvent and the volume occupied by our macromolecule. If the molecule has a higher density than solvent, M_b is positive, but for molecules with a lower density than solvent (like many lipoproteins in water), M_b is negative, and they will actually float toward the center of the rotor rather than sedimenting!

Rather than working with the density of the macromolecule, in AUC it is traditional instead to consider the *partial specific volume*, which is given the symbol $\overline{v}$, defined as the increase in the volume of the solution (in milliliters) per gram of added solute. The partial specific volume is essentially the inverse of the molecular density, but it also takes into account the fact that our macromolecule can alter the density of the water near its surface. With this definition, and for a solvent with density ρ, the buoyant mass can be related to the true mass through the formula $M_b = M \times (1 - \overline{v}\rho)$.

Sedimentation Coefficient

When a molecule is spun in the rotor, the centrifugal force F_c accelerates it in the radial direction. This motion is opposed by hydrodynamic friction, and very quickly the molecule reaches a final terminal velocity, v_t, where the centrifugal force is exactly balanced by the frictional force F_f. The sedimentation coefficient of the molecule, s, then is defined as the ratio of this terminal velocity to the applied centrifugal force.

Because M_b/N_A is the buoyant mass of one molecule (N_A is Avogadro's number), and the angular acceleration is given by $\omega^2 r$, where ω is the angular velocity (rad/s) and r is the radius, then from Newton's first law $F = ma$ we get $F_c = (M_b/N_A)\omega^2 r$. Also, from simple mechanics we have $F_f = -fv_t$, where f is the translational frictional coefficient. Therefore, from the definition above we find

$$s \equiv \frac{v_t}{\omega^2 r} = \frac{M_b}{N_A f} = \frac{M(1 - \bar{v}\rho)}{N_A f} \qquad (1)$$

Thus, the sedimentation coefficient depends on three intrinsic molecular properties: mass, shape (because the frictional coefficient depends on shape), and partial specific volume. It also depends on two less-interesting solvent properties: density and viscosity (f is proportional to viscosity). Therefore, in order to have a value that is related only to molecular properties, it is customary to convert the experimental value (often called a "raw" sedimentation coefficient) to a standard condition, the value that would be observed for that molecule in pure water at 20 °C. That standardized value is given the symbol $s_{20,w}$. These $s_{20,w}$ values are then absolute and can be compared directly (for different molecules, or the same molecule under different experimental conditions). This conversion is done through the formula

$$s_{20,w} = s_{\text{raw}} \frac{(1 - \bar{v}_{20,w}\rho_{20,w})}{(1 - \bar{v}_{T,b}\rho_{T,b})} \frac{\eta_{T,b}}{\eta_{20,w}} \qquad (2)$$

where η is the solvent viscosity, and the subscripts indicate values for either the experimental temperature and buffer or for water at 20 °C. The fundamental units for sedimentation coefficients are seconds, but because typical values lie in the range 10^{-13} to 10^{-12} seconds, the values are customarily given in units of Svedbergs, with $1\ \text{S} \equiv 10^{-13}$ seconds.

As will be discussed further in the section "Concentration Dependence and Solution Nonideality" below, at high protein concentrations the observed sedimentation coefficient will be smaller than in dilute solution. When sedimentation coefficients show a significant concentration dependence, it is customary to measure $s_{20,w}$ values at several concentrations and then extrapolate to zero concentration, and that extrapolated value is designated $s^0_{20,w}$. Often, this extrapolation is done by fitting to a simple linear dependence on the protein concentration, c, (in mg/mL) using the relation $s_{20,w}(c) = s^0_{20,w}/(1 + k_s c)$ (Rowe, 1992).

Boundary Motion and Shape

A sketch of a sedimentation velocity cell and a scan at a time when the boundary is near the midpoint of the cell are shown in Figure 1. The simulated boundary shapes and motion for a pure monoclonal antibody are shown in Figure 2. As shown by the dotted curves, if diffusion were absent, then the boundaries would be infinitely sharp. For a species of sedimentation coefficient s, the boundary positions r_b at any time t in the run are given by

$$\ln(r_b/r_m) = s\int_0^t \omega^2 t' dt' \tag{3}$$

where r_m is the meniscus position. The integral of ω^2 over time is necessary to account for the fact that the rotor does not accelerate instantaneously to the final speed; if it did, the right-hand term would simply be $s\omega^2 t$.

In a real experiment, diffusion of course spreads out the boundary, and for fairly homogeneous samples it is possible to use that boundary spreading to determine the diffusion coefficient, D, as well as the sedimentation coefficient. When the diffusion coefficient of a molecule is

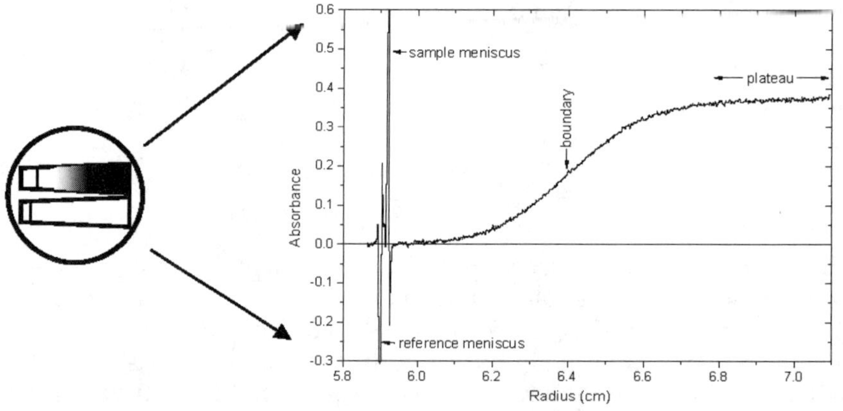

Figure 1 Sketch of a sedimentation velocity experiment. The dual-channel velocity cell is shown at left, with sample in the upper sector-shaped channel and the corresponding buffer in the lower (reference) channel. An absorbance scan taken near the middle of a run is shown at right. The off-scale positive and negative peaks at the left of the scan are caused by the air-liquid interface (meniscus) for the sample and reference channels, respectively. Sedimentation has removed the protein from the left-hand portion of the cell, giving a sedimentation boundary at ~6.45 cm.

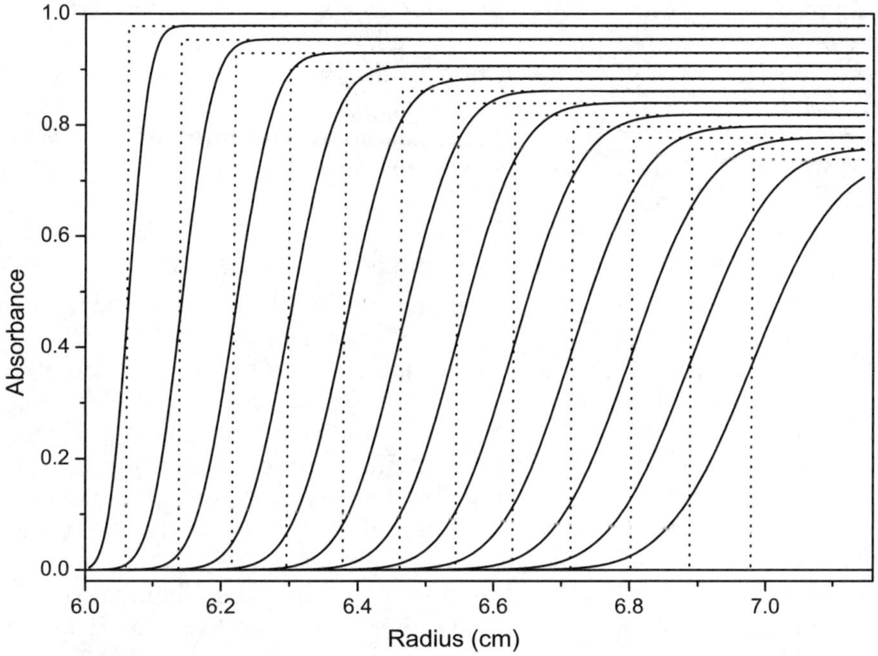

Figure 2 Simulated noise-free velocity data for a monoclonal antibody at 45,000 rpm ($s = 6.4$ S, 150 kDa, loading concentration giving 1 OD, meniscus at 6.0 cm) at an interval of 15 minutes between scans. The dotted lines indicate what would be observed if diffusion were absent; the solid lines include the simulated diffusion corresponding to the 150 kDa molecular mass.

known (from sedimentation velocity or dynamic light scattering) as well as its sedimentation coefficient, its molecular mass can be calculated by the relation

$$M(1 - \bar{v}\rho) = \frac{sRT}{D} \tag{4}$$

where R is the gas constant and T is the temperature (Kelvin). Through this relation, we see that only two of the three quantities s, D, and M are independent; the third can be calculated when any two are known.

One notable feature of the simulated curves in Figure 2 is the systematic decline of the concentration in the plateau region to the right of the boundary. This sample dilution is due to two effects: (1) the sector shape of the channel means that the molecules in each radial region of the cell move into a region of larger volume as they sediment, so conservation of mass requires the concentration to drop; (2) because the velocity is proportional to r, the molecules at higher radius always move slightly faster and pull away, diluting the sample in the radial dimension. The total dilution is about 25% when the leading edge of the boundary reaches the base of the cell.

There is no analytical solution to the differential equations governing the flow of molecules in a velocity experiment. However, highly accurate boundary shapes can be calculated by computer using numerical integration techniques, and various approximate analytical solutions are also useful.

Frictional Coefficient and Molecular Shape

Sedimentation (or diffusion) coefficients provide useful information about molecular shape or compactness when M is also known. One thing we must realize, however, is that proteins have a considerable amount of water bound to their surface. Typically the hydration, δ, is 0.3–0.4 g of water per gram of protein (which corresponds to a surface layer 1–2 molecules thick). When proteins sediment, this bound water adds to the hydrodynamic friction, but because it is neutrally buoyant, it contributes nothing to the net sedimentation force. Thus, the bound water adds only friction and slows sedimentation.

The first step in interpretation is calculating the frictional coefficient f using Equation 1. This value is then compared to f_0, the frictional coefficient for an anhydrous sphere of equivalent volume (a sphere has the minimum possible friction). From the Stokes equation, $f_0 = 6\pi\eta R_0$, where η is the viscosity and R_0 is the radius of the anhydrous sphere. Recognizing that the protein volume can be estimated as $M\bar{v}$, then $R_0 = (3M\bar{v}/4\pi N_A)^{1/3}$.

If we then calculate the ratio f/f_0, we get a quantity that depends only on molecular shape and hydration; this ratio is independent of solvent viscosity and also the absolute size of the molecule. Compact, globular proteins typically have $f/f_0 = 1.15$ to 1.3, more asymmetric and extended structures like antibodies have $f/f_0 \sim 1.6$, and a largely unfolded protein will have $f/f_0 \geq 2$.

Because f/f_0 depends on both shape and hydration, it is conceptually useful to separate the hydration expansion effect from that due to molecular shape:

$$\frac{f}{f_0} = \frac{f}{f_{\text{shape}}} \times \frac{f}{f_{\text{hyd}}} = \frac{f}{f_{\text{shape}}} \times \left(\frac{\bar{v} + \delta\bar{v}_s}{\bar{v}}\right)^{1/3} \tag{5}$$

where $\bar{v}_s$ is the partial specific volume of the solvent. The geometric f/f_{shape} factors can readily be calculated for simple shapes like ellipsoids or cylinders. For example, a prolate ellipsoid (cigar shape) with a major axis 5 times the minor axis has $f/f_{\text{shape}} = 1.25$. When the three-dimensional structure of a protein is known, its frictional coefficient can be calculated fairly accurately by modeling its detailed shape with a surface consisting of spherical beads (de la Torre, 2001).

Concentration Dependence and Solution Nonideality

As protein concentration increases and the protein molecules start to occupy a significant fraction of the total solution volume, the sedimentation coefficient for any particular species decreases from the value seen in dilute solution. Further, when multiple species are present, they no longer sediment independently, with each modifying the apparent sedimentation coefficients (and even the apparent concentration) of the others (van Holde, 1971). These "solution nonideality" or "molecular crowding" (Minton, 2001) effects are always quite significant at protein concentrations of 10 mg/mL or above. For globular proteins at 1 mg/mL and physiological ionic strength, the observed sedimentation coefficient will typically be only 0.9% below its value at infinite dilution (Cantor and Schimmel, 1990), with larger effects for highly extended conformations (Rowe, 1992).

For charged solutes like proteins, there is an electrostatic component to the nonideality effects, in addition to the excluded-volume contribution (Yphantis and Roark, 1972). This electrostatic effect is suppressed as the ionic strength increases, and for proteins at ionic strength ≥100 mM and not too far from their isoelectric point, the electrostatic effects are usually not significant. However, for example a peptide or protein dissolved only in dilute acid is likely to show strong nonideality even at 0.1 mg/mL.

Proteins undergoing reversible self-association will also, of course, exhibit concentration-dependent sedimentation. However, self-association always produces an *increase* in the overall weight-average sedimentation coefficient of the sample (van Holde, 1971). Consequently, an observation of an increasing sedimentation coefficient with concentration is definitive proof of reversible association.

Theory and Principles of Sedimentation Equilibrium

Although sedimentation velocity is excellent for characterizing molecular conformation and for separating mixtures, sedimentation equilibrium is a complementary approach whose strengths are determining solution molecular mass and characterizing the stoichiometry and equilibrium constants for reversible binding equilibria (self-association or binding interactions between two different molecules).

Rather than the very strong centrifugal force that is applied in sedimentation velocity experiments, for sedimentation equilibrium a much smaller force (lower rotor speed) is employed. Although the concentration of the protein does build up at the outside of the rotor, with this gentle force all the species present remain in solution (no pellet is formed). Eventually, the distribution of protein within the centrifuge cell reaches a steady-state, where sedimentation is exactly balanced by diffusion at every position in the cell. (Indeed, what today we usually abbreviate as "sedimentation equilibrium" should properly be called "sedimentation-diffusion equilibrium").

Under this equilibrium condition, two key properties are true: (1) the concentration gradient depends only on molecular mass (independent of molecular shape); and (2) the sample also maintains mass-action equilibrium for any reversible association reactions that may occur. Those two properties largely define the utility and power of this technique.

Equilibrium Distribution for a Single Ideal Species

Equilibrium is reached when the flux of solute due to sedimentation is canceled by the opposite flux due to diffusion. From that simple principle, one can derive the following expression for the concentration of a single solute species (van Holde, 1971), under conditions where the solution is "ideal" (solutes effectively occupy no volume, and have no long-range interactions, like an ideal gas):

$$c(r) = c(r_0)e^{M_b\omega^2/2RT\ (r^2 - r_0^2)} \tag{6}$$

where $c(r)$ is the concentration at radius r, and r_0 is a reference radial position (chosen at some arbitrary position within the sample). This expression shows that the curvature of the exponential function is directly proportional to the buoyant mass. Notably, because at equilibrium there is no net flow, this distribution is independent of the hydrodynamic friction coefficient and thus is independent of molecular shape.

Time to Reach Equilibrium

For a sample of column height (radial extent) h cm, the time (seconds) to reach a concentration distribution within ~0.1% of the final (infinite-time) distribution is given approximately by

$$t_{\text{equil}} = \frac{0.7h^2}{D} \tag{7}$$

where D is the diffusion coefficient (in cm^2/s). For globular proteins, D can be roughly estimated by $D = 3 \times 10^{-5} M^{-1/3}$ for M in daltons (van Holde and Baldwin, 1958).

Because of the strong dependence on sample size, sedimentation equilibrium samples typically have $h \cong 0.3$ cm rather than the $h \cong 1.2$ cm used in sedimentation velocity. At that column height, proteins of ~20 kDa will typically reach equilibrium in ~12 hours, whereas a ~200 kDa protein will take ~24 hours. Also note that D in Equation 7 depends on solvent viscosity; viscous additives like glycerol can greatly extend the time to reach equilibrium.

Self-Association

Consider a monomer that reversibly associates to form an N-mer. Each species will distribute according to its buoyant mass, so the total concentration will be given by the sum, or

$$c_{\text{tot}}(r) = c_1(r_0)e^{M_b\omega^2/2RT\,(r^2-r_0^2)} + c_N(r_0)e^{NM_b\omega^2/2RT\,(r^2-r_0^2)} \quad (8)$$

where c_1 and c_N are the concentration of monomer and N-mer, respectively. However, these species concentrations are interrelated by the equilibrium association constant, K_a (or equivalently the dissociation constant, K_d):

$$K_a = \frac{c_N}{c_1^N};\ K_d = \frac{1}{K_a} = \frac{c_1^N}{c_N} \quad (9)$$

Thus, we can rewrite Equation 9 as

$$c_{\text{tot}}(r) = c_1(r_0)e^{M_b\omega^2/2RT\,(r^2-r_0^2)} + K_a c_1^N(r_0)e^{NM_b\omega^2/2RT\,(r^2-r_0^2)} \quad (10)$$

Therefore, the equilibrium distribution for a self-associating protein is directly related to the quantities of interest: the monomer mass, the stoichiometry N, and the strength of the association. For a more complicated association scheme such as monomer ↔ dimer ↔ tetramer, we can simply extend the same approach, adding an additional term for each new species and its equilibrium constant with respect to the monomer.

Solution Nonideality

The theory and equations above apply to very dilute solutions, but when the protein occupies a significant fraction of the total volume or when strong electrostatic interactions are present, solution nonideality effects become important. When the nonideality is not too strong, its effects can be treated by allowing the *apparent* molecular mass to vary with concentration as

$$\frac{1}{M_{\text{app}}} = \frac{1}{M_{\text{true}}} + Bc \quad (11)$$

where B is a parameter called the second virial coefficient that measures the strength of the nonideality effects (an ideal solution corresponds to $B = 0$). The value of B is usually positive, which causes M_{app} to drop as the concentration increases. Although an increase in M_{app} with concentration due to self-association can be parameterized via a negative value of B (and this is often done for light-scattering data), customarily for AUC data the self-association is treated through specific association models as discussed above. When nonideality is considered, the concentration distribution for a single species (Equation 6 above) becomes (Haschemeyer and Bowers, 1970)

$$c(r) = c(r_0)e^{M_b\omega^2/2RT\,(r^2 - r_0^2) - BM[c(r) - c(r_0)]} \tag{12}$$

and the nonideality terms can be added to Equation 10 in a similar fashion.

Excluded volume and electrostatic effects both contribute to the value of B. The excluded volume contribution is always repulsive (positive values). For spherical particles, the excluded volume contribution is given by $4\bar{v}/M$ (Tanford, 1961). That value implies that for typical globular proteins, M_{app} will be about 3% below the true value at a protein concentration of 10 mg/mL. For highly asymmetric or unfolded proteins, the excluded volume contributions can be much larger.

For solutions containing only a single type of protein, the electrostatic effects also must always be repulsive. Electrostatic nonideality is suppressed as the ionic strength increases and generally is comparable to the excluded volume effects at physiological ionic strength and when the solution pH is within 1–2 units of the protein's isoelectric point (Yphantis and Roark, 1972).

Instrumentation and Data Analysis Software

Beckman Coulter (Fullerton, CA, USA) produces the only commercially available analytical ultracentrifuges, the ProteomeLab XL-A/XL-I Protein Characterization System. The XL-A model uses an absorbance optical system, which covers wavelengths from 190 to 800 nm. The XL-I model has in addition a Rayleigh interference optical system, which measures gradients in refractive index. Both optical systems measure the difference between the protein sample and a buffer reference (preferably a dialysate). One advantage of the interference optics is that it detects molecules like carbohydrates that lack distinctive absorbance bands. However, for protein concentrations below ~100 µg/mL, the signal/noise of the absorbance data is usually superior (assuming the buffer absorbance permits working at 230 nm or lower). The interference system has a very high linearity and dynamic range and can measure samples at considerably higher concentrations than the absorbance system, up to ~50 mg/mL. (Note, however, that the interpretation of data at such high concentrations can be quite difficult because the solution nonideality is so strong.) Data acquisition by the interference optics is much more rapid than the absorbance optics (about 5 seconds to scan each sample *vs.* ~80 seconds minimum for absorbance), which can be a significant advantage for sedimentation velocity experiments. A fluorescence optical system for the XL-A/I has been developed by Tom Laue's lab at the

University of New Hampshire, which will hopefully be available commercially in the near future as a third-party add-on.

Like most modern instruments, these instruments and their data acquisition are controlled by an external computer. Once the samples and rotor are loaded and the run started, little human intervention is required. The cleaning, assembly, and filling of sample cells is, however, a manual operation, as is data analysis (which takes some time to master), so overall this is still a technique that is fairly labor-intensive for highly trained personnel.

Rotors and Cells

Two types of analytical rotors are available. The AN60Ti rotor holds up to 3 measurement cells and has a top speed of 60,000 rpm. The AN50Ti rotor holds up to 7 cells and thus allows more than twice the sample throughput, but it has a top speed of only 50,000 rpm. For proteins smaller than ~50 kDa, limiting the rotor speed to 50,000 rpm will significantly reduce the resolving power of sedimentation velocity experiments, and the ability to distinguish monomer from dimer will be compromised. The relatively slow data acquisition for absorbance data also strongly limits the number of scans that can be obtained for each sample when seven cells are used in the AN50Ti. Thus, overall for velocity measurements, the higher throughput advantage of the AN50Ti may come with a trade-off of lower quality data.

The analytical cells all consist of a centerpiece that has two or more channels that hold the liquid. This centerpiece is then sandwiched between two optical windows, and this stack of components is squeezed together within a cylindrical aluminum housing to form a liquid-tight chamber. The windows are made from either quartz or synthetic sapphire. Sapphire is required for the interference optics and also works for absorbance measurements at 230 nm or above. Only quartz windows allow absorbance measurements below 230 nm.

For sedimentation velocity experiments, the centerpieces have two wedge-shaped channels about 1.4 cm long, as shown in Figure 1. The wedge shape allows the molecules to sediment without hitting the walls. The two-channel centerpiece that is most commonly used with proteins is made from Epon epoxy resin mixed with carbon. A version made from aluminum is also available, and officially this is the only version that can be used at rotor speeds above 44,000 rpm. However, these aluminum centerpieces have been found to induce aggregation of some proteins, so many protein labs routinely use the carbon-Epon centerpieces at speeds up to 60,000 rpm, accepting the fact that they will have to replace centerpieces more often. The standard centerpieces have an optical pathlength of 12 mm and use 400–430 μL of sample for a velocity experiment. A 3-mm pathlength carbon-Epon version is also available (fourfold lower sample volume). That 3-mm centerpiece allows fourfold higher concentrations to be measured and is quite useful for measurements over a large range of concentrations.

The two-channel centerpieces can also be used for sedimentation equilibrium, but then they are usually filled with only 100–110 μL of sample (about 3 mm of column height) so they come to equilibrium more rapidly (Equation 7). Because sedimentation equilibrium protocols usually involve measuring many aliquots of the same material loaded over a wide range of concentrations, specialized equilibrium cells holding larger numbers of small-volume samples are often used. The six-channel cells hold three sample-reference pairs with volumes up to ~120 μL maximum (3-mm column height). An eight-channel cell with sample volumes of only ~12 μL allows equilibrium to be reached in only 1–2 hours. However, these eight-channel cells require using interference optics, and the small number of data points limits the precision of the results. Thus, these eight-channel cells are used for quickly surveying many conditions or many samples, rather than for detailed or high-precision analyses.

Software for Data Analysis

First, it should be noted that a number of new innovations in data analysis have appeared in recent years, and continued development in this area may be expected. Although Beckman Coulter supplies analysis software that covers several methods, the more advanced software packages are in the public domain or marketed by third parties. Unfortunately, most of these programs, including the Beckman Coulter software, do not provide detailed reports, the ability to store a permanent record of how the analysis was done, or the ability to recall a previous analysis, as is particularly desirable in an industrial setting. There are far too many approaches and programs to discuss all of them here, so the focus will be on a few of highest utility from a biotechnology industry perspective.

Sedimentation Velocity Software

One important group of approaches derives a distribution of sedimentation coefficients, akin to a chromatogram, from the raw data, with each species producing a peak at its corresponding sedimentation coefficient. These approaches provide a way of summarizing and presenting the homogeneity or heterogeneity of the sample in a way that is easily understood by scientists who know nothing about AUC, as well as a good way to quantitate the various species that may be seen (from the area under the peaks).

The Time-Derivative, or $\partial c/\partial t$*, Method*: In this approach, originally developed by Walter Stafford (Stafford, 1994), a group of scans closely spaced in time are subtracted in pairs (giving concentration difference curves Δc). Those subtracted curves are then divided by the time difference between those scans Δt to approximate the partial derivative $\partial c/\partial t$ as a function of radius. Those $\partial c/\partial t$ curves can then be transformed from a radius scale to an apparent sedimentation coefficient scale using Equation 3, averaged together, and normalized to give the final distribution of apparent sedimentation coefficients, $g(s^*)$. The $\partial c/\partial t$ method is implemented in the Beckman Coulter software, a third-party commercial program called DCDT+, and in public domain software by Walter Stafford.

One important advantage of the $\partial c/\partial t$ method is that the pairwise subtraction of scans algebraically removes any time-independent noise in the data, such as that due to any dirt or scratches on the cell windows. This feature is particularly important for scans from the interference optics, where window nonuniformity and distortions cause the baseline readings (signal at zero protein concentration) to vary significantly across the cell. A second important advantage of this method is that it is essentially model-independent and does not rely on any assumptions about the properties of the sample. The major drawback of this method is that only a small fraction of the total scans, covering a limited range of boundary movement, can be used in the analysis. This restriction limits the signal/noise and also makes it difficult to analyze samples containing a very wide range of sedimentation coefficients.

Figure 3 illustrates the application of the $\partial c/\partial t$ method to data from a simulation for an antibody sample containing 20% of an irreversible dimer. This simulation is meant to simulate an optimum experiment, with data collected at approximately the maximum rate possible for absorbance scans in a run with three samples. The eight scans in the upper panel with heavy traces are the optimum ones for resolving dimer from monomer (the ones just before the dimer reaches the bottom of the cell). Eight scans are the *most* that should be used in this case; even with this few scans, the dimer peak in the resulting $g(s^*)$ distribution (lower panel) is already being broadened.

This example illustrates why the term *apparent* sedimentation coefficient and the s^* symbol are used here. Diffusion causes each species to produce a bell-shaped peak in the distribution rather than a single sharp spike. Only at the center of each such peak is s^* equal to the true sedimentation coefficient of that species—on the side of a peak the value of $g(s^*)$ is non-zero, but it only *appears* that material at that sedimentation coefficient is present because of the broadness of the peak. On the other hand, those peak widths do contain valuable information. Because they are related to diffusion, for well-resolved peaks it is possible to fit the peak widths to determine D and then to determine the molecular mass for that peak from Equation 4. Such mass estimates are generally only accurate to 5–10% (Philo, 2000), but that accuracy is more than sufficient to determine whether that species is a monomer, dimer, trimer, and so forth.

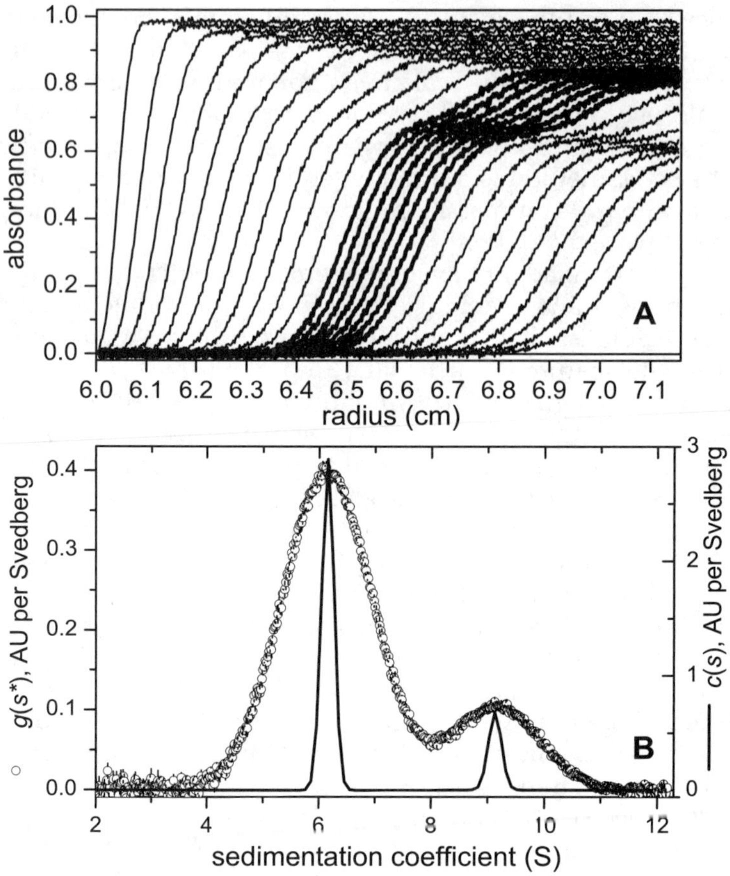

Figure 3 Simulated velocity data for 1 OD loading concentration of a 6.2 S antibody sample containing 20% dimer (at 9.2 S), with 0.005 OD r.m.s. random noise. Panel A shows the simulated data. Every other scan is shown, except that all scans are shown for the group of eight scans used for time-derivative analysis (heavier traces). Panel B shows the results of analyzing these simulated data using the time-derivative method (points with error bars, left axis) and using the $c(s)$ method (solid line, right axis).

The c(s) Method: A relatively new distribution method developed by Peter Schuck at the NIH and implemented in his program SEDFIT (Schuck, 2000) is a significant innovation. The so-called $c(s)$ distribution obtained by this approach has a much higher resolution than the $g(s^*)$ distribution because it explicitly removes the broadening caused by diffusion. It also achieves higher signal/noise and can cover a broader range of sedimentation coefficients because all of the scans from the beginning to the end of the run can be used.

In this method, a grid of possible sedimentation coefficients is defined (e.g., 100 points from the lowest to highest values to be covered by the analysis), and a diffusion coefficient is also assigned to each sedimentation coefficient (more about that later). From those values, a set of theoretical scans is constructed for each sedimentation coefficient in the grid, including the effects of diffusion (similar to those shown in Figure 2); those theoretical scans are calculated for the same times during the run as the experimental scans. Nonlinear least-squares fitting is then used to determine a set of concentrations for all 100 possible species that best fits the experimental scans, and those values then define the concentration as a function of sedimentation coefficient (the distribution).

A key question, though, is how to assign the diffusion coefficient corresponding to each sedimentation coefficient? (Remember, this is really equivalent to assigning a molecular mass to each possible species, through Equation 4.) This assignment is done through a simplifying assumption: all species are assumed to have the same hydrodynamic shape (as defined by the frictional coefficient ratio, f/f_0, as discussed above). Because we do not necessarily know in advance what f/f_0 ratio is correct, this ratio can be treated as a fitting parameter, allowing it to vary to find the ratio that gives the best fit of the raw data. If the sample consists of a single homogeneous species, that best-fit f/f_0 ratio will of course be the true ratio for that molecule; Dr. Schuck has shown that for heterogeneous samples, an f/f_0 ratio that corresponds to the weight-average value will give the best fit of the data. Another implicit assumption of the method is that the sample is a *mixture* of independent species rather than an interacting system of molecules that associate or dissociate during the course of the experiment.

An additional feature of this method is that sophisticated regularization routines can be applied to help insure all the peaks and features in the distribution are real and not just consequences of noise in the data. It also provides a means for removing the time-independent noise from the data (baseline distortions or scratches on the windows), although this is done through curve fitting (and is thus model-dependent) rather than the exact algebraic subtraction employed in the $\partial c/\partial t$ method.

The solid curve in Figure 3B shows the $c(s)$ distribution obtained from the simulated data shown in Figure 3A. The direct modeling of diffusion results in a much higher resolution relative to $g(s^*)$; in essence, the detailed information about diffusion and molecular mass is sacrificed in order to obtain that higher resolution. The higher signal/noise of $c(s)$ results from using the full set of scans rather than a limited subset. This distribution also gives an accurate description of the sample as 79.9% of a species at 6.200 S and 20.1% of a species at 9.203 S.

The principal drawback of this approach is that it may be difficult to know whether its assumptions are correct for any particular sample. Certainly, if it fails to give a good fit of the raw data, then the results are suspect (possible false peaks or significant distortions of peak positions and amplitudes). When the different species in a sample are well separated in size, the results tend to be fairly insensitive to the exact f/f_0 ratio that is used, and the reproducibility will be high. When the separation is fairly poor, as it inevitably is between monomers and dimers for monomers below ~50 kDa, then the positions and amplitudes of the minor peaks become more sensitive to details like the f/f_0 ratio and obtaining reliable results requires optimizing the experiment and extra care with the data analysis.

Discrete Mixture Analysis: For samples that appear to contain only a few non-interconverting species (e.g., by one of the distribution methods above), generally the best way to quantitate the properties of those species and the species fractions is to directly fit the raw scans using a discrete mixture model. In such a model, the theoretical scans for each species (defined by its concentration, sedimentation coefficient, and diffusion coefficient) are calculated (as in Figure 2) and then the species are summed. The c, s, and D of each species are then varied by a nonlinear least-squares algorithm to find the best overall fit of the raw data. Unlike the $c(s)$ distribution method, no assumptions are made about the hydrodynamic shapes of the different species, but like that method, scans covering essentially the full range of boundary movement can be included in the fitting, maximizing the resolution and signal/noise.

Generally, this approach will give the best accuracy for the molecular properties and the fraction of different species. Another significant advantage is that with this approach it is straightforward to estimate the statistical precision of those values (their confidence limits). Often, it is more convenient to use the molecular mass of each species as a fitting parameter rather than D; this is easily implemented, as D can be internally calculated by the software from s and M using Equation 4. That transformation of variables also makes it possible to constrain the mass of one species to be a multiple of another (e.g., to model a mixture of monomer, dimer, and trimer with an unknown monomer mass). Such mass constraints often make sense for highly pure pharmaceutical samples, as the only possible species are different oligomers (or different conformations with the same mass). Adding such constraints, based on other information about the sample, can then significantly increase the precision of the remaining fitting parameters.

Discrete mixture models are available in several software packages. Those include the public-domain programs LAMM by Joachim Behlke, SEDANAL by Walter Stafford, and SEDFIT by Peter Schuck, plus the third-party commercial programs SVEDBERG by John Philo and ULTRASCAN by Borries Demeler.

Sedimentation Equilibrium Software

Some simple graphical transforms of sedimentation equilibrium data can provide a quick semiquantitative interpretation of the data as well as useful diagnostics to guide further detailed analysis. The theoretical concentration distribution for a single ideal species (Equation 6) implies that a graph of the natural log of concentration *versus* $r^2/2$ will give a straight line, with a slope given by $M_b\omega^2/RT$. Figure 4A shows such a plot resulting from simulated data for a reversible monomer-trimer system. Because substantial amounts of trimer are present at the highest concentrations reached, it is clear that the slope of the plot rises with increasing radius (upward curvature), a diagnostic for aggregation or reversible association. Strong nonideality would manifest itself as a decreasing slope at higher concentrations.

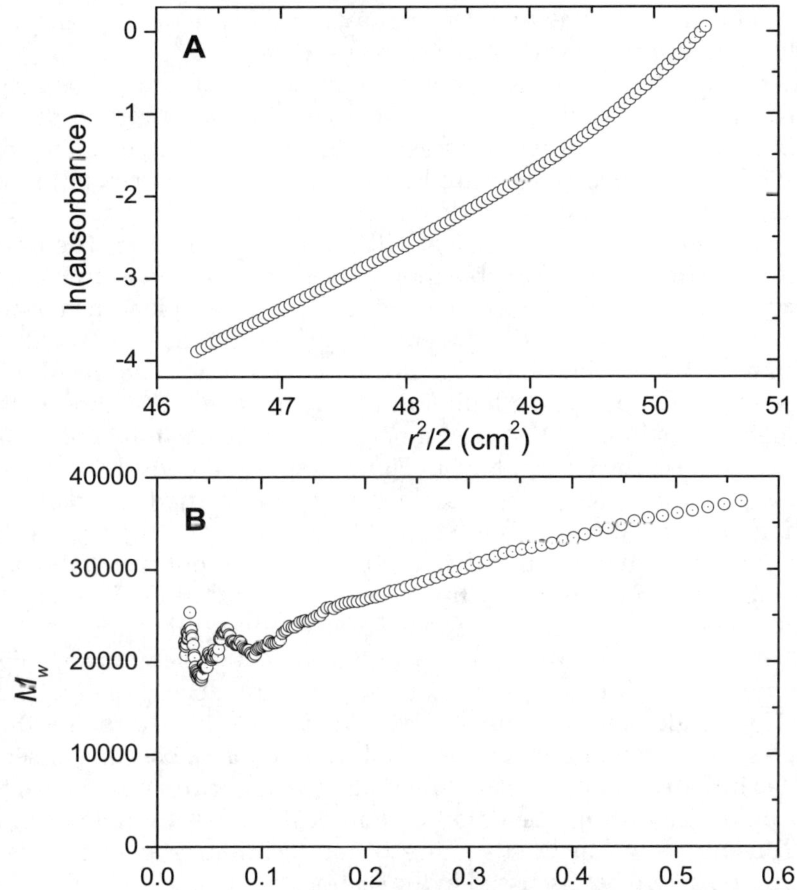

Figure 4 Simulated sedimentation equilibrium data (with random noise) at 25,000 rpm for a 20 kDa monomer exhibiting reversible association to trimer. Panel A shows the ln(*c*) *versus* $r^2/2$ plot. Panel B gives the weight-average molecular mass *versus* concentration plot from the same data. Each data point in B derives from the slope over 40 data points in A, with the 40-point window sliding through the dataset one point at a time.

Often, a more useful and sensitive plot can be constructed by taking the local slope through small regions of such a plot and calculating the local weight-average molecular mass and average concentration for that region, to construct an M_w *versus* concentration plot, as in panel B of Figure 4. This plot shows that the weight-average mass increases substantially over this range of concentration, rising from a value near that of monomer to nearly that of a dimer. Further, if this variation is due to reversible association, the fact that the M_w is close to dimer but still rising strongly with concentration suggests that oligomers larger than dimer are formed. One drawback of this type of plot, however, is that taking the slope through a limited number of data points tends to greatly magnify any noise in the data, as is particularly noticeable for the data at the lowest concentrations.

An important point about this example is that from this single sample, it is not possible to distinguish whether we are detecting reversible association or the presence of irreversible aggregates. That distinction requires measuring samples loaded at several different protein concentrations, and with such data in hand these M_w *versus* c plots can provide a very useful diagnostic, as illustrated by Figure 5. The upper panel shows the same reversible monomer-trimer system as in Figure 4 but now with data for samples over a fourfold range of loading concentrations. The key point is that where the concentrations overlap, all the data fall on a single curve (within error due to noise). That result is of course expected, because a hallmark of a reversible association is that the species fractions and M_w are functions only of the total concentration. In contrast, the lower panel in Figure 5 shows the results for a mixture of monomer with irreversible trimer. For a mixture, the three loading concentrations define separate, non-overlapping curves; the fraction of monomer and trimer in each sample is a function of the position within the cell, not a function of total concentration. Thus, an M_w *versus* c plot can be a useful diagnostic for whether it is appropriate to apply reversible association models, and if so what stoichiometry might be appropriate. The Beckman Coulter software does generate these plots for single samples, but unfortunately not directly for multiple samples (and not without prior manual subtraction of baseline offsets).

Quantitative analysis of sedimentation equilibrium data is usually done by directly fitting the data to theoretical models (Equations 6, 10, or 12) using nonlinear least-squares fitting techniques. An example of this approach will be discussed in the section "Case Studies" below. The Beckman Coulter equilibrium analysis software is quite user-friendly and provides models for both single species and self-associations with up to three different oligomers present and optional inclusion of nonideality effects through a second virial coefficient. For self-associating systems, this software has one major flaw: it cannot simultaneously analyze absorbance datasets recorded at different wavelengths or mix together absorbance and interference data. That flaw means this software

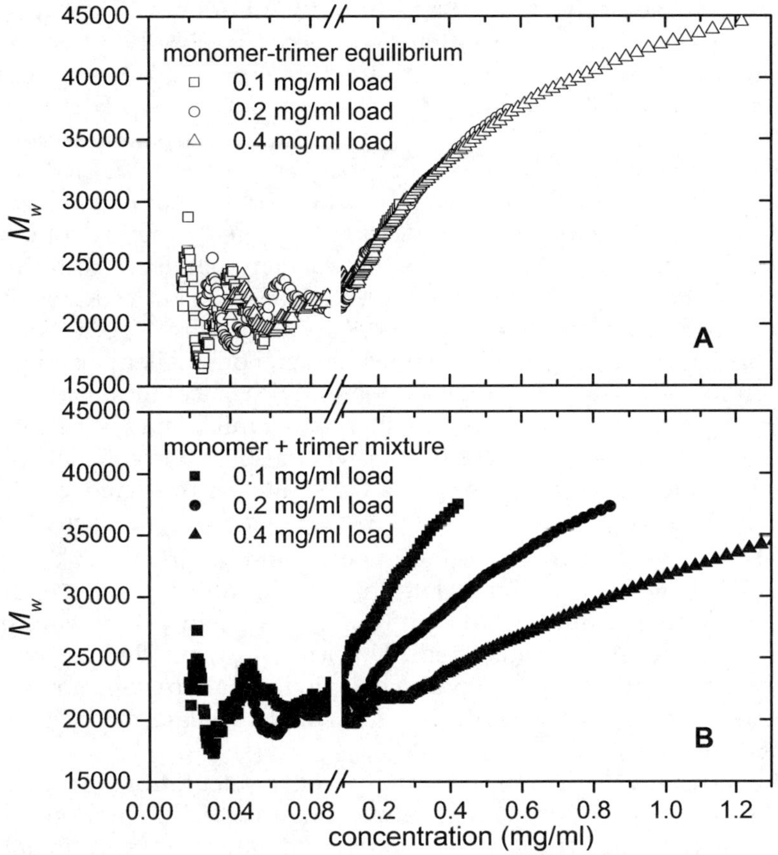

Figure 5 *M_w versus c* plots from simulations for a 20 kDa monomer. Panel A shows simulations for a reversible monomer-trimer association, with samples loaded at 0.1, 0.2, and 0.4 mg/mL. Panel B shows simulations for a 50:50 mixture of monomer with irreversible trimer, at the same loading concentrations.

cannot cover a large range of loading concentrations, which is really essential for distinguishing different possible stoichiometries and obtaining precisely determined dissociation constants. A public domain program called NONLIN (Johnson et al., 1981), developed largely at the University of Connecticut, overcomes that limitation and has other useful options. NONLIN is less user-friendly than the Beckman software but is nonetheless widely used as a proven "standard" in the field. The third-party package ULTRASCAN has similar capabilities to NONLIN plus some additional models.

Utility Software

Obtaining true molecular mass from the buoyant masses that are directly measured by AUC techniques requires knowing the partial specific volume of the protein and the solvent density. For sedimentation velocity data, we also often need to know the solvent viscosity. Although all these properties can directly be measured, and should be whenever possible, often it is impractical to do so. Fortunately, these properties can often be calculated, and those calculations are readily done via a public domain program called SEDNTERP developed by Tom Laue, David Hayes, and this author (Laue et al., 1992).

For partial specific volumes of nonconjugated proteins, all that is needed is their amino acid composition (usually obtained from the sequence). Generally, such calculated $\overline{v}$ values seem to be accurate within ±0.5%, which translates to about a ±2% uncertainty in M, but there are proteins where the deviations are much larger. For smaller peptides, the accuracy of calculated $\overline{v}$ values is poorer, with uncertainties of 1–2%. For glycosylated or PEGylated proteins, the overall $\overline{v}$ may be calculated based on the fraction by weight of polypeptide, X, the calculated $\overline{v}$ of the polypeptide, $\overline{v}_{\text{poly}}$, and the $\overline{v}_{\text{conj}}$ of the conjugate molecule (carbohydrate or PEG), as $\overline{v} = X\overline{v}_{\text{poly}} + (1 - X)\overline{v}_{\text{conj}}$. An excellent discussion of glycoprotein $\overline{v}$ values can be found in Lewis and Junghans (2000).

SEDNTERP also can calculate the density and viscosity of solutions containing common salts, buffering agents, and stabilizers such as sucrose, glycerol, or mannitol. Unfortunately, no density or viscosity data are available for certain excipients that are common in the biotech industry, such as various amino acids, trehalose, polysorbate 20 or 80, and Pluronic F68.

Finally, SEDNTERP handles many other calculations needed for AUC data interpretation, such as converting raw sedimentation coefficients to $s_{20,w}$ values and hydrodynamic calculations (f/f_0 and axial ratios for ellipsoid and cylinder shape models).

Case Studies

Sedimentation Velocity

Demonstrating Comparability of Conformation

It is often necessary to show that different preparations of a protein are similar enough to be "comparable." This need may arise, for example, following a change in the purification protocol, manufacturing site, or formulation, or after further manufacturing scale-up. After such changes, comparability data often must be submitted to regulatory agencies. There are many excellent tools for demonstrating comparability of chemical structure, purity, and so forth, but these are not sufficient. Because the bioactivity of the protein requires the correct conformation (secondary, tertiary, and quaternary structures), we must also show comparability of conformation.

Sedimentation velocity is an excellent tool for demonstrating comparability of conformation. Sedimentation coefficients are sensitive to small differences in protein structure (and also, of course, to larger differences such as change in oligomeric state). They can be measured with high precision and reproducibility and thus provide a more quantitative marker of conformation than spectroscopic approaches such as circular dichroism or infrared spectroscopy.

Thus, the basic approach is simply to compare the sedimentation coefficients observed for different protein preparations. Ideally, this is done in a direct side-by-side comparison of samples run simultaneously, and the expected precision of the derived sedimentation coefficients in such cases is ±0.2% or better if the experiments and analysis are done carefully. Direct comparison of data for samples run at different times (even years apart) can also be made; the reproducibility will be somewhat lower (due principally to differences in sample temperature), but it is still typically ±0.5% or better. In principle, any data analysis method that yields a sedimentation coefficient could be used. However, if the samples potentially contain multiple species (different sedimentation coefficients), then it may be important to distinguish changes in the distribution of species from changes in the sedimentation coefficient of individual species.

Figure 6 shows a comparison of samples of a monoclonal antibody made via two different purification processes, analyzed using the $\partial c/\partial t$ method. As shown by the vertical line, the sedimentation coefficients at the center of the main peak appear to be identical. By fitting a single Gaussian peak, the actual sedimentation coefficients returned for the peak centers are 6.518 ± 0.004 S and 6.513 ± 0.008 S, values that are clearly not significantly different. We also should note that the sedimentation coefficient of the main peak, together with the diffusion coefficient implied by the peak width, implies a molecular mass of 150 kDa (with an accuracy of ~5%), as expected for an antibody monomer.

If we did detect a small difference in sedimentation coefficient, that alone would not tell us whether this was due to a change in conformation or to a small change in molecular mass (perhaps clipping of terminal amino acids or a change in extent of glycosylation). Thus, we sometimes need to combine our biophysical data with results from other methods to get a unique interpretation.

In this study, a direct comparison of the raw sedimentation coefficients was possible because both samples were in the same formulation buffer and at the same concentration. Because of the high precision of these values, one must be careful that the buffers themselves are truly the same! Variations in salt or excipient concentrations that would cause density variations greater than ~0.03% or viscosity variations greater than ~0.1% can compromise the intrinsic precision. It is possible to compare data in different solvent conditions, but then we must convert the raw sedimentation coefficients to $s_{20,w}$ values (using Equation 2) before making the comparison. If the protein concentrations differ

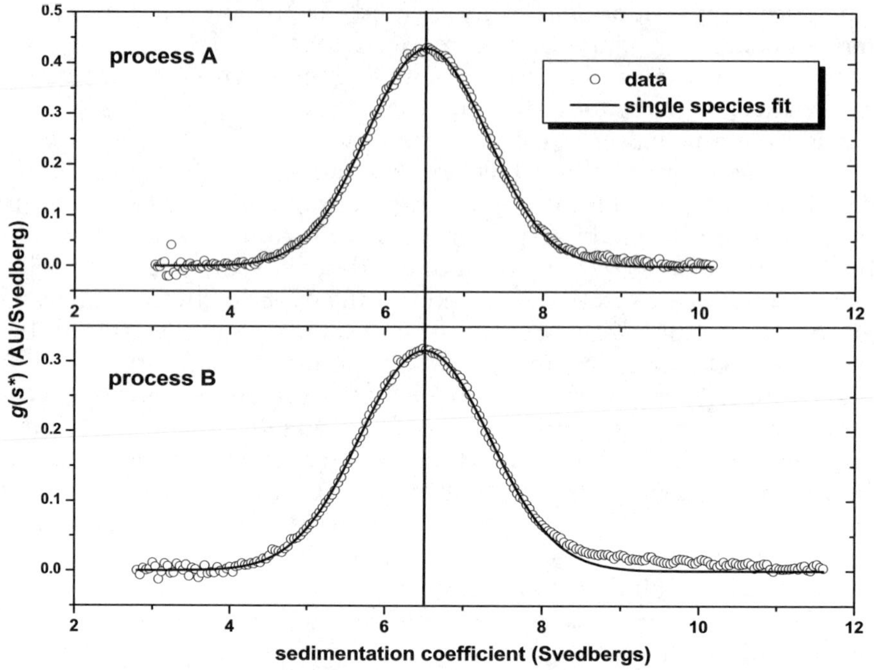

Figure 6 Comparability analysis of monoclonal antibody samples from two different manufacturing processes. The sedimentation coefficient distributions from time-derivative analysis are shown (points), along with a single-species fit for each sample (solid line). These data were measured on samples at 0.5 mg/mL using absorbance scans at 280 nm.

between samples, it may also be important to measure, and correct for, that concentration dependence.

Measuring Aggregate Content

Sedimentation velocity is an excellent tool to characterize aggregates in protein samples, and the data in Figure 6 from the case study above actually illustrates this also. A close inspection shows that the fitted single peak does not represent the data well for the region above about 8.2 S for either sample. This is due to the presence of aggregates in both samples, with significantly more aggregate for the process B sample. Taking the area under the curve for the actual data less the area for the main peak, these data imply 2.0% ± 0.5% aggregate for process A and 6.0% ± 0.5% for process B. Thus, the aggregate levels for these samples are *not* comparable. Running additional lots from each purification process showed a significant batch-to-batch variation in aggregate content for both processes (something that had not been detected by other methods).

It turns out that SEC analysis on these same samples in the same buffer gave 1.0% and 3.3% aggregate, respectively. Why did we detect more aggregate by sedimentation velocity? Although we did not explore this thoroughly in this case, we probably underestimated the aggregate content by SEC because the aggregates tend to stick to the frit or column resin (poor recovery) and/or because they are actually dissociated during chromatography (by dilution and/or shear forces). The absence of such losses is one major advantage of sedimentation velocity for measuring aggregates.

The *c*(*s*) distribution is now often the method of choice for analysis of aggregates, because its higher resolution allows better quantitation of individual species, because it can provide higher sensitivity, and because it can cover a wide range of sizes in one analysis. Figure 7 shows an example for a monoclonal antibody sample. The 100-fold amplified inset allows us to see that minor aggregate species can be detected at levels below 0.1%. Because sedimentation coefficients depend on conformation as well as mass, we cannot with certainty assign the state of oligomerization for these aggregate peaks, but the species at 10 S is almost certainly a dimer, the 13.6 and ~16 S peaks probably represent trimer and tetramer, and the ~23 S species may be hexamer or octamer.

Figure 8 shows the *c*(*s*) analysis for a highly aggregated sample of interferon-β, and clearly good resolution can be obtained even for

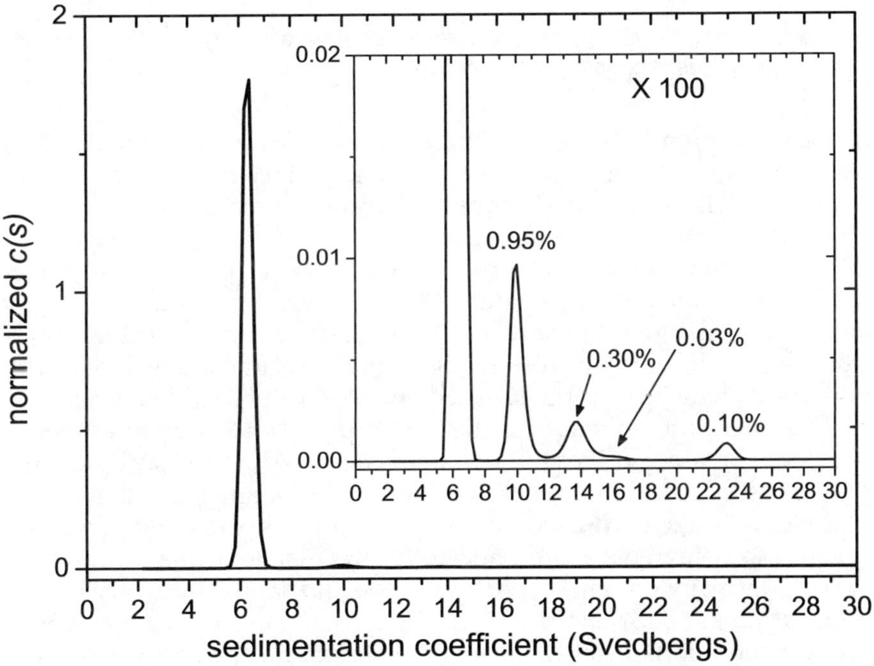

Figure 7 Detection of aggregates in a monoclonal antibody sample using the *c*(*s*) method. The inset shows a 100-fold vertical amplification of the same data so the minor peaks can be seen. These data were measured on a sample at 0.5 mg/mL using absorbance scans at 280 nm.

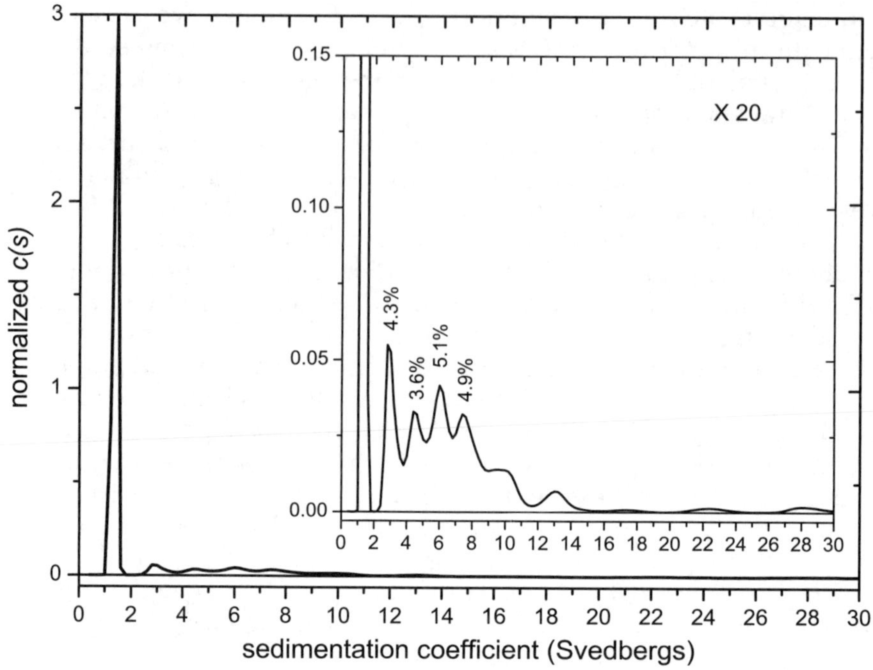

Figure 8 $c(s)$ analysis of an interferon-β sample with high levels of noncovalent aggregates, analyzed at 60,000 rpm.

such a small protein (monomer mass ~20 kDa) when it is run at 60,000 rpm. The largest aggregates detected here have masses of ~2 MDa (100-mer), demonstrating the very wide range of sizes that can be analyzed at one time (at one rotor speed).

Interferons as a class are prone to form both reversible and irreversible noncovalent aggregates. They are also very sticky proteins and are very difficult to analyze by SEC. Indeed, by the SEC methods typically used for interferon-β, this sample would analyze as 100% monomer! That happens because acetonitrile/water/trifluoroacetic acid mixtures are commonly used as the mobile phase for SEC of interferon-β in order to obtain good resolution and recovery. Although such solvents give robust and validated methods for *covalent* aggregates, they cause total dissociation of the noncovalent aggregates to monomer (even though those aggregates are often fully irreversible in the formulation buffer). For many proteins, good resolution and sample recovery in SEC requires high ionic strength and/or the addition of organic cosolvents, but such conditions are likely to alter the distribution of aggregates one is trying to measure. Thus, an important advantage of sedimentation velocity is that it can usually be run directly in the formulation buffer or physiologically relevant conditions.

One important application for sedimentation velocity is as an orthogonal method to cross-check whether SEC methods are detecting all the aggregates that are present. In the case of interferon-β, with the help of sedimentation velocity to serve as the gold standard, it was possible to develop a nondissociating SEC method that detects the noncovalent aggregates and correlates strongly with the velocity data. We should point out, however, that even when SEC is nondissociating, typically the larger aggregates elute at or near the exclusion limit and are not resolved (e.g., in Figure 8 all the species 6 S and larger would elute as one peak). That poor resolution of the larger aggregates means that SEC may not distinguish that different processes or different formulations have different distributions of large aggregates and therefore potentially different immunogenicity.

While Figures 7 and 8 hopefully illustrate that this approach can be quite powerful, and data of this quality have been obtained many times in the author's laboratory, the reader should not assume that the sensitivity to aggregates or other minor species will always be that high. The limits of detection will vary depending on the signal/noise and the degree of separation between species. The sedimentation coefficients for minor components, especially those present at levels below 1%, will also have much lower precision than that of the main component. As is true of all analytical methods, the true sensitivity and reproducibility can only be established by experiment.

Reversible Association or Irreversible Aggregation?

When multiple species are detected in a sedimentation velocity experiment, as in Figures 6 to 8, how do we know whether these are due to reversible self-association or irreversible aggregation? First, though, does this distinction even matter in the context of biopharmaceutical development? Yes, this can be an important distinction, for several reasons. Generally, the oligomers formed through rapidly reversible noncovalent association have fairly native-like tertiary structures and are less likely to be immunogenic than oligomers/aggregates that are irreversibly associated. Moreover, reversible oligomers that may form in the formulated product will likely dissociate when that product gets diluted after administration, which means they may have no impact on parameters such as pharmacokinetics or immunogenicity. Note, however, that the strong molecular crowding effects present *in vivo* will produce more reversible association than for a pure protein at the same concentration (Minton, 2001).

It is also important to realize that the distinction between "reversible" and "irreversible" is not clear-cut, and often whether a reaction appears irreversible depends on how long one is willing to wait. With proteins, it is not uncommon to see oligomers that are reversible but take many minutes to days to form and many minutes to days to dissociate (which we like to term *metastable* oligomers). Thus, whether an oligomer that may be present in our product has any significance, and whether it is detected by any particular analytical method, may really be most strongly related to its lifetime (the kinetics of association and dissociation) rather than simply whether it is "reversible."

In a sedimentation velocity experiment, we expect that irreversible aggregates and metastable oligomers with lifetimes comparable to or longer than the timescale of the separation (~1–2 hours) can be resolved as individual species. For a rapidly reversible monomer-dimer association, however, we should not resolve monomer and dimer but instead detect only one average species that will sediment at a rate intermediate between that of monomer and dimer. For rapidly reversible associations to trimers or higher oligomers, however, it is theoretically possible to resolve multiple species even when the association and dissociation rates are extremely rapid (Lee and Rajendran, 1994). In such cases, though, the peaks in the size distribution often do not represent individual, pure oligomers.

Figure 9 shows results for a protein exhibiting reversible self-association. Major changes in the size distribution are seen after a fourfold dilution. This protein probably exhibits a monomer ↔ dimer ↔ tetramer equilibrium, with only small amounts of dimer accumulating, although that was not definitely established. Note in particular that the sedimentation coefficient of pure tetramer is probably much higher than the ~6.7 S position of the major peak at the higher

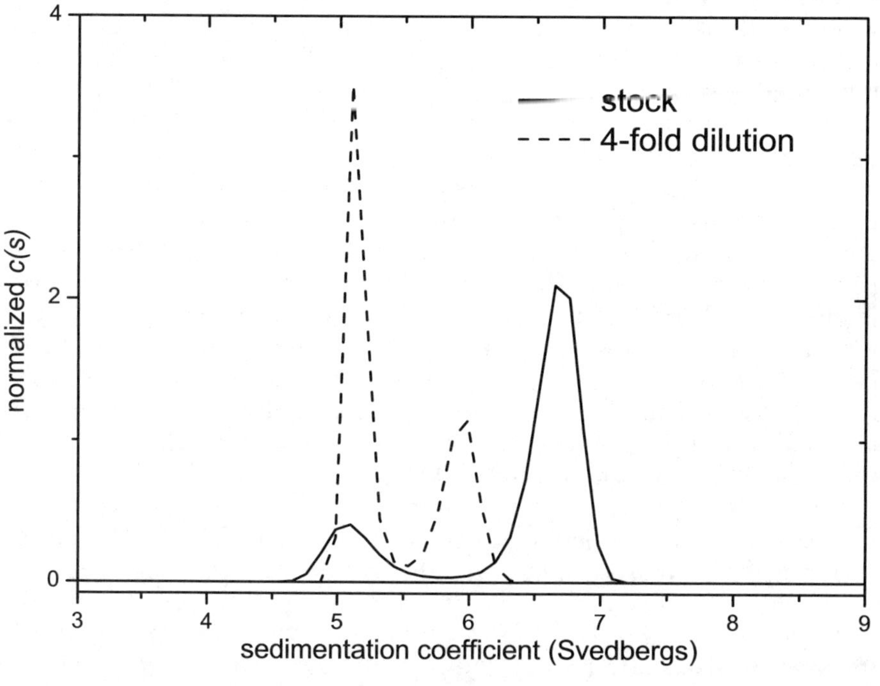

Figure 9 Sedimentation coefficient distributions for a reversibly associating protein at different loading concentrations.

concentration, and the intermediate peak at ~5.9 S at the lower concentration does not represent dimer—none of the peaks represent pure oligomers. In principle, then, we should not use the $c(s)$ distribution method, as that is really for mixtures rather than interacting systems, but we can calculate a valid weight-average sedimentation coefficient from these distributions, and that is a thermodynamically valid parameter to characterize interacting systems.

Overall, although it is often the case that the aggregate or oligomer peaks detected by sedimentation velocity represent fairly stable or long-lived species, only by running multiple concentrations is it possible to truly distinguish which species may be in reversible equilibrium.

Protein therapeutics such as monoclonal antibodies and Fc-fusion proteins are often formulated at concentrations higher than 10 mg/mL because the doses are large. This makes aggregation, or at least reversible association, very likely and also creates some special analytical challenges. To complicate further the situation, at high concentrations monoclonal antibodies seem to be prone to forming metastable oligomers (see, e.g., Moore et al., 1999). Trying to assess whether an antibody is still primarily monomeric at concentrations of 50 mg/mL or higher is a daunting challenge both experimentally and theoretically. One simple protocol we have found very useful in this situation is to dilute such samples to concentrations less than or equal to 1 mg/mL and then immediately run sedimentation velocity. Such a protocol detects all irreversible aggregates and metastable oligomers with lifetimes longer than about one hour that may have formed at the high concentration. Because these are the species that would also persist for some time *in vivo*, even for products rapidly diluted by intravenous injection, these are the aggregate species of highest concern. Although such a protocol will certainly not detect any rapidly reversible noncovalent oligomers that might be formed at the high concentration, it can be argued that such species are less significant because they are more native-like and transient. In one variant of this "dilute and shoot" protocol, the dilution is deliberately done into phosphate-buffered saline (PBS) in order to try to more closely mimic what may happen *in vivo*. It is also often useful to re-examine a diluted aliquot after several days to detect dissociation of any metastable oligomers.

Sedimentation Equilibrium

Determining Native-State Quaternary Structure

The task of determining whether a protein's native state in solution is a monomer, dimer, trimer, and so forth, would seem an easy one, but this author has seen numerous cases where such assignments were wrong and/or in dispute. Further, when the native state is a noncovalent oligomer, it may be important to know whether it dissociates substantially at lower concentrations. Sedimentation equilibrium is an excellent tool to address such issues.

The protein macrophage migration inhibitory factor (MIF), the first known cytokine, is one whose quaternary structure has been controversial. MIF has variously been reported as monomer, dimer, trimer, or irreversible mixtures of all three. MIF has several different biological activities, but those also seem to vary among different labs, and it was thought that might be related to differences in state of oligomerization (for details, see Philo et. al., 2004). Some sedimentation equilibrium data for three samples of highly purified MIF loaded at 39, 315, and 1050 μg/mL are summarized in Figure 10 as a weight-average mass *versus* concentration plot (described in the section "Software for Data Analysis," above). It is obvious from this plot that MIF is a trimer, with no evidence for significant dissociation into monomer at concentrations down to ~20 μg/mL.

Solution Mass Determination for Regulatory Submissions

A determination of solution mass is a normal part of a basic characterization package for regulatory filings on protein therapeutics. The regulatory agencies prefer to have an absolute method, independent of

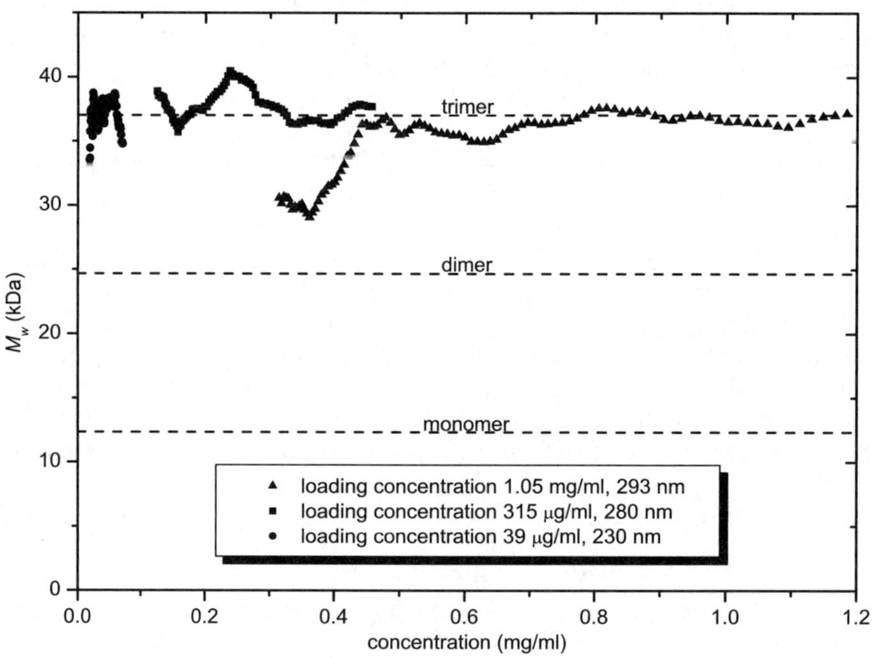

Figure 10 M_w *versus c* plot of sedimentation equilibrium data at 18,000 rpm for 3 samples of recombinant human MIF in PBS. The masses of monomer, dimer, and trimer (based on the sequence mass) are marked by horizontal dashed lines.

molecular conformation, like sedimentation equilibrium. For this purpose, we want a numerical value, preferably with a confidence limit, so the usual approach is global nonlinear least-squares fitting of multiple datasets. Figure 11 shows the data for protein X, a recombinant protein with sequence mass ~30 kDa expressed in *Escherichia coli*. Also shown is a global fit to a single ideal species model, which fits all fifteen datasets well. That fit returns a mass that is 99.98% of the sequence mass, with a 95% confidence interval of 99.56% to 100.41% of the sequence mass.

We should note that absolute accuracy depends on having the correct partial specific volume and solvent density. Generally, we

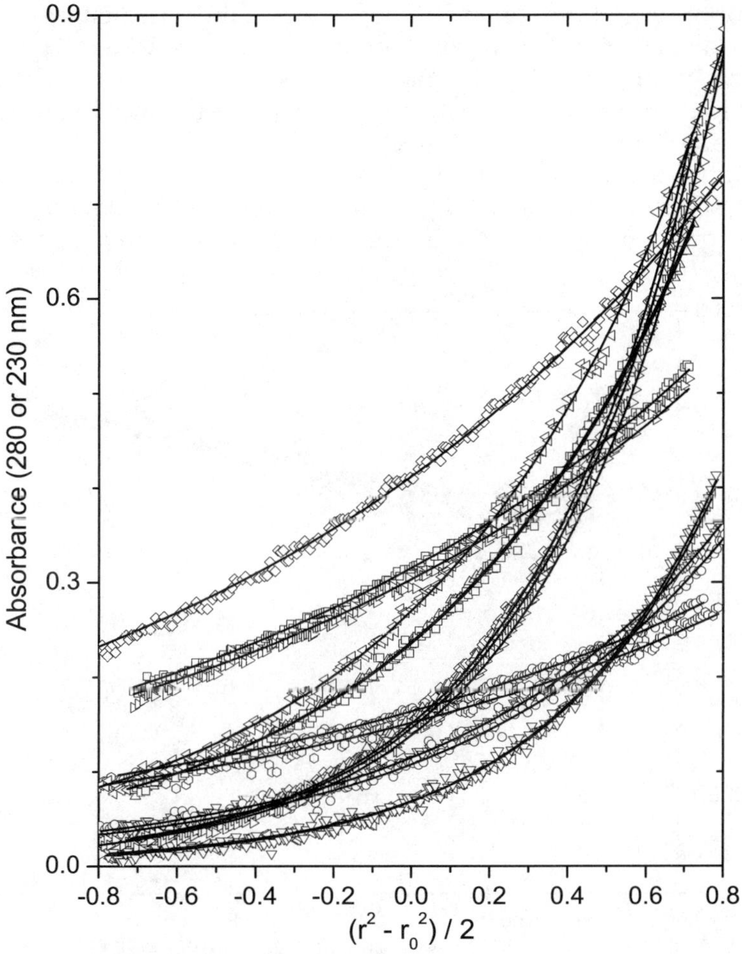

Figure 11 Overlay of sedimentation equilibrium data (points) for protein X and the fitted curves (lines) for a single ideal species model. These data were recorded at different wavelengths, so the actual range of concentration is much larger than the range of absorbance.

must rely on a calculated $\bar{v}$ value, and often this represents the biggest source of error. In our experience, with pharmaceutical-grade nonglycosylated proteins, we usually obtain a solution mass within ±2% of the expected value (based on the sequence mass) by using the calculated $\bar{v}$. However, that will not always be the case, and an experimental determination of $\bar{v}$ by precision density techniques is always best.

Quantifying Self-Association

For proteins that reversibly self-associate, we may need to determine the stoichiometry and equilibrium constants that characterize and quantify that association process. For example, we may want to know the proportions of monomer and oligomers at the formulation concentration or extrapolated to some other conditions. When significant amounts of different oligomerization states are present in the formulation, it may also be necessary to demonstrate comparability of association after a manufacturing change.

Figure 12 shows the M_w *versus* c plots from 18 datasets (9 loading concentrations at 2 different rotor speeds) for a recombinant glycoprotein, protein Y, with a monomer mass of 58 kDa (from mass

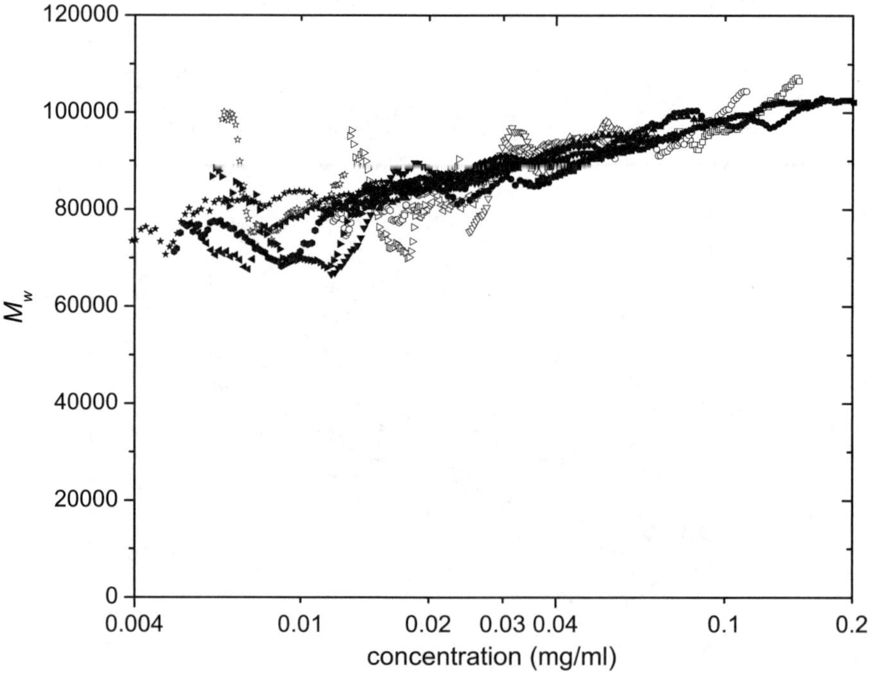

Figure 12 M_w *versus* c plot from eighteen sedimentation equilibrium datasets for protein Y that exhibits reversible association. Note the logarithmic concentration scale.

spectroscopy). Although there is some noise in the data, it is clear that there is a systematic variation with concentration, which proves this is a reversible association. Further, M_w is approaching the value for dimer at the highest concentrations, suggesting this protein may be in a monomer ↔ dimer equilibrium.

Global nonlinear least-squares analysis confirms that a monomer ↔ dimer model fits the data well (Figure 13); other stoichiometries provide significantly poorer fits. The monomer ↔ dimer fit returns a dimer → monomer dissociation constant of 518 nM with a 95% confidence

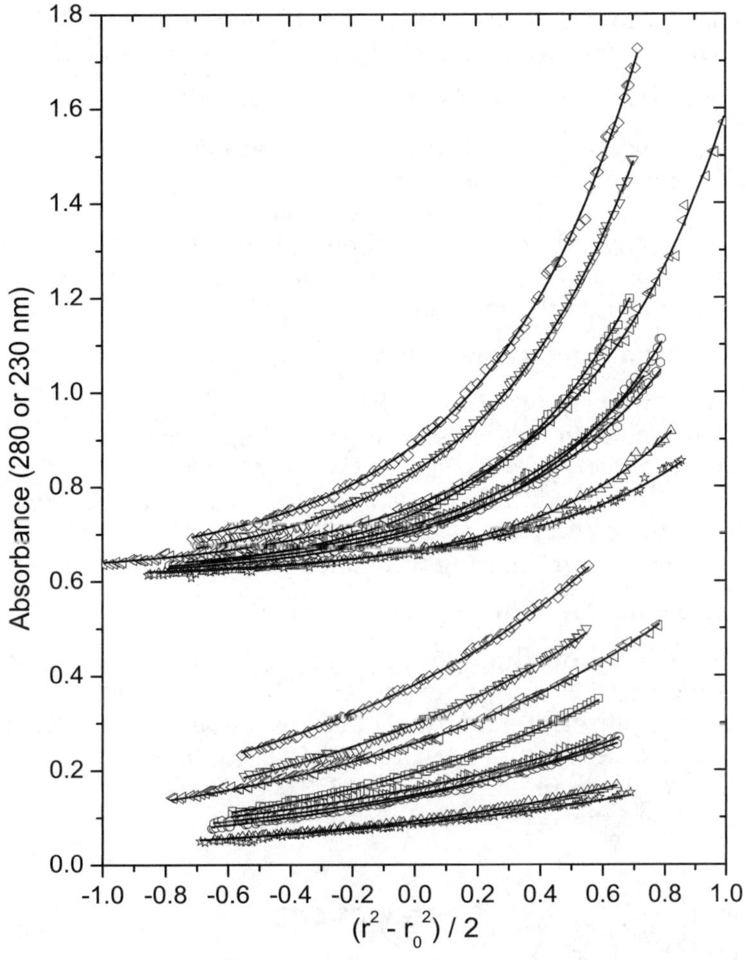

Figure 13 Overlay of sedimentation equilibrium data for protein Y (points) and fitted curves (lines) for a reversible monomer-dimer association model. The upper group of 9 datasets was taken at 13,000 rpm; they have been shifted upward by 0.6 OD for clarity. The lower group of 9 datasets was taken at 9000 rpm. The model returned a dimer-monomer dissociation constant of 520 nM.

interval of 490 to 548 nM. That confidence interval, however, assumes no uncertainty in the buoyant mass of the monomer. If we allow for the uncertainty in the $\bar{v}$ of the carbohydrate portion (Lewis and Junghans, 2000), the confidence interval widens to 410 to 720 nM.

Advantages and Limitations

Advantages common to both AUC methods are

1. They are "first principles," absolute methods that don't rely on protein standards.
2. They can usually be done in formulation buffers.
3. They have a strong theory that is very helpful for data analysis.

The particular advantages of sedimentation velocity are

1. It is an excellent quantitative measure of protein conformation.
2. It detects both covalent and noncovalent aggregates, with sensitivity below 1% for long-lived species.
3. It can have high resolution (better than SEC).
4. It covers a very large range of masses in a single experiment (much larger than SEC).

The particular advantages of sedimentation equilibrium are

1. It is a true equilibrium method.
2. It detects *all* forms of association (rapidly reversible, metastable, irreversible).
3. It is considered the gold standard for measuring the strength of self-association.
4. It is good at measuring weak interactions.

The primary drawbacks to both methods are

1. They have low throughput (3–7 samples/day for velocity, 9–21 samples/day for equilibrium).
2. The equipment and data analysis are not automated like HPLC; these techniques are fairly labor intensive.
3. The equipment is expensive ($250,000–$300,000).
4. Highly trained personnel are required.

Abbreviations

AUC	Analytical ultracentrifugation
MIF	Macrophage migration inhibitory factor
PBS	Phosphate buffered saline
SEC	Size-exclusion chromatography

References

1. Cantor CR, and Schimmel PR. *Biophysical Chemistry, Part II: Techniques for the Study of Biological Structure and Function*. San Francisco: W.H. Freeman; 1990, 365 pp.
2. de la Torre JG. Building Hydrodynamic Bead-Shell Models for Rigid Bioparticles of Arbitrary Shape. *Biophys Chem* 2001; 94:265–274
3. Haschemeyer RH, and Bowers WF. Exponential Analysis of Concentration or Concentration Difference Data for Discrete Molecular Weight Distributions in Sedimentation Equilibrium. *Biochemistry* 1970; 9:435–445
4. Johnson ML, Correia JJ, Yphantis DA, and Halvorson HR. Analysis of Data from the Analytical Ultracentrifuge by Nonlinear Least-Squares Techniques. *Biophys J* 1981; 36:575–588
5. Laue TM, Shah BD, Ridgeway TM, and Pelletier SL. Computer-Aided Interpretation of Analytical Sedimentation Data for Proteins. In: Harding SE, Rowe AJ, and Horton JC. *Analytical Ultracentrifugation in Biochemistry and Polymer Science*. Cambridge, UK: Royal Society of Chemistry; 1992:90–125
6. Lebowitz J, Lewis MS, and Schuck P. Modern Analytical Ultracentrifugation in Protein Science: A Tutorial Review. *Protein Sci* 2002; 11:2067–2079
7. Lee JC, and Rajendran S. Studies of Macromolecular Interaction by Sedimentation Velocity. In: Schuster TM, Laue T. *Modern Analytical Ultracentrifugation*. Boston: Birkhauser; 1994:138–155
8. Lewis MS, and Junghans RP. Ultracentrifugal Analysis of Molecular Mass of Glycoproteins of Unknown or Ill-Defined Carbohydrate Composition. *Methods Enzymol* 2000; 321:136–149
9. Minton AP. The influence of Macromolecular Crowding and Macromolecular Confinement on Biochemical Reactions in Physiological Media. *J Biol Chem* 2001; 276:10577–10580
10. Moore JM, Patapoff TW, and Cromwell ME. Kinetics and Thermodynamics of Dimer Formation and Dissociation for a Recombinant Humanized Monoclonal Antibody to Vascular Endothelial Growth Factor. *Biochemistry* 1999; 38:13960–13967
11. Philo JS. A Method for Directly Fitting the Time Derivative of Sedimentation Velocity Data and an Alternative Algorithm for Calculating Sedimentation Coefficient Distribution Functions. *Anal Biochem* 2000; 279:151–163
12. Philo JS, Yang TH, and LaBarre M. Re-examining the Oligomerization State of Macrophage Migration Inhibitory Factor (MIF) in Solution. *Biophys Chem* 2004; 108:77–87

13. Rowe AJ. The Concentration Dependence of Sedimentation. In: Harding SE, Rowe AJ, and Horton JC. *Analytical Ultracentrifugation in Biochemistry and Polymer Science*. Cambridge, UK: Royal Society of Chemistry; 1992, pp. 394–406

14. Schuck P. Size-Distribution Analysis of Macromolecules by Sedimentation Velocity Ultracentrifugation and Lamm Equation Modeling. *Biophys J* 2000; 78:1606–1619

15. Stafford WF III. Boundary Analysis in Sedimentation Velocity Experiments. *Methods Enzymol* 1994; 240:478–501

16. Tanford C. *Physical Chemistry of Macromolecules*. New York: John Wiley & Sons; 1961, 710 pp.

17. van Holde KE. *Physical Biochemistry*, 2nd ed. Englewood Cliffs, NJ: Prentice-Hall; 1985, 320 pp.

18. van Holde KE, and Baldwin RL. Rapid Attainment of Sedimentation Equilibrium. *J Phys Chem* 1958; 62:734–749

19. Yphantis DA, and Roark DE. Equilibrium centrifugation of Nonideal Systems. Molecular Weight Moments for Removing the Effects of Nonideality. *Biochemistry* 1972; 11:2925–2934

11

Field-Flow Fractionation

KARIN D. CALDWELL

Department of Surface Biotechnology, Uppsala University, Uppsala, Sweden

KARL-GUSTAV WAHLUND

Division of Technical Analytical Chemistry, Center for Chemistry and Chemical Engineering, Lund University, Lund, Sweden

Introduction

The recent and rapid evolution of protein-based pharmaceuticals has generated a growing need for quality control procedures that can assure the integrity of even the most delicate protein structures, whether they are present in a purified monomeric form or embedded in a liposome or other packaging matrix. From an analytical perspective, the two may represent extremes on a size scale that ranges from 1 to 1000 nm and beyond. Although there exists a wide variety of characterization methods that may reveal the size and homogeneity of a sample in any small interval of this scale, there are relatively few such methods that allow an examination of the entire size range in a reasonably short time period of just a few minutes.

An added discriminating condition that should be applied to potential characterization methods is that they must allow the analyst full freedom to select a liquid medium for the analysis that exerts no adverse effects on the structure of the sample components but leaves them in their biologically intact form. In addition to being chemically benign, the process must also be physically gentle, so that shear-related disturbances such as aggregation or breakup into subunits are kept to a minimum. The two conditions of range and environmental tolerance severely restrict the number of suitable analysis tools.

Among candidate techniques for use in protein characterization, the field-flow fractionation (FFF) methods have recently come into focus as being highly attractive for protein work (Arfvidsson and Wahlund, 2003; Litzen et al., 1993; Li and Giddings, 1996; Stevenson et al., 1999; Arfvidsson et al., 2004). This heightened interest is, in part, triggered by the emergence on the market of several newly designed FFF instruments that are easier to handle than the original prototypes. Because of these new developments, it appears warranted to review briefly the FFF family of separation techniques and their use in protein analysis. After a general survey of the workings of the different FFF subtechniques, we will concentrate on a more specific review of performance and instrumentation.

The FFF Principle of Operation

In virtually all its manifestations, FFF is an analytical technique, where accuracy is promoted by work with small, dilute samples. This condition is a direct consequence of the separation mechanism, which exposes the sample to the interplay between an externally applied field and the laminar flow of liquid through a capillary (Giddings, 1993). Although all ancillary equipment (i.e., pumps, detectors, monitors, etc.) is the same as that used with an ordinary high-performance liquid chromatography (HPLC) system, the FFF unit functions according to an entirely different principle, in that it performs the separation in an open, liquid-filled channel free of any stationary phase. This is the reason why even cells, viruses, and other delicate macromolecular assemblies can be processed in the separator with minimal shear-induced damage.

The differential migration along the length of the channel is due to the injected components' differential interactions with the field (Giddings, 1993), as seen in Figure 1. The figure depicts a flow channel with the rectangular cross-section that is most commonly employed, although cylindrical ducts can be used as well (Wijnhoven et al., 1996). As a result of its specific interaction with the field, each sample component will move to, and concentrate in, the quiescent flow-lines near the wall. This concentration will be somewhat offset by diffusion, and a number of more or less compact zones will thus form and be forced to move along the wall with velocities that directly reflect their zonal thickness. Because of the very steep velocity gradient that is established at the wall during laminar flow through a thin channel, even small differences in thickness translate into large differences in migration speed along the channel.

The entire process of zone formation and downstream transport can readily be modeled for thin channels with smooth, flat walls, where the *infinite parallel plate* geometry reduces the model to a mere two dimensions, with the parabolic velocity distribution in the x-direction, perpendicular to the walls, exclusively dependent on channel thickness,

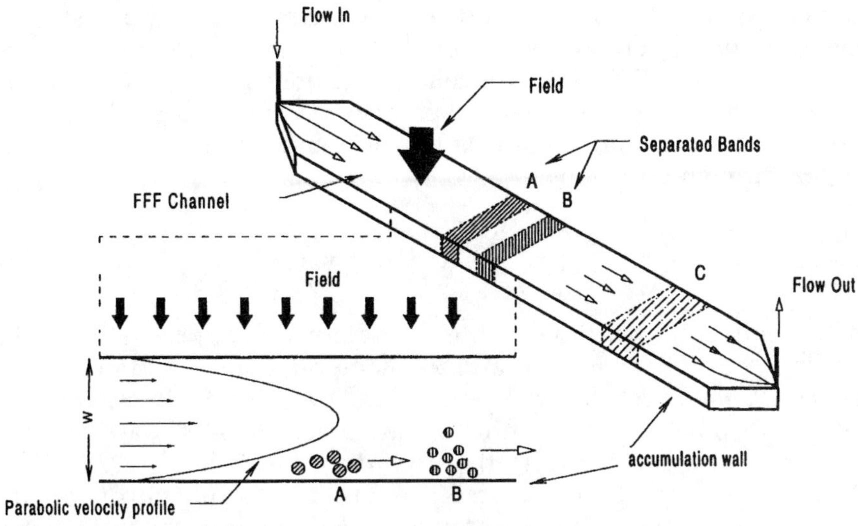

Figure 1 Schematic drawing of a conventional FFF channel with rectangular cross-section. The thin channel (dimension *w* is typically 250 μm or less) is bounded by parallel plates. The field is applied perpendicular to these plates, and the zones are transported through the channel with the laminar flow of the mobile phase. Analyte A has formed a more compact zone than analyte B and is therefore more retained. Reproduced from Schure et al. (2000) with permission by Wiley.

w, as shown in Equation 1 (Giddings, 1993).

$$v(x) = 6\langle v\rangle\left[\frac{x}{w} - \frac{x^2}{w^2}\right] \tag{1}$$

Here, $\langle v \rangle$ symbolizes the average velocity over the cross section. Because the field is applied perpendicular to the walls, the sample zones that are formed through the joint action of field and diffusion will likewise be distributed along the x-direction. Their concentration distribution $c(x)$ will be exponential and have its highest value $c(0)$ at the wall ($x = 0$), as shown in Equation 2:

$$c(x) = c(0)\exp\left(-\frac{x}{\ell}\right) \tag{2}$$

where parameter ℓ is the ratio of two important physical properties of the sample, namely its diffusion coefficient D and its field-induced velocity U, $\ell = D/U$. The magnitude of U, in turn, depends on the strength of the applied field and the component's susceptibility to the chosen field type.

The joint action of field and flow will result in the downstream migration of all zones at average velocities that are precisely determined by the product of the concentration and velocity profiles. It follows that a component's zonal velocity, measured through elution time t_e, is a direct function of its physical characteristics, as demonstrated through Equation 3:

$$R = \frac{\langle v \rangle_{zone}}{\langle v \rangle} = \frac{t^0}{t_e} = 6\lambda\left[\coth\left(\frac{1}{2\lambda}\right) - \frac{1}{2\lambda}\right] \approx 6\lambda \qquad (3)$$

In this equation, the dimensionless parameter λ represents the ratio ℓ/w, t^0 is the elution time for an inert compound, and parameter R is the *retention ratio*. At strong retention (i.e., long elution times, or small values of R), the function within square brackets rapidly approaches unity, so that R becomes directly proportional to λ (i.e., to D/U). The behavior described by Equation 3 is referred to as "the normal mode of FFF" and is applicable to sample particles whose dimensions are negligible compared to layer thickness ℓ.

Because the steady-state distribution of an injected sample plug tends to be just a few micrometers thick, (ℓ in Equation 2), a high sample concentration in the original injection will generate an utterly concentrated sample zone that, in turn, may give rise to unwanted intermolecular forces and other overloading phenomena when using the traditional channel geometry (Caldwell et al., 1988). It is for this reason that FFF, regardless of the choice of field, is best operated as an analytical technique with small samples at high dilution. Recently, efforts have been made to construct equipment allowing massively parallel runs, thereby increasing the amounts of sample that can be processed in one run and enabling semipreparative separations of particles (Maskos and Schupp, 2003).

Selecting the Field

To date, among the various subtechniques realizing the FFF principle, thermal fields have been used for polymer and particle separation (Gunderson and Giddings, 1986; Shiundu et al., 1995), dielectric and magnetic fields for cell separation (Huang et al., 1997; Zborowski et al., 1999), and electric fields for emulsion and particle characterization (Caldwell and Gao, 1993; Caldwell, 2000). If the purpose is protein separation, there are at present only two subtechniques that have shown significant promise, namely the *sedimentation FFF* (sdFFF) and the *flow FFF* (flFFF), the latter also referred to as the *universal FFF*. Of these, the sdFFF method has the highest size selectivity (Caldwell, 1988). Yet, the requirement that the mobile phase should flow from a stationary pump via a rotating seal into and through the spinning separation channel and then out again to a stationary detector sets an upper limit to the applicable spin rates, and hence a lower limit to the masses that might be retained in the system. This limitation is due to the tendency of the rotating seals to spring a leak at higher spin rates. The use of sdFFF is therefore typically limited to protein particles and similar structures with sizes in excess of 100 nm. For samples of higher density, such as Au-labeled proteins, smaller sizes are amenable to analysis, as seen from Equation 4 below. Despite the limitation, the many attractive features of this technique warrant a brief discussion.

Sedimentation FFF

Selectivity in sdFFF is generated by the application of a centrifugal field G to the thin channel, which for this application is coiled to fit inside a rotor basket. The sample is injected through a septum port directly into the channel while the rotor is at stand-still. Immediately following injection, the mobile phase flow is interrupted and the rotor is accelerated to its preselected rpm, where it stays spinning for the entire run if a constant field operation is selected. The field may also be programmed to decay in some specified manner to accelerate the elution of highly retained components. The flow of mobile phase is resumed after a suitable relaxation time of several minutes, the stop-flow period being designed to allow the components to assume their equilibrium distribution in the channel. Once this condition is at hand, the flow will not perturb the exponential distribution predicted by Equation 2.

For the sdFFF subtechnique, the λ parameter has the following form(s):

$$\lambda_{\text{sd}} = \frac{kT}{m\dfrac{\Delta\rho}{\rho_p}Gw} \tag{4a}$$

or equivalently, for spherical particles

$$\lambda_{\text{sd}} = \frac{6kT}{\pi d^3 \Delta\rho G w} \tag{4b}$$

In this case, the field susceptibility is the buoyant mass of the particle; that is, its mass m multiplied by the difference in density $\Delta\rho$ between the particle (ρ_p) and the mobile phase. If the particle can be treated as a sphere, its mass m may be replaced by the product of volume and density ($\pi d^3 \rho_p/6$), as done in Equation 4b. In this form, the retention parameter λ clearly shows the high size selectivity (diameter to the third power) of the sdFFF technique. This analytical power has been used extensively in the characterization of protein-coated latex particles (Caldwell et al., 1992; Langwost and Caldwell, 1992) for diagnostic and other uses.

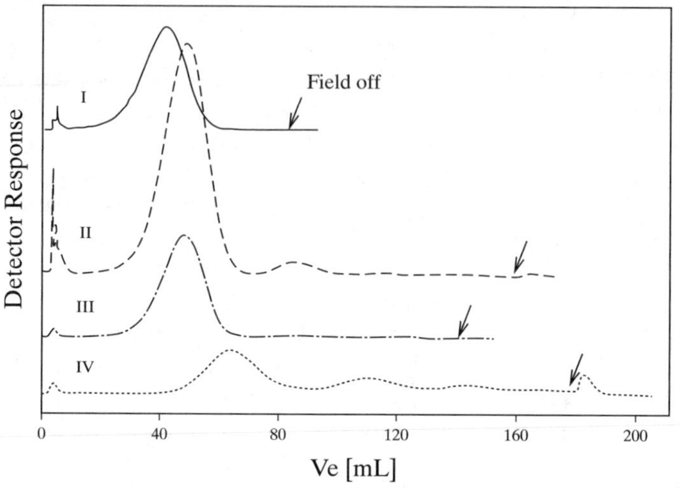

Figure 2 sdFFF fractograms of bare and protein-coated polystyrene latex particles with a diameter of 272 nm. The centrifugal field used in the analyses was 173 *g*. Trace I represents bare particles, trace II particles coated with human IgG (antigen in this case), trace III the particles from trace II "passivated" with human serum albumin (no protein uptake), and trace IV the particles from trace II after exposure to rabbit anti-human IgG. The shifts in elution volume of the first (monomer) peaks demonstrated that the antigen adsorbed with a surface concentration of 1.4 mg/m^2 while the antibody-antigen complex covered the surface to a concentration of 5.0 mg/m^2, or 2.6 anti-IgG per IgG. Reproduced from Langwost and Caldwell (1992) with permission.

Protein-Coated Latex Particles: An example of this type of analysis is shown in Figure 2, which illustrates the retention behavior of bare latex particles, particles coated with the protein human IgG, and the latter particles after having reacted with a goat-derived antibody against human IgG. From knowledge of the particle and protein densities, it was possible to determine the size of the core particle, the amount of protein adsorbed to each particle, and the number of antibodies bound to each human IgG molecule on the particle.

Protein Nanoparticles: At times, there is a need to characterize polydisperse populations of particles for intravenous use (e.g., as drug carriers or as carriers of radioactive marker substances for monitoring the status of the cardiovascular system in real-time). Suitable particle dispersion systems for such applications may be prepared through emulsion cross-linking of some plasma protein, such as serum albumin, which is nonimmunogenic and degradable in the liver. An early study (Caldwell et al., 1981) demonstrated the capability of sdFFF to determine both the average size and the polydispersity of a model colloid prepared from bovine serum albumin (BSA). In this case, the particle population had quite a broad size distribution, as can be seen in Figure 3.

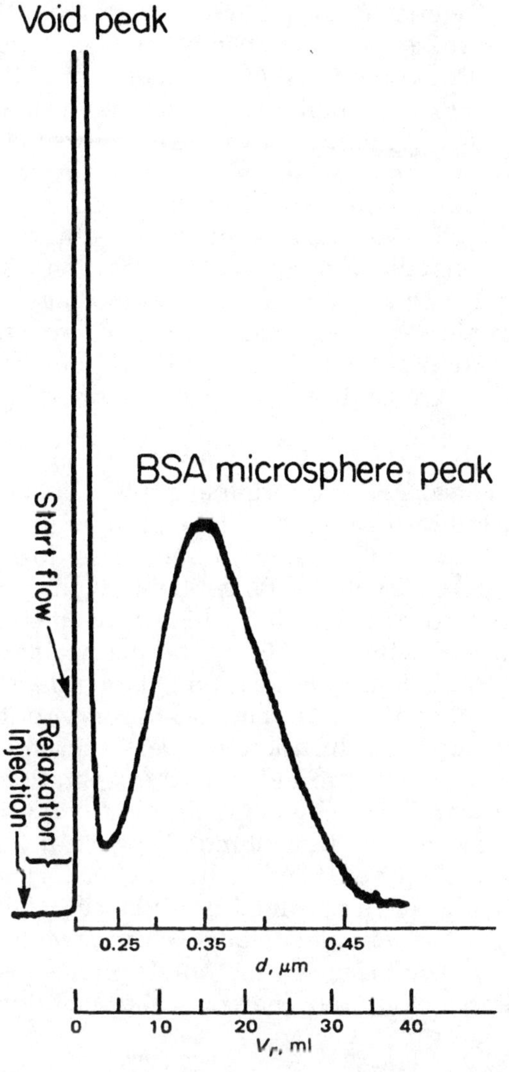

Figure 3 sdFFF elution diagram for a polydisperse sample of albumin microspheres. The analysis was performed under an applied field of 13.8 *g*. The diameter scale was calculated from Equation 4b, assuming a particle density of 1.36 g/cm^3. The FFF analysis was compared with an evaluation based on transmission electron microscopy whereby the average diameters and their standard deviations were determined as 0.349 ± 0.040 μm (FFF) and 0.354 ± 0.062 μm, respectively. Reproduced from Caldwell et al. (1981) with permission.

The average diameter, and its standard deviation within the population, were both corroborated by transmission electron microscopy.

Although both the above examples make use of the high size selectivity of the sdFFF system, they represent but a small subset of the protein characterization problems that are increasingly important to the biopharma community, whose needs primarily center around the quality control of singlet proteins with sizes around 1 nm and above. For these small molecules, the current sdFFF equipment is incapable of spinning fast enough to accomplish separation. The diffusion of these comparatively small molecules is far too rapid for the experimentally accessible centrifugal fields to accomplish the required sample concentration at the channel wall. Any FFF-based small protein characterization must therefore involve a different subtechnique.

Flow FFF

The flow, or universal, FFF in its original implementation is performed in the customary thin separation channel, although the walls of this channel are semipermeable to allow a transverse flow of carrier fluid to occur simultaneously with the tangential flow that moves the zones from injection port to channel exit. In this arrangement, the lower size limit of processable samples is set by the permeability of the membrane at the accumulation wall. Everything larger than the pore size of this membrane will remain in the channel to concentrate at the wall in a process counteracted by diffusion, whereas everything smaller will escape through the membrane and leave the separation process.

As in the other FFF techniques, the flFFF zones are propelled downstream by the laminar flow of mobile phase at velocities that reflect the thickness of the zonal layer. This thickness is a function of the concentrating field, in this case the linear velocity U of mobile phase through the membrane and the dispersive diffusive flux given by the sample's diffusion coefficient D, the only sample-specific characteristic that influences retention. Therefore, the retention parameter λ_{fl} has the following form:

$$\lambda_{fl} = \frac{D}{Uw} \tag{5}$$

Replacing U by the volumetric flow rate V_c divided by the area of the channel wall through which this flow is passing (void volume V^0 divided by channel thickness w), Equation 5 can be written as:

$$\lambda_{fl} = \frac{DV^0}{V_c w^2} = \frac{kT}{3\pi\eta V_c w^2}\frac{1}{d} \tag{5a}$$

From expression 5a, two things are clear; namely, first, that there is only one sample-specific parameter, D, that governs retention, and second, that the retention therefore depends on size *d* to the first order. This condition is demonstrated by Figure 4 (Li and Hansen, 2000), in which the observed retention (expressed as elution time) of a set of seven model proteins under constant cross-flow is shown to increase linearly with hydrodynamic size. It is worth noting that the depicted separation involves proteins ranging in molecular weight from 16.7 kDa to 1.5 MDa, all processed in a single run.

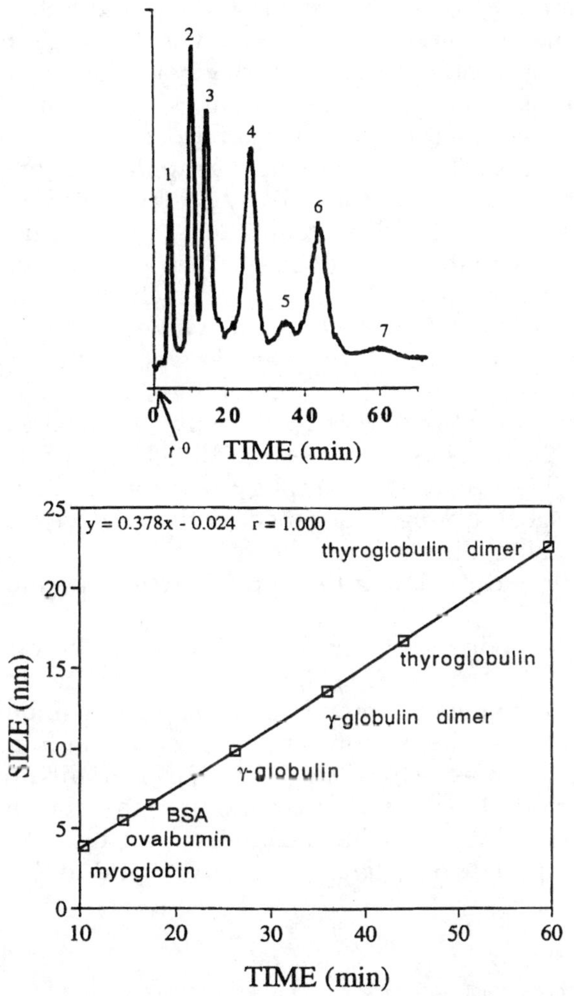

Figure 4 SyflFFF separation of proteins using a tangential flow of 0.8 mL/min and a cross-flow of 6.8 mL/mln. The retention times are seen to correlate well with hydrodynamic size for these model proteins, as predicted by Equation 5a. The peaks in the fractogram on top correspond to the following components: 1, vitamin B12; 2, myoglobin; 3, ovalbumin; 4, γ-globulin; 5, γ-globulin dimer; 6, thyroglobulin; 7, thyroglobulin dimer. Reproduced from Li and Hansen (2000) with permission.

In flFFF, contrary to SEC, the resolution between separated components can be effectively manipulated by the experimental conditions, in this case the cross-flow rate, whereas the separation time can be manipulated and optimized by the tangential flow rate through the channel. This flow optimization of resolution forms the basis for the development of rapid, high-resolution separations using high flow rates (Wahlund et al., 1986).

In recent years, the standard type of symmetrical flow FFF (syflFFF) channel, with its semipermeable top and bottom walls, has been complemented by an asymmetrical version (asyflFFF) in which the upper wall is a solid glass plate and only the lower wall has membrane characteristics (Wahlund and Giddings, 1987). The development of this technique was triggered by the need for higher resolution in flow FFF separations. It also represents a simplification of the construction of the channel. It was shown early on (Wahlund and Giddings, 1987) that in the asyflFFF system, the same retention equations (Equations 3 and 5a) that govern the symmetrical channel can be used to describe retention also in the asymmetrical channel.

Wahlund and Litzén (1989) invented a new feature that significantly improved the resolution; namely, an injection port a short distance downstream from the channel inlet, as seen in Figure 5. A sample injection at this port is then easily followed by a focusing cycle, in which buffer is pumped into the channel from both inlet and outlet. This operational strategy leads to a sharply focused zone at some position predetermined by the relative magnitudes of the two opposing flows. Once the focusing/relaxation is accomplished, the flow direction at the outlet is reversed, and the elution cycle can begin.

In a further refinement, Litzén and Wahlund (1991a) constructed a tapered asyflFFF channel (see Figure 5), in which the linear flow velocity tangential to the membrane can be made to maintain less variation from inlet to outlet, in contrast to the situation in a rectangular asymmetrical channel. The main advantages of this "trapezoidal" asyflFFF channel, however, are twofold: first, that the eluted sample zones are less diluted due to the lower flow rate at the channel outlet, as compared to rectangular asyflFFF and syflFFF channels,

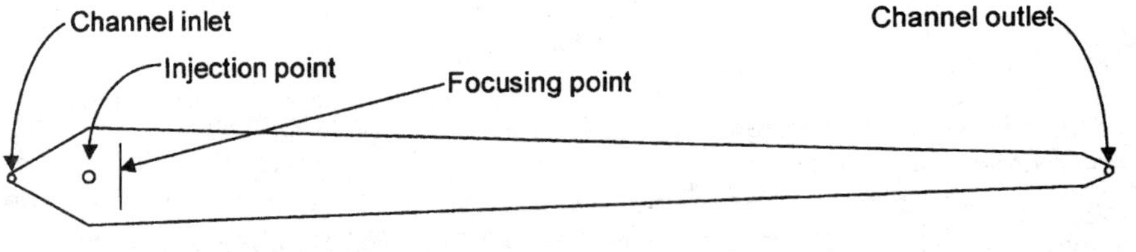

Figure 5 Illustration of the tapered trapezoidal asymmetrical flow FFF channel.

which decreases the detection limit; and second, that the volumetric cross-flow rate can be kept smaller due to the decreased area of the accumulation wall, thus decreasing the solvent consumption.

The high-speed performance characteristics of the tapered asyflFFF channel is demonstrated in Figure 6 for a mixture of proteins of a similar kind as in Figure 4. The same high resolution is obtained but in much less time (a factor of 16 smaller).

In recent years, the potential capability of protein analysis by flFFF has vastly been improved by using a multiangle light scattering (MALS) detector and a refractive index (RI) detector coupled on-line to the separation channel. This enables the direct determination of molecular weights (Wyatt, 1993) of the separated components. The flFFF-MALS-RI methodology as developed for water-soluble polymer characterization (Roessner and Kulicke, 1994; Wittgren and Wahlund, 1997) is therefore now also available for protein analysis, which of course greatly enhances the characterization possibilities for complex protein mixtures. Figure 7 illustrates such results for BSA.

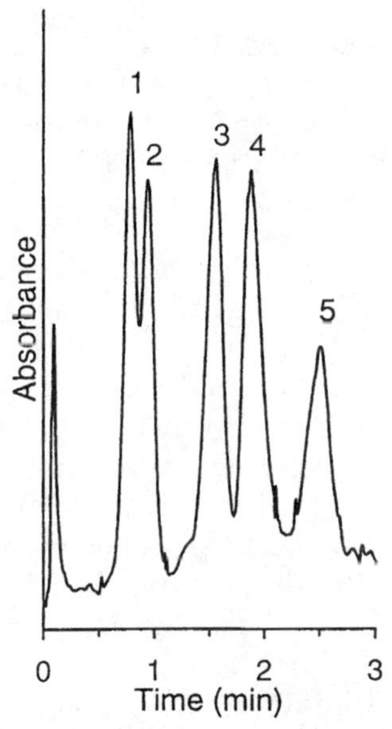

Figure 6 High-speed separation of five proteins of different molecular weights. 1, ovalbumin (43 kDa) at 0.25 mg/mL; 2, bovine serum albumin (67 kDa) at 0.27 mg/mL; 3, monoclonal antibody (156 kDa) at 0.20 mg/mL; 4, ferritin (440 kDa) at 0.050 mg/mL; 5, thyroglobulin (669 kDa) at 0.25 mg/mL. The injection volume was 7 μL. The carrier was a 0.050 M phosphate buffer at pH 6.5. V_{in} = 12 mL/min; V_{out} = 1.3 mL/min; V_C = 10.7 mL/min. Temperature: 24 °C. Channel thickness was 111 μm. Reproduced from Wahlund (2000) with permission by Wiley.

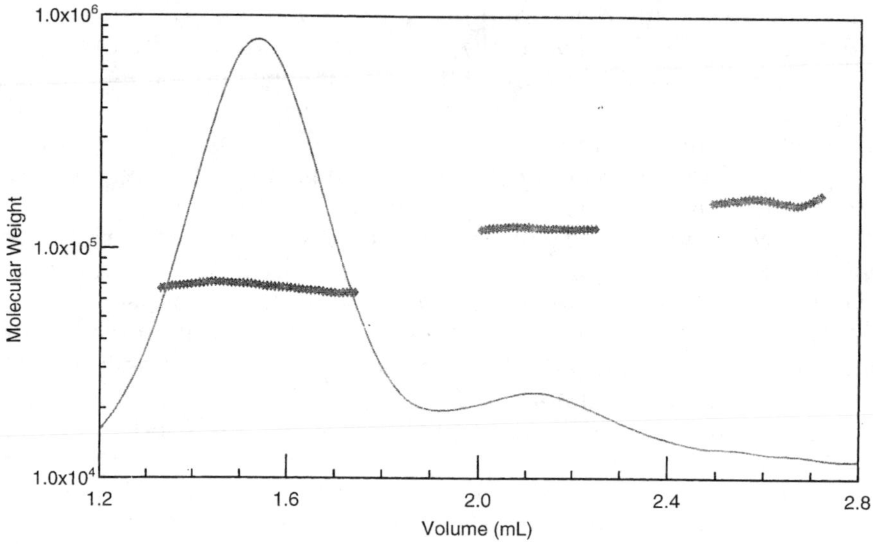

Figure 7 Molecular weights of bovine serum albumin monomer, dimer, and trimer determined on-line by a triple-angle light-scattering detector coupled to an asyflFFF channel. Molecular weights are on the ordinate and the elution volume on the abscissa. The continuous line is the protein concentration curve (a fractogram) registered by the RI detector as a function of elution time. The thick lines are the molecular weights obtained. Courtesy of D. Krunovar, B. Wittgren, and K-G. Wahlund.

Protein Analysis: Separation of proteins were among the first applications of flow FFF (Giddings et al., 1976a,b; Giddings et al., 1977; Wahlund and Giddings, 1987; Wahlund and Litzen, 1989; Wahlund et al., 1986), demonstrating the predictable relationship between protein molecular size and retention time and the ease by which the experimental conditions (cross-flow rate, channel flow rate) can be adjusted to fit proteins of different molecular weight ranges, as shown in Figures 4 and 6 above. The basic parameter behind these advantages is that the diffusion coefficient of a protein directly governs the retention time.

Protein Aggregates: A striking potential of flFFF is that it can be used to resolve protein aggregates (Wahlund and Litzen, 1989). In a 15-minute analysis time, complete resolution between human serum albumin monomer (68 kDa) and its dimer was obtained. After refinement of experimental conditions, the analysis time was brought down to 10 minutes for complete and high resolution of the monomer and dimer of, respectively, ferritin (440 kDa) and thyroglobulin (669 kDa), whereas for albumin the monomer and dimer were separated in 3 minutes (Litzen and Wahlund, 1989). Under stepwise flow gradient elution, even a trimer of albumin was observed.

In a separate study, a method was developed for characterizing the aggregation state of a pharmaceutical monoclonal antibody (Litzen et al., 1993). It is requested by regulation authorities that the absence of aggregates should be certifiable to some high level of probability, as these constitute impurities that can be suspected as potentially capable of causing immunoreactions. This requires that the purity of a pharmaceutical protein can be quantified in terms of its monomer content and the impurity content. The standard procedure is an SEC separation of the monomer from the dimer and all higher aggregates. This purity assessment requires only that there is a complete resolution between the monomer and the dimer but not among the higher aggregates.

Because the IgG antibody is a relatively large protein (165 kDa), it had been difficult to find good-enough resolution between the monomer and the dimer when SEC was applied. Figure 8 illustrates that complete monomer/dimer resolution of a monoclonal antibody was obtained by asyflFFF, as opposed to SEC. This was accomplished in much shorter time, requiring a mere three minutes by asyflFFF, as compared to fourteen minutes by SEC. It is remarkable that in the asyflFFF, the presence of higher order aggregates is clearly indicated, namely the trimer, tetramer, and pentamer forms of the protein, whereas these were not resolved at all by SEC. The aggregation was here induced by an acidic pH shift of the antibody solution.

In these IgG asyflFFF experiments, the carrier liquid was a 100 mM phosphate buffer, pH 6.5. It has been argued that flFFF will allow a higher degree of freedom in choosing the carrier composition than SEC due to the fact that the choice of mobile phase composition in SEC is restricted by the risk of adsorptive interactions between the protein and the particle support surface. Therefore, in SEC, one may under certain circumstances be forced to avoid using a mobile phase composition, which would otherwise have been ideal by virtue of its being identical to the composition of the pharmaceutical formulation. This makes the evaluation of protein aggregation by SEC unrepresentative of the conditions required in the formulation for the protein. This restriction is not operating in flFFF, which therefore enables the operator to choose solvent conditions that are relevant to the actual formulation conditions, thus avoiding buffer exchange. The only limitation is that the pH must be different from the pI of the protein (Litzén and Wahlund, 1991b) in order to avoid strong adsorption of protein to the accumulation wall.

Yet, even if today it is only required to differentiate between the monomer of a pharmaceutical protein, on one side, and all the aggregates on the other side, it may be expected that in the future, more detailed information on the aggregation properties could be requested. Moreover, investigations into the detailed molecular mechanisms of noncovalent protein aggregation may be necessary, as such knowledge is

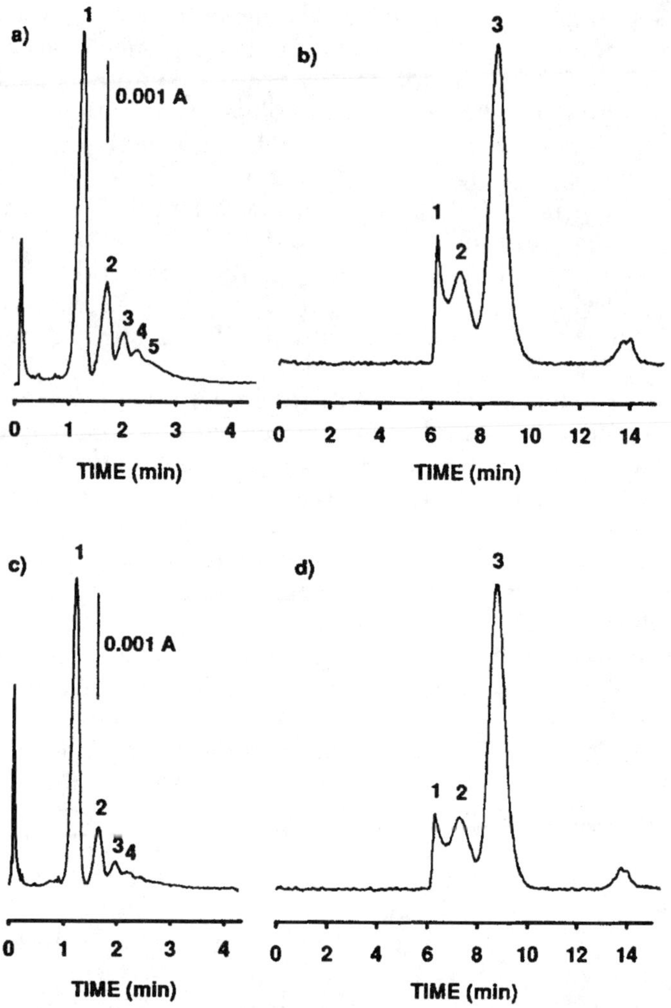

Figure 8 Comparison of asyflFFF and SEC for the separation of two different aggregated monoclonal antibodies (I and II). (a) Antibody I by asyflFFF; (b) antibody I by SEC; (c) antibody II by asyflFFF; (d) antibody II by SEC. asyflFFF: elution flow rates V_C = 8.19 mL/min; V_{out} = 1.81 mg/mL. $t°$ = 0.09 min. Samples: antibody I 0.2 mg/mL; antibody II 0.15 mg/mL. Injected volume: 20 µL. Peaks: 1, monomer; 2, dimer; 3–5, aggregates larger than dimers. SEC: Column, G 3000 SW. Flow rate: 1.0 mL/min. Sample concentrations: antibody I 0.41 mg/mL; antibody II 0.46 mg/mL. Peaks: 1, aggregates larger than dimers; 2, dimer; 3, monomer. Reproduced from Litzen et al. (1993) with permission by Academic Press.

scarce today. For example, the aggregation pattern (i.e., the size of aggregates and under which solvent conditions they form) may have to be characterized. This puts higher demand on a separation method necessitating that as many as possible of the formed aggregates can be detected.

The flFFF method has the potential to meet such demands as illustrated by Figure 9. It demonstrates the very high resolution between protein monomer, dimer, and trimer that can be obtained under optimized conditions for some model proteins. Determination of high-molecular-weight aggregates of pharmaceutical proteins has received recent interest (Fraunhofer et al., 2001).

Protein samples may sometimes contain insoluble matter. This is the case, for example, if a protein has been subjected to conditions that cause strong aggregation resulting in the formation of insoluble agglomerates in addition to smaller, soluble aggregates. Fortuitously, these samples can be analyzed by flFFF without prior filtration, because the agglomerates will be effectively separated from the dissolved protein (Litzen et al., 1993) during the analysis. This is contrary to SEC procedures, which require that such samples always be filtered to remove the particulates prior to their introduction to the SEC column.

Lipoproteins: The quantification of lipoproteins in human blood plasma is an essential task in drug development. The various lipoprotein fractions are important in evaluation of cardiovascular drugs. It was performed by symmetrical flFFF (Li and Hansen, 2000; Li et al., 1997) using both isocratic and programmed-field. Rapid quantification of the high-density lipoprotein (HDL), low-density lipoprotein (LDL), and very-low-density lipoprotein (VLDL) fractions of human

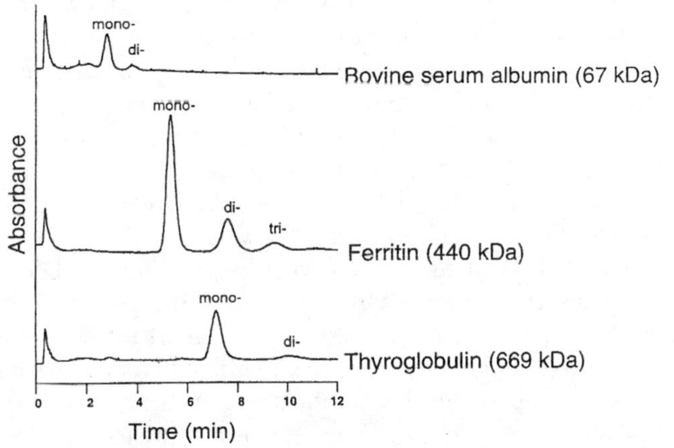

Figure 9 Test fractograms obtained in a trapezoidal asyflFFF channel using proteins that contain aggregates. For each protein, 20 μL of a 0.1 mg/mL solution in carrier was injected. The elution positions for monomers, dimers, and trimers are indicated. The carrier was 0.025 M phosphate buffer at pH 6.5. V_{in} = 7.9 mL/min; V_{out} = 0.5 mL/min; V_c = 7.4 mL/min. Temperature: 25 °C. The channel thickness was 122 μm. The ultrafiltration membrane was a Hoechst NADIR UF10-C regenerated cellulose membrane. The UV detector was set at 280 nm. Reproduced from Wahlund (2000) with permission by Wiley.

blood plasma was obtained. The lipoprotein fractions were clearly identified as different peaks. The plasma samples were measured without sample pretreatment, and differences in lipoprotein profiles could also be seen for different individuals. Not only could the HDL, LDL, and VLDL fractions be resolved and quantified, but with the use of a programmed field, even lipoprotein subspecies could be determined.

A specific form of flFFF, termed *membrane-selective flow FFF*, was demonstrated to be capable of isolating various colloidal constituents that may be present in blood plasma (Li and Giddings, 1996). It uses an accumulation wall membrane with large enough pores (e.g., a cutoff of 30 kDa or 100 kDa) to let smaller macromolecules leave the channel while retaining larger components such as colloids. The potential of the technique was demonstrated for the analysis of lipoprotein fractions found in human blood plasma.

Drug-Plasma Protein Binding: Flow FFF has demonstrated its applicability as a tool in the characterization of drug/plasma protein interactions (Madorin et al., 1997). The distribution of a lipophilic drug (*N*-benzoyl-staurosporine) within human plasma protein fractions was investigated with respect to high separation speed. This method was used to separate human plasma in its protein fractions. The drug distribution in the fractions was determined by monitoring the fractionated eluate for drug content using fluorescence spectroscopy. Human plasma was separated into human serum albumin (HSA), HDL, alpha(2)-macroglobulin, and LDL fractions in less than ten minutes. The influence of membrane type and composition of the carrier solution on the absolute recovery of *N*-benzoyl-staurosporine and fluorescein-isothiocyanate-albumin (FITC-albumin) was found to be quite significant. Both factors were optimized during the course of the investigations. *N*-Benzoyl-staurosporine was found to be enriched in the fraction containing HSA.

If experimental conditions are carefully selected and controlled in order to suppress drug and plasma protein adsorption at the separation membrane, particularly the asymmetrical flow FFF shows high recoveries and fast separation of, for example, human plasma proteins, and has proved to be a reliable tool for component quantification. For analytical purposes, it has the potential to rival established technologies like ultracentrifugation in terms of ease of use, precision, and separation time.

Protein Nanoparticles: Nanoparticles made from proteins are today explored as possible carrier systems for drugs, and their characterization is a major challenge. Drug-loaded nanoparticles may be used for parenteral delivery of DNA or for oral gene delivery in gene therapy. Fraunhofer et al. (2004) have carefully investigated the possibility of applying asyflFFF-UV and asyflFFF-MALS-RI or asyflFFF-MALS-UV for this purpose. A system of gelatin nanoparticles was studied, and it was found that the method was able to characterize comprehensively polymeric gelatin bulk materials as well as drug-loaded and unloaded nanoparticles in terms of particle size, size distribution, molecular weight, and loading efficiency. Figure 10 illustrates the results.

Instrumentation

The instrumental developments in FFF have resulted in the commercial availability of two subtechniques that are suitable for protein analysis: sdFFF and flFFF. The sdFFF has only one manufacturer, Postnova Analytics (Landsberg, Germany), which also manufactures a syflFFF system. The asyflFFF instrument, in turn, is manufactured by three companies: ConSenxus GmbH (Ober-Hilbersheim, Germany), Postnova Analytics, and Wyatt Technology Corp. (Santa Barbara, CA). In addition, a semipreparative flFFF instrument is manufactured by ConSenxus.

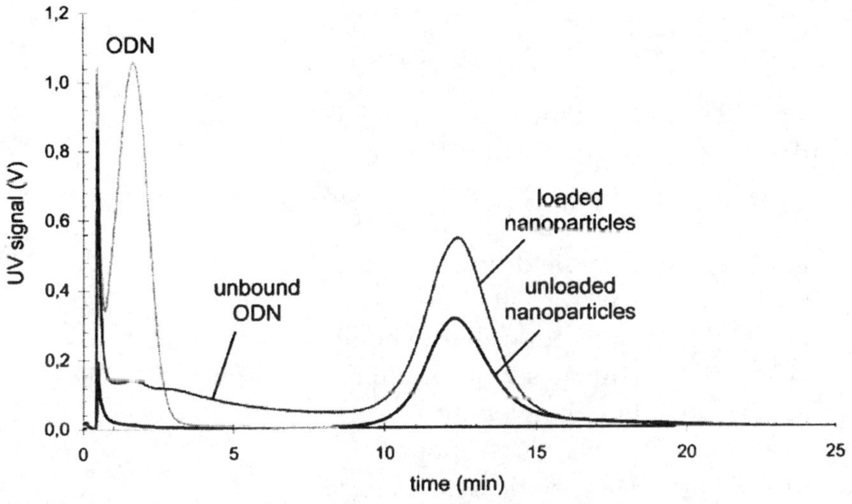

Figure 10 Fractograms from asymmetrical flow FFF-UV detection showing how a DNA specimen (ODN, oligonucleotide; 5.4 kDa) becomes loaded onto gelatin nanoparticles. The pale line (ODN) shows the response of the ODN before incubation with the nanoparticles. The black line shows the response from the nanoparticles alone before incubation (i.e., when they are unloaded). The gray line shows the result after incubation when the nanoparticles have become loaded with ODN so that it contributes to UV-light absorption by the particles while the response from the unbound ODN is much reduced. Reproduced from Fraunhofer et al. (2004) with permission by American Chemical Society.

Conclusions

The FFF techniques offer a unique resolution range for protein analysis, spanning molecular weights from a few kDa to multi MDa and beyond. The separation takes place in unobstructed capillary channels under low shear and is therefore minimally destructive on fragile molecules. Because the resolution is based on physical properties of the sample, as opposed to chemical affinity for most liquid chromatographic stationary phases, the operator has virtually complete freedom to choose a mobile phase that best suits a given sample (i.e., without having to pay attention to effects on matrix binding constants).

In recent years, these features have led to an increased market availability of FFF systems with different design features. This is likely to expand greatly the use of the technique for rapid protein analysis. From the fractogram, the operator can readily derive information regarding a sample's molecular size, in the case of flFFF, or mass, in the case of sdFFF. This feature has been made particularly attractive through the availability of powerful multi-angle laser light scattering detectors, which, in combination with the FFF separator, allow additional on-line evaluation of the molar mass spectrum of even highly complex samples in just a few minutes.

Abbreviations

asyflFFF	Asymmetrical flow field-flow fractionation
BSA	Bovine serum albumin
FITC	Fluorescein isothiocyanate
FFF	Field-flow fractionation
flFFF	Flow field-flow fractionation
HDL	High-density lipoprotein
HPLC	High-performance liquid chromatography
HSA	Human serum albumin
IgG	Immunoglobulin G
LDL	Low-density lipoprotein
MALS	Multi-angle light scattering
RI	Refractive index
sdFFF	Sedimentation field-flow fractionation
SEC	Size-exclusion chromatography
syflFFF	Symmetrical flow field-flow fractionation
VLDL	Very-low-density lipoprotein

Acknowledgment

Both authors are grateful for financial support from the Swedish Research Council.

References

1. Arfvidsson C, and Wahlund K-G. Mass Overloading in the Flow Field-Flow Fractionation Channel Studied by the Behaviour of the Ultra-Large Wheat Protein Glutenin. *J Chromatogr A* 2003; 1011:99–109
2. Arfvidsson C, Wahlund K-G, and Eliasson A-C. Direct molecular weight determination in the evaluation of dissolution methods for unreduced glutenin. *J Cereal Sci* 2004; 39:1–8
3. Caldwell KD. Field-Flow Fractionation. *Anal Chem* 1988; 60:959A–971A
4. Caldwell KD. Electrical Field-Flow Fractionation. In: Schimpf M, Caldwell KD, and Giddings JC. *Field-Flow Fractionation Handbook.* New York: Wiley Interscience, 2000:295–312
5. Caldwell KD, and Gao YS. Electrical Field-Flow Fractionation in Particle Separation. 1. Monodisperse Standards. *Anal Chem* 1993; 65:1764–1772
6. Caldwell KD, Karaiskakis G, Myers MN, and Giddings, JC. Characterization of Albumin Microspheres by Sedimentation Field-Flow Fractionation. *J Pharm Sci* 1981; 70:1350–1352
7. Caldwell KD, Brimhall SL, Gao YS, and Giddings JC. Sample Overloading Effects in Polymer Characterization by Field-Flow Fractionation. *J Appl Polymer Sci* 1988; 36:703–719
8. Caldwell KD, Li J-T, Li J, and Dalgleish DG. Adsorption Behavior of Milk Proteins on Polystyrene Latex – A Study Based on Sedimentation Field-Flow Fractionation. *J Chromatogr* 1992; 604:63–71
9. Fraunhofer W, Schroer C, Tschoepe M, and Winter G. Asymmetical Flow Field-Flow Fractionation: A New Approach to Analysis of High-Molecular Weight Protein Aggregates? In: *Proceedings - 28th International Symposium on Controlled Release of Bioactive Materials and 4th Consumer & Diversified Products Conference,* 2001; San Diego, CA, June 23–27, Vol 2, 900–901, Controlled Release Society, Minneapolis, MN
10. Fraunhofer W, Winter G, and Coester C. Asymmetrical Flow Field-Flow Fractionation and Multiangle Light Scattering for Analysis of Gelatin Nanoparticle Drug Carrier Systems. *Anal Chem* 2004; 76:1909–1920
11. Giddings JC. Field-Flow Fractionation: Analysis of Macromolecular, Colloidal, and Particulate Materials. *Science* 1993; 260:1456–1465
12. Giddings JC, Yang FJ, and Myers MN. Flow Field-Flow Fractionation: A Versatile New Separation Method. *Science* 1976a; 4259:1244–1245
13. Giddings JC, Yang FJ, and Myers MN. Theoretical and Experimental Characterization of Flow Field-Flow Fractionation. *Anal Chem* 1976b; 8:1126–1132
14. Giddings JC, Yang FJ, and Myers MN. Flow Field-Flow Fractionation as a Methodology for Protein Separation and Characterization. *Anal Biochem* 1977; 2:395–407

15. Gunderson JJ, and Giddings JC. Comparison of Polymer Resolution in Thermal Field-Flow Fractionation and Size-Exclusion Chromatography. *Anal Chim Acta* 1986; 189:1–15

16. Huang Y, Wang XB, Becker FF, and Gascoyne PRC. Introducing Dielectrophoresis as a New Force Field for Field-Flow Fractionation. *Biophys J* 1997; 73:1118–1129

17. Langwost B, and Caldwell KD. Solid Phase Immune Reactions as Monitored by Sedimentation Field-Flow Fractionation. *Chromatographia* 1992; 34:317–324

18. Li P, and Giddings JC. Isolation and Measurement of Colloids in Human Plasma by Membrane-Selective Flow Field-Flow Fractionation: Lipoproteins and Pharmaceutical Colloids. *J Pharm Sci* 1996; 8:895–898

19. Li P, and Hansen M. Protein Complexes and Lipoproteins. In: Schimpf M, Caldwell K, and Giddings JC. *Field-Flow Fractionation Handbook*. New York: Wiley-Interscience; 2000:433–470

20. Li P, Hansen M, and Giddings JC. Separation of Lipoproteins from Human Plasma by Flow Field-Flow Fractionation. *J Liq Chrom Rel Technol* 1997; 16-17:2777–2802

21. Litzen A, and Wahlund KG. Improved Separation Speed and Efficiency for Proteins, Nucleic-Acids and Viruses in Asymmetrical Flow Field Flow Fractionation. *J Chromatogr* 1989; 476:413–421

22. Litzén A, and Wahlund K-G. Zone Broadening and Dilution in Rectangular and Trapezoidal Asymmetrical Flow Field-Flow Fractionation Channels. *Anal Chem* 1991a; 63.1001–1007

23. Litzén A, and Wahlund KG. Effects of Temperature, Carrier Composition and Sample Load in Asymmetrical Flow Field-Flow Fractionation. *J Chromatogr* 1991b; 548:393–406

24. Litzen A, Walter JK, Krischollek H, and Wahlund KG. Separation and Quantitation of Monoclonal Antibody Aggregates by Asymmetrical Flow Field-Flow Fractionation and Comparison to Gel Permeation Chromatography. *Anal Biochem* 1993; 212:469–480

25. Madorin M, van Hoogevest P, Hilfiker R, Langwost B, Kresbach GM, Ehrat M, and Leuenberger H. Analysis of Drug Plasma Protein Interactions by Means of Asymmetrical Flow Field Flow Fractionation. *Pharm Res* 1997; 14:1706–1712

26. Maskos M, and Schupp W. Circular Asymmetrical Flow Field-Flow Fractionation for the Semipreparative Separation of Particles. *Anal Chem* 2003; 22:6105–6108

27. Roessner D, and Kulicke WM. Online Coupling of Flow Field-Flow Fractionation and Multi-Angle Laser-Light Scattering. *J Chromatogr A* 1994; 687:249–258

28. Schure MR, Schimpf ME, and Schettler PD. Retention - Normal Mode. In: Schimpf M, Caldwell K, and Giddings JC. *Field-Flow Fractionation Handbook*. New York: Wiley-Interscience; 2000:31–48

29. Shiundu P, Liu G, and Giddings JC. Separation of Particles in Non-Aqueous Suspensions by Thermal Field-Flow Fractionation. *Anal Chem* 1995; 67:2705–2713

30. Stevenson SG, Ueno T, and Preston KR. Automated Frit Inlet/Frit Outlet Flow Field-Flow Fractionation for Protein Characterization with Emphasis on Polymeric Wheat Proteins. *Anal Chem* 1999; 71:8–14

31. Wahlund K-G. Asymmetrical Flow Field-Flow Fractionation. In: Schimpf M, Caldwell K, and Giddings JC. *Field-Flow Fractionation Handbook*. New York: Wiley-Interscience; 2000:279–294

32. Wahlund KG, and Giddings JC. Properties of an Asymmetrical Flow Field-Flow Fractionation Channel Having One Permeable Wall. *Anal Chem* 1987; 59:1332–1339

33. Wahlund KG, and Litzen A. Application of an Asymmetrical Flow Field-Flow Fractionation Channel to the Separation and Characterization of Proteins, Plasmids, Plasmid Fragments, Polysaccharides and Unicellular Algae. *J Chromatogr* 1989; 461:73–87

34. Wahlund KG, Winegarner HS, Caldwell KD, and Giddings JC. Improved Flow Field-Flow Fractionation System Applied to Water-Soluble Polymers: Programming, Outlet Stream Splitting, and Flow Optimization. *Anal Chem* 1986; 58:573–578

35. Wijnhoven JEGJ, Koorn JP, Poppe H, and Kok WT. Influence of Injected Mass and Ionic Strength on Retention of Water-Soluble Polymers and Proteins in Hollow-Fibre Flow Field-Flow Fractionation. *J Chromatogr A* 1996; 732:307–315

36. Wittgren B, and Wahlund K-G. Fast Molecular Mass and Size Characterization of Polysaccharides Using Asymmetrical Flow Field-Flow Fractionation-Multiangle Light Scattering. *J Chromatogr A* 1997; 760:205–218

37. Wyatt PJ. Light Scattering and the Absolute Characterization of Macromolecules. *Anal Chim Acta* 1993; 272:1–40

38. Zborowski M, Sun L, Moore LR, and Chalmers JJ. Rapid Cell Isolation by Magnetic Flow Sorting for Application in Tissue Engineering. *ASAIO J* 1999; 45:127–130

12

Mass Spectrometry: Protein Conformational Analysis and Molecular Recognition

EWALD T.J. VAN DEN BREMER

OctoPlus Development BV Leiden, The Netherlands

ALBERT J.R. HECK

Department of Biomolecular Mass Spectrometry, Bijvoet Center for Biomolecular Research & Utrecht Institute for Pharmaceutical Sciences (UIPS), Utrecht University, Utrecht, The Netherlands

Introduction

Mass spectrometry is a technique that has been around for more than 100 years. However, only during the past decade has mass spectrometry (MS) developed into a mature tool to study larger biomolecules in general and proteins in particular. MS has proved to be extremely useful in determining the primary structure of proteins including possible post-translational modifications. Therefore, currently MS is a key technology in proteome research (see Chapter 13). More recently, MS has also been used in the analysis of secondary and tertiary protein structures, for instance in combination with hydrogen/deuterium exchange (e.g., Simmons and Konermann, 2002). Moreover, MS has been applied to study quaternary structures of proteins and protein complexes and their interactions with enzymes, cofactors, substrates, or other ligands. The use of MS in the analysis of protein higher order structures is the main focus of this chapter.

For many years, electron impact (EI) and chemical ionization (CI) were the only ionization methods available for analyzing molecules by MS; these methods were limited to volatile compounds. In the early 1980s, a new ionization technique was introduced, termed plasma desorption MS (PDMS) (Macfarlane and Torgerson, 1976; Sundqvist and Macfarlane, 1985). In PDMS, secondary gaseous ions are generated in a sputtering process by bombardment with high-energy primary ions (MeV). PDMS was one of the first ionization techniques that allowed the analysis of peptides. A real breakthrough for this application in protein chemistry at that time came after the observation in mass spectra of molecular ions of larger peptides such as insulin (Håkansson et al., 1982). Subsequently, Barber and co-workers developed around 1980 an ionization technique based on bombardment with neutral atoms (e.g., Xe, Ar) (Barber et al., 1981) termed fast atom bombardment (FAB) MS. FABMS, like PDMS, allowed observation of molecular ions of larger peptides (Barber et al., 1982; Dell and Morris, 1982). FAB involves the bombardment of a solid mixture of a matrix (small organic species, e.g., glycerol or thioglycerol) and analyte by a fast particle beam of neutral Ar or Xe atoms. When this high-energy atom beam strikes the matrix, kinetic energy is transferred to the surface molecules and is dissipated in various ways, some of which result in evaporation and ionization of both matrix and analyte molecules. These two techniques, PDMS and FAB, opened the development of new strategies in protein analysis and made it possible to analyze compounds of higher mass up to several tens of kDa. Disadvantages were still the low upper mass limit, the low ionization efficiency (Williams et al., 1982), and the high background due to matrix ion signals. Although PDMS and FAB were important developments, the current prominent role of mass spectrometry in the analysis of biological macromolecules is particularly due to two major technical breakthroughs in the late 1980s: the development of matrix-assisted laser desorption/ionization (MALDI)-MS and electrospray ionization (ESI)-MS. The inventors of these novel ionization methods, John Fenn and Koichi Tanaka, were awarded the 2002 Nobel Prize in chemistry, demonstrating the importance of these developments. The development of the "soft" ionization techniques MALDI and ESI was revolutionary, as it permits analysis of biomolecules in a mass range of up to and over 100 kDa. These techniques require less energy and thereby introduce the biomolecules intact into the gas-phase. Nowadays, these two techniques are routinely used for a wide variety of biomolecules, especially for peptide and protein analysis. These are high-sensitivity techniques, reaching down to the femtomole level. In the following sections, the principles of MALDI and ESI will be discussed in more detail.

Mass Spectrometric Techniques

Electrospray Ionization Mass Spectrometry

The principle of ESI found its origin in molecular beamwork at the end of the 1960s and was first described by Dole et al. (1968). The potential of ESI as a mass spectrometric technique was for the first time demonstrated by Fenn and co-workers in the late 1980s. They described that molecular ions (i.e., proteins) with a high charge-state could be obtained (Fenn et al., 1989). Electrospray is basically based on two processes: (1) the generation of a fine spray of charged droplets and (2) the removal of the solvent to produce molecular ions (e.g., proteins). As described in detail by Kebarle (Kebarle, 2000), in electrospray initially small charged droplets are formed at the tip of the capillary. These charged droplets are electrostatically attracted to the mass spectrometer inlet. Due to the action of the applied electric field on the solution at the capillary tip (1–4 kV), the liquid is enriched for positive ions (in positive ion mode).

Two mechanisms have been proposed to explain the final gas-phase ion formation, named the charge residue model and the ion evaporation model. The first mechanism proposes that ions detected in the mass spectrometer are the charged species that remain upon complete evaporation of the solvent from a droplet (Kebarle and Tang, 1993). Continuous evaporation and fission events would eventually lead to the formation of "droplets" containing a single highly charged ion. In the evaporation model, ions are emitted from the small, highly charged droplets when the repulsion between the charged ion and other charges of the droplet cause expulsion of the ions (Iribarne and Thomson, 1976). In ESI, the formation of ions takes place at atmospheric pressure under very gentle conditions (Wilm and Mann, 1994; Gaskell, 1997).

The introduction of nanoflow ESI, which is a low-flow variant, has played an important role in furthering the analysis of proteins. This allowed the analysis of very small amounts of analyte due to an introduction rate of nanoliters per minute (Wilm and Mann, 1994). Using nano-ESI-MS, typically less than 1 µL is introduced into the ion source through a gold-coated glass capillary with an orifice of 1–3 µm. The nano-ESI source disperses the liquid analyte purely by electrostatic means, and no assistance such as nebulizing gas is needed. It improves desolvation, ionization, and transfer efficiencies into the mass analyzer. The initial droplets have a diameter of less than 200 nm. Thus, their volume is about 100–1000 times smaller than that of the droplets generated using conventional ESI sources. Nano-ESI enables analysis of protein concentrations of less than 1 pmol/µL.

The evaporation process in ESI-MS requires that the solvents and buffers are volatile. This puts certain restrictions on its compatibility with wet biochemical procedures; ESI-MS is most suited for analysis of samples that are highly purified in terms of salts and low-molecular-weight contaminants. On the other hand, ESI-MS is well suited to study, for example, protein-protein, protein-ligand, and protein-metal interactions because in the ESI process weak noncovalent interactions are preserved. The use of ESI-MS for such applications is rapidly expanding and is even not solely restricted to investigate protein interactions at the intermolecular level but also at the intramolecular level (e.g., in protein folding).

Matrix-Assisted Laser Desorption Ionization Mass Spectrometry

The first results of laser/desorption ionization of high-weight molecules were presented in 1987 at the 2nd Japan-China Joint Symposium on Mass Spectrometry and published in 1988 (Tanaka, 1988). In the original work, Tanaka and co-workers used a sample preparation where the analyte is mixed with ultrafine cobalt powder and glycerol as vacuum-stable binding medium. When irradiated with a laser pulse, glycerol and intact analyte molecules could be desorbed into the gas phase. Nowadays, the original method is hardly used. Instead a closely related method, matrix-assisted laser desorption/ionization (MALDI), developed by Karas and Hillenkamp, has become the method of choice (Karas and Hillenkamp, 1988). This ionization technique requires a solid matrix with which the sample is co-crystallized. The three most frequently used matrices for the analysis of proteins and peptides are α-cyano-4-hydroxycinnamic acid, 2,5-dihydroxybenzoic acid, and sinapinic acid. The matrix α-cyano-4-hydroxycinnamic acid is mostly used for the analysis of peptides and small proteins and peptides. Sinapinic acid, in contrast, is mostly used for larger proteins. 2,5-Dihydroxybenzoic acid is a more generally applicable matrix for both peptides and proteins. The matrix absorbs most of the incident energy and is believed to facilitate the ionization process. The charged molecules are directed by electrostatic lenses from the ionization source into the mass analyzer. Subsequently, time-of-flight analysis is often used to separate the ions according to their mass-to-charge ratio. In contrast to ESI, the ionization process in MALDI mainly yields singly charged proteins or peptides (Karas et al., 2000), which makes analysis of mixtures of proteins or peptides relatively simple. MALDI has many attractive features. First, the sensitivity is remarkably high, with spectra of as little as 1 fmol of lysozyme being reported (Karas et al., 1989). Very little suppression is observed in mixture analysis (Beavis and Chait, 1990). The tolerance to the presence of salts and impurities of various kinds is high, which allows analysis of crude samples of biological origin (Goldberg et al., 1991). This high tolerance has been explained by co-crystallization of the protein with the matrix molecule with concomitant exclusion of impurities from the crystals (Karas et al., 1990). A drawback of MALDI is that protein complexes are often not preserved.

Structural Analysis of Proteins

Conformational Changes Based on Charge-State Distribution

Shortly after the introduction of ESI-MS, the potential of this technique to probe higher order structures of proteins was realized. Much of the early work with ESI was focused on the mechanistic aspects of desorption and ionization events. Solutions of pH 3–4 were commonly used for basic peptides and proteins because of their optimal ionization efficiency. The effect of higher pH toward the neutral, physiological regime and to the more basic end of the pH scale was studied for their effects on biological molecules (Guevremont et al., 1992; Kelly et al., 1992; Le Blanc et al., 1993). Early studies (Chowdhury et al., 1990; Loo et al., 1990) demonstrated that folded and unfolded proteins produced a different distribution of charge-states in their electrospray mass spectra. One obvious difference observed from high pH solutions was that the charge distribution typically shifted toward lower charging, that is, higher mass-to-charge, m/z. Fenn et al. have explained this phenomenon of electrospray ionization within the "ion-evaporation" model (Fenn, 1993). According to that theory, tightly folded protein molecules are expected to have significantly smaller projected areas compared to less-structured protein molecules, and would, therefore, accommodate fewer charges on the surface during the ion evaporation stage. As a result, two protein molecules whose geometries are very different from each other (e.g., compact folded states *vs.* less-folded states) will produce ions with significant differences in charge density. If both states are populated simultaneously in solution under certain conditions, then the resultant charge-state distribution in the ESI spectrum could exhibit a bimodal character. This effect has been used extensively during the past several years to investigate protein-unfolding reactions. Konermann and Douglas pioneered this approach to study acid-induced unfolding of several proteins (e.g., Konermann and Douglas, 1998). In Figure 1, ESI-MS spectra are shown of cytochrome c sprayed from solutions at different pH. Between pH 8.5 and 3.2, cytochrome c is folded and displays a "low" charge distribution. When the pH is further reduced, ESI-MS spectra reveal a "high" charge distribution indicative of a more unfolded protein.

More recently, ESI-MS was used to identify an interesting conformational feature of the antibacterial toxin colicin E9 DNase that was not detected in structural studies using NMR and x-ray crystallography (Figure 2) (Van den Bremer et al., 2002). In this case, ESI spectra showed already at neutral pH a clear bimodal charge-state distribution having maxima at 18+ (more unfolded) and 8+ (more folded) (Figure 2). The considerable distinction between the two m/z maxima in the ESI spectra indicates that even at neutral pH there are two protein populations in solution that have very distinct conformational properties. Upon acidification, the ion signals that represent the more folded protein population diminished. Related to this, Mohimen et al. reported a chemometric approach for the analysis of multiple overlapping charge-state distributions to resolve protein conformers in solution (Mohimen et al., 2003).

Konermann and co-workers developed an approach for studying protein folding, termed *time-resolved ESI.* This interesting approach

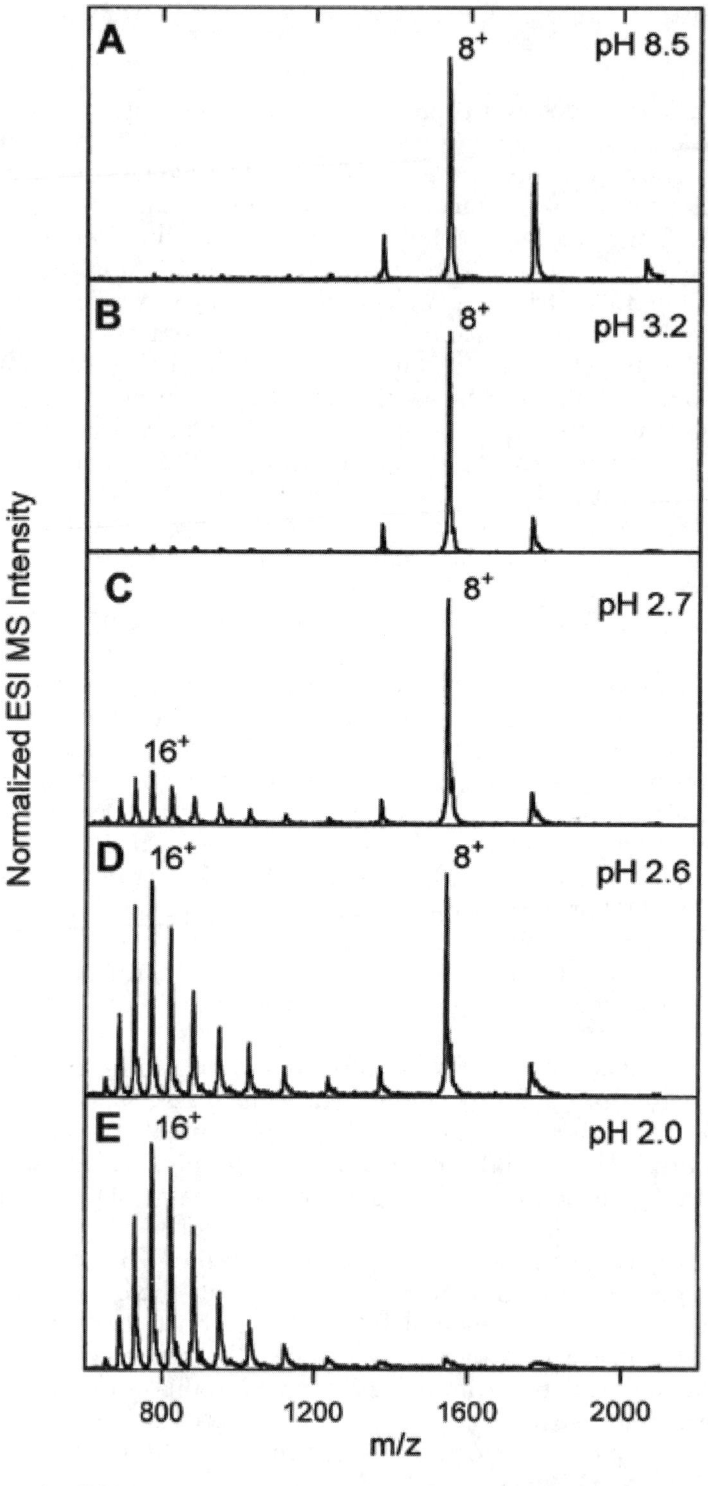

Figure 1 ESI mass spectra of cytochrome c recorded at different pH: (a) pH 8.5, (b) pH 3.2, (c) pH 2.7, (d) pH 2.6, and (e) pH 2.0. The pH was adjusted by addition of ammonium hydroxide and/or acetic acid. Reprinted from Konermann et al. (1998). Copyright (1998), with permission from John Wiley & Sons, Ltd.

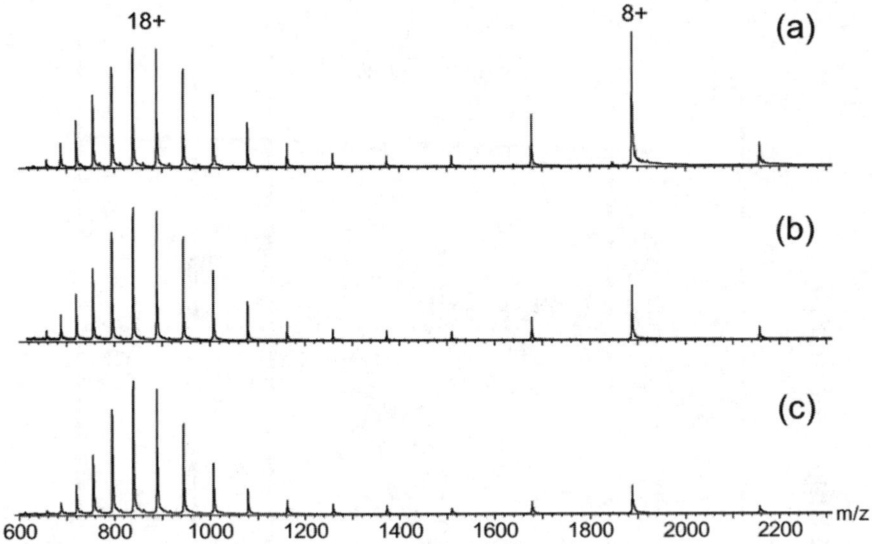

Figure 2 ESI mass spectra of the apo-E9 DNase recorded at different pH values: (a) pH 7.4, (b) pH 5.3, and (c) pH 4.1. Two distinctive charge-state distributions are observed, having maxima at 18+ and 8+ and representing the (more) unfolded and (more) folded protein conformers of colicin E9 DNase, respectively. Reprinted from van den Bremer et al. (2002). Copyright (2002), with permission from Cold Spring Harbor Laboratory Press.

has been used to study the folding of cytochrome c (Konermann et al., 1997a) and myoglobin (Konermann et al., 1997b). In the first of these studies, acid-denatured cytochrome c and a buffer (pH 3.0) were rapidly mixed using a continuous-flow mixing apparatus connected directly to the mass spectrometer. ESI was used to monitor changes in charge-state distributions with a time resolution of about 0.1 second. In this case, no conformational intermediate between folded and unfolded protein was detected. In contrast to these results with cytochrome c, a similar time-resolved ESI experiment with holo-myoglobin revealed the presence of an intermediate, partly unfolded holo-myoglobin species during its acid-induced denaturation into apo-myoglobin (Figure 3). The decay of the peak intensity of the charge-state distribution of the partly unfolded holo-myoglobin revealed that the lifetime of this intermediate was of the order 0.4 seconds, which correlated well with lifetime measurements obtained in related stopped-flow solution phase experiments. These time-resolved ESI experiments show that ESI mass spectrometry can be used to detect intermediates in protein-folding processes. Furthermore, ESI-MS has the advantage over other techniques such as circular dichroism (CD) and steady-state fluorescence in that subpopulations of protein conformers can be analyzed individually instead of the properties of the ensemble of the different subpopulations.

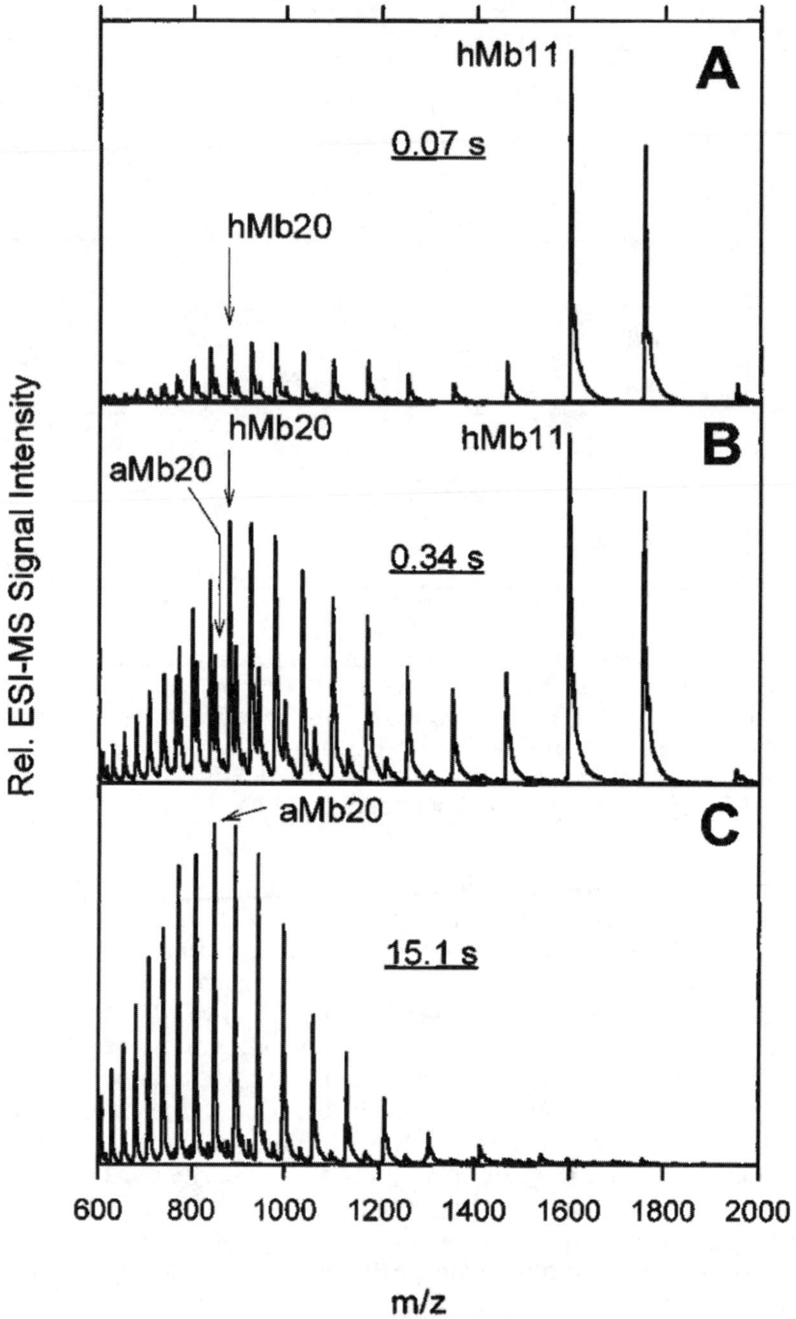

Figure 3 ESI mass spectra of Mb at different times after a pH-jump from 6.5 to 3.2; the spectra were recorded after 0.07 s (A), 0.34 s (B), and 15.1 s (C). Notation: hMb20, holo-myoglobin + 20 H^+; aMb20, apo-myoglobin + 20 H^+, and so forth. The decay of the peak intensity of the charge-state distribution of the partly unfolded holo-myoglobin revealed that the lifetime of this intermediate was of the order 0.4 seconds. Reprinted from Konermann et al. (1997). Copyright (1997), with permission from American Chemical Society.

Protein Structural Analysis by Hydrogen Deuterium Exchange

In the previous section, we showed applications of the use of charge-state distributions in protein conformational analysis and how this approach may be used to distinguish between folded and unfolded forms of a protein. Although charge-state distributions are highly informative about conformational states of a protein, an even more detailed view can be obtained by hydrogen/deuterium (H/D) exchange experiments. H/D exchange has long been used to obtain structural and dynamic information (Hvidt and Nielsen, 1966; Englander and Kallenbach, 1983), and mass spectrometry has simply extended these capabilities. The principle of H/D exchange is that some of the solvent-exposed hydrogens in proteins are capable to exchange with hydrogens from the solvent molecules surrounding the protein. If an isotope of hydrogen is used as the solvent, such as deuterium oxide (D_2O), incorporation into the protein leads to a mass increase that can be monitored. There are several different kinds of hydrogens in proteins. Hydrogens covalently bound to carbon essentially do not exchange, whereas the free amine hydrogens located at the side chains may exchange very rapidly. Backbone amide hydrogens exchange relatively slowly partly because they can be involved in the formation of hydrogen bonds in secondary structural elements (e.g., α-helices or β-sheets). The rate of exchange for these amide hydrogens is typically dependent on solvent accessibility (protein conformation), temperature, and pH. Amide hydrogens buried within the hydrophobic core of a protein or those involved in intramolecular hydrogen bonding can have exchange rates up to eight orders of magnitude slower than solvent-exposed amide hydrogens (Smith, 1998). The general assumption is that proteins with open structures have more hydrogens that are accessible to the solvent (D_2O), and therefore H/D exchange will proceed to a faster and greater extent. One of the first applications of ESI-MS to monitor H/D exchange was reported by Katta and Chait (1991, 1993). They detected the exchange of labile hydrogens with deuterium by monitoring the mass increase of the intact protein. For ubiquitin, they showed that only 62% of the exchangeable hydrogens in "native" ubiquitin exchanged with solvent deuterons, compared to 91% exchange for the unfolded molecule. Exchange rates can be highly informative about the state of a protein, and this has been used in a number of different proteins during the past decade including α-helical or β-sheet peptides (Wagner et al., 1994), lysozyme (Katta and Chait, 1993; Mao et al., 2003), GroEL-bound human dihydrofolate reductase (DHFR) (Gross et al., 1996), GroEL-bound α-lactalbumin (Robinson et al., 1994), myoglobin (Wang and Tang, 1996), cytochrome c (Wagner and Anderegg, 1994), and calmodulin (Nemirovskiy et al., 1999).

In one case, the role of a single amino acid mutation in the stability of cytochrome c_{553} was investigated (Guy et al., 1996). Four mutant proteins were constructed in which Tyr 64 was replaced with either Phe, Leu, Val, or Ala. Subsequently, the rates of H/D exchange were determined for each mutant (Figure 4) and were used to evaluate the relative stability of these four analogues in comparison to the wild-type: wild-type > Y64F = Y64L > Y64V > Y64A. Data obtained with the Phe, Leu, and Val mutants using spectrophotometric methods (Blanchard et al., 1994) indicated that only the Val mutation destabilized the protein, whereas the Phe and Leu mutations had no measurable effect on the stability of the protein.

The use of ESI-MS in combination with H/D exchange was extended by Demmers et al. to study membrane peptides in lipid bilayers (Demmers et al., 2000, 2001, 2002). In this approach, dispersions of fully hydrated lipid bilayers with reconstituted peptides of proteins were directly introduced into the mass spectrometer. Membrane peptides and proteins are characterized by an intrinsic hydrophobicity and insolubility in aqueous solvent, which make analysis by conventional ESI-MS difficult. Until recently, spraying such hydrophobic proteins was only performed from organic solvents, in which they are typically fully denatured. The twofold advantage of the method introduced by Demmers et al. is that (1) the lipid bilayer acts as "matrix" to bring hydrophobic, membrane-associated components from the aqueous buffer solution in the gas-phase, and (2) membrane peptides

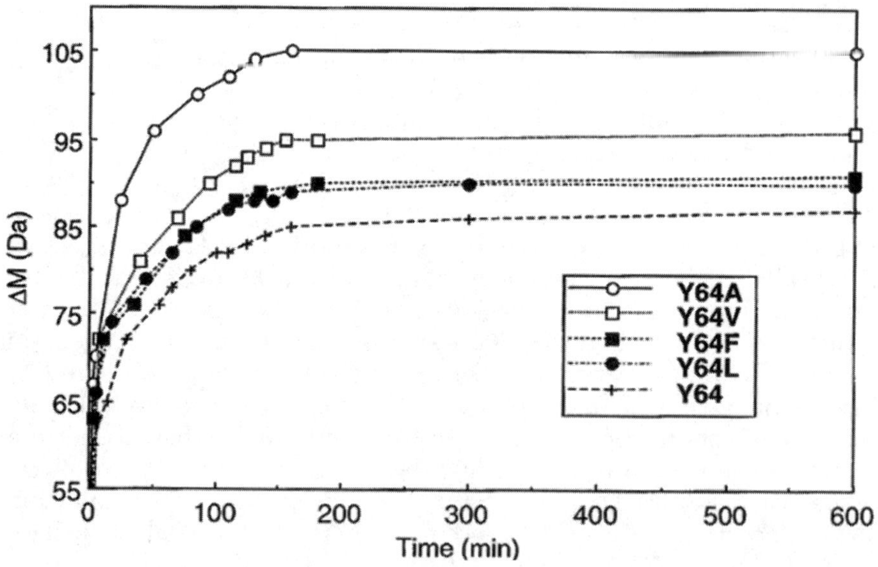

Figure 4 The time-course of H/D exchange in wild-type cytochrome c_{553} and four analogues. Based on the isotopic exchange rates, the relative stabilities of the mutants were found to be wild-type > Y64F = Y64L > Y64V > Y64A. Reprinted from Guy et al. (1996). Copyright (1996), with permission from Elsevier.

and proteins can be studied in their natural environment, the phospholipids bilayer membrane. In general, they measured the uptake of deuterium atoms in peptides while these were incorporated in lipid bilayers upon incubation with D_2O in time. It was shown that the exchanging hydrogens could be divided into several populations with different exchange rates. From this fact, one could derive what parts of the peptide are located within the membrane (aqueous solvent protected) and which parts are sticking out (less protected). This method showed that ESI-MS is capable to study a variety of properties of transmembrane peptides in lipid bilayers, such as the positioning in the bilayer interface and the conformational stability of an α-helix.

Another interesting strategy to monitor protein conformational dynamics was demonstrated by Simmons and Konermann by using time-resolved ESI (see previous section) in conjunction with H/D exchange (Simmons and Konermann, 2002). In Figure 5, the experimental setup of this approach is shown schematically, involving both the rapid mixing of first native heme with unfolded apo-myoglobin followed by mixing with D_2O. The reaction time for reconstitution of holo-myoglobin can be varied by the length of the mixing capillary (Wilson and Konermann, 2003). This approach combines structural

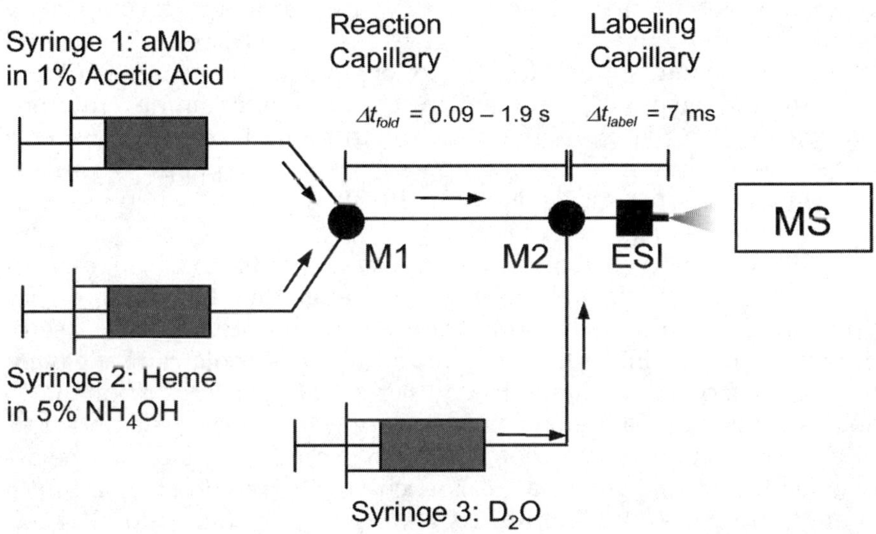

Figure 5 Scheme of the continuous-flow apparatus used to monitor the reconstitution of holo-myoglobin by isotopic pulse-labeling with on-line ESI-MS detection. Heme in basic solution and acid-denatured aMb are mixed at mixer M1, where reconstitution is initiated at pH 10.4. After various reaction times (90 milliseconds to 1.9 seconds), the mixture is diluted with D_2O in a 4:1 ratio (v:v) for isotopic pulse-labeling at mixer M2. On-line analysis by ESI-MS occurs 7 milliseconds after initiation of H/D exchange. Adapted from Simmons and Konermann (2002).

information from the ESI charge-state distribution and from the H/D exchange levels of individual protein states, while at the same time different ligand binding states can be monitored. This technique was used to study the reconstitution of holo-myoglobin (hMb) from unfolded apomyoglobin (aMb) and free heme. In this way, they observed a short-lived folding intermediate with two heme groups attached, representing a protein-bound heme dimer. This state appeared to have a compactness close to that of native hMb; however, H/D exchange results indicated a significantly perturbed structure.

A complication in on-line H/D exchange MS is that the final reaction mixture must be compatible with MS. Thus, the use of urea, guanidine salts, or other nonvolatile solvent additives can be difficult. Folding experiments triggered by a pH jump are relatively easy.

Most of the H/D exchange work in the examples given above characterized only the global behavior of proteins and polypeptides. Smith et al. introduced an elegant method to circumvent this problem by adapting a technique originally developed to probe 1H/3H exchange in a site-specific fashion (Rosa and Richards, 1979; Englander et al., 1985; Smith et al., 1997b). This methodology uses an enzymatic digestion of the protein under slow-back-exchange conditions, followed by measurement of the isotope incorporation in the proteolytic fragments. In the mass spectrometry–based strategy developed by Smith et al., protein amide exchange is quenched by reducing the pH to 2–3 and the temperature to 0 °C, followed by rapid proteolytic fragmentation of the protein with pepsin. The number of amide deuterons present in each peptic fragment can then be deduced directly from mass measurements, following the separation of peptic fragments with fast HPLC (Zhang and Smith, 1993). A general scheme of this methodology is presented in Figure 6.

Following several modifications (Wang and Smith, 2003), this approach to produce site-specific H/D exchange data has become very popular. One important modification of this procedure is the use of immobilized pepsin, so that the entire protein sample can be passed through a series of columns to digest separate peptides prior to MS analysis. Back exchange that occurs during the proteolysis and peptide separation steps does not appear to exceed 10% and may be accounted for by introducing a "back-exchange" correction factor (Smith et al., 1997a; Engen and Smith, 2001). Importantly, these "local" H/D exchange measurements can be carried out in parallel with "global" dynamics studies.

Although ESI has thus far been the most popular technique in combination with H/D exchange measurements, MALDI is now also gaining acceptance in the field. Early studies were carried out by Mandell et al. using the standard protocol (H/D exchange-quench-proteolysis) implemented with MALDI-MS to monitor deuterium incorporation in a site-specific fashion (Mandell et al., 1998).

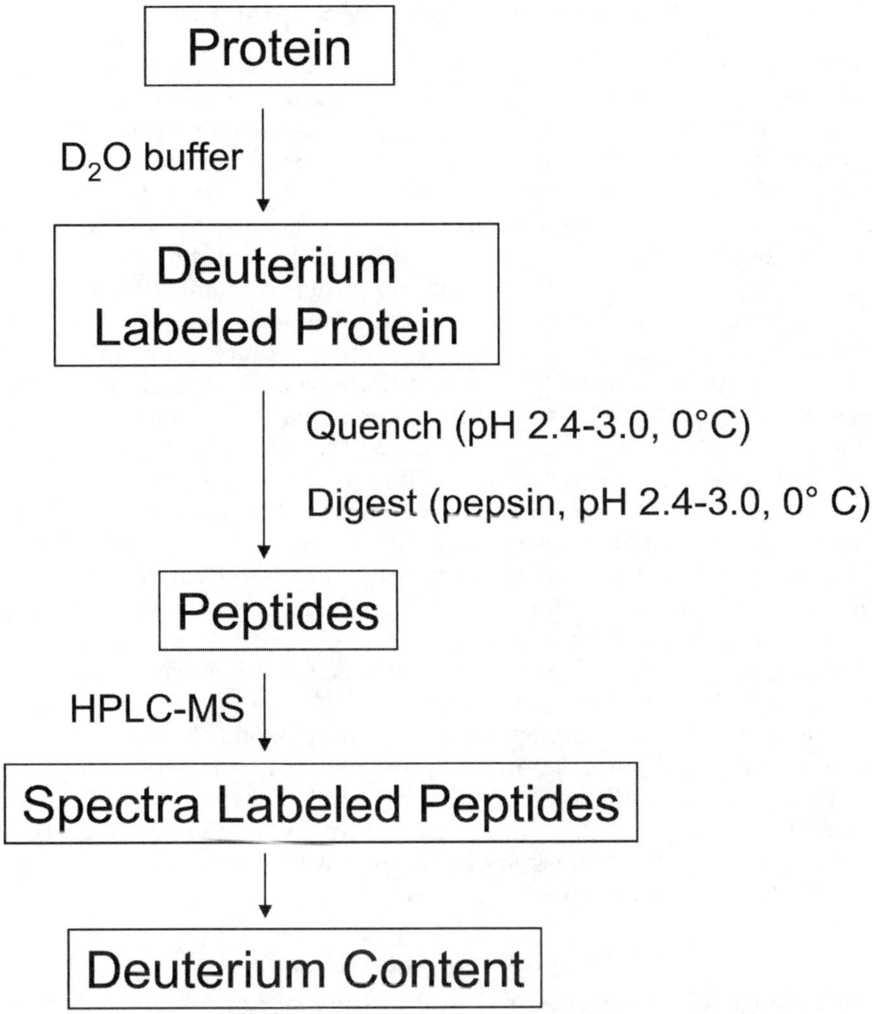

Figure 6 Procedure used to determine deuterium levels at peptide amide linkages in short segments of intact proteins incubated in D_2O.

Only minor modifications of established MALDI sample preparation protocols are required to carry out such experiments. For instance, quenching of amide and peptic digests after exchange are followed by chilled and rapid drying of samples on the MALDI target to ensure minimal loss of deuterons from peptide amide groups. It was found, however, that when the sample remains completely dry and is kept under high vacuum during the analysis, the deuterium content decreases gradually over time. Therefore, a correction for the back exchange prior to mass analysis had to be made (Mandell et al., 1998).

Noncovalent Protein Complexes

Noncovalent interactions of proteins are important in many biological processes, such as in signaling cascades, enzyme-substrate or enzyme-drug interactions, and multi-protein machineries including for instance viruses and proteasome/ribosomes. ESI-MS, with its unique ability to preserve weaker noncovalent bonds when transferring species from solution into the gas phase, has recently proved to be extremely useful for studying protein conformations and to characterize noncovalent protein complexes, including their dynamic behavior.

Although it is evident that the conformations and structures of multiple charged, desolvated proteins in the gas-phase are not identical to the corresponding structures in solution (Loo, 1997), there are pieces of indirect evidence that suggest that data obtained by mass spectrometry may sometimes be used to distill information about that species relevant to its behavior in solution. The most clear evidence lies in the specificity of the noncovalent interactions observed with ESI-MS in many well-documented cases, whereby the detected noncovalent interactions by MS are consistent with those measured using solution-phase–based methods. Starting from the early reports showing the ability to observe intact noncovalent complexes using ESI-MS (Ganem et al., 1991; Katta and Chait, 1991), numerous studies have now been described using this technique (Loo, 1997). A range of different types of interactions has been reported including protein-metal, protein-ligand, protein-oligonucleotide/DNA/RNA, and protein-protein interactions. In the following section, we will describe several recent studies without giving a comprehensive review of the vast literature on the application of mass spectrometry to investigate noncovalent protein interactions studies.

Protein-Metal Interactions

ESI-MS has become very useful in the study of noncovalent protein-metal interactions. ESI-MS can provide a direct and accurate measure of the metal-binding stoichiometry of a protein, based on an observed shift in the mass of the protein arising from the noncovalent binding of the metal ions (Lafitte et al., 1995; Veenstra et al., 1998; Loo, 2001). Here we describe two examples. First, ESI-MS has been used to evaluate the metal-binding stoichiometry of the Ca^{2+}-binding protein, calbindin D_{28K} (Veenstra et al., 1997). Calbindin is composed of six EF-hand Ca^{2+}-binding domains. Traditional methods suggested a stoichiometry anywhere from three to six Ca^{2+}-binding sites (Bredderman and Wasserman, 1974; Gross et al., 1987; Veenstra et al., 1995). ESI-MS conclusively showed that one calbindin D28K molecule binds four Ca^{2+} ions (Veenstra et al., 1997). This conclusion was based on a mass difference of 151 Da between the metal-free protein and the protein saturated with Ca^{2+} ($4 \times 38 = 152$). The mass of Ca^{2+} ion is 40 Da; however, a mass increase of 38 Da is observed for each Ca^{2+}. This result is observed because Ca^{2+} ions, and divalent metal ions in general, replace two protons (2 Da) upon binding. Metal ions are often involved in catalytic events and/or assist the protein to adopt a more stabilized conformation. An intriguing example was described by Van den Bremer et al. for the protein colicin E9 DNase (Van den Bremer et al., 2002). Solutions containing E9 DNase and a fourfold excess of Zn^{2+} ions in 50 mM ammonium acetate resulted in mass spectra that showed a dramatically different appearance compared to those spectra observed for the metal-free E9 DNase under identical solution conditions (Figure 7). The metal-free E9 DNase showed a bimodal charge-state distribution in which the charge-states centered around the 18-fold protonated ion peak were attributed to the loosely packed, conformational states of metal-free E9 DNase and those centered around the 8-fold protonated ion peak to more compact, folded states. The latter charge-state distribution was solely observed for the metal-bound state. The 1:1 metal-protein stoichiometry was in agreement with the x-ray structure (Kleanthous et al., 1999; Ko et al., 1999). These results

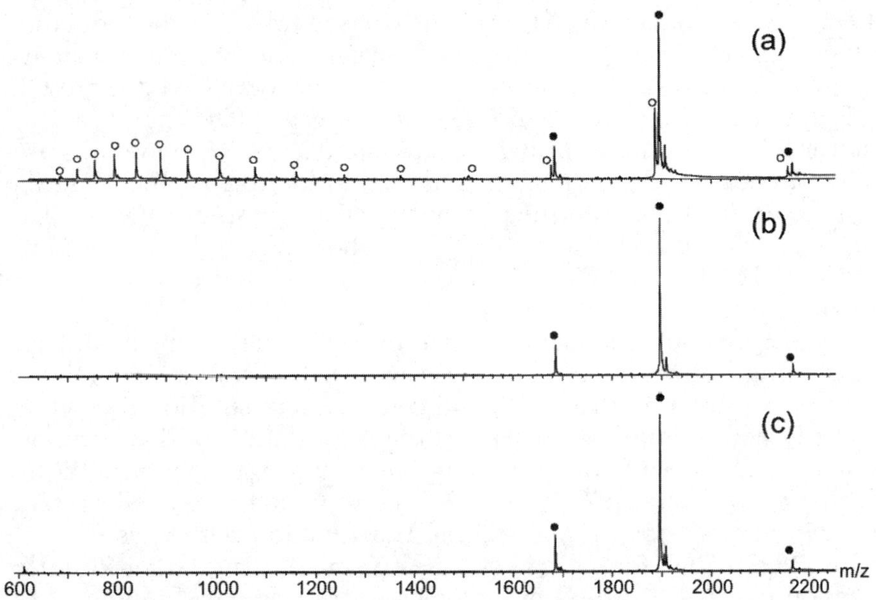

Figure 7 ESI mass spectra of E9 DNase at pH 7.4 in the presence of increasing amounts of Zn^{2+}. Molar ratios of zinc acetate to apo-E9 DNase: (a) 0.25:1, (b) 1:1, and (c) 4:1. Ion peaks of Zn^{2+} bound and metal-free E9 DNase are labeled with filled symbols (●) and open symbols (○), respectively. The 18+ charge-state distribution in (a) represents the more unfolded protein conformer of the E9 DNase and is solely observed for the nonbound metal-free protein. Reprinted from Van den Bremer et al. (2002). Copyright (2002), with permission from Cold Spring Harbor Laboratory Press.

have been interpreted as showing that the metal-free E9 DNase is significantly destabilized in solution with the ESI spectra able to reflect to the conformational states populated. In further agreement, it is known that the thermodynamic stability of E9 DNase increases substantially on metal ion binding (Pommer et al., 1999). Therefore, ESI-MS seems to be able not only to measure the stoichiometry of protein-metal complexes but also to probe metal-induced conformational changes in proteins.

Protein-Ligand Interactions

Some of the earliest studies showing noncovalent interactions by mass spectrometry between ligands and proteins were the observation of the intact heme-myoglobin complex (Katta and Chait, 1991) and an intact complex FK binding protein (FLBP) (Ganem et al., 1991), the latter being an immunosuppressive binding protein, bound to the immunosuppressive agents FK506 and rapamycin. Because noncovalent ESI-MS is able to measure the molecular mass of protein-ligand complexes, it is able to distinguish accurately two closely related ligands that bind to a common site. An illustrative example is the study of the ras protein that may be bound by GDP and GTP (Ganguly et al., 1992; Huang et al., 1993). The ras protein is involved in the regulation of cell growth and differentiation; however, the protein is inactive when bound to GDP. Therefore, in study of uncontrolled cell growth, it is important to be able to differentiate between GDP-bound ras protein and GTP-bound ras protein. GDP-bound and GTP-bound ras proteins are easily distinguished using ESI-MS due to the differing molecular mass between the two noncovalent complexes (MW GDP-ras = 19,295 Da; GTP-ras = 19,375 Da). The direct measurement provided by ESI-MS left no ambiguity to the identity of the ligand bound within the active site of the ras protein.

ESI-MS is not only useful to determine the mass of the ligand and its binding stoichiometry but can also be used to evaluate the relative strength of different protein-ligand complexes in solution (Loo et al., 1997; Jorgensen et al., 1998; Vollmerhaus et al., 2003). Loo et al. applied this to the SH2 domain on a tyrosine kinase protein from the Src family (Loo et al., 1997). The SH2 domain mediates crucial protein-protein interactions in many signal transduction pathways (Grucza et al., 1999). Therefore, it makes this a potential drug target for intervention in a wide variety of biological processes. The binding of tyrosyl phosphopeptides can mimic these interactions. The relative abundances of the multiple charged ions for the complex between Src SH2 protein and several nonphosphorylated and phosphorylated peptides were compared. The mass spectrometry data correlated well to the measured binding constants derived from solution-based methods, indicating that the mass spectrometry–based method can be used to assess the affinity of such interactions. ESI-MS of a solution containing Src SH2 with a mixture of phosphopeptides showed the expected protein-phosphopeptide complex as the dominant species in the mass spectrum, demonstrating the method's potential for screening mixtures from peptide libraries. Still, however, care must be taken when relating gas-phase findings with solution-phase binding properties because there are also reports that showed less correlation (e.g., Chung et al., 1999).

A study by Rostom et al. (2000) dealt with the 58 kDa periplasmic peptide receptor protein OppA that was allowed to interact with 11 peptides with diverse properties. Peptides with two, three, and five amino acid residues were added to OppA. ESI-MS showed that the highest affinity complexes were observed for OppA with tripeptide ligands. Lower affinity complexes were observed for OppA with dipeptide ligands, and no complex formation was detected with pentapeptides or a tripeptide in which the N-terminal amino acid group was acetylated. Tripeptides containing a single D amino acid residue were found not to bind to native OppA. Analysis of the proportions of peptide-bound and free proteins under low-energy MS conditions showed a reasonable correlation with solution-phase K_d measurements. Increasing the internal energy of the gas-phase complex led to dissociation of the complex. The ease of dissociation highlighted the stability of the complex. The results from this study correlate very well with the physiological role of this protein, that is, allowing into the cell di- and tripeptides containing naturally occurring amino acids, regardless of their sequence.

Protein-DNA Interactions

One of the first ESI-MS studies in which ds-DNA was complexed with a natural binding partner involved the DNA-binding domain of the eukaryotic transcription factor, PU.1 (Cheng et al., 1996). Mixtures of PU.1-DNA binding domain (DBD) protein and wild-type target DNA sequence yielded ESI-MS spectra showing only protein-ds-DNA complex ions of 1:1 stoichiometry and free ds-DNA. When PU.1-DBD protein, wild-type target DNA, and mutant target DNA lacking the consensus sequence were mixed, only the 1:1 complex with the wild-type DNA was observed, consistent with mobility shifts in a gel electrophoresis assay. These results provided strong evidence that the detected PU.1-protein DNA complex was not merely the result of a basic peptide nonspecifically interacting with the negative phosphate DNA backbone. Furthermore, the data highlighted the sequence-specific interactions as observed in solution.

In another report, Potier and co-workers used a similar approach to detect protein-DNA complexes (Potier et al., 1998). ESI-MS was used to study interactions among the trp apo-repressor (TrpR), tryptophan, and its specific operator DNA sequence. The trp operator/repressor is one of the most studied transcription systems in biochemistry (Bass et al., 1988). Binding of tryptophan to the repressor protein changes its conformational shape and allows it to bind to its operator DNA sequence, preventing transcription of genes by RNA polymerase. In 5 mM ammonium acetate, TrpR was detected as partially unfolded monomer. In the presence of a 21-base-pair DNA possessing the two symmetrically arranged CTAG consensus required for specific TrpR binding, a homodimer-dsDNA complex with a 1:1 stoichiometry was observed. Furthermore, to check the binding specificity of the TrpR for its target DNA, they performed a competition experiment. The protein was mixed with an equimolar amount of three different DNAs in which the two CTAG sequences were separated by 2, 4, and 6 bp, respectively. Only the DNA with the correct consensus spacing of 4 bp was able to form stable interactions with TrpR. This experiment demonstrates the potential of ESI-MS to test the sequence-specificity of protein-DNA complexes.

Although the application of ESI to protein-DNA complexes is relatively new, these two reports show that mass spectrometry can provide precise binding stoichiometries and can characterize relative affinities of a protein for a particular DNA sequence. However, one disadvantage in the analysis of protein-DNA complexes is the high polarity of the DNA phosphate backbone. Alkali metal cations have a tendency to bind very tightly to oligonucleotides, which may lead to broader peaks and to erroneous mass measurements. However, nowadays, the resolution of mass spectrometers is often sufficient to separate the sodium adducts so that a very precise assignment of the complex could be made.

Protein-Protein Interactions

Many proteins interact *in vivo* strongly and/or weakly with similar proteins to form oligomers or with other proteins to establish functional protein complexes and/or signaling pathways. ESI-MS may provide direct, unambiguous data showing the formation of a protein-protein complex based on the mass of the formed complex. This technique has a significant advantage compared to spectroscopic techniques, for instance, when examining cofactor-dependent protein-protein interactions (Fabris and Fenselau, 1999; Tahallah et al., 2002). The protein-protein and cofactor stoichiometry is readily determined within a single experiment, as the observed mass of the noncovalent complex should be the sum of the masses of all these identities.

ESI-MS on the quaternary structure of 4-Oxalocrotonate tautomerase was one of the first studies in which the oligomeric state of a protein was probed by MS (Fitzgerald et al., 1996). 4-Oxalocrotonate tautomerase was estimated to be a pentamer on the basis of size-exclusion chromatography and ultracentrifugation (Chen et al., 1992). X-ray crystallography and mass spectrometry clearly showed that the protein forms a hexamer of 41 kDa (Roper et al., 1994; Fitzgerald et al., 1996). Four single point-mutated 4-Oxalocrotonate tautomerases, on the other hand, only formed monomeric protein ions. Thus, these data provide information about the oligomerization behavior of wild-type protein and residues important in maintaining the hexameric structure.

Following the pioneering work of the research group of Standing, several other groups have used ESI-MS for the analysis of the stoichiometry of oligomeric protein complexes of known and unknown composition. Interesting studies are reported on complexes of proteasome activator proteins (Yao et al., 1999; Zhang et al., 1999), the chaperone complex GroEL (Rostom and Robinson, 1999), and many different variants of hemoglobin (Zal et al., 1997, 2000; Apostol, 1999; Lippincott et al., 2000; Versluis and Heck, 2001). ESI-MS experiments have confirmed that GroEL is composed of 14 noncovalently bound subunits arranged in two heptameric rings with a total molecular mass of 800 kDa (Rostom and Robinson, 1999). For its functional activity, GroEL requires the presence of a co-chaperone, GroES. The association of the mobile loop of GroES and the model substrate helix D from rhodandese to GroEL was also probed by mass spectrometry (Ashcroft et al., 2002). Together with the results from fluorescence binding studies, it was concluded that chaperone GroES and substrate proteins have, at least partially, distinct binding sites even in the intact GroEL tetradecamer.

Hemoglobin is in most mammalian systems present as a tetrameric complex of approximately 60 kDa possibly in equilibrium with dimers in lower abundance, as also revealed by mass spectrometry. Mass spectrometry has been used extensively to study different forms of hemoglobin, for example, the study of hemoglobin S and fetal hemoglobin in blood from material of patients with homozygous sickle cell disease (Ofori-Acquah et al., 2001). This study concentrated on the quantitative distribution of asymmetric hemoglobin hybrid and other tetrameric species in blood of patients with sickle cell disease. The noncovalent association of hemoglobin subunits was studied by ESI mass spectrometry. Mass spectra of both patient and fetal blood revealed intact hetero-hemoglobin tetramers of both $\alpha_2\beta_2$ and hybrid $\alpha_2\gamma\beta$. A unique tetrameric marker protein of average mass 64.6 kDa was identified in hemolysates from patients with sickle cell disease in accordance with the calculated mass of the asymmetric hemoglobin hybrid.

Another elegant study revealing the potential of mass spectrometry deals with the investigation of protein oligomerization on the transthyretin system (McCammon et al., 2002). The tetrameric human protein transthyretin transfers the hormone thyroxine and, through its noncovalent interactions with the retinol binding protein, retinol. Transthyretin is thought to be one of the human proteins involved in the formation of amyloid fibrils, which is associated with diseases such as Alzheimer disease, type II diabetes, and the transmissible spongiform encephalopathies. Under normal conditions, transthyretin transports thyroxine and retinol, but misfolding of wild-type and single point mutations lead to senile systematic amyloidosis and familial amyloid polyneuropathy, respectively. Dissociation of the tetrameric transthyretin complex is thought to be the first step in the formation of transthyretin amyloid fibrils (Miroy et al., 1996). McCammon et al. examined a series of eighteen possible inhibitors of this process by mass spectrometry (McCammon et al., 2002). The ligands were evaluated for their ability to bind to and stabilize the tetrameric structure, their cooperativity in binding to the transthyretin complex, and their ability to compete with the natural ligand thyroxine. The observation of the multifaceted ten-component complex containing six protein subunits (four transthyretin and two retinol binding protein), two retinol molecules, and two synthetic ligands allowed them to conclude that ligand binding does not inhibit association of transthyretin with retinol-bound retinol binding protein. Thyroxine and the probed synthetic ligands showed different cooperativity characteristics. Negative, positive, and noncooperativity mechanisms could be observed; among them the well established negative cooperativity for thyroxine binding.

Another interesting example of the use of mass spectrometry on very large protein complexes is the work of Pinkse et al. on the supramolecular assembly of *Helicobacter pylori* urease (Pinkse et al., 2003). *H. pylori* is a gram-negative spiral bacterium infecting approximately half of the world's population and an important etiologic agent in a variety of diseases (Covacci et al., 1999). The abundant protein complex urease is essential for survival in the acidic environment of the stomach. Urease catalyses the hydrolysis of urea, present in the gastric juice, to ammonium and carbamate. It has been suggested that the acid-resistant *H. pylori* urease is probably dependent on its supramolecular architecture and contributes to its ability to resist and survive in the harsh conditions in the stomach. Therefore, urease is considered as a potentially useful target for developing new therapeutic avenues for eradicating *H. pylori* infections. *H. pylori* urease consists of two subunits, a 25.5 kDa α-subunit and a 61.7 kDa β-subunit (Hu and Mobley, 1990; Evans et al., 1991). Nanoflow ESI mass spectrometry established that the urease complex has a molecular weight of 1,063,900 ± 600 Da, corresponding to a dodecameric $(\alpha\beta)_{12}$ assembly of α- and β-subunits (Figure 8). The observation that the dodecamer readily disassembles into $(\alpha\beta)_3$ subunits is in strong support for a $((\alpha\beta)_3)_4$ architecture. This structural design was consistent with the recently published x-ray structure, revealing a spherical assembly of twelve catalytic $(\alpha\beta)$ units (Ha et al., 2001).

Conclusions

In this chapter, we have described some of the prospectives of MS-based techniques for the investigation of protein folding and protein

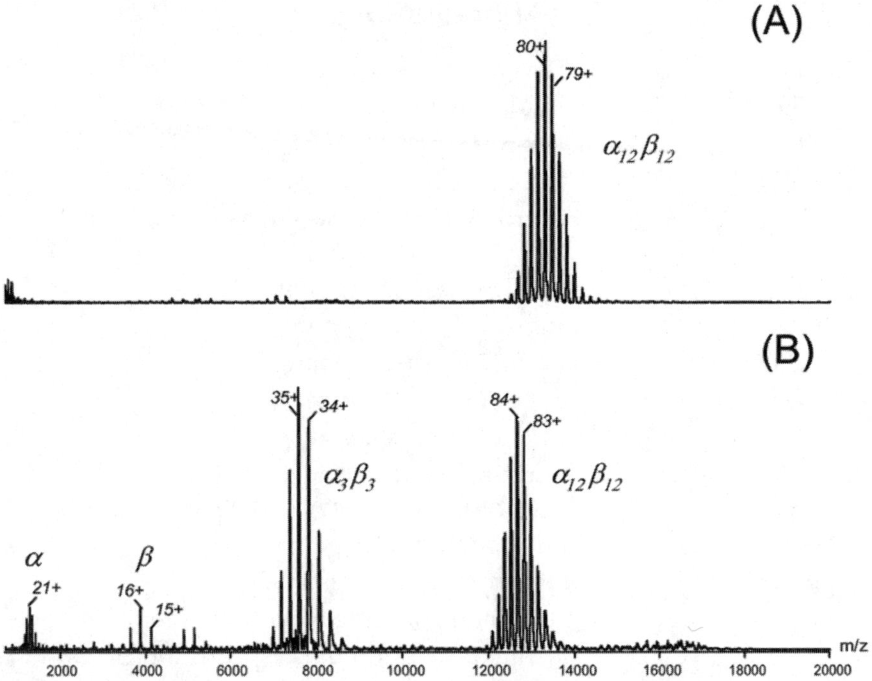

Figure 8 ESI mass spectra of *H. pylori* urease, electrosprayed from an aqueous 200 mM ammonium acetate solution, pH 6.4, at an estimated [$\alpha\beta$] concentration of ~20 μM, acquired in (A) positive ion mode with a source temperature of 40 °C; (B) with a source temperature of 70 °C. Adapted from Pinkse et al. (2003).

noncovalent interactions. The examples presented in this chapter indicate that in recent years, mass spectrometry has become an alternative tool in structural biology that may complement more established methods. Although mass spectrometry has its limitations, as it is still a gas-phase–based method, it also has its unique strengths. Mass spectrometry is highly sensitive, requiring very little sample, is very selective, and has nearly no limitations in size of the species under investigation. As ESI-MS allows the simultaneous monitoring of multicomponent species (differentiating by mass) or even conformations (differentiating by charge) present in solution, it allows the investigation of dynamic features, such as assembly of disassembly, and equilibria, and therefore opens up the possibility to study particular biological problems that cannot easily be addressed by alternative methods. The application of ESI-MS in structural biology is still in its infancy, but it is our belief that it will grow quite extensively in the near future in a direction from the more fundamental applications described here toward pharmaceutical applications. Beyond doubt, ESI-MS–based ligand/drug-screening of pharmaceutical proteins will be an important application.

Abbreviations

bp	Base pairs
CD	Circular dichroism
Da	Dalton, atomic mass unit
DBD	PU.1-DNA binding domain
ESI-MS	Electrospray ionization mass spectrometry
FAB	Fast atom bombardment
GDP	Guanosine diphosphate
GTP	Guanosine triphosphate
H/D	Hydrogen/deuterium
K_d	Dissociation constant
MALDI-MS	Matrix-assisted laser desorption/ionization mass spectrometry
MS	Mass spectrometry
NMR	Nuclear magnetic resonance
OppA	Periplasmic peptide receptor protein
PDMS	Plasma desorption mass spectrometry

Acknowledgments

We thank the Netherlands Organization for Scientific Research (NWO, grant no. 98034 to E.T.J.vdB.) and members past and present of the Biomolecular Mass Spectrometry Department at Utrecht University.

References

1. Apostol I. Assessing the Relative Stabilities of Engineered Hemoglobins Using Electrospray Mass Spectrometry. *Anal Biochem* 1999; 272:8–18
2. Ashcroft AE, Brinker A, Coyle JE, Weber F, Kaiser M, Moroder L, Parsons MR, Jager J, Hartl UF, Hayer-Hartl M, and Radford SE. Structural Plasticity and Noncovalent Substrate Binding in the GroEL Apical Domain. A Study Using Electrospray Ionization Mass Spectrometry and Fluorescence Binding Studies. *J Biol Chem* 2002; 277:33115–33126
3. Barber M, Bordoli RS, Sedwick RD, and Tyler AN. Fast Atom Bombardment of Solids (FAB): A New Ion Source for Mass Spectrometry. *J Chem Soc Chem Commun* 1981; 1981:325–327
4. Barber M, Bordoli RS, Elliot GJ, Sedgwick RD, and Tyler AN. Fast Atom Bombardment Mass Spectrometry of Bovine Insulin and Other Large Peptides. *J Chem Soc Chem Commun* 1982; 1982:936–938

5. Bass S, Sorrells V, and Youderian P. Mutant Trp Repressors with New DNA-Binding Specificities. *Science* 1988; 242:240–245

6. Beavis RC, and Chait BT. Rapid Sensitive Analysis of Protein Mixtures by Mass Spectrometry. *Proc Natl Acad Sci U S A* 1990; 87:6873–6877

7. Blanchard L, Dolla A, Bersch B, Forest E, Bianco P, Wall J, Marion D, and Guerlesquin F. Effects of the Tyr64 Substitution on the Stability of Cytochrome c553, a Low Oxidoreduction-Potential Cytochrome from Desulfovibrio vulgaris Hildenborough. *Eur J Biochem* 1994; 226:423–432

8. Bredderman PJ, and Wasserman RH. Chemical Composition Affinity for Calcium and Some Related Properties of the Vitamin D Dependent Calcium-Binding Protein. *Biochemistry* 1974; 13:1687–1694

9. Chen LH, Kenyon GL, Curtin F, Harayama S, Bembenek ME, Hajipour G, and Whitman C. P 4-Oxalocrotonate Tautomerase, an Enzyme Composed of 62 Amino-Acid-Residues Per Monomer. *J Biol Chem* 1992; 267:17716–17721

10. Cheng X, Morin PE, Harms AC, Bruce JE, Ben-David Y, and Smith RD. Mass Spectrometric Characterization of Sequence-Specific Complexes of DNA and Transcription Factor PU1 DNA Binding Domain. *Anal Biochem* 1996; 239:35–40

11. Chowdhury SK, Katta V, and Chait BT. Probing Conformational Changes in Proteins by Mass Spectrometry. *J Am Chem Soc* 1990; 112:9012–9013

12. Chung EW, Henriques DA, Renzoni D, Morton CJ, Mulhern TD, Pitkeathly MC, Ladbury JE, and Robinson CV. Probing the Nature of Interactions in SH2 Binding Interfaces—Evidence from Electrospray Ionization Mass Spectrometry. *Protein Sci* 1999; 8:1962–1970

13. Covacci A, Telford JL, Del Giudice G, Parsonnet J, and Rappuoli R. Helicobacter pylori Virulence and Genetic Geography. *Science* 1999; 284:1328–1333

14. Dell A, and Morris HR. Fast Atom Bombardment-High Field Magnetic Mass Spectrometry of 6000 Da Polypeptides. *Biochem Biophys Res Commun* 1982; 106:1456–1461

15. Demmers JA, Haverkamp J, Heck AJ, Koeppe RE 2nd, and Killian JA. Electrospray Ionization Mass Spectrometry as a Tool to Analyze Hydrogen/Deuterium Exchange Kinetics of Transmembrane Peptides in Lipid Bilayers. *Proc Natl Acad Sci U S A* 2000; 97:3189–3194

16. Demmers JA, van Duijn E, Haverkamp J, Greathouse DV, Koeppe RE 2nd, Heck AJ, and Killian JA. Interfacial Positioning and Stability of Transmembrane Peptides in Lipid Bilayers Studied by Combining Hydrogen/Deuterium Exchange and Mass Spectrometry. *J Biol Chem* 2001; 276:34501–34508

17. Demmers JA, Rijkers DT, Haverkamp J, Killian JA, and Heck AJ. Factors Affecting Gas-Phase Deuterium Scrambling in Peptide Ions and Their Implications for Protein Structure Determination. *J Am Chem Soc* 2002; 124:11191–11198
18. Dole M, Mack LL, Hines RL, Mobley RC, Ferguson LO, and Alice MB. Molecular Beams of Macroions. *J Chem Phys* 1968; 49:2240–2249
19. Engen JR, and Smith DL. Investigating Protein Structure and Dynamics by Hydrogen Exchange MS. *Anal Chem* 2001; 73:256A–265A
20. Englander SW, and Kallenbach NR. Hydrogen Exchange and Structural Dynamics of Proteins and Nucleic Acids. *Q Rev Biophys* 1983; 16:521–655
21. Englander JJ, Rogero JR, and Englander SW. Protein Hydrogen Exchange Studied by the Fragment Separation Method. *Anal Biochem* 1985; 147:234–244
22. Evans DJ Jr, Evans DG, Kirkpatrick SS, and Graham DY. Characterization of the Helicobacter pylori Urease and Purification of Its Subunits. *Microb Pathog* 1991; 10:15–26
23. Fabris D, and Fenselau C. Characterization of Allosteric Insulin Hexamers by Electrospray Ionization Mass Spectrometry. *Anal Chem* 1999; 71:384–387
24. Fenn JB. Ion Formation from Charged Droplets: Roles of Geometry Energy and Time. *J Am Soc Mass Spectrom* 1993; 4:524–535
25. Fenn JB, Mann M, Meng CK, Wong SF, and Whitehouse CM. Electrospray Ionization for Mass Spectrometry of Large Biomolecules. *Science* 1989; 246:64–71
26. Fitzgerald MC, Chernushevich I, Standing KG, Whitman CP, and Kent SB. Probing the Oligomeric Structure of an Enzyme by Electrospray Ionization Time-of-Flight Mass Spectrometry. *Proc Natl Acad Sci U S A* 1996; 93:6851–6856
27. Ganem B, Li YT, and Henion JD. Detection of Noncovalent Receptor Ligand Complexes by Mass-Spectrometry. *J Am Chem Soc* 1991; 113:6294–6296
28. Ganguly AK, Pramanik BN, Tsarbopoulos A, Covey TR, Huang E, and Fuhrman SA. Mass-Spectrometric Detection of the Noncovalent GDP-Bound Conformational State of the Human H-Ras Protein. *J Am Chem Soc* 1992; 114:6559–6560
29. Gaskell SJ. Electrospray: Principles and Practice. *J Mass Spectrom* 1997; 32:677–688
30. Goldberg DE, Slater AF, Beavis R, Chait B, Cerami A, and Henderson GB. Hemoglobin Degradation in the Human Malaria Pathogen Plasmodium falciparum: A Catabolic Pathway Initiated by a Specific Aspartic Protease. *J Exp Med* 1991; 173:961–969

31. Gross M, Robinson CV, Mayhew M, Hartl FU, and Radford SE. Significant Hydrogen Exchange Protection in GroEL-bound DHFR Is Maintained During Iterative Rounds of Substrate Cycling. *Protein Sci* 1996; 5:2506–2513

32. Gross MD, Nelsestuen GL, and Kumar R. Observations on the Binding of Lanthanides and Calcium to Vitamin D-Dependent Chick Intestinal Calcium-Binding Protein Implications Regarding Calcium-Binding Protein Function. *J Biol Chem* 1987; 262:6539–6545

33. Grucza RA, Bradshaw JM, Futterer K, and Waksman G. SH2 Domains: From Structure to Energetics, A Dual Approach to the Study of Structure-Function Relationships. *Med Res Rev* 1999; 19:273–293

34. Guevremont R, Siu KWM, Leblanc JCY, and Berman SS. Are the Electrospray Mass-Spectra of Proteins Related to Their Aqueous-Solution Chemistry. *J Am Soc Mass Spectrom* 1992; 3:216–224

35. Guy P, Remigy H, Jaquinod M, Bersch B, Blanchard L, Dolla A, and Forest E. Study of the New Stability Properties Induced by Amino Acid Replacement of Tyrosine 64 in Cytochrome C553 from Desulfovibrio vulgaris Hildenborough Using Electrospray Ionization Mass Spectrometry. *Biochem Biophys Res Commun* 1996; 218:97–103

36. Ha NC, Oh ST, Sung JY, Cha KA, Lee MH, and Oh BH. Supramolecular Assembly and Acid Resistance of Helicobacter pylori Urease. *Nat Struct Biol* 2001; 8:505–509

37. Håkansson P, Kamensky I, Sundqvist B, Fohlman J, Peterson P, McNeal CJ, and Macfarlane RD. 127-I Plasma Desorption Mass Spectrometry of Insuline. *J Am Chem Soc* 1982; 104:2948–2949

38. Hu LT, and Mobley HL. Purification and N-terminal Analysis of Urease from Helicobacter pylori. *Infect Immun* 1990; 58:992–998

39. Huang EC, Pramanik BN, Tsarbopoulos A, Reichert P, Ganguly AK, Trotta PP, Nagabhushan TL, and Covey TR. Application of Electrospray Mass-Spectrometry in Probing Protein Protein and Protein Ligand Noncovalent Interactions. *J Am Soc Mass Spectrom* 1993; 4:624–630

40. Hvidt A, and Nielsen SO. Hydrogen Exchange in Proteins. *Adv Protein Chem* 1966; 21:287–386

41. Iribarne JV, and Thomson BA. On the Evaporation of Small Ions from Charged Droplets. *J Chem Phys* 1976; 64:2287–2294

42. Jorgensen TJD, Roepstorff P, and Heck AJR. Direct Determination of Solution Binding Constants for Noncovalent Complexes Between Bacterial Cell Wall Peptide Analogues and Vancomycin Group Antibiotics by Electrospray Ionization Mass Spectrometry *Anal Chem* 1998; 70:4427–4432

43. Karas M, and Hillenkamp F. Laser Desorption Ionization of Proteins with Molecular Masses Exceeding 10000 daltons *Anal Chem* 1988; 60:2299–2301

44. Karas M, Ingendoh A, Bahr U, and Hillenkamp F. Ultraviolet-Laser Desorption Ionization Mass-Spectrometry of Femtomolar Amounts of Large Proteins. *Biomed Environ Mass Spectrom* 1989; 18:841–843

45. Karas M, Bahr U, Ingendoh A, Nordhoff E, Stahl B, Strupat K, and Hillenkamp F. Principles and Applications of Matrix-Assisted UV-Laser Desorption/Ionization Mass Spectrometry. *Anal Chim Acta* 1990; 1990:175–185

46. Karas M, Gluckmann M, and Schafer J. Ionization in Matrix-Assisted Laser Desorption/Ionization: Singly Charged Molecular Ions Are the Lucky Survivors. *J Mass Spectrom* 2000; 35:1–12

47. Katta V, and Chait BT. Conformational Changes in Proteins Probed by Hydrogen-Exchange Electrospray-Ionization Mass Spectrometry. *Rapid Commun Mass Spectrom* 1991; 5:214–217

48. Katta V, and Chait BT. Hydrogen-Deuterium Exchange Electrospray-Ionization Mass-Spectrometry—A Method for Probing Protein Conformational-Changes in Solution. *J Am Chem Soc* 1993; 115:6317–6321

49. Kebarle P. A Brief Overview of the Present Status of the Mechanisms Involved in Electrospray Mass Spectrometry. *J Mass Spectrom* 2000; 35:804–817

50. Kebarle P, and Tang L. From Ions in Solution to Ions in the Gas-Phase—The Mechanism of Electrospray Mass-Spectrometry *Anal Chem* 1993; 65:A972-A986

51. Kelly MA, Vestling MM, Fenselau CC, and Smith PB. Electrospray Analysis of Proteins—A Comparison of Positive-Ion and Negative-Ion Mass-Spectra at High and Low pH. *Org Mass Spectrom* 1992; 27:1143–1147

52. Kleanthous C, Kuhlmann UC, Pommer AJ, Ferguson N, Radford SE, Moore GR, James R, and Hemmings AM. Structural and Mechanistic Basis of Immunity Toward Endonuclease Colicins. *Nat Struct Biol* 1999; 6:243–252

53. Ko TP, Liao CC, Ku WY, Chak KF, and Yuan HS. The Crystal Structure of the DNase Domain of Colicin E7 in Complex with Its Inhibitor Im7 protein. *Structure Fold Des* 1999; 7:91–102

54. Konermann L, Collings BA, and Douglas DJ. Cytochrome c Folding Kinetics Studied by Time-Resolved Electrospray Ionization Mass Spectrometry. *Biochemistry* 1997a; 36:5554–5559

55. Konermann L, Rosell FI, Mauk AG, and Douglas DJ. Acid-Induced Denaturation of Myoglobin Studied by Time-Resolved Electrospray Ionization Mass Spectrometry. *Biochemistry* 1997b; 36:6448–6454

56. Konermann L, and Douglas DJ. Equilibrium Unfolding of Proteins Monitored by Electrospray Ionization Mass Spectrometry: Distinguishing Two-State from Multi-State Transitions. *Rapid Commun Mass Spectrom* 1998; 12:435–442

57. Lafitte D, Capony JP, Grassy G, Haiech J, and Calas B. Analysis of the Ion Binding Sites of Calmodulin by Electrospray Ionization Mass Spectrometry. *Biochemistry* 1995; 34:13825–13832

58. Le Blanc JCY, Guevremont R, and Siu KWM. Electrospray Mass-Spectrometry of Some Proteins and the Aqueous-Solution Acid-Base-Equilibrium Model in the Negative-Ion Detection Mode. *Int J Mass Spectrom Ion Process* 1993; 125:145–153

59. Lippincott J, Fattor TJ, Lemon DD, and Apostol I. Application of Native-State Electrospray Mass Spectrometry to Identify Zinc-Binding Sites on Engineered Hemoglobin. *Anal Biochem* 2000; 284:247–255

60. Loo JA. Studying Noncovalent Protein Complexes by Electrospray Ionization Mass Spectrometry. *Mass Spectrom Rev* 1997; 16:1–23

61. Loo JA. Probing Protein-Metal Ion Interactions by Electrospray Ionization Mass Spectrometry: Enolase and Nucleocapsid Protein. *Int J Mass Spectrometry* 2001; 204:113–123

62. Loo JA, Edmonds CG, Udseth HR, and Smith RD. Effect of Reducing Disulfide-Containing Proteins on Electrospray Ionization Mass Spectra. *Anal Chem* 1990; 62:693–698

63. Loo JA, Hu PF, McConnell P, Mueller WT, Sawyer TK, and Thanabal VA. Study of Src SH2 Domain Protein-Phosphopeptide Binding Interactions by Electrospray Ionization Mass Spectrometry. *J Am Soc Mass Spectrom* 1997; 8:234–243

64. Macfarlane RD, and Torgerson DF. ^{252}Cf-plasma desorption time-of-flight mass spectrometry. *Int J Mass Spectrom Ion Phys* 1976; 21:81–92

65. Mandell JG, Falick AM, and Komives EA. Measurement of Amide Hydrogen Exchange by MALDI-TOF Mass Spectrometry. *Anal Chem* 1998; 70:3987–3995

66. Mao D, Babu KR, Chen YL, and Douglas DJ. Conformations of Gas-Phase Lysozyme Ions Produced from Two Different Solution Conformations. *Anal Chem* 2003; 75:1325–1330

67. McCammon MG, Scott DJ, Keetch CA, Greene LH, Purkey HE, Petrassi HM, Kelly JW, and Robinson CV. Screening Transthyretin Amyloid Fibril Inhibitors: Characterization of Novel Multiprotein Multiligand Complexes by Mass Spectrometry. *Structure* 2002; 10:851–863

68. Miroy GJ, Lai Z, Lashuel HA, Peterson SA, Strang C, and Kelly JW. Inhibiting Transthyretin Amyloid Fibril Formation via Protein Stabilization. *Proc Natl Acad Sci U S A* 1996; 93:15051–15056

69. Mohimen A, Dobo A, Hoerner JK, and Kaltashov IA. A Chemometric Approach to Detection and Characterization of Multiple Protein Conformers in Solution Using Electrospray Ionization Mass Spectrometry. *Anal Chem* 2003; 75:4139–4147

70. Nemirovskiy O, Giblin DE, and Gross ML. Electrospray Ionization Mass Spectrometry and Hydrogen/Deuterium Exchange for Probing the Interaction of Calmodulin with Calcium. *J Am Soc Mass Spectrom* 1999; 10:711–718

71. Ofori-Acquah SF, Green BN, Davies SC, Nicolaides KH, Serjeant GR, and Layton DM. Mass Spectral Analysis of Asymmetric Hemoglobin Hybrids: Demonstration of Hb FS (alpha2gammabetaS) in Sickle Cell Disease. *Anal Biochem* 2001; 298:76–82

72. Pinkse MW, Maier CS, Kim JI, Oh BH, and Heck AJ. Macromolecular Assembly of Helicobacter pylori Urease Investigated by Mass Spectrometry. *J Mass Spectrom* 2003; 38:315–320

73. Pommer AJ, Kuhlmann UC, Cooper A, Hemmings AM, Moore GR, James R, and Kleanthous C. Homing in on the Role of Transition Metals in the HNH Motif of Colicin Endonucleases. *J Biol Chem* 1999; 274:27153–27160

74. Potier N, Donald LJ, Chernushevich I, Ayed A, Ens W, Arrowsmith CH, Standing KG, and Duckworth HW. Study of a Noncovalent Trp Repressor: DNA Operator Complex by Electrospray Ionization Time-of-Flight Mass Spectrometry. *Protein Sci* 1998; 7:1388–1395

75. Robinson CV, Gross M, Eyles SJ, Ewbank JJ, Mayhew M, Hartl FU, Dobson CM, and Radford SE. Conformation of Groel-Bound Alpha-Lactalbumin Probed by Mass-Spectrometry *Nature* 1994; 372:646–651

76. Roper DI, Subramanya HS, Shingler V, and Wigley DB, Preliminary Crystallographic Analysis of 4-Oxalocrotonate Tautomerase Reveals the Oligomeric Structure of the Enzyme. *J Mol Biol* 1994; 243:799–801

77. Rosa JJ, and Richards FM. An Experimental Procedure for Increasing the Structural Resolution of Chemical Hydrogen-Exchange Measurements on Proteins: Application to Ribonuclease S Peptide. *J Mol Biol* 1979; 133:399–416

78. Rostom AA, and Robinson CV, Disassembly of Intact Multiprotein Complexes in the Gas Phase. *Curr Opin Struct Biol* 1999; 9:135–141

79. Rostom AA, Tame JR, Ladbury JE, and Robinson CV. Specificity and Interactions of the Protein Oppa: Partitioning Solvent Binding Effects Using Mass Spectrometry. *J Mol Biol* 2000; 296:269–279

80. Simmons DA, and Konermann L. Characterization of Transient Protein Folding Intermediates During Myoglobin Reconstitution by Time-Resolved Electrospray Mass Spectrometry with On-line Isotopic Pulse Labeling. *Biochemistry* 2002; 41:1906–1914

81. Smith DL. Local Structure and Dynamics in Proteins Characterized by Hydrogen Exchange and Mass Spectrometry. *Biochemistry* 1998; 63:338–348

82. Smith DL, Deng Y, and Zhang Z. Probing the Non-covalent Structure of Proteins by Amide Hydrogen Exchange and Mass Spectrometry. *J Mass Spectrom* 1997a; 32:135–146

83. Smith RD, Bruce JE, Wu QY, and Lei QP. New Mass Spectrometric Methods for the Study of Noncovalent Associations of Biopolymers. *Chem Soc Rev* 1997b; 26:191–202

84. Sundqvist B, and Macfarlane RD. Cf-252 Plasma Desorption Mass-spectrometry. *Mass Spectrom Rev* 1985; 4:421–460

85. Tahallah N, Van Den Heuvel RH, Van Den Berg WA, Maier CS, Van Berkel WJ, and Heck AJ. Cofactor-Dependent Assembly of the Flavoenzyme Vanillyl-Alcohol Oxidase. *J Biol Chem* 2002; 277:36425–36432

86. Tanaka K. Protein and Polymer Analysis Up to m/z 100000 by Laser Ionization Time-of-Flight Mass Spectrometry. *Rapid Commun Mass Spectrom* 1988; 2:151–153

87. Van den Bremer ETJ, Jiskoot W, James R, Moore GR, Kleanthous C, Heck AJR, and Maier CS. Probing Metal Ion Binding and Conformational Properties of the Colicin E9 Endonuclease by Electrospray Ionization Time-of-Flight Mass Spectrometry. *Protein Sci* 2002; 11:1738–1752

88. Veenstra TD, Gross MD, Hunziker W, and Kumar R. Identification of Metal-Binding Sites in Rat Brain Calcium-Binding Protein. *J Biol Chem* 1995; 270:30353–30358

89. Veenstra TD, Johnson KL, Tomlinson AJ, Naylor S, and Kumar R. Determination of Calcium-Binding Sites in Rat Brain Calbindin D28K by Electrospray Ionization Mass Spectrometry. *Biochemistry* 1997; 36:3535–3542

90. Veenstra TD, Johnson KL, Tomlinson AJ, Craig TA, Kumar R, and Naylor S. Zinc-Induced Conformational Changes in the DNA-Binding Domain of the Vitamin D Receptor Determined by Electrospray Ionization Mass Spectrometry. *J Am Soc Mass Spectrom* 1998; 9:8–14

91. Versluis C, and Heck AJR. On the Gas-Phase Dissociation of Hemoglobin. *Int J Mass Spectrom* 2001; 210:637–649

92. Vollmerhaus PJ, Breukink E, and Heck AJR. Getting Closer to the Real Bacterial Cell Wall Target: Biomolecular Interactions of Water-Soluble Lipid II with Glycopeptide Antibiotics. *Chem-Eur J* 2003; 9:1556–1565

93. Wagner DS, and Anderegg RJ. Conformation of Cytochrome c studied by Deuterium Exchange-Electrospray Ionization Mass Spectrometry. *Anal Chem* 1994; 66:706–711

94. Wagner DS, Melton LG, Yan Y, Erickson BW, and Anderegg RJ. Deuterium Exchange of Alpha-Helices and Beta-Sheets as Monitored by Electrospray Ionization Mass Spectrometry. *Protein Sci* 1994; 3:1305–1314
95. Wang L, and Smith DL. Downsizing Improves Sensitivity 100-Fold for Hydrogen Exchange-Mass Spectrometry. *Anal Biochem* 2003; 314:46–53
96. Wang F, and Tang X. Conformational Heterogeneity of Stability of Apomyoglobin Studied by Hydrogen/Deuterium Exchange and Electrospray Ionization Mass Spectrometry. *Biochemistry* 1996; 35:4069–4078
97. Williams DH, Bradley CV, Santikarn S, and Bojesen G. Fast-Atom-Bombardment Mass Spectrometry: A New Technique for the Determination of Molecular Weights and Amino Acid Sequences of Peptides. *Biochem J* 1982; 201:105–117
98. Wilm MS, and Mann M. Electrospray and Taylor-Cone Theory Doles Beam of Macromolecules at Last. *Int J Mass Spectrom Ion Process* 1994; 136:167–180
99. Wilson DJ, and Konermann L. A Capillary Mixer with Adjustable Reaction Chamber Volume for Millisecond Time-Resolved Studies by Electrospray Mass Spectrometry. *Anal Chem* 2003; 75:6408–6414
100. Yao Y, Huang L, Krutchinsky A, Wong ML, Standing KG, Burlingame AL, and Wang CC. Structural and Functional Characterizations of the Proteasome-Activating Protein PA26 from Trypanosoma brucei. *J Biol Chem* 1999; 274:33921–33930
101. Zal F, Green BN, Lallier FH, and Toulmond A. Investigation by Electrospray Ionization Mass Spectrometry of the Extracellular Hemoglobin from the Polychaete Annelid Alvinella pompejana: An Unusual Hexagonal Bilayer Hemoglobin. *Biochemistry* 1997; 36:11777–11786
102. Zal F, Green BN, Martineu P, Lallier FH, Toulmond A, Vinogradov SN, and Childress JJ. Polypeptide Chain Composition Diversity of Hexagonal-Bilayer Haemoglobins Within a Single Family of Annelids, The Alvinellidae. *Eur J Biochem* 2000; 267:5227–5236
103. Zhang ZQ, and Smith DL. Determination of Amide Hydrogen-Exchange by Mass-Spectrometry - A New Tool for Protein-Structure Elucidation. *Protein Sci* 1993; 2:522–531
104. Zhang ZG, Krutchinsky A, Endicott S, Realini C, Rechsteiner M, and Standing KG. Proteasome activator 11S REG or PA28: Recombinant REG alpha/REG beta Hetero-Oligomers Are Heptamers. *Biochemistry* 1999; 38:5651–5658

13

Mass Spectrometry: Proteomics

EDWIN P. ROMIJN, JEROEN KRIJGSVELD, and ALBERT J.R. HECK

Department of Biomolecular Mass Spectrometry, Bijvoet Center for Biomolecular Research, and Utrecht Institute for Pharmaceutical Sciences, Utrecht University, Utrecht, The Netherlands

Introduction

With the DNA sequences of more than 90 genomes completed, as well as a draft sequence of the human genome (McPherson et al., 2001; Venter et al., 2001), a major challenge in modern biology is to understand the expression, function, and regulation of the entire set of proteins encoded by an organism. These are the aims of the new field of proteomics. This information will be invaluable for understanding how complex biological processes occur at a molecular level, how they differ in various cell types, and how they are altered in disease states. The technological tools are now becoming available to study properties of proteins at a more comprehensive scale, including not only their expression patterns but also the protein networks they form (Gavin et al., 2002; Ho et al., 2002; Ozier et al., 2003; Vazquez et al., 2003; Wilhelm et al., 2003), the way they fold (Dobson, 1999; Fersht and Daggett, 2002; Kazmirski et al., 2002; Mayor et al., 2003), and the way they are post-translationally modified (Haynes and Aebersold, 2000; Geng et al., 2001; MacCoss and Yates, 2001; McLachlin and Chait, 2001; Vosseller et al., 2001; Mann et al., 2002; Shou et al., 2002; Kaji et al., 2003; Zhang et al., 2003). The proteome can be defined as the protein constituents of a cell characterized in terms of their localization, post-translational modification, interactions, and turnover. The proteome is of an intrinsically dynamic nature, with alterations in protein expression, activation, and/or modification occurring in response to signals or alterations in the microenvironment. Therefore, it has an inherent complexity that surpasses that of the genome or the transcriptome, the mRNA complement of a cell. The total number of protein species in a mammalian cell is not known. An estimate of the total number of transcripts in a mammalian cell is 5000 (Kettman et al., 2002). Beyond this are the post-translational modifications that generate a further degree of complexity, such that 100,000 or more different protein forms potentially exist. Each protein may be compartmentalized in discrete organelles or shuttle between them, their location often governing functional activity. Regulation of protein shuttling is a mechanism for governance of function; thus, a systematic knowledge of the subcellular localization of proteins, and the other proteins with which they interact, is just as significant as knowledge of their expression.

One of the core components of proteomics is the ability to quantify and identify systematically every protein expressed in a biological cell or tissue. The technology for such proteome analysis involves separation science for the separation of proteins and peptides, analytical science for the identification and quantitation of these biomolecules, and bioinformatics for data management, including linking the proteomics data to data obtained via other genome-wide approaches. Proteomics is a multidisciplinary research activity wherein separation science and mass spectrometry play pivotal roles. In this chapter, we will give an overview of conventional and some of the newest analytical strategies in mass spectrometry–based proteomic research.

Conventional Strategies in Proteomics

Because of the much more heterogeneous nature of the proteins compared to nucleic acids, they are much more difficult to isolate, separate, and analyze. Additionally, the protein world lacks a PCR equivalent that could be used for amplification. The conventional strategy for proteome analysis is outlined in Figure 1. In general, proteomics starts with the isolation of the proteins from the biological matrix, such as fluids, cells, or tissue. The procedures for these steps are highly dependent on the particular sample and/or organism of origin. It is important to separate the proteins from nonprotein material, such as lipids, DNA, and RNA. For the separation and analysis of the proteins conventionally, two analytical methods are being used; namely, two-dimensional (2D) electrophoresis and mass spectrometry.

2D Gel Electrophoresis

Two-dimensional polyacrylamide gel electrophoresis (2D-PAGE) is currently the core technology for proteomics (Gorg et al., 1999; Rabilloud, 1999; Gorg et al., 2000; Rabilloud, 2000; Rabilloud, 2002), and this 2D gel can be considered as the "proteome map" of a sample (Figure 2). It involves the separation of proteins from a complex mixture according to two independent parameters in two distinct steps. Proteins are separated according to their isoelectric point on a pH gradient; this gradient then is applied to a sodium dodecyl sulfate SDS-polyacrylamide gel matrix where the proteins are separated

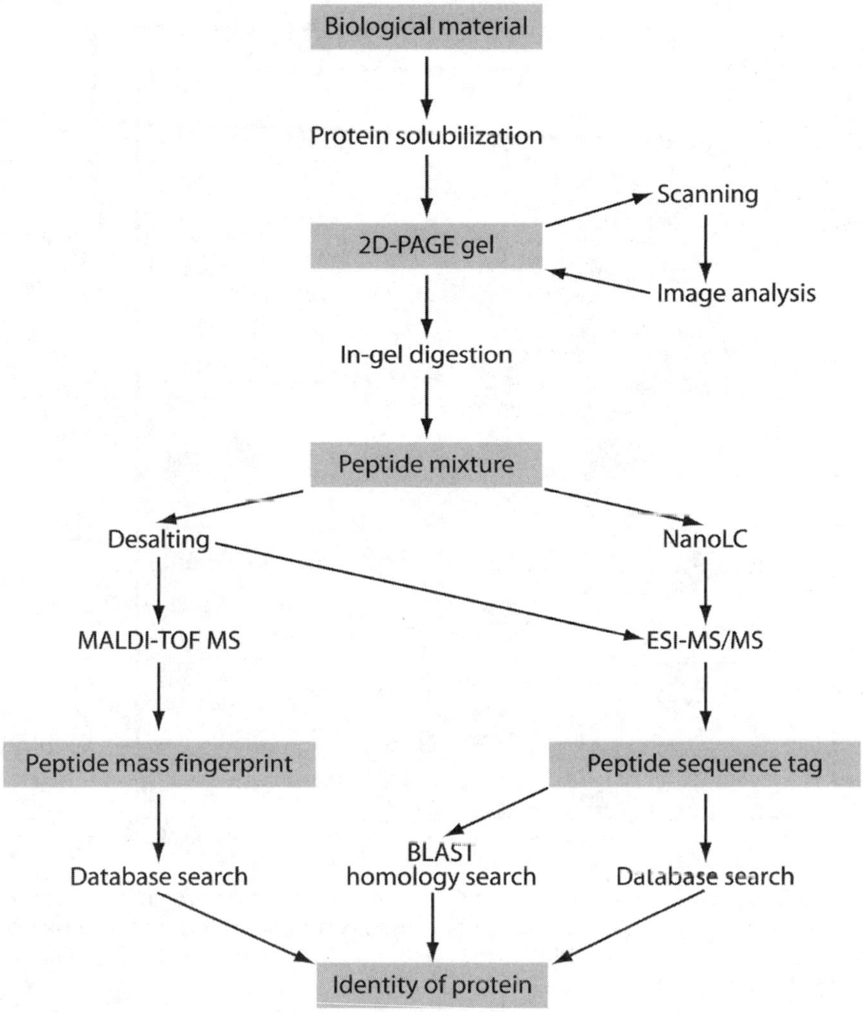

Figure 1 Schematic overview of the conventional approach in proteome analysis based on 2D gel electrophoresis, mass spectrometry, and database searches.

according to their size perpendicular to the first dimension. Spot patterns can be visualized by stains such as Coomassie blue, silver, and, more recently, staining by fluorescent dyes (Martin et al., 2002; Patton, 2002; Hart et al., 2003a; Hart et al., 2003b) and analyzed by specialist computer software to detect differences in protein expression between samples. For expression proteomics, a disadvantage of most staining techniques is that the magnitude of staining often depends on the

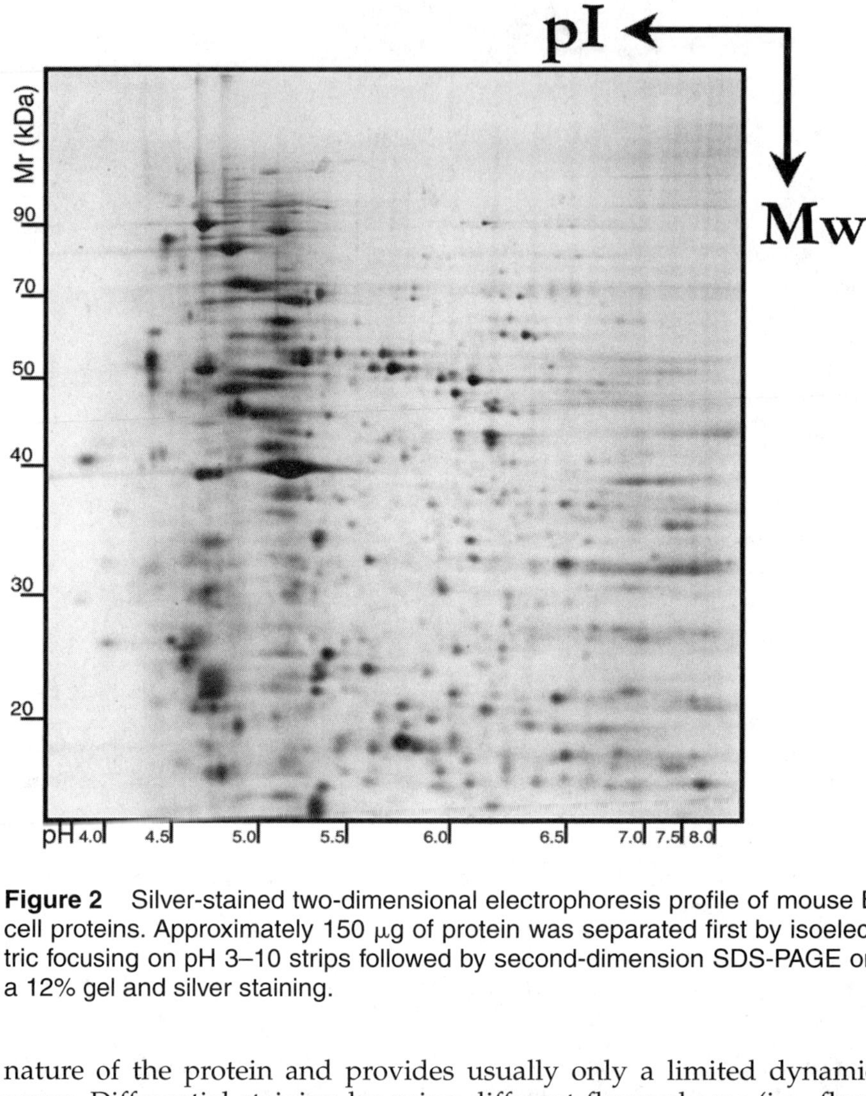

Figure 2 Silver-stained two-dimensional electrophoresis profile of mouse B cell proteins. Approximately 150 μg of protein was separated first by isoelectric focusing on pH 3–10 strips followed by second-dimension SDS-PAGE on a 12% gel and silver staining.

nature of the protein and provides usually only a limited dynamic range. Differential staining by using different fluorophores (i.e., fluorescent dyes) is a recent advancement, which increases the dynamic range of image profiling of protein expression patterns (Unlu et al., 1997; Patton, 2002; Yan et al., 2002; Alban et al., 2003; Hu et al., 2003; Lee et al., 2003; Van den Bergh et al., 2003a; Van Den Bergh et al., 2003b). Imaging of the stained two-dimensional gels provides a very nice way to compare two proteomes, for instance for a stressed and normal state of a cell culture, or, more generally, a normal and an abnormal cellular state or phenotype. Proteins of interest then can be excised and identified by MS techniques, such as peptide mass fingerprinting, or peptide fragmentation to obtain amino acid sequence data.

Typically, 2D gel electrophoresis is used in expression proteomic studies where the focus is on studying alterations in protein expression profiles due to the appearance of a new protein spot or changes in the intensity of an existing protein spot. In summary, 2D gel electrophoresis is still one of the best techniques for separating and visualizing a large number of proteins. Up to 10,000 different spots may be resolved on the best gels, especially when magnified zoom-gels are used (Gorg et al., 2000). Although having a superior resolution compared to other techniques, current 2D gel–based methods still have several technical drawbacks that limit their widespread application. One of the drawbacks of these gel-based technologies is that they require relatively large amounts of sample (e.g., greater than 200 μg total protein lysate per narrow pI-range gel). It also is difficult to obtain sufficient material to identify low-abundance proteins of interest by in-gel tryptic digestion following a total protein stain such as silver staining. This limits the utility of "conventional" 2D-PAGE for analysis of low-abundance samples such as patient materials.

Mass Spectrometry

The desired endpoint of proteomics studies is the characterization of all cellular proteins in terms of their abundance, modification, localization, and function. Such a comprehensive database of cellular proteins could serve as a tool for studying the proteomic response of cells or tissues in response to a variety of conditions (Gygi and Aebersold, 2000) and depends to a great extent on the methods available for protein identification and characterization. Since the late 1980s, mass spectrometry (MS) has, due to its sensitivity, evolved into an indispensable technique in protein analysis. MS, in conjunction with the development of comprehensive protein databases (Keough et al., 2000) and advances in computational methods (Lahm and Langen, 2000), is also easily adaptable to high-throughput formats, which has made it the method of choice for protein identification and characterization (Lahm and Langen, 2000, Pandey and Mann, 2000; Yates, 2000) (Figure 3).

Nowadays, two distinct mass spectrometric techniques are employed for protein identification at low concentrations (attomole to femtomole) (Yates, 1998; Keough et al., 2000; Lahm and Langen, 2000; Yates, 2000; Mann et al., 2001). In the first method, a peptide fingerprint of a protein is recorded, usually by matrix assisted laser desorption ionization–time of flight–mass spectrometry (MALDI-ToF-MS). In the second method, short amino acid sequences, so-called sequence tags, are generated usually by electrospray ionization (ESI)-tandem MS. Before discussing the two methods described above (peptide fingerprinting and sequence tagging), we will first briefly describe the two ionization methods (MALDI and ESI) and commonly used mass analyzers.

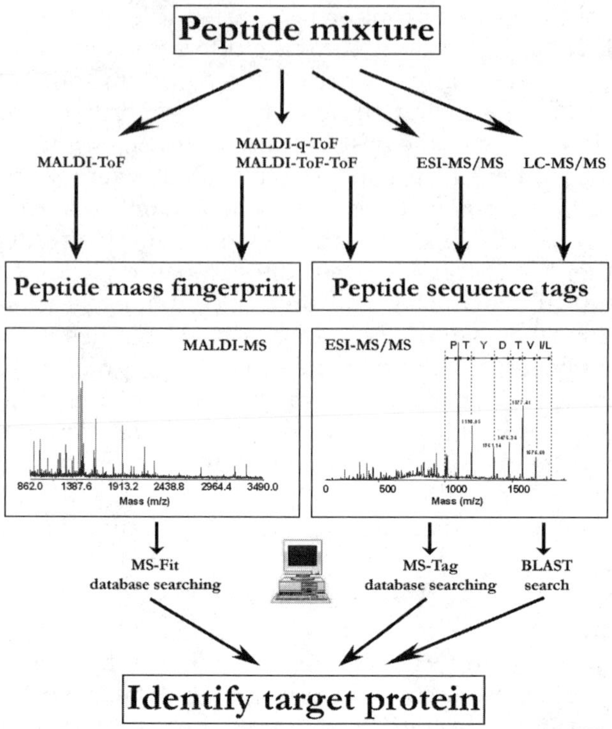

Figure 3 Typical mass spectrometry scheme for protein identification using peptide mass mapping and tandem mass spectrometry. Protein spots from 2D gel are digested, and the peptides are analyzed using either MALDI-ToF or a tandem MS. The peptide mass fingerprint or amino acid sequence is then compared against databases to identify the proteins.

Ionization

MALDI and ESI are the two main technologies that are used to transfer larger biomolecules from the liquid or solid phase into the gas phase and ionize them while keeping them intact (Karas and Hillenkamp, 1988; Fenn et al., 1989). The creation of intact molecular ions enables accurate measurement of its molecular weight. Nowadays, MALDI and ESI are routine ionization techniques to determine molecular masses of different types of biomolecules. For proteins and specifically for peptides, the sensitivity is considered extremely good. Although applicable to different types of biomolecules and other macromolecules, the focus will here be on proteins and peptides.

In ESI-MS (Fenn et al., 1989), the protein sample is in solution, and an electrical potential (1–3 kV) is applied to create a fine mist of charged droplets that are subsequently further desolvated until the bare ions are introduced into the mass analyzer (Lahm and Langen, 2000). The ESI process can be used to produce negative or positive ions, but typically peptides and proteins are analyzed as positive ions using the capillary as an anode and the MS inlet as the cathode. ESI uses a steady stream of solvent to produce a continuous beam of ions and therefore can readily be coupled to high-pressure liquid chromatography (HPLC). At low flow rates, electrospray can be induced solely by an electrical potential, without any nebulization gas that is needed at high solvent flow rates (0.1–1 mL/min). Two operating regimes have been defined at low flow rates. The first is microelectrospray, where the flow rate is approximately 0.1–0.5 μL/min; and the second is nanoelectrospray, where a flow rate of 100 nL/min or less is used (Emmett and Caprioli, 1994; Wilm and Mann, 1994). These lower flow rates create smaller droplets and a much reduced Taylor cone, allowing the bulk of the spray to be directed into the mass spectrometer's inlet (Wilm and Mann, 1994). As a result, much less sample is required than at higher flow rates because of the concentration sensitive nature of electrospray. In contrast to MALDI, ESI produces multiply charged ions, usually without fragmentation of the ions in the gas phase (Mann et al., 2001). Proteins, oligonucleotides, sugars, and polar lipids of all molecular weights can be analyzed with ESI-MS (Mann et al., 2001) as multiple charges can be added to the molecules lowering the mass-to-charge (*m/z*) ratio to a range detectable by mass analyzers (Yates, 2000; Mann et al., 2001). Furthermore, multiple protonation of peptides and proteins promotes more facile amide bond fragmentation when the ions are activated for dissociation. These features of ESI have led to its widespread application for proteomics research.

MALDI uses energy from laser photons to ionize biomolecules. In MALDI (Karas and Hillenkamp, 1988), the protein samples to be analyzed are incorporated in an energy-absorbing matrix (Hillenkamp et al., 1991; Lahm and Langen, 2000). Ionization occurs when these matrix molecules absorb the light provided by a laser (usually 337 nm). Release of the energy causes a rapid thermal expansion of matrix and analyte into the gas phase. Proton transfer from matrix to analyte may result in formation of singly charged ion observed in the gas phase (Kruger et al., 2001). The matrix is typically a small molecule such as dihydrobenzoic acid, 4-hydroxycinnamic acid (for peptides less than 5 kDa), or sinapinic acid (for proteins) that is deposited onto a metal substrate along with the sample (Mann et al., 2001), and ions are generated when the matrix is struck and excited by a laser. MALDI generates, in contrast to ESI, mostly single charged ion species and is less sensitive to salts in the buffer than ESI, although salt and matrix adducts of analyte ions can form. Because MALDI depends on pulsed laser radiation, ions are created in bunches or packets. Consequently, mass analyzers capable of analyzing ions created by MALDI are required. The most common mass analyzer used with MALDI is the time-of-flight mass analyzer.

Mass Analyzers

The time-of-flight (ToF) analyzer is commonly used in combination with ESI and MALDI, although ESI is often coupled to other mass analyzers as well (quadrupoles or ion traps, for example). The ToF analyzer is probably one of the simplest mass analyzers and is based on accelerating a set of ions to a detector with the same amount of kinetic energy (U_k). Because $U_k = {}^1/_2 mv^2$, ions with different masses (m) will get different velocities (v), resulting in different times of flight. The time of flight is not only dependent on the masses of the ions but also on the charges (z) of the ions. Thus, the m/z values of the ions can be determined from the ions' time of arrival at the detector. To increase the resolution of the ToF analyzer, reflectron ToF analyzers have been developed. The addition of the reflectron reduces the kinetic energy distribution of the ions that reach the detector, and consequently the resolution increases. This gain in resolution, however, comes at the expense of sensitivity and mass range.

Time-of-flight analysis can be coupled to quadrupole (Q) mass analyzers, resulting in a quadrupole-ToF mass spectrometer. A quadrupole mass analyzer separates ions on the basis of stability of the ion trajectory through four parallel rods, in which opposing rods carry charges of the same sign. By applying an alternating radio frequency (RF) voltage on top of a direct current (DC) voltage, ions travelling at relatively low velocity with different m/z values follow different trajectories through the rods. At a certain voltage, ions with one m/z value traverse the quadrupole, while others collide with the rods and are lost. By scanning the voltages while keeping the ratio between DC and RF voltages constant, ions with different m/z values pass successively through the mass analyzer. Subsequently, ions are pushed into the orthogonal ToF tube for mass analysis based on the time of flight of a particular ion.

Soon after their introduction eight years ago, Q-ToF mass spectrometers rapidly conquered their place in the (bio-) analytical community. This is because they are powerful and robust instruments with unique capabilities. In Q-ToF mass spectrometers, ions are extracted orthogonally into the ToF tube to correct for spatial or velocity spread of the ions in the direction of the ion source and so to improve resolution. The quadrupole acts (when it operates in RF-only mode) as a wide bandpass filter transmitting ions across the entire m/z range. In tandem MS (MS/MS) mode, the quadrupole acts as an m/z filter for selection of the precursor ion for collision-induced dissociation in the collision cell. [For an extensive review on Q-ToF-MS, see Chernushevich et al., (2001).] The Q-ToF mass spectrometer is still one of the most powerful instruments in the bioanalytical field, although recently new instruments have been introduced. These new instruments include MALDI-Q-ToF and MALDI-ToF/ToF mass spectrometers (Medzihradszky et al., 2000; Bienvenut et al., 2002; Huang et al., 2002; Yergey et al., 2002), linear ion traps (Hager and Yves Le Blanc, 2003; Le Blanc et al., 2003; Londry and

Hager, 2003), and (linear ion trap-) Fourier transform ion cyclotron resonance (FTICR) mass spectrometers (Harkewicz et al., 2002; Tang et al., 2002; Belov et al., 2003). [For an extensive review on these recent trends in mass spectrometer development, see Hager, (2003).]

Protein Identification

In the first of the two approaches to identify proteins, a protein spot to be identified is cut out of the gel and digested (in-gel) with a specific protease (see Figure 1), mostly trypsin, which specifically cleaves at arginine or lysine residues (Olsen et al., 2004). The resulting peptide mixture is eluted from the gel and analyzed by MALDI-ToF-MS or alternatively by using ESI-MS. Collectively, the peptide masses resulting from in-gel digestion form a fingerprint, which is indicative for the protein concerned. This fingerprint is then compared to theoretically expected tryptic peptide masses for each protein entry in the database. The hits are generally ranked according to the number of peptides that match. Unique identification does not require the whole protein to be covered by the tryptic peptides. Usually, 10–20% coverage is already sufficient. After a hit, a second search is usually done to correlate the remaining peptides in the mass spectrum with the data obtained from the protein hit, taking into account possible post-translational modifications. The great precision of mass determination, which may be achieved by modern mass spectrometers, allows the discrimination of even highly homologous proteins, or proteins that only differ by a single amino acid. Generally, peptide fingerprinting is still the most rapid and efficient method for protein identification. As identification occurs via consultation of protein and genome databases, it may be apparent their increasing comprehensiveness greatly aids in protein identification.

If the peptide mass fingerprint does not allow identification of the protein (e.g., because insufficient peptides have been generated and/or the fingerprint is not unique), then it is possible to determine a small piece of amino acid sequence from one or more of the peptides by tandem MS. Tandem mass spectrometry of peptides is done using ESI-Q-ToF-MS/MS, ESI-triple quadrupole-MS/MS, or ESI-ion-trap-MS/MS, although nowadays tandem MS is also possible with MALDI-Q-ToF and MALDI-ToF/ToF instruments (Medzihradszky et al., 2000; Bienvenut et al., 2002; Huang et al., 2002; Yergey et al., 2002). A small piece of amino acid sequence from a peptide is generated by fragmentation of the peptide in the mass spectrometer and mass analysis of resulting fragments. The fragment ions produced in this process can be separated into two classes. One class retains the charge on the N-terminal region while cleavage is observed from the C-terminal region, resulting in a_n, b_b, and c_n type ions [nomenclature according to Roepstorff et al., (1984)]. The second class retains the charge on the C-terminal region while cleavage is observed from the N-terminal region, resulting in x_n, y_b, and z_n type ions (Figure 4). Although these fragmentation patterns

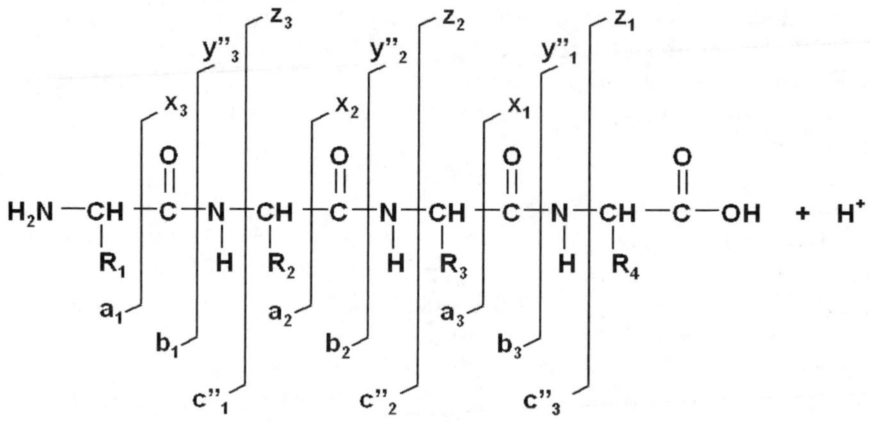

Figure 4 Roepstorff nomenclature for peptide fragmentation. Fragmentation is observed predominantly at the amide bond. N-terminal fragments are named a_n, b_b, and c_n type ions and C-terminal fragments are named x_n, y_b, and z_n type ions.

may be complicated, they generally allow the determination of partial sequences. With this partial sequence, possibly in combination with the peptide-fingerprint already obtained, the chance of a unique hit in the database is considerably enhanced. With one or two of these short sequence tags (often no more than five amino acids), it is often possible to unambiguously identify a protein. Should a match not be found in a database, additional peptides have to be mass spectrometrically sequenced in this manner. Determination of the total sequence of a protein still requires considerably more effort.

The above-described conventional proteomic strategies, accomplished by the combination of two-dimensional gel electrophoresis and MALDI-ToF-MS or ESI-ToF-MS/MS, are nowadays more and more replaced by methods that involve protein or peptide preseparation by using high-efficiency capillary separations techniques, such as liquid chromatography and/or electrophoresis.

Liquid Chromatography–Based Approaches in Proteomics

The benefits of on-line coupling liquid chromatography (LC) with MS include an enhanced ability to handle the most complex mixtures, improved dynamic range, and increased sensitivity arising from the concentration of each mixture component into a small (nanoliter range) volume delivered from the chromatographic column to the mass spectrometer ion source. The implementation of in particular nanoflow LC-MS offers unique opportunities for speed, sensitivity, and automation of proteomics research (Meiring et al., 2002). In addition, the LC-MS combination allows facile desalting of samples prior to MS analysis. In some circumstances, off-line combination of LC and MS may be advantageous in allowing multiple MS analyses of individual fractions, but the advantage of sample concentration is lost.

Nano-LC columns with 50–100 μm diameters are nowadays routinely used for the analysis of protein digests, requiring the delivery of solvent gradients at nanoliter per minute flow rates. This has typically been achieved using generic HPLC pumping systems for the delivery of microliter per minute gradients that are either flow-split or sampled. At the end of the capillary, often a tip is pulled with an inner diameter of approximately 10 μm. The total peptide mixture is typically first loaded onto a nanocolumn (usually 75 μm internal diameter) containing reverse-phase C18 material and then eluted by using a gradient directly into a tandem mass spectrometer. The peptides are eluted at a low flow rate, typically 100–200 nL per minute, and the elution time of each peptide is approximately 10–30 seconds. The peptides elute off the column into the ionization source of the mass spectrometer. First, a mass spectrum in the survey scan mode of the intact peptides can be obtained. In the data-dependent acquisition mode, the instrument can be set to fragment and collect automatically MS/MS data on any number of peptides observed in the MS spectrum based on their intensity, m/z, value or charge state.

Using LC-MS/MS in proteomics, quite a few different strategies may be taken. In one approach, close to the conventional one, 2D gel electrophoresis is still employed for the separation of the intact proteins. Digests resulting from proteolysis of the cut-out spots are analyzed by LC-MS or LC-MS/MS. This approach may be complementary to a conventional peptide mapping approach, as usually different sets of peptides may be more abundant/present in mass spectra obtained by MALDI and ESI. Such a parallel MALDI peptide mapping and ESI LC-MS approach may therefore enhance the protein coverage.

To overcome some of the inherent disadvantages of 2D gels, several 2D gel–independent LC-MS/MS approaches have been introduced. In this approach, the proteins in the total proteome are separated and resolved by molecular weight using one-dimensional (1D) gels. This 1D gel is cut into pieces, all proteins in such a band are digested, and the mixture of peptides are analyzed by LC-MS and/or LC-MS/MS (Rappsilber et al., 2000; Gavin et al., 2002; Ho et al., 2002; Wu et al., 2002; De Boer et al., 2003). This approach provides an intermediate form between analyzing the very complex large peptide mixture obtained when digesting all proteins of a lysate and the single protein digested when using a 2D gel. As compared to the 2D gel, a 1D gel–based approach is less elaborate. Additionally, very large and basic proteins (known to be difficult to analyze using 2D gels) are easier to handle using single-dimensional gels.

In another approach, the whole cell lysate is digested chemically or by a protease and analyzed using multidimensional protein identification technology (MudPIT) (Yates et al., 1999; Washburn et al., 2002). This generates a very large and complex set of peptides, beyond the separation capacity of one-dimensional separation techniques. For the analysis of such complex mixtures, various multidimensional separation techniques have been introduced. In one approach, peptides are first separated using a strong cation exchange column and a stepped solvent gradient. They then are introduced into a reverse-phase column that is coupled to a tandem mass spectrometer. In an off-line system, the fractions from the cation exchange column are retained and can be applied to the reverse-phase LC-MS/MS. The direct coupling of the two stages of separation is convenient, and especially the pioneering work by the group of Yates has revealed the power of this procedure. They introduced the rapid and large-scale proteome analysis of total cell lysates by multidimensional protein identification technology, termed by them MudPIT (Yates et al., 1999; Washburn et al., 2002) (Figure 5). These technologies employ for instance a biphasic

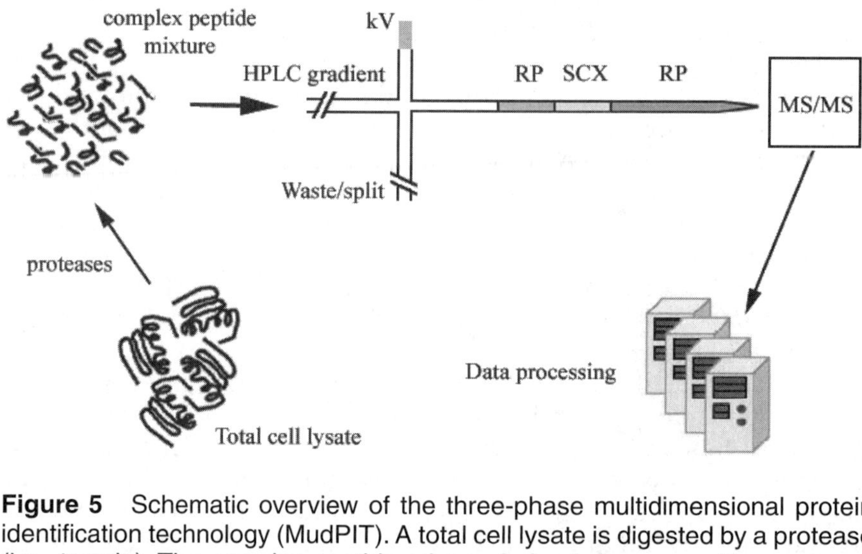

Figure 5 Schematic overview of the three-phase multidimensional protein identification technology (MudPIT). A total cell lysate is digested by a protease (i.e., trypsin). The complex peptide mixture is loaded on a capillary column. The HPLC gradient is delivered to a junction, and part of the flow is split off through the column to get a flow of a couple hundreds of nanoliters per minute. The peptides are separated by a three-phasic column, with in-line a reverse phase–strong cation exchange and another reverse phase, before they enter the ionization source of the mass spectrometer. Peptide elutes in a time window of several seconds, allowing the mass measurements and the determination of peptide sequences by tandem mass spectrometry. This data is used for protein identification using database searches.

column with a section of reverse-phase material flanked by strong cation exchange resin. They applied this MudPIT technology to the proteome analysis of *Saccharomyces cerevisiae*, which yielded, together with the work of Peng et al. (Peng et al., 2003a), one of the largest proteome analyses to date, as far as the number of detected/identified proteins is concerned (Washburn et al., 2001). A total of 1500 proteins were detected and identified (Washburn et al., 2001; Peng et al., 2003a). Categorization of these hits demonstrated the ability of this technology to detect and identify proteins rarely seen in proteome analysis, including low-abundance proteins like transcription factors and protein kinases. Even more recently, McDonald et al., (McDonald et al., 2002) reported the use of a third dimension/phase adding an additional section of RP material behind the first RP and SCX (Figure 5). Although this extra RP dimension of separation indeed allowed the identification of even more proteins, the whole data collection time had to be significantly extended.

On the other hand, an "off-line" approach may provide a more versatile approach. In an "off-line" study of the yeast proteome (Peng et al., 2003a), 1 mg of yeast lysate was separated into 80 fractions on a strong cation exchange column, and each of these fractions was individually analyzed by reverse-phase LC-MS/MS, resulting in the identification of 1504 proteins. This exceeds the number of proteins that can readily be identified using gel-based approaches using this amount of material, and methods such as the use of different/multiple proteases (Choudhary et al., 2003) may further extend the identification of proteins. The very large volume of data generated during LC-MS/MS analyses of complex peptide mixtures may be moderated when more targeted analyses (e.g., after subcellular fractionation) are performed.

This is nicely illustrated by a report on the analysis of proteins found in yeast mitochondria (Pflieger et al., 2002). Here, 40 μg of protein from a yeast mitochondrial extract was first separated on a 1D gel, which was then cut into 27 slices (approximately 2 mm length). Proteins in each slice were digested using trypsin and analyzed by LC-MS/MS using a reverse-phase column. The analysis was performed twice. During the first LC-MS/MS run, precursor ions between m/z 400 and 720 were selected for fragmentation to derive peptide sequence information; precursors between m/z 680 and 1300 were selected in the second run. This strategy identified 179 mitochondrial proteins, out of around 500 predicted from the yeast genome sequence. Although these numbers may appear limited in terms of the numbers of protein spots that appear on a 2D polyacrylamide (PA) gel, it is important to note the small amount of starting material. A disadvantage of the retention of the gel-based protein separation is that protein losses can occur either during the electrophoresis experiment or during recovery of tryptic peptides from the gel.

Compared to the conventional 2D gel–based approach, multidimensional LC has also proved to be profitable for the analysis of membrane proteins. Membrane proteins are usually under-represented on conventional 2D gels. Using MudPIT strategies, the group of Yates identified 131 proteins in yeast with three or more predicted transmembrane domains (Link et al., 1999; Washburn et al., 2001), and more recently Schirmer et al., used MudPIT technology to identify 13 known nuclear envelope integral membrane proteins and 67 uncharacterized open reading frames with predicted membrane-spanning regions of which all of the 8 proteins tested targeted to the nuclear envelope, indicating that there are substantially more integral proteins of the nuclear envelope than previously thought (Schirmer et al., 2003). MudPIT is useful for proteome analysis in general and may specifically be applied to membrane proteins to obtain detailed information on this class of proteins.

The last approach is based on Fourier transform ion cyclotron resonance MS (FTICR/MS). FTICR/MS is a relatively new tool for the analysis of proteins and offers a number of advantages. Ions are trapped within a high magnetic field and process with frequencies that are directly related to mass/charge. Accurate and precise determination of cyclotron frequency allows the achievement of high accuracy and precision of m/z determination. Thus, accuracies of less than 1 ppm are readily achieved (compared with 5–50 ppm on a ToF instrument) (Bruce et al., 1999), and it is possible to differentiate between two peptides that differ by only 0.0045 Da (He et al., 2001). This facilitates protein identification based on mass determination of one or a few tryptic peptides without fragmentation and *de novo* sequencing. The exceptional sensitivity and resolving power have been used to great advantage in a number of studies on whole proteomes. For example, analysis of 40 μg of a tryptic digest of a yeast cytosolic fraction using FTICR on-line reverse-phase HPLC column yielded approximately 110,000 putative polypeptide signals that could be used as accurate mass tags for database searching. These matched to greater than 9000 peptides from around 1000 proteins (Shen et al., 2001). It is expected that LC-FTICR/MS will become a very powerful additional tool in proteomics research in the near future. Altogether, it is expected that nanoflow LC-MS/MS–based strategies combined with database searching will soon dominate the field of protein identification.

Quantitative Proteomics

As outlined so far, the large-scale identification of proteins is one of the key aspects of proteomics. However, a second and increasingly important aspect is the quantitation of protein expression levels. This is of great importance when samples of different origin (e.g., healthy *versus* diseased tissue) are compared with the aim to find proteins responsible for or involved in this difference. The traditional approach has been to separate proteins in such samples on 2D gels and compare spot densities of corresponding spots. One of the drawbacks in this approach is that results critically depend on sample processing and staining efficiency of individual gels. To overcome some of these problems, the use of fluorescent dyes emitting at different wavelengths has gained popularity (Unlu et al., 1997; Tonge et al., 2001; Patton, 2002). In this setup, individual samples can be derivatized with a specific dye, and several samples can be run simultaneously on a single gel. This greatly avoids inter-gel variation and facilitates matching of spots.

Mass Spectrometry–Based Protein Quantitation

Apart from gel-based protein quantitation, a more recent approach is quantitation via MS. The principle of such an approach is based on the linear relationship between the amount of peptide/protein in a sample and its peak height in a mass spectrum (Chelius et al., 2002). It should be noted that direct quantitation of proteins and peptides from a single mass spectrum is virtually impossible because the absolute intensity of the detected ion is determined by its physicochemical properties. However, mass spectrometry–based quantitation can be adopted very well to determine relative protein abundance (i.e., to uncover the up- or downregulation of a protein in one sample compared to another). The basic principle of all examples discussed below is that for every peptide of interest, an internal standard is added for relative quantitation. A key aspect in any of these approaches is that the chemical properties of the internal standard are equal to those of the analyte. In this respect, the application of stable isotopes has proved highly suitable. The general procedure implies the spiking of the sample with the analyte (i.e., peptide) modified with a stable isotope (^{2}H, ^{13}C, ^{15}N). The nonderivatized and mass-labeled peptide will behave (i.e., ionize, dissociate) identically but can be distinguished because of the difference in mass. These combined characteristics permit quantitation based on the intensity ratio of the internal standard to the analyzed compound. To adopt this principle for proteome-wide studies, ideally all peptides in a sample should be derivatized with a mass label, and indeed several approaches have been developed to achieve this. The derivatized sample (e.g., collected under normal condition) is then mixed with an unlabeled sample (e.g., collected from a diseased or stressed situation). The strength of this methodology is that every labeled peptide in the mixture serves as an internal standard for the corresponding unlabeled peptide, and that in principle every protein in a sample can be quantified (Figure 6).

On the technical side, several labeling strategies have been developed during the past few years to obtain global incorporation of stable isotopes. Basically, four approaches can be distinguished as outlined in Figure 7 and as detailed below. In each of these methods, the mass label is introduced at a different stage of sample preparation, thereby dictating the moment that labeled and unlabeled samples can be mixed

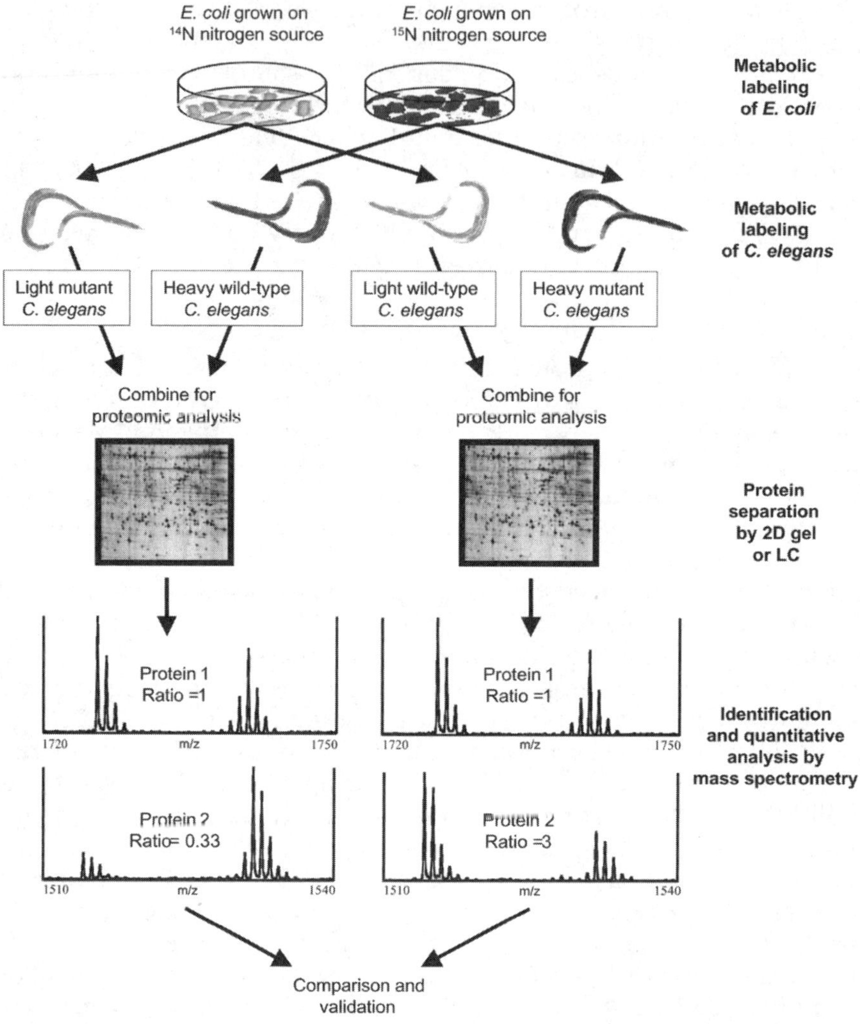

Figure 6 Schematic overview of the experimental approach for a *C. elegans* quantitative proteomics study using metabolic labeling. *E. coli* was grown in two separate batches on identical medium, except that one batch was 98% enriched in ^{15}N. Subsequently, wild-type and mutant nematodes were grown using the labeled and unlabeled batch of *E. coli*, respectively, as the only food source. Next, labeled wild-type and unlabeled mutant worms were pooled in 1:1 ratio. Protein was extracted and separated by 2D gel electrophoresis, protein spots were excised, and tryptic digests were analyzed by MS. In the bottom panels, mass spectra are shown displaying peaks of "light" and "heavy" peptides. The relative intensity of these peaks is used as direct measure for protein abundance. As a validation of the quantitation, this whole procedure was repeated in a reciprocal arrangement by labeling *glp-4* mutants instead of wild-type nematodes.

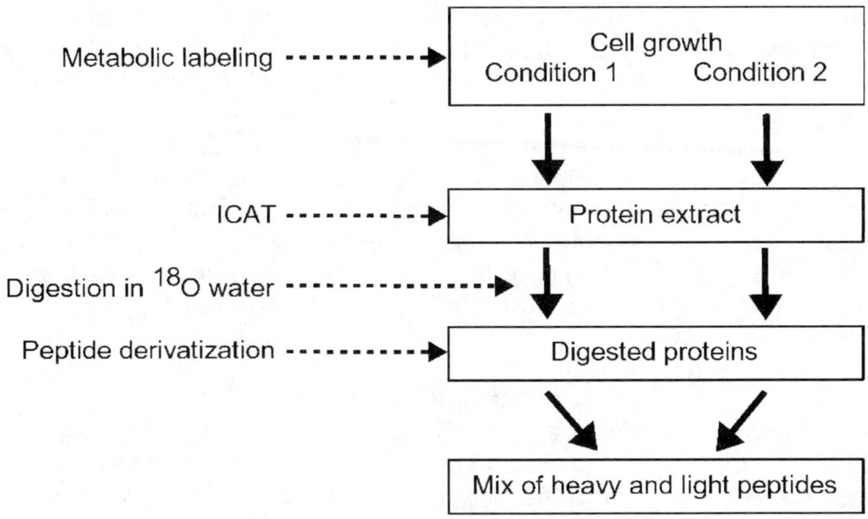

Figure 7 Various strategies for isotopic labeling of proteins and peptides. Stable isotope labels can be introduced in proteins at various stages of sample preparation. In a typical experiment, labeled and unlabeled samples should be combined directly after the labeling step. Metabolic labeling is the earliest possible point of entry and allows combination of samples directly after cell harvesting, thereby minimizing errors due to sample handling. Later entry points are provided by ICAT, digestion in ^{18}O water, and labeling after proteolytic digestion.

(i.e., directly after the labeling step). Derivatization downstream in the procedure necessitates parallel sample preparation, while early introduction of the isotope, such as in metabolic labeling, offers the advantage that samples can be combined even prior to protein extraction.

Labeling of Peptides

Complete derivatization of all proteins or peptides in a sample is a crucial part of any labeling strategy in comparative proteomics. After tryptic digestion, virtually all peptides contain a primary amine and a carboxyl group at the N- and C-termini, respectively, and both are available for chemical derivatization. Differential isotope labeling has been performed by acylation of N-terminal amines using *N*-hydroxysuccinimide or its trideuterioacetylated analogue (Geng et al., 2000; Ji et al., 2000). Nicotinyl-*N*-hydroxysuccinimide and its deuterated form have been used for the selective modification of N-termini, without affecting ε-amines in lysine (Munchbach et al., 2000).

Labeling with ^{18}O During Proteolysis

In virtually all applications in proteomics, proteins are proteolyzed with trypsin. This step can be used to incorporate a mass label in a single step by performing the reaction in $H_2{}^{18}O$ (Schnolzer et al., 1996; Mirgorodskaya et al., 2000; Wang et al., 2001; Yao et al., 2001). Under these conditions, one ^{18}O atom is added to the C-terminal carboxilic group at every cleavage site. As a result, all peptides get a mass increase of 2 Da compared to the nonderivatized peptide. This approach can be very powerful because labeling is universal, and high labeling efficiencies can be obtained. However, potential complications include the exchange of a second ^{18}O atom into the peptide (Schnolzer et al., 1996; Stewart et al., 2001) and decreased labeling efficiency due to hydrolysis, variation in pH, and relative concentration of $H_2{}^{18}O/H_2{}^{16}O$ (Stewart et al., 2001).

Isotope Coded Affinity Tag Labeling

A different approach was taken by Gygi et al. (Gygi et al., 1999), who designed a chemical tag for protein derivatization (isotope coded affinity tag, or ICAT). Apart from the fact that these tags were synthesized with and without heavy isotopes (deuterium or ^{13}C), they consist of different entities. First, the ICAT reagent contains a thiol-specific group for specific derivatization of cysteines. Next, there is a linker region containing either eight hydrogen atoms (d0, light) or eight deuterium atoms (d8, heavy). Finally, a biotin group allows the specific recovery of tagged peptides. For differential screening, proteins from two different sources are reacted with the d0 and d8 version of the tag, respectively. Next, these samples are combined, digested with a proteolytic enzyme, and derivatized peptides are selectively captured using avidin-chromatography (Figure 8). One of the advantages of this method is its specific enrichment for cysteine-containing peptides. This preference for cysteine-free peptides results in a substantial reduction in sample complexity and a reduced number of peptides to be analyzed.

One of the complications with the original ICAT tag was caused by an isotope effect during chromatography causing slightly different elution profiles for unlabeled and deuterated peptides. This was overcome by using a ^{13}C-labeled tag instead of a deuterated label (Hansen et al., 2003). Another modification included the introduction of a cleavable linker allowing the removal of the relatively large biotin moiety. This resulted in improved peptide fragmentation during MS and more straightforward data interpretation (Li et al., 2003). A more recent modification includes the linkage of the ICAT reagent to a solid support via a photo-cleavable linker, resulting in a more sensitive probing of low-abundance proteins (Zhou et al., 2002).

So far, a limited number of large-scale studies applying ICAT have been published. One example of the global analysis of protein expression profiles is the differential analysis of microsomal fractions of human cells. This has resulted in the identification and quantitation of 491 proteins (Han et al., 2001). A recent and powerful application of ICAT labeling is to determine specificity of protein-protein

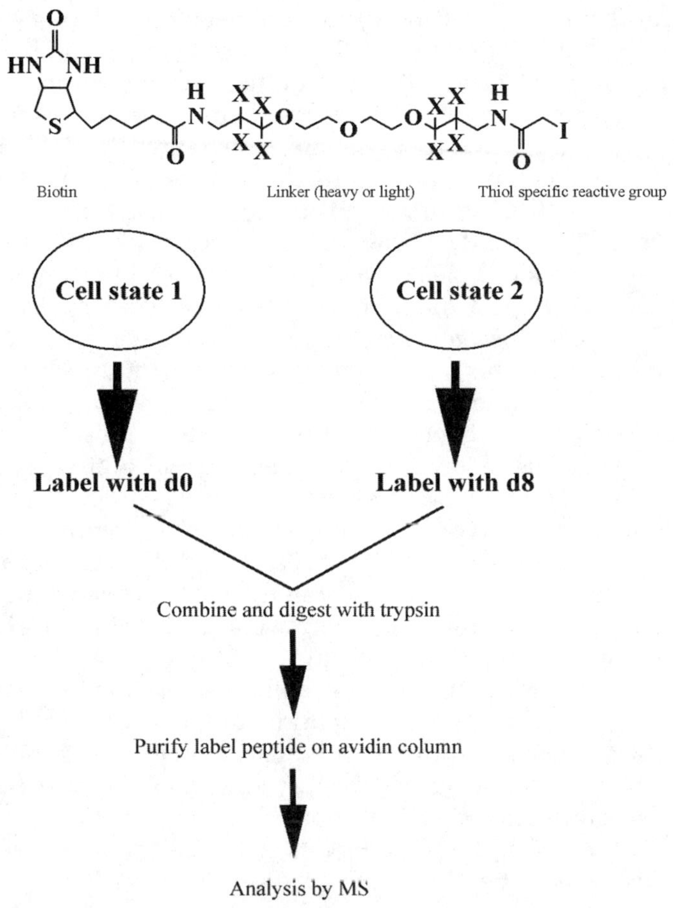

Figure 8 Application of isotope coded affinity tags (ICAT). Two protein pools are reacted with the heavy (X = D) and light form of the label (X = H), respectively. Both pools are mixed, proteolyzed, and labeled peptides are isolated by avidin affinity chromatography. The resulting fraction is then analyzed by reverse-phase LC-MS/MS.

interactions. When looking for binding partners of a specific protein, nonspecifically binding proteins will be captured in equal abundance (i.e., ratio of 1) irrespective of the pool of proteins that is searched. The hallmark of specific binding, however, is the capture of proteins in a ratio drastically deviating from 1 (Ranish et al., 2003). In a similar way, dynamic changes in the composition of protein complexes can be monitored (Brand et al., 2004).

Metabolic Labeling

Metabolic labeling offers the most comprehensive way of proteome coverage because no chemical derivatization steps are involved. Also, it is the earliest time point to introduce labels (see Figure 6). The advantage is that cells/organisms to be analyzed can be combined even before protein extraction, thereby eliminating any variation due to subsequent sample handling. The biggest limitation of this method is that it can only be applied to organisms able to grow in defined media to which the desired label can be added. So far, applications have been shown for yeast (Oda et al., 1999), bacteria (Pasa-Tolic et al., 1999; Conrads et al., 2001), and cell cultures (Conrads et al., 2001; Ong et al., 2002). Recently, this has been extended to multicellular organisms by the complete labeling of the nematode *Caenorhabditis elegans* and the fruit fly *Drosophila melanogaster* (Krijgsveld et al., 2003). This latter study showed that this approach is sensitive enough to quantify changes in expression levels even below a factor of 1.5. In all these organisms, there is basically no restriction in the type of isotope used: labeling has been achieved by complete labeling by ^{15}N (Oda et al., 1999; Conrads et al., 2001) but could be done using ^{13}C equally well.

A variation on this type of labeling introduced recently is stable isotope labeling with amino acids in cell culture (SILAC) (Ong et al., 2002). This approach makes use of deuterated or ^{13}C-labeled amino acids added to the growth media of separately cultured cell lines, giving rise to cells containing "light" and "heavy" proteins, respectively. Upon mixing of cell lysates, proteins are usually resolved by SDS-PAGE followed by in-gel trypsin digestion. Peptides are then separated based on hydrophobic properties from microcapillary columns and ionized into the mass spectrometer using electrospray ionization. Because there is no difference in chromatographic behavior between peptides containing ^{12}C and ^{13}C (or ^{14}N and ^{15}N), peptides with the same amino acid sequence will co-elute. Proteins can then be simultaneously identified and relatively quantified using mass spectrometry (Ong et al., 2002; Blagoev et al., 2003; Ong et al., 2003a; Ong et al., 2003b). Compared to the ICAT approach, no reduction in sample complexity is obtained. This can also be regarded as an advantage, as a higher protein coverage can be achieved with no bias against cysteine-free proteins.

Absolute Quantitation

All strategies discussed above were based on the relative quantitation of an unlabeled *versus* a labeled analyte. If the absolute concentration of the standard is known, then the absolute quantity of the analyte can be determined accurately. This is shown in recent studies (Stemmann et al., 2001; Gerber et al., 2003) in which proteins as well as their phosphorylation status were quantified in a complex mixture. Such an approach would be of special interest when one focuses on a single protein in a complex mixture: after synthesizing a mass-tagged peptide, the protein of interest can be followed across many samples. Proteome-wide absolute quantitation seems out of reach with the current stage of technology, as this would require the synthesis of a huge set of peptides each uniquely identifying a protein in the proteome and each containing a mass label. However, it has been suggested that this may become one of the long-term goals in quantitative proteomics (Aebersold, 2003).

Post-translational Modifications

Most of the above technologies have concentrated on the profiling of the total protein content of a cell type, compartment, or complex of interest in terms of the proteins present and their relative abundance. However, an additional important facet of proteomics is the ability to study post-translational modification, which can affect activity and binding of a protein and alter its role within the cell (Pandey and Mann, 2000; Choudhary et al., 2001). One of the unique features of proteomics is that post-translational modifications can be investigated at a more global level (Pandey and Mann, 2000; Aebersold and Goodlett, 2001). The expressed products of a single gene represent a protein population that may contain large amounts of microheterogeneity (Figure 9), each different state (e.g., another phosphorylation, glycosylation, ubiquitination, acetylation, lipidation, etc., state) adding a large amount of diversity to the expression profile of that protein. More than 100 modification types are described, and very likely more will be found in the future (O'Donovan et al., 2001). Obviously, MS can in principle readily distinguish between analogous peptides or proteins that incorporate, or not, a specific modification,

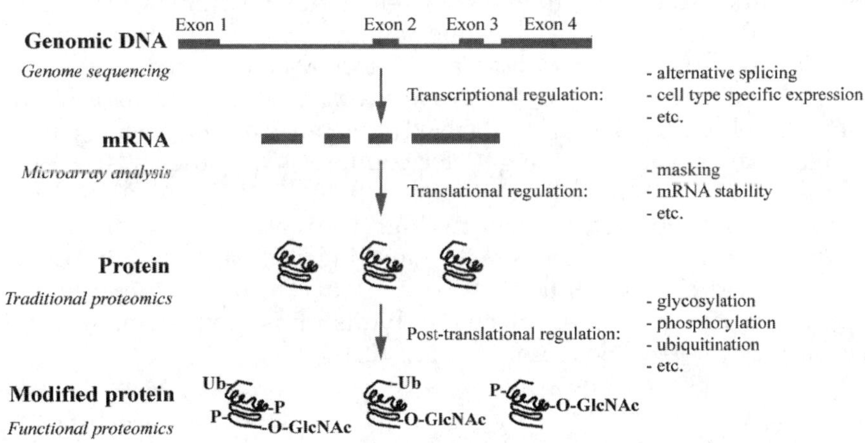

Figure 9 Schematic overview showing that a single gene can give rise to a variety of different gene products, illustrating that the complexity increases dramatically from genomics to proteomics.

but these changes are frequently not observed or identified in the global studies outlined earlier. In order to study a specific post-translational modification, it may be helpful to enrich a sample for that modification, usually by some form of affinity purification. To study phosphorylation, for example, it may mean using either a phosphopeptide-specific antibody or immobilized metal affinity chromatography (IMAC) technology. Clearly, post-translational modification of proteins is an event that has a drastic effect on the complexity of the proteome and possibly hampers the development of recombinant therapeutic proteins. Because of space limitations, we will focus here on phosphorylation, one of the most important post-translational modifications that has been studied in global proteomics studies using LC-MS/MS–based technologies.

The study of protein phosphorylation has grown exponentially in recent years as researchers from various disciplines have come to realize that key cellular functions are regulated by the reversible phosphorylation and dephosphorylation of proteins on serine, threonine, and tyrosine residues (Hunter, 1991; Sefton, 1991; Cohen, 1992; Hubbard et al., 1993). Approximately 2000 genes encoding protein kinases and probably half as many genes encoding protein phosphatases are estimated to exist in the human genome (Hunter, 1995). At present, only a small subset of eukaryote protein kinases and protein phosphatases has been characterized with respect to biological function and protein substrate specificity. To understand more about protein phosphorylation and dephosphorylation, it is necessary to identify the specific amino acid residues that become phosphorylated, because identification of these sites in proteins may reveal which protein kinase regulates the protein and thereby help elucidate the biological function and significance of novel phosphoproteins. A conventional method used to study protein phosphorylation employs radiolabeling with ^{32}P inorganic phosphate ($^{32}P_i$). To measure differences in relative abundances of phosphorylation, ^{32}P-labeled proteomes are resolved by two-dimensional polyacrylamide gel electrophoresis and the relative spot intensities are compared (Mason et al., 1998). The use of $^{32}P_i$ to label proteins does not lend itself to high-throughput proteome-wide analysis due to safety issues with handling radioactive compounds and the associated contamination of analytical instrumentation. Recently, also a stain that selectively binds phosphoproteins in a gel has been developed (Schulenberg et al., 2003).

However, several new methods depend on MS for the analysis of phosphoproteins and fall into two general categories. In a first approach, enrichment of phosphoproteins/peptides is performed prior to either electrophoretic separation at the protein level or LC-MS of digestions of enriched phosphoproteins. A second strategy involves precursor ion scanning during tandem MS to detect all components that fragment to yield a product ion characteristic of a phosphopeptide.

Thus, a product ion of m/z 98 is characteristic of serine and threonine phosphorylation (Qin et al., 1997), whereas the presence of a phosphotyrosine residue is indicated by an immonium ion at m/z 216.043 (the differentiation of this ion from those of the same nominal, but different, exact mass is essential to selective detection) (Steen et al., 2001).

Phosphorylation is a dynamic process, and usually only a fraction of the protein is phosphorylated (at a certain time point). Moreover, the negatively charged phosphogroup suppresses ionization (in positive ion mode) dramatically. Therefore, to be able to analyze protein phosphorylation, the first strategy involving enrichment of phosphoproteins/peptides has been most successful. However, enrichment of phosphoproteins by antibody-based methods is problematic if the enriched population is to be analyzed directly by MS because of the presence of large amounts of antibody and salt contamination. Fortunately, several new methods have been developed recently that provide tools for the specific enrichment of phosphopeptides from complex mixtures (Goshe et al., 2001; Oda et al., 2001; Zhou et al., 2001; Ficarro et al., 2002; Goshe et al., 2002). The enrichment step involves modifying phosphoproteins with affinity tags. These chemical modifications are often performed in combination with stable isotope incorporation that allows for their subsequent identification and quantitative analysis by MS (Oda et al., 1999; Oda et al., 2001; Zhou et al., 2001; Goshe et al., 2002). Below we will describe these strategies and their applications in more detail.

A first method was developed concurrently, and independently, by two groups, and are essentially identical (Goshe et al., 2001; Oda et al., 2001; Goshe et al., 2002). This approach to phosphoprotein enrichment and mapping is based on site-specific modification of phosphoseryl and/or phosphothreonyl residues. The phosphates are chemically replaced by biotinylated moieties, allowing purification of phosphoproteins by avidin affinity chromatography. After elution of phosphoproteins and proteolysis with trypsin, enrichment is carried out by a second round of avidin purification (Oda et al., 2001). Essentially, the same strategy has been used with a labeled version of 1,2-ethanedithiol (EDT) (d0 and d4) for the quantitation of phosphorylation and is described later (Goshe et al., 2001) (Figure 10). It is basically the same as the ICAT strategy described earlier (Figure 8) but specifically applicable to protein phosphorylation. Because the reactions are performed in a single tube, this procedure is easier to perform than the method described below, and losses resulting from multiple purification steps are minimized. A second strategy, developed by Zhou et al. (Zhou et al., 2001), is applicable to phosphotyrosine-containing peptides in addition to those containing phosphoserine and phosphothreonine residues. Several chemical reactions and purification steps are needed using this approach, which possibly lead to substantial losses. As is generally the case with chemical modification–based approaches, both the above

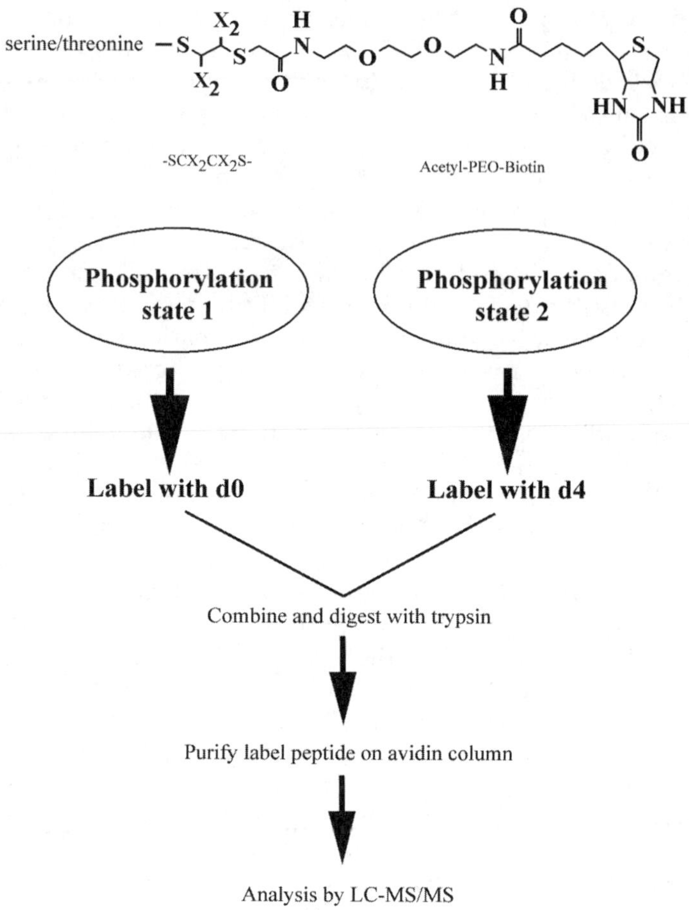

Figure 10 Phosphoproteome analysis using phosphoprotein isotope-coded affinity tags. The first phase involves labeling of the phosphoseryl and phosphothreonyl residues. In the second phase, the proteins are digested, and the sample is enriched for the labeled peptides by avidin affinity chromatography. This enriched mixture is then analyzed by capillary reverse-phase LC-MS/MS.

methods require substantial amounts of sample with the result that only abundant proteins are easily identified. Nevertheless, these approaches are promising and could be coupled to other fractionation steps to improve the overall recovery of low-abundance proteins.

As an alternative, IMAC can be used. Here, phosphoproteins (or peptides) are bound under acidic conditions to an immobilized metal such as gallium (III) or iron (III), unbound peptides are washed off, and bound peptides are recovered by raising the pH (Posewitz et al., 1999). Several studies have used this in conjunction with LC-MS/MS to identify either phosphorylation sites on a specific protein (Neville et al., 1997) or phosphorylated proteins in a complex mixture. IMAC-LC-MS/MS analysis of 100 μg of a yeast lysate tryptic digest has been shown to detect greater than 1000 phosphoproteins. A total of 216 peptide sequences were determined, containing 383 sites of phosphorylation (Ficarro et al., 2002). This technique can be modified to allow relative quantitation of phosphoproteins by using stable isotope labeling techniques such as ICAT, but there remains some way to go before routine determinations of changes in protein phosphorylation can be achieved.

Of the more than 100 modification types that have been described (O'Donovan et al., 2001), only phosphorylation has been illustrated above. LC-MS or LC-MS/MS have been used to study other modifications, such as glycosylation (Haynes and Aebersold, 2000; Kaji et al., 2003; Zhang et al., 2003) ubiquitination (Hitchcock et al., 2003; Peng et al., 2003b), palmitolation (Gomez et al., 2002), lipidation (Wu et al., 2000), and sulfonation. As LC-MS/MS is at the moment probably one of the most appropriate tools for the comprehensive analysis of protein post-translational modifications, we expect many more reports addressing post-translational modifications by LC-MS/MS in the near future.

Conclusions

During the past decade, the developments in MS-based proteomics have been enormous. With the current tools, extremely complex samples such as total cell lysates, treated with a protease, become directly amenable for analysis. Speed, sensitivity, and selectivity in proteome analysis has risen substantially and will expand even further in the near future. It has become increasingly evident that such tools are necessary to get to understand the full complexity of the proteome. As outlined in Figure 9, a single gene may lead to various gene products (i.e., largely heterogeneous proteins) via alternative splicing, masking at mRNA level, and post-translational modifications. The tools developed in proteomics research will have a significant impact on the development of therapeutic proteins. First, more drug targets will be discovered; second, these drug targets can be much better characterized including their functional micro-heterogeneities; and finally, better structural validation of therapeutic recombinant proteins brought to the market will be possible by fast, sensitive, and selective MS-based analysis.

Abbreviations

2D	Two-dimensional
DC	Direct current
ESI	Electrospray ionization

EDT	1,2-Ethanedithiol
FTICR/MS	Fourier transform ion cyclotron resonance MS
HPLC	High-pressure liquid chromatography
ICAT	Isotope coded affinity tag
IEF	Isoelectric focusing
IMAC	Immobilized metal affinity chromatography
LC	Liquid chromatography
MALDI	Matrix assisted laser desorption ionization
MS	Mass spectrometry
MS/MS	Tandem mass spectrometry
MudPIT	Multidimensional protein identification technology
PA	Polyacrylamide
PTM	Post-translational modification
RF	Radio frequency
SDS-PAGE	Sodium dodecyl sulfate polyacrylamide gel electrophoresis
ToF	Time of flight

References

1. Aebersold R. Constellations in A Cellular Universe. *Nature* 2003; 422(6928):115–116

2. Aebersold R, and Goodlett DR. Mass Spectrometry in Proteomics. *Chem Rev* 2001; 101(2):269–295

3. Alban A, David SO, Bjorkesten L, Andersson C, Sloge E, Lewis S, and Currie I. A Novel Experimental Design for Comparative Two-Dimensional Gel Analysis: Two-dimensional Difference Gel Electrophoresis Incorporating a Pooled Internal Standard. *Proteomics* 2003; 3(1):36–44

4. Belov ME, Zhang R, Strittmatter EF, Prior DC, Tang K, and Smith RD. Automated Gain Control and Internal Calibration with External Ion Accumulation Capillary Liquid Chromatography-Electrospray Ionization Fourier Transform Ion Cyclotron Resonance. *Anal Chem* 2003; 75(16):4195–4205

5. Bienvenut WV, Deon C, Pasquarello C, Campbell JM, Sanchez JC, Vestal ML, and Hochstrasser DF. Matrix-assisted Laser Desorption/ionization-tandem Mass Spectrometry with High Resolution and Sensitivity for Identification and Characterization of Proteins. *Proteomics* 2002; 2(7):868–876

6. Blagoev B, Kratchmarova I, Ong SE, Nielsen M, Foster LJ, and Mann M. A Proteomics Strategy to Elucidate Functional Protein-protein Interactions Applied to EGF Signaling. *Nat Biotechnol* 2003; 21(3):315–8

7. Brand M, Ranish JA, Kummer NT, Hamilton J, Igarashi K, Francastel C, Chi TH, Crabtree GR, Aebersold R, and Groudine M. Dynamic Changes in Transcription Factor Complexes during Erythroid Differentiation Revealed by Quantitative Proteomics. *Nat Struct Mol Biol* 2004; 11(1):73–80

8. Bruce JE, Anderson GA, Wen J, Harkewicz R, and Smith RD. High-Mass-measurement Accuracy and 100% Sequence Coverage of Enzymatically Digested Bovine Serum Albumin from an ESI-FTICR Mass Spectrum. *Anal Chem* 1999; 71(14):2595–2599

9. Chelius D, and Bondarenko PV. Quantitative Profiling of Proteins in Complex Mixtures Using Liquid Chromatography and Mass Spectrometry. *J Proteome Res* 2002; 1(4):317–323

10. Chernushevich IV, Loboda AV, and Thomson BA. An Introduction to Quadrupole-time-of-flight Mass Spectrometry. *J Mass Spectrom* 2001; 36(8):849–865

11. Choudhary G, Wu SL, Shieh P, and Hancock WS. Multiple Enzymatic Digestion for Enhanced Sequence Coverage of Proteins in Complex Proteomic Mixtures Using Capillary LC with Ion Trap MS/MS. *J Proteome Res* 2003; 2(1):59–67

12. Choudhary JS, Blackstock WP, Creasy DM, and Cottrell JS. Matching Peptide Mass Spectra to EST and Genomic DNA Databases. *Trends Biotechnol* 2001; 19(10 Suppl):S17–22

13. Cohen P. Signal Integration at the Level of Protein Kinases, Protein Phosphatases and their Substrates. *Trends Biochem Sci* 1992; 17(10):408–413

14. Conrads TP, Alving K, Veenstra TD, Belov ME, Anderson GA, Anderson DJ, Lipton MS, Pasa-Tolic L, Udseth HR, Chrisler WB, et al. Quantitative Analysis of Bacterial and Mammalian Proteomes Using a Combination of Cysteine Affinity Tags and 15N-metabolic Labeling. *Anal Chem* 2001; 73(9):2132–2139

15. De Boer E, Rodriguez P, Bonte E, Krijgsveld J, Katsantoni E, Heck A, Grosveld F, and Strouboulis J. Efficient Biotinylation and Single-step Purification of Tagged Transcription Factors in Mammalian Cells and Transgenic Mice. *Proc Natl Acad Sci USA* 2003; 100(13): 7480–7485

16. Dobson CM. Protein Misfolding, Evolution and Disease. *Trends. Biochem Sci* 1999; 24(9):329–332

17. Emmett MR, and Caprioli RM. Micro-electrospray Mass Spectrometry: Ultra-high-sensitivity Analysis of Peptides and Proteins. *Journal of the American Society for Mass Spectrometry* 1994; 5(7):605–613

18. Fenn JB, Mann M, Meng CK, Wong SF, and Whitehouse CM. Electrospray Ionization for Mass Spectrometry of Large Biomolecules. *Science* 1989; 246(4926):64–71
19. Fersht AR, and Daggett V. Protein Folding and Unfolding at Atomic Resolution. *Cell* 2002; 108(4):573–582
20. Ficarro SB, McCleland ML, Stukenberg PT, Burke DJ, Ross MM, Shabanowitz J, Hunt DF, and White FM. Phosphoproteome Analysis by Mass Spectrometry and its Application to *Saccharomyces Cerevisiae*. *Nat Biotechnol* 2002; 20(3):301–305
21. Gavin AC, Bosche M, Krause R, Grandi P, Marzioch M, Bauer A, Schultz J, Rick JM, Michon AM, Cruciat CM, et al. Functional Organization of the Yeast Proteome by Systematic Analysis of Protein Complexes. *Nature* 2002; 415(6868):141–147
22. Geng M, Ji J, and Regnier FE. Signature-peptide Approach to Detecting Proteins in Complex Mixtures. *J Chromatogr A* 2000; 870(1–2):295–313
23. Geng M, Zhang X, Bina M, and Regnier F. Proteomics of Glycoproteins Based on Affinity Selection of Glycopeptides from Tryptic Digests. *J Chromatogr B Biomed Sci Appl* 2001; 752(2):293–306
24. Gerber SA, Rush J, Stemman O, Kirschner MW, and Gygi SP. Absolute Quantification of Proteins and Phosphoproteins from Cell Lysates by Tandem MS. *Proc Natl Acad Sci U S A* 2003; 100(12): 6940–6945
25. Gomez SM, Nishio JN, Faull KF, and Whitelegge JP. The Chloroplast Grana Proteome Defined by Intact Mass Measurements from Liquid Chromatography Mass Spectrometry. *Mol Cell Proteomics* 2002; 1(1): 46–59
26. Gorg A, Obermaier C, Boguth G, Harder A, Scheibe B, Wildgruber R, and Weiss W. The Current State of Two-dimensional Electrophoresis with Immobilized pH Gradients. *Electrophoresis* 2000; 21(6):1037–1053
27. Gorg A, Obermaier C, Boguth G, and Weiss W. Recent Developments in Two-dimensional Gel Electrophoresis with Immobilized pH Gradients: Wide pH Gradients up to pH 12, Longer Separation Distances and Simplified Procedures. *Electrophoresis* 1999; 20(4–5):712–717
28. Goshe MB, Conrads TP, Panisko EA, Angell NH, VeenstraTD, and Smith RD. Phosphoprotein Isotope-coded Affinity Tag Approach for Isolating and Quantitating Phosphopeptides in Proteome-wide Analyses. *Anal Chem* 2001; 73(11):2578–2586
29. Goshe MB, Veenstra TD, Panisko EA, Conrads TP, Angell NH, and Smith RD. Phosphoprotein Isotope-coded Affinity Tags: Application to the Enrichment and Identification of Low-abundance Phosphoproteins. *Anal Chem* 2002; 74(3):607–616
30. Gygi SP, and Aebersold R. Mass Spectrometry and Proteomics. *Curr Opin Chem Biol* 2000; 4(5):489–494

31. Gygi SP, Rist B, Gerber SA, Turecek F, Gelb MH, and Aebersold R. Quantitative Analysis of Complex Protein Mixtures Using Isotope-coded Affinity Tags. *Nat Biotechnol* 1999; 17(10):994–999
32. Hager JW. Recent Trends in Mass Spectrometer Development. *Anal Bioanal Chem* 2003; 378(4):845–850
33. Hager JW, and Yves Le Blanc JC. Product ion Scanning Using a Q-q-Q Linear Ion Trap (Q TRAP) Mass Spectrometer. *Rapid Commun Mass Spectrom* 2003; 17(10):1056–1064
34. Han DK, Eng J, Zhou H, and Aebersold R. Quantitative Profiling of Differentiation-induced Microsomal Proteins Using Isotope-coded Affinity Tags and Mass Spectrometry. *Nat Biotechnol* 2001; 19(10):946–951
35. Hansen KC, Schmitt-Ulms G, Chalkley RJ, Hirsch J, Baldwin MA, and Burlingame AL. Mass Spectrometric Analysis of Protein Mixtures at Low Levels Using Cleavable 13C-ICAT and Multi-Dimensional Chromatography. *Mol Cell Proteomics* 2003; 2(5):299–314
36. Harkewicz R, Belov ME, Anderson GA, Pasa-Tolic L, Masselon CD, Prior DC, Udseth HR, and Smith RD. ESI-FTICR Mass Spectrometry Employing Data-dependent External Ion Selection and Accumulation. *J Am Soc Mass Spectrom* 2002; 13(2):144–154
37. Hart C, Schulenberg B, Diwu Z, Leung WY, and Patton WF. Fluorescence Detection and Quantitation of Recombinant Proteins Containing Oligohistidine Tag Sequences Directly in Sodium Dodecyl Sulfate-polyacrylamide Gels. *Electrophoresis* 2003a; 24(4):599–610
38. Hart C, Schulenberg B, Steinberg TH, Leung WY, and Patton WF. Detection of Glycoproteins in Polyacrylamide Gels and on Electroblots Using Pro-Q Emerald 488 dye, a Fluorescent Periodate Schiff-base Stain. *Electrophoresis* 2003b; 24(4):588–598
39. Haynes PA, and Aebersold R. Simultaneous Detection and Identification of O-GlcNAc-modified Glycoproteins Using Liquid Chromatography-tandem Mass Spectrometry. *Anal Chem* 2000; 72(21):5402–5410
40. He F, Hendrickson CL, and Marshall AG. Baseline Mass Resolution of Peptide Isobars: a Record for Molecular Mass Resolution. *Anal Chem* 2001; 73(3):647–650
41. Hillenkamp F, Karas M, Beavis RC, and Chait BT. Matrix-Assisted Laser Desorption/ionization Mass Spectrometry of Biopolymers. *Anal Chem* 1991; 63(24):1193A–1203A
42. Hitchcock AL, Auld K, Gygi SP, and Silver PA. A Subset of Membrane-associated Proteins is Ubiquitinated in Response to Mutations in the Endoplasmic Reticulum Degradation Machinery. *Proc Natl Acad Sci U S A* 2003; 100(22):12735–12740
43. Ho Y, Gruhler A, Heilbut A, Bader GD, Moore L, Adams SL, Millar A, Taylor P, Bennett K, Boutilier K, et al. Systematic Identification of

Protein Complexes in Saccharomyces Cerevisiae by Mass Spectrometry. *Nature* 2002; 415(6868):180–183

44. Hu Y, Wang G, Chen GY, Fu X, and Yao SQ. Proteome Analysis of *Saccharomyces Cerevisiae* Under Metal Stress by Two-dimensional Differential Gel Electrophoresis. *Electrophoresis* 2003; 24(9):1458–1470

45. Huang L, Baldwin MA, Maltby DA, Medzihradszky KF, Baker PR, Allen N, Rexach M, Edmondson RD, Campbell J, Juhasz P, et al. The Identification of Protein-protein Interactions of the Nuclear Pore Complex of *Saccharomyces Cerevisiae* Using High Throughput Matrix-assisted Laser Desorption Ionization Time-of-flight Tandem Mass Spectrometry. *Mol Cell Proteomics* 2002; 1(6):434–450

46. Hubbard MJ, and Cohen P. On target with a New Mechanism for the Regulation of Protein Phosphorylation. *Trends Biochem Sci* 1993; 18(5):172–177

47. Hunter T. Protein Kinase Classification. *Methods Enzymol* 1991; 200:3–37

48. Hunter T. Protein Kinases and Phosphatases: the Y in and Yang of Protein Phosphorylation and Signaling. *Cell* 1995; 80(2):225–236

49. Ji J, Chakraborty A, Geng M, Zhang X, Amini A, Bina M, and Regnier F. Strategy for Qualitative and Quantitative Analysis in Proteomics based on Signature Peptides. *J Chromatogr B Biomed Sci Appl* 2000; 745(1):197–210

50. Kaji H, Saito H, Yamauchi Y, Shinkawa T, Taoka M, Hirabayashi J, Kasai K, Takahashi N, and Isobe T. Lectin Affinity Capture, Isotope-coded Tagging and Mass Spectrometry to Identify N-linked Glycoproteins. *Nat Biotechnol* 2003; 21(6):667–672

51. Karas M, and Hillenkamp F. Laser Desorption Ionization of Proteins with Molecular Masses Exceeding 10,000 Daltons. *Anal Chem* 1988; 60(20):2299–2301

52. Kazmirski SL, Isaacson RL, An C, Buckle A, Johnson CM, Daggett V, and Fersht AR. Loss of a Metal-binding Site in Gelsolin Leads to Familial Amyloidosis-Finnish Type. *Nat Struct Biol* 2002; 9(2):112–116

53. Keough T, Lacey MP, Fieno AM, Grant RA, Sun Y, Bauer MD, and Begley KB. Tandem Mass Spectrometry Methods for Definitive Protein Identification in Proteomics Research. *Electrophoresis* 2000; 21(11):2252–2265

54. Kettman JR, Coleclough C, Frey JR, and Lefkovits I. Clonal Proteomics: One Gene-family of Proteins. *Proteomics* 2002; 2(6):624–631

55. Krijgsveld J, Ketting RF, Mahmoudi T, Johansen J, Artal-Sanz M, Verrijzer CP, Plasterk RH, and Heck AJ. Metabolic Labeling of C. Elegans and D. melanogaster for Quantitative Proteomics. *Nat Biotechnol* 2003; 21(8):927–931

56. Kruger R, Pfenninger A, Fournier I, Gluckmann M, and Karas M. Analyte Incorporation and Ionization in Matrix-assisted Laser Desorption/ionization Visualized by pH Indicator Molecular Probes. *Anal Chem* 2001; 73(24):5812–5821

57. Lahm HW, and Langen H. Mass Spectrometry: A Tool for the Identification of Proteins Separated by Gels. *Electrophoresis* 2000; 21(11):2105–2014

58. Le Blanc JC, Hager JW, Ilisiu AM, Hunter C, Zhong F, and Chu I. Unique Scanning Capabilities of a New Hybrid Linear Ion Trap Mass Spectrometer (Q TRAP) Used for High Sensitivity Proteomics Applications. *Proteomics* 2003; 3(6):859–269

59. Lee JR, Baxter TM, Yamaguchi H, Wang TC, Goldenring JR, and MGAnderson. Differential Protein Analysis of Spasomolytic Polypeptide Expressing Metaplasia Using Laser Capture Microdissection and Two-dimensional Difference Gel Electrophoresis. *Appl Immunohistochem Mol Morphol* 2003; 11(2):188–193

60. Li J, Steen H, and Gygi SP. Protein Profiling with Cleavable Isotope-coded Affinity Tag (cICAT) Reagents: The Yeast Salinity Stress Response. *Mol Cell Proteomics* 2003; 2(11):1198–1204

61. Link AJ, Eng J, Schieltz DM, Carmack E, Mize GJ, Morris DR, Garvik BM, and Yates JR, 3rd. Direct Analysis of Protein Complexes Using Mass Spectrometry. *Nat Biotechnol* 1999; 17(7):676–682

62. Londry FA, and Hager JW. Mass Selective Axial Ion Ejection From a Linear Quadrupole Ion Trap. *J Am Soc Mass Spectrom* 2003; 14(10):1130–47

63. MacCoss MJ, and Yates JR, 3rd. Proteomics: Analytical Tools and Techniques. *Curr Opin Clin Nutr Metab Care* 2001; 4(5):369–375

64. Mann M, Hendrickson RC, and Pandey A. Analysis of Proteins and Proteomes by Mass Spectrometry. *Annu Rev Biochem* 2001; 70:437–473

65. Mann M, Ong SE, Gronborg M, Steen H, Jensen ON, and Pandey A. Analysis of Protein Phosphorylation Using Mass Spectrometry: Deciphering the Phosphoproteome. *Trends Biotechnol* 2002; 20(6):261–268

66. Martin K, Hart C, Schulenberg B, Jones L, and Patton WF. Simultaneous red/green Dual Fluorescence Detection on Electroblots Using BODIPY TR-X Succinimidyl Ester and ELF 39 Phosphate. *Proteomics* 2002; 2(5):499–512

67. Mason GG, Murray RZ, Pappin D, and Rivett AJ. Phosphorylation of ATPase Subunits of the 26S Proteasome. *FEBS Lett* 1998; 430(3):269–274

68. Mayor U, Guydosh NR, Johnson CM, Grossmann JG, Sato S, Jas GS, Freund SM, Alonso DO, Daggett V, and Fersht AR. The Complete Folding Pathway of a Protein from Nanoseconds to Microseconds. *Nature* 2003; 421(6925):863-867

69. McDonald WH, Ohi R, Miyamoto DT, Mitchison TJ, Yates I, and John R. Comparison of Three Directly Coupled HPLC MS/MS Strategies for Identification of Proteins from Complex Mixtures: Single-dimension LC-MS/MS, 2-phase MudPIT, and 3-phase MudPIT. *Int J Mass Spectrom* 2002; 219(1):245–251

70. McLachlin DT, and Chait BT. Analysis of Phosphorylated Proteins and Peptides by Mass Spectrometry. *Curr Opin Chem Biol* 2001; 5(5):591–602

71. McPherson JD, Marra M, Hillier L, Waterston RH, Chinwalla A, Wallis J, Sekhon M, Wylie K, Mardis ER, Wilson RK, et al. A Physical Map of the Human Genome. *Nature* 2001; 409(6822):934–941

72. Medzihradszky KF, Campbell JM, Baldwin MA, Falick AM, Juhasz P, Vestal ML, and Burlingame AL. The Characteristics of Peptide Collision-induced Dissociation Using a High-performance MALDI-TOF/TOF Tandem Mass Spectrometer. *Anal Chem* 2000; 72(3):552–558

73. Meiring HDH, van der; Hove E, ten; Jong GJ, de APJM. Nanoscale LC-MS(n): Technical Design and Applications to Peptide and Protein Analysis. *J Sep Sci* 2002; 25(9):557–568

74. Mirgorodskaya OA, Kozmin YP, Titov MI, Korner R, Sonksen CP, and Roepstorff P. Quantitation of Peptides and Proteins by Matrix-assisted Laser Desorption/ionization Mass Spectrometry using (18)O-labeled Internal Standards. *Rapid Commun Mass Spectrom* 2000; 14(14):1226–1232

75. Munchbach M, Quadroni M, Miotto G, and James P. Quantitation and Facilitated De Novo Sequencing of Proteins by Isotopic N-terminal Labeling of Peptides with a Fragmentation-directing Moiety. *Anal Chem* 2000; 72(17):4047–4057

76. Neville DC, Rozanas CR, Price EM, Gruis DB, Verkman AS, and Townsend RR. Evidence for Phosphorylation of Serine 753 in CFTR Using a Novel Metal-ion Affinity Resin and Matrix-Assisted Laser Desorption Mass Spectrometry. *Protein Sci* 1997; 6(11):2436–2445

77. Oda Y, Huang K, Cross FR, Cowburn D, and Chait BT. Accurate Quantitation of Protein Expression and Site-specific Phosphorylation. *Proc Natl Acad Sci USA* 1999; 96(12):6591–6596

78. Oda Y, Nagasu T, and Chait BT. Enrichment Analysis of Phospho-rylated Proteins as a Tool for Probing the Phosphoproteome. *Nat Biotechnol* 2001; 19(4):379–382

79. O'Donovan C, Apweiler R, and Bairoch A. The Human Proteomics Initiative (HPI). *Trends Biotechnol* 2001; 19(5):178–181

80. Olsen JV, Ong SE, and Mann M. Trypsin Eleaves Exclusively C-terminal to Arginine and Lysine Residues. *Mol Cell Proteomics* 2004; 3(6):608–614

81. Ong SE, Blagoev B, Kratchmarova I, Kristensen DB, Steen H, Pandey A, and Mann M. Stable Isotope Labeling by Amino Acids in Cell Culture, SILAC, as a Simple and Accurate Approach to Expression Proteomics. *Mol Cell Proteomics* 2002; 1(5):376–386

82. Ong SE, Foster LJ, and Mann M. Mass Spectrometric-based Approaches in Quantitative Proteomics. *Methods* 2003a; 29(2):124–130

83. Ong SE, Kratchmarova I, and Mann M. Properties of 13C-Substituted Arginine in Stable Isotope Labeling by Amino Acids in Cell Culture (SILAC). *J Proteome Res* 2003b; 2(2):173–181

84. Ozier O, Amin N, and Ideker T. Global Architecture of Genetic Interactions on the Protein Network. *Nat Biotechnol* 2003; 21(5):490–491

85. Pandey A, and Mann M. Proteomics to Study Genes and Genomes. *Nature* 2000; 405(6788):837–846

86. Pasa-Tolic L, Jensen PK, Anderson GA, Lipton MS, Peden KK, Martinovic S, Tolic N, Bruce JE, and Smith RD. High Throughput Proteome-wide Precision Measurements of Protein Expression Using Mass Spectrometry. *J Am Chem Soc* 1999; 121:7949–7950

87. Patton WF. Detection Technologies in Proteome Analysis. *J Chromatogr B Analyt Technol Biomed Life Sci* 2002; 771(1–2):3–31

88. Peng J, Elias JE, Thoreen CC, Licklider LJ, and Gygi SP. Evaluation of Multidimensional Chromatography Coupled with Tandem Mass Spectrometry (LC/LC-MS/MS) for Large-scale Protein Analysis: the Yeast Proteome. *J Proteome Res* 2003a; 2(1):43–50

89. Peng J, Schwartz D, Elias JE, Thoreen CC, Cheng D, Marsischky G, Roelofs J, Finley D, and Gygi SP. A Proteomics Approach to Understanding Protein Ubiquitination. *Nat Biotechnol* 2003b; 21(8):921–926

90. Pflieger D, Le Caer JP, Lemaire C, Bernard BA, Dujardin G, and Rossier J. Systematic Identification of Mitochondrial Proteins by LC-MS/MS. *Anal Chem* 2002; 74(10):2400–2406

91. Posewitz MC, and Tempst P. Immobilized Gallium(III) Affinity Chromatography of Phosphopeptides. *Anal Chem* 1999; 71(14): 2883–2892

92. Qin J, and Chait BT. Identification and Characterization of Posttranslational Modifications of Proteins by MALDI Ion Trap Mass Spectrometry. *Anal Chem* 1997; 69(19):4002–4009

93. Rabilloud T. Silver Staining of 2-D Electrophoresis Gels. *Methods Mol Biol* 1999; 112:297–305

94. Rabilloud T. Detecting Proteins Separated by 2-D Gel Electrophoresis. *Anal Chem* 2000; 72(1):48A–55A

95. Rabilloud T. Two-dimensional Gel Electrophoresis in Proteomics: Old, Old Fashioned, But It Still Climbs Up the Mountains. *Proteomics* 2002; 2(1):3–10

96. Ranish JA, Yi EC, Leslie DM, Purvine SO, Goodlett DR, Eng J, and Aebersold R. The Study of Macromolecular Complexes by Quantitative Proteomics. *Nat Genet* 2003; 33(3):349–355

97. Rappsilber J, Siniossoglou S, Hurt EC, and Mann M. A Generic Strategy to Analyze the Spatial Organization of Multi-protein Complexes by Cross-linking and Mass Spectrometry. *Anal Chem* 2000; 72(2):267–275

98. Roepstorff P, and Fohlman J. Proposal for a Common Nomenclature for Sequence Ions in Mass Spectra of Peptides. *Biomed Mass Spectrom* 1984; 11(11):601

99. Schirmer EC, Florens L, Guan T, Yates JR, 3rd and Gerace L. Nuclear Membrane Proteins with Potential Disease Links Found by Subtractive Proteomics. *Science* 2003; 301(5638):1380–1382

100. Schnolzer M, Jedrzejewski P, and Lehmann WD. Protease-Catalyzed Incorporation of 18O into Peptide Fragments and its Application for Protein Sequencing by Electrospray and Matrix-assisted Laser Desorption/ionization Mass Spectrometry. *Electrophoresis* 1996; 17(5):945–953

101. Schulenberg B, Aggeler R, Beechem JM, Capaldi RA, and Patton WF. Analysis of Steady-state Protein Phosphorylation in Mitochondria Using a Novel Fluorescent Phosphosensor Dye. *J Biol Chem* 2003; 278(29):27251–27255

102. Sefton BM. Measurement of Stoichiometry of Protein Phosphorylation by Biosynthetic Labeling. *Methods Enzymol* 1991; 201:245–251

103. Shen Y, Tolic N, Zhao R, Pasa-Tolic L, Li L, Berger SJ, Harkewicz R, Anderson GA, Belov ME, and Smith RD. High-throughput Proteomics Using High-efficiency Multiple-capillary Liquid Chromatography with On-line High-performance ESI FTICR Mass Spectrometry. *Anal Chem* 2001; 73(13):3011–3021

104. Shou W, Verma R, Annan RS, Huddleston MJ, Chen SL, Carr SA, and Deshaies RJ. Mapping Phosphorylation Sites in Proteins by Mass Spectrometry. *Methods Enzymol* 2002; 351:279–296

105. Steen H, Kuster B, Fernandez M, Pandey A, and Mann M. Detection of Tyrosine Phosphorylated Peptides by Precursor Ion Scanning Quadrupole TOF Mass Spectrometry in Positive Ion Mode. *Anal Chem* 2001; 73(7):1440–1448

106. Stemmann O, Zou H, Gerber SA, Gygi SP, and Kirschner MW. Dual Inhibition of Sister Chromatid Separation at Metaphase. *Cell* 2001; 107(6):715–726

107. Stewart II, Thomson T, and Figeys D. 18O L:abeling: a Tool for Proteomics. *Rapid Commun Mass Spectrom* 2001; 15(24):2456–2465

108. Tang K, Tolmachev AV, Nikolaev E, Zhang R, Belov ME, Udseth HR, and Smith RD. Independent Control of Ion Transmission in a Jet Disrupter Dual-channel Ion Funnel Electrospray Ionization MS Interface. *Anal Chem* 2002; 74(20):5431–5437

109. Tonge R, Shaw J, Middleton B, Rowlinson R, Rayner S, Young J, Pognan F, Hawkins E, Currie I, and Davison M. Validation and Development of Fluorescence Two-dimensional Differential Gel Electrophoresis Proteomics Technology. *Proteomics* 2001; 1(3):377–396

110. Unlu M, Morgan ME, and Minden JS. Difference Gel Electrophoresis: a Single Gel Method for Detecting Changes in Protein Extracts. *Electrophoresis* 1997; 18(11):2071–2077

111. Van den Bergh G, Clerens S, Cnops L, Vandesande F, and Arckens L. Fluorescent Two-dimensional Difference Gel Electrophoresis and Mass Spectrometry Identify Age-related Protein Expression Differences

for the Primary Visual Cortex of Kitten and Adult Cat. *J Neurochem* 2003a; 85(1):193–205

112. Van Den Bergh G, Clerens S, Vandesande F, and Arckens L. Reversed-Phase High-performance Liquid Chromatography Prefractionation Prior to Two-dimensional Difference Gel Electrophoresis and Mass Spectrometry Identifies New Differentially Expressed Proteins Between Striate Cortex of Kitten and Adult Cat. *Electrophoresis* 2003b; 24(9): 1471–1481
113. Vazquez A, Flammini A, Maritan A, and Vespignani A. Global Protein Function Prediction From Protein-protein Interaction Networks. *Nat Biotechnol* 2003; 21(6):697–700
114. Venter JC, Adams MD, Myers EW, Li PW, Mural RJ, Sutton GG, Smith HO, Yandell M, Evans CA, Holt RA, et al. The Sequence of the Human Genome. *Science* 2001; 291(5507):1304–1351
115. Vosseller K, Wells L, and Hart GW. Nucleocytoplasmic O-glycosylation: O-GlcNAc and Functional Proteomics. *Biochimie* 2001; 83(7):575–581
116. Wang YK, Ma Z, Quinn DF, and Fu EW. Inverse 18O Labeling Mass Spectrometry for the Rapid Identification of Marker/target Proteins. *Anal Chem* 2001; 73(15):3742–3750
117. Washburn MP, Ulaszek R, Deciu C, Schieltz DM, and Yates JR, 3rd. Analysis of Quantitative Proteomic Data Generated Via Multi-dimensional Protein Identification Technology. *Anal Chem* 2002; 74(7):1650–1657
118. Washburn MP, Wolters D, and Yates JR, 3rd. Large-scale Analysis of the Yeast Proteome by Multidimensional Protein Identification Technology. *Nat Biotechnol* 2001; 19(3):242–247
119. Wilhelm T, Nasheuer HP, and Huang S. Physical and Functional Modularity of the Protein Network in Yeast. *Mol Cell Proteomics* 2003; 2:292–298
120. Wilm MS, and Mann M. Electrospray and Taylor-Cone theory, Dole's Beam of Macromolecules at Last?. *International Journal of Mass Spectrometry and Ion Processes* 1994; 136(2–3):167–180
121. Wu SL, Amato H, Biringer R, Choudhary G, Shieh P, and Hancock WS. Targeted Proteomics of Low-level Proteins in Human Plasma by LC/MSn: Using Human Growth Hormone as a Model System. *J Proteome Res* 2002; 1(5):459–465
122. Wu XD, Kircher RA, McVerry PH, and Malinzak DA. Characterization of Post-translationally Modified Recombinant Protein Using Liquid Chromatography/mass Spectrometry. *Dev Biol (Basel)* 2000; 103:61–67
123. Yan JX, Devenish AT, Wait R, Stone T, Lewis S, and Fowler S. Fluorescence Two-dimensional Difference Gel Electrophoresis and Mass Spectrometry Based Proteomic Analysis of *Escherichia Coli*. *Proteomics* 2002; 2(12):1682–1698

124. Yao X, Freas A, Ramirez J, Demirev PA, and Fenselau C. Proteolytic 18O Labeling for Comparative Proteomics: Model Studies with Two Serotypes of Adenovirus. *Anal Chem* 2001; 73(13):2836–2842

125. Yates JR, 3rd. Mass Spectrometry and the Age of the Proteome. *J Mass Spectrom* 1998; 33(1):1–19

126. Yates JR, 3rd. Mass Spectrometry. From Genomics to Proteomics. *Trends Genet* 2000; 16(1):5–8

127. Yates JR, 3rd, ECarmack, LHays, AJLink, and JKEng. Automated Protein Identification Using Microcolumn Liquid Chromatography-Tandem Mass Spectrometry. *Methods Mol Biol* 1999; 112:553–569

128. Yergey AL, Coorssen JR, Backlund PS, Jr., Blank PS, Humphrey GA, Zimmerberg J, Campbell JM, and Vestal ML. De Novo Sequencing of Peptides Using MALDI/TOF-TOF. *J Am Soc Mass Spectrom* 2002; 13(7):784–791

129. Zhang H, Li XJ, Martin DB, and Aebersold R. Identification and Quantification of N-linked Glycoproteins Using Hydrazide Chemistry, Stable Isotope Labeling and Mass Spectrometry. *Nat Biotechnol* 2003; 21(6):660–666

130. Zhou H, Ranish JA, Watts JD, and Aebersold R. Quantitative Proteome Analysis by Solid-phase Isotope Tagging and Mass Spectrometry. *Nat Biotechnol* 2002; 20(5):512–515

131. Zhou H, Watts JD, and Aebersold R. A Systematic Approach to the Analysis of Protein Phosphorylation. *Nat Biotechnol* 2001; 19(4): 375–378

14

Immunoassays

GIDEON KERSTEN and JANNY WESTDIJK

Netherlands Vaccine Institute, Unit Research and Development, Bilthoven, The Netherlands

Introduction

An immunoassay is an analytical technique that uses (1) antibodies or antibody-related molecules to characterize or quantify antigens of interest or (2) antigens to characterize or quantify antibodies or antibody responses. When antibodies are used to detect the antigen, the antibodies may also be called the reagent, and in that case the antigen is the analyte. On the other hand, when antigens are used to detect antibodies, the antigen is called reagent, and the antibodies are the analyte. In this chapter, descriptions of immunoassays will be based on the detection of antigens (= analyte) with aid of antibodies (= reagent) unless indicated otherwise.

Immunoassays are specific and sensitive. As the number of different antibodies is very large (see the section "Antibodies," below), immunoassays can be very specific, provided that the antibodies are properly selected. As the binding of antibody to antigen can be very strong, the sensitivity of immunoassays is often high. Moreover, many immunoassays are easy to perform. These are some of the reasons why immunoassays have expanded enormously since the introduction of the technique by Yalow and Berson (1959).

Apart from widespread use in clinical research and diagnosis, immunoassays are also used on a regular basis in the pharmaceutical industry. In the field of drug development, the technique contributes to, for example, toxicological, pharmacological, pharmacokinetic, and immunogenicity studies. Vaccine development would not be that well advanced without immunoassays. Some applications have even found their way into the household. The home pregnancy test is an example of this.

The expansion of applications during the past forty years is accompanied by an increase of assay designs and a wide variety of signal reagents and detection systems. These days, especially in laboratories where many samples are handled, immunoassays are fully automated.

This chapter deals mostly with the principles of the assay and its applications with an emphasis on applications in the pharmaceutical field. It is impossible to give a full insight into the subject in one chapter. The interested reader is referred to excellent handbooks on the subject of immunoassays (Wild, 1994; Price and Newman, 1997).

Components of Immunoassays: Antibodies and Antisera

Antibodies

Most antibodies or immunoglobulins consist of two identical heavy and two identical light chains interconnected by disulfide bonds forming a Y-shaped molecule (Figure 1). One end of the light and heavy chain forms the antigen-binding site, or paratope. Each immunoglobulin contains two identical antigen-binding sites. The antigen-binding site is located in an area called the Fab (fragment antigen binding) region. The Fab region consists of the light chain plus a part of the heavy chain. The Fab region is highly variable in composition and determines the specificity of an antibody. It is estimated that mammalian immune systems are able to generate more than 10 billion different antibody molecules. On a genetic level, this is done economically by gene rearrangements of a limited number of gene segments.

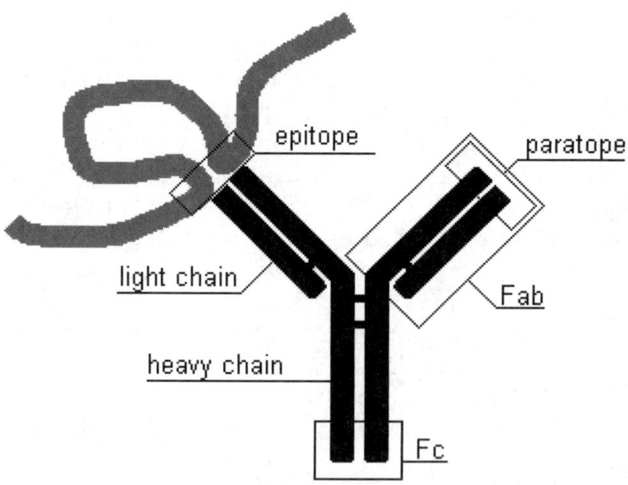

Figure 1 Schematic representation of an antibody-antigen complex. Paratope and epitope interact. In this case, the epitope is discontinuous. The different parts of the antibody are indicated.

Mammalian antibodies exist in five isotypes: IgG, IgM, IgA, IgD, and IgE. The isotype is determined by the amino acid sequence of the heavy chain constant region. For immunoassays, the IgG isotypes are most relevant. This is mainly because IgGs are most abundant. Serum can contain up to 13 mg/mL immunoglobulin, about 75% of which is IgG. IgA and IgM are also present in considerable amounts; IgD and IgE only in traces. IgG molecules are monomeric molecules, in contrast to IgM, which forms pentamers, with a molecular weight of about 150,000 g/mol (150 kDa). They are divided into four subclasses: IgG1, IgG2a, IgG2b, and IgG3 in mice (IgG1, 2, 3, and 4 in man). The subclasses differ in heavy chain structure and the number of disulfide bridges between the heavy chains.

Antibody-Antigen Interactions

As mentioned above, antibodies recognize antigens with the Fab region. The region that actually binds to antigens, the paratope, consists of six hypervariable loops containing a total of about 50 amino acids. The hypervariable loops are called complementarity determining regions (CDRs). The paratope is complementary to a part of about 3 × 3 nm of an antigen. This part of the antigen is an epitope. Electrostatic as well as hydrophobic interactions play a role in the interaction between paratope and epitope (Price, 1998). Of the 50 amino acids in the hypervariable part, about 15 are in direct contact with the epitope (i.e., the distance between antibody and antigen is 0.4 nm or less).

Epitopes on proteins consist of 15 to 20 amino acids, although only about 3 to 5 contribute significantly to the binding constant. This low number of binding amino acids does not imply that epitopes are linear. It is assumed that most protein epitopes are to some extent conformation dependent. In addition, epitopes can be continuous or discontinuous (Figure 1). This seems contradictory to the observation that blotted proteins denatured by sodium dodecyl sulfate are in many cases detectable with antibodies in immunoblot procedures. It is thought, however, that proteins retain (or regain) a considerable part of their conformation under these "denaturing" conditions. Only when protein epitopes are present on small peptides (degraded protein or synthesized peptides) can it be demonstrated that many antibodies do not bind anymore.

Antibodies as Reagents in Immunoassays

Antibodies used as reagents for immunoassays can be polyclonal or monoclonal. Polyclonal antibodies are easily generated, and they often have a higher affinity (or, better, avidity; i.e., the binding force of the combined interactions) as compared to monoclonal antibodies. To generate antiserum as an assay reagent, small experimental animals are less suitable because of low serum yields. Often rabbit, goat, sheep, or even cow or horse are used. In these species, enough serum can be obtained for routine testing. Antibodies from chicken, IgY, may also be convenient because they are present in large amounts in egg yolk.

Monoclonal antibodies, on the other hand, have the advantage of ultimate specificity and, once the hybridoma is available, unlimited supply. Production of hybridomas and monoclonal antibodies requires certain expertise, but the methods are well established and straightforward.

Polyclonal antibodies have the advantage that destruction or shielding of an epitope due to a label or binding of a monoclonal antibody in a sandwich assay (see the section "Excess Reagent Format," below) does in general not lead to complete inhibition of the binding because antibodies to other epitopes are present.

For the generation of polyclonal antibodies, antigen is needed for immunization. For the generation of monoclonal antibodies, antigen is needed for both immunization and selection of the hybridomas. Special attention must be paid to the quality of the antigen preparation. The presence of contaminating proteins may influence the specificity of the antiserum dramatically, especially if the contaminants are highly immunogenic and the target antigen only weakly. Therefore, the specificity of the antiserum should be tested thoroughly. Not only should this be done by immunoblotting of the target antigen but preferably also with the proper negative control (i.e., the matrix from which the antigen was purified minus the antigen). For recombinant proteins, this can be a lysate of the parent cell line. A second quality aspect about antigen preparations used to generate or select antibodies is the antigen itself. Preferably, this should be identical with respect to structural integrity, aggregation state, and conformation to the protein that has to be detected or quantified in the future assay. Although this sounds obvious, it may be difficult to achieve because, for instance, the production cell lines may not be available yet. The adjuvants should be chosen with care also. These substances are added to the immunizing formulation in order to boost the immune response. They often adsorb the antigen or are surface active or oily. This may affect antigen stability and, as a result, the specificity of the response.

For the generation of monoclonal antibodies, the quality of the immunizing antigen is less important because a selection procedure will follow. A problem may occur, however, during this phase. The selection of antigen-specific hybridomas is done by screening antibody-containing culture supernates, often in antigen-coated ELISA plates. Impure antigen will lead to the selection of more nonrelevant hybridomas. By Western blot analysis, these nonspecific antibodies can be identified and discarded.

In addition to using structurally and conformationally intact antigen, it is important to check whether the structure of the antigen does not change in the selection phase. Upon coating of ELISA plates with antigen, the antigen may undergo dramatic conformational changes, resulting in loss of epitopes and exposure of hidden ones (see the section "Heterogeneous Immunoassays," below). To prevent this, catching antibodies coated on the solid support can be used. This can of course only be done if antibodies are available already.

Quantity, Antigenicity, Immunogenicity

Because they are often used incorrectly, a few words are spent on the difference between antigenicity and immunogenicity. The ability of an antigen to bind antibodies is called antigenicity. The ability of an antigen to induce an immune response or to generate antibodies is called immunogenicity. One could say that antigenicity is "passive" and immunogenicity is "active." Immunoassays typically measure the antigenicity of a protein, unless of course the immunoassay uses serum as analyte with the purpose to measure whether a protein pharmaceutical induces undesired antibody responses (i.e., whether the protein is immunogenic) (see the section "Immunogenicity of Protein Pharmaceuticals," below).

Immunoassays are often designed as quantification assays. It should be kept in mind, however, that they actually are antigenicity assays. They measure the concentration of intact epitopes accessible to antibodies. This may or may not correlate with protein concentration. Aggregation, interaction with matrix components, denaturation, and glycosylation can change the ratio between protein concentration and antigenicity. It is therefore important that factors that may affect the antigenicity of a particular protein are identified during assay development and validation.

Classification of Immunoassays

Immunoassays can be classified in several ways. For example, on the basis of the kind of analyte or reagent used in the assay or on basis of assay conditions. If the sensitivity and the precision of the assay are taken into consideration, there are three major criteria on the basis of which immunoassays can be ordered (Price and Newman, 1997): (1) the use of limited reagent ("competitive" assay) or excess reagent (sandwich or immunometric assay), (2) the use of a homogeneous (liquid-phase assay) or heterogeneous assay (solid phase) format, (3) not using a label or using a label and then the choice of the label.

Analyte to Reagent Ratio

Limited Reagent Format

In the limited reagent format, nonlabeled antigen in a sample and labeled antigen compete with each other for a limited amount of antibodies (Figure 2a).

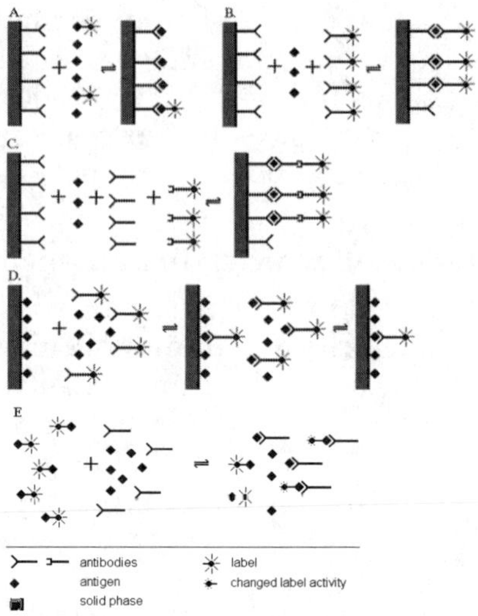

Figure 2 Immunoassay formats. (a) Heterogeneous, competitive (or limited reagent), labeled antigen. (b) Heterogeneous, sandwich (or excess reagent). (c) Heterogeneous, sandwich, indirect detection. (d) Heterogeneous, competitive, labeled antibody. (e) Homogeneous, competitive.

Erroneously, this assay type is also being called a competitive assay. Competition, however, only occurs if antibody-binding sites will be fully saturated. This is not true for the limited reagent assay format. The fact is that the use of a label in this assay format is only a means to distinguish the amount of bound antigen from the amount of free antigen. However, because the phrase "competitive assay" has general currency in scientific literature, it will also be used in this chapter.

The antibody-bound antigen is separated from the free antigen. The amount of label in one of the two fractions is measured. The concentration of antigen in the sample is then calculated by interpolation from the calibration curve of a series of standards with known antigen concentrations.

Immunoassays in which the antigen is labeled have some practical and theoretical disadvantages: (1) the avidity of the antibody determines the assay sensitivity, (2) the assay has a relatively small dynamic range because of the limited amount of antigen binding sites, and (3) antigen labeling may cause a conformational change in an epitope or may expose a hidden epitope. Antibodies might recognize the antigen barely or not at all.

Excess Reagent Format

Many proteins have several different or identical epitopes with enough distance between them to bind two antibodies at the same time. The excess reagent immunoassay format (also known as the two-site immunometric assay or sandwich assay) is based on this principle.

In this approach, the first (capturing) antibody is coupled to a solid phase (Figure 2b). Next, the sample with antigen is added. Usually, a dilution series is prepared, but if the approximate concentration is known, one dilution (containing subsaturating amounts of analyte) suffices. The capturing antibody binds the antigen to the solid phase through a first epitope. Unbound components are washed away. Subsequently, excess amount of a labeled (detecting) antibody binds the antibody-antigen complex through a second epitope. The excess of free detecting antibody is washed away, and the label activity, which is bound to the antibody-antigen-antibody complex, is measured. The amount of labeled antibody bound through the antigen to the solid phase is directly proportional to the amount of antigen in the sample.

This method has a few advantages compared to the limited reagent immunoassay format. In theory, the assay is more sensitive because in a sandwich assay, the signal rises with increasing antigen concentration. In general, it is far better to measure a large signal against a small background signal than it is to measure the difference between two large signals in case of a competitive assay. Also, the dynamic range of the assay is larger, and the antigen does not need to be labeled. Drawbacks are the larger consumption of antibodies and the need to use two antibodies in the assay.

A variation on this format is the indirect sandwich assay (Figure 2c). Instead of a labeled detecting antibody, a nonlabeled one is used. For detection, a labeled anti-immunoglobulin antibody is used. These are commercially available for many species; for example, anti-mouse (or -rat or -rabbit) IgG. The advantage of this format becomes apparent when the detecting antibody is a monoclonal. In that case, it is possible to use panels of monoclonals without the need to label them. A disadvantage is that the catching antibody must come from a different species than the detecting antibody because otherwise the labeled anti-species IgG will cause a high background by binding to the catching antibodies. For example, a format could be the following: catching with polyclonal anti-serum from rabbits, antigen, detecting the antigen with a mouse monoclonal antibody, followed by labeled anti-mouse IgG.

The excess reagent format is especially used for the quantification of large molecules (like proteins) because small molecules (e.g., haptens) simply cannot bind two large antibody molecules.

Heterogeneous or Homogeneous Assays

An immunoassay is heterogeneous or homogeneous depending on the need to carry out a separation step. A heterogeneous system includes one or more separation steps to remove excess of unbound reagent. In a homogeneous system, all components are mixed in a reaction vessel and measured without previous separation steps. Because the activity of the label changes as soon as antibody binding occurs, a selective detection of the label (and antigen) is possible.

Heterogeneous Immunoassays

In a competitive heterogeneous assay (Figures 2a and 2d), generally the unbound label (and analyte) is removed so that only the antibody-bound fraction is left. In a sandwich assay (Figure 2b), the free, labeled antibodies are removed. In both assay formats, the efficiency of the separation defines the quality of the assay. In the past, classical separation techniques like centrifugation, chromatography, and precipitation have been used. Nowadays, separation is almost exclusively achieved by coupling antibody (or antigen) to solid-phase supports. The number of supports that are available is large (Table 1). Microtiter plates with 96 wells and an 8 × 12 format are frequently used after a hesitant introduction. The slow introduction was caused by the absence of plate washers and dedicated spectrophotometers. Reasons for this success are the low consumption of reagents and the ease of use. Microtiter plate readers often have accompanying software to calculate calibration curves, sample values, and quality control parameters.

Disadvantages of microtiter plates are the relatively low binding capacity compared to suspended particles and problems with reproducibility. The former is caused by a low specific surface area and the latter by temperature gradients during incubation (edges *vs.* plate center), difference in incubation time due to manual filling, and batch to batch or even plate to plate differences introduced during plate manufacturing. Many qualities of surfaces exist. They are made of polystyrene or polyvinylchloride. Because of environmental reasons, polystyrene is preferred nowadays. The binding characteristics can be changed by UV, gamma, or electron irradiation or chemical pretreatment. This decreases the hydrophobicity, resulting in increased binding capacity for some proteins. Problems during assay development such as high background, low sensitivity, or low reproducibility may disappear by using another plate type. Finding the best plate type for a particular assay is mostly a matter of trial and error.

In competitive assays, a less efficient separation might be caused by (1) nonspecific binding of label (the remaining label can cause high background signals), (2) too stringent separation procedures, resulting in removal of bound label, (3) interference in the reaction between antibody and antigen, or (4) a variable sample matrix.

In competitive assays (heterogeneous or homogeneous), nonspecific binding is a frequent problem. A high background signal causes the assay to be less precise, especially at high concentrations. Therefore, the use of solid phases or second antibody separation systems (e.g., affinity chromatography with monoclonal antibodies) might reduce the high background signal and improve the assay precision.

Type	Examples	Advantages	Disadvantages
Microparticles (less than 20 μm)	Latex, some magnetic particles, glass	Dispensing as for liquids. Agitation not required. High antibody binding capacity.	Centrifugation required. Long magnetic separation.
Particles (less than 1 mm)	Sepharose, sephacryl, sephadex beads	Centrifugation not required. Short magnetic separation.	Agitation required. Slower reaction kinetics. Moderate antibody-binding capacity.
Macroparticles (larger than 1 mm)	Polystyrene, nylon	Centrifugation not required. Agitation not required.	Some variability in antibody coupling. Lower antibody-binding capacity. Difficulty in dispensing. Poor reaction kinetics.
Fibers	Membranes, glass fiber, nylon, silicone rubber	Centrifugation not required. Agitation not required. No dispensing of reagent. Simple to use.	Medium antibody-binding capacity.
Solid surface	Tubes, dipsticks, microtiter plates	Centrifugation not required. Agitation rare. No dispensing of reagent. Simple to use.	Variability in antibody coupling. Lowest antibody-binding capacity. Slowest reaction kinetics.

Table 1 Solid-Phase Supports Used in Immunoassays

In sandwich assays, a major problem arises if not all labeled antibodies can be removed. This could have a negative effect on the assay sensitivity, especially at low concentrations. However, one of the advantages of solid-phase separations is that thorough washing is possible. When optimized properly, the washing procedure results in less background, increased sensitivity, and higher precision.

The solid phases have to be coated with antigen (or antibody). This can be done by:

1. Adsorption: This procedure is simple. Because of the orientation of adsorbed molecules, the random binding capacity is relatively low. Butler et al. (1992) have shown that less than 10% of the binding sites may be available after random coating. This low amount is not only due to steric factors; adsorption of proteins to a surface often results in denaturation. The interaction between protein and nontreated plastic surfaces is predominately hydrophobic.
2. Indirect immobilization: The above-mentioned problem of decreased binding capacity of randomly coated antibodies can be circumvented by using a solid phase with anti-Fc antibodies. This forces the antibody in the optimal orientation and prevents solid-phase–induced denaturation. This approach may increase availability of binding sites from less than 10% to more than 70% (Butler et al., 1992). Other capture molecules are protein A, protein G (both for antibodies), and streptavidin. The latter requires biotinylation of the coating protein. The problem of random orientation remains in this case.
3. Direct covalent coupling: Many coupling chemistries exist, often using amino groups of proteins. Coupled immunoreagents are often more stable than the reagent in solution. Sometimes, coated ELISA plates can be stored dried and frozen for many years.

Homogeneous Immunoassays with a Label

These assays require no separation steps. Pipetting, incubating, and measuring are the only steps required. The reactions occur in solution generally without beads or solid-phase attachments to interfere with low-affinity interactions. In any homogeneous assay, all the components of the assay are present during measurement.

Homogeneous assays can be divided in assays with direct modulation, indirect modulation, or coupled modulation (Miyai, 1985).

Direct modulation means that the binding of an antibody to a labeled antigen modifies a property of the label (e.g., enzyme activity) in a way that the signal decreases or increases (Figure 2e). If a specific antibody binds the antigen-enzyme conjugate for instance, the activity of the enzyme could be sterically inhibited. With the antibody and conjugate present in limited amounts and the concentration of antigen varied, the more antigen present, the greater the antigen-antibody concentration and the smaller the concentration of antibody-antigen-enzyme complex. Consequently, the activity of the conjugate increases. In indirect modulation, the label is a cofactor or inhibitor: the binding of antibody to labeled antigen affects (inhibits mostly) the capacity of the cofactor to activate an apoenzyme. Coupled modulation stands for an assay in which a signal is changed when binding of antibody-antigen occurs. An example is measurement of fluorescence polarization (or anisotropy, see Chapter 2). Antibody-antigen complexes rotate slower in solution than free antigen. This difference in mobility can be detected by using fluorescently labeled antigen and measurement of the polarization of the fluorescence. The slower rotating complex results in increased polarization signals.

Homogeneous Assays Without a Label: Immune Complex Formation These types of immunoassays are based on the formation of immune complexes: an antigen with multiple identical epitopes or different and distinct epitopes can produce a large immune complex of many antigen and antibody molecules (Heidelberger and Kendall, 1935). Figure 3 is an illustration of the antigen-antibody reaction, resulting in

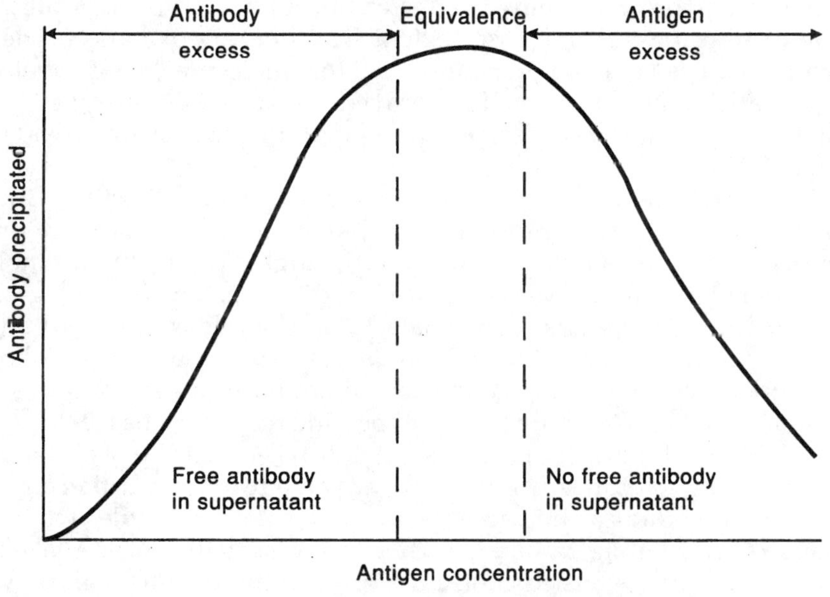

Figure 3 The Heidelberger-Kendall curve, showing the relation between free antibody and antigen and immune complexes. At equivalence, the immune complex formation is maximal and with it the amount of precipitate and level of turbidity.

the precipitin reaction (the shown curve is often referred to as the Heidelberger-Kendall curve): as increasing concentrations of antigen are added to a constant amount of antibody, the amount of immune complex precipitated rises and then falls. The precipitin curve generated in this way has three regions: (1) the antibody excess region in which the immune complexes are small, (2) the equivalence region, in which there is an optimum ratio of antibody bridging in relation to antigen concentration, and (3) the antigen excess region in which there is a reduced supply of bridging antibody molecules.

Immune complex formation can be monitored by agglutination assays. These assays depend on a visual inspection (e.g., the blood-type assay is based on agglutination of erythrocytes and antibodies). Instrumental detection of immune complexes is typically done by light scattering.

Signal Generation and Detection Systems

Radioisotopes and Labeling Methods

The most commonly used label for antibodies or antigens is iodine-125 (^{125}I). ^{125}I has a high activity and a relatively long half-life (60 days), by which the reagent shelf life is acceptable.

Iodine is easily incorporated in many molecules. A common labeling method for proteins is substitution of ^{125}I onto the aromatic ring of tyrosine (or to a lesser extent histidine) by oxidation. These methods have a damaging effect on the molecule being labeled. Enzymatic radioiodination has proved to be less damaging. The method relies on the incorporation of iodine by lactoperoxidase (David, 1972).

Some protein molecules cannot be radioiodinated directly because they do not have the appropriate group (e.g., tyrosine). For these proteins, a suitable method is conjugation labeling. In this method, a molecule containing an appropriate group is radioiodinated and conjugated to the molecule that has to be labeled. An example is iodinated 3-(*p*-hydroxy-phenyl) propionic acid *N*-hydroxysuccinimide ester, or Bolton-Hunter reagent, and conjugation to the amino groups in any lysine residue or at the N-terminus of a polypeptide that has to be labeled.

Tritium (^{3}H) is also used as a label. It has a half-life of well over twelve years. Tritium labeling has the advantage that the tritiated molecule is structurally the same as the unlabeled molecule, whereas many other labeling techniques change the structure of the labeled molecule. Titrium-labeled molecules are produced by chemical synthesis or exchange reactions. Many radioactive labels are commercially available.

Edwards (1997) provides the interested reader with a more extensive explanation of radiolabeled immunoassays.

Enzyme Immunoassay

Van Weeman and Schuurs (1971) were the first to introduce enzymes in an immunoassay as an alternative to radioisotopes. Since then, the enzyme immunoassay (EIA) has become the most widely used immunoassay.

By making use of the catalytic properties of the enzyme to generate colored, fluorescent, or luminescent products, very low concentrations of labeled antigens (or antibodies) can be measured. A single enzyme molecule is able to convert many molecules of substrate by which the signal increases and consequently adds to the sensitivity of the assay. Enzyme labels work equally well in competitive assays and in sandwich assay formats.

One of the drawbacks of using enzymes is that all enzymes have high molecular weights. As a consequence, they diffuse slowly and are more sensitive to bind nonspecific to reaction vessels than some other labels. Another drawback of enzymes is their vulnerability to inactivation by external factors despite their stability and the long shelf life of enzyme conjugates.

In radioimmunoassays, the incubation of antibody and antigen is the only crucial step that may be optimized during the development of the assay. In EIAs, however, there are two incubation stages to be optimized and kept free from interference. The first incubation involves the binding of antibody to antigen and the second concerns the signal generation step.

Enzyme Labels

The enzymes commonly used in EIA are horseradish peroxidase (HRP) and alkaline phosphatase (AP). Of many other enzymes that have been used over the years, β-D-galactosidase is the only one that has gained a widespread use in commercial products.

HRP is a 44-kDa glycoprotein that reacts with the substrate hydrogen peroxide and becomes oxidized. Oxidized HRP then oxidizes another substrate, producing a colored, fluorescent, or luminescent product, depending on the substrate that is used (Table 2).

AP is a dimeric 140-kDa glycoprotein that catalyzes the dephosphorylation of phosphate esters of primary alcohols, phenols, and amines.

Enzymes are covalently coupled to antigen or antibody, depending on which assay is used. To make conjugates of enzymes with antibodies, antigens, or haptens, the conjugation method should be as mild as possible in order to maintain the functional activity of the conjugate. Moreover, the conjugates should be stable.

Many conjugation methods are described in literature. HRP and AP may be conjugated through glutaraldehyde for instance (Avrameas et al., 1978). HRP may also be conjugated through carbohydrate residues in

Enzyme	Assay Method	Substrate
Horseradish peroxidase	Colorimetric	H_2O_2 and *o*-phenylene-diamine (OPD)
		H_2O_2 and tetramethyl-benzidine (TMB)
		H_2O_2, 3-methyl-2-benzothia zolinone hydrazone (MBTH), and 3-(dimethylamino) benzoic acid (DMAB)
	Fluorometric	3-*p*-Hydroxyphenyl-propionic acid
	Luminometric	Luminol
Alkaline phosphatase	Colorimetric	4-Nitrophenyl-phosphate
	Fluorometric	4-Methyl-umbelliferyl phosphate
	Luminometric	Adamantyl-1,2-dioxyethane
		Phenylphosphate-substituted dioxetane
	Time-resolved (TR) Fluorometric	5-Fluorosalicyl phosphate
β-D-galactosidase	Colorimetric	2-Nitrophenyl-*β*-D-galactoside
	Fluorometric	4-Methyl-umbelliferyl-*β*-D-galactoside (MUG)
	Luminometric	Lactose with enzyme-coupled system

[a]Adapted from Gosling (1997).

Table 2 Most Used Enzyme-Substrate Combinations[a]

accordance with the periodate method (Tijssen and Kurstak, 1984). Gosling (1997) extensively describes other conjugation methods such as:

- the heterobifunctional reagent method: reagents that can react with two different functional groups, one of each of the two proteins to be conjugated; an example of this is the preparation of Fab′-enzyme conjugates.
- chimeric proteins as conjugates: recombinant DNA technology is used to prepare protein-protein conjugates; for instance, conjugates of single-chain Fv fragment and enzyme).
- enzyme-labeled haptens.

Colorimetry

Enzyme labels can generate different signals, depending on the substrate that is used (see Table 2). The simplest signal to detect is color. A colorimetric method has many advantages:

1. After the reaction has stopped, the intensity of the color remains stable.
2. The use of simple and relative low-cost spectrophotometers that are able to measure very fast (2–5 seconds per microtiter plate).
3. The possibility to evaluate visually a large number of samples in a large-scale experiment.

Colorimetric endpoint assays are limited by the working range of the spectrophotometer. Typically, optical absorbance readings in the range of ~0.1 to ~1.5 are used. As the relationship between optical absorbance and the intensity of transmitted light is logarithmic, the dynamic range of the assay would be about 15. This is sufficient for competitive assays, but sandwich assays are capable of a wider working range. However, the range can be extended by reading high levels of absorbance at a wavelength offset from the center of the main peak.

A commonly used peroxidase substrate in colorimetry is TMB (3,3',5,5'-tetramethylbenzidine). TMB gives (in the presence of hydrogen peroxide) colored products with high absorbance values, low backgrounds, and is not mutagenic. The substrate *p*-nitrophenyl phosphate is often used in combination with alkaline phosphatase.

Fluoroimmunoassay

As mentioned in the section "Colorimetry" above, enzyme labels can generate different signals, depending on the substrate that is used. Fluorescent probes are one of the labels used in enzyme-linked immunoassays (see also Chapter 2).

There are two types of fluoroimmunoassays (FIAs). The first type uses directly a fluorophore label such as fluorescein or rhodamine. The second type is an enzyme immunoassay in which substrates are converted into fluorescent end products. Based on theoretical models and certain assumptions regarding the mechanism of reactions, colorimetric assays are equally sensitive as radioimmunoassays. Ishikawa (1987), however, showed that fluorometric enzyme immunoassays manage to reach a factor of 5 to 1000 times higher sensitivities (dependent on the type of enzyme) with the same enzymes as used in colorimetric assays. A commonly used substrate in FIAs is 4-methylumbelliferyl phosphate (4-MUP) with alkaline phosphatase as the enzyme label (Table 2).

Conventional fluorophores have some disadvantages. Because of their relatively short decay times ($\tau \approx 1$ to 16 nanoseconds, dependent on the fluorophore), interference from light scattering ($\tau = 0$ nanoseconds), background fluorescence ($\tau = 1$ to 10 nanoseconds), and quenching may reduce the sensitivity of fluorescent measurements by factors between 100 and more than 1000 (Soini and Hemmila, 1979). This interference can be avoided by using time-resolved fluorescence in combination with a probe with a relatively long lifetime. Fluorescent lanthanides like europium and terbium, chelated with light-adsorbing ligands, are very suitable for time-resolved fluorescence measurements. The much longer lived fluorescence of the europium ion (decay time is 714 microseconds or 925 microseconds, depending on the ligand used) is measured during a time window (e.g., 400–800 microseconds) chosen so that there is minimal interference from background fluorescence (Wood and Barnard, 1997). The dissociation enhanced lanthanide fluorescence immunoassay (DELFIA) uses antigen or antibody labeled with a stable europium chelate. In this form, the europium ion has only weak fluorescence. The DELFIA system improves lanthanide fluorescence by dissociation of the europium ion from the hydrophilic chelator to a more hydrophobic one.

There are several applications of FIAs besides the DELFIA system. Wood and Barnard (1997) have described some heterogeneous and homogeneous methods and their appropriate fluorometers.

Chemiluminescence Immunoassay

Chemiluminescence is the emission of light resulting from chemical reactions. Such reactions generate large amounts of energy that are converted in photons if the excited molecules fall back to their stable ground state. Unlike fluorescence measurement, there is no incident light. Consequently, measurements of chemiluminescence emission are possible against lower backgrounds.

Chemiluminescence can normally be found in nature. Most examples are seen in the form of bioluminescence. A well-known example of bioluminescence is the firefly. The fly has an enzyme (luciferase) that catalyzes the oxidation of a specific substrate (luciferin) in the presence of adenosine triphosphate (ATP).

The most widely known chemiluminescent substance is luminol (5-amino-2,3-dihydrophthalazine-1,4-dione). In the presence of hydrogen peroxide, horseradish peroxidase causes oxidation of luminol by which light is emitted. Acridinium salts are also employed as immunoassay label, forming the base of many commercial immunoassay kits (Weeks et al., 1983).

Most enzyme-based luminescent reactions have low quantum yields and generate a weak light emission that rapidly decays. However, the intensity of light may increase substantially when another chemical compound is added. The amount of light generated by peroxidase-assisted oxidation of luminol increases more than a factor of 1000 in the presence of certain substituted phenols and naphthols. Background light emission of luminol and peroxide is also minimized, improving the signal-to-noise ratio (Wild, 1994). When the emitted light is rather a glow than a flash of light that may prolong for several hours, it is called enhanced chemiluminescence, although signal reading normally occurs within 20 minutes.

Chemiluminescence can be detected and quantified with great sensitivity. The most widely used detector is a photomultiplier tube (PMT). If less sensitivity is demanded, a low-cost luminometer with a photodiode as a detector may be used.

In recent years, imaging luminometers based on the use of charge-coupled devices (CCDs) have been developed. Weeks (1997) discusses in detail light-measurement methods. He also goes extensively into the design of heterogeneous and homogeneous chemiluminescence immunoassays.

Streptavidin/Avidin-Biotin

Biotin is a small vitamin with molecular weight of 244 g/mol. It has a high affinity to the egg-white protein avidin. Avidin is extremely basic (pI = 10.5) through which the risk of nonspecific binding by electrostatic forces is present. That is why in immunoassay, streptavidin is preferable. Streptavidin, a protein isolated from *Streptomyces*, has a much less basic isoelectric point, thereby reducing the risk of nonspecific binding. Each streptavidin molecule has four binding sites for biotin molecules. This makes the streptavidin-biotin system an excellent enhancement technique for immunoassays. Biotin in this case serves as the label instead of a radioisotope or enzyme. Biotinylation of antibodies, for instance, is a gentle process that does not reduce the activity of the antibody.

At the end of the immunoassay, a conjugate of streptavidin and a signal-generating substrate is added. Examples of conjugates are streptavidin-horseradish peroxidase, strepavidin-^{125}I, and streptavidin-fluorescein. The conjugate and biotin bind together with high affinity through which the bound fraction is labeled with the signal generator. Because in an excess reagent format assay antibodies may be labeled with many biotin molecules, many extra streptavidin molecules can bind, thus amplifying the signal. If the conjugate is made up of two or three molecules of streptavidin, further enhancement of the signal may be achieved.

Assay Sensitivity

Immunoassays have been so successful that literally thousands of publications dealing with these techniques appeared during the past thirty years. But when comparing the performance characteristics of various assays and in particular estimates of their sensitivity, a lack of standardization is evident. Differences in volume, incubation time, integration time of signal detection, sample matrix, and so forth, can all lead to considerable differences in reported sensitivity for even the same assay. Therefore, it is difficult to pronounce sensitivities upon various kinds of immunoassays. However, some theoretical considerations are give here, because they have formed the base of the development of novel assay designs.

All assays, regardless of which component is labeled, can be described in one or two ways: either they measure unoccupied sites (limited reagent format) or they measure occupied sites (excess reagent format). Ekins (1991) has developed mathemathical arguments to estimate sensitivities, based on this antibody occupancy principle.

Briefly, Ekins shows that the sensitivity of competitive assays is governed by three factors: the equilibrium constant, the precision of signal measurement, and the level of nonspecific binding. The converse is true when occupied sites are measured directly as in sandwich assays. The major determining factors for assay sensitivity are the level and imprecision of the nonspecific binding and the error in measurement of the signal. A high sensitivity assay therefore requires high concentrations of antibody in relation to the antigen, a low specific binding, and a high specific activity label.

On theoretical grounds, competitive assays can reach sensitivities ranging from 0.2 pM to 20 pM, assuming equilibrium constants of 10^9 to 10^{11} L/mol, high specific labeled antigens, and 1% experimental errors (Jackson and Ekins, 1983). The maximum sensitivity of a sandwich assay format should be of the order 0.20 fM to 20 fM, assuming an antibody affinity of 10^9 to 10^{11} L/mol, a nonspecific binding less than 0.1%, and an error in the signal response of 1%.

Applications

Immunogenicity of Protein Pharmaceuticals

Foreign macromolecules have the ability to induce immune responses after repeated contact: they are immunogenic. This is, except of course for proteins in vaccines, unwanted because it can affect their therapeutic efficacy. It is therefore crucial that immunogenicity of therapeutic proteins is assessed during clinical and sometimes preclinical evaluation [see Schellekens (2002) for a review]. Currently, it is not predictable how immunogenic a particular protein will be. This depends on many factors. Minor sequence variation or differences in glycosylation pattern may render recombinant proteins immunogenic even when their origin (i.e., the gene) is human. But also the dose, duration of treatment, immune status, and genetic background of the patient, route of administration, presence of impurities, stability, and formulation may have a profound effect on the immunogenicity (Schellekens, 2002). Often, the presence of antibodies has no measurable influence on the therapeutic effect, but for more than a few biopharmaceuticals loss of efficacy has been reported (Wadhwa, 2003). Among them are preparations of insulin, several lymphokines, Factor

VIII, and human chorionic gonadotropin (HCG). Severe adverse effects may occur when neutralizing antibodies are generated that cross-react with the native protein produced by the patient. This was demonstrated with erythropoietin (EPO) (Casadevall et al., 2002). The anti-EPO response in some patients led to a complete standstill of erythrocyte development. Only frequent blood transfusions could save these patients.

This unpredictable behavior stresses the importance of testing for the induction of antibodies. For product registration, this is mandatory. One should distinguish between binding antibodies and neutralizing antibodies. "Neutralizing" is here defined as inhibition of the biological effect (*in vitro*). This would imply that an immunoassay is not sufficient to assess immunogenicity (note that an immunoassay measures antigenicity not immunogenicity; see the section "Quantity, Antigenicity, Immunogenicity" above). Indeed, after establishing the presence of antibodies with an immunoassay, some kind of bioassay should be done in order to estimate the clinical relevance of the finding.

Detecting and quantifying antibodies induced by biopharmaceuticals bear some intrinsic difficulties. First, availability of positive controls and reference sera is often a problem. This hampers assay validation or qualification and inter-laboratory comparison. Second, patients may have antibodies before the treatment started. To correct for that, pre-sera are needed. In clinical trials this can be done, but when immunogenicity problems arise after registration, pre-titers are often not available. Third, low-affinity antibodies may go undetected, although they may have an impact on the efficacy of the product. A bioassay or biosensor analysis (see Chapter 15) may reveal the effect of these undetectable antibodies or the presence of the antibodies, respectively. Fourth, the presence of immune complexes (i.e., product plus bound antibody) may result in an underestimation of the antibody titer. Fifth, conformational differences between the material used in the assay and the product *in vivo* may cause flawed results.

To conclude, qualitative detection of anti-biopharmaceutical antibodies can be relatively uncomplicated, but to quantify the response and to establish the effects on efficacy is difficult.

Quantification of Host Cell Proteins by Immunoassays

Almost as important as characterization and quantification of pharmaceutical proteins is the level of contaminating proteins. For recombinant DNA products, the regulatory agencies such as the European Agency for the Evaluation of Medicinal Products (EMEA) require that manufacturers monitor absence of host cell proteins (HCPs). The level of contaminations is low, so it is not possible to measure total protein and subtract the amount of active component (besides this, a reference with established purity is needed). An immunoassay is the method of choice, but this poses another problem: the impurities are mostly unknown and may consist of numerous proteins. Monoclonal antibodies are clearly not suitable for this purpose. In cases like this, the high specificity of antibodies is a disadvantage. The hardest part in developing a HCP immunoassay is the generation of a representative anti-HCP serum. The use of crude cell lysates (free of product) to generate an antiserum has some disadvantages. First, the composition of such a lysate is probably very different compared to the pattern of impurities that remain after downstream processing (DSP). Second, although the lysate contains almost all HCPs, the resulting antiserum may contain antibodies to only a few immunogenic proteins. A protein that is only weakly immunogenic may thus go undetected even if it accounts for half the HCP contamination. The first problem is difficult to circumvent. Theoretically, one should perform a purification procedure with the cell lysate similar to the DSP of the product. By definition, this will result in very low yields. As an alternative, one can perform the first one or two purification steps. These are often filtration or centrifugation, resulting in removal of the cell debris. In any case, it is important to take into account the production process of the active compound.

The second problem (presence of immunodominant components or components in excess) can be avoided by cascade immunizations (Anicetti et al., 1992). First, an antiserum is generated with a cell lysate. Next, the antiserum obtained is used to remove the immunodominant antigens from this lysate. This is done by immunoaffinity chromatography, coupling the antiserum to a column matrix. The flow-through of the chromatographed lysate contains proteins that are not recognized by the antibodies from the antiserum. These proteins are used for re-immunization. This will lead to a more evenly distributed immune response. With the resulting antiserum containing antibodies to most proteins in the lysate, an immunoassay can be developed.

Development and Validation of a CAMPATH-1H Quantification Assay

This example will show that assay development and validation are time consuming, laborious processes. CAMPATH-1H is a humanized monoclonal antibody against CD52. CD52 is a glycoprotein expressed on the surface of lymphocytes. CAMPATH-1H is used for the treatment of leukemia, autoimmune disease, and prevention of transplant rejection. For pharmacokinetic studies and dose finding, an assay is needed for CAMPATH levels in serum (Rebello and Hale, 2002). The assay must detect low concentrations of immunoglobulin in the presence of large amounts of natural immunoglobulins. This limits the choice of the method to biological assays or immunoassays. The authors start by reviewing existing methodology and available reagents.

Based on availability, sensitivity, and specificity, continuous cell lines expressing CD52 were chosen as the antigen source used to bind CAMPATH-1H.

Apart from two bioassays, two immunoassays were considered: ELISA and FIA. Both are of the excess reagent type (i.e., noncompetitive). With ELISA, plates are coated with target cells, and sample is added followed by a secondary labeled antibody. With FIA, the target cells are incubated with sample, and CAMPATH bound to the cells is labeled with a secondary fluorescent antibody. Fluorescence is measured with a flow cytometer. Although technically more complicated, the FIA method was selected for further development and validation because of expected higher sensitivity and dynamic range.

In the development phase, the FIA method was compared with the most promising bioassay: a complement-mediated cytotoxicity assay. Three cell lines were tested in these two assays. Not the most sensitive cell line was selected but the one with the widest working range: HUT-78, a human T-cell line, had a working range of a factor of 170.

The reference, one chromatographically purified lot of CAMPATH, was stored in single-use aliquots at –70 °C. Literature data were available about the stability (minimal 6 years at –70 °C). Protein content was determined by UV absorption. The quality of the target cells was maintained by using a master seed bank, used for the preparation of working cell banks. This ensures defined cell passage numbers.

In general, validation parameters for an assay are as follows:

- Accuracy: How close is the measured value to the actual value?
- Precision: How variable are results obtained:
  1. under identical conditions ("within run variance"),
  2. from day to day ("between run variance" or intermediate precision),
  3. under nonidentical conditions (different operators, equipment, etc.); this is also called reproducibility.
- Sensitivity: Limit of quantification (LOQ) and limit of detection (LOD).
- Robustness: How sensitive is the test for disturbances (matrix effects, temperature, differences in incubation times, etc.). The specificity of a test can also be considered as a robustness parameter.

In this example, numerous parameters were investigated:

Accuracy, precision, and reproducibility: The accuracy was 109% and the coefficient of variation was 12–13%. These latter values were not only obtained within a single test (precision) but also on different days and on different flow cytometers (reproducibility).

Sensitivity: The LOQ was found to be 0.25 µg/mL. This was determined by producing 20 calibration curves ranging from 0.025 to 100 µg/mL CAMPATH. Subsequently, the accuracy and precision for each dilution is determined. The lowest concentration with acceptable precision and accuracy was 0.25 µg/mL. The LOD was between 0.05 and 0.1 µg/mL.

Robustness: This encloses all variables that may occur in the test. In this case, the following:

- Specificity: For this purpose, a nonbinding CAMPATH mutant and Fab fragments of CAMPATH were used. The latter is an important control because the assay must detect only intact CAMPATH. The antibody has to be intact in order to activate complement or effector cells.
- Cell quantity and quality (culture time, storage time before cytometry).
- Assay temperature: Room temperature and on ice.
- Lot and concentration of the fluorescent detection reagent.
- Stability of standards and detection reagent. This was done in retrospect after the two-year study.
- Matrix effects: Human serum was spiked with CAMPATH.
- Sample collection and storage: Human blood was spiked with CAMPATH, and the effect of clotting conditions and repeated freeze thawing on the recovery were established. The authors also studied effects of heat inactivation (routine procedure to inactivate the complement system) and storage time of processed serum.
- Finally, more than 2500 samples from 90 patients receiving CAMPATH were measured. The overall coefficient of variation was 11%.

New Developments

Although the basics of immunoassays have hardly changed in the past decades, there are a number of interesting developments. The use of immunosensors is an important one, both for research and routine applications. Chapter 15 of this book is devoted to this. Some other developments with relevance for protein analysis follow.

Antibody Engineering

A notable exception in immunoglobulin structure are the immunoglobulins of camels and lamas. A fraction of these antibodies lack the light chain. This facilitates antibody engineering. Stable antibody fragments can be expressed by yeast in large quantities.

Phage display technologies enable expression of "antibodies" on or by bacteriophages by recombinant DNA techniques. It is possible to prepare random or specific peptide libraries (the latter from mRNA obtained from B-cells). Phages expressing these peptides on their surface or excreted soluble fragments can be used as a reagent in immunoassays (Nissim et al., 1994). Applications are still limited because of problems related to either affinity (too low because antibody maturation does not occur) or specificity (the antibody fragment for a certain antigen is not present in the phage or peptide bank), which are inversely related. Phage banks containing high-affinity phages or antibodies are often designed for certain antigens, (i.e., they are obtained from mRNA from immunized animals). Random, synthetic banks can contain many specificities, but many antibodies have a low affinity, making them not or less suitable for immunoassays.

Signal Detection and Amplification

Attempts to increase the sensitivity of immunoassays have resulted in some interesting approaches. Liposomes, bilayer vesicles composed of phospholipid, can be used to amplify signals. The liposome aqueous interior contains marker molecules or enzymes and the bilayer contains streptavadin in order to bind to biotinylated secondary antibodies. Bound liposomes are lysed with a detergent, releasing the marker (Rongen et al., 1997). Another example of signal amplification is the use of immuno polymerase chain reaction (PCR) (Hendrickson et al., 1995). Detecting antibodies are labeled with DNA. The DNA is amplified by PCR. PCR products are quantified measuring the intensity of DNA bands after agarose gel electrophoresis.

Multianalyte Assays and Antibody Arrays

The era of proteomics has resulted in the accelerated development of protein and antibody arrays. Most applications are in the fields of (immuno)diagnostics, proteomics, and high-throughput screening. Some are useful for the analysis of protein pharmaceuticals, for instance for the evaluation of immune modulating proteins and vaccines. Particle arrays are among them. An example is the use of fluorescent beads or nanoparticles. Mixtures of beads are used as solid phase for antibodies (or other capture molecules). The beads are distinguishable by their fluorescence. Up to 100 different beads can be identified by coloring the beads with accurate mixtures of fluorophores. Each type of bead is then coated with a different capture molecule (e.g., one type of antibody for one type of bead). Bound analytes are detected with fluorescent antibodies. Subsequently, beads are analyzed with a fluorimeter similar to fluorescence activated cell sorters [reviewed by Vignali (2000)]. With these systems (commercially available as Luminex), it is possible to determine the properties of immunomodulating pharmaceuticals by measuring the concentration of fifteen or twenty cytokines in one assay, saving sample and time. Also, it is possible to measure antibodies against numerous antigens simultaneously.

PCR-based immunoassays (see "Signal Detection and Amplification," above) are also suitable for detection and quantification of several antigens. This is achieved by labeling each antigen-specific antibody with a DNA fragment of a different length. Each PCR product will appear as a separate band on the gel.

Conclusions

Immunoassays have existed for several decades. They are indispensable tools for the characterization and quantification of pharmaceutical proteins. Immunoassays possess what many physicochemical characterization methods lack: high specificity and high sensitivity. The other side of the test spectrum for proteins is occupied by bioassays. Immunoassays often outperform them with regard to robustness. This ensures that immunoassays are here to stay for many years to come.

Abbreviations

AP	Alkaline phosphatase
ATP	Adenosine triphosphate
CCD	Charge-coupled device
CDR	Complementarity determining regions
DELFIA	Dissociation enhanced lanthanide fluorescence immunoassay
EIA	Enzyme immunoassay
ELISA	Enzyme-linked immunosorbent assay
EPO	Erythropoietin
Fab	Fragment antigen binding
FIA	Fluorescence immunoassay
Fv	Fragment variable
HCG	Human chorionic gonadotropin
HRP	Horseradish peroxidase
kDa	Kilodalton
MUP	Methylumbelliferyl phosphate
PMT	Photomultiplier tube
TMB	3,3′,5,5′-Tetramethylbenzidine

References

1. Anicetti VR, Simonetti MA, Blackwood LL, Jones AJ, and Chen AB. Immunization Procedures for E. coli Proteins. *Appl Biochem Biotechnol* 1989; 22:151–168

2. Avrameas S. Amplification Systems in Immunoenzymatic Techniques. *J Immunol Meth* 1992; 150:23–32

3. Butler JE, Ni L, Nessler R, Joshi KS, Suter M, Rosenberg B, Chang J, Brown WR, and Cantanero LA. The Physical and Functional Behavior of Capture Antibodies Adsorbed to Polystyrene. *J Immunol Meth* 1992; 150:77–90
4. Casadevall N, Nataf J, Viron B, Kolta A, Kiladjian JJ, Martin-Dupont P, Michaud P, Papo T, Ugo V, Teyssandier I, et al. Pure Red Cell Aplasia and Anti-erythropoietin in Patients Treated with Recombinant Erythropoietin. *N Engl J Med* 2002; 346:469–475
5. David GS. Solid-State Lactoperoxidase: A Highly Stable Enzyme for Simple, Gentle Iodination of Proteins. *Biochem Biophys Res Commun* 1972; 48:464
6. Edwards, R. Radiolabeled Immunoassays. In: Price CP, and Newman DJ. *Principles and Practice of Immunoassay.* New York: Stockton Press; 1997:325–348
7. Ekins RP. Immunoassay Design and Optimisation. In: Price CP, and Newman DJ. *Principles and Practice of Immunoassay.* London: Macmillan; 1991:96–153
8. Gosling JP. Enzyme Immunoassay: With and Without Separation. In: Price CP, and Newman DJ. *Principles and Practice of Immunoassay.* New York: Stockton Press; 1997:349–388
9. Heidelberger M, and Kendall F. Quantitative Theory of the Precipitin Reaction: Study of Azoprotein-Antibody System. *J Exp Med* 1935; 61:467–483
10. Hendrickson ER, Truby TM, Joerger RD, Majarian WR, and Ebersole RC. High Sensitivity Multianalyte Immunoassay Using Covalent DNA-Labeled Antibodies and Polymerase Chain Reaction. *Nucleic Acid Res* 1995; 23:522–529
11. Ishikawa E. Development and Clinical Application of Sensitive Enzyme Immunoassay for Macromolecular Antigens: A Review. *Clin Biochem* 1987; 20:375–385
12. Jackson TM, and Ekins RP. Theoretical Limitations on Immunoassay Sensitivity: Current Practice and Potential Advantages of Fluorescent Eu^{3+} Chelates as Nonradioactive Tracers. *J Immunol Methods* 1983; 87:13–20
13. Miyai K. Advances in Non-isotopic Immunoassay. *Adv Clin Chem* 1985; 24:61–110
14. Nissim A, Hoogenboom HR, Tomlinson IM, Flynn G, Midgley C, Lane D, and Winter G. Antibody Fragments from a 'Single Pot' Phage Display Library as Immunochemical Reagents. *EMBO J* 1994; 13:692–698
15. Price CP. Progress in Immunoassay Technology. *Clin Chem Lab Med* 1998; 36:341–347

16. Price CP, and Newman DJ. Classification of Immunoassays. In: Price CP, and Newman DJ. *Principles and Practice of Immunoassay.* New York: Stockton Press; 1997:4–10
17. Rebello P, and Hale G. Pharmacokinetics of CAMPATH-1H: Assay Development and Validation. *J Immunol Methods* 2002; 260:285–302
18. Rongen HAH, Bult A, and van Bennekom WP. Liposomes and Immunoassays. *J Immunol Methods* 1997; 204:105–133
19. Schellekens H. Immunogenicity of Therapeutic Proteins: Clinical Implications and Future Prospects. *Clin Therapeutics* 2002; 24:1720–1740
20. Soini E, and Hemmilä J. Fluoroimmunoassay: Present Status and Key Problems. *Clin Chem* 1979; 25:353–361
21. Tijssen P, and Kurstak E. Highly Efficient and Simple Methods for the Preparation of Peroxidase and Active Peroxidase-Antibody Conjugates for Enzyme Immunoassays. *Anal Biochem* 1984; 136:451–457
22. Van Weeman BK, and Schuurs AHWM. Immunoassay Using Antigen-Enzyme Conjugates. *FEBS Lett* 1971; 15:232–236
23. Vignali DA. Multiplexed Particle-Based Flow Cytometric Assays. *J Immunol Methods* 2000; 243:243–255
24. Wadhwa M, Bird C, Dilger P, Gaines-Das R, and Thorpe R. Strategies for Detection, Measurement and Characterization of Unwanted Antibodies Induced by Therapeutic Biologicals. *J Immunol Methods* 2003; 278:1–17
25. Weeks I. Chemiluminescence Immunoassay. In: Price CP, and Newman DJ. *Principles and Practice of Immunoassay*. New York: Stockton Press; 1997:425–442
26. Weeks I, Beheshti I, McCapra F, Campbell AK, and Woodhead JS. Acridinium Esters as High Specific Activity Labels in Immunoassay. *Clin Chem* 1983; 29:1474–1479
27. Wild D (ed). *The Immunoassay Handbook.* New York: Stockton Press; 1994. 618p.
28. Wood PJ, and Barnard G. Fluoroimmunoassay. In: Price CP, and Newman DJ. *Principles and Practice of Immunoassay.* New York: Stockton Press; 1997:389–424
29. Yalow RS, and Berson SA. Assay of Plasma Insulin in Human Subjects by Immunological Methods. *Nature* 1959; 184:1648–1649

15

Surface Plasmon Resonance

MICHELLE J. CANNON and DAVID G. MYSZKA

Center for Biomolecular Interaction Analysis, University of Utah, Salt Lake City, Utah

Introduction

Surface plasmon resonance (SPR) biosensors have had and continue to have a significant impact on life science research, drug discovery, and pharmaceutical development (Myszka and Rich, 2000; Cooper, 2002). The technology can be used to obtain detailed information about the kinetics and binding affinities of nearly any biomolecular interaction. In an SPR experiment, one molecule is immobilized on a surface and the other is passed over the tethered ligand to analyze binding events between the two reactants. When the biosensor was originally engineered in the early 1990s, it was primarily used as an immunosensor—a biosensor that relies on an antibody as one of the biological recognition elements (Jönsson, 1991; Karlsson et al., 1991, 1993). In the ensuing years, the technology has evolved and has been applied to study a myriad of interactions involving proteins, oligomers, oligosaccharides, lipids, small molecules, and even membrane receptors (Rich and Myszka, 2001, 2002, 2003). One of the recent trends in SPR technology is the application of biosensors to investigate and develop therapeutic proteins (Morton and Myszka, 1998; Faggioni et al., 2001; Darling et al., 2002; Mason et al., 2003).

Although biosensor assays are similar to conventional immunoassay approaches that measure analyte binding to an immobilized ligand, SPR offers many advantages. An SPR assay does not require modification of the analyte with reporter molecules, as is commonly done with fluorescence-based or radioligand assays. This allows for rapid experimental setup. More importantly, the label-free methodology permits analytes to be examined in their native states. Additionally, reactions can be monitored in real time to determine binding rate and affinity constants using biosensor technology. In contrast, an approach like enzyme-linked immunosorbent assay (ELISA) only provides a snapshot of the final complex distribution to yield an equilibrium constant. Furthermore, a wide range of affinities (1 mM–1 pM) can be measured using SPR, enabling the analysis of both transient interactions and very tight complexes. These biosensor analyses can often be accomplished using very little sample. For example, to immobilize a ligand, 1–2 μg of sample is needed for a single surface. The amount of analyte required, however, is dependent on the affinity of the complex being studied, but sample volumes are low, typically 100 μL per assay. Finally, fully automated systems such as Biacore 2000, 3000, and C (the platform designed specifically for GLP/GMP applications) afford unattended operation, which is beneficial for high-throughput studies, and also improves data quality.

Although biosensor use has expanded to study a range of biomolecular interactions, recently the technology has returned to its original application as an immunosensor (Rich and Myszka, 2001, 2002, 2003; Swanson et al., 2002; Wilson and Howell, 2002; Mason et al., 2003; Swanson, 2003; Drake et al., 2004; Canziani et al., 2004). This is due to the tremendous interest in creating antibodies for therapeutic applications. Although many different biosensor technologies are available commercially (Baird and Myszka, 2001), our discussion will primarily be focused on Biacore technology applications, as they are the most routinely used. Here we describe how biosensors are ideal for characterizing therapeutically important antibodies because of the technology's sensitivity range and high-throughput capabilities. We discuss in detail how to screen antibodies with SPR biosensors and obtain high-resolution data to determine the binding kinetics and affinity of antibody/antigen interactions. Finally, we will illustrate how biosensors are being applied to support clinical studies and the production of biopharmaceutical products.

SPR Technology

Detector

All SPR biosensors are composed of three main components: the detector, interaction surface, and sample delivery system. The most commonly used biosensor is manufactured by Biacore AB in Uppsala, Sweden, and is based on the Kretschmann configuration, shown in Figure 1A (Kretschmann 1968). Polarized light is coupled through a prism that is placed on a biosensor chip. At a critical angle, the incoming light is totally internally reflected and the resulting light intensity is detected by a photodiode array. Biacore instruments use a gold surface located at the interface of reflection to induce the formation of plasmon waves. The plasmons are sensitive to the refractive index of the medium, so as an analyte binds to a ligand immobilized on the surface of a biosensor chip, the refractive index at the chip

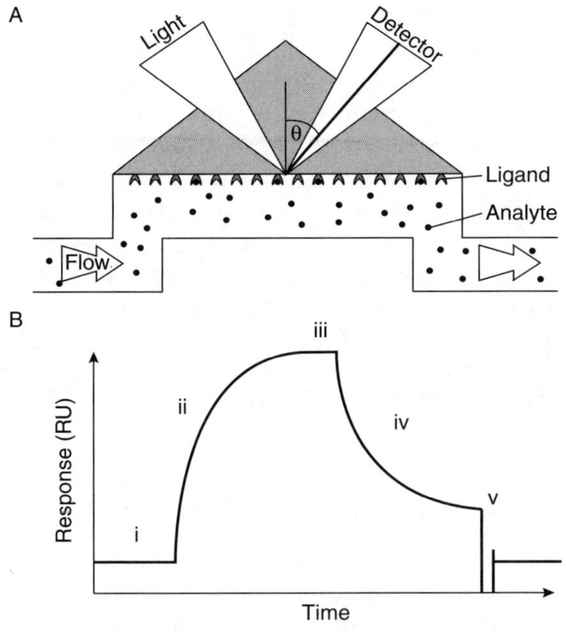

Figure 1 Surface plasmon resonance detection. (A) Schematic depicting the Kretschmann configuration of a Biacore SPR biosensor. The ligand is immobilized on the biosensor chip surface. An analyte sample is introduced through a microfluidic flow cell and binds to the immobilized ligand. Light is directed through a prism and the angle intensity changes as molecules bind at the surface. The angle intensity is monitored over time. (B) A schematic of a binding response for a typical binding cycle depicting the five steps of a biosensor reaction: (i) baseline response, (ii) association phase, (iii) equilibrium, (iv) dissociation phase, and (v) a regeneration step. The reactions are measured in time and recorded as a response.

surface changes, and the detector monitors the binding events in real time (Homola et al., 2002).

Interaction Surface

To evaluate binding interactions using an SPR biosensor, it is necessary to tether one of the molecules of interest to the surface of the sensor chip. On many sensor chips, a dextran layer provides a highly fluid, water-like environment that helps enable an immobilized ligand to remain active. There are a range of chemistries available to immobilize the ligand, including noncovalent attachment using capturing systems such as antibodies, or direct immobilization through amine groups, carbohydrates, and thiols (Johnsson et al., 1991). These coupling methods can be used to control the amount of ligand immobilized and typically produce a highly stable reaction surface. Tethering the ligand to the surface of a chip is routine and straightforward, with the process taking roughly 20 minutes.

Delivery System

Biacore 2000 and 3000 instruments use four independent flow cells that deliver analyte to the sensor surface. Typically, three flow cells are used for immobilization of three different ligands, and the remaining flow cell is left underivatized to serve as an internal reference. The four flow cells can be addressed individually or in a series through the microfluidics delivery system. A continuous flow system enables analyte to be held at a constant concentration as it is injected over the reaction surfaces during an interaction analysis. Additionally, this configuration allows buffer to pass continually over the sensor chip surface between injections of analyte. This feature can be used to wash rapidly the surfaces to ensure that residual analyte does not remain in the flow cell as complex dissociation is monitored and allows an aqueous environment to be maintained at the chip surface at all times.

How to Use an SPR Biosensor

The data collected by a SPR instrument is often referred to as a sensorgram and is plotted as a change in binding response *versus* time. The binding response units (RU) are proportional to the amount of mass accumulated on the chip surface that has bound the immobilized target. A sensorgram can be broken into the five steps illustrated in Figure 1B. In step (i) of the binding cycle, a stable baseline is established by flowing buffer over the biosensor chip. With an injection of analyte in step (ii), the binding response increases as analyte binds to the immobilized ligand. This is commonly referred to as the association phase. The sensorgram plateaus in step (iii) and indicates that the concentrations of free and bound analyte are in equilibrium. During the dissociation phase, shown in step (iv), the surface is washed with buffer, and the binding response returns to baseline as complexes dissociate. Weak interactions (those having a fast k_d) may decay back to baseline within minutes. However, if a tightly bound complex is formed and the analyte does not completely dissociate from the immobilized ligand, a mild acid or base can be used to disrupt the complexes and regenerate a fully active surface. Following the regeneration step (v), the signal returns to baseline as buffer is flowed over the sensor chip. This enables the reaction surface to be used for multiple binding cycles in a screening assay or to learn more about a given interaction by running an analyte concentration series over the same surface. The number of binding cycles that can be studied on a single surface is dependent upon the stability of the immobilized ligand but typically ranges from 100 to 200 cycles.

Assay Design

When studying biomolecular interactions using SPR, it is important to have the proper assay design. The orientation of each molecule can

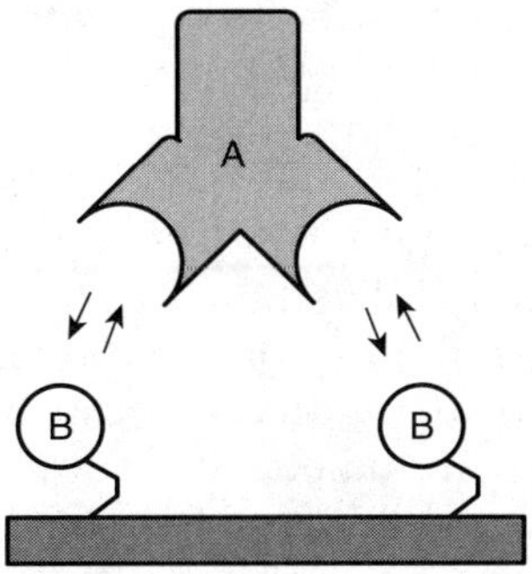

Figure 2 The binding of (A) a bivalent analyte to (B) an immobilized monovalent ligand. This orientation allows the bivalent molecule to cross-link with ligands tethered to the biosensor chip.

influence the reactions measured and sometimes complicate analysis of the binding data. To avoid avidity effects, bivalent molecules such as antibodies should be coupled to the biosensor chip surface (rather than injected across immobilized antigen). Unfortunately, there are numerous examples in the literature of binding interactions mistakenly performed with multivalent analytes in solution analyzed with an immobilized monovalent ligand. When the antibody is in solution, it has the potential to cross-link with multiple antigen molecules on the surface, as shown in Figure 2. This produces an erroneously high apparent affinity for the antibody/antigen interaction. Coupling to the sensor surface may be done by either directly immobilizing the antibody or by capturing it with a ligand that has been tethered to the surface. This capturing approach is useful when screening large panels of antibodies.

SPR Applications

Kinetic Screening of Antibodies

Through genetic engineering, the ability to generate large quantities of monoclonal antibodies against different targets has become routine (Glaser et al., 1992; Hoogenboom et al., 1992). Currently, the bottleneck for therapeutic antibody development has shifted from producing antibodies to identifying and characterizing those of interest for further analysis. Biosensors can improve the pace at which antibodies are screened by providing a medium-throughput level of studying roughly 100–400 antibodies a day per instrument, while returning information-rich data (Myszka and Rich, 2000; Canziani et al., 2004). For example, in contrast to ELISA-based equilibrium assays, both association and dissociation rate constants for an entire panel of antibodies can be derived using a biosensor screening approach. This allows antibodies to be classified based not only on their affinities, but also on the kinetic rate constants, to ultimately select candidates for further characterization and development.

To illustrate how a Biacore instrument can be adapted for a screening process, Figure 3A shows the assay approach used to select antibodies that bind to a target of interest. A ligand such as protein A or an anti-Fc antibody is immobilized on the surface of a sensor chip and used to capture out an antibody from a crude hybridoma supernatant or from other sources. Three antibodies can be captured in separate flow cells and analyzed simultaneously for antigen binding. Antigen is injected over the surfaces and interacts with the captured antibodies. This method (Canziani et al., 2004) has several benefits. First, it enables the amount of antibody expressed in the supernatant to be determined directly as the biosensor monitors how well the antibody is captured onto the surface. Second, kinetic information for an antibody/antigen interaction can be obtained without having to purify the antibody. With this approach, the antibodies of greatest interest can be selected for further purification. Third, all of the antibodies captured onto the surface are oriented in the same way so that the antigen binding sites are homogeneously presented. Using a direct immobilization method, the antibodies might be tethered to the chip in a heterogeneous manner. Finally, the same antigen sample can be used to test multiple antibody surfaces. This allows direct comparison of the relative antigen binding responses to various antibodies.

Recently, Canziani and colleagues demonstrated a screening approach that was used to characterize twenty-four crude monoclonal antibody samples (Figure 3B) (Canziani et al., 2004). In this example, anti-IgG was immobilized on the four flow cells of a biosensor chip, and three different monoclonal antibodies were captured in three separate flow cells. The remaining flow cell that contained anti-IgG served as an internal reference to correct the data for instrumental drift and also to assess the level of nonspecific binding (Morton and Myszka, 1998). To test for binding, antigen was injected through all four flow cells of the chip. After monitoring the dissociation phase, the entire antibody/antigen complex was stripped from the biosensor chip to regenerate the anti-IgG surfaces. Three new monoclonal antibodies were then assayed in a subsequent screening cycle. A total of eight binding cycles were done with the same concentration of antigen to complete analysis of the antibody panel (Canziani et al., 2004).

The antigen binding responses were processed to remove systematic artifacts and bulk refractive index changes (Myszka, 1999) and then normalized by dividing each antigen response curve by the corresponding antibody capture level. Normalizing the sensorgrams makes it possible to fit globally the antigen binding kinetics of many antibodies captured at variable surface densities. To obtain the reaction kinetics for each antibody using the single antigen binding responses, all the sensorgrams were globally fit with a simple bimolecular interaction model (Morton and Myszka, 1998). This simple model $A + B \underset{k_d}{\overset{k_a}{\rightleftharpoons}} AB$

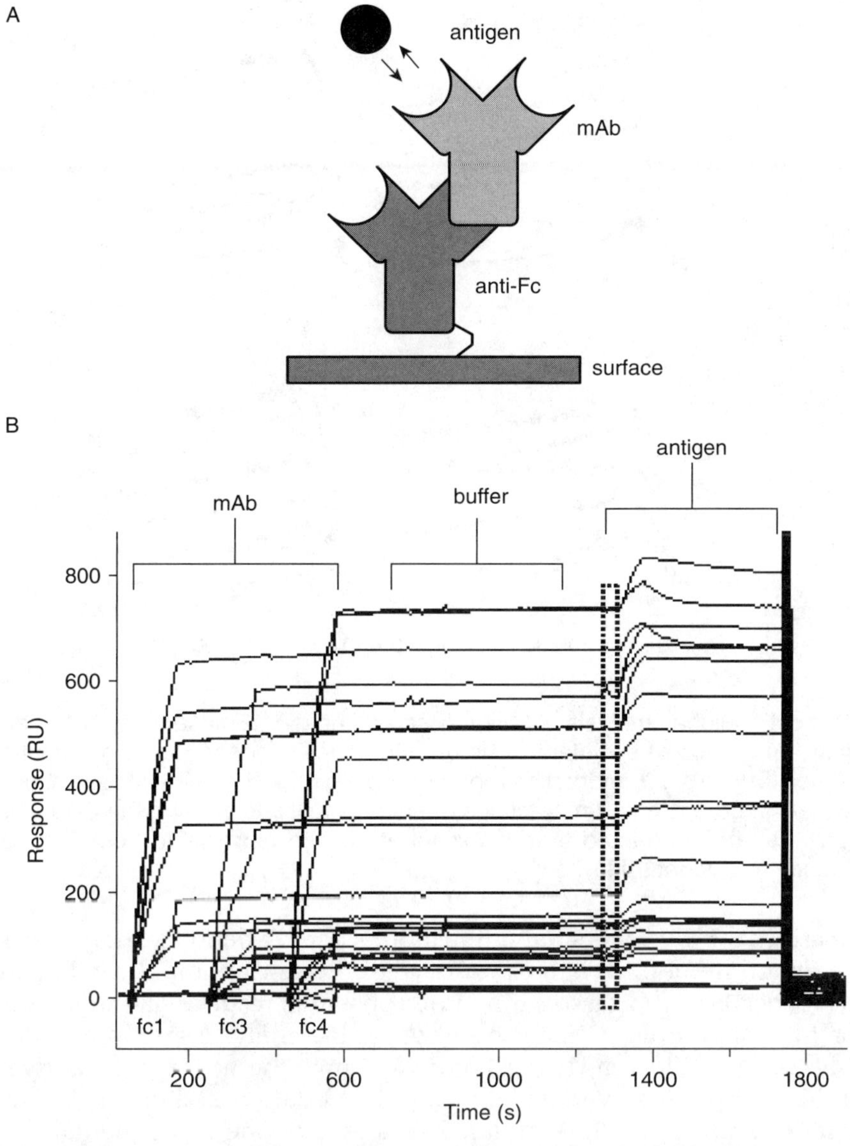

Figure 3 Kinetic antibody screening. (A) Assay design for antibody screening. Using a capture approach, an Fc-specific antibody is directly immobilized on the biosensor chip. The anti-Fc antibody is used to capture the therapeutic antibody from a crude mixture. An antigen is then injected to collect kinetic binding data. (B) Anti-IgG was immobilized on the surface of the biosensor chip. Twenty-four supernatants were diluted and injected over each flow cell (Fc1, Fc3, and Fc4), followed by an injection of buffer. Antigen was then injected over all four surfaces (Fc1, Fc2, Fc3, and Fc4). The anti-IgG surfaces were regenerated, and the signal returned to baseline. The dotted box denotes the region used to determine the average antibody capture response level. This figure has been adapted from Canziani et al. with permission from Elsevier Science (copyright 2004).

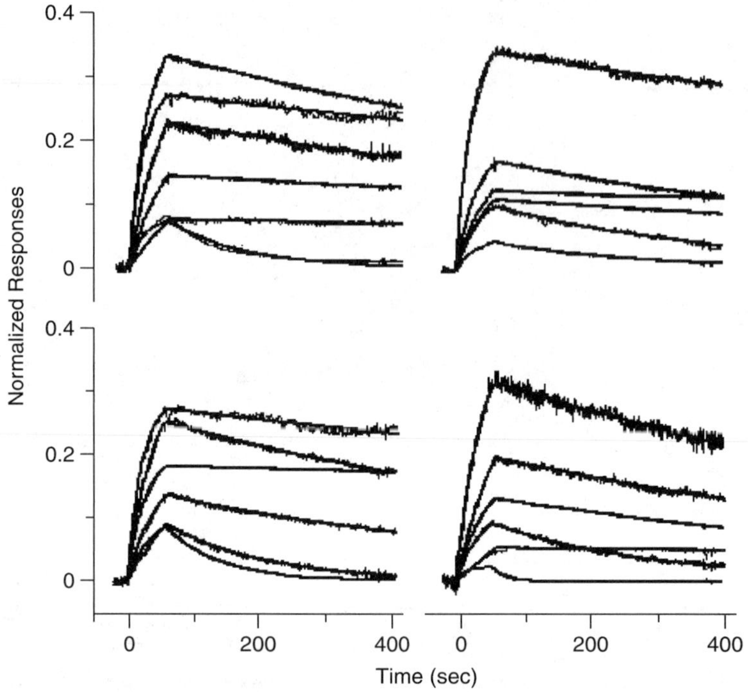

Figure 4 Global analysis of antibody screening data. Antigen responses are shown in groups of six interactions per plot. The responses were all simultaneously fit with a 1:1 interaction model using a global R_{max}. The fits are shown overlaid in groups of six interactions per plot to facilitate viewing the data. This figure has been adapted from Canziani et al. with permission from Elsevier Science (copyright 2004).

describes a fully reversible interaction where A denotes analyte, B is the ligand immobilized at the surface of the sensor chip, k_a and k_d are the association and dissociation rate constants, respectively, and AB is the formed complex. Data were fit using the analysis software Clamp (Myszka and Morton, 1998). Figure 4 shows the normalized antigen binding responses overlaid with simulated data based on the best-fit kinetic parameters. The simulated data overlay the experimental data well, which indicates that the kinetic model is in good agreement with all of the data (Canziani et al., 2004).

The reaction kinetics and binding equilibrium for the 24 antibodies were compared in a two-dimensional plot (Figure 5). To show the distribution of the kinetics, the on- and off-rates of each antibody were plotted on the x-axis and the y-axis, respectively. The diagonal lines in the graph represent the equilibrium isotherms to help visualize the affinity distributions of each of the interactions (Canziani et al., 2004). As the graph shows, two antibodies can have the same affinity, but it is derived from the ratios of different kinetic rates. Additionally, it is worth noting that the plot also illustrates the antibodies with the

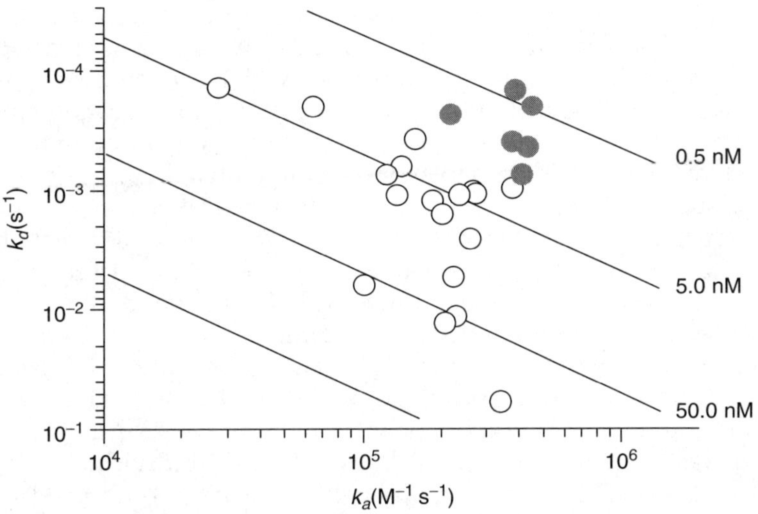

Figure 5 Kinetic distribution plot. The dissociation rate constant *versus* the association rate constant for each antibody-antigen complex studied was plotted. The equilibrium dissociation constants of each of the complexes are also shown. The affinities of complexes were distributed over a range of 0.5–500 nM. The highest affinity complexes are clustered in the upper-right corner of the plot. Because of their high affinity, we would select the antibodies denoted by the gray circles to be analyzed in higher-resolution experiments. This figure has been adapted from Canziani et al. with permission from Elsevier Science (copyright 2004).

slowest off-rates do not necessarily have the highest affinity for the antigen. An interaction's affinity is dependent upon both the association and dissociation rate constants. When the k_a is very fast, it can be an important factor in the affinity of an interaction.

Coupling SPR technology with a screening assay provides a fairly rapid and easy approach to study a panel of antibodies. In general, when using current Biacore 2000 and 3000 platforms, 100–150 antibody samples can be analyzed in a 24-hour period. Because the instruments are automated, the experimental setup is not labor intensive. The amounts of material required for studying this number of antibodies depends on the expression level in the crude hybridoma supernatants. In general, by using a flow-through capturing system, antibodies that are expressed at a level of 1 μg/mL or higher can be captured out of a crude supernatant. The amount of antigen required to complete the analysis is dependent upon the interaction.

Array-Based SPR Biosensors

Until recently, commercially available biosensors have been limited in throughput because of the low number of surfaces (typically up to four) that could be analyzed concurrently. To meet the demand for a higher throughput platform, HTS Biosystems recently introduced the Flexchip Kinetic Analysis system. This instrument allows up to 400 reactions to be monitored and compared simultaneously with the injection of a single concentration of analyte. The instrument uses grating-coupled SPR to monitor binding interactions within a 1-cm^2 surface. In this case, the system relies on a disposable plastic chip with a fine grating molded onto its surface. The grating is coated with a thin layer of gold to which biomolecules are tethered. Incident light is reflected and detected by a charge-coupled device (CCD) camera. As analyte is passed over the gold-coated sensor, the angle of reflected light changes and the binding events that occur are monitored by the camera.

Proteins are spotted onto the chip using solid-pin spotting similar to the platforms used to produce nucleic acid arrays. Using typical spotting methods, antibody concentrations of 25–100 μg/mL are required. The biosensor accommodates up to 400 individual sampling spots within the chip and regions next to each spot are used for internal referencing of the binding signals. In some assays, the spotted proteins can also be regenerated to enable replicate analysis of each spot. A challenge, however, with using the array platform is preparing the antibodies for immobilization on the chip.

Figure 6 shows an example of data that can be collected with the Flexchip Kinetic Analysis System. A uniform concentration of protein A/G was spotted onto a sensor chip in a 12 × 13 array, and serial dilutions of mouse IgGκ were passed over the surface. The binding responses show that the mouse IgGκ interacted with protein A/G in a concentration-dependent manner, where higher concentrations of the IgG antibody produced larger signals. The data collected can be globally fit, in an identical manner to how Biacore data are analyzed, to determine the best-fit kinetic parameters that describe the reaction. The experimental data are shown overlaid with simulated data generated from a simple 1:1 interaction model. In this example, IgGκ is shown binding to 156 spots of protein A/G and the data will illustrate the large amount of information that is generated by an array instrument. For these binding responses, the spot-to-spot variability in the binding kinetics was less than 8%. This enables the instrument to be used in both a high-throughput and a high-resolution mode.

High-Resolution Characterization

Biosensors are often used as biophysical tools to evaluate antibody/antigen complex formation in detail. When filing an investigational new drug sheet (IND), two of the necessary characteristics to include about a therapeutic monoclonal antibody are its kinetics and binding affinity for the target. Although reaction kinetics are derived using the screening approaches described above, characterizing high-affinity antibodies in more detail often requires further analysis to identify clearly the binding mechanism and rate constants. The assay

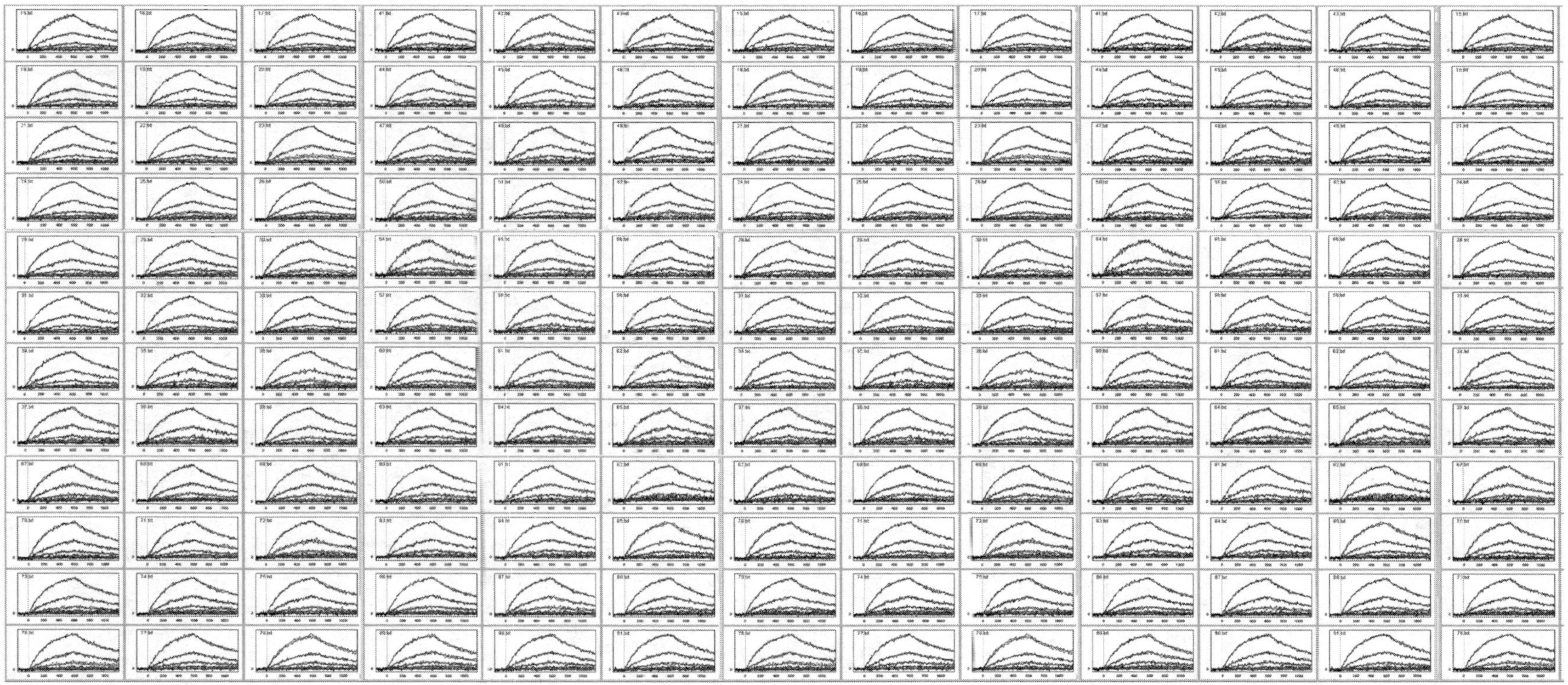

Figure 6 Flexchip Kinetic Analysis data. A uniform concentration of protein A/G was spotted in a 12 $\times$ 13 array on the biosensor chip. Serial dilutions of mouse IgGκ were passed over the chip to analyze binding. All of the experimental data were globally fit to a simple bimolecular interaction model.

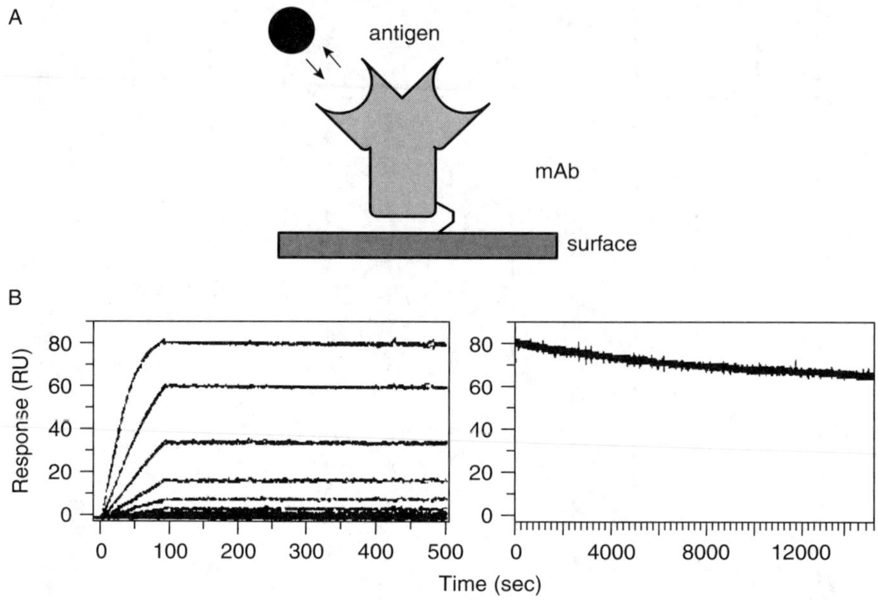

Figure 7 High-resolution analysis of antibody/antigen interaction. (A) The assay design for high-resolution analysis of an antibody/antigen interaction. Using direct immobilization, an antibody is coupled to the surface of a biosensor chip. An antigen is injected over the immobilized antibody to analyze the interaction and collect kinetic binding data. (B) High-resolution SPR data of an antibody/antigen interaction. The left panel shows six concentrations of antigen injected over the immobilized antibody. The dissociation phases were monitored for ten minutes. The right panel of data shows analysis of the antibody/antigen interaction with an extended dissociation time of four hours. The experimental data were globally fit to determine the kinetic rates, $k_a = 2.7 \times 10^6\ M^{-1}\ s^{-1}$; $k_d = 1.6 \times 10^{-5}\ s^{-1}$; $K_D = 6.1$ pM. This figure has been adapted from Drake et al. with permission from Elsevier Science (copyright 2004).

design that is typically used to obtain high-resolution data for an antibody/target interaction is illustrated in Figure 7A. An antibody is directly immobilized on the sensor chip to provide a very stable surface for monitoring antigen binding. Two primary options are available for immobilizing an antibody: (1) amine coupling through the amino terminus or surface lysine residues on the immunoglobulin, and (2) carbohydrate coupling through the sugars on the Fc region of the antibody. Although both methods can be used, carbohydrate coupling has an advantage over amine coupling because it orients the antibodies on the sensor surface. This typically improves the activity of the immobilized antibody by avoiding modification of residues within the antigen binding site. To evaluate complex formation in a high-resolution manner, serial dilutions of the target antigen are injected over the surface.

With improved antibody selection and maturation technology, today's antibodies often display very high affinity for their antigens. It is not uncommon to encounter antibodies having affinities approaching 1 picomolar. By adjusting the experimental design, biosensors can be used to determine robustly the affinities and, importantly, the dissociation rate constants for such complexes. As an example, Figure 7B shows sensorgrams obtained from an extremely high-affinity antibody/antigen interaction (Drake et al., 2004). The antibody was immobilized on the surface of a biosensor chip, and six concentrations of antigen were injected over the tethered immunoglobulin. The binding responses are concentration dependent, with higher concentrations of antigen producing larger signals (Figure 7B left panel). Little decay is observed over the 10-minute time period the dissociation phase was monitored and shows that the complex dissociates very slowly. Accurately determining a dissociation rate in this case requires that the time of the dissociation phase must be extended. The right panel of Figure 7B shows an experiment where the antibody/antigen interaction was again analyzed by injecting the highest concentration of antigen over the surface and monitoring complex dissociation for four hours. As the surface is monitored for an extended period of time, a discernable decay is observed. In many cases it is necessary to monitor complex dissociation for a longer period of time than shown in this example, and when using a Biacore instrument, the dissociation phase can be extended up to twenty-four hours, if necessary. To establish that the decay was reliable, this binding cycle was repeated three times. Samples of buffer were also injected over the surface, and the signal was monitored for four hours to show that the baseline is stable. Reproducibility and buffer injections are important to test because repetition and baseline stability lend confidence to interpreting the data.

To analyze the high-resolution data, the binding responses from both experiments are combined and globally fit to a simple bimolecular interaction model. When combined with the concentration-dependent data (left panel Figure 7B), data from the long dissociation analysis (right panel Figure 7B) yield information about the dissociation constant. The red lines represent a simulation of the data set based on best-fit parameter values when fitting the binding responses and they overlay the experimental data very well. This indicates that this simple interaction model is likely appropriate for the antibody/antigen reaction. The association rate constant (k_a) obtained by fitting the data with the simple model is $2.7 \pm 0.6 \times 10^6\ \mathrm{M}^{-1}\ \mathrm{s}^{-1}$, and the interaction has an off-rate (k_d) of $1.6 \pm 0.2 \times 10^{-5}\ \mathrm{s}^{-1}$. This corresponds to a half-life of 11.9 hours and shows that the interaction is very stable. The ratio of the kinetics, k_d/k_a, predicts an affinity of $K_D = 6.2 \pm 0.8$ pM.

One criticism of SPR technology, because it is a surface-based approach, has been that the thermodynamics and binding kinetics derived from an analysis may be altered because one of the reactants is tethered to a surface. To address this concern, the results from studying the antibody/antigen interaction described above were compared with results obtained from a solution-based method called KinExA. Using this fluorescence-based method, Drake and colleagues measured association and dissociation rate constants of $2.8 \pm 0.5 \times 10^6\ M^{-1}\ s^{-1}$ and $1.1 \pm 0.2 \times 10^{-5}\ s^{-1}$, respectively. This solution-based approach also predicted an affinity of $K_D = 4 \pm 2$ pM (Drake et al., 2004). Within experimental error, there was no difference in the rate constants or the affinity of the antibody/antigen complex obtained using the surface-based Biacore or the solution-based KinExA. This is consistent with other studies in which solution-based methods such as calorimetry and stopped-flow spectroscopy produced rate and thermodynamic constants that agreed well with those measured using the biosensor (Day et al., 2002; Deinum et al., 2002; Myszka et al., 2003). Together these studies further validate the SPR technology as a biophysical tool and demonstrate its ability to collect reliable equilibrium, kinetic, and thermodynamic information with a single platform.

Clinical Applications

Once a potential biopharmaceutical agent has been identified, an important aspect of product development is to establish whether or not it is immunogenic in patients. If a therapeutic is recognized as foreign by the host immune system, the patient will likely generate an adverse immune response to the biopharmaceutical. By validating a Biacore immunoassay for serological analysis of antibody titers, Swanson and colleagues have shown that SPR technology can be applied to characterize the humoral immune response of a therapeutic agent (Swanson et al., 1999; Mason et al., 2003). With their approach, the protein-based drug is immobilized on the surface of a Biacore chip, and multiple human serum samples can be screened for drug binding. A serum sample is determined to have anti–drug-specific activity by comparing the sample's maximum binding response to a threshold value. If the sample's binding response is above the threshold value, its anti–drug-specific activity is then confirmed to be an antibody-binding signal by injecting a confirmatory reagent, such as an anti-human IgG. The assay was shown to be robust and highly reproducible, for the variability of results obtained from multiple instruments and different analysts did not exceed a coefficient of variation of 20%. This validation provides a benchmark for assay and instrument performance during the analysis of serological samples.

Another application of Biacore technology in studying the immunogenicity of a protein-based drug is to identify the isotype of serum antibodies that bind the drug (Swanson et al., 2002). As an example of this, Figure 8 shows an antibody captured out of serum that binds to the surface-immobilized therapeutic protein. The resulting complex establishes a new, stable baseline (baseline 2). Anti–species-specific IgE, IgA, IgM, and IgG are sequentially injected

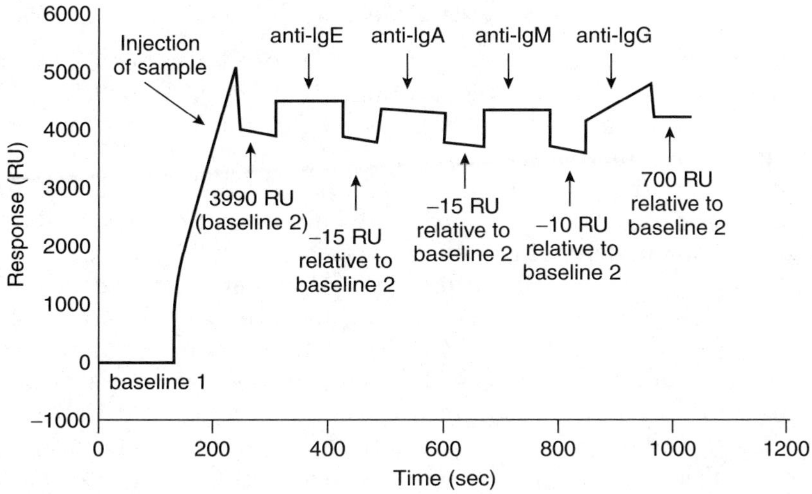

Figure 8 Isotyping of a therapeutic antibody. The therapeutic antibody is injected over the antigen surface. In this example, 3990 RU were detected relative to baseline. This new reference point is called baseline 2. Next, isotyping reagents (anti-IgE, anti-IgA, anti-IgM, and anti-IgG) are added to examine additional increases in RU. The response depicted in this figure is predominantly IgG, because an increase of 700 RU was detected relative to baseline 2, whereas no increase in RU was evident with anti-Ig3, IgA, or IgM. This figure has been adapted from Swanson et al. with permission from Karger (copyright 2002).

over the surface to determine the isotype of the captured immunoglobulin. In this example, an increase of nearly 700 RU (relative to baseline 2) is observed after the injection of anti-IgG. Increases in the binding signals were not observed after injecting anti-IgE, anti-IgA, or anti-IgM, which suggests that the antibody captured on the biosensor surface was a serum IgG.

While developing a therapeutic agent, it is necessary for companies to adhere strictly to the regulations of Good Laboratory Practice (GLP) and Good Manufacturing Practice (GMP). Biacore has recently released an instrument, Biacore C, which is designed to comply with GLP/GMP regulations. Among its applications, the instrument allows for rapid assay development and can be used for production by monitoring antibody expression levels. It is also designed to perform concentration analyses of pharmaceuticals and vaccines within a validated GMP/GLP environment. The software used to support Biacore C and, most recently, the new compliant software for the Biacore 3000 platform conforms to FDA regulations by maintaining data integrity and allowing multiple user access levels through increased security. Given the expanding use of pharmaceutical agents, the application of sensor technology to support clinical trials and production is expected to increase significantly.

Conclusions

Surface plasmon resonance optical biosensors have been used for more than a decade to analyze biomolecular interactions from pure samples or complex mixtures found in serum, urine, supernatants, and other crude media. Improved hardware and novel experimental design now permit SPR biosensor technology to be used in nearly every stage of developing a therapeutic antibody or a protein-based drug. Biacore instrumentation is used to screen kinetically large panels of antibodies from crude supernatants and select candidates for further characterization. The candidates are analyzed in a higher resolution mode by injecting multiple antigen concentrations over the selected antibody surface. This is done to determine accurately the reaction rate constants and affinity and can be used to identify mechanistic differences in antigen binding to multiple antibodies. At the clinical level, SPR technology has been used to analyze serum samples within a validated immunoassay format as well as in quality control and manufacturing. Through continued developments in instrument hardware, software, and applications, the role of SPR technology in biopharmaceutical research and development will continue to expand.

Abbreviations

SPR	Surface plasmon resonance
ELISA	Enzyme-linked immunosorbent assay
RU	Response unit
CCD	Charge-coupled device
IND	Investigational new drug
GLP	Good laboratory practice
GMP	Good manufacturing practice
Fc	Flow cell

References

1. Baird CL, and Myszka DG. Current and Emerging Optical Biosensors. *J Mol Recognit* 2001; 14:261–268

2. Cooper MA. Optical Biosensors in Drug Discovery. *Nature Rev* 2002; 1:515–526

3. Canziani G, Klakamp S, and Myszka DG. Kinetic Screening of Antibodies from Crude Samples. *Anal Biochem* 2004; 325:301–307

4. Darling RJ, Kuchibhotla U, Glaesner W, Micanovic R, Witcher DR, and Beals JM. Glycosylation of Erythropoietin Affects Receptor Binding Kinetics: Role of Electrostatic Interactions. *Biochemistry* 2002; 41:14524–14531

5. Day YS, Baird CL, Rich RL, and Myszka DG. Direct Comparison of Binding Equilibrium, Thermodynamic, and Rate Constants Determined by Surface- and Solution-based Biophysical Methods. *Protein Sci* 2002; 11:1017–1025
6. Deinum J, Gustavsson L, Gyzander E, Kullman-Magnusson M, Edstrom A, and Karlsson R. A Thermodynamic Characterization of the Binding of Thrombin Inhibitors to Human Thrombin Combining Biosensor Technology, Stopped-flow Spectrophotometry, and Microcalorimetry. *Anal Biochem* 2002; 300:152–162
7. Drake AW, Myszka DG, and Klakamp SL. Characterizing High Affinity Antigen/Antibody Complexes by Kinetic- and Equilibrium-Based Methods. *Anal Biochem* 2004; 328:35–43
8. Faggioni R, Cattley RC, Guo J, Flores S, Brown H, Qi M, Yin S, Hill D, Scully S, Chen C, et al. IL-18-binding Protein Protects Against Lipopolysaccharide-induced Lethality and Prevents the Development of Fas/Fas Ligand-mediated Models of Liver Disease in Mice. *J Immunol* 2001; 167:5913–5920
9. Glaser SM, Yelton DE, and Huse DH. Antibody Engineering by Codon-based Mutagenesis in a Filamentous Phage Vector System. *J Immunol* 1992; 149:3903–3913
10. Homola J, Yee SS, and Myszka DG. Surface Plasmon Resonance. In: Rowe Taitt CA. *Optical Biosensors: Present and Future.* Amsterdam: Elsevier Science B.V.; 2002:207–251
11. Hoogenboom HR, Marks JD, Griffiths AD, and Winter G. Building Antibodies from Their Genes. *Immunol Rev* 1992; 130:41–68
12. Johnsson B, Löfås S, and Lindquist G. Immobilizations of Proteins to a Carboxymethyldextran-modified Gold Surface for Biospecific Interaction Analysis in Surface Plasmon Resonance Sensors. *Anal Biochem* 1991; 198:269–277
13. Jönsson U. Real-time Biospecific Interaction Analysis Using Surface Plasmon Resonance and a Sensor Chip Technology. *Bio Techniques* 1991; 11:620–627
14. Karlsson R, Michaelsson A, and Mattsson L. Kinetic Analysis of Monoclonal Antibody-Antigen Interactions with a New Biosensor Based Analytical System. *J Immunol Methods* 1991; 145:229–240
15. Karlsson R, Fägerstam L, Nilshans H, and Persson B. Analysis of Active Antibody Concentration. Separation of Affinity and Concentration Parameters. *J Immunol Methods* 1993; 166:75–84
16. Kretschmann EZ, and Raether H. Radiative Decay of Non-radioative Surface Plasmons Excited by Light. *Z Naturforsch* 1968; 23O:2135–2136
17. Mason S, La S, Mytych D, Swanson SJ, and Ferbas J. Validation of the Biacore 3000 Platform for Detection of Antibodies Against

Erythropoietic Agents in Human Serum Samples. *Curr Med Res Opin* 2003; 19:651–659

18. Morton TA, and Myszka DG. Kinetic Analysis of Macromolecular Interactions Using Surface Plasmon Resonance Biosensors. *Methods Enzymol* 1998; 295:268–294
19. Myszka DG. Improving Biosensor Analysis. *J Mol Recognit* 1999; 12:279–284
20. Myszka DG. Kinetic, Equilibrium, and Thermodynamic Analysis of Macromolecular Interactions with Biacore. *Methods Enzymol* 2000; 323:325–340
21. Myszka DG, and Morton TA. CLAMP: A Biosensor Kinetic Data Analysis Program. *Trends Biochem Sci* 1998; 23:149–150
22. Myszka DG, and Rich RL. Implementing Surface Plasmon Resonance Biosensors in Drug Discovery. *Pharm Sci Technol Today* 2000; 3:310–317
23. Myszka DG, Abdiche YN, Arisaka F, Byron O, Eisenstein E, Hensley P, Thomson JA, Lombardo CR, Schwarz F, Stafford W, and Doyle ML. ABRF-MIRG'02 Study: Assembly State, Thermodynamic and Kinetic Analysis of an Enzyme/Inhibitor Interaction. *J Biomol Tech* 2003; 14:247–269
24. Rich RL, and Myszka DG. Survey of the 2000 Commercial Optical Biosensor Literature. *J Mol Recognit* 2001; 14:273–294
25. Rich RL, and Myszka DG. Survey of the 2001 Commercial Optical Biosensor Literature. *J Mol Recognit* 2002; 15:352–376
26. Rich RL, and Myszka DG. Survey of the 2002 Commercial Optical Biosensor Literature. *J Mol Recognit* 2003; 16:351–382
27. Swanson SJ. New Technologies for the Detection of Antibodies to Therapeutic Proteins. *Dev Biol (Basel)* 2003; 112:127–133
28. Swanson SJ, Jacobs SJ, Mytych D, Shah C, Indelicato SR, and Bordens RW. Applications for the New Electrochemiluminescent (ECL) and Biosensor Technologies. *Dev Biol Stand* 1999; 97:135–147
29. Swanson SJ, Mytych D, and Ferbas J. Use of Biosensors to Monitor the Immune Response. *Dev Biol (Basel)* 2002; 109:71–78
30. Wilson S, and Howel S. High-Throughput Screening in the Diagnostic Industry. *Biochem Soc Trans* 2002; 30:794–797

16

Microarray-Based Methods

DANA AULT-RICHÉ, BRUCE ATKINSON, and
KRISHNANAND D. KUMBLE

Pointilliste, Inc., Mountain View, CA

Introduction

Along with advances made in processing speeds in the computer industry during the past two decades, the drug industry has experienced an ever-increasing appetite for higher data throughput. The high-throughput screening industry emerged in the 1980s as robotics and manufacturing principles were applied to chemical testing. A second leap forward in data throughput followed when massive parallel testing of nucleic acids became enabled by DNA microarrays. DNA microarray technology allows tens of thousands of DNA fragments and messenger RNA (mRNA) transcripts to be simultaneously tested and in the process is transforming genetics research (Gerhold et al., 2002). More than 60,000 tests in a single chip have now become routine, and commercial DNA microarray products are now widely available.

The emergence of DNA microarray technology created an industry of instruments and manufacturing tools (Holloway et al., 2002). Many of the instruments originally developed for DNA microarray technology, such as printing robotics and imagers, can be readily adapted for use with proteins, and several companies that specialize in selling antibody products have begun selling antibody microarrays.

The technology for DNA microarrays preceded that of protein microarrays, primarily due to the relative robustness of nucleic acid chemistry in comparison to protein chemistry. DNA can be chemically synthesized, and binding (hybridization) based on Watson-Crick base pairing rules is very specific. Most proteins are inherently less stable than nucleic acids, and the chemistry of protein binding is much more complex than nucleic acid hybridization. Additionally, proteins cannot be directly chemically synthesized, and molecules that specifically bind proteins are not easily obtained. As an example, antibodies are one of the most specific protein binders, yet high-quality (high affinity and high specificity) antibodies are difficult to produce. This problem (the lack of specific protein binders) has been dubbed "the protein biochip content problem."

The usefulness of information gained from DNA microarray technology has turned out to be much more limited than was originally suspected. This is primarily because proteins, rather than DNA or RNA, are more directly involved in diseases and are the primary targets of drugs. Further, the amount and activity of proteins cannot be directly inferred from determining the amount of mRNA (Griffin et al., 2002). Despite the "protein biochip content problem," the market for protein biochips was $70 million in 2001, $100 million in 2002, and is expected to grow to $400 million by 2007 (Protein Biochips 2003, Select Biosciences). These market numbers highlight the thirst for technologies that can provide information about proteins with the same degree of informational throughput that DNA microarray technology has provided for genes.

Microarrays can be divided into two main categories: planar and liquid. A planar microarray is a collection of 100–150 μm diameter spots arranged in an array pattern on a flat surface. Different capture agents, such as antibodies, are deposited at known locations in the array, resulting in the spots. A test sample is then applied to the array, loosely bound material is washed away, and the remaining bound material is detected. Sample binding is then correlated to a capture agent located at a specific physical address in the array.

A liquid array is created using a collection of encoded beads, which can be decoded after binding to the sample. The most common type of liquid array is composed of a collection of dye-encoded latex beads. The beads are decoded, or read, by a device, such as a fluorescence-activated cell sorter (FACS), and sample binding is then correlated with a capture agent by indirect correlation with a specific dye.

Hardware

Arrayers

Several types of devices have been developed for the manufacture of microarrays. These devices, commonly referred to as "arrayers" or "spotters," were originally developed to create DNA microarrays. Most arrayers can be adapted for use with proteins. Arrayers can generally be categorized as either contact or noncontact, depending on whether the device makes physical contact with the surface of the substrate upon which the microarray is being created.

The most common type of contact arrayer uses a series of pins to deposit droplets onto the substrate surface. Pin-style arrayers have been adapted for use in making protein microarrays (Haab et al., 2000). The pins are typically manufactured from stainless steel and can either be solid or split, like a quill, to contain a reservoir of a few microliters at their tip. The pins are manually loaded into a holding block located at the end of a robotic arm. The arm, whose movements are driven by precise stepper-motors, is used to physically maneuver the pins from sample reagent source plates to the target substrates. The sample-filled pin tips are lowered to a point where (ideally) only the meniscus of the protein reagent, at the tip of the pin, contacts the substrate and deposits a defined volume of reagent (typically 50–300 nL) at the point of contact to create a spot in the microarray. Pin-dispensed spot volumes generally fall within a 20% coefficient of variation. To prevent sample-to-sample contamination, the pins are brought to a wash station prior to loading with the next sample.

Pins require a high level of maintenance to prevent reservoir tip clogging and are also susceptible to physical damage should they be brought into contact with the target substrate with excessive force. Clogged or damaged pins will result in a failure to correctly print a spot. Failures to print are referred to as "dropouts" in the array. Washing accounts for a majority of the time involved in arraying. To increase microarray manufacturing speed, arrayers have been formatted to contain multiple pins. However, a major source of variation in pin-style arrayers is due to variation between different pins.

Ink-jet–style arrayers evolved directly from computer ink-jet printing technology. These noncontact devices typically employ a piezoelectric motor, located within the print head, which causes the emission of precise, small-volume (nanoliter) reagent droplets that are then deposited onto the target substrate (Delahanty and Ligler, 2003). Depending on the configuration, either the print head or the substrate platform is robotically driven into position prior to droplet ejection. Reagent source well contents are fed to the print head via tubing, and the system is sealed to maintain correct liquid pressure to the print head. Ink-jet arrayers have the advantage of greater control in the volumes dispensed (typically 5% coefficient of variation). Dropouts are primarily the result of clogged components. Some ink-jet arrayers have both an external and an internal wash system to minimize contamination and motor clogging. Some systems also optically monitor dispensing, providing data on droplet formation and dropouts, which can be used to "fill-in" dropouts. Ink-jet arrayers are typically faster than pin-style arrayers but are typically more expensive and require greater maintenance.

Acoustic dispensers are the newest arrayer technology platform. These employ a precisely controlled microtransducer that emits a focused acoustic energy wavelength to drive a sample from the source plate to the target substrate (Ellson, 2002). The transducer is held in a fixed position, while the sample plate and target substrate platforms are robotically moved over the transducer. Droplets are "shot" from the sample plate upward, onto the substrate. Sample volumes can accurately be delivered in the picoliter range; more typically 5–50 nL volumes are delivered with a 5% coefficient of variation. For these systems to work, the acoustic properties of the sample must first be assessed using the transducer. These systems have the significant advantages of avoiding the need for wash steps between samples and the problems associated with clogging. Table 1 lists various types of commercially available arrayers.

Detectors

Detection of a specific event on the array, such as protein binding to an array, can be performed using a variety of different methods. The most common methods of detection rely on labeling the sample either before or after binding to the array. For example, a relative measurement of

Type of Printer	Features	Manufacturer URL
Microcontact pin and ring system	Spot size of 150 to 200 μm with 50 pL delivery volume	www.affymetrix .com
Microcontact solid or quill pins	Spot size range dependent upon pin type (60 to 110 μm with 50 to 110 pL delivery volume)	www.cartesiantech.com
Microcontact solid or quill pins	Spot size range dependent upon pin type; HEPA filtration for dust-free printing	www.genemachines.com
Microcontact solid pins	Spot size of 250 μm; linked database for storing source plate information	www.genetix.co.uk
Noncontact piezo print-head	Spot size approximately 200 μm; can deliver either 24 or 96 spots simultaneously	www.genescan-europe.com
Noncontact piezo print-head	Spot size range dependent upon pin type (75 to 200 μm)	www.packardbioscience.com

Table 1 Printing Systems for Microarrays.

protein abundance can be obtained by comparing two samples—experimental sample and reference—by independently labeling the samples with different fluorophores and then mixing them before binding to the microarray. The two fluorophores most commonly used are the dyes Cy-3 and Cy-5 (Haab et al., 2001). In addition, binding can be detected via label-independent processes using a variety of more sophisticated detectors. To date, the vast majority of protein microarray–derived data found in literature has been generated using label-dependent detection. The choice of a detection system is ultimately driven by the assay format employed and the cost. Table 2 lists some of the various types of scanners currently available in the market.

Label-Dependent Detection

In parallel with the adaptation of DNA arrayers for use in protein microarray manufacture, fluorescence-based scanning systems originally developed for use with DNA microarrays have also been adapted for use in protein microarray signal detection and imaging. Microarray "scanners" use lasers to excite a fluorophore, which is typically coupled to a specific secondary detection reagent. An example is fluorophore-labeled streptavidin, which can bind to a biotin-labeled sample. The sample is biotinylated prior to exposure to the array. After the sample is bound to the array, the labeled streptavidin is added, the array washed, and the remaining streptavidin detected (Valeur, 2002). Scanner optics focus emitted light through an emission filter, which then strikes a photomultiplier tube (PMT). The scanner outputs a color-coded array image for subsequent analysis by the user. These instruments typically enable a range of user-defined scanning resolutions that correspond to the feature sizes of the array elements, adjustable detection sensitivity through variable PMT settings, and postimage signal analysis options.

Fluorescence scanners, when paired with the wide range of commercially available fluorescence chemistries, are by far the most flexible systems for use in protein microarray image detection. Many scanners carry laser capacity for four-color scanning, thus enabling simultaneous multifluorophore detection. Fluorescence-based detection generally provides a higher range of sensitivity and dynamic range than those of chemiluminescence-based systems. The greater number of photons generated by a fluorophore (per unit time) when excited by a laser wavelength centered on, or near, its peak excitation spectrum forms the basis for greater sensitivity (Agbaria et al., 2002). The choice of a particular fluorescence-based system does limit the user to a given set of fluorophores.

Chemiluminescence, a detection chemistry widely used within the protein microarray field, is a defined chemical process whose product is broad-spectrum white light. Chemiluminescence-based detection instrumentation relies on a charge-coupled device (CCD),

Type of Scanner	Features	Manufacturer URL
Laser scanning microscope	Sensitivity of less than 1 fluor/μm^2 Two excitation filters: 635 nm and 532 nm	www.affymetrix.com
Scanning laser	Sensitivity of less than 1 fluor/μm^2 Two excitation filters: 635 nm and 532 nm; scanner focus adjusted continually	www.agilent.com
Scanning laser	Sensitivity of less than 0.1 fluor/μm^2 for Cy3 and Cy5; multiple filter options	www.axon.com
Laser confocal optics	Sensitivity of less than 0.1 fluor/μm^2; multiple laser options	lifesciences.perkinelmer.com
Scanning laser	Sensitivity of less than 0.1 fluor/μm^2; multiple laser options; can read 96-well plates	www.tecan.com
CCD digital imager	Broad band excitation source with up to 8 selectable excitation and emission filters; slide or 96-well plate formats	www.alphainnotech.com
CCD digital imager	Broad band excitation source with up to 4 selectable excitation and emission filters; luminescence, fluorescence, absorbance (transmission/detection), and radiographic detection	www.kodak.com

Table 2 Scanning Systems for Microarrays.

together with conventional camera optics. For image detection, the camera lens shutter is opened, exposing the CCD to light generated from the chemiluminescent reaction, which is then converted by software to a black-and-white image. These instruments typically offer extensive user-defined exposure times and camera settings. Postimage signal analysis and image resolution options are also possible using certain scanners.

Chemiluminescence is extremely flexible for applications in a wide range of assay formats. As such, it can be considered more readily adaptable for a given assay readout. CCD-based systems have become increasingly sophisticated, both in terms of their hardware (number of CCD pixels, camera optic quality), and their software (user-defined options for camera settings, postacquisition signal and image manipulation). In addition, many CCD-based systems contain an internal ultraviolet light source and a number of emission filters, which permit the user to perform some fluorescence-based detection as well.

Detection Systems for Liquid Arrays

Liquid arrays are dependent on the ability to detect different sets of optically distinct beads. Beads of approximately 5 μm in diameter, produced from latex or other similar materials, are embedded with fluorescent dyes that partition the emission wavelengths of the beads into optically distinguishable signatures. A user-provided, specific binding molecule, such as an antibody, is coupled to each of the unique bead sets. Conventional solution-phase binding assays are then performed by incubating coupled beads with analyte-containing samples. A secondary fluorescent dye, coupled to an adapter molecule, is used as a reporter for specific analyte binding to the coupled beads. These bound complexes are then passed through a capillary flow system containing excitation lasers, emission filters, and detectors. The system rapidly analyzes many members of each of the optically unique bead sets, providing a readout for each bead set (Earley et al., 2002).

Such systems are refered to as "on-the-fly," as binding events are deconvoluted while the beads pass in front of the detector. These systems have successfully been applied to a wide range of multiplexed, multi-analyte analyses of autoimmune and cancer markers (Prabhakar et al., 2002; Keyes et al., 2003) and cytokine profiles (Carson and Vignali, 1999).

Nanoparticle-based bio-bar codes combine the ability to increase assay sensitivity using polymerase chain recation (PCR)-based DNA amplification, similar to rolling circle amplification, with a multiplexed decoding strategy using DNA tags and DNA microarrays, as an alternative to optical tags and FACS detectors (Nam et al., 2003). This method remains based around the concept of an immunosandwich assay. In this case, the capture antibody is immobilized on a magnetic particle and the detection antibody is immobilized to a nanoparticle containing specific DNA tags. Combinations of magnetic particles containing different capture antibodies and nanoparticles containing different detection antibodies can be mixed with a sample. The analyte in the sample initiates the formation of a cross-linked complex containing the magnetic particle, the antigen, and the nanoparticle. These formed complexes are then separated from the sample by use of a magnetic field. After separation, the DNA tags are released from the complexes and then amplified using PCR. Multiple PCR products can simultaneously be detected and "decoded" using a DNA microarray. The number of potentially different DNA signatures is much higher than the number of distinct optical signatures currently separable by FACS. However, the usefulness of this approach has yet to be determined, as the primary limitation with protein arrays is in the number of compatible sandwich antibody pairs, rather than the ability to array them either by spatial location or by optical decoding.

Label-Independent Detection

A major limitation of label-dependent methods is the variation caused by inconsistencies in labeling chemistry. Variation can occur both between and within samples. Label-independent methods such as surface plasmon resonance (SPR) avoid this shortcoming. SPR-based assays are commonly used to measure antibody affinity for an antigen by flowing antigens over antibodies bound to a thin film in a flow cell (see Chapter 15). These systems are highly sensitive for such analyses, typically requiring sub-microgram quantities of material. In addition, samples analyzed for specific binding can be crude, provided the immobilized probe is specific for a given analyte.

Surface Chemistries

The following factors need to be considered for selecting a surface chemistry to print protein microarrays: (i) high protein binding capacity with retention of protein function, (ii) applicability to a wide range of protein types, (iii) low background with complex samples, and (iv) ease of manufacture with minimum intra- and inter-batch variability.

Derivatized glass is the most common surface used for printing protein microarrays (Schäeferling et al., 2002) either through physical adsorption or by covalent binding. The most commonly used chemicals for coating glass to adsorb proteins include nitrocellulose (Stillman and Tonkinson, 2002) and poly-L-lysine (Haab et al., 2000). Nitrocellulose has higher protein binding capacity, stability, and performance as compared to most other derivatized glass surfaces (Stillman and Tonkinson, 2002).

Chemistries used for covalent binding of proteins to surfaces can be either generic or specialized to bind only certain types of proteins.

Generally, chemicals are deposited on glass either by vapor deposition or by dipping into a solution; both methods often result in an uneven surface coating, giving rise to significant micro-heterogeneity within the surface, causing uneven protein binding on the surface.

Generic surfaces have been created by derivatization of glass with epoxy- or aldehyde-silane compounds (MacBeath and Schreiber, 2000; Tam et al., 2002), succinimidyl esters (Shriver-Lake et al., 1997; Kusnezow et al., 2003), and maleimides (Kusnezow et al., 2003). Surfaces such as those coated with salicylhydroxamic acid created for covalent linking of proteins modified with phenyl(di)boronic acid based on reversible complexation chemistry have also been described (Stolowitz et al., 2001). Surface hydrophobicity caused by silanization results in protein denaturation, requiring addition of stabilizing agents such as glycerol in the spotting buffer (Kusnezov and Hoheisel, 2003).

Surfaces have been created to capture specific types of proteins: these include nickel-coated and streptavidin-coated glass for capture of hexahistidine-tagged proteins (Zhu et al., 2000) and biotinylated proteins (Mooney et al., 1996), respectively. Avidin was coated on glass treated with thiolsilane and was used for attachment of Fab fragments biotinylated at the hinge region (Rowe et al., 1999). Avidin interaction with biotin, with a binding affinity constant of about 10^{15} M, is commonly used for immunoassays (Schetters, 1999). Antibodies have also been attached to surfaces by their Fc portion using protein A (Kanno et al., 2000) or via carbohydrate residues using amino-modified surfaces (Nisnevitch and Firer, 2001).

To preserve functionality and increase the amount of bound protein, three-dimensional surfaces, such as hydrated gels, have been used. There are a limited number of such surfaces available due to challenges in the manufacturing process, and these surfaces create a barrier for diffusion, requiring longer incubation times to achieve equilibrium at low analyte concentrations (Arenkov et al., 2000). Slides coated with gels such as agarose (Afanassiev et al., 2000) and polyacrylamide (Arenkov et al., 2000) have been used for performing various types of immunoassays and enzymatic reactions, as well as assays with live cells (Barsky et al., 2002).

Polystyrene has routinely been used for conventional ELISA assays. This surface has the advantage of using commercially available 96-well plates that allow simultaneous and automated processing of 96 samples using standard robotic tools at relatively low cost (Moody et al., 2001). This method for preparation and use of microarrays will advance the technology by allowing analysis of larger numbers of samples on focused arrays with fewer elements (Kumble, 2003). Polystyrene can be easily modified by radiation and other techniques to alter the surface chemistry in order to create functional groups on the surface (Brash and Horbett, 1997). Polystyrene has also been modified with a variety of reactive groups such as amine and maleimide to

Type of Substrate	Chemistry for Protein Attachment	Manufacturer URL
Glass	Aldehyde	www.arrayit.com
Glass	Amine	www.arrayit.com
Glass	Aminosilane	www.corning.com
Glass	Epoxysilane	www.genescan-europe.com
Glass	Hydrogel	www.packardbioscience.com
Glass	Nitrocellulose	www.s-and-s.com
Glass	Nylon	www.pall.com
Glass	*N*-hydroxysuccinimide ester	www.amershambiosciences.com
Polystyrene	Aminosilane	www.nuncbrand.com
Polystyrene	Aldehyde	www.nuncbrand.com
Polystyrene	Streptavidin	www.piercenet.com
Polystyrene	Maleimide	www.immunosurface.com
Polymer	MaxiSorp	www.nuncbrand.com

Table 3 Surface Chemistries for Printing Protein Microarrays.

allow covalent attachment of proteins on the surface (Rasmussen, 1997; Smythe, 1964). Polystyrene plates coated with concanavalin A, protein A, and streptavidin are commercially available and can be used for binding specific types of proteins. Table 3 describes various types of surfaces along with the chemistries that are commercially available for printing protein microarrays.

Software

An array image must be interpreted and transformed into a dataset. Several commercially available software packages are available to do this. As a first step in image processing, background-corrected intensity values for each of the spots are obtained using software programs that grid the elements of the array. Background can be corrected using one of several options: local (area around individual spots) or global (average signal in area outside of the grid) background corrections or user-defined values such as those from negative control data points contained within the array.

Once background-corrected intensity values for each of the spots has been computed, the data is normalized with respect to differences in sample quality, sample labeling, and other sources of systematic and biological variation (Bilban et al., 2002). The method used for normalization is a crucial choice that can impact efforts to compare accuracy and precision between arrays, samples, and experiments.

Several methods for statistical analysis of microarray data are available depending upon the experimental setup and the kind of biological question that needs to be addressed. Hierarchical clustering of the data is the most commonly used analysis due to its simplicity of data organization (Slonim, 2002); several clustering packages are available in the public domain and can freely be used by academic researchers (Holloway et al., 2002). More advanced machine-learning techniques such as neural networks (Xu et al., 2002) and support vector machines (Brown et al., 2002) have also been used for data analyses. The data generated from protein microarrays will have to be integrated with data from other high-throughput assays including transcript profiling in order to enable a comprehensive analysis of the information gathered. This can be done by adopting common standards for data generation and representation such as those recommended for DNA microarrays (Brazma, 2001; Brazma et al., 2001).

Applications

Antibody Arrays

The ability to test the levels of proteins in a sample is a fundamental practice in nearly all of the approximately 40,000 life science research laboratories worldwide. The two predominant methods for determining protein levels are the ELISA (enzyme-linked immunosorbent assay) and the Western blot technique. The market for products related to ELISA and Western blotting is several hundred million dollars a year (Deutsche Bank Securities Report, 2002).

The multi-analyte microspot immunoassay was originally described by Roger Ekins (Ekins, 1989). This method uses surface-immobilized antibodies to detect the abundance of analytes in samples via a second interaction using another labeled antibody as the detector (Silzel et al., 1998; Joos et al., 2000). The simplest extension of this format was printing antibodies into arrays in the wells of 96-well plates to simultaneously determine the abundance of multiple analytes in samples (Moody et al., 2001; Tam et al., 2002).

Antibodies are the most common protein class arrayed on surfaces due to their similar structures, stability, and specificity of antigen binding. With the potential for greater sensitivity and the advantages of multiplexing, antibody microarrays could one day replace both ELISA and Western blots. Profiling entire proteomes of organisms using sandwich ELISA would require thousands of well-characterized antibody pairs to be available for capture and detection. In theory, an array containing antibodies to all human proteins could be developed given the appropriate collection of antibodies. However, a collection of antibodies capable of quantitative analysis of all human proteins in an array format does not currently exist.

Microarray-based immunoassays can simultaneously measure multiple analytes from low sample amounts (less than 50 μL) with a high degree of sensitivity (pg/mL) and wide dynamic range (3 logs) (Schweitzer et al., 2002). In order to measure the abundance and modification of proteins in biological samples, capture agents that will specifically bind proteins from samples are needed. An initial test of this approach was performed by screening a small number of polyclonal and monoclonal antibodies against 5,000 different yeast proteins deposited on a glass slide. The results showed that in addition to recognizing their cognate proteins, the antibodies cross-reacted with other proteins to varying degrees (Michaud et al., 2003), highlighting the importance of using highly specific antibodies as capture agents.

Many different monoclonal antibodies have been created over the years to react with various epitopes on cytokines, thus creating a diverse source of potential cytokine capture antibodies. These antibodies have been used in antibody microarrays to measure cytokine concentrations (Huang et al., 2002; Schweitzer et al., 2002; Patterson and Aebersold, 2003). The first antibody arrays that have become commercially available products contain sets of antibodies to human cytokines. These products have been introduced to the market by companies that also sell other antibody products; for example, BD Biosciences (http://www.bdbiosciences.com), Pierce (http://www.piercenet.com), and Biosource (http://www.biosource.com).

Initial ELISAs with anticytokine antibodies printed at the bottom of 96-well polystyrene plates could measure 1 to 10 pg/mL of 9 different cytokines using less than 50 μL of sample (Moody et al., 2001). Cytokine expression levels in connexin43-transfected cells determined using an array of 43 anticytokine antibodies showed a significant decrease in monocyte chemotactic factor-1 (MCP-1) expression (Huang et al., 2002). These types of arrays have been multiplexed to have 50 or more anticytokine antibodies for which ELISAs are already being performed. An array of 75 anticytokine antibodies segregated into subarrays of 37 and 38 antibodies in quadruplicate spots was used to measure cytokine levels in supernatants from lipopolysaccharide- or tumor necrosis factor-α–induced dendritic cells (Schweitzer et al., 2002). Printing the antibodies in two subarrays allowed for switching capture antibodies between subarrays to minimize cross-reactivity within each subarray. The assay sensitivity was increased to femtomolar levels with a 3 log dynamic range of detection using rolling circle amplification in conjunction with universal antibodies, which is sufficient to measure most biologically relevant changes in cytokine secretion (Figeys, 1999).

One impediment to the widespread adoption of antibody microarrays as replacements for ELISAs has been the time and expense required to optimize a large collection of antibodies such that cross-reactivity of antigens between antibodies in the array does not interfere with the ability of each arrayed antibody to produce a quantitative signal. This challenge was highlighted in studies using monoclonal antibodies to probe expressed proteins. These studies show significant cross-reactivity of antibodies to proteins other than the intended targets (Walter et al., 2000; Cahill, 2001).

The size of the market for human cytokine antibodies in an array format justified the commercial development required. However, it is unclear how many other array products will follow, as most research laboratories are involved in highly specialized projects, so their ELISA needs are also highly specialized. Laboratories usually have distinct collections of antibodies discovered over the history of the laboratory. Creating antibody microarray products for the needs of each laboratory is commercially impractical. It remains to be seen what collections of antibodies, other than cytokine antibodies, will have enough broad-scale utility to become commercially successful products in microarray formats.

Shifting the burden of array optimization to the individual laboratories by delivering a flexible, universal tool may be the solution to widespread use of antibody microarrays as a common research tool. Pointilliste (http://www.pointilliste.com) has developed a collection of antibodies to short epitope tags, which act as linkers to enable different sets of antibodies and proteins to be formatted into microarrays. This *universal antibody microarray* enables researchers to format their own antibodies into a microarray. It also enables researchers to test rapidly various combinations of antibodies to define antibody sets that are compatible in an array format.

Similar to the Pointilliste universal antibody microarray are the FACS systems offered by Luminex (http://www.luminex.com) and BD Biosciences. These instruments can quantify proteins present in a sample that have been captured by antibodies linked to dye-encoded microbeads. Kits for attaching antibodies to microbeads are currently being sold by several companies including BioRad Laboratories (http://www.biorad.com), Linco Research (http://www.lincoresearch.com), and Upstate Group (http://www.upstate.com). BD Biosciences sells sets of beads preloaded with antibodies, and kits of beads to attach antibodies of interest, as well as an instrument to read the bead assays.

It has been suggested that if a large enough density of representative antibody fragments ("perhaps 100,000 to 1,000,000 spots") could be formatted into a microarray, the concept of specificity would become irrelevant. The entire array would then be analyzed as a diverse heterogeneous binding surface to analyze differential proteome-wide changes in proteins. In this way, "patterns-of-interaction" become the predominant readout, rather than quantitative signals from any one particular spot. In a step toward enabling this approach, microarray displays of *in vitro* synthesized antibody libraries have been screened for binding to specific antigens (de Wildt et al., 2000). A significant advantage of this format is the demonstration that immobilization of bacterial cells expressing recombinant antibodies rather than purified proteins can be used for screening binders.

Large diversity arrays of recombinant antibody fragments have also been suggested (Holt et al., 2000; Wingren et al., 2003), but the difficulty of this approach is in economically creating an array with such a large diversity of spots. Pointilliste is attempting to develop a unique approach to solving this problem through the use of its antibody-epitope tagging system. Pointilliste can genetically distribute defined epitopes across a recombinant antibody library and then sort the resulting recombinant proteins onto sets of different arrays using its universal antibody microarray (Figure 1). Multiple arrays can be combined to create ultimately highly diverse collection of arrays that, taken together, display more than a million different antibody fragments. The

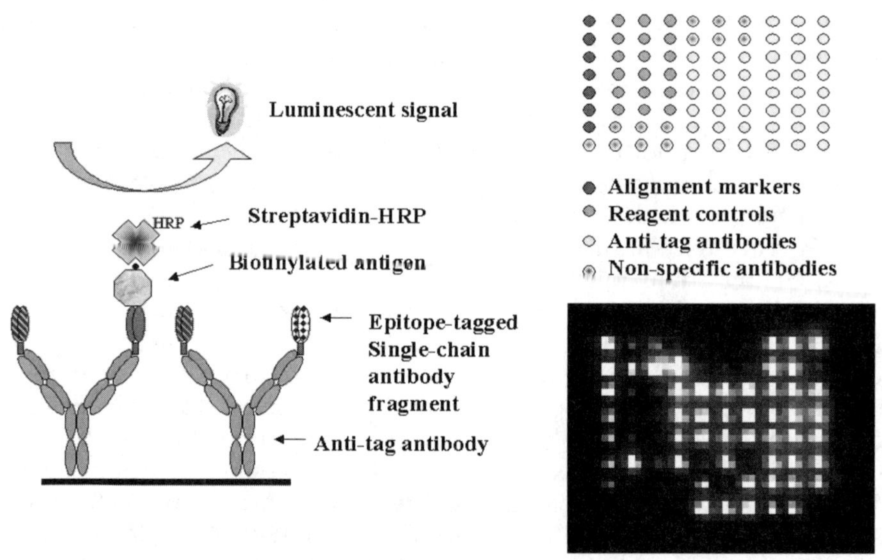

Figure 1 Recombinant histidine-tagged single-chain antibodies (ScFv) binding human fibronectin were expressed with various epitope tags and sorted on an array of antitag antibodies specific for a particular tag sequence. The displayed ScFvs captured biotinylated human fibronectin, which was then detected with streptavidin-conjugated horseradish peroxidase using luminol as the substrate. The array had mouse antibody conjugated to horseradish peroxidase for grid alignment during signal quantification. Reagent controls printed on the array included a biotinylated mouse antibody, antihuman fibronectin antibody, and antihexahistidine antibody. Nonspecific mouse antibodies served as negative controls.

key question of how many antibodies need to be displayed to generate unique "pattern-of-interaction" fingerprints, however, still remains to be answered.

Antibody microarrays have also been applied to the analysis of serum samples. Profiling of serum to identify and characterize all proteins is complicated by the fact that concentrations of serum proteins can vary by greater than 10 orders of magnitude (Patterson and Aebersold, 2003). There is also a necessity to establish standard protocols for serum collection and handling to enable data comparison from different studies (Brazma, 2001).

Although cell lysates have been analyzed using antibody arrays for determination of the presence of specific antigens (Knezevic et al., 2001; Sreekumar et al., 2001), their analysis presents a formidable challenge. Concentrations of various proteins in cells vary as much as 10^{14}, requiring systems that can simultaneously detect both low- and high-abundance proteins in a single array. In order to detect accurately the entire range of protein levels, reduction of sample complexity using liquid-phase protein separation systems such as preparative isoelectric focusing has been suggested (Gurpide-Madoz et al., 2001). Another strategy would be to separate cell organelles and analyze the proteome in each compartment separately. However, these approaches limit the number of samples that can be processed simultaneously.

Proteins in cell lysates can also be labeled with fluorescent dyes or other haptens, using reagents such as *N*-hydroxysuccinimidyl (NHS)-conjugated Cy dyes (Haab et al., 2000) and NHS-biotin (Espejo et al., 2002) prior to capture on antibody arrays. Inclusion of internal calibration proteins for assessment of labeling efficiency and sample proteolysis during processing is crucial for these experiments. Accurate quantification can be achieved by inclusion of sample-specific and reagent-specific controls to account for inter-array and inter-experiment variations. Samples can also be labeled with fluorescent tags followed by direct detection after capture; other haptens can also be used as label, followed by detection with an antihapten antibody conjugated to a fluorescent dye or to an enzyme such as horseradish peroxidase (Haab et al., 2001; Knezevic et al., 2001; Sreekumar et al., 2001).

Transcript profiles using DNA arrays have been used for identification of differences between normal and disease tissues to reveal patterns of expression characteristic of particular disease states (Alizadeh et al., 2001; Khan et al., 2001). However, differences in transcript profiles did not translate into corresponding changes in protein expression patterns. Identification of biomarkers that correlate with disease is a rapidly growing part of the new proteomics landscape. New biomarkers have the promise of allowing much greater predictive capabilities in disease diagnosis and prognosis, drug toxicity, and pharmacogenomics. Serum-based proteomic profiling has identified patterns of protein changes in ovarian cancer, which would never have been identified if individual proteins were measured (Petricoin et al., 2002). Discovery of new protein targets identified directly from diseased human tissue is much more valuable than targets identified from tissue-culture cells and model systems, as these may not represent the cellular microenvironment found in diseased human tissue.

Once disease-specific proteins have been identified through screening for global protein changes, arrays containing the "diagnostic" subset of disease-specific proteins can then be used for high-throughput, large-sample applications. These arrays can be manufactured in a low-density format such that replicate arrays can be assayed in parallel on the same slide, using multiple patient samples, thereby increasing assay throughput and simplifying data analysis.

Distinctive patterns of cell binding were obtained for normal peripheral blood leukocytes and various types of leukemic cells after application of cell suspensions on a cluster of differentiation (CD) array with anti-CD antibodies (Belov et al., 2001). Proteomic studies have also been used to determine toxicity profiles of various drugs by analysis of liver proteins (Steiner et al., 1996) and serum proteins (Eberini et al., 1999).

Antigen Arrays

Antigen arrays have been shown to be four- to eight-fold greater in sensitivity than conventional ELISAs with a linear 3-log range down to nanogram per milliliter limits of detection (Robinson et al., 2002). In addition to increased sensitivity, antigen arrays enable the same sample to be tested against a variety of antigens. Several studies have analyzed serum to determine the presence of antibodies to clinically relevant antigens, leading to disease identification (Bacarese-Hamilton et al., 2002; Mezzasoma et al., 2002; Robinson et al., 2002, 2003). Arrays of 18 different common autoantigen diagnostic markers for autoimmune diseases were used to measure autoantibody titers in patient sera, demonstrating the potential for microarrays in routine diagnostics and screening applications (Joos et al., 2000). Autoantibody responses were also measured from autoimmune patient sera against 196 arrayed antigens, providing a tool for studying specificity of autoantibody responses and for defining relevant autoantigens in human diseases (Robinson et al., 2002). A similar format was used to measure antibodies against 5 pathogen-derived antigens in 60 human sera with greater than 80% concordance with conventional ELISA (Mezzasoma et al., 2002).

An example of the use of protein antigen arrays as an assay to guide drug development was demonstrated by a group from Stanford University School of Medicine who took advantage of the multiplexing feature to look at epitope spreading in an animal model of multiple sclerosis (MS) (Robinson et al., 2003). In MS, B cells develop autoreactive antibody responses to myelin proteins. The Stanford group used an array of 232 distinct antigens, covering multiple epitopes on several myelin-associated proteins, to track the diversity of the B cell response as a function of disease progression. Using *myelin proteome arrays* the group was able to identify sets of overlapping, yet distinct, epitopes associated with disease relapse and could demonstrate a correlation between reduced epitope spreading and improved disease outcomes. The group then went on to use the information gain from the arrays to guide the development of a cocktail of tolerizing DNA vaccines, which they hoped would limit epitope spreading and improve the disease. Early results seem encouraging (Robinson et al., 2003).

Protein Arrays

Arrays have the ability to perform rapidly functional screens using proteins generated from high-throughput expression and purification, as small amounts of material is sufficient for printing a large number of arrays. Binding interactions using an array of the entire yeast proteome was used to identify proteins with calmodulin and phospholipid-binding motifs (Zhu et al., 2001). This is the first report of cloning, expression, and purification of nearly all proteins from a cell, followed by arraying of these proteins to perform functional assays. After cloning 5800 yeast open reading frames in a vector that contained two tags—glutathione-*S*-transferase and hexa-histidine—the proteins were purified in a single affinity purification step and arrayed on nickel-coated glass slides. The proteins were then probed for binding to various phospholipids, resulting in the identification of more than 150 proteins with previously unreported phospholipid binding activity.

Functional properties of kinases (Zhu et al., 2000; Houseman et al., 2002) and proteases (Gosalia and Diamond, 2003) have successfully been measured using arrays. The former uses immobilized kinases that are then incubated with peptide substrates as samples, whereas the latter are arrays of fluorogenic substrates used for profiling protease specificity.

Arrays of all yeast protein kinases were used to characterize substrate specificities using 17 different substrates, identifying novel aspects of protein kinase selectivity (Zhu et al., 2000). Protein kinase arrays have been used for the identification of small-molecule binders and substrates (MacBeath and Schreiber, 2000). This assay demonstrates the ability to speed-up identification of protein kinase inhibitors in the human genome, where ~500 distinct kinases have been identified (Mitchell, 2002). Enzyme arrays have also been used to screen for inhibitors of activity (MacBeath et al., 1999), demonstrating the use of this technology to perform high-throughput screening of multiple enzymes to identify small-molecule inhibitors. Other functional measurements such as protein-protein, protein-DNA, and protein-lipid interactions have also been described (Ge, 2000; Zhu et al., 2001; Espejo et al., 2002; Newman and Keating, 2003).

Microarrays can be used to screen the specificity of a drug to its protein target while simultaneously analyzing its binding to other structurally and functionally related cellular proteins. Protein targets can be arrayed, and then labeled small-molecule compounds are applied to the arrays to screen for binders.

G-protein coupled receptors (GPCRs) make up the largest class of receptors in the human genome and are also the largest class of therapeutic targets. The finding that GPCRs immobilized using carbohydrate-specific chemistry remained stable through several activation/deactivation cycles upon ligand binding enables the use of microarray-based GPCR screening assays (Bieri et al., 1999). A β2 adrenergic receptor immobilized on arrays was used for monitoring of activation with an agonist by measuring changes in fluorescence intensity (Neumann et al., 2002). Competitive binding assays were also performed on GPCR arrays using mixtures of fluorescently labeled ligands and unlabeled inhibitors to demonstrate dose dependency and selectivity of binding to compounds by different receptor subtypes (Fang et al., 2002).

Other Types of Arrays

Peptide arrays have been used for identification of peptides that act as antagonists to proteins that are likely drug targets (Reineke et al., 1996). Specificities of proteases were identified using peptidyl coumarin substrate arrays (Salisbury et al., 2002). These substrates were created such that the highly fluorescent coumarin leaving group was linked via the C-terminus to the solid support, enabling the synthesis of peptide libraries at the N-terminal end via solid phase synthesis. Positional scanning of these peptide libraries can help in the rapid identification of substrate specificities. Synthetic peptides have also been arrayed either by directly depositing on the surface (Falsey et al., 2001; Houseman et al., 2002) or by *in situ* synthesis (Kramer and Schneider-Mergener, 1998). These peptide arrays have been used for mapping epitopes (Reineke et al., 1999), measuring enzyme activity (Houseman et al., 2002), and for capturing whole cells (Falsey et al., 2001).

Small-molecule ligands for proteins can be used as tools for understanding protein function as well as for the development of novel pharmaceuticals. The introduction of split-pool strategy for synthesis (Furka et al., 1991) and the appropriate tagging technology (Lam et al., 1991) have helped in synthesizing large compound collections that are a rich source for the discovery of protein ligands. The feasibility of screening small-molecule ligands for protein binders was demonstrated by immobilizing three well-characterized organic molecules on modified glass slides, followed by measurement of binding to fluorescently labeled proteins (MacBeath et al., 1999). Winssinger et al. (2002) tethered a library of small molecules to peptide-nucleic acid tags, providing an address for each small molecule while using the nucleic acid for immobilization on chips. Fluorescently labeled proteins have been used for screening small-molecule libraries directly printed on surfaces to identify new ligands for these proteins (MacBeath et al., 1999; Kuruvilla et al., 2002).

Proteins in the form of covalent mRNA-protein fusions were displayed on single-stranded DNA arrays through nucleic acid hybridization of the attached mRNA (Weng et al., 2002). This format can orient proteins in a manner that would allow efficient interaction with analytes in the sample. Similarly, alternative sets of capture agents such as nucleic acid aptamers have also been proposed (Petach and Gold, 2002).

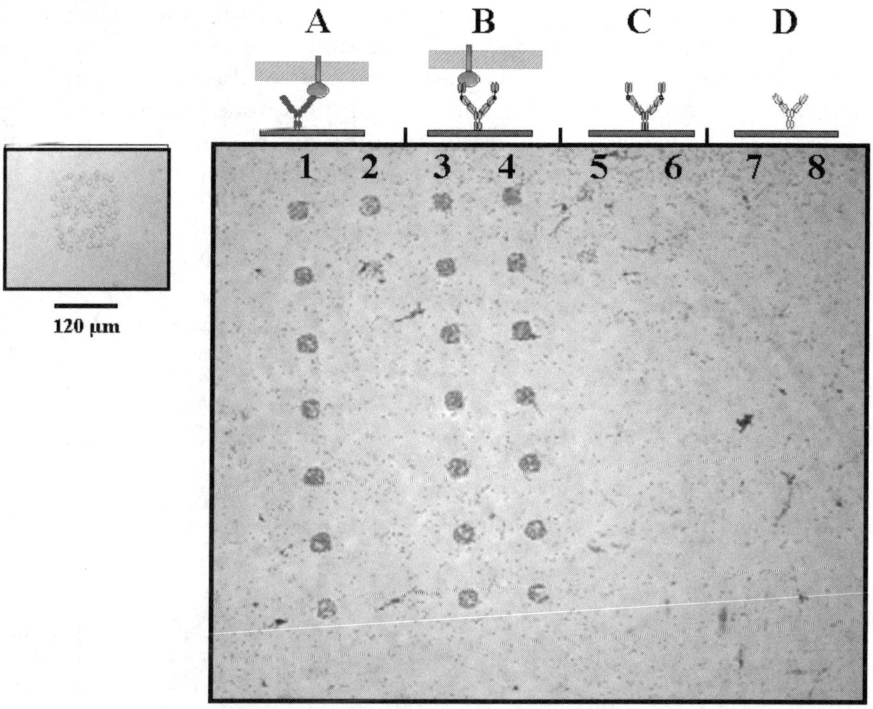

Figure 2 Capture of B cells on an array. An array of antibodies was printed to demonstrate capture of specific cell type for enabling array-based functional assays. Columns 1 and 2 were printed with a B-cell-specific antibody that recognizes a surface IgM; columns 3 and 4 were spotted with an anti-epitope tag antibody that captures a B-cell-specific single-chain antibody expressed with the epitope tag; columns 5 and 6 had a different anti-epitope tag antibody printed that was used to display a nonspecific single-chain antibody; columns 7 and 8 were printed with an antibody specific for the T-cell receptor. A, B, C, and D represent the assay formats for columns 1 and 2, 3 and 4, 5 and 6, and 7 and 8, respectively. Each spot captured 40 to 50 B cells (inset).

Binding specificities of lectins were profiled on an array of monosaccharides and the interactions shown to be specific using competition experiments and surface plasmon resonance (Houseman and Mrksich, 2002). The specificity of anticarbohydrate antibodies in serum samples were identified on arrayed carbohydrates, demonstrating their use in serum profiling (Wang et al., 2002).

Future Directions

In perhaps a sign of things to come, Pointilliste is using protein microarrays to display and test the effects of mixtures of potential drug molecules on living cells. In this application of protein microarray technology, antibodies that bind to defined cell surface receptors are included in each spot in the microarray. These antibodies function to capture and "anchor" living cells to the spots in the microarray and prolong contact between the cells and other molecules being displayed in the spot (Figure 2). The effect on cells of the interaction with displayed molecules in the microarray spot is determined by fixing the cells to the surface and then staining with specific secondary stains. As an example, Pointilliste captured malignant human T cells

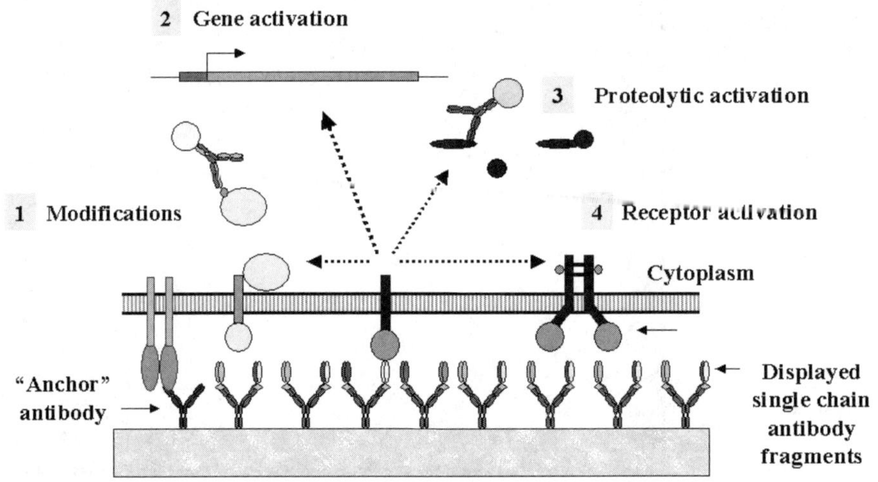

Figure 3 Drug libraries displayed on arrays can be used for cell-based assays. Cells are captured on arrays followed by measuring changes in cell physiology using various readouts. (1) Fixed cells can be stained with antibodies specific for post-translational modifications such as phosphorylation of signaling proteins. (2) Cells transfected with reporter genes can be used to identify changes in reporter gene activity such as that of green fluorescent protein. (3) Proteolytic processing of cellular proteins as a result of modification by arrayed drug can be identified using antibodies specific for modified proteins such as caspase modification on apoptosis. (4) Cell activation can be measured by staining for cytokines.

using an antibody to the T cell receptor and monitored the induction of apoptosis by staining with an antibody that specifically recognizes the activated form of caspase 3 as an indication of apoptosis in the cells. By monitoring the functional consequence of the cell's interaction with the microarray spot, the microarray can be used to discover functional receptor agonists without prior knowledge of the agonist's target receptor. This use of the protein microarray is significant in its transformation of the microarray from a purely analytical tool to measure the levels of analytes in a sample to an active screening tool (Figure 3). It will be fascinating to watch how many more transformations this exciting new research tool will undergo as it is more widely adopted in the research community.

References

1. Afanassiev V, Hanemann V, and Wolfl S. Preparation of DNA and Protein Micro Arrays on Glass Slides Coated with an Agarose Film. *Nucleic Acids Res* 2000; 28:E66
2. Agbaria RA, Oldham PB, McCarroll ME, McGown LB, and Warner IM. Molecular Fluorescence, Phosphorescence, and Chemiluminescence Spectrometry. *Anal Chem* 2002; 74:3952–3962
3. Alizadeh AA, Ross DT, Perou CM, and van de Rijn M. Towards a Novel Classification of Human Malignancies Based on Gene Expression Patterns. *J Pathol* 2001; 195:41–52
4. Arenkov PA, Kukhtin A, Gemmell S, Voloshchuk V, Chupeeva V, and Mirzabekov AD. Protein Microchips: Use for Immunoassay and Enzymatic Reactions. *Anal Biochem* 2000; 278:123–131
5. Bacarese-Hamilton T, Mezzasoma L, Ingham C, Ardizzoni A, Rossi R, Bistoni F, and Crisanti A. Detection of Allergen-Specific IgE on Microarrays by Use of Signal Amplification Techniques. *Clin Chem* 2002; 48:1367–1370
6. Barsky VE, Kolchinsky AM, Lysov YP, and Mirzabekov AD. Biological Microchips with Hydrogel-Immobilized Nucleic Acids, Proteins, and Other Compounds: Properties and Applications in Genomics *Mol Biol (Mosk)* 2002; 36:437–455
7. Belov L, de la Vega O, dos Remedios CG, Mulligan SP, and Christopherson RI. Immunophenotyping of Leukemias Using a Cluster of Differentiation Antibody Microarray. *Cancer Res* 2001; 61:4483–4489
8. Bieri C, Ernst OP, Heyse S, Hofmann KP, and Vogel H. Micropatterned immobilization of a G Protein-Coupled Receptor and Direct Detection of G Protein Activation. *Nat Biotechnol* 1999; 17:1105–1108
9. Bilban M, Buehler LK, Head S, Desoye G, and Quaranta V. Normalizing DNA Microarray Data. *Curr Issues Mol Biol* 2002; 4:57–64

10. Brash JL, and Horbett TA. *Proteins at Interfaces:* ACS Symposium Series. Washington DC: American Chemical Society; 1987
11. Brazma A. On the Importance of Standardization in Life Sciences. Bioinformatics 2001; 17:113–114
12. Brazma A, Hingamp P, Quackenbush J, Sherlock G, Spellman P, Stoeckert C, Aach J, Ansorge W, Ball CA, Causton HC, et al. Minimum Information About a Microarray Experiment (MIAME)—Toward Standards for Microarray Data. Nat Genet 2001; 29:365–371
13. Brown MPS, Grundy WN, Lin D, Cristianini N, Sugnet CW, Furey TS, Manuel Ares M, and Haussler D. Knowledge-Based Analysis of Microarray Gene Expression Data by Using Support Vector Machines. *Proc Natl Acad Sci U S A* 2000; 95:14863–14868
14. Cahill DJ. Protein and Antibody Arrays and Their Medical Applications *J Immunol Methods* 2001; 250:81–91
15. Carson RT, and Vignali DAA. Simultaneous Quantitation of 15 Cytokines Using a Multiplexed Flow Cytometric Assay. *J Immunol Methods* 1999; 227:41–52
16. Delahanty JB, and Ligler FS. Method for Printing Functional Protein Microarrays. *Biotechniques* 2003; 34:380–385
17. de Wildt RM, Mundy CR, Gorick BD, and Tomlinson IM. Antibody Arrays for High-Throughput Screening of Antibody-Antigen Interactions. *Nat Biotechnol* 2000; 18:989–994
18. Earley MC, Vogt RF, Shapiro HM, Mandy FF, Kellar KL, Bellisario R, Pass KA, Marti GE, Stewart CC, and Hannon WH. Report from a Workshop on Multianalyte Microsphere Assays. *Cytometry* 2002; 50:239–242
19. Eberini I, Miller I, Zancan V, Bolego C, Puglisi L, Gemeiner M, and Gianazza E. Proteins of Rat Serum IV. Time-Course of Acute-Phase Protein Expression and Its Modulation by Indomethacine. *Electrophoresis* 1999; 20:846–853
20. Ekins RP. Multianalyte Immunoassay. *J Pharm Biomed Anal* 1989; 7:155–168
21. Ellson RN. Picoliter: Enabling Precise Transfer of Nanoliter and Picoliter Volumes *Drug Discov Today* 2002; 7:S32-S34
22. Espejo A, Cote J, Bednarek A, Richard S, and Bedford MT. A Protein-Domain Microarray Identifies Novel Protein-Protein Interactions. *Biochem J* 2002; 367:697–702
23. Falsey JR, Renil M, Park S, Li S, and Lam KS. Peptide and Small Molecule Microarray for High Throughput Cell Adhesion and Functional Assays. *Bioconjugate Chem* 2001; 12:346–353
24. Fang Y, Frutos AG, and Lahiri J. G-Protein-Coupled Receptor Microarrays. *ChemBioChem* 2002; 3:987–991

for the temperature of interest according to

$$\Delta H(T) = (6.7 \pm 1.5) + (0.12 \pm 0.03) \times (T_m - 60) \quad (11)$$

It should be noted, however, that the error increases the further T_m is from the reference temperature of 60°C. In addition, the degree of folding can vary for some proteins and can deviate markedly from that of a standard protein as in the case of gp120, described in the section "Example: Determination of Active Fraction of gp120," below.

Determination of Concentration of Binding Sites by ITC

ITC is very useful for the determination of the active protein concentration as long as the protein has a well-defined site for a ligand, preferably a small-molecular-weight molecule. Such an active site titration requires the ligand concentration to be known with high accuracy. The concentration and purity of a stock ligand solution can accurately be determined by an appropriate analytical method. Once the concentration of the ligand is determined, the binding to the protein from a batch of unknown activity is studied by ITC for determination of the stoichiometry. In general, the apparent stoichiometry, n, is equal to the fraction of active protein, F_a, present in the sample:

$$n = F_a = \frac{[P]_{\text{active}}}{[P]_{\text{tot}}} \quad (12)$$

where $[P]_{\text{active}}$ and $[P]_{\text{tot}}$ are the active and total concentrations of protein in the calorimetric cell, respectively. The resulting apparent stoichiometry is consequently equal to one if the protein presents 100% activity.

As long as the concentration of the ligand in the syringe is known with high accuracy, the enthalpy and affinity are always exactly determined, and the uncertainty in protein concentration due to the presence of an inactive protein fraction will only be reflected in the value for the apparent stoichiometry. Because the enthalpy value obtained from the analysis depends only on the concentration of the titrating reactant in the syringe, it can be used as an additional independent control parameter for the reactant in the syringe in the active protein concentration determination assay. In that way, even if the concentration of reactant in the syringe is unknown, if the expected binding enthalpy is known, it is possible to use two independent adjustable parameters, stoichiometry and binding enthalpy, during the analysis of the ITC experiment to determine simultaneously both the concentration of reactant in the syringe (from the estimated enthalpy) and the concentration of reactant in the cell (from the estimated stoichiometry).

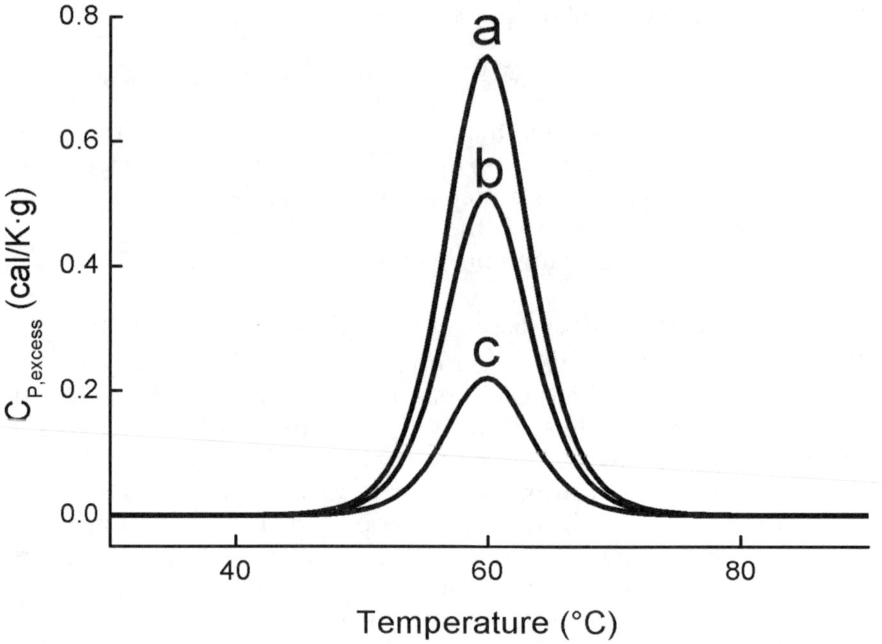

Figure 5 Simulated temperature scans of a protein from a batch of (a) 100%, (b) 70%, and (c) 30% activity. The denaturation enthalpy for the protein of 100% activity was 6.5 cal/g, which serves as the standard enthalpy, ΔH^o, in Figure 6.

Example: Determination of Active Fraction of gp120

The presence of inactive protein fractions poses a serious problem in studies where both the reactants are proteins. In the case of proteins with measurable enzymatic activity, the activity can usually be determined by spectrophotometric or fluorometric assays. However, in the absence of enzymatic activity, the active fraction of the reactants must be determined by other means (e.g., as described in this section).

The glycoprotein gp120, found on the surface of the viral envelope of the HIV-1 particle, binds to the soluble form of the cell receptor CD4 (sCD4) with an affinity of 5 nM and an enthalpy of –35 kcal/mol (Myszka et al., 2000) at 25°C. Before any meaningful information can be obtained from a titration of gp120 with sCD4, the active protein fraction must be determined for at least one of the proteins. DSC, as suggested in the section "DSC and the Determination of Active Protein Fraction" above, can obtain a measure of the degree of activity. Earlier studies have shown that the thermal denaturation of fully active gp120 is characterized by an enthalpy of 2.3 cal/(g × K) (Leavitt, 2003), which serves as our reference enthalpy according to Equation 10. The ratio between the enthalpy from the unknown and standardized protein batch according to Equation 10 (see also Figures 5 and 6) is used to adjust the concentration of the gp120 prior to any ITC

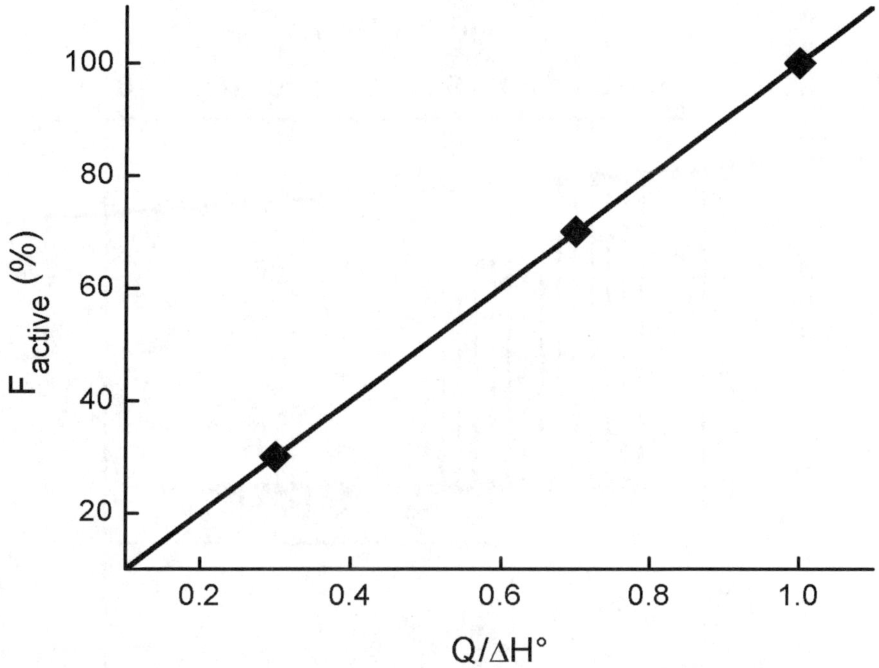

Figure 6 Fraction active protein as a function of the ratio between the experimental enthalpy Q and the standardized enthalpy ΔH^o according to Equation 10. The plot was based on the areas derived from the simulated curves in Figure 5. The value for ΔH° was 6.5 cal/g.

experiment with sCD4. Any deviation in the apparent stoichiometry obtained in subsequent ITC experiment can thus be attributed to an uncertainty in the concentration of sCD4.

Example: Determination of Active Fraction of HIV-1 Protease

HIV-1 protease is an aspartic protease crucial in the maturation steps in the HIV-1 virus life cycle. Biochemical studies on this enzyme require the assessment of the concentration of active protein. As mentioned above, in the case of proteins with measurable enzymatic activity, the activity (determined by spectrophotometric or fluorometric assays with appropriate substrates) can be used to monitor active fraction concentration. However, ITC provides an excellent methodology for performing active site titrations using both moderate and high-affinity ligands ($K_a \sim 10^6$ to 10^{10} M^{-1}).

Calorimetric studies with acetyl-pepstatin, a universal aspartic-protease inhibitor, have shown that it binds with moderate affinity and endothermically to HIV-1 protease ($K_a = 2.3 \times 10^6$ M^{-1} and $\Delta H =$ 7.7 kcal/mol, in acetate buffer 10 mM, pH 5, DMSO 2%, 25°C) (Todd and Freire, 1999; Velazquez-Campoy et al., 2001). Figure 7 shows the result from an ITC experiment in which the inhibitor is injected into a

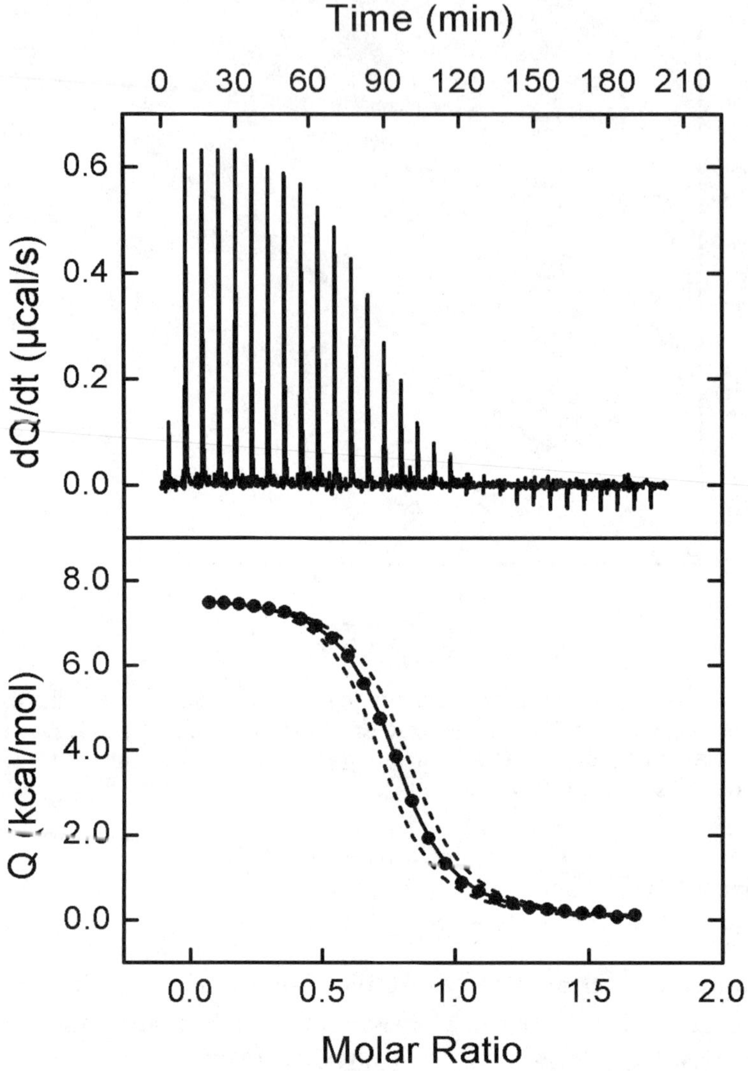

Figure 7 Results from a titration experiment in which 230 μM acetyl-pepstatin is injected into a solution of 29 μM HIV-1 protease. The regression analysis provides the following results: $K_a = 2.3 \pm 0.2 \times 10^6\ M^{-1}$, $\Delta H = 7.7 \pm 0.1$ kcal/mol, and $n = 0.76 \pm 0.01$. The apparent stoichiometry of 0.76 indicates that only 76% of the total protein concentration, as determined spectrophotometrically, is active. If the analysis is repeated using 22 μM (corrected for the inactive fraction) as the concentration of protein, the analysis yields a value of 1 for the stoichiometry (100% active). The two dashed lines represent the titration plot with an n value of 0.76 ± 0.05, in order to observe the precision obtained in this parameter. The value obtained for the binding enthalpy corresponds to the expected one, indicating that the concentration of acetyl-pepstatin was well determined.

solution of HIV-1 protease. An apparent stoichiometry of less than one obtained from the regression analyses using the known inhibitor concentration and the nominal protein concentration determined from its extinction coefficient at 280 nm reflects the presence of impurities or inactive protein.

It is interesting to point out that for an active site titration to be accurate, it is not necessary to use highly potent ligands. That is not the case in traditional enzymatic active site titrations where (sub)nanomolar inhibitors are employed in order to minimize the error in determining, usually by graphic extrapolation, the concentration of active fraction. Besides, as mentioned before, the enthalpy value obtained through the analysis of the experiments can be used to check the concentration of titrant. Any deviation from the expected value will reflect an error in the determination of such concentration.

Abbreviations

DSC	Differential scanning calorimetry
DMSO	Dimethyl sulfoxide
gp120	Glycoprotein 120
HIV-1	Human immunodeficiency virus type 1
ITC	Isothermal titration calorimetry
NMR	Nuclear magnetic resonance spectroscopy
sCD4	Soluble form of the cell surface receptor, CD4

Acknowledgments

Arne Schön and Adrian Velazquez-Campoy as well as the research work presented in this paper were supported by grants GM 57144 and GM 56550 from the National Institutes of Health and grant MCB-0131241 from the National Science Foundation to Ernesto Freire.

References

1. Boudker O, Todd MJ, and Freire E. The Structural Stability of the Co-chaperonin GroES. *J Mol Biol* 1997; 272:770–779
2. Freire E. Differential Scanning Calorimetry. *Methods Mol Biol* 1995a; 40:191–218
3. Freire E. Thermodynamics of Partly Folded Intermediates in Proteins. *Annu Rev Biophys Biomol Struct* 1995b; 24:141–165
4. Freire E, and Murphy KP. Molecular Basis of Co-operativity in Protein Folding. *J Mol Biol* 1991; 222:687–698

5. Freire E, Mayorga OL, and Straume M. Isothermal Titration Calorimetry. *Anal Chem* 1990a; 62:950A–959A
6. Freire E, van Osdol WW, Mayorga OL, and Sanchez-Ruiz JM. Calorimetrically Determined Dynamics of Complex Unfolding Transitions in Proteins. *Annu Rev Biophys Biophys Chem* 1990b; 19:159–188
7. Johnson CR, Morin PE, Arrowsmith CH, and Freire E. Thermodynamic Analysis of the Structural Stability of the Tetrameric Oligomerization Domain of p53 Tumor Suppressor. *Biochemistry* 1995; 34:5309–5316
8. Kasimova MR, Milstein SJ, and Freire E. The Conformational Equilibrium of Human Growth Hormone. *J Mol Biol* 1998; 277:409–418
9. Leavitt S, and Freire E. Direct Measurement of Protein Binding Energetics by Isothermal Titration Calorimetry. *Curr Opin Struct Biol* 2001; 11:560–566
10. Leavitt SA, Schön A, Klein JC, Manjappara U, Chaiken IM, and Freire E. Interactions of HIV-1 Proteins gp120 and Nef with Cellular Partners Define a Novel Allosteric Paradigm. *Curr Protein Peptide Sci* 2004; 5:1–8
11. Manning MC, Patel K, and Borchardt RT. Stability of Protein Pharmaceuticals. *Pharm Res* 1989; 6:903–918
12. Myszka DG, Sweet RW, Hensley P, Brigham-Burke M, Kwong PD, Hendrickson WA, Wyatt R, Sodroski J, and Doyle ML. Energetics of the HIV gp120-CD4 Binding Reaction. *Proc Natl Acad Sci U S A* 2000; 97:9026–9031
13. Privalov PL, and Potekhin SA. Scanning Microcalorimetry in Studying Temperature-Induced Changes in Proteins. *Methods Enzymol* 1986; 131:4–51
14. Remmele RL Jr, and Gombotz WR. Differential Scanning Calorimetry: A Practical Tool for Elucidating Stability of Liquid Pharmaceuticals. *Biopharm Eur* 2000; 56:58–60
15. Remmele RL Jr, Nightlinger NS, Srinivasan S, and Gombotz WR. Interleukin-1 Receptor (Il-1R) Liquid Formulation Development Using Differential Scanning Calorimetry. *Pharm Res* 1998; 15:200–208
16. Robertson AD, and Murphy KP. Protein Structure and the Energetics of Protein Stability. *Chem Rev* 1997; 97:1251–1268
17. Todd MJ, and Freire E. The Effect of Inhibitor Binding on the Structural Stability and Cooperativity of the HIV-1 Protease. *Proteins* 1999; 36:147–156
18. Todd MJ, Semo N, and Freire E. The Structural Stability of the HIV-1 Protease. *J Mol Biol* 1998; 283:475–488
19. Velazquez-Campoy A, Kiso Y, and Freire E. The Binding Energetics of First- and Second-Generation HIV-1 Protease Inhibitors: Implications for Drug Design. *Arch Biochem Biophys* 2001; 390:169–175

20. Velazquez-Campoy A, Leavitt SA, and Freire E. Characterization of Protein-Protein Interactions by Isothermal Titration Calorimetry. In: Fu H. *Protein-Protein Interactions. Methods and Applications.* Totowa, NJ: Humana Press; 2004:35–54
21. Wiseman T, Williston S, Brandts JF, and Lin LN. Rapid Measurement of Binding Constants and Heats of Binding Using a New Titration Calorimeter. *Anal Biochem* 1989; 179:131–137

18

Atomic Force Microscopy

ANDREAS ENGEL

Maurice E. Müller Institute for Microscopy, Biozentrum, University of Basel, Basel, Switzerland

Introduction

Membrane proteins compose more than 30% of the proteome of higher organisms. With characteristic dimensions of 5–10 nm, they are membrane-embedded nanomachines that fulfill key functions such as energy conversion, solute transport, secretion, and signal transduction. Their central role in a wide range of diseases and in cell-cell communication may explain the fact that 70% of all drug targets are membrane proteins. Although the structure of more than 10,000 soluble proteins has been solved, the number of membrane protein structures solved is smaller than 100. This lack of structural information is related to the instability of membrane proteins in a detergent-solubilized state, making the growth of three-dimensional (3-D) crystals difficult. Because a large majority of therapeutics acts on membrane proteins, knowledge of their structure is urgently required. Two-dimensional (2-D) crystals of purified membrane proteins reconstituted in the presence of lipids provides a close to native environment and allows the structure and function of membrane proteins to be assessed.

The surface structure of such membrane protein arrays can be determined in buffer solutions by atomic force microscopy (AFM) (Binnig et al., 1986) at a lateral resolution of 0.4–1 nm and a vertical resolution of 0.1–0.2 nm (Czajkowsky et al., 1998; Engel and Müller, 2000). In addition, single proteins can be addressed directly with the tip and unfolded, revealing the forces that dictate the protein's fold (Oesterhelt et al., 2000). The unique capability of the atomic force microscope to observe and manipulate single proteins in a physiological environment allows their function and dynamics to be directly monitored. Thus, information about the surface structure of membrane proteins can be obtained by AFM from tightly packed but disordered membranes. In contrast, electron microscopy relies on highly ordered

2-D crystals, from which the 3-D structure of a native membrane protein can be elucidated to a resolution of 0.3 nm by crystallographic averaging (Fujiyoshi, 1998).

Although direct drug-protein interactions have not been characterized as yet by AFM imaging methods, the potential possibility to achieve this by force mapping has been proposed (Hinterdorfer et al., 1996; Klein et al., 2003; Riener, 2003). Alternatively, the inhibition of conformational changes could possibly be observed by AFM, indicating the drug's ability to bind tightly and specifically to its target. In addition, changes of cellular properties (shape, dimensions, morphology, micromechanics) may be monitored, which would provide indirect evidence for the effect of a drug. Drug-protein interactions are expected to result in additional forces that should be measurable when the target is unfolded. On the other hand, cantilever arrays are now becoming sufficiently sensitive to allow highly specific detection of compounds that bind to a lever-immobilized substrate (Lang et al., 2002; McKendry et al., 2002; Arntz et al. 2003). Although these approaches are all promising, the atomic force microscope is not as yet a standard tool in drug discovery and it is not a high-throughput technique, although there are great opportunities to push this technology further. In this chapter, we describe the crucial technical aspects of the atomic force microscope, methods for preparing membrane samples, and discuss several applications.

Atomic Force Microscope Instrumentation

Imaging Modes

In the atomic force microscope, a topograph is recorded by raster scanning the sample below the stylus, which is attached to the end of a flexible cantilever. The atomically sharp stylus thus follows the contours of the sample, while an optical system tracks the cantilever deflections to a resolution of about 0.1 nm. In contact mode imaging (Figure 1A) the deflection signal is coupled to a feedback loop that displaces the sample or the cantilever vertically to maintain a constant cantilever deflection. To react, the servo needs a difference signal, which is pronounced at sharp edges (Figure 1B). Thus, the error signal provides an image that reveals the edges of the surface topography. With modern instruments and an appropriate cantilever, stable contact mode imaging is possible at forces of about 50 pN, provided that the sample is in an aqueous solution. Alternatively, a cantilever with a magnetic end is oscillated vertically during scanning by activating it with an oscillating magnetic field. The oscillation of the deflection signal is reduced (Figure 1C) and a phase difference between the cantilever oscillation and the magnetic oscillation arises (Figure 1D) when the tip approaches the sample. Both signals provide different kinds of information about the sample properties. The liquid cell shown in Figure 1E is required for scanning samples in buffer solution.

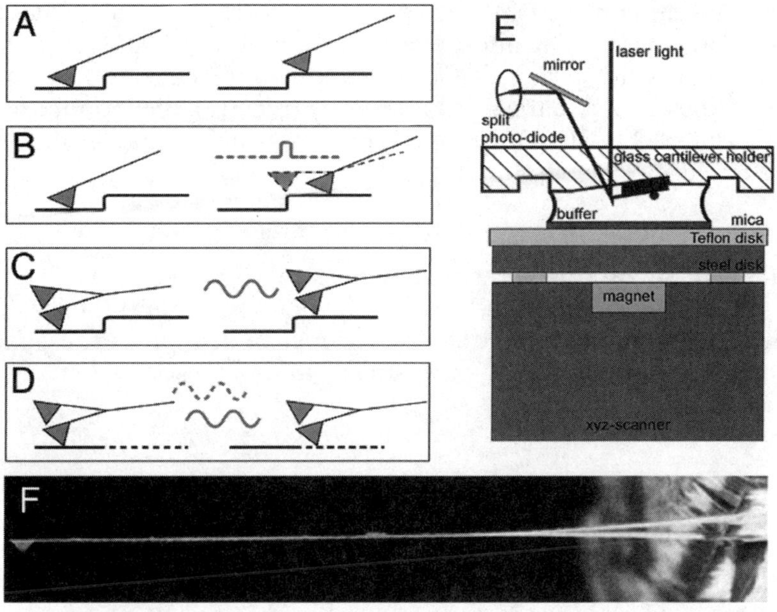

Figure 1 AFM instrumentation and modes of operation. (A) In the contact mode, the servo system moves the tip up over elevations and down over depressions while the sample is raster scanned below the stylus. (B) The servo needs a difference signal to react for keeping the cantilever deflection constant even at sharp edges. This error signal provides an image that reveals the edges of the surface topography. (C) Approaching the sample with an oscillating cantilever results in tapping of the sample by the stylus. This reduces the oscillation amplitude, providing the signal to activate the servo. Because the tip-sample contact is disrupted periodically, the friction forces are eliminated in the tapping mode. (D) The phase difference between the measured oscillation (solid wave) and the oscillation driving the cantilever depends on the mechanical and chemical properties of the sample. This phase signal thus produces a sensitive material contrast. (E) Key elements of an atomic force microscope are the cantilever with a pyramidal stylus that touches the sample, an optical lever consisting of a laser and a photo diode to measure the cantilever's deflection, a piezoelectric translator to displace the sample in *x*, *y*, *z*, and a computer to control these movements and store the surface contours. The instrument is working in buffer solution under ambient condition. Different liquids can be injected, and a Peltier-element allows precise adjustment of the temperature. (F) The cantilever (typical dimensions ~200 x 20 x 0.5 μm, spring constant 0.1–0.01 N/m) is deflected upwards when the tip is pushed toward the sample surface (repulsive forces) or downwards when the tip is retracted from the sample surface (attractive forces). (By courtesy of F. Frederix.)

Force-Distance Curves

With a typical cantilever (see Figure 1F), forces from slightly below 50 pN to 2000 pN can be measured. For biomolecules that are tethered to support and stylus, force-distance spectra can be recorded while increasing the tip-support distance (Rief et al., 2000; Marszalek et al.,

2001; Janovjak et al., 2003). To record such force-distance curves, the tip is approached vertically toward the sample until contact is made and subsequently retracted. Occasionally a molecule grafted to the support does attach to the stylus either by (mostly) hydrophobic or by specific interaction (such as thiol-gold cross-links) and is unfolded while tip-support separation increases.

Cantilevers and Tips

Cantilevers are produced by microfabrication. The geometry (length l, width w, and thickness t and shape), the density (ρ), and the elasticity module (E) of the material dictate their mechanical properties, generally expressed by the spring constant (k_L) and the resonance frequency (f_0) in vacuum:

$$k_L = \frac{1}{4}\frac{E \cdot w \cdot l^3}{l^3} \tag{1}$$

$$f_0 = \frac{ct\sqrt{E/\rho}}{l^2} \tag{2}$$

Here, c is a geometry constant that amounts to $c = 0.164$ for a rectangular cantilever. For imaging of biological samples in contact mode, the spring constant of the cantilever should be taken between 0.01 N/m and 0.1 N/m. The resonance frequency for these cantilevers ranges from 5 kHz to 50 kHz in vacuum and is slightly lower in air. When operating the cantilever in fluid, it is reduced by a factor of 2–5. For example, a conventional rectangular silicon nitride cantilever [E = 129 GPa, $\rho = 3440$ kg/m^3 (Yasumura et al., 2000)] with $l = 100$ μm, w = 40 μm, and $t = 0.4$ μm has a spring constant of $k_L = 0.08$ N/m and a resonance frequency in vacuum of $f_0 = 40$ kHz.

Another parameter characterizing the cantilever is the quality factor Q, which is defined by the stored energy W_0 and the total energy loss per oscillation cycle ΔW (Yasumura et al., 2000):

$$Q = 2\pi\frac{W_0}{\Delta W} \tag{3}$$

In high vacuum, Q is a true cantilever parameter, because the energy losses result from internal damping mechanisms (Yasumura et al., 2000; Yang et al., 2002). However, in air or in fluid, the hydrodynamic damping by the surrounding medium dominates the energy losses (Sader et al., 1999; Yasumura et al., 2000; Yang et al., 2002). Although suitable contact mode cantilevers still have a quality factor of $Q = 10$ to 100 in air, typical values drop down to $Q = 1$ to 5 when they are operated in fluid. The resonance frequency and quality factor can be determined from the noise in the deflection signal (Chon et al., 2000), even when no oscillation mode is available. The thermal noise spectrum of the cantilever in water or air can be obtained by recording the deflection signal without a sample in close proximity and at a sampling rate that is at least twice the resonance frequency of the cantilever.

For most applications, it is important to know the mechanical properties of the cantilever. Unfortunately, the microfabrication process is not accurate enough to warrant a cantilever thickness variation below 20%. As indicated by Equation 1, this introduces large variations of the spring constant, making the routine characterization of cantilevers an imperative exercise. The spring constant can be calculated from the amplitude (in nm) of the thermal deflection noise using the fluctuation method (Hutter and Bechhöfer, 1993). To this end, the deflection signal (usually a voltage) needs first to be calibrated simply by pushing the cantilever upwards by a known amount and measuring the deflection voltage. Alternatively, the spring constant can be derived from the resonance frequency and quality factor of the cantilever, together with its top view dimensions (Sader, 1998). Interestingly, calibration of the deflection signal is not required here. The method assumes that the energy losses are dominated by the hydrodynamic damping, implying that the density and viscosity of the surrounding medium (air or liquid) are known. The given method (Sader, 1998) is derived for rectangular cantilevers but can be extended to different shapes as well if a reference cantilever is available (Sader, 1998; Gibson et al., 2004). Both methods (Hutter and Bechhöfer, 1993; Sader, 1998) give an uncertainty of <20%.

The minimal force ($F_{\min}$) that a cantilever can measure is ultimately limited by the thermal noise of the cantilever. This can be expressed as a function of the spring constant, resonance frequency, and quality factor, or in the case of vacuum also as function of the dimensions of the cantilever and the quality factor:

$$F_{\min} = \sqrt{\frac{4k_B T B k_L}{2\pi f_0 Q}} \tag{4}$$

where k_B is the Boltzmann constant, T the absolute temperature, and B the measurement bandwidth (Yasumura et al., 2000). For a cantilever as mentioned above (k_L = 0.08 N/m and assuming f_0 = 10 kHz and Q = 2 in fluid) that is operated at room temperature with a sampling rate equal to the resonance frequency (i.e., B = 10 kHz), Equation 4 gives a minimal detectable force of about 10 pN. This equation helps to optimize cantilever dimensions to achieve high force sensitivity, within the boundary conditions determined by the spring constant (0.01–0.1 N/m) and the limitations of microfabrication. In addition, a high resonance frequency is desirable to allow high recording speeds for observing real-time dynamics. Both goals (i.e., a better force sensitivity and a higher measurement speed) can be achieved by reducing the cantilever dimensions (Viani et al., 1999a,b; Yang et al., 2004). For example, a rectangular cantilever with $l = 20\ \mu m$, $w = 5\ \mu m$, and $t = 0.16\ \mu m$ has a 10 times higher resonance frequency in vacuum compared to the 100-μm-long, rectangular cantilever mentioned above, while exhibiting the same spring constant (0.08 N/m). Thus, in contact mode the sensitivity is enhanced by a factor of three when operated at the same bandwidth B as a conventional cantilever or the measurement speed can be increased by a factor of 10 without an increase of the thermal noise.

Another main feature of the cantilever is the tip sharpness. Although suppliers specify tip radii not better than 10–50 nm, topographs of flat biological surfaces that exhibit a resolution better than 1 nm have been acquired routinely (Schabert and Engel, 1994; Schabert et al., 1995; Czajkowsky et al., 1998; Müller et al., 1999; Engel and Müller, 2000). Provided that long-range interactions (e.g., electrostatic forces) are compensated for, only the short-range forces determine the tip-sample interaction. Consequently, only the tip atoms nearest to the sample determine the lateral resolution, which is considerably smaller than would be expected based on the tip radius given by the manufacturer. The tips often have a small asperity that protrudes sufficiently to contour the finest surface structures. Such a small asperity can exert a prohibitively high pressure on the underlying structure, inducing its deformation. Therefore, pH and ion strength are adjusted to balance the tip-sample interactions (Müller et al., 1999).

Sample Preparation

Sample preparation methods for raster-scanning samples in buffer solutions all have the objective to attach the specimen firmly to a support. Immobilization is required for the contact-imaging mode, as even small lateral forces exerted by the tip tend to push away the structure to be imaged. Lateral forces are largely eliminated by the dynamic imaging mode, where the tip is oscillated vertically, lifting it away from the sample surface most of the time. For small particles and filaments, the contact area between sample and support is small, hence such structures can be immobilized for contact mode imaging by chemical fixation (Karrasch et al., 1993). However, the interaction area is much larger with membranes, which can simply be adsorbed to freshly cleaved mica, as prepared simply by lifting off the top layer using adhesive tape. Adjusting the electrolyte concentration and the pH of the buffer solution facilitates this: the pH dictates the surface charge of the membrane, while the electrolyte concentration determines the thickness of the Debye layer (Müller et al., 1997a,b). As detailed below for the tip-sample interactions, adsorption results from attractive van der Waals forces that are relatively short-ranged, whereas the decay length of attractive or repulsive electrostatic forces corresponds to the thickness of the Debye layer. Because most biological membranes have a negative surface charge, the repulsive electrostatic force resulting from the negative surface charge of mica prevents adsorption unless the Debye layer is thin compared to the decay length of van der Waals forces. As illustrated by Figure 2, the critical concentration of a KCl solution for adsorbing purple membranes to mica is about 50 mM. Once a membrane is absorbed, the gap between sample and support will decrease until van der Waals attraction and osmotic pressure resulting from the electrolytes in the gap are in equilibrium. Because that gap is small, solutes cannot quickly diffuse away, and it is possible to exchange the adsorption buffer with a buffer optimized for imaging without desorbing the membranes. Interestingly, a Nernst potential is generated in this case that needs to be considered when investigating membrane channels.

Imaging

Derjaguin, Landau, Verwey, Overbeck Forces

To achieve high resolution, topographs are recorded in buffer solution. In this case, mainly electrostatic and van der Waals forces govern the tip-sample interactions. In biological systems measured in aqueous solution, van der Waals interactions do not depend on the ionic strength, decay rapidly, and are always attractive. Electrostatic interactions

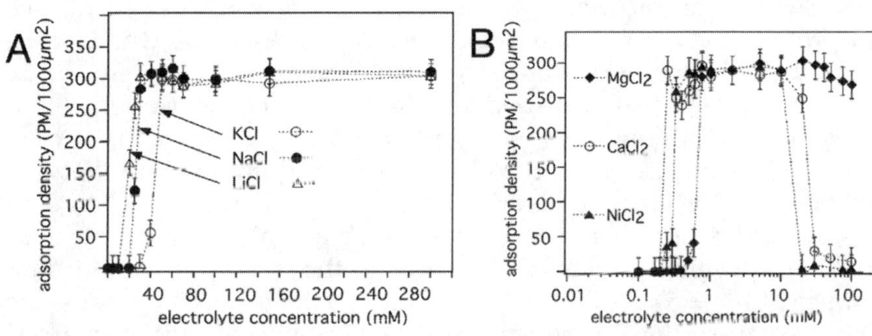

Figure 2 Adsorption of purple membranes to freshly cleaved mica. (A) Purple membrane adsorbed in monovalent electrolytes. With increasing electrolyte concentrations, the maximum adsorption density is first reached with LiCl, followed by NaCl and KCl. (B) Purple membrane adsorbed in divalent electrolytes. With increasing electrolyte concentrations, the maximum adsorption is first reached with $CaCl_2$, followed by $NiCl_2$ and $MgCl_2$. At salt concentrations above 10 mM ($CaCl_2$ and $NiCl_2$) and 50 mM ($MgCl_2$), the purple membranes aggregate and the ratio of single sheets adsorbed to mica decreases. Purple membranes were dissolved in electrolyte (40 µg/mL protein, clean Millipore water) and deposited (50 µL) onto freshly cleaved mica disks (radii 6 mm) at room temperature. After an adsorption time of 5 min, samples were gently rinsed with the appropriate electrolyte and were ready to be inspected with the atomic force microscope. (By courtesy of D.J. Müller.)

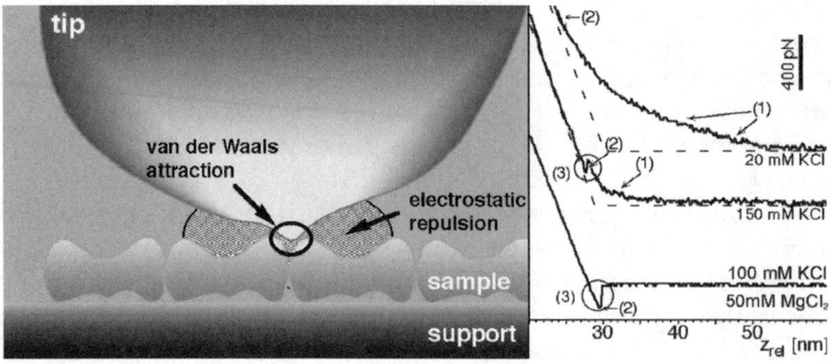

Figure 3 DLVO forces. Force-distance curves recorded on the extracellular purple membrane surface. Data were obtained for different electrolyte concentrations at constant pH (7.6). Force-distance curves were recorded during the approach of sample and AFM tip. The dotted lines (....) represent force-distance curves without electrostatic repulsion. Scan frequency 1.97 Hz; scan range 50 nm (512 pixel). Arrows (1) mark the onset of measurable electrostatic repulsion, whereas arrows (2) indicate the point of contact between tip and sample. (By courtesy of D.J. Müller.)

result from hydrophilic surfaces that are often charged in water. They are long-ranged and can be attractive or repulsive, depending on the sign of the surface charges, which, in turn, depend on the pH. Because the stylus (silicon nitride, Si_3N_4) is negatively charged at neutral pH, and protein layers are often negatively charged as well, the electrostatic forces are in most instances repulsive. The surface charges can be screened with electrolytes, and the electrolyte concentration determines the decay of the electrostatic interaction. In this way, the electrostatic interactions can be controlled. The Derjaguin, Landau, Verwey, Overbeek (DLVO) theory describes these forces quantitatively and allows the interactions between a spherical tip and a planar sample to be modeled, providing clues to optimize the recording conditions (Müller et al., 1999). As illustrated in Figure 3 the tip surfs on a cushion of electrostatic repulsion under optimal conditions, while the small asperity discussed above is in contact with the sample without inducing deformation (Müller et al., 1999). Figure 3 also demonstrates how the repulsive electrostatic forces between stylus and sample depend on the ionic strength.

Imaging Purple Membrane

When operating a commercial atomic force microscope under optimal recording conditions, the surfaces of biomolecules are contoured routinely at a lateral resolution better than 1 nm and a vertical resolution of around 0.1 nm. Figure 4 shows an image of the extracellular surface of a purple membrane crystal (Fotiadis and Engel, 2004). Purple membrane

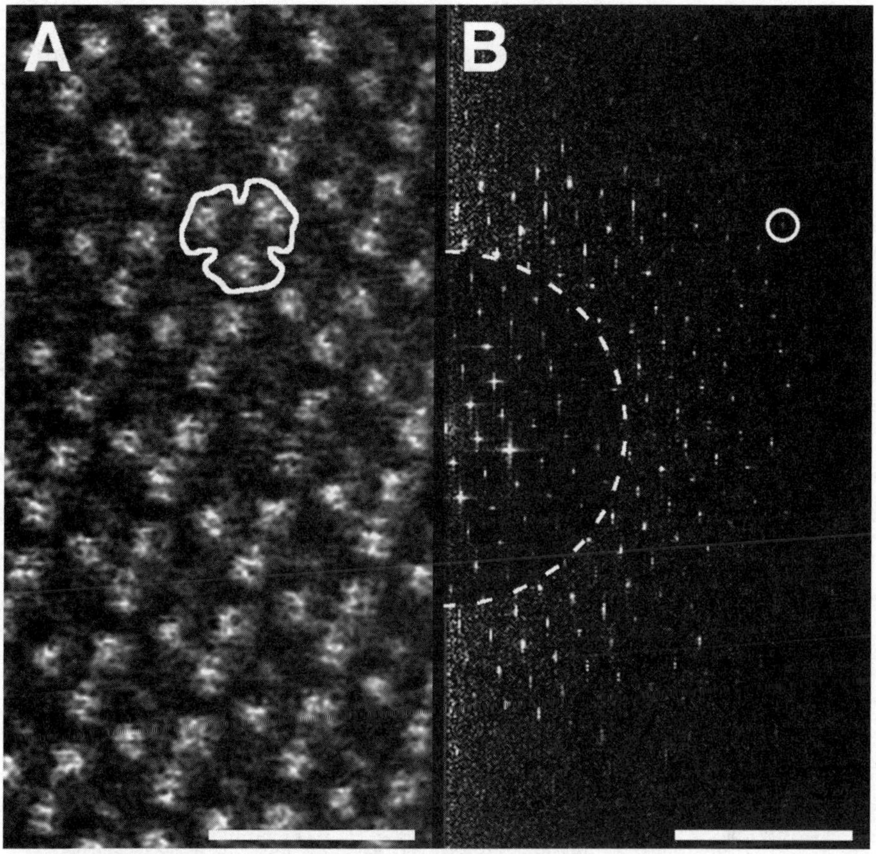

Figure 4 High-resolution topography of the extracellular surface of bacteriorhodopsin (bR). (A) AFM topograph recorded in buffer solution at a force of about 100 pN (10 mM Tris-HCl, pH 7.5, 150 mM KCl) (scale bar: 10 nm; full gray scale: 0.5 nm). (B) Calculated power spectrum from image shown in (A). Half-circle indicates a resolution of 1 nm, small full circles surround diffraction spots at a resolution beyond 0.5 nm. (By courtesy of D. Fotiadis and P. Frederix.)

is a native two-dimensional crystal composed of bacteriorhodopsin (bR) trimers (see contour line in Figure 4A) and lipids (Blaurock and Stoeckenius, 1971). The topograph shows the conformation of the extracellular surface of bR in its most native state. In the trimer, structural details can be clearly distinguished. The 2-D power spectrum of this topograph (Figure 4B) documents the lateral resolution obtained in such an image. Many diffraction spots are visible beyond $(1\ \text{nm})^{-1}$ (dashed half circle) in the power spectrum with spots corresponding to a resolution of 0.46 nm and beyond (e.g., solid circle). Because many gray levels are resolved in this image of purple membrane, whose extracellular surface exhibits a corrugation of 0.6 nm, the vertical resolution achieved in this case is better than 0.1 nm.

Imaging Native Membranes

The best way to study membrane proteins under native or near-native conditions is to prepare directly and observe native membranes in buffer solution. The atomic force microscope is currently the only instrument that allows these conditions in combination with a resolution better than 2 nm. Recently, topographs of the native bacterial photosynthetic membrane of *Rhodopseudomonas viridis* have revealed the molecular organization of the photosynthetic core complex (Scheuring et al., 2003). Moreover, AFM images of the disk membranes of murine retina have elucidated, in contrast to early optical studies, the native conformation of rhodopsin molecules, which are arranged in rows of dimers (Fotiadis et al., 2003) (Figure 5). The importance of this observation is

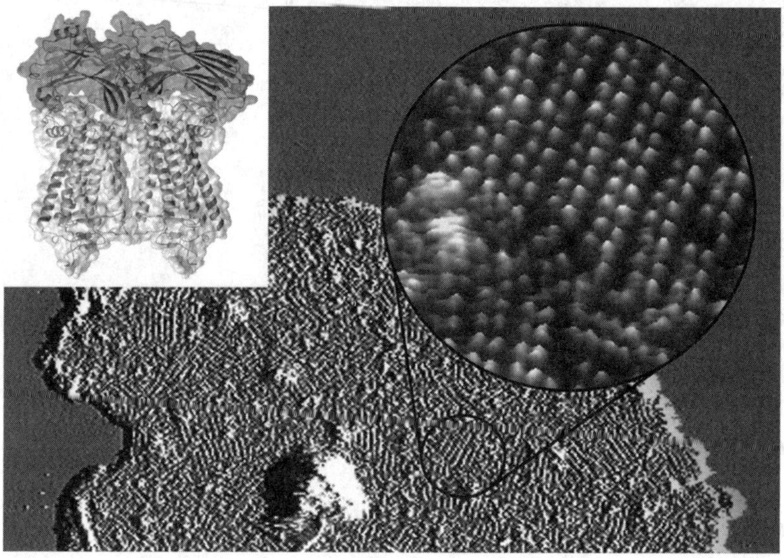

Figure 5 Topography of native eye-disk membrane adsorbed on mica and recorded by AFM. The overview shows part of a disk membrane adsorbed with its extracellular surface facing the mica surface. The line pattern in the overview image recorded in error signal mode indicates tight packing with rhodopsin. Dense lines are seen as rows of rhodopsin dimers in the circular inset recorded in contact mode at high magnification (Fotiadis et al., 2003). The disk membrane has a diameter of about 1 μm, while the double rows represent a paracrystalline array with a slightly skewed unit cell (a = 8.4 nm, b = 3.8 nm, γ = 85°) (Liang et al., 2003). After adsorption of osmotically shocked disk membranes on mica, their topography was measured in buffer solution (20 mM Tris-HCl, pH 7.8, 150 mM KCl, and 25 mM $MgCl_2$). Inset top left: Model of the arrestin-rhodopsin dimer complex. The dimer packing arrangement was elucidated from the AFM data. The theoretical model reflects the interaction of one arrestin molecule with the rhodopsin dimer. (By courtesy of D. Fotiadis, Y. Liang, and K. Palczewski.)

remarkable: for a long time, cross-linking studies and pharmacokinetics have suggested that G-protein coupled receptors (GPCR) work as dimers (Rios et al., 2001; Angers et al., 2002; Mercier et al., 2002; Cheng et al., 2003; Lee et al., 2003; Overton et al., 2003; Soyer et al., 2003; Trettel et al., 2003; Bai 2004). Because rhodopsin is the only GPCR with a known structure (Palczewski et al., 2000), AFM data now provide a solid basis to model a GCPR dimer (Figure 5, inset) and to discuss possible interaction with the cognate proteins, such as arrestin and the G-protein heterotrimer (Liang et al., 2003).

Image Processing

After investing a significant effort to acquire quantitative topographic information of a native biological membrane at sub-nanometer resolution, it is worthwhile to scrutinize this information with the tools provided by digital image processing. Figure 6 shows the cytoplasmic surface of bR. To record this image, the force applied to the stylus was approximately 50 pN, preventing a force-induced conformational change of the loop connecting helices E and F, which is known to deform at a force greater than 100 pN, thereby changing the conformation of the bR surface (Müller et al., 1995). Interestingly, the variation of this region between different atomic models from x-ray crystallography is pronounced: the conformation of the flexible EF loop, which is involved in the contacts leading to the 3-D crystals, is dictated by the 3-D packing arrangement of the bR molecules (Heymann et al., 1999). Although all bR trimers are identical in the 2-D crystal displayed in Figure 6A, their surface structure varies significantly. These changes are related to the flexibility of the bR surface rather than any noise introduced by the atomic force microscope. Although the flexibility of the surface lowers the resolution of the ensemble average (Figure 6B), the structural details of each trimer are still discernable.

Image processing now allows a variety of crucial parameters to be extracted. To this end, the individual molecular images need to be aligned laterally and angularly. Although this is not required for highly ordered 2-D crystals, it can be accomplished to sub-nanometer accuracy with randomly oriented single particles as a result of the high signal-to-noise (S/N) ratio of topographs recorded by AFM. From aligned single molecular topographs, their average and the related standard deviation of the height measurement are calculated pixel by pixel, yielding the maps displayed in Figures 6B and 6C. The vertical scale of Figure 6C documents that vertical measurements can be better than 0.1 nm, provided that the sample is sufficiently stiff. Calculating two independent averages from all odd- and all even-numbered single particles then allows the resolution to be determined by comparing the Fourier coefficients at a given resolution of these two averages. This yields the phase residual (Figure 6D) and the spectral S/N ratio (Figure 6E) as function of resolution. Both analyses indicate the lateral

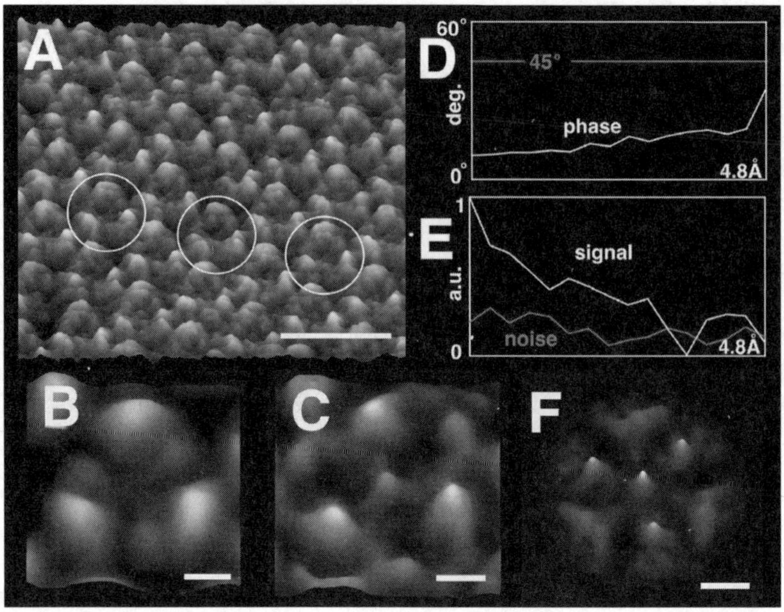

Figure 6 Image analysis of the cytoplasmic bacteriorhodopsin surface. Topographs were recorded in 10 mM Tris-HCl, pH 7.8, 100 mM KCl. (A) At low force (~100 pN), the cytoplasmic surface is dominated by three protrusions at the periphery of each bacteriorhodopsin trimer (circles) resulting from the loop connecting helices E and F (Müller et al., 2000). (B) Average topograph calculated from aligned trimers. (C) The standard deviation map reveals variations of the vertical signal. Large variations are related to the flexibility of the structure (Schabert et al., 1994). The vertical brightness range is 0.01–0.03 nm. (D) Phase residual analysis from 1403 topographs of the cytoplasmic bR surface. The Nyquist frequency is 1/(0.48 nm). (E) Spectral signal-to-noise ratio analysis from the same dataset. (F) The protrusion position probability map calculated from 1403 aligned trimers (Scheuring et al., 2002a) reveals prominent peaks representing protrusions that exhibit only a small lateral flexibility, whereas flexible structures such as the loop EF are rather delocalized. Scale bars: (A) 10 nm; (B), (C), and (F) 1 nm. Brightness range in (A), (B): 1 nm. (By courtesy of S. Scheuring.)

resolution to be 0.5 nm. Clearly, vertical fluctuations will emerge from lateral displacements of protruding loops that connect the helices of bR. Therefore, it is advantageous to characterize variations in the trimer structure by calculating the probability $P_d(x,y)$ to find a certain loop (or domain) at a certain position (x,y) by mapping the corresponding peak positions of all individual bR trimers, as illustrated by Figure 6F (Scheuring et al., 2002a). This map is readily converted to a free energy landscape G_d using Boltzmann's law:

$$G_d = -k_B T \ln[p_d(x,y)] \quad (5)$$

where k_BT is the thermal energy, which is 4.1×10^{-21} J at room temperature.

Nanodissection

The examples shown were all recorded at minimal forces (less than 100 pN) to prevent deformation of the molecules, which would cause a reduction of resolution. However, at higher forces the stylus may be used as a nanoscalpel to disrupt supramolecular assemblies (Fotiadis et al., 2002; Schabert et al., 1995). In this case, the force applied to the tip is increased to 0.75–10 nN, depending on the damage to be achieved. Quite small forces (typically 1 nN) are sufficient to separate stacked layers of membranes or 2-D crystals (Schabert et al., 1995; Fotiadis et al., 2000; Müller et al., 2002a;). Even smaller forces and repeated scanning at high magnification suffice to push away extrinsic proteins that are specifically complexed to an integral membrane protein (Fotiadis et al., 1998; Scheuring et al., 2003; Fotiadis et al., 2004; Siebert et al., 2004). Such nanodissection is of interest to get access to the structure of interfaces, and it also can be used to estimate the interaction forces between subunits in a complex.

Single-Molecule Force Spectroscopy

"Unzipping" Membrane Proteins

A modern commercial atomic force microscope offers both a high spatial resolution and force sensitivity down to some 10 pN. Such instruments allow the forces of single molecular interactions to be measured. For single-molecule force spectroscopy using the atomic force microscope, molecules must be tethered to both the support and the tip in some way. Although a variety of chemical surface activation techniques have been used, large 2-D assemblies such as reconstituted or native membranes (Oesterhelt et al., 2000) and bacterial S-layers (Müller et al., 1999; Scheuring et al., 2002b) are simply physisorbed to mica as described above. The AFM stylus is then attached to an individual protein of such an assembly by pressing the stylus down with a force of ~1 nN for approximately 1 s, which induces the denaturation of some protruding domain that is thus stuck to the silicon nitride tip. Specifically designed mutant proteins that bear a cysteine on a surface-exposed domain allow specific attachment of the protein to a gold-coated stylus (Oesterhelt et al., 2000). Alternatively, a ligand may be covalently immobilized on the stylus through a suitable spacer, while the receptor is tethered to the support (Rief et al., 2000). When approaching the tip to the support, a bond may form between ligand and receptor. In all cases, the change in cantilever deflection upon tip retraction is then recorded, yielding a force-extension curve (Figure 7B). By scanning the sample before (Figure 7A) and after (Figure 7C) such molecular "unzipping"

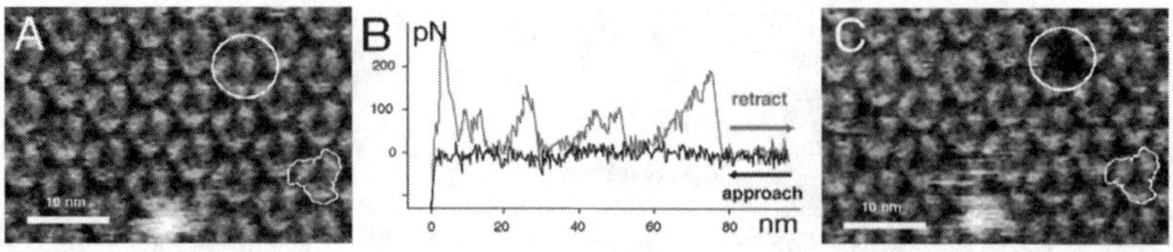

Figure 7 Assessing topology and secondary structure of membrane proteins by imaging and single-molecule force spectroscopy. (A) Patch of intact purple membrane with a contamination at the bottom. (B) Typical force-extension curve recorded while unfolding an individual bacteriorhodopsin molecule (circle) from the native purple membrane shown in (A). The force curve can be divided into major peaks, which can be fitted using the worm-like chain (WLC) model and into minor peaks (Oesterhelt et al., 2000). The major peaks document the pairwise unfolding of transmembrane α-helices. (C) The same patch imaged after the force experiment reveals the site of the unzipped molecule. Scale bars: 10 nm. (By courtesy of D.J. Müller.)

events, the structural damage can directly be assessed and correlated with the respective force-extension curve (Figure 7B).

Dynamic Force Spectroscopy

The atomic force microscope also allows biomolecular bonds (e.g., ligand/receptor) to be disrupted in a quantitative manner. The dissociation of biomolecular bonds is a temperature-driven, dynamic process. To probe it at different timescales, the acquisition of force-extension curves is achieved at different extension rates, a method that is referred to as dynamic force spectroscopy (DFS). Quantitative ensemble measurements can only provide macroscopic thermodynamical quantities (e.g., the free energy of complex formation and/or dissociation). From DFS measurements, however, the virtual energy landscape of a single molecular interaction can be mapped, giving a detailed insight into the reaction pathway of single molecular processes. Thermodynamical models of the rupture of a single bond are described in Evans (2001), Galligan et al. (2001), and Heymann and Grubmüller (2000).

A liner increase of unbinding forces with the logarithm of the loading rate has been observed (Merkel et al., 1999). A similar behavior has been demonstrated for the unfolding of proteins, at least in the limit of small forces. This is a direct consequence of an exponential decrease of the bond lifetime with an applied force, which can be explained by assuming that the dissociation barrier is linearly decreased by applying a force F (Evans, 2001):

$$\Delta E^0 - xF \tag{6}$$

and that the bond lifetime $t(F)$ is given by the Arrhenius expression:

$$t(F) = t^0\, exp[-xF/k_B T] \tag{7}$$

where t^0 is bond lifetime without applied force. The bond length x describes the geometry of the energy landscape. It is interpreted as the distance between the ground state and an energy barrier along the dissociation path (the reaction coordinate) projected onto the direction of the applied force. In principle, many reaction pathways exist, but the applied force is expected to select a specific reaction coordinate in the multidimensional energy landscape. In other words, the system will follow the reaction pathway corresponding to the lowest energy barriers. From Equations 6 and 7, the thermal off-rate $k_{off}(F)$ of a bond under an applied force can be written as:

$$k_{off}(F) = K_{off} \exp[F/F^0] \tag{8}$$

where K_{off} is the natural thermal off-rate for dissociation and F^0 is a force-scale factor ($F^0 = k_BT/x$). Although the bond lifetime is a relevant parameter to describe the kinetics of bond rupture, the quantity directly measured in a typical AFM experiment is the unbinding force. However, because the rupture of a bond is a stochastic process, we cannot expect to measure only one specific unbinding force. Rather, the measured unbinding force should follow a distribution whose width is mainly determined by the force-scale factor. In this case, it can be shown that the most probable unbinding force (F^*) depends on the applied loading rate r as:

$$F^* = F^0 \ln[r/(F^0 K_{off})] \tag{9}$$

DFS measurements can be performed with an unmodified commercial atomic force microscope or using external data acquisition capabilities that enhance sensitivity and flexibility of the instrument. To achieve reproducible measurements, the spring constants of all cantilevers used must be calibrated as described above. Because the temperature is a crucial parameter in these experiments, it should be controlled, for example using a regulated Peltier element, which is in contact with the buffer solution (Schumakovitch et al., 2002). Because unbinding events are caused by thermal fluctuations rather than by mechanical instability, unbinding forces show a distribution whose width σ is mainly determined by the force scale factor $F^0 = k_BT/x$ (Merkel et al., 1999).

When approaching the tip to the surface, nonspecific attachments may occur even if the support has been passivated or is a pure polymer surface. Thus, it is imperative to test the specificity of the interaction (see inset, Figure 8). Unspecific interactions can be minimized using linkers [e.g., poly(ethylene)glycol (PEG) linkers] that shift the region where unbinding takes place away from the surface. Finally, to quantify the most probable value for the unbinding force of a single complex, conditions must be chosen in which the probability that two or more duplexes are attached to the tip is very low. This is

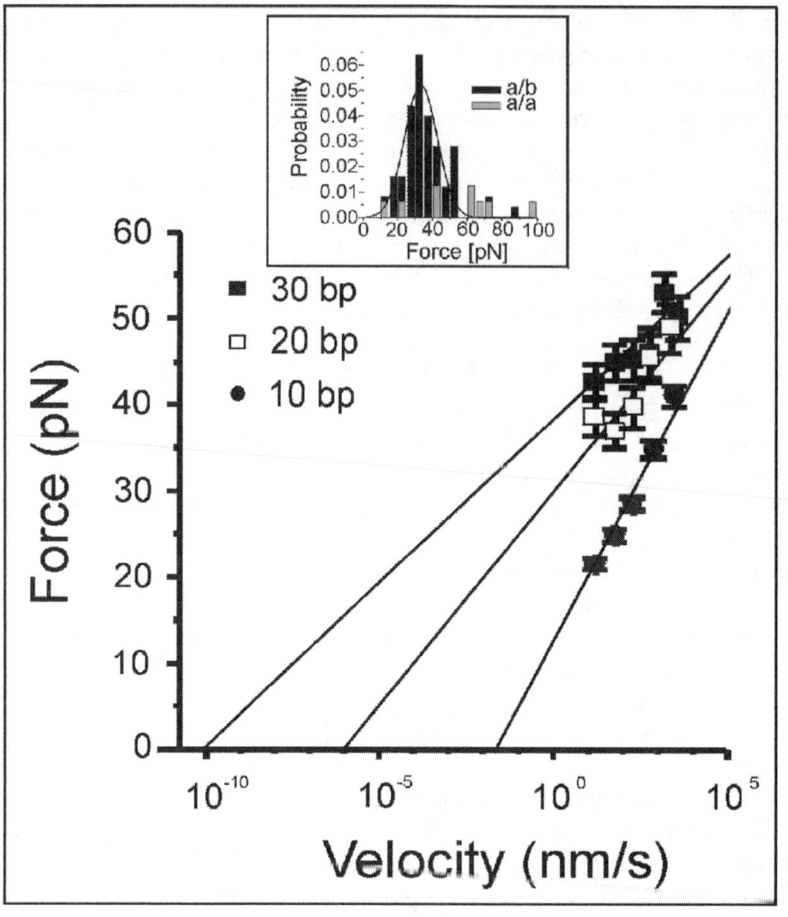

Figure 8 Rupturing dsDNA molecules in dynamic force spectroscopy experiments. The graph shows the velocity dependence of the most probable unbinding force (Hegner and Engel, 2002). Black squares, 30 bp; empty squares, 20 bp; circles, 10 bp. From a linear fit, both the force-scales $F^0 = k_BT/\Delta x$ and thermal off-rates can be determined. Inset specificity of the rupturing force: A typical probability distribution for the rupture force (about 500 approach/retract cycles, retract velocity 100 nm/s). For this experiment, an oligomer **a** (see text) was attached to the tip of the atomic force microscope-cantilever and its complement **b** was immobilized on the surface (complements were pulled apart at their opposite 5-ends). Gray rectangles (**a** against **a**), black rectangles (**a** against **b**). To minimize unspecific interactions (e.g., **a** against **a**) and multiple unbinding events, 30-nm-long PEG linkers were attached to the 5-ends of oligonucleotides. (By courtesy of M. Hegner.)

achieved at low concentration and using linkers that have a length that is comparable to the diameter of the AFM-tip (about 50 nm). Nevertheless, rupture event series may still be observed. Therefore, it is the last rupture event that should be considered to measure the single molecule rupture force F^*.

An illustrative example is the DFS measurements on complementary DNA strands. The strands had different length [10, 20, and 30 base pairs (bp)] and were pulled apart at their opposite 5′-ends. The base sequences of the oligonucleotides were designed to favor the binding to their complementary oligonucleotides in the ground state rather than in an intermediate state in which the strand is shifted relative to its complement. Oligomer **a** (5′-G-G-C-T-C-C-C-T-T-C-T-A-C-C-A-C-T-G-A-C-A-T-C-G-C-A-A-C-G-G-3′), which contains 30 bases and in which every three-base motive occurs only once in the sequence, was chosen as a bait. For this sequence, self-complementarities are avoided because the complement of each three-base motive is not contained in the sequence. Oligomer **a** was tested against its complement **b** (30 bp) and against truncated components **c** (20 bp) and **d** (10 bp), respectively. As expected, an F^* *versus* $\ln(r)$ plot shows a linear behavior for each duplex (Equation 9, see Figure 8). For each duplex, the distance Δx from the ground state to the energy barrier and the thermal off rates under load, k_{off}, were determined according to Equation 9. The Δx distance was found to follow the linear relation: $\Delta x = [(0.7 \pm 0.3) + (0.07 \pm 0.03) \times n]$ nm, where n is the number of base pairs. This increase of Δx with n clearly indicates cooperativity in the unbinding process. Measurements of k_{off} can be described by: $k_{\text{off}} \approx 10^{\alpha - \beta n}\ \text{s}^{-1}$, where $\alpha = 3 \pm 1$ and $\beta = 0.05 \pm 0.1$. The k_{off} values obtained by DFS are in good agreement with thermodynamical data (Strunz et al., 1999). It should be noted that an exponential decrease of the thermal off-rate with the number of base pairs is expected due to the increase of the activation energy for dissociation.

The temperature dependence of duplex DNA rupture has also been studied (Schumakovitch et al., 2002). Sequence **e** (5′-T-A-T-T-A-A-T-A-T-C-A-A-G-T-T-G-3′) was attached to the tip and its complement **f** immobilized on the surface. As previously, PEG linkers were used, and DNA strands were pulled apart at their opposite 5′-ends. The specificity of the interaction was comparable to the one obtained in base-pair dependent measurements (Figure 8). However, one important property of the energy landscape becomes visible when the temperature is varied. The slope of the F^* *versus* $\ln(r)$ plots changes as a function of temperature, which documents a strong temperature dependence of Δx. This result emphasizes the fact that the energy landscape for a DNA-duplex is likely to be more complicated than that of ligand-receptor bonds. The decrease of Δx upon increasing the temperature, which is not expected for bonds with a one-dimensional energy landscape exhibiting a sharp energy barrier, suggests that entropic contributions need to be considered to describe fully the unbinding of DNA and the unfolding of RNA or proteins (Moy et al., 1994).

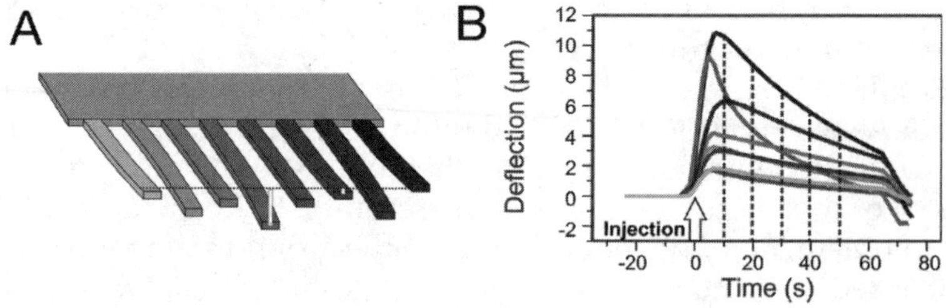

Figure 9 Nanoanalytics using cantilever arrays (Lang et al., 2002). (A) Cantilever sensor array for detection of gases or biomolecules in liquid. Specifically functionalized cantilevers bend differently upon interaction with a gas or mixture of biomolecules. (B) The characteristic cantilever deflection pattern allows gases to be identified. (By courtesy of Ch. Gerber and H.P. Lang.)

Cantilever Arrays

To detect biomolecules with high specificity, cantilevers are functionalized to exhibit a chemically active and a chemically inactive surface. The specific biomolecule will bind to the active cantilever surface, which can either be detected by cantilever bending or resonance frequency shift. Bending is induced by the electrostatic repulsion between adsorbed molecules or simply by their tight packing. This has been exploited to detect specific gas molecules (Lang et al., 2002), proteins (Arntz et al., 2003), or picomolar concentrations of oligonucleotides (McKendry et al., 2002). The mass increase resulting from the bound biomolecules can also be detected by the resonance frequency shift, allowing the detection of less than 1 pg (1 pg = 10^{-12} g) (Lang et al., 2002). Pairs of cantilevers have been operated in parallel, one functionalized on one surface and the other having both surfaces passivated to serve as reference. This differential measurement increases the sensitivity and makes the method robust against changes of the environment. Parallelized detection using cantilever arrays combined with diode arrays has been demonstrated, allowing quantitative DNA-binding assays (Arntz et al., 2003) and analyses of gas mixtures (Lang et al., 2002) (Figure 9). With the millipede concept, which allows thousands of cantilevers to be monitored simultaneously (Vettiger et al., 2002), cantilever based DNA or protein sensors will probably play an important role in the future, because they allow unlabeled biomolecules to be detected at very low concentrations.

Conclusions

Progress in instrumentation and sample preparation methods have opened avenues to image single-membrane proteins and to assess the nanomechanics of single molecules. Such novel experimental possibilities find a wide range of applications in the study of structure and function of biomolecules, the building blocks of life, as well as their interaction with cognate ligands. Although the amazing S/N-ratio and resolution of the atomic force microscope allows the conformational states of single protein surfaces to be probed at a lateral resolution of 0.4–1 nm and 0.1–0.2 nm vertical resolution, the force sensitivity (greater than 20 pN) of this instrument also opens the door to measure the forces required to unfold a protein. Combining single-molecule imaging and force spectroscopy gives an unprecedented insight into the nature of intra- and intermolecular interactions in and between biomolecules.

Using DFS measurements, the energy landscape of molecular bonds can be mapped. Moreover, relevant parameters such as the location and height of the barriers and the thermal off-rates can be determined. To overcome the limited range of loading rates available in an atomic force microscope, DFS experiments may be achieved with the biomembrane force probe (Merkel et al., 1999). An alternative solution is to use small cantilevers, which allow faster pulling and exhibit less thermal noise, so smaller unbinding forces can be detected. These small cantilevers are still experimental (Viani et al., 1999a), and great efforts are being made to commercialize them in the future. These developments will also require instrument development, so it is expected to be a few years before they can be widely used. It is likely that DFS will be applied in the future to assess the binding affinity of ligands to biomolecules on large arrays (Vettiger et al., 2002), but significant technical developments need to be implemented before such experiments are feasible on a routine basis.

Abbreviations

AFM	Atomic force microscopy
bR	Bacteriorhodopsin
DFS	Dynamic force spectroscopy
GPCR	G-protein coupled receptors

Acknowledgments

The research discussed here was supported by the Swiss National Research Foundation, the M.E. Müller Foundation, and the Swiss National Center of Competence in Research (NCCR) "Structural Biology" and the NCCR "Nanoscale Science." The author is indebted to his creative and motivated colleagues Dimitrios Fotiadis, Patrick Frederix, Daniel J. Müller, Christoph Gerber, Martin Hegner, and Simon Scheuring, who have all contributed to the work summarized in this review. He also would like to thank Kris Palczewski for most fruitful and inspiring collaborations and for his friendship.

References

1. Angers S, Salahpour A, and Bouvier M. Dimerization: An Emerging Concept for G Protein-Coupled Receptor Ontogeny and Function. *Annu Rev Pharmacol Toxicol* 2002; 42:409–435
2. Arntz Y, Seelig J, Lang H, Zhang J, Hunziker P, Ramseyer J, Meyer E, Hegner M, and Gerber C. Label-Free Protein Assay Based on a Nanomechanical Cantilever Array. *Nanotechnology* 2003; 14:86–90
3. Bai M. Dimerization of G-Protein-Coupled Receptors: Roles in Signal Transduction. *Cell Signal* 2004; 16:175–186
4. Binnig G, Quate CF, and Gerber C. Atomic Force Microscope. *Phys Rev Lett* 1986; 56:930–933
5. Blaurock AE, and Stoeckenius W. Structure of the Purple Membrane. *Nat New Biol* 1971; 233:152–155
6. Cheng ZJ, Harikumar KG, Holicky EL, and Miller LJ. Heterodimerization of Type A and B Cholecystokinin Receptors Enhance Signaling and Promote Cell Growth. *J Biol Chem* 2003; 278:52972–52979
7. Chon JWM, Mulvaney P, and Sader JE. Experimental Validation of Theoretical Models for the Frequency Domain Response of Atomic Force Microscope Cantilever Beams Immersed in Fluid. *J Appl Phys* 2000; 87:3978–3988
8. Czajkowsky DM, Sheng S, and Shao Z. Staphylococcal a-Hemolysin Can Form Hexamers in Phospholipid Bilayers. *J Mol Biol* 1998; 276:325–330
9. Engel A, and Müller DJ. Observing Single Biomolecules at Work with the Atomic Force Microscope. *Nat Struct Biol* 2000; 7:715–718
10. Evans E. Probing the Relation Between Force - Lifetime - and Chemistry in Single Molecular Bonds. *Ann Rev Biophys Biomol Struct* 2001; 30:105–128
11. Fotiadis D, and Engel A. High-Resolution Imaging of Bacteriorhodopsin by Atomic Force Microscopy. *Methods Mol Biol* 2004; 242:291–303
12. Fotiadis D, Müller DJ, Tsiotis G, Hasler L, Tittmann P, Mini T, Jenö P, Gross H, and Engel A . Surface Analysis of the Photosystem I Complex by Electron and Atomic Force Microscopy. *J Mol Biol* 1998; 283:83–94
13. Fotiadis D, Hasler L, Müller DJ, Stahlberg H, Kistler J, and Engel A. Surface Tongue-and-Groove Contours on Lens MIP Facilitate Cell-to-Cell Adherence. *J Mol Biol* 2000; 300:779–789
14. Fotiadis D, Scheuring S, Müller SA, Engel A, and Müller DJ. Imaging and Manipulation of Biological Structures with the AFM. *Micron* 2002; 33:385–397
15. Fotiadis D, Liang Y, Filipek S, Saperstein DA, Engel A, and Palczewski K. Atomic-Force Microscopy: Rhodopsin Dimers in Native Disc Membranes. *Nature* 2003; 421:127–128

16. Fotiadis D, Qian P, Philippsen A, Bullough PA, Engel A, and Hunter CN. Structural Analysis of the Reaction Center Light-Harvesting Complex I Photosynthetic Core Complex of Rhodospirillum rubrum Using Atomic Force Microscopy. *J Biol Chem* 2004; 279:2063–2068

17. Fujiyoshi Y. The Structural Study of Membrane Proteins by Electron Crystallography. *Adv Biophys* 1998; 35:25–80

18. Galligan E, Roberts CJ, Davies MC, Tendler SJB, and Williams PM. Simulating the Dynamic Strength of Molecular Interactions. *J Chem Phys* 2001; 114:3208–3214

19. Gibson CT, Johnson DJ, Anderson C, Abell C, and Rayment T. Method to Determine the Spring Constant of Atomic Force Microscope Cantilevers. *Rev Sci Instrum* 2004; 75:565–567

20. Hegner M, and Engel A. Single Molecule Imaging and Manipulation. *Chimia* 2002; 56:506–514

21. Heymann B, and Grubmüller H. Dynamic Force Spectroscopy of Molecular Adhesion Bonds. *Phys Rev Lett* 2000; 84:6126–6129

22. Heymann JB, Müller JD, Landau EM, Rosenbusch JP, Pebay-Peyroula E, Büldt G, and Engel A. Charting the Surfaces of the Purple Membrane. *J Struct Biol* 1999; 128:243–249

23. Hinterdorfer P, Baumgartner W, Gruber HJ, Schilcher K, and Schindler H. Detection and Localization of Individual Antibody-Antigen Recognition Events by Atomic Force Microscopy. *Proc Natl Acad Sci U S A* 1996; 93:3477–3481

24. Hutter JL, and Bechhöfer J. Calibration of Atomic-Force Microscope Tips. *Rev Sci Instrum* 1993; 64:3342

25. Janovjak H, Kessler M, Oesterhelt D, Gaub H, and Müller DJ. Unfolding Pathways of Native Bacteriorhodopsin Depend on Temperature. *EMBO J* 20003; 22:5220–5229

26. Karrasch S, Dolder M, Schabert F, Ramsden J, and Engel A. Covalent Binding of Biological Samples to Solid Supports for Scanning Probe Microscopy in Buffer Solution. *Biophys J* 1993; 65:2437–2446

27. Klein DC, Stroh CM, Jensenius H, van Es M, Kamruzzahan AS, Stamouli A, Gruber HJ, Oosterkamp TH, and Hinterdorfer P. Covalent Immobilization of Single Proteins on Mica for Molecular Recognition Force Microscopy. *Chemphyschem* 2003; 4:1367–1371

28. Lang H, Hegner M, Meyer E, and Gerber C. Nanomechanics from Atomic Resolution to Molecular Recognition Based on Atomic Force Microscopy Technology. *Nanotechnology* 2002; 13:R29–R36

29. Lee SP, O'Dowd BF, Rajaram RD, Nguyen T, and George SR. D2 Dopamine Receptor Homodimerization Is Mediated by Multiple Sites of Interaction, Including an Intermolecular Interaction Involving Transmembrane Domain 4. *Biochemistry* 2003; 42:11023–11031

30. Liang Y, Fotiadis D, Filipek S, Saperstein DA, Palczewski K, and Engel A. Organization of the G Protein-Coupled Receptors Rhodopsin and Opsin in Native Membranes. *J Biol Chem* 2003; 278:21655–21662
31. Marszalek PE, Li H, and Fernandez JM. Fingerprinting Polysaccharides with Single-Molecule Atomic Force Microscopy. *Nat Biotechnol* 2001; 19:258–262
32. McKendry R, Zhang J, Arntz Y, Strunz T, Hegner M, Lang HP, Baller MK, Certa U, Meyer E, Güntherodt HJ, et al. Multiple Label-Free Biodetection and Quantitative DNA-Binding Assays on a Nanomechanical Cantilever Array. *Proc Natl Acad Sci U S A* 2002; 99:9783–9788
33. Mercier JF, Salahpour A, Angers S, Breit A, and Bouvier M. Quantitative Assessment of Beta 1- and Beta 2-Adrenergic Receptor Homo- and Heterodimerization by Bioluminescence Resonance Energy Transfer. *J Biol Chem* 2002; 277:44925–44931
34. Merkel R, Nassoy P, Leung A, Ritchie K, and Evans E. Energy Landscapes of Receptor-Ligand Bonds Explored with Dynamic Force Spectroscopy. *Nature* 1999; 397:50–53
35. Möller C, Fotiadis D, Suda K, Engel A, Kessler M, and Müller DJ. Determining Molecular Forces That Stabilize Human Aquaporin-1. *J Struct Biol* 2003; 142:369–378
36. Moy VT, Florin EL, and Gaub HE. Iintermolecular Forces and Energies Between Ligands and Receptors. *Science* 1994; 266:257–259
37. Müller DJ, Büldt G, and Engel A. Force-Induced Conformational Change of Bacteriorhodopsin. *J Mol Biol* 1995; 249:239–243
38. Müller DJ, Amrein M, and Engel A. Adsorption of Biological Molecules to a Solid Support for Scanning Probe Microscopy. *J Struct Biol* 1997a; 119:172–188
39. Müller DJ, Engel A, and Amrein M. Preparation Techniques for the Observation of Native Biological Systems with the Atomic Force Microscope. *Bioelectrochem Bioenerg* 1997b; 12:867–877
40. Müller DJ, Fotiadis D, Scheuring S, Müller SA, and Engel A. Electrostatically Balanced Subnanometer Imaging of Biological Specimens by Atomic Force Microscopy. *Biophys J* 1999; 76:1101–1111
41. Müller DJ, Heymann JB, Oesterhelt F, Möller C, Gaub H, Büldt G, and Engel A. Atomic Force Microscopy of Native Purple Membrane. *Biochim Biophys Acta* 2000; 1460:27–38
42. Müller DJ, Hand GM, Engel A, and Sosinsky GE Conformational Changes in Surface Structures of Isolated Connexin 26 Gap Junctions. *EMBO J* 2002a; 21:3598–3607
43. Müller DJ, Kessler M, Oesterhelt F, Möller C, Oesterhelt D, and Gaub H. Stability of Bacteriorhodopsin Alpha-Helices and Loops Analyzed by Single-Molecule Force Spectroscopy. *Biophys J* 2002b; 83:3578–3588

44. Oesterhelt F, Oesterhelt D, Pfeiffer M, Engel A, Gaub HE, and Müller DJ. Unfolding Pathways of Individual Bacteriorhodopsins. *Science* 2000; 288:143–146
45. Overton MC, Chinault SL, and Blumer KJ. Oligomerization, Biogenesis, and Signaling Is Promoted by a Glycophorin A-like Dimerization Motif in Transmembrane Domain 1 of a Yeast G Protein-Coupled Receptor. *J Biol Chem* 2003; 278:49369–49377
46. Palczewski K, Kumasaka T, Hori T, Behnke CA, Motoshima H, Fox BA, Le Trong I, Teller DC, Okada T, Stenkamp RE, et al. Crystal Structure of Rhodopsin: a G Protein-Coupled Receptor. *Science* 2000; 289:739–745
47. Rief M, Gautel M, and Gaub HE. Unfolding Forces of Titin and Fibronectin Domains Directly Measured by AFM. *Adv Exp Med Biol* 2000; 481:129-136; discussion 137–141
48. Riener CK, Stroh CM, Ebner A, Klampfl C, Gall AA, Romanin C, Lyubchenko Y, Hinterdorfer P, and Gruber HJ. Simple Test System for Single Molecule Recognition Force Microscopy. *Anal Chim Acta* 2003; 479:59–75
49. Rios CD, Jordan BA, Gomes I, and Devi LA. G-Protein-Coupled Receptor Dimerization: Modulation of Receptor Function. *Pharmacol Ther* 2001; 92:71–87
50. Sader JE. Frequency Response of Cantilever Beams Immersed in Viscous Fluids with Applications to the Atomic Force Microscope. *J Appl Phys* 1998; 84:64–76
51. Sader JE, Chon JWM, and Mulvaney P. Calibration of Rectangular Atomic Force Microscope Cantilevers. *Rev Sci Instrum* 1999; 70:3967–3969
52. Schabert FA, and Engel A. Reproducible Acquisition of Escherichia coli Porin Surface Topographs by Atomic Force Microscopy. *Biophys J* 1994; 67:2394–2403
53. Schabert FA, Henn C, and Engel A. Native Escherichia coli OmpF Porin Surfaces Probed by Atomic Force Microscopy. *Science* 1995; 268:92–94
54. Scheuring S, Müller DJ, Stahlberg H, Engel HA, and Engel A. Sampling the Conformational Space of Membrane Protein Surfaces with the AFM. *Eur Biophys J* 2002a; 31:172–178
55. Scheuring S, Stahlberg H, Chami M, Houssin C, Rigaud JL, and Engel A. Charting and Unzipping the Surface Layer of Corynebacterium glutamicum with the Atomic Force Microscope. *Mol Microbiol* 2002b; 44:675–684
56. Scheuring S, Seguin J, Marco S, Levy D, Robert B, and Rigaud JL. Nanodissection and High-Resolution Imaging of the Rhodopseudomonas viridis Photosynthetic Core Complex in Native Membranes by AFM. Atomic Force Microscopy. *Proc Natl Acad Sci U S A* 2003; 100:1690–1693

57. Schumakovitch I, Grange W, Strunz T, Bertoncini P, Güntherodt H-J, and Hegner M. Temperature Dependence of Unbinding Forces Between Complementary DNA Strands. *Biophys J* 2002; 82:517–521

58. Siebert CA, Qian P, Fotiadis D, Engel A, Hunter CN, and Bullough PA. Molecular Architecture of Photosynthetic Membranes in Rhodobacter sphaeroides: The Role of PufX. *EMBO J* 2004; 23:690–700

59. Soyer OS, Dimmic MW, Neubig RR, Goldstein RA. Dimerization in Aminergic G-Protein-Coupled Receptors: Application of a Hidden-Site Class Model of Evolution. *Biochemistry* 2003; 42:14522–14531

60. Strunz T, Oroszlan K, Schäfer R, and Güntherodt H-J. Dynamic Force Spectroscopy of Single DNA Molecules. *Proc Natl Acad Sci U S A* 1999; 96:11277–11282

61. Trettel F, Di Bartolomeo S, Lauro C, Catalano M, Ciotti TM, and Limatola C. Ligand-Independent CXCR2 Dimerization. *J Biol Chem* 2003; 278:40980–40988

62. Vettiger P, Cross G, Despont M, Drechsler U, Dürig U, Gotsmann B, Häberle W, Lantz M, Rothuizen H, Stutz R, et al. The "Millipede"-Nanotechnology Entering Data Storage. *IEEE Trans Nanotechnol* 2002; 1:39–55

63. Viani MB, Schäfer TE, Chand A, Rief M, Gaub H, and Hansma PK. Small Cantilevers for Force Spectroscopy of Single Molecules. *J Appl Phys* 1999a; 86:2258–2262

64. Viani MB, Schäffer TE, Paloczi GT, Pietrasanta LI, Smith BL, Thompson JB, Richter M, Rief M, Gaub HE, Plaxco KW, et al. Fast Imaging and Fast Force Spectroscopy of Single Biopolymers with a New Atomic Force Microscope Designed for Small Cantilevers. *Rev Sci Instrum* 1999b; 70:4300–4303

65. Yang J, Ono T, and Esashi M Energy Dissipation in Submicrometer Thick Single-Crystal Silicon Cantilevers. *J Microelectromech Sys* 2002; 11:775–783

66. Yang JL, Despont M, Hoogenboom BW, Drechsler U, Frederix PLTM, Martin S, Hug HJ, Vettiger P, and Engel A. Ultra-small Single Crystal Silicon Cantilevers for Scanning Force Microscopy. *J Microelectromech Sys* 2004; 560–563

67. Yasumura KY, Stowe TD, Chow EM, Pfafman T, Kenny TW, Stipe BC, and Rugar D. Quality Factors in Micron- and Submicron-Thick Cantilevers. *J Microelectromech Sys* 2000; 9:117–125

19

Molecular Modeling

ED E. MORET and DAVID GESTEL

Department of Medicinal Chemistry and Department of Pharmaceutics, Utrecht Institute for Pharmaceutical Sciences, Utrecht University, Utrecht, The Netherlands

Introduction

Both the pharmacodynamics and the pharmacokinetics of protein pharmaceuticals ultimately depend on their three-dimensional *structure* and interaction *properties*.

The three-dimensional (3-D) structure can be determined experimentally with electron microscopy, x-ray, and NMR, and coordinates are normally deposited in the Protein Data Bank (PDB). Theoretical structure models can be made as well, based on comparative (homology) modeling or fold recognition (threading). Furthermore, molecular force field simulations are increasingly being used to predict protein structure *ab initio*.

The most important pharmacodynamic properties of pharmaceutical proteins are their interactions with other proteins, which can be simulated by electrostatics and docking calculations. Their pharmacokinetic behavior is largely determined by their stability with respect to refolding, their chemical and post-translational modification, and mutational variability. Furthermore, protein immunogenicity can roughly be predicted. In 2002, 40% of the New Molecular Entities registered at the U.S. Food and Drug Administration (FDA) were pharmaceutical proteins (Frantz and Smith, 2003). Unlike small drug molecules, there is not much known about the structure of these biotech drugs. Furthermore, in light of the newly defined conformational diseases (Lomas and Carrell, 2002), where protein conformation is related to pathology, protein modeling can be used to study the transitions between conformations.

This chapter will guide the reader through molecular modeling and the information avalanche of structural bioinformatics. Databases and software for prediction, visualization, and simulation have become freely available on user-friendly platforms and are listed in Tables 1 and 2.

Protein Data Bank	http://www.rcsb.org/pdb/
Research Collaboratory for Structural Bioinformatics (RCSB)	
Structural Classification of Proteins	
SCOP	http://scop.mrc-lmb.cam.ac.uk/scop/
MRC Laboratory of Molecular Biology, University of Cambridge, UK	
CATH	http://www.biochem.ucl.ac.uk/bsm/cath/
Biomolecular Structure and Modelling Group, University College London, UK	
DALI	http://www.ebi.ac.uk/dali/
European Bioinformatics Institute, Hinxton, UK (Structural similarities of proteins already in the PDB are in the FSSP database)	
Protein Homology and Similarity	
PFAM	http://pfam.wustl.edu/
The Pfam Consortium, Washington University, St. Louis, MO, USA	
SMART	http://smart.embl-heidelberg.de/
European Molecular Biology Laboratory, Heidelberg, Germany	
BLAST	http://www.ncbi.nlm.nih.gov/BLAST/
National Center for Biotechnology Information, National Institutes of Health, Bethesda, MD, USA	
Fold Recognition	
HOMSTRAD	http://www-cryst.bioc.cam.ac.uk/~homstrad/
Crystallography and Bioinformatics, University of Cambridge, UK	
3D-PSSM	http://www.sbg.bio.ic.ac.uk/~3dpssm/
Structural Bioinformatics Group, Imperial College London, UK	
3D-Jury	http://bioinfo.pl/Meta/
Bioinformatics Laboratory, BioInfoBank Institute, Poznan, Poland (Structure prediction meta server)	

Table 1 Tools for Structure Prediction.

(continued on following page)

(continued from previous page)

Threading	
PROSPECT	http://compbio.ornl.gov/PROSPECT/
Oak Ridge National Laboratory, Oak Ridge, TN, USA	
Comparative Protein Structure Modeling	
WHAT IF Web Interface	http://www.cmbi.kun.nl/gv/servers/WIWWWI/
Centre for Molecular and Biomolecular Informatics, University of Nijmegen, Netherlands	
3D-JIGSAW	http://www.bmm.icnet.uk/servers/3djigsaw/
Biomolecular Modelling Laboratory, Cancer Research UK London Research Institute, UK	
ModWeb	http://alto.compbio.ucsf.edu/modweb-cgi/main.cgi
Andrej Sali Lab, University of California, San Francisco, CA, USA	
***Ab Iinitio* Structure Prediction**	
ROSETTA	http://depts.washington.edu/bakerpg/
Baker Laboratory, University of Washington, Seattle, WA, USA	
Molecular Visualization Programs	
Cn3D	http://www.ncbi.nih.gov/Structure/CN3D/cn3d.shtml
National Center for Biotechnology Information, National Institutes of Health, Bethesda, MD, USA	
(Cn3D is a visualization tool for biomolecular structures, sequences, and sequence alignments)	
Protein Explorer, Chime & Rasmol	http://www.umass.edu/microbio/rasmol/
Department of Microbiology, University of Massachusetts, Amherst, MA, USA	
Pymol	http://pymol.sourceforge.net/
DeLano Scientific LLC, San Carlos, CA, USA	
(PyMOL is a Python-enhanced molecular graphics program for the 3-D visualization of proteins, small molecules, electron density, surfaces, and trajectories)	

(continued on following page)

(continued from previous page)

DeepView (Swiss-PdbViewer)	http://www.expasy.org/spdbv/
Swiss Institute of Bioinformatics, Biozentrum - Basel University, Switzerland	
(DeepView is a tool for viewing and manipulating protein structures and models. The comparative modeling server SWISS-MODEL is accessible from the program)	
MOLMOL	http://www.mol.biol.ethz.ch/wuthrich/software/molmol/
Wütrich Laboratory, Swiss Federal Institute of Technology, Zürich, Switzerland	
(MOLMOL is a molecular graphics program for display, analysis, and manipulation of 3-D structures of biological macromolecules)	
GRASP2	http://trantor.bioc.columbia.edu
Honig Laboratory, Columbia University, New York, NY, USA	
(GRASP2 is an updated version of the GRASP program used for macromolecular structure and surface visualization)	
Protein Structure Validation	
Procheck	http://www.biochem.ucl.ac.uk/~roman/
Biomolecular Structure and Modelling Group, University College London, UK	
WhatCheck	http://www.cmbi.kun.nl/gv/servers/WIWWWI/
Centre for Molecular and Biomolecular Informatics, University of Nijmegen, The Netherlands	
Biotech Validation Suite	http://biotech.ebi.ac.uk:8400/
European Bioinformatics Institute, Hinxton, UK	
(Server providing PROCHECK V3.5 and WHAT IF V4.99)	

Protein Structure

The year 1962 was remarkable. The Nobel Prize for Medicine was given to Watson, Crick, and Wilkins for the elucidation of the structure and function of DNA nine years earlier. This enabled the beginning of molecular biology in the 1970s. In 1962 as well, Perutz and Kendrew were awarded the Nobel Prize for Chemistry for elucidating the structure of

Epitope Prediction	
ProtScale	http://www.expasy.org/cgi-bin/protscale.pl
Swiss Institute of Bioinformatics, Biozentrum - Basel University, Switzerland	
(ProtScale allows you to compute and represent the profile produced by any amino acid scale like Hopp and Woods, and Kyte and Doolitle)	
SYFPEITHI	http://syfpeithi.bmi-heidelberg.com/
Biomedical Informatics, Heidelberg, Germany	
(Prediction of MHC class I and II binding peptides)	
MHCPred	http://www.jenner.ac.uk/MHCPred/
Edward Jenner Institute for Vaccine Research, Compton, UK	
(Prediction of MHC class I and II binding peptides)	
NetMHC	http://www.cbs.dtu.dk/services/NetMHC/
Center for Biological Sequence Analysis, Technical University of Denmark, Lyngby, Denmark	
(Prediction of MHC class I binding peptides)	
NetChop	http://www.cbs.dtu.dk/services/NetChop/
Center for Biological Sequence Analysis, Technical University of Denmark, Lyngby, Denmark	
(Prediction of cleavage sites of the human proteasome)	
Docking Programs	
AUTODOCK	http://www.scripps.edu/pub/olson-web/doc/autodock/
Molecular Graphics Laboratory, Scripps Research Institute, La Jolla, CA, USA	
DOCK	http://dock.compbio.ucsf.edu/
Molecular Design Institute, University of California, San Francisco, CA, USA	
ICM-DISCO	http://abagyan.scripps.edu/~jfrecio/ICMprotdock/
Abagyan Group, Scripps Research Institute, La Jolla, CA, USA	

Table 2 Tools for Protein-Protein Interaction.

(continued on following page)

(continued from previous page)

Protein Mutation Databases
MutaProt http://atlantis.weizmann.ac.il/mutaprot/ Weizmann Institute of Science, Rehovot, Israel (Comparison of PDB files that differ by point mutations)
Protein Mutant Database http://pmd.ddbj.nig.ac.jp/ Center for Information Biology and DNA Databank of Japan, National Institute of Genetics, Mishima, Japan (Information on functional and/or structural influences brought about by amino acid mutations)

other biomacromolecules: the proteins myoglobin and hemoglobin. This discovery was the start of structural biology. These colleagues from the Cavendish Laboratory witnessed the birth of biology as a molecular discipline, and the following decades have given us an atomic understanding of life, as seen in eukaryotes, prokaryotes, and archaea. And now that the techniques are mature enough to expect high-throughput structural biology and structural genomics, people have started yearning for an understanding of physiology on a more integrated level, in systems biology. Still, the atomic detail differentiates between inactive and active proteins and will always be needed for understanding life.

The first protein structure led to some disappointment. In contrast to the beautiful symmetry of DNA, the first protein structures resembled spaghetti on a plate. However, currently around 800 folds (Andreeva et al., 2004) have been found, and some of them are strikingly beautiful.

Experimental Protein Structure

The 3-D structure of proteins can be elucidated experimentally by electron microscopy, x-ray and neutron diffraction, and multidimensional NMR methods. Despite predictions from the 1920s, theoretical computation based on quantum mechanics is still impossible. Only recently, classical force-field-based mechanics and dynamics simulations have been successful in predicting the structure of a very small protein. So, current knowledge of protein structure seems to be sufficient for solving Levinthal's paradox (*vide infra*). The center of this knowledge is, without doubt, the Protein Data Bank.

Protein Data Bank

Originally set up in 1971 at the Brookhaven National Laboratory (Bernstein et al., 1977), the Protein Data Bank (PDB) has been curated by the Research Collaboratory for Structural Bioinformatics since 1999 (Berman et al., 2000, 2003). As of December 2004, it holds 22,500 x-ray structures and 3500 NMR structures of biomacromolecules like nucleic acids, proteins, and viruses and can be queried in various ways. Because text queries depend heavily on the names given to the protein by the depositors, the most reliable way of searching the PDB is by aligning a protein sequence to all PDB sequences using BLAST (*vide infra*). A major part of our understanding of biochemistry and (patho)physiology is based on these structures, some of which have become very rich sources of knowledge. Public and free access to the coordinates is one of the hallmarks of the PDB initiative. Biology has followed the example of high-energy physics rather than that of chemistry in giving information freely to the entire scientific community. But this trend is threatened by commercial initiatives like the Celera database version of the Human Genome. Fortunately, all major journals require authors of structural biology papers to disseminate their results to the PDB before they are accepted. However, especially in case of pharmaceutical targets, delayed disclosure is often allowed for up to a maximum of one year (on-hold status). Although there is an initiative to introduce a standard format for both small molecules and biomacromolecules (Crystallographic Information File; CIF), the current PDB text-file format is widely accepted and the *de facto* standard of the modeling community.

A PDB file consists of an information header followed by the atomic X,Y,Z-coordinates, one atom per line. If the coordinates come from crystallography, the atomic occupancy and the anisotropic temperature factors follow the coordinates. In contrast to the isotropic temperature factors, which reflect the random variation of atomic positions in the crystal cells, the anisotropic factors reflect the mobility that all comparable atoms were undergoing in the course of the experiment, because of the dynamic nature of the protein. In NMR structures, the same information can be gleaned by overlaying all solutions to the nuclear Overhauser effect (NOE; see Chapter 6) constraints, as all these conformations might be minima for the protein under study.

Unfortunately, only one minimum is selected for further analysis by the biochemist and the biologist, neglecting a lot of information about the protein dynamics. In addition, the structural models follow from refining the coordinates to fit the distance constraints (NMR) or the observed electron density (crystallography), using force fields from earlier work, which leads to the current paradigm of structural biology; where x-ray structures are the foundation of force fields that are used to refine new x-ray structures. This resembles bootstrapping, and it would be very hard to find a molecular structure that deviates strongly from our current paradigm.

Secondary Databases

Several secondary databases automatically analyze the newly released PDB structures for further classification. These secondary databases typically hold structures or parts of structures (domains), superimposed structures, structure-based sequence alignments, and intra- and intermolecular interactions. Fortunately, these databases are freely accessible as well.

All entries in the Protein Data Bank are categorized in structural classes in the SCOP (structure classification of proteins; Murzin et al., 1995), DALI (Holm and Sander, 1996), and CATH (class, architecture, topology, homology; Orengo et al., 1997; Pearl et al., 2003) databases. Proteins are classified in alpha, beta, and mixed alpha/beta classes. Further subclassification leads to highly graphic fold descriptions as horseshoe, barrel, propeller, and sandwich. Fold families are represented by one member, creating a nonredundant knowledge base of structural biology. This sometimes leads to confusion, if, for example, an epidermal growth factor (EGF) domain or an immunoglobulin (Ig) fold is found in proteins without EGF or Ig properties. Classification databases rely on structural superimposition and comparison. These procedures have recently been validated (Novotny et al., 2004), and all perform equally well in classification.

Protein structure superposition is far from trivial. Comparison of two conformations of the same molecule is done by a simple least-squares fit of identical atoms, leading to a root mean square (RMS) deviation for all atoms (Cohen and Sternberg, 1980). Sometimes only the heavy backbone atoms or even only the C_α-atoms are used for these superpositions. Comparing two different molecules requires a definition of comparable residues, which makes this procedure dependent on the sequence alignment. Most methods, however, superimpose secondary structure motifs first, and some use simple vector representations.

Structure-based multiple sequence alignments like HOMSTRAD (Homologous Structure Alignment Database; Mizuguchi et al., 1998) perform better than normal sequence alignments, as they are also based on structural superposition. Another family database is Protein Families (Pfam; Bateman et al., 2004), which consists of hidden Markov models (probabilistic alignments). The families are manually checked and can be enriched by homologous sequences from molecules for which no structure is known (de Bakker et al., 2001).

Predicting Protein Structure

In the 1960s, Anfinsen proved that ribonuclease refolds to an active enzyme after denaturation. Levinthal, who started computer modeling (Levinthal, 1966), computed that a protein of 100 residues, with only 3 conformations per residue, could have 3^{100} conformations but manages to find its native fold in less than a millisecond. *Levinthal's paradox* is still the impetus for much research effort in structure prediction. In the age of genomics sequencing, and with the knowledge that structure rather than sequence determines function, an arms race of structure prediction has been initiated. Some researchers try to tackle Levinthal's paradox with modeling; others apply fold recognition or fold prediction techniques. Some modelers dedicate their entire career to folding proteins entirely by computer simulation with molecular dynamics. Progress is impressive, and the community is fueling progress by stimulating a structure prediction competition called CASP (Moult et al., 2003).

Comparative Modeling

The first critical assessment of structure prediction (CASP) meeting was held in Asilomar in 1994. This biannual event depends first on brave modelers who want to publicly deposit their predictions, second on altruistic structural biologists who are solving structures and do not publish the results before the CASP deadline, and third on impartial jurors who have to develop new tools for structure comparison, preferably automated routines, because the number of predictions is increasing enormously every two years. It is good to see that the field of comparative modeling has reached maturity since the first comparative model of α-lactalbumin, based on lysozyme (Browne et al., 1969). There has not been much progress in this field where models are made based on structures with considerable homology (more than 30%). The only large source for errors is the alignment of template and model sequence, which should be done with more sophisticated techniques than simply aligning the two sequences. Following the alignment, all or part of the template structure is mutated to the model sequence. The alignment insertions and deletions have to be fixed. After some refinement, the structure is checked for flaws.

Comparative modeling became a competitive discipline in the drug discovery field. G-protein coupled receptors (GPCRs) are abundant in the human genome, with more than 950 estimated members. These receptors function as the cell's sensors for light, smell, neurotransmitters, hormones, proteins, and so forth, and therefore were a grateful target to 45% of our current drug repertoire. This explains why all innovative pharmaceutical companies made comparative models for these receptors (Figure 1), based on a low-resolution microscopy of bacterial rhodopsin that was published in 1990 (Henderson et al., 1990). Even though this was not G-protein coupled, it did have the anticipated seven transmembrane (TM) helix topology of this class (only very few researchers were of the opinion that there were eight TM helices; they still could be right). In 2000, an x-ray structure of bovine rhodopsin was elucidated (Palczewski et al., 2000) and it became obvious that most predictions defined the wrong sequence in the first TM helix. In Figure 2, one can see that a misalignment of even 1

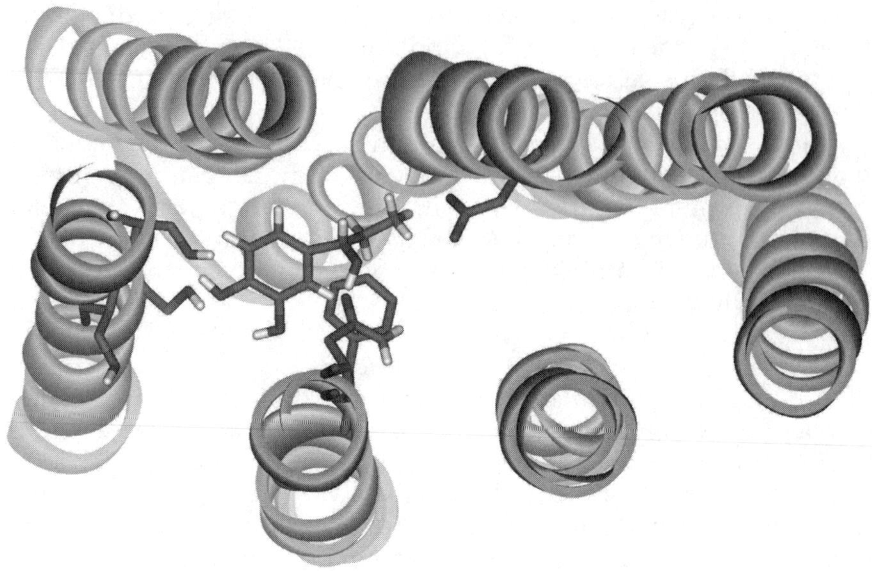

Figure 1 Noradrenalin modeled in a $\beta 2$-adrenoceptor model, which was taken from the GPCRDB (Horn, 2003). A shift of one residue in the sequence alignment would lead to a 120° shift for the most important receptor residues, preventing the model to be useful.

residue would prevent the essential amino acids from forming a binding pocket for the ligand. It is best to remember this quote by George Box: "All models are wrong; some are useful" (Box, 1976).

Alignment So, a comparative model is as good as the sequence alignment with the template structure. The first algorithms for global alignments (Needleman and Wunsch, 1970) and local alignments (Smith and Waterman, 1981) were capable of aligning identical nucleotides or amino acids. The first-generation algorithms used a unitary matrix with a score of 1 for identical residues and 0 for nonidentical residues (example 1 in Figure 2). A problem was the allowance of gap insertions. If you allow gaps of infinite size, an alignment will always succeed but reflect nonsense. Therefore, gap insertion was penalized. Second-generation alignment programs started using substitution matrices, appointing scores for amino acid substitution (example 2 in Figure 2). The newer matrices are based on observed mutation rates in multiple alignments. The PAM250 matrix refers to an evolutionary distance of 250 accepted point mutations per 100 amino acids, which is an extrapolation from a multiple alignment of closely related sequences to evolutionary very distant sequences. The BLOSUM62 matrix, default in BLAST (Basic Local Alignment Search Tool; Altschul et al., 1990), is derived from a multiple gapless alignment of proteins that are

1. Pairwise sequence alignment using an unitary matrix

```
seqA                VIVIALRLKIA
seqB                VIVS-LRL--A
                    ***  ***  *
```

2. Pairwise sequence alignment using a substitution matrix

```
seqA                VIVIALRLKIA
seqB                VI-VSLRL--A
                    ** ::***  *
```

3. Multiple sequence alignment

```
seqA                VIVIALRLKIA
seqB                VI--VSSLRLA
seqC                VI--VLRLKLA
seqD                VI-IALSLKIA
                    **  .  *::*
```

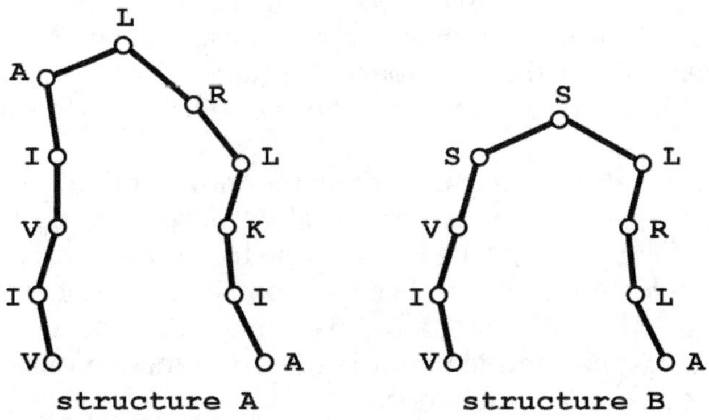

4. Structural alignment

```
strA                VIVIALRLKIA
strB                VIVS--SLRLA
                    ***    *::*
```

5. Multiple sequence alignment using structural information

```
strA                VIVIALRLKIA
strB                VIVS--SLRLA
seqC                VIVL--RLKLA
seqD                VIIA-LSLKIA
                    **:    *::*
```

Figure 2 Various types of alignments between sequences and structures. See the text for explanations.

less than 62% identical. The high diagonal scores of substituting a residue by itself reflect important function (e.g., cysteine) or less common occurrence in biology (e.g., tryptophan). Substitutions that are more common than expected are called positives and reflect the physicochemical similarity of residues (e.g., valine, isoleucine, leucine, and methionine) and/or similarity at codon level (e.g., serine is coded by AGC and arginine by AGA). Apart from scoring positive and negative substitutions, possible gap sizes are tried as well but at a penalty. The length of the total alignment is taken into account as well in the final score. The expectancy value reflects the chance of finding the alignment by chance alone and should be close to zero.

Because aligning two sequences is often unreliable, it is better to align the model to a multiple sequence alignment (example 3 in Figure 2). The best multiple alignments are based on probabilistic methods or based on structures (see the section "Secondary Databases," above). If the structures are known, it can be wise to insert gaps manually at positions where no alignment program would put them (see alignments 4 and 5 in Figure 2). The best alignments are finished manually.

BLAST is "the Microsoft Word program" of the field of bioinformatics; BLAST and other sequence comparison routines are also essential to gene finding, as most programs translate genomic or copy DNA sequences and then compare the translations to protein databases to find homologues and clues for the function of the gene.

Modeling The oldest approach in comparative modeling relies on separating the sequence in structurally conserved regions (SCR), which form the closely packed core, and structurally variable regions (SVR), which form random coils on the outside. In Ig heavy chains, it is obvious that the immunoglobulin framework is sequentially conserved, whereas the complementarity determining regions (CDRs) are not; they are hypervariable loops.

The coordinates for the backbone atoms of the SCR are copied from the template(s) to the model structure and the side chains of identical residues as well. Then, the side chains of substituted residues are added, using a rotamer library. Some programs first mutate the residues, whose local sequence can be found in the PDB. If the template has the sequence ILVEF and the model ILDEF, the program tries to find other occurrences of LDE in the hope of finding realistic rotamer conformations of the new amino acid (Chinea et al., 1995; Rodriguez et al., 1998). Another algorithm copies the Cα-atoms from a homologue and tries to match segments from other protein structures onto this scaffold. Yet another method does not copy coordinates, but rather optimizes the model to fit within a set of spatial restraints, taken from aligned residues in the template and the PDB, and geometrical restraints from a force field (Fiser et al., 2002). Most algorithms do not adhere to the one-on-one relationship between template and model but use many templates for the various SCR parts of the sequence to be modeled, creating a hybrid model (Schwede et al., 2003). Here, the danger lies within the fact that the protein fold does not resemble any observed structure anymore and is an extrapolation of our current knowledge, just like the mythical hybrids. The same holds for multidomain modeling. If we model an Ig light chain based on template X and an Ig heavy chain based on template Y, combining the two model chains might lead to overlapping residues, incompatible elbow angles, and so forth.

Because coordinates and secondary structures are often taken from widely varying proteins, there might not always be an evolutionary relationship between template and model. Therefore, the term *homology modeling* is replaced by *comparative modeling*. In the recent CASP5 meeting, it was shown that all programs for comparative modeling perform quite well, and not much progress has been observed since CASP4 (Tramontano and Morea, 2003). The chance of finding homologous and analogous structures is increasing with every new entry to the PDB.

Fold Recognition

If no structure can be found with more than 30% sequence homology to the model sequence, the number of SCR residues becomes too small for reliable predictions and the chance increases that hits are found in entirely different structural classes, making the initial choice impossible. Especially if no further biochemical data can support the choice for a certain homologue, comparative modeling becomes impossible. Fold recognition (Ponder and Richards, 1987; Bowie et al., 1991; Jones et al., 1992) is used more and more for these instances. Why are these methods called threading? A library of protein folds is selected, for example representative structures from SCOP or CATH. Then the model sequence is threaded through the structure, and all residues from the sequence visit all positions in the structure. At all positions, the residues are scored with a simplified energy function. In order to speed up the energy evaluations, atom-to-atom energies are replaced by potentials of mean force, reflecting the interaction between two new centers, for example centered at the C_β atom of the residues. These interaction parameters are taken from the PDB (knowledge-based) and should reflect observations such as hydrophobic leucines are not favored on the outside of proteins, and prolines are not favored in helices.

Fold recognition methods do not assume homology, but success is dependent on the availability of suitable folds, thereby limiting their usefulness, just like comparative modeling. Currently, new programs combine (multiple) alignment with threading. Other programs focus on local secondary structure prediction first, most notably the Rosetta program from the Baker group (Simons et al., 1997), a procedure that can be applied to new folds as well.

Ab Initio Methods

Because our knowledge base depends entirely on the folds that are currently known, structural genomics initiatives have been started to elucidate those structures first that cannot yet be classified into a known fold family based on their sequence. The sooner all protein folds are present in the PDB, the better comparative modeling and fold recognition will work. The Rosetta approach (Simons et al., 1997) from the Baker group seemed to be ahead of the competition in a past CASP meeting (Aloy et al., 2003) in the prediction of new folds. In their approach, the model sequence is split up in nine residue segments, which rapidly convert between secondary structures, in a stochastic approach. The secondary structures are taken from PDB files. A simplified energy function is used to evaluate the ensemble of locally folded segments.

It is possible, however, that some folds will never lead to protein crystals for x-ray diffraction, will never be amenable to NMR spectroscopy, are transient states of naturally unordered and floppy proteins, or only show order in a particular environment like membranes. For these instances, total *ab initio* methods may prove their worth. These methods do not rely on the PDB as a knowledge base but rather on atomic force fields.

Force Field Of course, theoretical physical chemistry gives an alternative for structure elucidation. If the laws of physics and chemistry are valid and can be applied, no experimental structures would be needed any more. Dirac, one of the pioneers of quantum mechanics, hinted at this in 1929. However, the Schrödinger equation can only be analytically solved for two-particle systems. Approximate methods were introduced and put in computer algorithms in the 1960s by Pople (Hartree-Fock) and Kohn (density functional theory; both shared the Nobel Prize in 1998), but even with the computer speed increasing according to Moore's law (doubling every two years), we still can only perform one self-consistent field calculation on one conformation of a protein. Fortunately, the classical mechanics laws can be applied successfully, and molecular mechanics and even molecular dynamics of proteins have been studied since the 1970s. A typical classical mechanics force field uses databases of empirical geometric and electronic parameters for as many atom types as needed. These parameters are input in the force-field equations to compute bonded energies for bond distance (between two atoms), bond angle (between three atoms), and torsion angle (between four angles or two planes). Harmonic potentials are used for these bonded energies, giving a minimum value for every possible bond and angle. A cosine function is used to reflect periodic minima for the torsion angles.

The nonbonded interactions are simulated by Coulomb electrostatic and Lennard-Jones steric attractions and repulsions. Because the Coulomb energy scales with $1/r$, it is the longest range nonbonded force for molecular recognition and a strong force for formally charged systems, only masked by the dielectric properties of the medium in between. In closer proximity, the steric attraction is added to the electrostatic attraction, scaled as $1/r^6$, and at shortest distance, van der Waals overlap is prevented by a repulsion term that scales as $1/r^{12}$. Sometimes, additional terms have been added to improve simulation of hydrophobicity and hydrogen bonds, to account for cross-terms (e.g., the effect of bond stretching on bond angles), polarization, and solvation, but no addition has found widespread use over the terms mentioned here.

Simulations The evaluation of the energies of several conformations of a molecule is known as conformational analysis. If an optimization routine is applied, all new conformational states are found by changing the atomic positions toward lower energy (minimization). We could also compute the energy of all conformations, while changing one conformational variable, for example by changing a torsion angle with 30° increments or by random changes (stochastic methods, e.g., Monte Carlo). Both procedures, however, fail to sample the entire conformational space of a typical protein due to Levinthal's paradox (*vide supra*).

Molecular dynamics is a conformational analysis where time connects the conformations (deterministic rather than stochastic). The molecular mechanics force field is applied to compute the energy of a molecular system, and this results in forces on the surrounding atoms. Based on their mass, they move with a certain speed in a certain direction and the energy is evaluated once again, intermittently like a stroboscopic light flash. Then, the direction and size of the force and acceleration of the atoms will change, and the procedure starts again. It is important to evaluate the energies often enough, to prevent atoms from colliding and eventually the system from exploding, so typically the energy is evaluated every femtosecond. Furthermore, minimization should ensure that most of the strain is out of the system before dynamics is started, without starting in a deep minimum. After gradual heating, even physically unrealistic temperatures can be applied to enhance the atomic speeds and overcome high energy barriers between conformational minima. In order to simulate coupling to a heat bath of, for example, 300 K, all velocities are rescaled periodically (Berendsen et al., 1984). In molecular dynamics, solvent water molecules are often taken into account as well, and the experts are certain that their algorithms reflect the molecular interactions well enough to simulate the protein folding process. Much time is spent, however, on the computation of nonbonded energies of every possible nonbonded atom pair. In a protein of 5000 atoms, we would have to evaluate 25 million nonbonded atom pairs. Because every femtosecond the energies have to be evaluated between thousands of atoms and millions of nonbonded atom pairs, it takes quite some computer time to simulate the folding process that is assumed to take milliseconds.

Still, the first successful molecular dynamics simulation of protein folding was reported in 1998 by Duan and Kollman for a very small 36-residue villin headpiece subdomain. However, this does not mean that only our limited computer speed prevents total *ab initio* folding of all proteins, because many proteins have disulfide-bridged intermediates in their protein folding pathway, and currently molecular dynamics cannot make and break covalent bonds. Still, it is good to know that our understanding of molecular behavior and our current algorithms are able to mimic biochemical reality. It also means that most modelers are constantly trying to come up with approximations to speed up calculations, like larger time steps between energy evaluations (SHAKE; van Gunsteren and Berendsen, 1977), implicit solvation (see the section "Protein Protonation and Solvation," below), and a cutoff distance for the nonbonded atom pairs (sometimes a larger cutoff distance for electrostatic interactions than steric interactions), introducing new artificial boundaries and simulation artifacts.

Partial Atomic Charges In quantum mechanics, a molecular orbital (MO) calculation is performed to self-consistent field (SCF; when the ingoing and outgoing atomic orbital contributions to the molecular orbitals are the same), sometimes followed by geometrical changes of the nuclei in an optimization algorithm, again followed by another MO-SCF calculation, and so forth. At all times, the charge distribution of the system reflects its current geometry. This is not the case in molecular mechanics, where the partial atomic charge initially given to an atom stays unchanged during the entire simulation. This leads to erroneous results, and therefore most force-field developers have used the concept of atom types, reflecting an atom's context. So, an amide oxygen atom has another charge and size than a hydroxyl oxygen atom, but further polarization during the simulation is not allowed. In some force fields, well over 200 atom types have been defined. It is quite difficult for the user to parameterize new atom types in case of nonpeptides like drug-like molecules, cofactors, or post-translational modifications. This has led to quantum dynamics and quantum/molecular mechanics hybrid force fields, which are currently being fine-tuned (Titmuss et al., 2002; Vreven et al., 2003).

Prediction of Disorder and Membrane Binding

Intrinsically disordered proteins and most membrane-bound proteins have resisted structural elucidation. A new CASP contest is dealing with predicting disordered proteins (Melamud and Moult, 2003). Normally, these proteins have high polarity and refuse to form secondary structures. However, it has been suggested that most proteins can occur in at least two states: ordered and disordered (Ptitsyn and Uversky, 1994; Radijovac et al., 2004). This would lead to a refinement of Anfinsen's postulate, which states that the amino acids completely determine the native fold. With a two-state model, we might be able to shed light on what has been called conformational diseases (Lomas and Carrell, 2002).

There are several prediction programs for transmembrane segments (Melen et al., 2003). They are all based on hydropathy plots or empirical observations. The main problem in predicting the topology is that stretches of around 20 hydrophobic amino acids favor helix as well as sheet formation. Combined with secondary structure propensity values, this leads to predicted topologies of transmembrane sheets (e.g., porins) or helices (e.g., 7TM GPCRs). With the use of a database of experimental data on transmembrane segments (Ikeda et al., 2003), the algorithms can be trained to predict optimally transmembrane propensities. A recent method, BPROMPT, uses a Bayesian network to find consensus between all the other transmembrane prediction servers (Taylor et al., 2003). Reliable structural models for transmembrane proteins have not been made yet, unfortunately, although there have been molecular dynamics simulations of GPCRs in membrane environments (Mehler et al., 2002). Solid-phase NMR studies might offer further help.

Checking the Quality of the Model

After modeling a protein, the resulting structure should be checked for biochemical and structural integrity. Procheck (Laskowski et al., 1993) and WhatCheck (EU 3-D Validation Network, 1998) are popular programs. They measure the internal geometrical parameters and compare these with the mean values of nonredundant subsets of structures of a comparable resolution. Especially, the phi/psi torsion angles are important. Ramachandran et al. (1963) plotted the phi/psi angles of protein residues, immediately showing that the majority of the residues in a protein are part of a secondary structure.

In 1996, Hooft et al. reported 1 million errors in the PDB (Hooft, 1996a). Although structural biologists strongly disagreed with their conclusions, there are many pitfalls in structure elucidation (Kleywegt, 2000). In some structures, residues are missing, or wrongly assigned, or in conflict with the SwissProt sequence database (Gasteiger et al., 2003). Often atoms are missing as well, or disordered in the electron density. In general, only if the structure is resolved to less than 2 Å resolution has strong electron density been observed for all side chain atoms.

The electron density for isosteric groups like CH_2, NH, and O is similar, so it is possible that side chains are rotated in the wrong way. An exhaustive investigation of the possible hydrogen bond networks helps in finding the best rotamers for residues like asparagine, glutamine, and histidine. WhatCheck computes a hydrogen bond interaction network and suggests side chain flips and warns for unpaired polar groups, which would destabilize the protein (Hooft, 1996b).

Protonation and ionization of the amino acids is another important modeling aspect. This will be discussed further on. At this point, it is relevant to mention that in case the PDB depositors did not use IUPAC conventions for atom naming, the atoms are considered as heteroatoms, and modeling programs cannot automatically add hydrogen atoms or partial charges to the atoms. The pK_a values of the isolated amino acids are well established, and modeling programs automatically charge the acids to glutamate and aspartate, as well as the bases arginine and lysine. The pH value of the crystallization conditions is never mentioned in the PDB file, however. Histidine with a pK_a value of 6 is either protonated or neutral, depending on its local environment. In enzymes, like serine proteases, this property of histidine is crucial to the enzymatic mechanism. Other titratable groups are cysteine and tyrosine. The N and C terminus of all protein chains has to be charged as well by the protein modeler, unless biochemical experiments have shown acetylation or amidation modifications.

The topology of prosthetic groups is another point of attention, as modeling programs only recognize the twenty proteinogenic amino acids and four nucleotides. Ligands, cofactors, modified amino acids, and other prosthetic groups should be manually corrected for atom and bond types before the modeler can perform a computation. The structure of noncovalent ligands is often based on very poor electron density (Davis et al., 2003).

Protein Design

In contrast to the problems of predicting 3-D structure from sequence, predicting sequence from structure is much easier and often applied by protein designers. In order to make an all α-helix protein, one just has to use residues that have strong α-helix propensity, like alanine, leucine, and glutamate. In a very interesting study, Dahiyat and Mayo (1997) fully designed a sequence to mimic a zinc finger structure. The zinc finger Zif268 is one of the shortest sequences with a strong preferential fold, mainly because of two histidine and two cysteine residues that are ligand to a Zn^{2+} ion. The authors classified 18 of the 28 positions as surface exposed, where 10 amino acids could occur; 7 boundary positions where 16 amino acids could occur; one core position with 7 possible amino acids; and 2 positions where they needed a glycine. With these choices, they reduced the search space from 20^{28} (all amino acids possible at all positions) to 10^{27}. They also wanted to evaluate side chain rotameric positions to prevent steric clashes, thereby increasing their search space again to 10^{62} possibilities. With a dead-end–elimination algorithm (Desmet et al., 1992), they came up with a list of hits in a reasonable time. The top hit had no sequence resemblance to Zif268, or any other protein, known at that time. It also lacked cysteine and histidine residues. NMR structural elucidation, however, did prove that their sequence gave rise to the exact same structure as Zif268.

Even more striking is the full design of a 93-residue protein with a completely novel fold (Kuhlman et al., 2003). X-ray and NMR analysis showed that the model was correct.

Programs for Protein Structure Predictions

In Table 1, an overview of programs for structure prediction is given. It is also indicated whether or not the user can submit his own sequences. In this table, free tools and servers are listed for making, validating, and visualizing protein models.

Protein Interaction Properties

Proteins do not function in vacuum, but their fate is determined by their surroundings. This could be temperature, light, water, ions, pH, inorganic and organic molecules, peptides, proteins, and nucleic acids. This causes the variation in, and the highly dynamic nature of the human proteome, where a possible 100,000 proteins have different interactions in 100 trillion cells in 6 billion people.

The pharmacodynamics of pharmaceutical proteins is determined by interactions with other proteins, insulin with the insulin receptor, omalizumab with IgE, etanercept with TNF-α, and so forth. The pharmacokinetics is determined by interactions with water, temperature, pH, ions, endocytosis receptors, modifying enzymes, immune system components, and so forth.

We will first focus on the covalent modifications that a protein can undergo and then focus on the noncovalent interactions.

Protein Modifications

Proteins have many disguises after ribosomal synthesis. Even during translation, changes can already occur. Most proteins have to be trafficked, signal peptides have to be cleaved off, proteolytic activations and inactivations can occur, covalent groups are attached, and mutations can occur within individuals (e.g., in tumors) and between individuals. All these but the last are known as post-translational modifications (PTM).

Post-translational Modifications

Studying PTM at an individual level will become increasingly important in medicinal research, as PTM determines the phenotype. In the past, PTM was outside the reach of genomics and generalized proteomics studies. The use of mass spectrometry (MS) has changed the face and pace of proteomics, however (see Chapter 13). Proteins are separated by two-dimensional (2-D) gel electrophoresis, based on their pI value and their molecular size. The individual spots are taken from the gel and characterized by MS, after trypsin digestion. The mass spectrum is compared to the masses of theoretically digested proteins in the SwissProt database. The mass changes caused by all currently known PTM are checked as well.

Deamidation of glutamines and asparagines is an interesting PTM, as it does not require enzymatic help. Because the electron density of glutamine and glutamate, and of asparagine and aspartate, is almost identical, crystal structures often show conflicts with the SwissProt sequence databases. It is very difficult to establish if deamidation has already occurred. Robinson and Robinson (2001) have proposed that deamidation functions as a molecular clock, a timer for starting or stopping protein action. If deamidation occurs in long-lived proteins, the effects of introducing a negative charge are quite destabilizing and may lead to changes in conformation and behavior of these proteins. The deamidation propensity of asparagine and glutamine in different pentapeptide contexts was measured (Robinson and Robinson, 2001), and the deamidation of asparagines in most PDB proteins was predicted (Robinson, 2002).

Pharmaceutical proteins can become immunogenic after PTM. Koning and his group have shown that deamidation of glutamine residues enhances T cell response in celiac disease models (van de Wal et al., 1998). Deamination of arginine leads to citrulline. In multiple sclerosis patients, myelin basic proteins with high citrulline content have been observed (de Seze et al., 2001). In rheumatoid arthritis, citrulline is so common that anticitrulline antibodies are excellent diagnostic markers (Rantapaa-Dahlqvist et al., 2003). Molecular modeling could help in predicting these PTM.

Protein Mutations

One of the early payoffs from genomics is the mapping of single nucleotide polymorphisms, or SNPs. A polymorphism is a mutation that occurs in more than 1% of the population. SNPs do not have to be located in genes to be useful; they can always be used as diagnostic markers in pathology and pharmacogenetics (Roses, 2000). Although it was at first estimated that one in every thousand bases could be a SNP, more recently it was shown that many SNPs are inherited simultaneously and that the human population can be subdivided in a limited number of SNP haplotypes (The International HapMap Consortium, 2003). Some SNPs could occur in genes and some of these could indeed result in the change of an amino acid, the nonsynonymous SNPs. But what does this point mutation mean for the protein? Does it change protein conformation, stability, interactions? Much attention has been paid recently to the effect of point mutations (e.g., Herrgard et al., 2003).

In early work on human growth hormone (HGH), Wells and collaborators at Genentech have shown that affinity can be determined by just a few residues (Wells, 1996; Figure 3). After mutating a tryptophan in the HGH receptor and a threonine in HGH, activity was lost and the affinity decreased by a 1000-fold (Guo et al., 2000). Interestingly enough, they were able to restore the affinity and activity by a small compound that could fit the gap that was created by the mutations. This implies that even some pathological mutations could be corrected with a drug-like molecule.

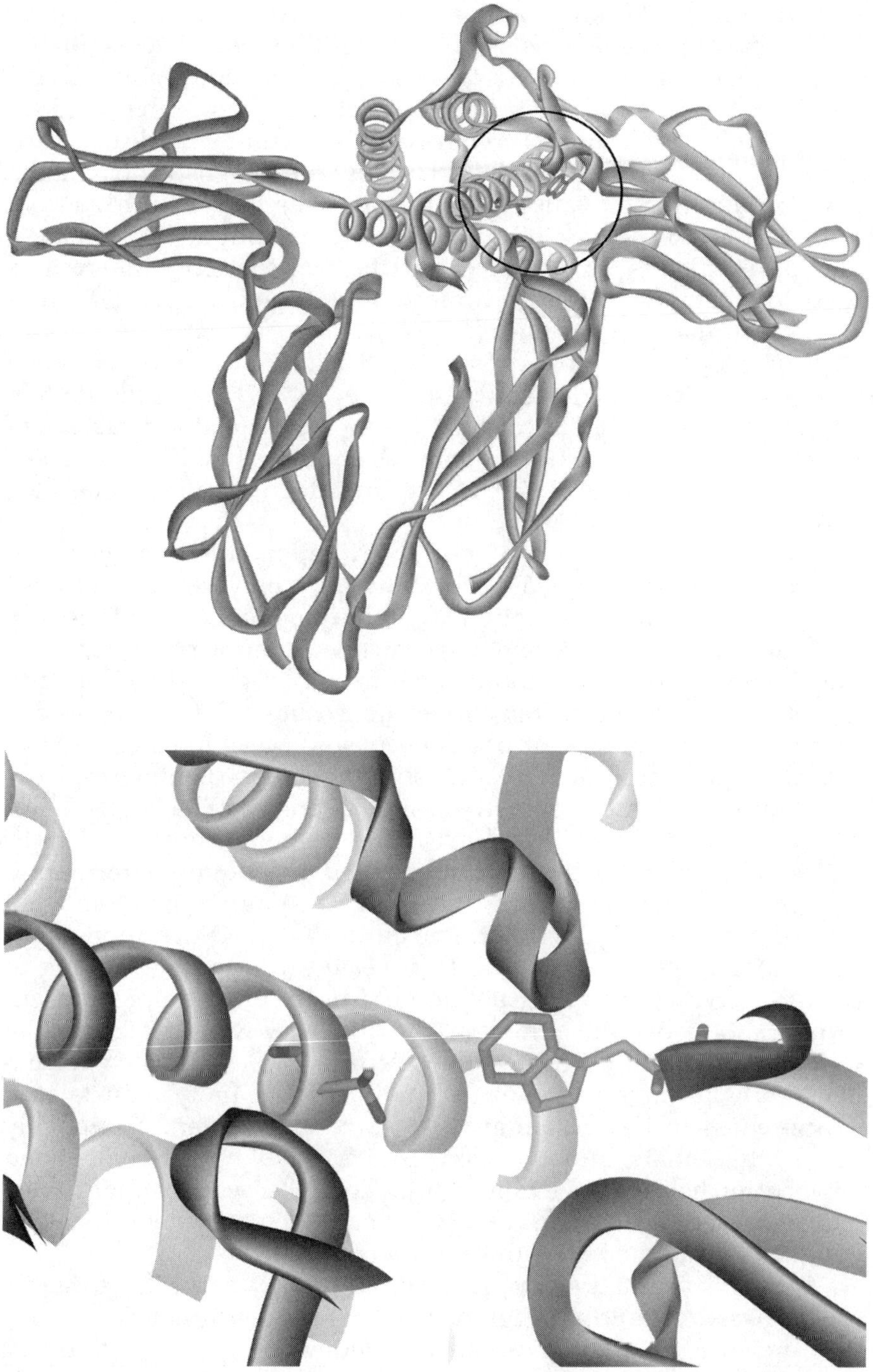

Figure 3 HGH binding to the receptor dimer (top, PDB-file 3hhr, de Vos, 1992). The important tryptophan and threonine residues are shown after zooming in (bottom).

Elaborate mutation studies involve T4 lysozyme. At a surface-exposed position, Ser44 was mutated to all other amino acids (Blaber et al., 1993). Fifteen x-ray structures were determined, enabling a correlation between structure and protein stability, as represented by the melting temperature. Never was the structure significantly different from the wild type, except a slight helix distortion in the Ser44Pro mutant. This caused a drop in melting temperature as well. Unfortunately, the structural changes are too subtle to be correlated to or be predictive for stability. But the fact that some mutations failed to crystallize might indicate that these mutations did cause major conformational changes.

Several databases hold energetic and calorimetric information on mutations, for example, MutaProt (Eyal et al., 2001) and the Protein Mutant Database (Nishikawa et al., 1994). There are also databases of observed mutations in HIV proteins, as parts of the HIV drug resistance database (Rhee et al., 2003), and intended mutations in GPCRs, GRAP (Beukers et al., 1999).

The effect of mutations is not always apparent from structural analyses. Germ-line antibodies normally have only weak affinity for antigens. After somatic mutation, an affinity mature antibody is produced by plasma cells. Wedemayer et al. (1997) performed a structural comparison between an affinity mature and a germ-line antibody against a small hapten, a transition-state isostere. The antibodies differed in nine positions, but only one position was in direct contact with the hapten, according to x-ray structures of the complexes. How could this explain the observed affinity difference of 30,000? The answer came after solving the structure of the uncomplexed antibodies. The affinity mature antibody had the same conformation with and without the hapten, but the germ-line antibody showed signs of substantial induced fit. The authors argued that induced fit means that more energy had to be invested for binding, leading to affinity decrease. If one breaks down affinity in enthalpic and entropic components, the affinity mature antibody was rigidized in the right conformation, leading to less entropy loss and less conformational enthalpy investment upon binding. The affinity mature antibody seemed to be preconfigured to bind the hapten, presumably because the somatic mutations led to a very stable hydrogen-bonded network in the antibody framework. This observation might have wider applicability: the lock-and-key model would by definition lead to higher affinity than the induced fit model.

Sometimes mutations appear to be correlated, according to statistical analysis of multiple alignments. This suggests that these positions are in a functional interaction, under the same evolutionary pressure. The correlated mutation analysis was applied to find possible interacting residues in GPCRs (Oliveira et al., 2002).

Protein-Protein Interactions

The principles of protein-protein interaction are universal, but the goal of these interactions gives rise to some subtle differences. Antibodies and antigens show large complementarity, whereas complexes between T cell receptors (TCR) and major histocompatibility complex (MHC) proteins do not, because these complexes have to be transient; the TCR has to probe more MHC-peptide complexes on more antigen-presenting cells. Most complementarity, due to evolutionary refinement, can be found in enzyme-substrate complexes. Still, serine proteases in blood coagulation have to be very specific, whereas serine proteases in digestion cleave without much specificity.

Surface plasmon resonance (see Chapter 15) has become a standard tool for measuring protein-protein interaction. Yeast two-hybrid screening has become the standard for rapid screening for protein-protein interactions (Mendelsohn and Brent, 1999). Modelers can predict protein-protein interactions by docking methods (*vide infra*) and sometimes even by simple sequence analysis. Marcotte et al. (1999) predicted that domains of different proteins would interact, if these domains are also found to be part of one single protein in any genome. The possibility of intramolecular interaction is predictive for intermolecular interaction. Despite many false positives, this method was the first contribution of bioinformatics to interaction proteomics.

If we have two protein structures, we do not need to dock to predict the binding interfaces. Sometimes steric complementarity is obvious, if we use a surface display of the proteins. McCoy et al. (1997) have used finite difference Poisson-Boltzmann calculations to prove that protein-protein interfaces show considerable electrostatic complementarity. Electrostatic complementarity becomes obvious if we color the surface according to the molecular electrostatic potential. This is also indicative of the large hydrophobic surface patches, which are believed to be the driving force for protein interaction. Still, a visual inspection of complexes of homodimers revealed that hydrophobic contact surface is only apparent in a minority of cases (Larsen et al., 1998).

Here we will focus on those interactions that are relevant to pharmaceutical proteins. We will discuss the interaction of proteins with water, with the immune system, and end with protein-protein docking.

Protein Protonation and Solvation

Interaction with water and hydrophobicity are important protein properties. In 1971, Lee and Richards defined the solvent accessible surface of a protein by rolling a probe of, for example, 1.4 Å, comparable to the radius of water, over a protein model without penetration of the van der Waals radii, and computing the contact surface, traced by the center of the probe. Later, in 1983, Connolly computed the contact surface with a probe of a given radius. The solvent accessible surface area

(SASA) turned out to be related to experimentally observed energetic and calorimetric parameters (see Chapter 17). The contact surface of two proteins, as computed by subtracting the SASA from two interacting proteins, was correlated to binding affinity. Especially, the apolar contact area between carbon and apolar hydrogen atoms was predictive.

Formal charges determine the polarity of a protein. As discussed earlier, the modeler has to make decisions about the charged state. HIV proteases are aspartyl proteases, where two aspartic acids in a homodimer form the active site. From physical chemistry, we know that it would be quite unlikely if both residues were negatively charged. But all possible protonation states occur, depending on the binding partner (Figure 4). This leaves us with the humble notion that the hydrogen atoms that are invisible to x-rays play the decisive role in biochemistry. Before we can predict interaction, we need to know the protonation states at the interaction surfaces. A recent overview was given by Nielsen and McCammon (2003).

High-quality diffraction studies of small molecules have allowed an experimental determination of electronic structures; for larger molecules, we have to use protein modeling to compute electronic properties. What does a molecule look like? Most likely, molecular electrostatic potentials are more relevant for molecular recognition than atomic topology.

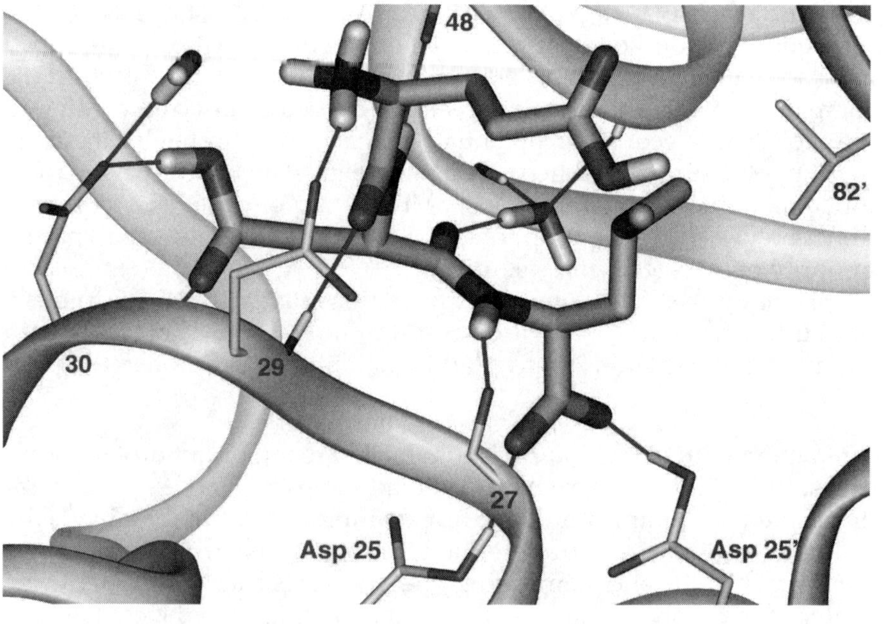

Figure 4 The tripeptide EDL bound to HIV-1 protease (PDB-file 1a30, Louis, 1998) with both catalytic Asp25-residues protonated (figure by M.B. Gillies).

Protein electrostatics was pioneered by the Honig group (Gilson and Honig, 1987). They compute the electrostatic potential by the finite difference Poisson-Boltzmann equation, with the protein placed in a grid. These authors did not use explicit water molecules but treated solvation implicitly with continuum electrostatics to speed up computation.

Molecular dynamics aims to simulate real biological processes, even though it reflects life at a concentration of 10^{-23} M. Reality is helped by the explicit addition of solvent molecules and counterions. In order to prevent the outer molecules from evaporating, periodic boundary conditions are often applied. In its simplest form, the molecular ensemble is copied twenty-six times, to generate twenty-seven identical boxes with the simulation box in the middle. If a water molecule leaves the box in the upper-left corner, it will reappear in the bottom-right corner, keeping the constant volume and constant pressure conditions intact.

Because of the computational cost of explicit water molecules, some implicit methods of solvation have been developed. These methods are based on continuum electrostatics or solvent accessible hydrophobic areas or a combination thereof, like the generalized Born/surface-area method (Still et al., 1990). Effective dielectric models were the first trick to simulate the fact that at longer distances, bulk water has a dielectric constant of 80 (at room temperature), thereby preventing charges from sensing each other clearly, but if two approaching molecules have totally desolvated and are in close proximity, a dielectric constant of 1 would apply. By replacing the dielectric constant by the distance r, the Coulomb electrostatic energy scales to $1/r^2$ and better reflects this solvent discontinuity (Figure 5). This is known as the distance-dependent dielectric constant model.

Other modelers have chosen the opposite approach and assume a continuous dielectric phase in the solvent (typically 80), but a different one in the protein (typically 1, 2, or 4). This means that charged residues have long-range electrostatic effects within proteins and short-range interactions outside. So, most charged amino acids could influence the electrostatics of an active site, whereas two equally charged residues on two separate arms of the proteins would not immediately lead to electrostatic collapse (Figure 6). It is good to realize that by modifying the electrostatics, protein modelers completely determine the outcome of their simulations.

Antigenicity

Many pharmaceutical proteins face the problem of immunogenicity, leading to suboptimal pharmacokinetics. Immunogenicity is not a problem but essential to vaccines and antibodies. What is immunogenicity and can it be predicted? Foreign organisms are detected by Toll-like receptors on dendritic cells, but all proteins, self and non-self

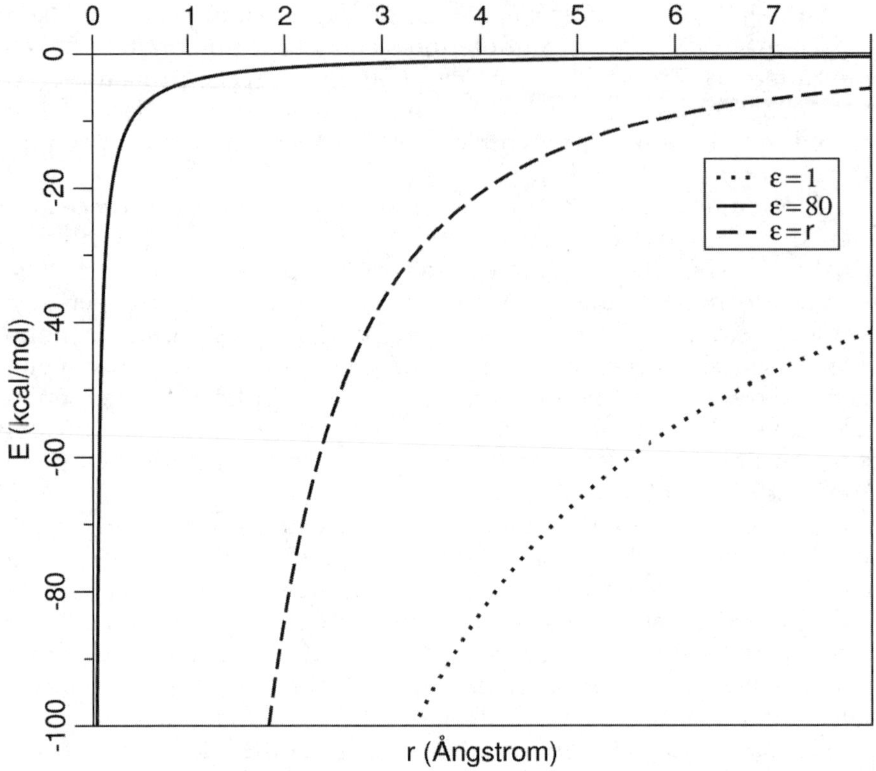

Figure 5 Electrostatic energies between opposing charges, modeled by -1/r, -1/80r, and -1/r2. The latter one resembles experiment best.

can be detected by antibodies from B cells and by T cell receptors (TCRs) on T cells. Normally, discrimination between B cell epitopes and T cell epitopes is made in vaccine studies. Both epitopes could be part of the same immunogenic protein or come from different parts of a microorganism. B cell epitopes have to be recognized in their protein 3-D context by antibodies, whereas T cell epitopes are processed by antigen-presenting cells and presented to TCRs in a standard fashion by molecules of the MHC. This immediately suggests that B cell epitopes are always on the outside and T cell epitopes could be on the inside of proteins, microorganisms, or cells. Much effort has been directed to the prediction of these epitopes.

B Cell Epitopes One of the first prediction algorithms was by Hopp and Woods (1981), based on a hydrophilicity scale like the one from Kyte and Doolittle (1982). Later, Jameson and Wolf (1988) defined an antigenicity index as a weighted linear combination of several properties. Apart from hydrophilicity, they compute secondary structure propensities (Chou and Fasman, 1974, 1978; Koehl and Levitt, 1999),

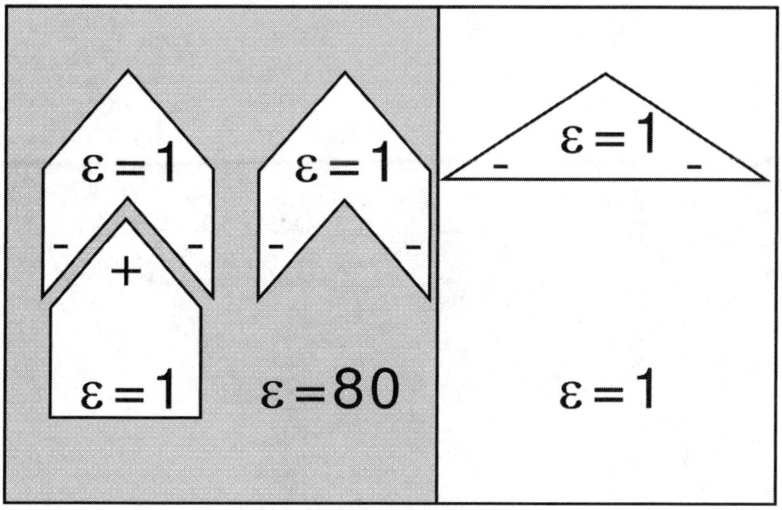

Figure 6 Two anions in a protein interact with a cation in the other protein. Without this protein, water should mask the anions from causing an electrostatic repulsion, forcing the protein to collapse in the normal vacuum modeling situation.

surface probability (Emini et al., 1985), and flexibility (Karplus and Schulz, 1985). All properties should be high, except secondary structure content. This corresponds well with the observation that secondary structures like α-helices and β-sheets and hydrophobic areas in general are found on the inside of proteins. There is also a statistical antigenicity propensity on a residue level; for example, cysteine, leucine, and valine are quite common in epitopes (Welling, 1985).

In an investigation of the known epitopes of foot-and-mouth disease virus, all well-known predictors performed worse than simply measuring the distance to the center of the virion (unpublished results), thereby proving the obvious (Thornton, 1986), that antibodies bind the most accessible sites. From an immunological perspective, we know that just a subset of antigens is immunogenic. If hydrophilic surface loops are sufficient for binding antibodies, our immune system would not be able to differentiate between self and non-self. We have investigated this aspect and found a good correlation between those positions that are variable in multiple alignments of orthologous sequences, as measured by sequence entropy, with observed antibody binding. Antibodies have some affinity for most exposed hydrophilic loops but high affinity for those that possess non-self residues (Figure 7).

A disadvantage of all these programs is their inability to predict discontinuous epitopes. Furthermore, it is hard to differentiate between all these predictors, as they strongly intercorrelate. Hydrophilic residues are always on the outside of proteins and so are loops.

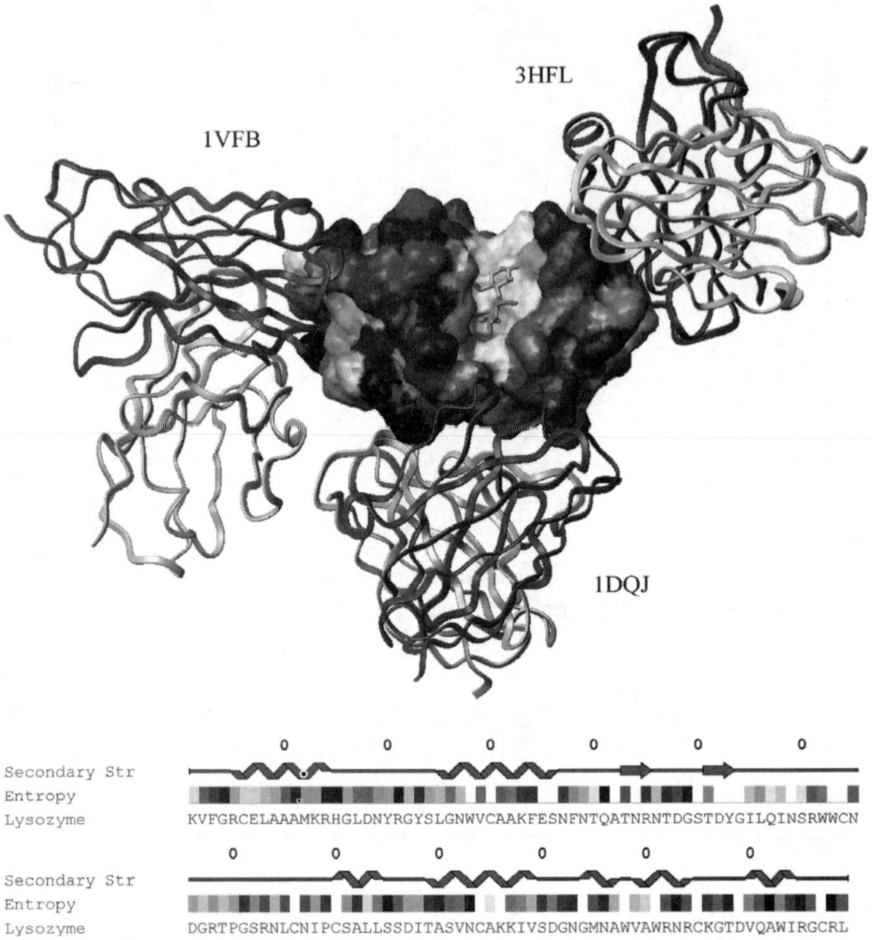

Figure 7 The antibody fragments (Fv) shown are D1.3 (PDB-file 1vfb, Bhat 1994), HyHEL-5 (PDB-file 3hfl, Cohen, 1996), and HyHEL-63 (PDB-file 1dqj, Li, 2000). Sequence, secondary structure, and sequence entropy are given as well, where high entropy is darker. Hen egg lysosome is shown with a substrate (PDB-file 1h6m, Vocadlo, 2001) and its molecular surface is color coded by sequence entropy; light near the substrate and dark near the Fv domains.

Sequence variation between orthologs is often strongest on the outside, if we assume that the function of most proteins is conserved in some interior residues. Viruses select for mutations on exposed positions to escape the immune system.

There are more than 200 antibody-antigen complexes in the PDB. Analyzing these complexes also shows a preference for loops and turns (Figure 8) in the antigenic epitopes (Dyson and Wright, 1995). Despite reasonable success in epitope prediction based on sequence, a better way of finding epitopes is by docking antibodies to antigens

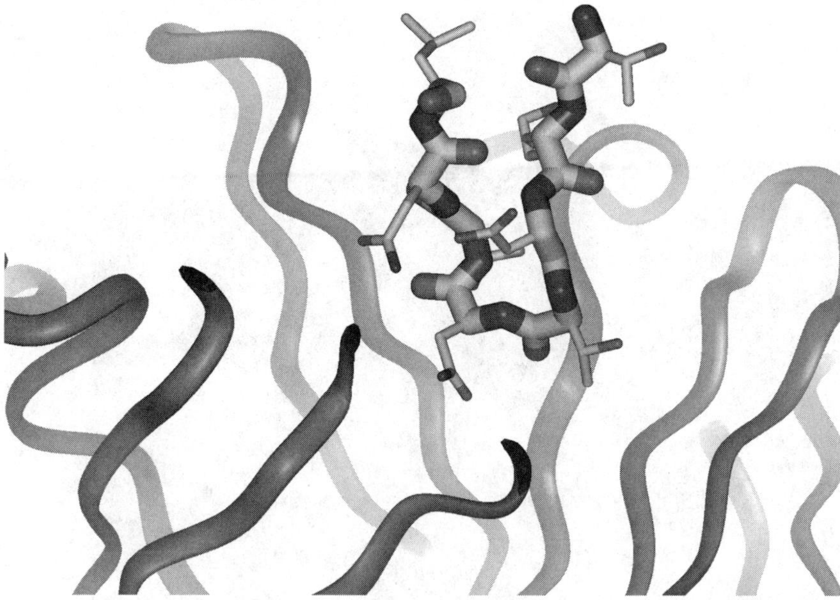

Figure 8 A peptide from *N. meningitidis* outer porin complexed as a type I β-turn to MN12 Mab (PDB-file 1mpa, van den Elsen, 1997). The CDRs of the light chain (front) have been removed for clarity.

(see the section "Protein-Protein Docking," below), in case the structures are known.

T Cell Epitopes All cells present internal proteins as eight to ten amino acid peptides via MHC class I molecules to TCRs on cytotoxic T cells, enabling swift response to, for example, viral infection of these cells. Many specialized immune cells are able to process proteins from the surroundings as well and present nine to twenty-five amino acid peptides with MHC class II molecules to helper T cells, which could respond by proliferation and the expression of cytokine messengers in order to help B cells and preserve immunological memory. The MHC molecules stretch the peptides using widely spaced hydrogen bonds with well conserved MHC residues for optimal presentation to the probing TCRs (Figure 9).

Humans express six class I and ten class II MHC molecules. These MHC molecules are polymorphous. This polymorphism underlies graft rejection after transplantation, as well as the genetic associations to autoimmune diseases like insulin-dependent diabetes mellitus, rheumatoid arthritis, and multiple sclerosis. The polymorphic residues of the MHC molecules line the pockets for peptide anchor residues and determine the specificity of the MHC molecule for a certain peptide repertoire (Figure 10). HLA-DR1, for example, preferentially presents

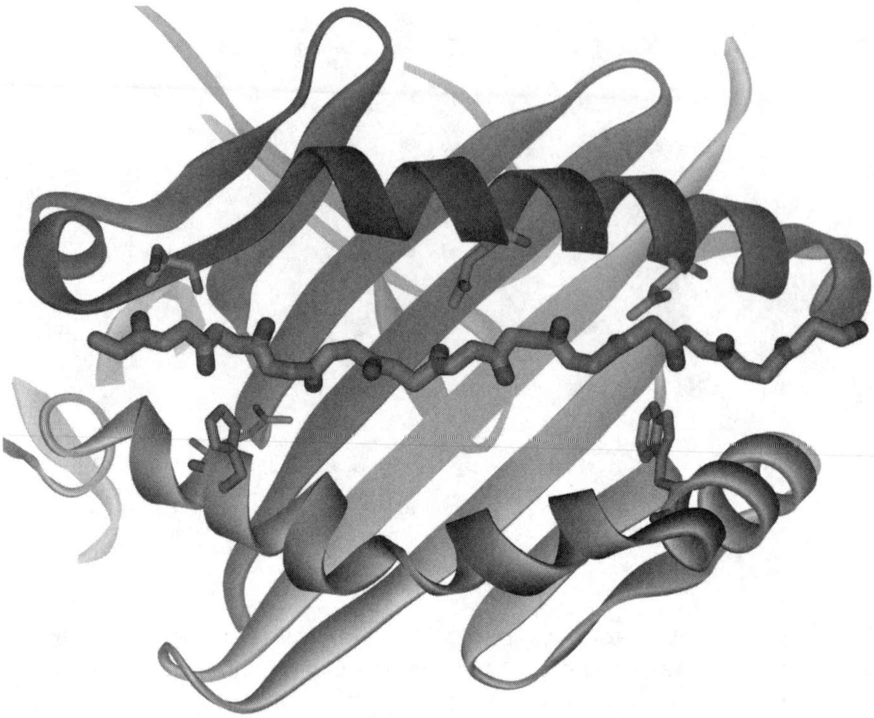

Figure 9 Conserved hydrogen bonds between the residues 53, 62, and 69 of the α-chain (top), as well as 61, 81, and 82 of the β-chain (bottom) to the backbone of the influenza hemagglutinin peptide (PDB-file 1dlh, Stern, 1994).

peptides of the XX[ILVY]XXXXXXX[VAILF]XX motif (Kropshofer et al., 1992), where the brackets denote sequence ambiguity at a single position. With only two or three anchor positions, MHC molecules are quite promiscuous binders, and every MHC molecule is believed to present thousands of different peptides. This also means that the TCR is responsible for the real self–non-self recognition. It is also obvious that we would need several T cell epitopes to have a vaccine that works in every individual, regardless of their MHC repertoire. The first structure of a peptide-MHC-TCR complex showed that only one peptide residue is in actual contact with the TCR (Garboczi et al., 1996). It is unlikely that TCRs "base their judgment" on only one residue, so the other residues must change the MHC conformation, making it more antigenic for the TCR.

T cell epitope prediction programs align sequences to motifs, which have been generated by peptide elution assays (Falk et al., 1991) or other binding studies and stored in databases like SYFPEITHI (Rammensee et al., 1999) and MHCPEP (Brusic et al., 1998). Furthermore, in principle, it is possible to obtain motifs from MHC 3-D structure (de Haan et al., 2002). The prediction programs are based on comparison to these motifs or they are trained artificial

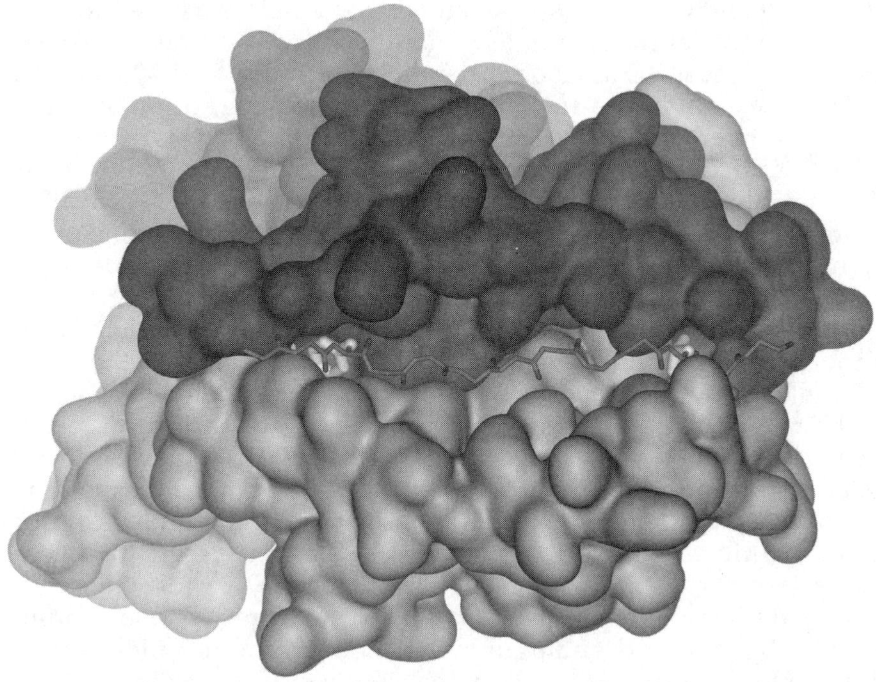

Figure 10 Top view of HLA-DR1 and the hemagglutinin peptide with only the white side-chains of the major anchor residues tyrosine and leucine shown (PDB-file 1dlh, Stern, 1994).

neural networks (Bisset and Fierz, 1993) to recognize new binders. In order to prevent many false positives, it is worthwhile to predict first which peptides are processed by the proteasome, as only every 1 in 200 peptides is presented by MHC class I (Kesmir et al., 2002). T cell epitopes are important for understanding cellular autoimmunity, for improving subunit vaccines, and they are candidates for removal in nonhuman pharmaceutical proteins. Their prediction has reached a high level of sophistication (Flower, 2003).

Protein-Protein Docking

Although molecular dynamics simulations, in principle, could be used to simulate the process of two proteins docking to one another, the computational cost would be enormous. Therefore, specialized docking algorithms have been developed that force the docking process in a certain direction, rather than simulate realistic motion. Levinthal was pioneering again (Levinthal et al., 1975). The first specialized program, DOCK, was developed to dock small molecules to proteins in 1986 by the Kuntz group (DesJarlais et al., 1986). In 1990, the AutoDock program used a combination of Monte Carlo steps and simulated annealing to dock small flexible ligands to proteins

(Goodsell and Olson, 1990). An atomic force field was used to evaluate the new translated, rotated, and orientated positions at discrete points in space, using Goodford's atomic affinity grids (Goodford, 1985). The first successful protein-protein docking, of an antibody and an antigen, did not use an atomic force field, but was completely based on surface complementarity (Walls and Sternberg, 1992). Later on, tabulated inter-residue contacts (Miyazawa and Jernigan, 1985) were used for scoring docking orientations. As a further simplification, only rigid body docking is performed. Flexibility and atomic force fields have only partially been applied by the Baker group, in an extension of the Rosetta fold predictor (Gray et al., 2003). Brooijmans and Kuntz have recently reviewed the field (2003).

In docking, new complex orientations are generated by different techniques: random placement, Metropolis Monte Carlo moves, molecular dynamics, genetic algorithms, and so forth. Because docking is an optimization procedure, rather than a general conformational analysis, prediction of the affinity is crucial to finding the best solution. Affinity prediction or scoring can be performed in several ways. Nonpartitioning methods, like free energy perturbation molecular dynamics (Bash et al., 1987), try to compute the binding affinity by slowly changing one molecule to another molecule, in the context of solvent and in the context of a binding partner (Figure 11).

These simulations require enormous computations. Partitioning methods partition the free energy of association into different components. Often, a master equation is used, combining separate calculations

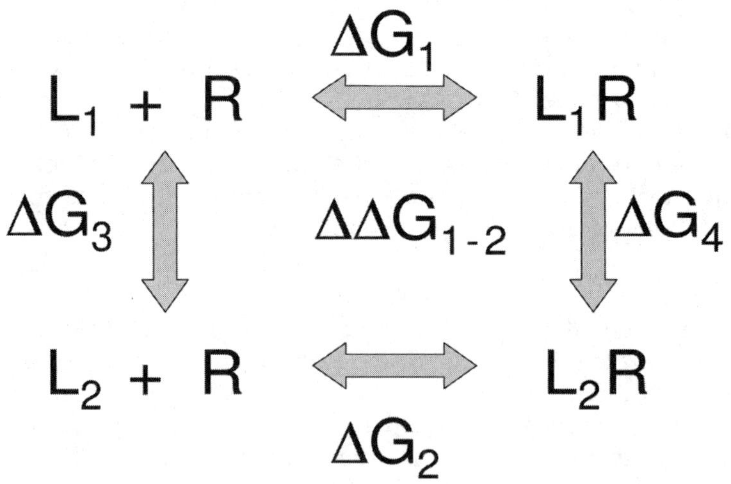

Figure 11 Free energy cycle where ligand 1 is slowly changed into ligand 2, before and after binding to binding partner R, to give the relative binding free energies.

of steric, electrostatic, hydrophobic, and solvation energy. Steric and hydrophobic contact surfaces are relatively easy to compute. For the other energies, more complicated electrostatics free energy cycles have to be computed (Figure 12). It is also possible to use empirical data, like the geometrical descriptors of an atomic interaction or molecular descriptors (Moret et al., 1999).

The quality of a new docking algorithm is always validated by redocking PDB complexes. All docking programs find the native complex conformation within the best 100 hits. In order to classify the native conformation as hit number one, atomic force fields would have to be used, leading to extremely long docking times. We have recently been able to address this problem by using atomic force field energy evaluations on proteins, which were reduced to their binding surface. The CDRs were extracted from antibody structures and subsequently docked to antigens. In order to prevent the new artificial surface of the antibody CDRs (i.e., the backside) from binding, dummy atoms were placed at strategic positions. During redocking experiments, by

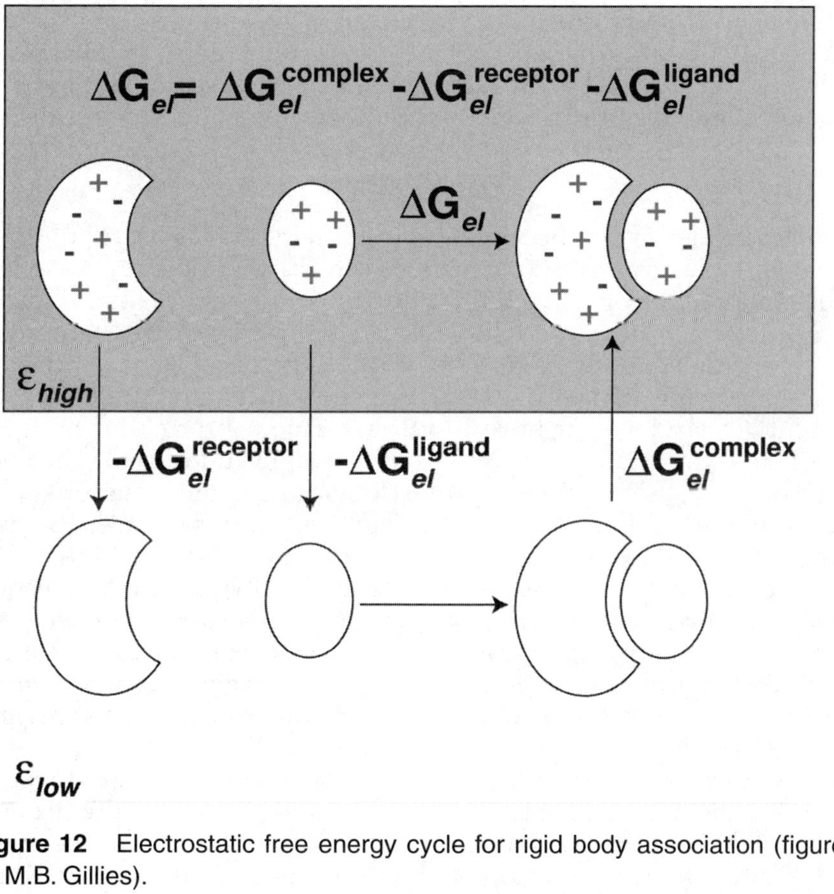

Figure 12 Electrostatic free energy cycle for rigid body association (figure by M.B. Gillies).

generating hundreds of thousands of generations with a Lamarckian genetic algorithm (meaning that optimization in one generation is allowed, as implemented in AutoDock 3; Morris et al., 1998), the native complex structure was scored as hit number 1 in 100% of the cases.

By analogy to CASP, the critical assessment of predicted interactions (CAPRI) was launched in 2001. Accurate predictions were given for four to five out of seven complexes (Janin et al., 2003). Five antibodies were initially given in their binding conformation, as the free antibodies had not been crystallized yet. This was not considered as bias by the jurors, as antibodies hardly induce their fit to antigens (Mendez et al., 2003). Most participants reranked their best hits manually. It would be a good suggestion to use an atomic force field for the end-stage docking, like was just described for AutoDock 3.

Programs for Prediction of Protein-Protein Interactions

In Table 2, an overview is given of programs for protein-protein docking, as well as some interaction and mutation databases that have recently been initiated. Furthermore, some epitope predictors are listed. The Human Proteome Organization has recently published a standard for publishing protein interaction data in databases (Hermjakob et al., 2004). In Table 2, free tools and servers for the prediction of epitopes, mutations, and protein-protein docking are listed.

Conclusions

Protein modeling has become a mature field. It is as useful to functional genomics as knockout studies are. Furthermore, because it is not hindered by physical limitations, it could help in studying the unseen. With protein modeling we can look at the electrostatics that governs molecular recognition. With dynamics, we can look at those molecules that resist structure elucidation and are either inherently disordered or have multiple stable conformations, which can lead to multiple phenotypes (e.g., soluble and fibrils). The benefit for molecular medicine is most obvious in conformational diseases, like Alzheimer, Creutzfeldt-Jakob, and serpinopathies (Lomas, 2002), where a protein conformational change is associated with the onset and gravity of disease. The instability of the protein in question cannot easily be investigated by experimental structural techniques. Docking and molecular dynamics might offer help here. And simple physicochemical properties like hydrophobicity and secondary structure propensity can explain the effects of point mutations on amyloid formation (Chiti et al., 2003). Even scientists without modeling experience can benefit from the Web servers. According to the latest results of the critical assessment of fully automated structure predictions, CAFASP3 (Fischer et al., 2003), these procedures can compete with the best CASP predictions, as long as homologous structures are known.

Although the pharmacist should guarantee the quality of drugs, in case of protein pharmaceuticals this would be a problem. For small organic drugs, all structural properties, which help in quality assurance, are well documented. For most pharmaceutical proteins, the sequence is proprietary information. Hopefully, sequence and structure for all pharmaceutical proteins will become readily available soon. Analogous to rational drug design, this would enable rational design of protein modifications to optimize their stability, solubility, antigenicity, and post-translational modifications (Marshall et al., 2003). Good examples are Tenekteplase and Insulin Lispro. By improving the pharmacodynamics and the pharmacokinetics, the patients will benefit most.

Abbreviations

BLAST	Basic Local Alignment Search Tool
BLOSUM62	Block substitution matrix 62%
CAFASP	Critical assessment of fully automated structure predictions
CAPRI	Critical assessment of prediction of interactions
CASP	Critical assessment of structure predictions
CDR	Complementarity determining regions
CIF	Crystallographic information file
EGF	Epidermal growth factor
FDA	Food and Drug Administration
GPCRs	G-protein coupled receptors
HGH	Human growth hormone
HIV	Human immunodeficiency virus
HLA	Human leukocyte antigen
Ig	Immunoglobulin
IUPAC	International Union of Pure and Applied Chemistry
MHC	Major histocompatibility complex
MO-SCF	Molecular orbital–self-consistent field
NOE	Nuclear Overhauser effect
PAM250	250 accepted point mutations per 100 amino acids
PDB	Protein Data Bank
PTM	Post-translational modifications
RMS	Root mean square
SASA	Solvent accessible surface area

SCR	Structurally conserved regions
SNP	Single nucleotide polymorphism
SVR	Structurally variable regions
TCR	T cell receptor
(7)TM	(7) transmembrane (helices)
TNF-α	Tumor necrosis factor α

Acknowledgment

Malcolm B. Gillies is gratefully acknowledged for some of the figures. Color figures can be found at http://wwwcmc.pharm.uu.nl/moret/pub/msapp/.

References

1. Aloy P, Stark A, Hadley C, and Russell RB. Predictions Without Templates: New Folds, Secondary Structure, and Contacts in CASP5. *Proteins* 2003; 53(Suppl 6);436–456
2. Altschul SF, Gish W, Miller W, Myers EW, and Lipman DJ. Basic Local Alignment Search Tool. *J Mol Biol* 1990; 215:403–410
3. Andreeva A, Howorth D, Brenner SE, Hubbard TJ, Chothia C, and Murzin AG. SCOP Database in 2004: Refinements Integrate Structure and Sequence Family Data. *Nucleic Acids Res* 2004; 32 Database issue:D226–229
4. Bash PA, Singh UC, Langridge R, and Kollman PA. Free Energy Calculations by Computer Simulation. *Science* 1987; 236:564–568
5. Bateman A, Coin L, Durbin R, Finn RD, Hollich V, Griffiths-Jones S, Khanna A, Marshall M, Moxon S, Sonnhammer EL, et al. The Pfam Protein Families Database. *Nucleic Acids Res* 2004; 32 Database issue:D138–141
6. Berendsen HJC, Postma JPM, Van Gunsteren WF, DiNola A, and Haak JR. Molecular Dynamics with Coupling to an External Bath. *J Chem Phys* 1984; 81:3684–3690
7. Berman HM, Westbrook J, Feng Z, Gilliland G, Bhat TN, Weissig H, Shindyalov IN, and Bourne PE. The Protein Data Bank. *Nucleic Acids Res* 2000; 28:235–242
8. Berman H, Henrick K, and Nakamura H. Announcing the Worldwide Protein Data Bank. *Nat Struct Biol* 2003; 10:980
9. Bernstein FC, Koetzle TF, Williams GJ, Meyer EF Jr, Brice MD, Rodgers JR, Kennard O, Shimanouchi T, and Tasumi M. The Protein Data Bank: A Computer-Based Archival File for Macromolecular Structures. *J Mol Biol* 1977; 112:535–542

10. Beukers MW, Kristiansen I, IJzerman AP, and Edvardsen I. TinyGRAP Database: A Bioinformatics Tool to Mine G-Protein-Coupled Receptor Mutant Data. *Trends Pharmacol Sci* 1999; 20:475–477
11. Bhat TN, Bentley GA, Boulot G, Greene MI, Tello D, Dall'Acqua W, Souchon H, Schwarz FP, Mariuzza RA, and Poljak RJ. Bound Water Molecules and Conformational Stabilization Help Mediate an Antigen-Antibody Association. *Proc Natl Acad Sci U S A* 1994; 91:1089–1093
12. Bisset LR, and Fierz W. Using a Neural Network to Identify Potential HLA-DR1 Binding Sites Within Proteins. *J Mol Recognit* 1993; 6:41–48
13. Blaber M, Zhang XJ, and Matthews BW. Structural Basis of Amino Acid Alpha Helix Propensity. *Science* 1993; 260:1637–1640
14. Bowie JU, Luthy R, and Eisenberg D. A Method to Identify Protein Sequences That Fold into a Known Three-Dimensional Structure. *Science* 1991; 253:164–170
15. Box GEP. Science and Statistics. *J Am Statistical Assoc* 1976; 71:791–799
16. Brooijmans N, and Kuntz ID. Molecular Recognition and Docking Algorithms. *Annu Rev Biophys Biomol Struct* 2003; 32:335–373
17. Browne WJ, North AC, Phillips DC, Brew K, Vanaman TC, and Hill RL. A Possible Three-Dimensional Structure of Bovine Alpha-Lactalbumin Based on That of Hen's Egg-White Lysozyme. *J Mol Biol* 1969; 42:65–86
18. Brusic V, Rudy G, and Harrison LC. MHCPEP, a Database of MHC-binding Peptides: Update 1997. *Nucleic Acids Res* 1998; 261:368–371
19. Chinea G, Padron G, Hooft RW, Sander C, and Vriend G. The Use of Position-Specific Rotamers in Model Building by Homology. *Proteins* 1995; 23:415–421
20. Chiti F, Stefani M, Taddei N, Ramponi G, and Dobson CM. Rationalization of the Effects of Mutations on Peptide and Protein Aggregation Rates. *Nature* 2003; 424:805–808
21. Chou PY, and Fasman GD. Conformational Parameters for Amino Acids in Helical, Beta-Sheet, and Random Coil Regions Calculated From Proteins. *Biochemistry* 1974; 13:211–222
22. Chou PY, and Fasman GD. Empirical Predictions of Protein Conformation. *Annu Rev Biochem* 1978; 47:251–276
23. Cohen FE, and Sternberg MJ. On the Prediction of Protein Structure: The Significance of the Root-Mean-Square Deviation. *J Mol Biol* 1980; 138:321–333
24. Cohen GH, Sheriff S, and Davies DR. Refined Structure of the Monoclonal Antibody HyHEL-5 with Its Antigen Hen Egg-White Lysozyme. *Acta Crystallogr D Biol Crystallogr* 1996; 52:315–326
25. Connolly ML. Solvent-Accessible Surfaces of Proteins and Nucleic Acids. *Science* 1983; 221:709–713

26. Dahiyat BI, and Mayo SL. De novo Protein Design: Fully Automated Sequence Selection. *Science* 1997; 278:82–87

27. Davis AM, Teague SJ, and Kleywegt GJ. Application and Limitations of X-Ray Crystallographic Data in Structure-Based Ligand and Drug Design. *Angew Chem Int Ed Engl* 2003; 42:2718–2736

28. de Bakker PI, Bateman A, Burke DF, Miguel RN, Mizuguchi K, Shi J, Shirai H, and Blundell TL. HOMSTRAD: Adding Sequence Information to Structure-Based Alignments of Homologous Protein Families. *Bioinformatics* 2001; 17:748–749

29. de Haan EC, Wauben MH, Grosfeld-Stulemeyer MC, and Moret EE. Structure-based Design and Evaluation of MHC Class II Binding Peptides. *Biologicals* 2001; 29:289–292

30. de Seze J, Dubucquoi S, Lefranc D, Virecoulon F, Nuez I, Dutoit V, Vermersch P, and Prin L. IgG Reactivity Against Citrullinated Myelin Basic Protein in Multiple Sclerosis. *J Neuroimmunol* 2001; 117:149–155

31. DesJarlais RL, Sheridan RP, Dixon JS, Kuntz ID, and Venkataraghavan R. Docking Flexible Ligands to Macromolecular Receptors by Molecular Shape. *J Med Chem* 1986; 29:2149–2153

32. Desmet J, De Maeyer M, Hazes B, and Lasters I. The Dead-End Elimination Theorem and Its Use in Protein Sidechain Positioning. *Nature* 1992; 356:539–542

33. de Vos AM, Ultsch M, and Kossiakoff AA. Human Growth Hormone and Extracellular Domain of Its Receptor: Crystal Structure of the Complex. *Science* 1992; 255:306–312

34. Duan Y, and Kollman PA. Pathways to a Protein Folding Intermediate Observed in a 1-Microsecond Simulation in Aqueous Solution. *Science* 1998; 282:740–744

35. Dyson HJ, and Wright PE. Antigenic Peptides. *FASEB J* 1995; 9:37–42

36. Emini EA, Hughes JV, Perlow DS, and Boger J. Induction of Hepatitis A Virus-Neutralizing Antibody by a Virus-Specific Synthetic Peptide. *J Virol* 1985; 55:836–839

37. EU 3-D Validation Network. Who Checks the Checkers? Four Validation Tools Applied to Eight Atomic Resolution Structures. *J Mol Biol* 1998; 276:417–436

38. Eyal E, Najmanovich R, Sobolev V, and Edelman M. MutaProt: a Web Interface for Structural Analysis of Point Mutations. *Bioinformatics* 2001; 17:381–382

39. Falk K, Rotzschke O, Stevanovic S, Jung G, and Rammensee HG. Allele-Specific Motifs Revealed by Sequencing of Self-Peptides Eluted from MHC Molecules. *Nature* 1991; 351:290–296

40. Fischer D, Rychlewski L, Dunbrack RL Jr, Ortiz AR, and Elofsson A. CAFASP3: the Third Critical Assessment of Fully Automated Structure Prediction Methods. *Proteins* 2003; 53(Suppl 6):503–516

41. Fiser A, Feig M, Brooks CL 3rd, and Sali A. Evolution and Physics in Comparative Protein Structure Modeling. *Acc Chem Res* 2002; 35:413–421

42. Flower DR. Towards in Silico Prediction of Immunogenic Epitopes. *Trends Immunol* 2003; 24:667–674

43. Frantz S, and Smith A. New Drug Approvals for 2002. *Nat Rev Drug Discov* 2003; 2:95–96

44. Garboczi DN, Ghosh P, Utz U, Fan QR, Biddison WE, and Wiley DC. Structure of the Complex Between Human T-Cell Receptor, Viral Peptide and HLA-A2. *Nature* 1996; 384:134–141

45. Gasteiger E, Gattiker A, Hoogland C, Ivanyi I, Appel RD, and Bairoch A. ExPASy: The Proteomics Server for In-depth Protein Knowledge and Analysis. *Nucleic Acids Res* 2003; 31:3784–3788

46. Gilson MK, and Honig BH. Calculation of Electrostatic Potentials in an Enzyme Active Site. *Nature* 1987; 330:84–86

47. Goodford PJ. A Computational Procedure for Determining Energetically Favorable Binding Sites on Biologically Important Macromolecules. *J Med Chem* 1985; 28:849–857

48. Goodsell DS, and Olson AJ. Automated Docking of Substrates to Proteins by Simulated Annealing. *Proteins* 1990; 8:195–202

49. Gray JJ, Moughon S, Wang C, Schueler-Furman O, Kuhlman B, Rohl CA, and Baker D. Protein-Protein Docking With Simultaneous Optimization of Rigid-Body Displacement and Side-Chain Conformations. *J Mol Biol* 2003; 331:281–299

50. Guo Z, Zhou D, and Schultz PG. Designing Small-Molecule Switches for Protein-Protein Interactions. *Science* 2000; 288:2042–2045

51. Henderson R, Baldwin JM, Ceska TA, Zemlin F, Beckmann E, and Downing KH. Model for the Structure of Bacteriorhodopsin Based on High-Resolution Electron Cryo-Microscopy. *J Mol Biol* 1990; 213:899–929

52. Hermjakob H, Montecchi-Palazzi L, Bader G, Wojcik J, Salwinski L, Ceol A, Moore S, Orchard S, Sarkans U, Von Mering C, et al. The HUPO PSI's Molecular Interaction Format—a Community Standard for the Representation of Protein Interaction Data. *Nat Biotechnol* 2004; 22:177–183

53. Herrgard S, Cammer SA, Hoffman BT, Knutson S, Gallina M, Speir JA, Fetrow JS, and Baxter SM. Prediction of Deleterious Functional

Effects of Amino Acid Mutations Using a Library of Structure-Based Function Descriptors. *Proteins* 2003; 53:806–816

54. Holm L, and Sander C. The FSSP Database: Fold Classification Based on Structure-Structure Alignment of Proteins. *Nucleic Acids Res* 1996; 24:206–209
55. Hooft RW, Vriend G, Sander C, and Abola EE. Errors in Protein Structures. *Nature* 1996a; 381:272
56. Hooft RW, Sander C, and Vriend G. Positioning Hydrogen Atoms by Optimizing Hydrogen-Bond Networks in Protein Structures. *Proteins* 1996b; 26:363–376
57. Hopp TP, and Woods KR. Prediction of Protein Antigenic Determinants from Amino Acid Sequences. *Proc Natl Acad Sci U S A* 1981; 78:3824–3828
58. Horn F, Bettler E, Oliveira L, Campagne F, Cohen FE, and Vriend G. GPCRDB Information System for G Protein-Coupled Receptors. *Nucleic Acids Res* 2003; 31:294–297
59. Ikeda M, Arai M, Okuno T, and Shimizu T. TMPDB: a Database of Experimentally-Characterized Transmembrane Topologies. *Nucleic Acids Res* 2003; 31:406–409
60. Jameson BA, and Wolf H. The Antigenic Index: a Novel Algorithm for Predicting Antigenic Determinants. *Comput Appl Biosci* 1988; 4:181–186
61. Janin J, Henrick K, Moult J, Eyck LT, Sternberg MJ, Vajda S, Vakser I, and Wodak SJ. CAPRI: a Critical Assessment of PRedicted Interactions. *Proteins* 2003; 52:2 9
62. Jones DT, Taylor WR, and Thornton JM. A New Approach to Protein Fold Recognition. *Nature* 1992; 358:86–89
63. Karplus PA, and Schultz GE. Prediction of Chain Flexibility in Proteins. *Naturwissenschaften* 1985; 72:212–213
64. Kesmir C, Nussbaum AK, Schild H, Detours V, and Brunak S. Prediction of Proteasome Cleavage Motifs by Neural Networks. *Protein Eng* 2002; 15:287–296
65. Kleywegt GJ. Validation of Protein Crystal Structures. *Acta Crystallogr D Biol Crystallogr* 2000; 56 (Pt 3):249–265
66. Koehl P, and Levitt M. Structure-based Conformational Preferences of Amino Acids. *Proc Natl Acad Sci U S A* 1999; 96:12524–12529
67. Kropshofer H, Max H, Muller CA, Hesse F, Stevanovic S, Jung G, and Kalbacher H. Self-peptide Released from Class II HLA-DR1 Exhibits a Hydrophobic Two-Residue Contact Motif. *J Exp Med* 1992; 175:1799–1803
68. Kuhlman B, Dantas G, Ireton GC, Varani G, Stoddard BL, and Baker D. Design of a Novel Globular Protein Fold with Atomic-Level Accuracy. *Science* 2003; 302:1364–1368

69. Kyte J, and Doolittle RF. A Simple Method for Displaying the Hydropathic Character of a Protein. *J Mol Biol* 1982; 157:105–132

70. Larsen TA, Olson AJ, and Goodsell DS. Morphology of Protein-Protein Interfaces. *Structure* 1998; 6:421–427

71. Laskowski RA, Moss DS, and Thornton JM. Main-Chain Bond Lengths and Bond Angles in Protein Structures. *J Mol Biol* 1993; 231:1049–1067

72. Lee B, and Richards FM. The Interpretation of Protein Structures: Estimation of Static Accessibility. *J Mol Biol* 1971; 55:379–400

73. Levinthal C. Molecular Model-Building by Computer. *Sci Am* 1966; 214:42–52

74. Levinthal C, Wodak SJ, Kahn P, and Dadivanian AK. Hemoglobin Interaction in Sickle Cell Fibers. I: Theoretical Approaches to the Molecular Contacts. *Proc Natl Acad Sci U S A* 1975; 72:1330–1334

75. Li Y, Li H, Smith-Gill SJ, and Mariuzza RA. Three-Dimensional Structures of the Free and Antigen-Bound Fab from Monoclonal Antilysozyme Antibody HyHEL-63(,). *Biochemistry* 2000; 39:6296–6309

76. Lomas DA, and Carrell RW. Serpinopathies and the Conformational Dementias. *Nat Rev Genet* 2002; 3:759–768

77. Louis JM, Dyda F, Nashed NT, Kimmel AR, and Davies DR. Hydrophilic Peptides Derived from the Transframe Region of Gag-Pol Inhibit the HIV-1 Protease. *Biochemistry* 1998; 37:2105–2110

78. Marcotte EM, Pellegrini M, Thompson MJ, Yeates TO, and Eisenberg D. A combined algorithm for genome-wide prediction of protein function. *Nature* 1999; 402:83–86

79. Marshall SA, Lazar GA, Chirino AJ, and Desjarlais JR. Rational Design and Engineering of Therapeutic Proteins. *Drug Discov Today* 2003; 8:212–221

80. McCoy AJ, Chandana Epa V, and Colman PM. Electrostatic Complementarity at Protein/Protein Interfaces. *J Mol Biol* 1997; 268:570–584

81. Mehler EL, Periole X, Hassan SA, and Weinstein H. Key Issues in the Computational Simulation of GPCR Function: Representation of Loop Domains. *J Comput Aided Mol Des* 2002; 16:841–853

82. Melamud E, and Moult J. Evaluation of Disorder Predictions in CASP5. *Proteins* 2003; 53(Suppl 6):561–565

83. Melen K, Krogh A, and von Heijne G. Reliability Measures for Membrane Protein Topology Prediction Algorithms. *J Mol Biol* 2003; 327:735–744

84. Mendelsohn AR, and Brent R. Protein Interaction Methods—Toward An Endgame. *Science* 1999; 284:1948–1950

85. Mendez R, Leplae R, De Maria L, and Wodak SJ. Assessment of Blind Predictions of Protein-Protein Interactions: Current Status of Docking Methods. *Proteins* 2003; 52:51–67

86. Miyazawa S, and Jernigan RL. Estimation of Effective Interresidue Contact Energies from Protein Crystal Structures: Quasi-Chemical Approximation. *Macromolecules* 1985; 18:534–552

87. Mizuguchi K, Deane CM, Blundell TL, and Overington JP. HOMSTRAD: a Database of Protein Structure Alignments for Homologous Families. *Protein Sci* 1998; 7:2469–2471

88. Moret EE, van Wijk MC, Kostense AS, and Gillies MB. Scoring Peptide(mimetic)-Protein Interactions. *Med Chem Res* 1999; 9:604–620

89. Morris GM, Goodsell DS, Halliday RS, Huey R, Hart WE, Belew RK, and Olson AJ. Automated Docking Using a Lamarckian Genetic Algorithm and Empirical Binding Free Energy Function. *J Comput Chem* 1998; 19:1639–1662

90. Moult J, Fidelis K, Zemla A, and Hubbard T. Critical Assessment of Methods of Protein Structure Prediction (CASP)—Round V. *Proteins* 2003; 53(Suppl 6):334–339

91. Murzin AG, Brenner SE, Hubbard T, and Chothia C. SCOP: a Structural Classification of Proteins Database for the Investigation of Sequences and Structures. *J Mol Biol* 1995; 247:536–540

92. Needleman SB, and Wunsch CD. A General Method Applicable to the Search for Similarities in the Amino Acid Sequence of Two Proteins. *J Mol Biol* 1970; 48:443–453

93. Nielsen JE, and McCammon JA. Calculating pKa Values in Enzyme Active Sites. *Protein Sci* 2003; 12:1894–1901

94. Nishikawa K, Ishino S, Takenaka H, Norioka N, Hirai T, Yao T, and Seto Y. Constructing a Protein Mutant Database. *Protein Eng* 1994; 7:733

95. Novotny M, Madsen D, and Kleywegt GJ. Evaluation of Protein Fold Comparison Servers. *Proteins* 2004; 54:260–270

96. Oliveira L, Paiva AC, and Vriend G. Correlated Mutation Analyses on Very Large Sequence Families. *Chembiochem* 2002; 3:1010–1017

97. Orengo CA, Michie AD, Jones S, Jones DT, Swindells MB, and Thornton JM. CATH—a Hierarchic Classification of Protein Domain Structures. *Structure* 1997; 5:1093–1108

98. Palczewski K, Kumasaka T, Hori T, Behnke CA, Motoshima H, Fox BA, Le Trong I, Teller DC, Okada T, Stenkamp RE, et al. Crystal Structure of Rhodopsin: a G Protein-Coupled Receptor. *Science* 2000; 289:739–745

99. Pearl FM, Bennett CF, Bray JE, Harrison AP, Martin N, Shepherd A, Sillitoe I, Thornton J, and Orengo CA. The CATH Database: an

Extended Protein Family Resource for Structural and Functional Genomics. *Nucleic Acids Res* 2003; 31:452–455

100. Ponder JW, and Richards FM. Tertiary Templates For Proteins. Use of Packing Criteria in the Enumeration of Allowed Sequences for Different Structural Classes. *J Mol Biol* 1987; 193:775–791

101. Ptitsyn OB, and Uversky VN. The Molten Globule Is a Third Thermodynamical State of Protein Molecules. *FEBS Lett* 1994; 341:15–18

102. Radivojac P, Obradovic Z, Smith DK, Zhu G, Vucetic S, Brown CJ, Lawson JD, and Dunker AK. Protein Flexibility and Intrinsic Disorder. *Protein Sci* 2004; 13:71–80

103. Ramachandran GN, Ramakrishnan C, and Sasisekharan V. Stereochemistry of Polypeptide Chain Configurations. *J Mol Biol* 1963; 7:95–99

104. Rammensee H, Bachmann J, Emmerich NP, Bachor OA, and Stevanovic S. SYFPEITHI: Database for MHC Ligands and Peptide Motifs. *Immunogenetics* 1999; 50:213–219

105. Rantapaa-Dahlqvist S, de Jong BA, Berglin E, Hallmans G, Wadell G, Stenlund H, Sundin U, and van Venrooij WJ. Antibodies Against Cyclic Citrullinated Peptide and IgA Rheumatoid Factor Predict the Development of Rheumatoid Arthritis. *Arthritis Rheum* 2003; 48: 2741–2749

106. Rhee SY, Gonzales MJ, Kantor R, Betts BJ, Ravela J, and Shafer RW. Human Immunodeficiency Virus Reverse Transcriptase and Protease Sequence Database. *Nucleic Acids Res* 2003; 31:298–303

107. Robinson NE. Protein Deamidation. *Proc Natl Acad Sci U S A* 2002; 99:5283–5288

108. Robinson NE, and Robinson AB. Molecular Clocks. *Proc Natl Acad Sci U S A* 2001; 98:944–949

109. Rodriguez R, Chinea G, Lopez N, Pons T, and Vriend G. Homology Modeling, Model and Software Evaluation: Three Related Resources. *Bioinformatics* 1998; 14:523–528

110. Roses AD. Pharmacogenetics and the Practice of Medicine. *Nature* 2000; 405:857–865

111. Schwede T, Kopp J, Guex N, and Peitsch MC. SWISS-MODEL: an Automated Protein Homology-Modeling Server. *Nucleic Acids Res* 2003; 31:3381–3385

112. Simons KT, Kooperberg C, Huang E, and Baker D. Assembly of Protein Tertiary Structures from Fragments with Similar Local Sequences Using Simulated Annealing and Bayesian Scoring Functions. *J Mol Biol* 1997; 268:209–225

113. Smith TF, and Waterman MS. Identification of Common Molecular Subsequences. *J Mol Biol* 1981; 147:195–197

114. Stern LJ, Brown JH, Jardetzky TS, Gorga JC, Urban RG, Strominger JL and Wiley DC. Crystal Structure of the Human Class II MHC Protein HLA-DR1 Complexed with an Influenza Virus Peptide. *Nature* 1994; 368:215–221

115. Still WC, Tempczyk A, Hawley RC, and Hendrickson T. Semianalytical Treatment of Solvation for Molecular Mechanics and Dynamics. *J Am Chem Soc* 1990; 112:6127–6129

116. Taylor PD, Attwood TK, and Flower DR. BPROMPT: a Consensus Server for Membrane Protein Prediction. *Nucleic Acids Res* 2003; 31:3698–3700

117. The International HapMap Consortium. The International HapMap Project. *Nature* 2003; 426:789–796

118. Thornton JM, Edwards MS, Taylor WR, and Barlow DJ. Location of 'continuous' Antigenic Determinants in the Protruding Regions of Proteins. *Embo J* 1986; 5:409–413

119. Titmuss SJ, Cummins PL, Rendell AP, Bliznyuk AA, and Gready JE. Comparison of Linear-Scaling Semiempirical Methods and Combined Quantum Mechanical/Molecular Mechanical Methods for Enzymic Reactions. II. An Energy Decomposition Analysis. *J Comput Chem* 2002; 23:1314–1322

120. Tramontano A, and Morea V. Assessment of Homology-based Predictions in CASP5. *Proteins* 2003; 53(Suppl 6):352–368

121. van den Elsen JM, Herron JN, Hoogerhout P, Poolman JT, Boel E, Logtenberg T, Wilting J, Crommelin DJ, Kroon J, and Gros P. Bactericidal Antibody Recognition of a Pora Epitope of Neisseria Meningitidis: Crystal Structure of a Fab Fragment in Complex with a Fluorescein-Conjugated Peptide. *Proteins* 1997; 29:113–125

122. van de Wal Y, Kooy Y, van Veelen P, Pena S, Mearin L, Papadopoulos G, and Koning F. Selective Deamidation by Tissue Transglutaminase Strongly Enhances Gliadin-Specific T Cell Reactivity. *J Immunol* 1998; 161:1585–1588

123. van Gunsteren WF, and Berendsen HJC. Algorithms for Macromolecular Dynamics and Constraint Dynamics. *Mol Phys* 1977; 34:1311–1327

124. Vocadlo DJ, Davies GJ, Laine R, and Withers SG. Catalysis by Hen Egg-White Lysozyme Proceeds via a Covalent Intermediate. *Nature* 2001; 412:835–838

125. Vreven T, Morokuma K, Farkas O, Schlegel HB, and Frisch MJ. Geometry Optimization with QM/MM, ONIOM, and Other Combined Methods. I. Microiterations and Constraints. *J Comput Chem* 2003; 24:760–769

126. Walls PH, and Sternberg MJ. New Algorithm to Model Protein-Protein Recognition Based on Surface Complementarity. Applications to Antibody-Antigen Docking. *J Mol Biol* 1992; 228:277–297

127. Wedemayer GJ, Patten PA, Wang LH, Schultz PG, and Stevens RC. Structural Insights into the Evolution of an Antibody Combining Site. *Science* 1997; 276:1665–1669

128. Welling GW, Weijer WJ, van der Zee R, and Welling-Wester S. Prediction of Sequential Antigenic Regions in Proteins. *FEBS Lett* 1985; 188:215–218

129. Wells JA. Binding in the Growth Hormone Receptor Complex. *Proc Natl Acad Sci U S A* 1996; 93:1–6

20

Concluding Remarks: Analytical Methods for Protein Formulations

TUDOR ARVINTE

School of Pharmacy, Department of Pharmaceutics and Biopharmaceutics, University of Geneva, Geneva, Switzerland

Structural analysis of protein pharmaceuticals is the first and most important step in the development of a new therapeutic protein product. The methods presented in this book are fundamental for the characterization of a protein drug substance and for the assessment of the chemical and physical stability of a protein. In this final chapter, some aspects of the complexity and challenges in the analysis of protein formulations will be discussed.

The analysis of protein properties in formulations and protein drug delivery systems requires specific, "tailor-made" analytical methods. These methods should be adapted to the necessities of the formulation and not to the needs of the analytical techniques. Analytical methods used to characterize formulated proteins should be able to detect small differences against a complex background and, if necessary, provide information on the protein structure in heterogeneous formulations such as aggregated slow-release formulations, microparticles, or proteins adsorbed onto adjuvants in vaccines.

A summary of some analytical techniques used for characterizing higher order protein structures and formulations is presented in Figure 1. Some techniques give information on large molecular aggregates (e.g., light microscopy) or on the properties of protein populations (e.g., DSC, chromatography; see column "Macroscopic" in Figure 1). Microscopic techniques such as NMR and x-ray provide very detailed information at atomic level resolution. The classification in Figure 1 is not strict and is used to highlight the techniques in the middle column, which can provide information at the molecular level as well as at the macroscopic level. With specific adaptations, some of these techniques can operate over a wide range of protein concentrations in both homogeneous and heterogeneous states (e.g., protein solutions from 0.05 mg/mL up to 50–150 mg/mL, or depot formulations). In specific cases, techniques from the "Macroscopic" and "Microscopic" groups can also be particularly useful for the characterization of protein structure and formulations. For example, NMR can be used to characterize the batch-to-batch consistency of disulfide bonds in the production of recombinant proteins, as was illustrated for transforming growth factor $\beta 3$ (TGF-$\beta 3$) (Blommers and Cerletti, 1997).

Examples of protein formulations that require specific analytical methods include slow-release formulations [e.g., human calcitonin fibrils (Arvinte et al., 1992); zinc hirudin complexes (Arvinte, 1994)], protein crystals (e.g., Mittl et al., 1996), gels, microparticles, liposomes, or formulations with very low or very high protein concentrations [e.g., 1–50 μg/mL TGF-$\beta 3$ formulations (Arvinte et al., 2000); 100–250 mg/mL antibody liquid formulations (Arvinte and Fauquex, 2002)].

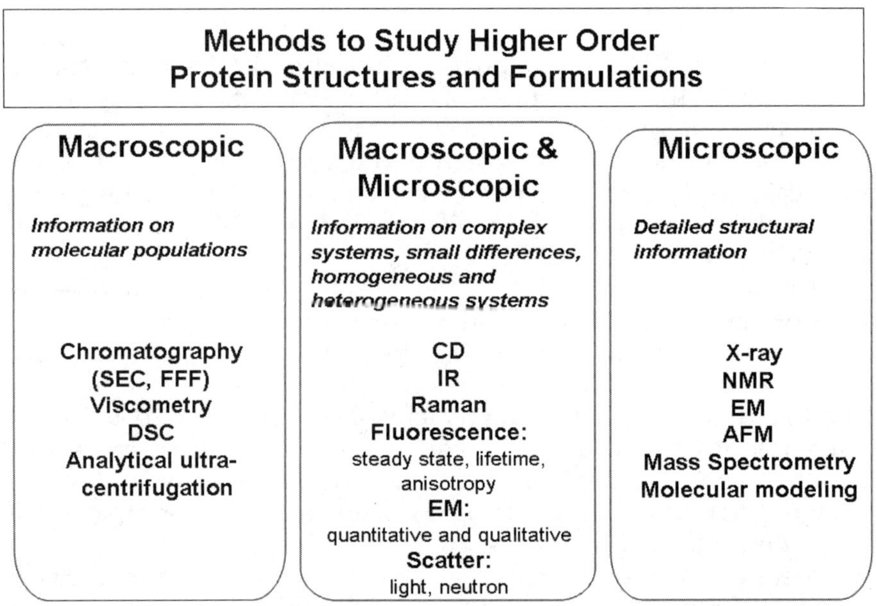

Figure 1 Some of the methods used to characterize protein structures and formulations. The methods in the box "Macroscopic" provide information on the properties of a protein population: SEC, size exclusion chromatography; FFF, free flow fractionation; DSC differential scanning calorimetry. The methods in the middle box, by providing information about the macroscopic and well as microscopic levels, are particularly useful for the characterization of protein formulations. The analytical techniques in the box "Microscopic" provide very detailed information on the protein structure on the Angstrom and nanometer scales.

Human calcitonin (hCT) has a strong tendency to aggregate and form fibrils. The hCT fibrils were shown to have beneficial properties as slow-release formulations for hCT (Arvinte et al., 1992) and for the release of TGF-β3 included in hCT fibrils (Arvinte et al., 2000). Different analytical methods were adapted to characterize the large protein fibrils formed during hCT fibrillation. These methods included fluorescence spectroscopy, circular dichroism (CD), infrared (IR) spectroscopy, electron microscopy (EM), and atomic force microscopy (AFM) (Arvinte and Drake, 1993; Arvinte et al., 1993; Bauer et al., 1995; Arvinte, 1996; Goldsbury et al., 1999). The CD method developed to study the structure of hCT in fibrils (concentrations of 30 mg/mL hCT) required very narrow cuvettes (0.01 mm) with the cuvette positioned to minimize the influence of light scattering (Arvinte et al., 1993). The CD spectroscopy measurements showed the presence of an α-helix structure for hCT in the fibrillated state. On the other hand, infrared spectroscopy measurements of hCT fibrils showed the presence of β-sheet conformation (Arvinte et al., 1993). Thus, the spectroscopic data provided a good illustration of the complementary nature of CD and infrared spectroscopy, with the former being more sensitive to α-helix and the latter to β-sheet (Arvinte et al., 1993).

The antibody vaccine MMA 383 loses immunogenic properties *in vivo* after lyophilization (Taschner et al., 2001). This loss in biological activity occurs without degradation of the protein or aggregation detected by standard analytical methods such as SDS/polyacrylamide gel electrophoresis, size exclusion chromatography, and capillary zone electrophoresis (Taschner et al., 2001). This suggested that the loss of functionality could be related to morphological changes in the antibody structure. Quantitative and qualitative EM, fluorescence and CD spectroscopy techniques showed that lyophilization induced subtle changes in the MMA 383 antibody structure associated with reduction in the flexibility of the Fab unit in the hinge region. These changes in antibody flexibility may be related to the reduced immunogenicity observed after lyophilization (Taschner et al., 2001). The combination of electron microscopy, fluorescence and CD spectroscopies described for the MMA 383 antibody provides a novel analytical approach to study the subtle changes in the morphology and flexibility of antibodies induced by lyophilization.

Many new biotechnology drugs such as monoclonal antibodies are commonly formulated at high concentrations (between 50 mg/mL and 200 mg/mL). Often in these cases, protein aggregation is a major impediment in development because the aggregates may decrease the bioavailability of the drug and/or may contribute to side effects as well as the generation of immunological responses (Schellekens, 2002). The term *protein aggregate* covers a broad range of protein complexes. At one extreme, protein aggregates are rigid structures formed by strong electrostatic and hydrophobic interactions as well as by covalent bonds. These rigid aggregates can be well characterized by standard methods such as light scattering and size exclusion chromatography methods. At the other extreme, protein aggregates are loose, transient structures formed by weaker interactions. Standard light scattering and size exclusion chromatography methods do not readily detect these loose aggregates, and their existence often remains unnoticed. However, accumulating indirect evidence indicates that these loose aggregates play a major role in toxicity and diseases (e.g., Bucciantini et al., 2002). The characterization of loose protein aggregates that may form in highly concentrated protein samples (e.g., 100 mg/mL) by analyzing diluted samples (e.g., 0.5 mg/mL) with standard size exclusion and light scattering methods is not appropriate and may be misleading. New analytical methods to study protein aggregation have to be developed that do not require dilution of the sample prior to the measurements. Along these lines, new methods are being developed that involve the staining of protein aggregates with fluorescence dyes such as the hydrophobic probe Nile red in procedures that do not require dilution of the formulations (Demeule et al., 2003; Arvinte et al., manuscript in preparation). The Nile red staining method has been used together with viscosity measurements in the development of E25 antibody formulations with high concentrations up to 300 mg/mL (Arvinte and Fauquex, 2002).

On the other hand, for formulations with very low protein concentrations (0.1–50 µg/mL), analysis of protein aggregation or protein binding to surfaces is also cumbersome. One example is TGF-β3 (therapeutic dose in the range of 0.1–20 µg/mL), a protein that has a strong tendency to adsorb to walls and aggregate in the physiological pH region (Pellaud et al, 1999). To characterize the adsorption and the aggregation of TGF-β3 at concentrations as low as 0.1 µg/mL, fluorescence and light scattering methods were developed (Arvinte, 2000). For example, when TGF-β3 adsorbs to the cuvette walls, the concentration of the protein in the solution decreases. This change in protein concentration can be monitored by measuring the reduction in fluorescence intensity (experimental details are presented in Arvinte, 2000). Fluorescence and light scattering methods were used for screening different excipients and had a crucial contribution to the development of physically and chemically stable TGF-β3 formulations (Arvinte, 2000).

The examples presented in this chapter show that for many protein biopharmaceuticals, analytical tools need to be adapted to the formulation. The use of complementary analytical methods strengthens the interpretation of the data and simplifies further refinement of the formulations. In the near future, approaches that are today regarded as less conventional, such as NMR or electron microscopy, will become common practice, adding to the array of techniques available for adequate characterization of protein formulations.

Abbreviations

AFM	Atomic force microscopy
CD	Circular dichroism
DSC	Differential scanning calorimetry
EM	Electron microscopy
hCT	Human calcitonin
IR	Infrared
NMR	Nuclear magnetic resonance
TGF-β3	Transforming growth factor β3

References

1. Arvinte T. Patent on Aqueous Depot Formulation Comprising Water, Hirudin, and Calcium, Magnesium or Zinc Ions in the Form of Water Insoluble Salt. European Patent Application, Publication No. 624 375 A2; 1994
2. Arvinte T. Human Calcitonin Fibrillogenesis. In: *The Nature and Origin of Amyloid Fibrils* (Ciba Foundation Symposium No. 199, 23-25 October 1995, Oeiras, Portugal). Chichester: John Wiley & Sons; 1996:90–97
3. Arvinte T, and Drake AF. Comparative Study of Human and Salmon Calcitonin Secondary Structure in Solutions with Low Dielectric Constants. *J Biol Chem* 1993; 268:6408–6414
4. Arvinte T, and Fauquex PF. Stable liquid Formulation of Antibodies with High Protein Concentrations. Patent Application, Publication No. WO 02/096457 A2; 2002
5. Arvinte T, Cudd A, and Phillips J. Use of Fibrillated Calcitonin in the Treatment of Calcium Deficiency Diseases. Ciba-Geigy Patent. European Patent Application, Publication No. 0510549; 1992
6. Arvinte T, Cudd A, and Drake AF. The Structure and Mechanism of Formation of Human Calcitonin Fibrils. *J Biol Chem* 1993; 268: 6415–6422
7. Arvinte T, Schote UT, and Sigg J. Pharmaceutical Compositions Comprising TGF-β3. Patent Application, International Publication No. WO 00/547797; 2000
8. Bauer HH, Aebi U, Haener M, Hermann R, Müller M, Arvinte T, and Merkle HP. Electron Microscopical Study on the Structure and Polymorphism of Fibrillar Supramolecular Assemblies Produced by in Vitro Aggregation of Human Calcitonin. *J Struct Biol* 1995; 115:1–15
9. Blommers MJJ, and Cerletti N. High Resolution NMR, a Useful Tool to Control Batch-to-Batch Consistency of Disulphide Bonds in

Biopharmaceuticals: Application to Transforming Growth Factor-β3. *Pharm Sci* 1997; 3:29–36

10. Bucciantini M, Giannoni E, Chiti F, Baroni F, Formigli L, Zurdo J, Taddei N, Ramponi G, Dobson CM, and Stefani M. Inherent Toxicity of Aggregates Implies a Common Mechanism for Protein Misfolding Diseases. *Nature* 2002; 416:507–511
11. Demeule B, Arvinte T, and Gurny R. Detection and Characterization of IgG1 Aggregation: Spectroscopy and Light Microscopy Methods. *Papers of the Controlled Release Society*, German chapter, Annual meeting, April 4, 2003, Munich, Germany
12. Goldsbury C, Kistler J, Aebi U, Arvinte T, and Cooper GJS. Watching Amyloid Fibrils Grow by Time-Lapse Atomic Force Microscopy. *J Mol Biol* 1999; 285:33–39
13. Mittl P, Grütter M, and Arvinte T. Use of Novel TGF-β3 Crystals for Slow-Release Formulations. International Patent Application, Patent No. WO9705166-A 97.02.13 (9709); 1996
14. Pellaud J, Schote U, Arvinte T, and Seelig J. Conformation and Self-Association of Human Recombinant Transforming Growth Factor-β3 in Aqueous Solutions. *J Biol Chem* 1999; 274:7699–7704
15. Schellekens H. Bioequivalence and the Immunogenicity of Biopharmaceuticals. *Nat Rev Drug Discovery* 2002; 1:457–462
16. Taschner N, Venkateshwar RA, Goldie KN, Müller SA, Drake AF, Aebi U, and Arvinte T. Modulation of Antigenicity Correlated to Changes in Antibody Flexibility Upon Lyophilization. *J Mol Biol* 2001; 310:169–179

INDEX